Molecular Mechanisms
of Functional Food

Molecular Mechanisms of Functional Food

Edited by

Rocio Campos-Vega
Postgraduate Program in Food of the Center of the Republic (PROPAC), Research and Graduate Studies in Food Science, School of Chemistry, Universidad Autónoma de Querétaro, Santiago de Querétaro, Mexico

B. Dave Oomah
(Retired) Formerly with Summerland Research and Development Centre, Agriculture and Agri-Food Canada, Summerland, BC, Canada

WILEY

This edition first published 2023

Registered Office(s)
John Wiley & Sons, Inc., 111 River Street, Hoboken, NJ 07030, USA
John Wiley & Sons Ltd, The Atrium, Southern Gate, Chichester, West Sussex, PO19 8SQ, UK

Editorial Office
The Atrium, Southern Gate, Chichester, West Sussex, PO19 8SQ, UK

For details of our global editorial offices, customer services, and more information about Wiley products visit us at www.wiley.com.

Wiley also publishes its books in a variety of electronic formats and by print-on-demand. Some content that appears in standard print versions of this book may not be available in other formats.

Library of Congress Cataloging-in-Publication Data Applied for
Hardback ISBN: 9781119804024

Cover Design: Wiley
Cover Image: © Piyaset/Shutterstock; avoferten/Shutterstock; gresei/Shutterstock; Dr Feelgood®/Getty Images; VladKK/Shutterstock

Set in 9.5/12.5pt STIXTwoText by Straive, Pondicherry, India

Contents

List of Contributors

Cristóbal N. Aguilar
Food Research Department
School of Chemistry
Universidad Autónoma de Coahuila
Saltillo, Mexico

Clara Albano
Institute of Sciences of Food Production
(ISPA), CNR,
Lecce, Italy

Priscila Zaczuk Bassinello
Embrapa Rice and Beans
Santo Antônio de Goiás,
Goiás, Brazil

Juliana Aparecida Correira Bento
School of Agronomy
Federal University of Goiás (UFG)
Goiânia, Brazil

Irshad Ul Haq Bhat
Department of Applied Sciences,
Chemistry Section
University of Technology and Applied
Sciences
Al-Khuwair, Muscat, Oman

Rajeev Bhat
ERA-Chair for Food (By-) Products
Valorisation Technologies (VALORTECH),
Estonian University of Life Sciences, Tartu,
Estonia, EU

Federica Blando
Institute of Sciences of Food Production
(ISPA), CNR,
Lecce, Italy

Cheyenne Braojos
Department of Agricultural Chemistry and
Food Science
Universidad Autónoma de Madrid,
Madrid, Spain

Martín Calderón-Juárez
Plan de Estudios Combinados en Medicina,
National Autonomous University
of Mexico,
Faculty of Medicine,
Mexico City, Mexico

Pedro de Souza Freitas Campos
Bioprocess Laboratory
Food Science and Nutrition Department,
School of Engineering,
University of Campinas,
Campinas, SP, Brazil

Rocio Campos-Vega
Programa de Posgrado en Alimentos del
Centro de la Republica (PROPAC),
Research and Graduate Studies in
Food Science,
School of Chemistry,
Universidad Autónoma de Queretaro (UAQ),
Queretaro, Mexico

Silvia Cañas
Department of Agricultural Chemistry and Food Science,
Universidad Autónoma de Madrid,
Madrid, Spain

Paz Cano-Muñoz
Department of Agricultural Chemistry and Food Science,
Universidad Autónoma de Madrid,
Madrid, Spain

Anaberta Cardador-Martínez
Escuela de Ingenieria y Ciencias,
Departamento de Bioingenierias,
Tecnologico de Monterrey,
Queretaro, Mexico

Eugenia Lugo-Cervantes
Tecnologia Alimentaria,
Centro de Investigación y Asistencia en Technologia y Diseño del Estado de Jalisco A.C., CIATEJ, Zapopan
Jalisco, Mexico

Mónica L. Chávez-González
Food Research Department,
School of Chemistry,
Universidad Autónoma de Coahuila
Saltillo, Mexico

Marisol Cruz-Requena
School of Chemistry
Autonomous University of Coahuila
Saltillo, Coahuila, Mexico

Behrokh Daei-Hasani
Department of Biology
Payame Noor University
Tehran, Iran

Maria Dolores del Castillo
Instituto de Investigación en Ciencias de la Alimentación (CIAL), CSIC-UAM
Calle Nicolás Cabrera
Madrid, Spain

Laura Guspí Domènech
Instituto de Investigación en Ciencias de la Alimentación (CIAL), CSIC-UAM
Calle Nicolás Cabrera
Madrid, Spain

Elisa Dufoo-Hurtado
Programa de Posgrado en Alimentos del Centro de la Republica (PROPAC)
Research and Graduate Studies in Food Science
School of Chemistry
Universidad Autónoma de Queretaro (UAQ)
Queretaro, Mexico

Adriana Carolina Flores-Gallegos
School of Chemistry
Autonomous University of Coahuila
Saltillo, Coahuila, Mexico

Alessandra Gambero
Bioprocess Laboratory
Food Science and Nutrition Department
School of Engineering
University of Campinas
Campinas, SP, Brazil

Jesus David García-Ortiz
School of Chemistry
Autonomous University of Coahuila
Saltillo, Coahuila, Mexico

Vanize M. Genova
Bioprocess Laboratory
Food Science and Nutrition Department
School of Engineering
University of Campinas
Campinas, SP, Brazil

Mariana Juste Contin Gomes
Department of Nutrition and Health
Federal University of Viçosa (UFV)
Viçosa, Brazil

Ena Gupta
Department of Home Science
University of Allahabad
Allahabad, India

David Fonseca-Hernandez
Tecnologia Alimentaria
Centro de Investigación y Asistencia en Technologia y Diseño del Estado de Jalisco
A.C., CIATEJ, Zapopan
Jalisco, Mexico

Alan Javier Hernández-Álvarez
School of Food Science and Nutrition
University of Leeds
Leeds, UK

Tana Hernandez-Barrueta
Department of Food Science
University of California
Davis, USA

Amaia Iriondo-DeHond
Instituto de Investigación en Ciencias de la Alimentación (CIAL), CSIC-UAM
Calle Nicolás Cabrera
Madrid, Spain

Cristian Jiménez-Martínez
Escuela Nacional de Ciencias Biológicas
Instituto Politécnica Nacional
Mexico City, Mexico

Maria Bibiana Zapata Londoño
School of Nutrition and Dietetics
University of Antioquia
Medellin, Colombia

Ivan Luzardo-Ocampo
Institute of Neurobiology
Universidad Nacional Autónoma de México (UNAM)
Queretaro, Mexico

Gabriela A. Macedo
Bioprocess Laboratory
Food Science and Nutrition Department
School of Engineering
University of Campinas
Campinas, SP, Brazil

Maria Elena Maldonado-Celis
Faculty of Health Sciences
University of Quindio
Armenia, Colombia

Maria A. Martín-Cabrejas
Institute of Food Science Research (CIAL) (UAM-CSIC)
Madrid, Spain

Hércia Stampini Duarte Martino
Department of Nutrition and Health
Federal University of Viçosa (UFV)
Viçosa, Brazil

Guiomar Melgar-Lalanne
Centro de Investigaciones Biomédicas
Universidad Veracruzana
Xalapa, Mexico

Martin Mondor
St-Hyacinthe Research and Development Centre
Agriculture and Agri-Food Canada
Saint-Hyacinthe, Quebec City
Canada & Institute of Nutrition and Functional Foods (INAF)
Université Laval,
Quebec City Canada

Luis Mojica
Tecnologia Alimentaria, Centro de Investigación y Asistencia en Technologia y Diseño del Estado de Jalisco
A.C., CIATEJ, Zapopan
Jalisco, Mexico

Jose Edgar Zapata Montoya
Universidad de Antioquin (UdeA)
Medellin
Antioquia, Colombia

Sendar Daniel Nery Flores
School of Chemistry
Autonomous University of Coahuila
Saltillo, Coahuila, Mexico

Menandes Alves de Souza Neto
School of Agronomy
Federal University of Goiás (UFG)
Goiânia, Brazil

B. Dave Oomah
(Retired) Formerly with Summerland
Research and Development Centre
Agriculture and Agri-Food Canada
Summerland, BC, Canada

Johanny Aguillón Osma
Faculty of Exact and Applied Sciences
Instituto Tecnológico Metropolitano
Medellin, Colombia

Sandra Pacios-Michelena
Food Research Department
School of Chemistry
Universidad Autónoma de Coahuila
Saltillo, Mexico

A. Sócrates Palacios-Ponce
Facultad de Ingenieria en Mecánica y
Ciencias de la Producción
Escuela Superior Politécnica del
Litoral (ESPOL)
Polytechnic University
Guayaquil, Ecuador

Lissethe Palomo-Ligas
School of Chemistry
Autonomous University of Coahuila
Saltillo, Coahuila, Mexico

Shazia Parveen
Chemistry Department
Faculty of Science
Taibah University
Yanbu, Saudi Arabia

Claudia Mariana Pérez-Juárez
School of Chemistry
Autonomous University of Coahuila
Saltillo, Coahuila, Mexico

Yousef Rasmi
Department of Biochemistry
Faculty of Medicine & Cellular and
Molecular Research Center
Urmia University of Medical Sciences,
Urmia, Iran

Miguel Rebollo-Hernanz
Department of Agricultural Chemistry and
Food Science
Universidad Autónoma de Madrid
Madrid, Spain

Mario E. Rodríguez-García
Departamento de Nanotechnologia,
Centro de Fisica Aplicada y Technologia
Avanzada
Universidad Nacional Autónoma de
México (UNAM)-Campus Juriquilla,
Queretaro, Mexico

Raúl Rodriguez-Herrera
School of Chemistry
Autonomous University of Coahuila
Saltillo, Coahuila, Mexico

Elihud Salazar
Centro Universitario de la Costa
University of Guadalajara
Puerto Vallarta, Mexico

Maria del Pilar Salazar
Universidad de Antioquin (UdeA)
Medellín
Antioquia, Colombia

Salvador A. Saldaña-Mendoza
Food Research Department
School of Chemistry
Universidad Autónoma de Coahuila
Saltillo, Mexico

Oscar Abel Sánchez-Velázquez
Facultad de Ciencias Químico-Biológicas
Universidad Autónoma de Sinaloa
Culiacan, Mexico

Olga Lucia Torres-Vargas
Research Group on Agroindustrial Sciences
Interdisciplinary Science Institute
Universidad del Quindio
Armenia, Colombia

Maria Lilibeth Manzanilla Valdez
School of Food Science and Nutrition
University of Leeds
Leeds, UK

Leidy Johana Valencia-Hernández
Food Research Department
School of Chemistry
Universidad Autónoma de Coahuila
Saltillo, Mexico

Montserrat Alcazar-Valle
Tecnologia Alimentaria
Centro de Investigación y Asistencia en
Technologia y Diseño del Estado de Jalisco
A.C., CIATEJ, Zapopan
Jalisco, Mexico

Abraham Wall-Medrano
Instituto de Ciencias Biomédicas
Universidad Autónoma de Ciudad Juárez
Chihuahua, Mexico

Yvonne V. Yuan
School of Nutrition
Toronto Metropolitan University,
Toronto, Ontario, Canada

Diego Fernando Uribe Yunda
School of Nutrition and Dietetics
University of Antioquia
Medellín, Colombia

Preface

The global functional food ingredients market valued at US $98.9 billion in 2021 is forecast to reach US $137.1 billion by 2026 due to increased consumption of nutritive convenience and fortified foods according to markets. The key functional food ingredients include amino acids, antioxidants, carotenoids, hydrocolloids, minerals, plant extracts, and prebiotics. However, plant sources of these food ingredients (proteins, omega-3 fatty acids, fibers, phytochemicals, and plant extracts) is the fastest growing segment because of their low carbon footprint, increasing consumer awareness of their health benefits and industry drive toward complete utilization to reduce by-product waste. The SARS-CoV-2 (COVID-19) pandemic and current boom in functional foods have led to the development of entirely novel ingredient sectors including probiotic and prebiotic components, foods that enhance immunological tolerance, improve physical and mental well-being, and protect against oxidative stress and chronic diseases, among other health benefits.

The first chapter illustrates the importance of phenolic compounds in improving human health by exerting considerable antiglycation potential to reduce disease progression (particularly diabetes and ageing) by different mechanisms. Many antiglycating phenolic compounds can be recovered from industrial by-products (e.g. citrus residues or other fruit pulps, grape marc, and peanut processing), millets, algaes, cinnamon, and herbal plants, providing a roadmap for further large-scale development of natural products to reduce the risk and/or prevent various pathologies.

In a global consumer survey, 'mental wellbeing' ranked top of health concerns according to Euromonitor market analysis. The survey also revealed that 57% of global respondents suffer at least 'moderate' levels of stress, suggesting an opportunity for innovation. Psychobiotics can be considered one of the potential consumer products that can alleviate stress. Psychobiotics are a class of probiotics that, when ingested in adequate amounts, induce health benefits in patients with psychiatric illness. Some psychobiotics, for example, *Lactobacillus plantarum* PS128™, reduce anxiety and depression and improve visceral hypersensitivity, exerting their psychotropic effects by altering monoamine neurotransmitter levels and by modulating glucocorticoid, anti-inflammatory and antioxidative responses, and sleep structure among highly stressed individuals. Chapter 2 highlights two main mechanisms of action for potential psychobiotic activity. The first is the production of gamma-aminobutyric acid (GABA), the major inhibitory neurotransmitter in the human central nervous system, and the second is modulation of host serotonin production. The benefits of psychobiotic interventions are also discussed, along with limitations in this research area. The South Korean Health Functional Food has commercialized a probiotic mixture NV-1703 (*Lactobacillus reuteri* NK33 and *Bifidobacterium adolescentis* NK98) representing an example of a good psychobiotic. NV-1703 improved sleep quality (significant insomnia reduction) and gut microbiota and decreased depressive and anxiety symptoms in healthy adults in a recent randomized, double-blind, placebo-controlled trial.

Sleep disturbances are a common concern that has been further accelerated with the COVID-19 pandemic. Natural circadian rhythm is adversely impacted by chronic sleep disturbances, particularly in the world's most sleep-deprived countries, Japan and China, where 41% of adults surveyed were unsatisfied with their sleep quality. Impaired sleep has been associated with several health consequences, including cardiovascular disease, chronic pain, neurodegenerative and cognitive disorders, depression, obesity, and all-cause mortality. Sleep quality is influenced by eating patterns, diets, and food components (amino acids, fiber, polyphenolics, lipids, and alkaloids) that affect circadian rhythm. Chronobiotics (Chapter 3) is an emerging discipline with clinical and therapeutic potential to regulate circadian rhythm, modify dietary patterns, and promote chronobiotics intake. The chapter explores chronobiotics as functional foods, particularly natural products/compounds that synchronize and restore the circadian clocks and mechanisms that control clock-controlled genes preventing circadian misalignment and associated disorders. Ongoing research shows that *in vitro* gastrointestinal digestion of pistachio increases phyto-melatonin bioaccessibility and colonic metabolites (i.e. short-chain fatty acids [SCFAs]) biosynthesis with chronobiotic potential to mitigate chronodisruption associated diseases.

The common bean (*Phaseolus vulgaris* L.), one of the most widely cultivated legumes, forms part of the human daily diet in the Americas, Europe, Africa, and Asia. Two consecutive chapters (Chapters 4 and 5) demonstrate its health benefits; the first is related to bioactive components that modulate markers of noncommunicable diseases, particularly type 2 diabetes and obesity. These bioactive compounds (carbohydrates, carotenoids, dietary fiber, phytosterols, polyphenolics, proteins and peptides, saponins, and tocopherols) are associated with the antidiabetic and antiobesity potential of common beans. Carioca is the most favored bean type by consumers and growers in Brazil due to indigenous influence and early cultivation by Brazilian natives. Its evolution in the late 1960s restructured the Brazilian bean market due to its high productivity in low fertility soils, disease resistance, and fast cooking time, dominating the market to date. This is, to our knowledge, the first chapter focusing on carioca beans as functional foods and/ingredients. Chapter 5 describes the multitude of bioactive compounds present in carioca beans (peptides, phenolic compounds including anthocyanins, phytate, saponins, complex carbohydrates with resistant starch as prebiotics, minerals, and vitamins) responsible for its myriad health benefits. Carioca bean peptides, previously known only for their antioxidant and dipeptidyl peptidase-IV inhibition, are now considered multifunctional (i.e. stomach mucosal membrane regulation, ubiquitin-mediated proteolysis, neuroprotection, and memory enhancement and satiety [appropriate for weight management]). The modulation of gut bacteria by carioca bean is of prime importance, particularly in regard to studies demonstrating increased *in vitro* iron bioavailability of Fe-biofortified compared to Fe-standard carioca beans. These gut microbiome changes are primarily due to increased abundance of bacteria associated with phenolic catabolism and beneficial SCFAs-producing bacteria. The first Brazilian plant-based product uses carioca bean as an ingredient in vegetable chicken (shredded, strips, and cubes) extruded together with soy and pea proteins. Carioca beans as a cost-competitive abundant raw material has a small environmental footprint with regional socioeconomic impact. Current research is aimed at producing carioca bean protein isolates and concentrates to formulate alternative "meat-like" convenience products (e.g. hamburgers and meatballs) and promote further meat reduction consumption (~50%) in Brazil. Plant-based products made with carioca bean protein are expected to provide quality, diversity, and accessibility of alternative protein sources to the Brazilian consumer.

The pandemic exacerbated food insecurity, highlighting the natural susceptibility to conventional farming that promoted local production by fostering geographical diversification. Sustainable, regenerative, and resilient farming systems and practices require biodiversity restoration simultaneously limiting water use. The investment in climate-smart plants to integrate them and their products in the functional food sector is illustrated in two subsequent chapters. The bioactive compounds of the Mexican semi-desert area endemic plants such as *Agave*, *Opuntia*, *Prosopis*, *Jatropha*, and others (Chapter 6) are described along with their physiological benefits, which can be exploited for functional food and nutraceutical ingredients. In addition to their traditional use, these plants and their bioactives offer health benefits such as antibacterial, anticancer, antioxidant, antiparasitic, hepatoprotective, and neurological effects, and vasodilator activities. *Opuntia*, one of these endemic plants, and its physiological health benefits is the focus of Chapter 7, since it grows either wild or cultivated in many countries. The antidiabetic activity of *Opuntia* cladode is of special interest to functional foods and nutraceuticals development based on results from clinical trials as well as its prebiotic effects in improving gut health. Recent studies have demonstrated the wound-healing, hypocholesterolemic, and anti-inflammatory effects of *Opuntia*, with potential applications in functional foods and other health-related industries. Interestingly, many *Opuntia*-derived products are already in the market.

Chapter 8 is dedicated to edible algae due to its chronic disease risk reduction in Asia and the Pacific where it is part of the daily diet that has spread (or taken root) in Western cultures. Several studies indicate the bioactive roles of edible macro- and microalgal constituents as functional foods and nutraceuticals. The biological and molecular mechanisms ascribed to these bioactives include anticarcinogenic, antidiabetic, antioxidant, antiviral, and cardioprotective effects as well as modulation of blood glucose and lipoprotein profiles. The food industry has benefited from algal research and introduced several products such as the docosahexaenoic acid (DHA)-enriched European Union approved algal oil, protein with meat-like properties, and antiviral products.

Hempseed (Chapter 9), an old world crop, is being recognized as "green, sustainable and environmentally friendly" due to its agronomic characteristics. However, regulation regarding cannabidiol (CBD) limits its production in many parts of the world, although industrial hempseed is distinct from other *Cannabis* (marihuana or drug type hemp). New regulation now enables production of industrial hemp (low tetrahydrocannabinol [THC]) that is already being considered for animal feed. In spite of this setback, research on industrial hemp has revealed its potential for human health benefits due to the high quality of its oil and the multifunctionality of its protein and peptides, as well as the lignanamides present in only a few crops. The chapter delves into detailed description of the composition of hempseed and its products (industrial coproducts), followed by extensive cellular, animal, and human studies of hempseed and/or its components. Interestingly, hempseed oil, protein, and lignanamides share a common mechanism of action in providing human health benefits. Notable are the convergence of studies demonstrating that several hempseed constituents can suppress or alleviate severe acute respiratory syndrome coronavirus-2 (SARS-CoV-2).

We have dedicated three chapters to coffee, one of the highest globally traded food commodities. Chapter 10 reviews coffee's bioactive compounds (cafestol, caffeine, chlorogenic acids, kahweol, sterols, and trigonelline), their extraction, bioactivity, and mechanisms of their physiological health benefits. The chapter also describes commercial new coffee-based food products, nutraceuticals, and supplements particularly from by-products and coproducts for complete circular sustainability of the coffee industry. Chapter 11 deals exclusively with coffee as an alternative protein source to diversify, upcycle, reduce environmental impact, and enhance sustainability of the coffee industry. Therefore, characteristics, quality, and peptides and biological parameters have been reviewed for proteins extracted from coffee by-products, including cascara, green and defective coffee beans, silverskin, and spent coffee grounds. Some coffee protein products are already commercially available to formulate foods for human consumption. Previous studies and reviews have focused on coffee proteins mostly based on their association with coffee flavor, aroma, and quality. Chapter 12 reviews the molecular mechanisms of action of coffee on human health along with those of ginger and cinnamon. These mechanisms involve the prevention and/or regulation of chronic noncommunicable diseases associated with the antidiabetic, anti-inflammatory, antioxidant, cardioprotective, and hepatoprotective effects of coffee and the abovementioned spices. The regulation and/or modulation of these mechanisms are ascribed to the major bioactive compounds of these natural products: caffeine and gallic acid from coffee; gingerols, shogaols, and paradols from ginger; and cinnamaldehyde, cinnamate, and cinnamic acid from cinnamon.

The last section of this book is dedicated to emerging (chontaduro, saffron, and cocoa shell) or established (pseudocereals and berries) crops or products. These emerging natural products, although elevating/promoting regional biodiversity, have previously not been considered as functional foods, presumably due to lack of investigations and/or interest. Chontaduro (*Bactris gasipaes* Kunth) or peach palm (Chapter 13), consumed mostly in Latin America, is a rich source of bioactive compounds (dietary fiber, carotenoids, and vitamins): like many Amazonian fruits, exhibiting antioxidant activity with health benefits. Chontaduro polysaccharides have high bifidogenic effect and its protein is highly digestible; the antioxidant and therapeutic effects are ascribed to bioavailable carotenoids with high retinol activity equivalents. Interest in saffron bioactives has increased recently because of their potential preventive and therapeutic applications due to their numerous and varied activities (antioxidant, anti-inflammatory, anticarcinogenic, antigenotoxic, antimutagenic, antitumoural, cytotoxic, antihypertensive, antiatherogenic, antidiabetic, antihyperlipidemic, anticonvulsive, antidepressive, and neuroprotective). Chapter 14 describes the bioactivity of the three main saffron compounds: crocetin/crocins, picrocrocin, and safranal; saffron's beneficial effects have been demonstrated in attenuating depression by regulating the endocannabinoid system in adults with major depressive disorders. Furthermore, clinical trials demonstrate the neuroprotective effects of saffron. The phenolic-rich tepals, saffron's main by-product, exert hepatoprotective effects and improve renal function. Another industrial by-product, cocoa shell (Chapter 15) is reviewed in regard to its biological activity and molecular mechanisms associated with the prevention of obesity, diabetes, and cardiovascular disease. These cardiometabolic disorders are attenuated by methylxanthines (theobromine and caffeine), phenolic compounds, and dietary fibers acting interactively or synergistically, thereby modulating various signaling pathways. Consumption of cocoa shell has potential human health benefits including cognitive and anticancer effects mediated by colonic-derived metabolites from its bioactive compounds, according to ongoing research. Cocoa shells are already being used in snacks and its flour for production of muffins and biscuits. Moreover, cocoa shells are a source material for the production of extracts rich in fibers, polyphenols, and antioxidants for extensive applications in the functional food industry.

Pseudocereals (amaranth, buckwheat, chia, and quinoa) fostering regional crop biodiversity are regarded as "super-foods" or alternative grains for celiac patients and/or gluten-free diets. These ancient grains are increasingly popular in Western countries, form part of many diverse food products, and can be used to develop products and/or formulations to treat dysbiosis, celiac disease, lactose intolerance, inflammatory bowel diseases, and inflammation-mediated disorders. Chapter 16 reviews pseudocereals bioactives in regard to *in vitro*, *in vivo*, and clinical data associating biological markers of diseases (cancer, cardiovascular diseases, diabetes, inflammation, and kidney diseases) at the subcellular, cellular, organ, and biological system levels. Berries have been the most valuable produce category in Europe due to increased consumption and research promoting the positive and profound impact of its intake on human health. It is interesting to note that two Latin American countries (Peru and Chile) are now major exporters of blueberries to Europe and other countries. The wide range of berry phytochemicals regulates the activities of metabolizing enzymes, modulates gene expression and signaling pathways, and repairs DNA oxidative damage. Chapter 17 recapitulates the multimechanistic actions of berry phytochemicals based on their important bioactivities (anti-inflammatory, antidiabetic, antioxidant, cardioprotective, neuroprotective, anticancer properties). The chapter also includes studies demonstrating human health benefits of berry consumption to those with metabolic diseases and reducing type 2 diabetic risk in older populations.

This book provides useful up-to-date information to researchers, nutritionists, health, and industry professionals to advance the field of functional foods based on their mechanism of actions.

Rocio Campos-Vega
B. Dave Oomah

Part I

Functional Food Bioactives

1

Polyphenolic Compounds Mechanisms as Inhibitors of Advanced Glycation End Products and Their Relationship to Health and Disease

Vanize Martins Genova, Alessandra Gambero, Pedro de Souza Freitas Campos, and Gabriela A. Macedo

Bioprocesses Laboratory, Food Science and Nutrition Department, School of Food Engineering, University of Campinas, UNICAMP, Campinas, SP, Brazil

1.1 Introduction

The glycation reaction occurs when an amino group of a protein reacts with a carbonyl group of a reducing sugar, which through several cascades forms glycated proteins, which can undergo oxidation generating advanced glycation end products (AGEs) (Ahmed, 2005). These highly toxic products can be formed both endogenously and exogenously causing health problems.

Natural products have been used since antiquity by different societies to fight diseases. Currently, this field presents a promising path to develop formulations based on natural products on a large scale, through biotechnology (Sun et al., 2020), in order to provide sources of treatment and/or prevention for various pathologies through a product, as a functional ingredient or as a natural intervention (e.g. supplements). Phenolic compounds are widely studied for their high antioxidant capacity; these compounds are metabolites that have a regulatory role and response to infections in plants. Their structure consists of a benzene ring attached to at least one hydroxyl group (Moreno & Peinado, 2012), and its antiglycant capacity has been based on a possible direct relationship with its chemical structures (Xie & Chen, 2013).

Thus, phenolic compounds have demonstrated to be efficient as different inhibition mechanisms in AGEs formation and have been shown to be an intervention of great importance for health promotion and reducing complications and/or disease progression.

1.2 Glycation Reaction and Formation of Advanced Glycation End Products Formation

1.2.1 Advanced Glycation End Products—AGEs

AGEs are the final components of a reaction that takes place both *in vivo* (defined as an endogenous glycation reaction) and in food preparation processes (which can be defined as an exogenous Maillard reaction). In the human body, it occurs in the presence of proteins with a

Molecular Mechanisms of Functional Food, First Edition. Edited by Rocio Campos-Vega and B. Dave Oomah.

sugar reducer during the natural life process, common mostly in people with diabetes mellitus where high blood glucose levels induce this reaction. In foods, they are formed through the Maillard reaction, defined mainly as acrylamides ingested as a component of the diet.

The Maillard reaction, also known as the nonenzymatic browning reaction, consists of a series of reactions that modify color, aroma, and nutritional properties of foods. It begins with the formation of the Schiff base from the condensation via nucleophilic addition of carbonyls of reducing sugars with amino groups of lipids, nucleic acids, free amino acids, or their residues in peptides, aminophospholipids, or proteins (Jakus & Rietbrock, 2004). The Schiff base undergoes dehydration and rearrangements to a more stable and reversible form, called Amadori products. Amadori products, in turn, undergo dehydration, oxidation, and rearrangements, generating AGEs/acrylamides, or oxidize or hydrolyze to form carbonyl compounds. In addition to reducing sugars, other agents and/or processes produce the reactive carbonyl compounds known as α-dicarbonyls or oxoaldehydes, such as the intermediates of the glycolytic pathway, the oxidation products of lipids, amino acids, and ascorbic acid by fragmenting phosphate trioses and catabolizing threonine and acetone (Monnier, 2003).

The α-dicarbonyls have potential for glycation and react with ε-amino groups of lysine, cysteine (cys) thiol (-SH), arginine guanidino, and N-terminal amino acids, leading to the formation of AGEs that have high thermal stability and, therefore, constitute the final phase of the Maillard reaction (Akillioglu & Gokmen, 2016; Barbosa, Souza, Santana, & Goulart, 2016; Wu, Huang, Lin, & Yen, 2011). As seen at the figure 1.1.

AGEs can have harmful health effects when accumulated in the body. These effects arise from their interactions with proteins and/or cellular receptors, which lead to increased expression of inflammatory mediators, morpho-functional changes, and oxidative stress (Ahmed, 2005; Bierhaus, Hofmann, & Nawroth, 1998). Thus, the glycation process is the main spontaneous cause of damage to cellular and extracellular proteins, mainly of long half-life, with the loss of essential amino acids and formation of substances potentially harmful to health (Paul & Bailey, 1999; Thornalley et al., 2003).

Studies have shown an association between the formation and accumulation of AGEs with the progression of diseases such as diabetes mellitus, Alzheimer and Parkinson's

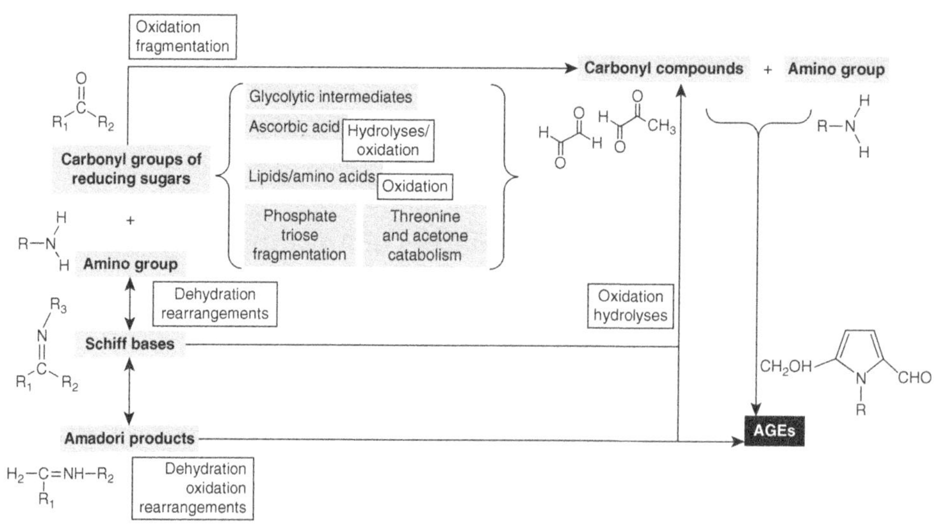

Figure 1.1 Overview of of AGEs formation. *Source:* Adapted from Melo (2018).

(Li, Liu, Lu, & Zhang, 2012), atherosclerosis (Wang et al., 2012), renal insufficiency (Kalousova et al., 2006), neuropathy, retinopathy and cataracts (Ahmed, 2005), rheumatoid arthritis (Degroot, 2001), and liver cirrhosis (Sebekova, Kupcová, Schinzel, & Heiland, 2002), in addition to the aging process (Uribarri et al., 2007), development and/or complications related to aging, various metabolic and neoplastic diseases, among others (Ahmed, 2005).

AGEs are, therefore, a heterogeneous class of molecules that can be classified into two large groups based on the fluorescence property and crosslinking structure: (a) AGEs with fluorescence property and crosslinking structure and (b) AGEs without fluorescence property and without crosslinking structure (Wu, Huang, Lin, & Yen, 2011) and in six different classes, according to their origin (Takeuchi & Yamagishi, 2008) (Figure 1.2).

The reactive carbonyl compounds precursor to AGEs, called glyoxal (GO), methylglyoxal (MGO), and 3-deoxyglucosone (3-DG), are represented in Figure 1.2. GO and MGO present glycation potential around 200 times higher than glucose (Zeng, Dunlop, Rodgers, & Davies, 2006).

As explained, the deleterious effects on health are due to both the endogenous formation and exogenous sources of AGEs, as well as through a diet rich in these products (Koschinsky et al., 1997; Vlassara et al., 2002) and smoking. Studies show that the total AGEs consumed in common Western diet is greater than the amount present in biological systems, such as plasma and tissues (Henle, 2003). The urinary excretion of AGEs also demonstrates that 90% of pyrralin

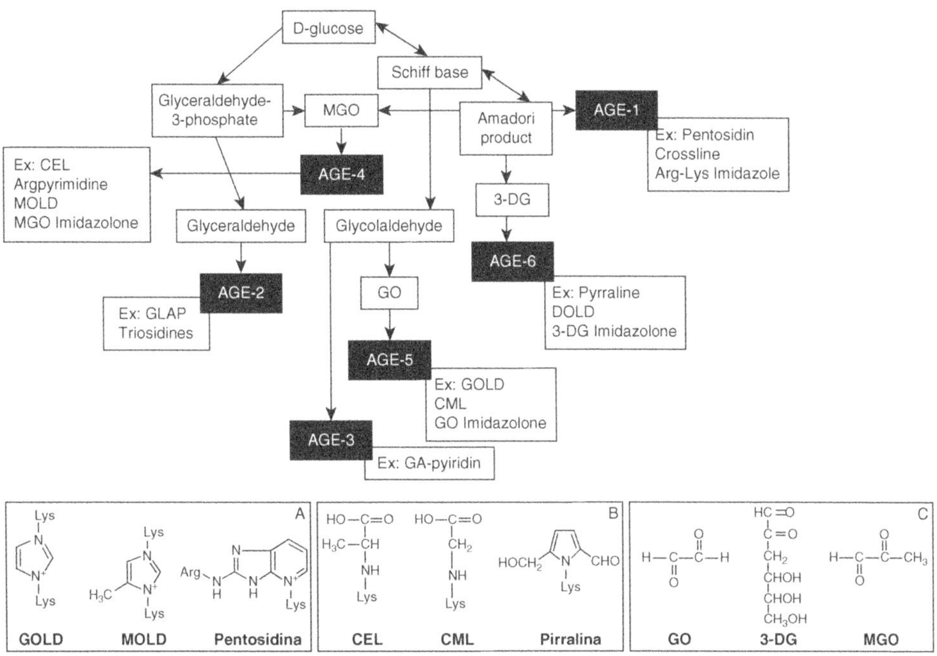

Figure 1.2 Origin of AGEs classification. AGEs with fluorescence properties and crosslinking structure (A); AGEs without fluorescence properties and without crosslinking structure (B); precursors in the formation of AGEs (C). Abbreviations: AGE-1: glucose-derived AGEs; AGE-2: glyceraldehyde derivatives; AGE-3: alpha-dicarbonyl derivatives; AGE-4: methyl glyoxal derivatives; AGE-5: glyoxal derivatives; AGE-6: 3-deoxyglucosone derivatives. AGE: advanced glycation end products; GOLD: glyoxal-lysine dimer; MOLD: methyl glyoxal-lysine dimer; DOLD: lysine dimer and 3-deoxyglucosone; GA: glycoaldehyde; CML: carboxymethyllsine; CEL: carboxyethyllisine; Arg-Lys: arginine-lysine; GLAP: pyridinium derived from glyceraldehyde; GO: glyoxal; DG: deoxyglucosone; 3-DG: 3-deoxyglucosone. *Source:* Adapted from Melo (2018).

and fructoselysine are from dietary origin (Förster, Kühne, & Henle, 2005). The high chronic consumption of AGEs correlates with oxidative stress markers, which can overlap the body's antioxidant and detoxifying capacity, contributing to the pathophysiology of several diseases.

The content of AGEs in food is influenced by the distribution of macronutrients, mainly proteins and fats, in addition to numerous factors such as pH, cooking time, and temperature, the relationship between amino acid and carbohydrate, protein origin, carbohydrate reduction properties (mono-, di-, oligo-, and polysaccharides), and glycoconjugate structure (Sanmartin, Arboleya, Villamiel, & Moreno, 2009). The binomial time and temperature of food processing contribute to a greater or lesser AGEs production, since cooking at high temperatures for extended period in limited water enables greater formation of these products (Uribarri et al., 2010). Among the different processing methods, those that showed the highest AGEs formation in decreasing order are grilling (225 °C), frying (177 °C), roasting (177 °C), and boiling (100 °C) (Goldberg et al., 2004).

Glycation also results in physical–chemical modifications of food proteins, directly influencing their functional properties and molecular weight (Liu, Ru, & Ding, 2012). These modifications include increased thermal stability (Sato, Sawabe, & Saeki, 2005), sparkling properties (Chobert, Gaudin, Dalgalarrondo, & Haertlé, 2006), and antioxidant activity (Lertittikul, Benjakul, & Tanaka, 2007), improved texture and emulsifying properties, and antimicrobial or bactericidal activity (Diftis & Kiosseoglou, 2004; Gerrard, Brown, & Fayle, 2003).

AGEs content is generally high in red meats, butter, cheese, margarine, mayonnaise, oils, nuts, tofu, fish, and eggs, but relatively low in carbohydrates-rich foods particularly processed items, grains, milk, vegetables, starchy vegetables, and breads (Goldberg et al., 2004). Henle (2003) estimated, using chromatographic methods, the ingestion of glucotoxins in a conventional diet (containing heated milk, confectionery, and coffee) at around 1,500–4,000 $\mu mol\, day^{-1}$ of derivatives from the Maillard reaction and around 100 to 300 $\mu mol\, day^{-1}$ of AGEs [such as pyrralin and carboxymethyllysine (CML)]. AGEs consumption has been estimated at approximately $16{,}000 \pm 5{,}000$ $kU\, day^{-1}$ based on the analysis of the 3-day food record of 90 healthy North American individuals. However, information about the quantity and presence of reactive carbonyl compounds in foods is scarce due to the high reactivity and the possible polymerization or formation of adducts with other components during analyses (Uribarri et al., 2015).

Pyrraline and pentosidine AGEs have the highest absorption rate (60–80%). Most of the absorbed AGEs are simple amino acids, low-molecular-weight peptides, or high-molecular-weight compounds. Nε- CML is absorbed by passive diffusion, while pyrraline is absorbed into intestinal tissue via peptide transport. The path of protein digestion is the first step toward AGEs absorption, but the proteins in AGEs are modified and there is an impediment to the natural process to digestion and absorption. After digestion, fructoselysine and CML are associated with low-molecular-weight peptides (less than 1000 Da) making them easily absorbed. The absorption and assimilation of Nε-1-carboxyethyllysine (CEL) and CML were studied in rats by intravenous administration of these glycated compounds (Bergmann et al., 2001). AGEs reached and accumulated in the liver temporarily.

1.2.2 Detoxification Process

In addition to the metabolism itself, some resources such as drugs and food can also be used to protect cells and systems against glycation damage. Chemically, AGEs are products derived from the condensation reaction between the aldehyde group of sugars and the amino group of proteins. This reaction is more efficient and intense at pH 7.0 (Figure 1.2). The mechanisms of detoxification that begin after the aldimines formed at the initial phase are rearranged in covalent bonds to generate Amadori products. Up to this point, the reactions are reversible, stage when drugs or phytotherapies can act in order to prevent AGEs formation. After this phase, the numerous chemical rearrangements that Amadori compounds undergo to form AGEs are irreversible reactions (Figure 1.2). Several intermediate products are formed in this reaction cascade with different physiological effects. AGEs can be found in food in molecular forms as follows: CML, CEL, pyrraline, GO, MGO, acrylamide, furan, and derivatives of bis(lysyl)imidazolium, such as deoxyglucosone-derived lysine dimer[1,3-di(Nε-lysino)-4(2,3,4-trihydroxybutyl)-imidazolium salt] (DOLD) and GO-derived lysine dimer[1,3-di(Nε-lysino) imidazolium salt] (GOLD).

Inhibitory mechanisms of formation and/or accumulation of advanced glycation products *in vivo* include reducing AGEs absorption, inhibiting Amadori products formation, decreasing carbonyl stress, preventing the progression of Amadori products to AGEs, and interrupting biochemical pathways and detoxification of dicarbonyl intermediates (Huebschmann, Regensteiner, Vlassara, & Reusch, 2006). The inhibitory mechanisms used are generally based on blocking the attack of proteins by carbonyls, preventing oxidative stress, or breaking crosslinks formed by AGEs (Peng, Ma, Chen, & Wang, 2011).

AGEs formed in tissues can be degraded by proteolysis or phagocytosed by macrophages mediated by scavenger receptors, which release them as small, soluble peptides excreted by the kidneys (Monnier, 2003). In addition, several systems are in place to limit tissue damage caused by glycation, which include reducing agents (glutathione), antioxidant and enzymatic pathways, and detoxifying systems (glyoxalase, aldose reductase, and aldehyde dehydrogenase) (Shinohara et al., 1998). The glyoxalase system present in the cytosol of all mammalian cells acts in the presence of glutathione, converting GO to glycolate and MGO to D-lactate. This system uses the enzymes glyoxylase I and glyoxylase II, in addition to glutathione (Shangari & O'brien, 2004). Antioxidant pathways possibly act by attenuating oxidative stress by chelating metal ions and reducing free radicals (Peng et al., 2011).

1.2.3 Mechanisms of Detoxification and Intracellular Accumulation

1.2.3.1 Receptor for Advanced Glycation End Products (RAGE)

AGEs are capable of inducing cellular responses through signaling at receptors. The detoxification mechanism is related to receptor for advanced glycation end products (RAGE) initially described in 1992 as a transmembrane protein of the immunoglobulin superfamily (Neeper et al., 1992; Schmidt et al., 1992). Subsequently, numerous splicing variants have been described at mRNA and protein level. The detailed structure of the integral protein (full-length RAGE protein), the membrane RAGE (mRAGE), is composed of three immunoglobulin domains: the extracellular one that consists of a type V and two types C (C1 and C2), a transmembrane domain, and a short C terminal cytoplasmic domain (Koch et al., 2010; Park, Adsit, & Boyington, 2010). The analysis of the amino acid composition showed that the V region has the unusual characteristic of containing many residues of arginine and lysine, which gives a positive charge on neutral pH on the VC1 surface.

In contrast, the C2 domain has more acidic residues on the surface and is negatively charged. These characteristics indicate that the extracellular domain of RAGE is composed of two subdomains that reflect on the ligand-binding properties (Fritz, 2011). Most ligands bind to the VC1 domain (Allmen, Koch, Fritz, & Legler, 2008; Ostendorp et al., 2007; Sturchler, Galichet, Weibel, Leclerc, & Heizmann, 2008) with the C1 region as the recognition function of ligand, which has negative charges as standard (Xue et al., 2011), in addition to participating in the stabilization of region V (Dattilo et al., 2007).

The extracellular domain without the cytoplasmic domain and the transmembrane domain is called soluble RAGE (sRAGE), which is found extracellularly in the circulation (Kierdorf & Fritz, 2013), being a product of the cleavage of mRAGE by metalloproteinases ADAM10 (Raucci et al., 2008) and MMP-9 (Zhang et al., 2008). Cleavage (shedding) is observed after the connection of high-mobility group box (HMGB)1 in mRAGE, for example, and represents an elegant model of negative feedback loop that attenuates signaling via RAGE, as sRAGE competes for ligands (Kierdorf & Fritz, 2013). Endogenous secretion RAGE (endogenous secretory RAGE, esRAGE) is formed from splicing of full-length RAGE mRNA capable of AGEs binding without activating intracellular responses (Yonekura et al., 2003). Tissue analyses showed that baseline mRAGE expression is low under physiological conditions in all tissues except the lung (Marinakis, Bagkos, Piperi, Roussou, & Diamanti-Kandarakis, 2014), but it increases in response to aging and various diseases (Chen, Gregersen, & Zhao, 2015; Lue et al., 2001; Lue, Walker, Jacobson, & Sabbagh, 2009).

RAGEs are multi-ligand receptors that participate in numerous pathophysiological processes, including inflammatory diseases, neurodegenerative diseases, and cardiovascular and neoplastic diseases (Neeper et al., 1992; Schmidt et al., 1992). RAGE is activated by AGEs; however, it can also be generated by a wide repertoire of ligands, including β2 integrin/Mac-1, amyloid β-peptides, damage-associated pattern molecules (DAMPs), HMGB1/amphoterin, complement 3a (C3a), heat shock protein 70 (HSP70), and proteins of the S100 group/calgranulins (Fritz, 2011; Hori et al., 1995; Leclerc, Fritz, Vetter, & Heizmann, 2009; Ruan et al., 2010; Somensi et al., 2017; Yan et al., 1996). This leads to the recruitment of a formalin molecule (formin molecule), mammalian diaphanous-1, or DIAPH1 by the C-terminal cytoplasmic domain (Hudson et al., 2008). Other adapter molecules have also been reported such as ERK-1/2 (Ishihara, Tsutsumi, Kawane, Nakajima, & Kasaoka, 2003), TIRAP and MYD88 (Sakaguchi et al., 2011), the latter being adapter proteins used by members of toll-like receptor (TLR) family, as well as RAGE can be phosphorylated by PKC determining the activation of important signaling pathways such as Ras/MAPK/NF-κB, JAK-2/STAT-1, and Rac1/Cdc42 (Huang et al., 2001; Hudson et al., 2008; Lander et al., 1997) varying according to the ligand, cell type, and tissue microenvironment. The activation of these signaling pathways results in ROS production, the generation of pro-inflammatory mediators such as interleukin (IL-6) and tumor necrosis factor (TNF)-α (Yamamoto et al., 2011), and the generation of growth factors and profibrotic mediators such as tumor growth factor (TGF)-β (Iwamoto et al., 2008). Additionally, RAGE does not act in isolation and can form heterodimers with members of the TLR family, particularly with TLR4 and TLR2 as well as sharing ligands in common with them, such as HMGB1 (Qin et al., 2009).

In contrast to the findings, reductions in the circulating levels of soluble forms of RAGE (sRAGE) have been described in some diseases, such as hypertension (Geroldi et al., 2005), coronary artery diseases (Falcone et al., 2005), chronic obstructive pulmonary disease (Smith et al., 2011), rheumatic arthritis (Pullerits, Bokarewa, Dahlberg, & Tarkowski, 2005), and Alzheimer's disease (Emanuele et al., 2005). However, circulating sRAGE levels are high in diabetes (Challier, Jacqueminet, Benabdesselam, Grimaldi, & Beaudeux, 2005; Fujisawa et al., 2013) and in advanced kidney disease (Kalousova et al., 2006), providing evidence for potential use of sRAGE as a biomarker for AGEs-associated diseases.

Phenolic compounds, in order to promote health through their antioxidant capacity, can inhibit the glycation process by two general mechanisms: indirectly through oxidative stress reduction (Lunceford & Gugliucci, 2005) and directly through the interception of reactive carbonyl species, inhibition of AGE precursor formation and reduction of RAGE expression (Khangholi, Majid, Berwary, Ahmad, & Aziz, 2016), reduction of AGE absorption, inhibition of the formation of Amadori products, reduction of carbonyl stress, preventing the progression of Amadori products to AGEs, interrupting biochemical pathways, and detoxifying dicarbonyl intermediates (Huebschmann et al., 2006). They can also induce antioxidant enzymes, such as glutathione peroxidase, catalase, and superoxide dismutase, inhibit the expression of enzymes such as xanthine oxidase, and modulate pro-inflammatory genes expression such as cyclooxygenase, lipoxygenase, nitric oxide synthase, and various cytokines, acting mainly through the inhibition of NF-kB-activated protein kinase signaling (Ahmed et al., 2002; Bhardwaj et al., 2007).

1.2.4 Polyphenolic Compounds Mechanisms as Antiglycation Agents: Proposed Mechanisms

1.2.4.1 Role of Phenolic Antioxidant and Antiglycation Capacity

The antioxidant capacity of a compound can be directly related to its antiglycation ability; compounds with high antioxidant capacity have high antiglycation potential. The antiglycation capacity of a phenolic compound is generally related to its molecular structure, presenting different mechanisms. Some of them can be described as: (a) decrease of hydroxyl glycosylation in flavonoids: decreases antiglycation capacity as the degree varies; (b) 5,7-dihydroxy structure in isoflavones increased antiglycation activity; (c) hydroxylated structure in ring C: compounds with this structure reduces activity, whereas hydroxylation in rings A and B increases antiglycant activity; (d) methylation of stilbenes and hydroxylation in ring B decrease antiglycant activity; (e) methylation of flavonols generally reduce its antiglycating capacity, except for 3-O-methylation; (f) flavones with C2=C3 double-bond hydrogenation reduce capacity; (g) ellagitannins and ellagic acids have good antiglycant capacity that increases in the presence of hydroxylation or methylation; (h) catechins decrease antiglycant activity in relation to proanthocyanidin trimers or dimers; (i) anthraquinone with metastructure or ortho dihydroxyl in the benzene ring exhibit high antiglycation capacity, as well as phenolic acids with multiple hydroxyls; and (j) catechins with gallo groups showed antiglycant activity, just as hydroxyl at C-3 was much more effective than the β-hydroxyl group at C-3 (Xie & Chen, 2013).

Phenolic compounds can be considered as antiglycation agents, since studies have been elucidating their mechanisms as AGE inhibitors. Some compounds show simple mechanisms, and others have more advanced mechanisms attaining greater degree of antiglycation potential. Briefly, these mechanisms are as follows: (a) interruption of the Maillard reaction, interfering in hydroxymethylfurfural (HMF) formation, complexing intermediates, and thus controlling AGEs formation (black/green pepper extract) (Favre et al., 2020); (b) reduction of carbonyl stress, blood glucose levels, RAGE expression, and AGE accumulation in rats by epigallocatechin-3-gallate (green tea *Camellia sinensis* extract) (Sampath, Rashid, Sang, & Ahmedna, 2017); (c) albumin protection, restricting post-glycation and bovine serum albumin (BSA) fibrillation, masking arginine and lysine residues, reducing the formation of fluorescent AGEs (vanilla extract, with vanillin as the main phenolic) (Awasthi & Saraswathi, 2016); (d) MGO capture and blocking cell damage (Chilean gooseberry extract, rich in anthocyanins, hydroxycinnamic acid, and flavonoids) (Jiménez-Aspee et al., 2016); (e) MGO capture forming mono- and di-adducts, decreasing the formation of fluorescent AGEs (quercetin-rich Chi *Camellia nitidissima*) (Wang et al., 2016); (f) antioxidant action reducing the formation of fluorescent AGEs (of phlorotannins-rich brown and red algae extract) (Kuda et al., 2016); (g) reduction of RAGE activity and formation of AGEs and fluorescent AGEs and *in vitro* RAGE activity (chlorogenic and 3,5-di-*O*-caffeoylquinic acids-rich *Aster koraiensis* extract) (Kim, Jo, Lee, Kim, & Kim, 2016); (h) multiple antioxidant action, reducing protein glycation and eliminating carbonyl intermediates (*p*-coumaric and chlorogenic acids-rich barnyard millet *Echinochloa frumentacea* extract) (Anis & Sreerama 2020); (i) GO and MGO capture, inhibiting CML formation (corn silk extract, rich in robinin, procyanidins, luteolin, and quercetin) (Zhang, Wang, & Liu, 2020); (i) glycosylation glycation and BSA inhibition (5-*O*-caffeoylquinic *Vaccinium cylindraceum SM* and *Vaccinium padifolium* extracts) (Spínola 2018); (j) BSA protection against charge modification and crosslinking reactions in the advanced stage (skin extract from white grapes rich in quercetin, kaempferol, and heterosides) (Harsha, Lavelli, & Scarafoni, 2014); (k) elimination of ROS and MGO capture by conjugation reaction and delay in carbohydrate-hydrolyzing enzymes (α-glucosidase and α-amylase) activities (resveratrol) (Shen, Xu, & Sheng, 2017); (l) inhibition of AGE accumulation in plasma, liver, adipose tissue, and kidney and expression about three times (epigallocatechin-3-gallate) (Sampath et al., 2017); (m) MGO capture during milk storage (naringenin) (Zhu, Poojary, Andersen, & Lund, 2019); (n) MGO reaction forming adducts that inhibit inflammation and ROS production, demonstrating high capacity to eliminate MGO in human umbilical vein endothelial cells (HUVECs) through the negative regulation of p-ERK 1-2, RAGE, and NADPH oxidase activity of NF-κB expression (apigenin [4′, 5,7-trihydroxyflavone]) (Zhou, Cheng, Gong, Li, & Wang, 2019); (o) MGO capture through the antioxidant, antioxidative, and anti-carbonyl capacities of the mono-MGO adduct of PG (MM-PG) generated during pork roasting (propyl gallate) (Hengqing et al., 2018); and (p) suppressing activities of enzymes involved in carbohydrates metabolism and insulin resistance reduction (Farias et al., 2021). The data discussed in this section are summarized in Table 1.1.

1.3 Current and Future Perspectives

Foods such as tea, cinnamon, rosemary, mate, and other herbal plants rich in polyphenols (primarily phenylpropanoids and flavonoids) efficiently trap Maillard reaction intermediates that are active glycating agents. Several aqueous and/or ethanol extracts have antiglycant capacity; these extracts were obtained from different natural sources (*Allium cepa, Allium sativum, Curcuma longa, Mentha piperita, Murraya koenigii* Spreng, *Petroselinum crispum,*

Table 1.1 Mechanisms of antiglycant action of extracts rich in phenolic compounds and pure patterns.

Extract (s)	Main active phenolic compound (s)	Antiglycant mechanism (s)	References
Green/black pepper	Epigallocatechin-3-gallate	Control of the Maillard reaction and the formation of AGEs in food, interfering in HMF formation through active agents of the β-CD complex, as well as complexation of Maillard reaction intermediates interrupting the reaction.	Favre et al. (2020)
Green tea	(*Camellia sinensis*)	In rats: inhibition of AGE accumulation and RAGE expression, activation of nuclear factor erythroid 2-related factor 2 (Nrf2), decreased blood glucose levels, reduced insulin resistance, and increased glutathione ratio in reduced form (GSH)/glutathione in oxidized form (GSSG) that leads to carbonyl stress reduction.	Sampath et al. (2017)
Vanilla	Vanillin	Reduction of the formation of fructosamine and fluorescent AGEs, masking lysine and arginine residues, albumin protection by preventing the conformational transition from its native helical to the beta leaf structure, and restriction of post-glycation aggregation and BSA fibrillation.	Awasthi and Saraswathi (2016)
Chilean gooseberries	Anthocyanins, hydroxycinnamic acids, and flavonoids	Protection against oxidative cell damage induced by hydrogen peroxide and MGO capture.	Jiménez-Aspee et al. (2016)

(Continued)

Table 1.1 (Continued)

Extract (s)	Main active phenolic compound (s)	Antiglycant mechanism (s)	References
Camellia nitidissima Chi	Quercetin	Reduction of the formation of fluorescent AGEs, MGO capture, and formation of mono- and di-MGO adducts.	Wang et al. (2016)
Brown and red algae	Phlorotannins	Reduction of the formation of fluorescent AGEs and performance as antioxidants.	Kuda et al. (2016)
Aster koraiensis	Chlorogenic and 3-5-di-O-caffeoylquinic acids	Reduction of the formation of fluorescent AGEs and RAGE activity *in vitro*. In diabetic rats, reduction of AGE formation and inhibition of NF-kB activation and iNOS expression.	Kim et al. (2016)
Barnyard millet (*Echinochloa frumentacea*)	*p*-Coumaric and chlorogenic acids	Elimination of carbonyl intermediates, decreased protein glycation, in addition to multiple antioxidant activities.	Anis and Sreerama (2020)
Corn silk extract	Robinin, procyanidins, luteolin, and quercetin	Inhibition of the formation of Nε-carboxymethyllysine capturing glyoxal and methylglyoxal in a casein glucose-fatty acid model system.	Zhang et al. (2020)
Vaccinium cylindraceum SM and Vaccinium padifolium	5-*O*-caffeoylquinic	Ability to inhibit glycation of BSA and glycosidases.	Spínola (2018)
Garcinia mangostana L.	Epicatequina, aromadendrina-8-C-glicopiranosídeo, 2,3′,4,5′,6-pentahidroxibenzofenona e garcimangosona	Inhibition of the formation of fluorescent and nonfluorescent AGEs, fructosamine, β-amyloid structure, and protein thiols.	Abdallah et al. (2016)

Pomegranate	Ellagic acid, punicalin, and punicalagin	Reduction of the formation of fluorescent AGEs *in vitro*, glycated hemoglobin, glycoalbumin, and levels of blood lipid peroxidation *in vivo*.	Kumagai et al. (2015)
Chinese olive	Gallic acid, ferulic acid, and rutin	Induced by hydrogen peroxide of the thiol groups in the BSA.	Kuo, Liu, Hsu, Lin, and Chen (2015)
White grapes peel	Quercetin, kaempferol, and heterosides	Reduction of the formation of fluorescent AGEs and protection of BSA against the variation in load that occurs at the beginning of glycation and against crosslinking reactions in the advanced stage.	Harsha, Lavelli and Scarafoni (2014)
Bark extracts of *Cinamomum sp. (Lauraceae)*	Procyanidin B2	Eighty percent of capacity to inhibit glycation by protective effects on glucose and MGO levels.	Mana, Das, Ghosh, and Sil, (2010)
Cladodes polysaccharide aqueous extract	-	Reduction of formation of AGEs and antioxidant activity	
Ethyl acetate extract of ginger	-	Effects on scavenging activity of DPPH; antiglycation activity by reducing glucose levels	Rani, Krishna, Padmakumari, Raghu, & Sundaresan, (2012)
Biotransformed soybean extracts	Daidzein, glycitein, and genistein	Reduction of AGES formation and antioxidant activity.	Genova et al. (2021)
Wild fruits	Anthocyanin	Reduction of the formation of fluorescent AGEs and elimination of free radicals.	Harris et al. (2014)
Naringenin		Ability to capture MGO during milk storage.	Zhu et al. (2019)

(*Continued*)

Table 1.1 (Continued)

Extract (s)	Main active phenolic compound (s)	Antiglycant mechanism (s)	References
Apigenin (4′,5,7-trihydroxyflavone)		High capacity for eliminating MGO in HUVECs (human umbilical vein endothelial cells) by its reaction with MGO-forming adducts that inhibit inflammation and ROS production, through the negative regulation of p-ERK ½, RAGE, NADPH oxidase activity, NF expression-κB; decrease in pro-inflammatory mediators MCP-1 and IL-6; positive regulation of Nrf2 and antioxidant profile.	Zhou, Cheng, Xiao, and Wang (2020))
Propilgalate		Imprisonment of MGO through the antioxidant, antioxidative, and anti-carbonyl capacity of the mono-MGO adduct of PG (MM-PG) generated during pork roast.	Hengqing et al. (2018)
Resveratrol		Reduction of the formation of fluorescent AGEs, delay in the activities of carbohydrate-hydrolyzing enzymes (α-glucosity and α-amylase), elimination of EROs and capture of MGO by conjugation reaction.	Shen, Xue, and Sheng (2017)
Epigallocatechin-3-gallate		Increase in the GSH/GSSG ratio in the liver, kidney, and adipose tissue of rats that received a diet with a high lipid content; inhibition of the accumulation of AGEs in plasma, liver, adipose tissue, and kidney; inhibition of RAGE expression by about three times; increased translocation of Nrf2 from the cytoplasm to the nucleus and increased expression of heme oxygenase 1.	Sampath et al. (2017)
			Sang et al. (2007)

Quercetin	Capture of methylglyoxal and glyoxal, dicarbonyl species through the B ring and formation of mono-, di-, and tri-MGO adducts.	Liu et al. (2017)
Gallic acid	Reduction of the formation of fructosamine and carbonyl proteins.	Adisakwattana et al. (2017)
Rutin	Inhibition of the formation of fructosamine, carbonyl proteins, and proteins ions.	Dubey, Ganeshpurkar, Ganeshpurkar, Bansal, and Dubey, (2017)
Curcumin	Inhibition of the aggregation of glycosylated and native bovine liver by decreasing the exposed hydrophobic surface of the catalase, preventing thermal aggregation caused by glycation, decreasing the exposed lysine residue, and eliminating EROs.	Najjar (2017)
	Reduction of the formation of fluorescent AGEs, inhibition of the activity of α-glycosidade and α-amilase enzymes, elimination of free radicals, reduction of LDL oxidation, and reduction of cellular oxidative stress.	Lekshmi, Arimboor, Nisha, Menon, and Raghu (2014)

Piper nigrum, Thymus vulgaris, and Zingiber officinale) (Buera et al., 2021). The antiglycant capacity was investigated for industrial by-products (orange juice [citrus residues containing hesperidin, narirutin, and tangeretin], pectin, peanut processing, and grape marc from wine production (rich in catechin and procyanidin B2) in glycation models: BSA/fructose, BSA/MGO, and arginine/MGO (Fernandes, Santana, et al., 2020). Peanut skin (50, 100, and 150 $\mu g\,mL^{-1}$) ethanol extract exerted the highest antiglycant activity (80–94% glycation inhibition) and protected RAW264.7 cells against oxidative damage and inflammatory stress mediators (TNF-α and IL-6) and NO levels induced by AGEs. Hesperidin from citrus waste/residue is recognized for its capacity to inhibit AGEs formation; it inhibits AGEs [(>60%) at 10, 15, and 20 $mg\,mL^{-1}$)] strongly dependent on polyphenolic content and antioxidant capacity, as well as inhibiting pancreatic lipase and α-glucosidase (76 and 98%, respectively). Removal of pectin from the citrus waste significantly increases the phenolic compounds (hesperidin from 25 to 88 $mg\,g^{-1}$ lyophilized extract), increasing thereby its antiglycant capacity (Fernandes, Santana, et al. 2020; Fernandes, Vieira, et al., 2020).

Pomegranate phenolics (punicalagin, ellagic acid, and urolithin) at 200 μM inhibit fructose-induced collagen glycation (64, 15, and 17%, respectively), MGO-induced collagen (71, 18, and 14%, respectively), and the formation of Amadori products (46, 32, and 40%, respectively). Punicalagin can reverse collagen glycation and crosslinking relating to skin damage. Pomella® (Geelawson.com) is a patented and proprietary *Punica granatum* extract standardized to punicalagins (Kheradia & Cropper, 2018). Amazon cherry (*Eugenia punicifolia* Kunth) (traditionally used to treat diabetes mellitus) pulp methanol extracts have high antiglycant and antioxidant activities due to the presence of *O*-glycosylated flavonoids (quercitrin and myricitrin) (Ramos et al., 2019). Antiglycant by non-oxidative pathway used fructose instead of GO and BSA for analysis. The methanolic extracts of *Eugenia punicifolia* (yellow pulp) exhibited antiglycant activity by the non-oxidative pathway higher (>three-folds) than aminoguanidine standard (IC_{50} 7.9 vs. 28.6 $\mu g\,mL^{-1}$) that correlated highly with total phenolic content.

Flavones (luteolin, apigenin, and chrysin) and flavonols (myricetin and quercetin) inhibit *in vitro* AGE production; luteolin and apigenin have similar AGE inhibitory effect, higher than the reference aminoguanidine (prototype AGE inhibitor) (60 vs. 40%) presumably due to the presence of hydroxyl groups in the B ring. These flavonoids also inhibited metalloproteinases (MMP-1 and MMP-13) demonstrating chelation as their fundamental mechanism of action of antiglycation. Polyphenol-rich (predominantly chlorogenic and dicaffeoylquinic acids) aqueous leaf extract of *Antirhea borbonica* (Rubiaceae), a tree endemic to Mascarene and Madagascar, exerts antiglycation effects against MGO-induced damage (hyperglycemia) *in vivo* in zebrafish (Delveaux et al., 2020). The extract prevents albumin from glycation and oxidation and β-amyloid-type aggregation and partially reduces hemolysis from MGO-induced injuries in erythrocytes and protects erythrocyte morphology from glycation impairment (preserves the deformability capacity of erythrocyte membrane). Moreover, the extract preserves and protects erythrocytes from MGO-induced damages *in vitro* and *in vivo* in a zebrafish model.

Antiglycant activity has also been investigated for food products and nutraceuticals. For example, biscuits containing antioxidant dietary fiber (5 g/30 g serving) extracted from spent coffee grounds reduced AGEs (up to sixfold) compared to traditional biscuits due to the bioaccessibility of phenolic compounds (gallic acid and catechin) after *in vitro* digestion

(Vázquez-Sánchez et al., 2018). Moreover, the fortified biscuit has antidiabetic potential demonstrated by its α-glucosidase inhibitory and antiglycative activities and high ascorbic acid bioavailability. Aqueous ethanol (70% EtOH) extract of *Cassia grandis* (Cesalpinaceae) fruit pulp in tablet form (granulated with microcrystalline cellulose) exhibited strong antiglycant effect by the oxidative and non-oxidative (via chelation) pathways and produced hypoglycemic effect comparable or similar to glibenclamide in diabetic-induced female Wistar rats (Prada et al., 2020). The presence of phenolics, flavonoids, and other components in the extract presumably account for its potent antiglycation effect (via breaking the crosslinking structures in the formed AGEs, blocking the carbonyl groups, and inhibiting the formation of late-stage Amadori products). The tablet can potentially be used to control blood glucose levels of diabetics and prevent diabetic complications.

Recently, two patents claim antiglycant activity of phenolic-rich extracts. Polyphenolic-rich extracts from *Osmanthus fragrans* (10% verbascoside), *P. granatum* (10% punicalagin), and *Olea europaea* (3% hydroxytyrosol) that synergistically inhibit AGEs generation can be used in beverages, skincare products, and food and healthcare products (Wang, Zhu, Luo, & Tang, 2021). Hydrolyzable tannins from green mango (*Mangifera indica*) aqueous extract protect and suppress glycation and is proposed for use in skincare (UV protection, skin whitening, and suppress collagen glycation) and healthcare (Lin, Yu, & Chung, 2021). Many food/crops containing polyphenolic compounds exert antiglycant activity, and their mechanisms of action have been reported (Fernandes, Vieira, et al., 2020). These food includes *Ribes nigrum* and other *Ribes species, Camellia nitidissima, Canarium album,* several *Rubus and Vaccinium* species, *cranberry, Silybum marianum, Garcinia mangostana, Scutellaria alpina and altissima, Carpobrotus edulis, Moringa oleifera, Annona crassiflora Mart,* propolis (ethanol extract), and pomegranate fruit extract.

1.4 Conclusion

The knowledge about AGEs metabolism and their presence in food and implications in health has increased over the decades. However, due to their complexity and wide distribution in biological systems, this topic still has some gaps to be addressed. Several studies demonstrate phenolic compounds as strong glycation-inhibiting agents, proposing several mechanisms of action against AGEs. This antiglycant activity is directly related to antioxidant capacity that depends on the phenolic chemical structure. For example, phenolics with hydroxylation in rings A and B (ellagitannins and ellagic acids) have high antiglycant activity because of hydroxylation and decrease due to methylation. Similarly, anthraquinone with meta structure or ortho dihydroxyl in the benzene ring, as well as phenolic acids with multiple hydroxyls, catechins with gallo groups, just as hydroxyl at C-3 are more effective in inhibiting glycation compared to the other forms (Xie & Chen, 2013). Glycation is inhibited by several mechanisms, such as oxidative stress reduction, reactive carbonyl species interception, inhibition of AGE precursor formation, reduction of RAGE expression, and AGEs capture (Khangholi et al., 2016; Lunceford & Gugliucci, 2005).

The studies to date indicate the potential use of extracts and phenolic compounds to reduce the progression of disease complications with pathophysiology correlated with the presence of AGEs and their carbonyl precursors. In this context, the use of combined phenolic compounds, in pure form or combined in food, can presumably synergize with their antioxidant capacities and, thus, increase their antiglycation potential, simultaneously acting through several mechanisms, thus contributing to greater relationship with health promotion and treatment.

References

Abdallah, H. M., El-Bassossy, H., Mohamed, G. A., El-Halawany, A. M., Alshali, K. Z., & Banjar, Z. M. (2016). Phenolics from garcinia mangostana inhibit advanced glycation endproducts formation: Effect on amadori products, cross-linked structures and protein thiols. *Molecules, 21*(2), 251. https://doi.org/10.3390/molecules21020251

Adisakwattana, S., Thilavech, T., Sompong, W., & Pasukamonset, P. (2017). Interaction between ascorbic acid and gallic acid in a model of fructose-mediated protein glycation and oxidation. *Electronic Journal of Biotechnology, 27*, 32–36. doi.org/10.1016/j.ejbt.2017.03.004

Ahmed, N. (2005). Advanced glycation endproducts—Role in pathology of diabetic complications. *Diabetes Research and Clinical Practice, 67*(1), 3–21. https://doi.org/10.1016/j.diabres.2004.09.004

Ahmed, S., Rahman, A., Hasanain, A., Lalonde, M., Goldberg, V. M., & Haqqi, T. M. (2002). Green tea polyphenol epigallocatechin-3-gallate inhibits the IL-1β-induced activity and expression of cyclooxygenase-2 and nitric oxide synthase-2 in human chondrocytes. *Free Radical Biology and Medicine, 33*(8), 1097–1105. https://doi.org/10.1016/s0891-5849(02)01004-3

Akillioglu, H. G., & Gokmen, V. (2016). Kinetic evaluation of the inhibition of protein glycation during heating. *Food Chemistry, 196*, 1117–1124. https://doi.org/10.1016/j.foodchem.2015.10.066

Allmen, E. U., Koch, M., Fritz, G., & Legler, D. F. (2008). V domain of RAGE interacts with AGEs on prostate carcinoma cells. *Prostate, 68*(7), 748–758. https://doi.org/10.1002/pros.20736

Anis, M. A., & Sreerama, Y. N. (2020). Inhibition of protein glycoxidation and advanced glycation end-product formation by barnyard millet (*Echinochloa frumentacea*) phenolics. *Food Chemistry, 315*, 126265. https://doi.org/10.1016/j.foodchem.2020.126265

Awasthi, S., & Saraswathi, N. T. (2016). Sinigrin, a major glucosinolate from cruciferous vegetables restrains non-enzymatic glycation of albumin. *International Journal of Biological Macromolecules, 83*, 410–415. https://doi.org/10.1016/j.ijbiomac.2015.11.019

Barbosa, J. H. P., Souza, I. T., Santana, A. E. G., & Goulart, M. O. F. (2016). Determination of advanced glycation (ages) and lipoxidation (ales) end products in foods and biological systems: advances, challenges and perspectives. *Química Nova, 2016*(15), 1–13.

Bergmann, R., Helling, R., Heichert, C., Scheunemann, M., Mäding, P., Wittrisch, H., . . . Henle, T. (2001). Radio fluorination and positron emission tomography (PET) as a new approach to study the in vivo distribution and elimination of the advanced glycation endproducts N-epsilon-carboxymethyllysine (CML) and N-epsilon-carboxyethyllysine (CEL) and N epsilon-carboxyethyllysine (CEL). *Nahrung, 45*, 182–188.

Bhardwaj, A., Sethi, G., Vadhan-Raj, S., Bueso-Ramps, C., Takada, Y., Gaur, U., . . . Aggarwal, B. B. (2007). Resveratrol inhibits proliferation, induces apoptosis, and overcomes chemoresistance through down-regulation of STAT3 and nuclear factor-kB–regulated antiapoptotic and cell survival gene products in human multiple myeloma cells. *Blood, 109*(6), 2293–2302. doi.org/10.1002/1521-3803(20010601)45:3<182::AID-FOOD182>3.0.CO;2-Q

Bierhaus, A., Hofmann, A. M., & Nawroth, P. P. (1998). AGEs and their interaction with AGE-receptors in vascular disease and diabetes mellitus. I. The AGE concept. *Cardiovascular Research, 37*(3), 586–600.

Buera, P., Abirached, C., Alamilla-Beltrán, L., Busch, V. M., dosSantos, C. I., . . . Vasile, F. E. (2021). Ibero-American network as a collaborative strategy to provide tools for the development of phytopharmaceuticals and nutraceuticals. In D. N. Chauhan & K. Shah (Eds.), *Phytopharmaceuticals: Potential therapeutic applications* (pp. 19–57). Beverly, MA, USA: Scrivener Publishing LLC.

Challier, M., Jacqueminet, S., Benabdesselam, O., Grimaldi, A., & Beaudeux, J. L. (2005). Increased serum concentrations of soluble receptor for advanced glycation endproducts in patients with type 1 diabetes. *Clinical Chemistry, 51*(9), 1749–1750. https://doi.org/10.1373/clinchem.2005.051961

Chen, P. M., Gregersen, H., & Zhao, J. B. (2015). Advanced glycation end-product expression is upregulated in the gastrointestinal tract of type 2 diabetic rats. *World Journal of Diabetes, 6*(4), 662–672. https://doi.org/10.4239/wjd.v6.i4.662

Chobert, J. M., Gaudin, J.-C., Dalgalarrondo, M., & Haertlé, T. (2006). Impact of Maillard type glycation on properties of beta-lactoglobulin. *Biotechnology Advances, 24*(6), 629–632. https://doi.org/10.1016/j.biotechadv.2006.07.004

Dattilo, B. M., Fritz, G., Leclerc, E., Kooi, C. W., Heizmann, C. W., & Chazin, W. J. (2007). The extracellular region of the receptor for advanced glycation end products is composed of two independent structural units. *Biochemistry, 46*(23), 6957–6970. https://doi.org/10.1021/bi7003735

Degroot, J. (2001). Age-related decrease in susceptibility of human articular cartilage to matrix metalloproteinase mediated degradation: The role of advanced glycation end products. *Arthritis and Rheumatism, 44*(11), 2562–2571. https://doi.org/10.1002/1529-0131(200111)44:11<2562::aid-art437>3.0.co;2-1

Delveaux, J., Turpin, C., Veeren, B., Diotel, N., Bravo, S. B., Begue, F., . . . Rondeau, P. (2020). *Antirhea borbonica* aqueous extract protects albumin and erythrocytes from glycoxidative damages. *Antioxidants, 9*(5), 415. http://dx.doi.org/10.3390/antiox9050415

Diftis, N., & Kiosseoglou, V. (2004). Competitive adsorption between a dry-heated soy protein-dextran mixture and surface-active materials in oil-in-water emulsions. *Food Hydrocolloids, 18*(4), 639–646. doi.org/10.1016/j.foodhyd.2003.11.007

Dubey, S., Ganeshpurkar, A., Ganeshpurkar, A., Bansal, D., & Dubey, N. (2017). Glycolytic enzyme inhibitory and antiglycation potential of rutin. *Future Journal of Pharmaceutical Sciences, 3*(2), 158–162. doi.org/10.1016/j.fjps.2017.05.005

Emanuele, E., D'Angelo, A., Tomaino, C., Binetti, G., Ghidoni, R., Politi, P., . . . Geroldi, D. (2005). Circulating levels of soluble receptor for advanced glycation end products in Alzheimer disease and vascular dementia. *Archives of Neurology, 62*(11), 1734–1736. https://doi.org/10.1001/archneur.62.11.1734

Falcone, C., Emanuele, E., D'Angelo, A., Buzzi, M. P., Belvito, C., Cuccia, M., & Geroldi, D. (2005). Plasma levels of soluble receptor for advanced glycation end products and coronary artery disease in nondiabetic men. *teriosclerosis Thrombosis, and Vascular Biology*, *25*(5), 1032–1037. https://doi.org/10.1161/01.ATV.0000160342.20342.00

Farias, D. P., Araújo, F. F., Neri-Numa, I. A., & Pastore, G. M. (2021). Anidiabetic potential of dietary polyphenols: A mechanistic review. *Food Research International*, *145*, 11038. https://doi.org/10.1016/j.foodres.2021.110383

Favre, L. C., Rolandelli, G., Mshicileli, N., Vhangani, L. N., Ferreira, C. S., Wyk, J. V., & Buera, P. (2020). Antioxidant and anti-glycation potential of green pepper (Piper nigrum): Optimization of β-cyclodextrin-based extraction by response surface methodology. *Food Chemistry*, *316*, 126280.

Fernandes, A. C. F., Santana, A. L., Martins, I. M., Moreira, D. K. T., Macedo, J. A., & Macedo, G. A. (2020). Anti-glycation effect and the α-amylase, lipase, and α-glycosidase inhibition properties of a polyphenolic fraction derived from citrus wastes. *Preparative Biochemistry and Biotechnology*, *8*(50), 794–802.

Fernandes, A. C. F., Vieira, N. C., Santana, A. L., Gandra, R. L. P., Rubia, C., Castro-Gamboa, I., . . . Macedo, G. A. (2020). Peanut skin polyphenols inhibit toxicity induced by advanced glycation endproducts in RAW264.7 macrophages. *Food and Chemical Toxicology*, *145*, 111619. https://doi.org/10.1016/j.fct.2020.111619

Förster, A., Kühne, Y., & Henle, T. (2005). Studies on absorption and elimination of dietary Maillard reaction products. *New York Academy of Sciences*, *1043*, 474–481. https://doi.org/10.1016/j.foodchem.2020.126280

Fritz, G. (2011). RAGE: a single receptor fits multiple ligands. *Trends in Biochemical Sciences*, *36*(12), 625–632. https://doi.org/10.1016/j.tibs.2011.08.008

Fujisawa, K., Katakami, N., Kaneto, H., Naka, T., Takahara, M., Sakamoto, F., . . . Shimomura, I. (2013). Circulating soluble RAGE as a predictive biomarker of cardiovascular event risk in patients with type 2 diabetes. *Atherosclerosis*, *227*(2), 425–428. https://doi.org/10.1016/j.atherosclerosis.2013.01.016

Genova, V. M., Fernandes, A. C. F., Hiramatsu, E., Queirós, L. D., Macedo, J. A., & Macedo, G. A. (2021). Biotransformed antioxidant isoflavone extracts present high-capacity to attenuate the in vitro formation of advanced glycation end products. *Food Biotechnology*, *35*(1), 50–66. https://doi.org/10.1080/08905436.2020.1869564

Geroldi, D., Falcone, C., Emanuele, E., D'Angelo, A., Calcagnino, M., Buzzi, M. P., . . . Fogari, R. (2005). Decreased plasma levels of soluble receptor for advanced glycation end-products in patients with essential hypertension. *Journal of Hypertension*, *23*(9), 1725–1729. https://doi.org/10.1097/01.hjh.0000177535.45785.64

Gerrard, J. A., Brown, P. K., & Fayle, S. E. (2003). Maillard crosslinking of food proteins III: the effects of glutaraldehyde, formaldehyde and glyceraldehyde upon bread and croissants. *Food Chemistry*, *80*(1), 45–50. https://doi.org/10.1016/S0308-8146(02)00233-9

Goldberg, T., Cai, W., Peppa, M., Dardaine, V., Suresh, B., Baliga, B. S., . . . Vlassara, H. (2004). Advanced glycoxidation end products in commonly consumed foods. *Journal of the American Dietetic Association*, *104*, 1287–1291. https://doi.org/10.1016/j.jada.2004.05.214

Harris, C. S., Cuerrier, A., & Lamont, E. (2014). Investigating wild berries as a dietary approach to reducing the formation of advanced glycation endproducts: Chemical correlates of *in vitro* antiglycation activity. *Plant Foods for Human Nutrition*, *69*(1), 71–77.

Harsha, P., Lavelli, V., & Scarafoni, A. (2014). Protective ability of phenolics from white grape vinification by-products against structural damage of bovine serum albumin induced by glycation. *Food Chemistry, 156*, 220–226. https://doi.org/10.1016/j.foodchem.2014.01.104

Hengqing, C., Fei Tao, Y., Hou, Y. L., Tiesong, Z., Shengmin, S., & Lishuang, L. (2018). Dual effects of propyl gallate and its methylglyoxal adduct on carbonyl stress and oxidative stress. *Food Chemistry, 265*, 227–232. https://doi.org/10.1016/j.foodchem.2018.04.045

Henle, T. (2003). AGEs in foods: Do they play a role in uremia? *Kidney International, 63*(84), S145–S147. https://doi.org/10.1016/j.jada.2004.05.214

Hori, O., Brett, J., Slattery, T., Cao, R., Zhang, J., Chen, J. X., . . . Schmidt, A. M. (1995). The receptor for advanced glycation end products (RAGE) is a cellular binding site for amphoterin. Mediation of neurite outgrowth and co-expression of rage and amphoterin in the developing nervous system. *Journal of Biological Chemistry, 270*(43), 25752–25761. https://doi.org/10.1074/jbc.270.43.25752

Huang, J. S., Guh, J. Y., Chen, H. C., Hung, W. C., Lai, Y. H., & Chuang, L. Y. (2001). Role of receptor for advanced glycation end-product (RAGE) and the JAK/STAT-signaling pathway in AGE-induced collagen production in NRK-49F cells. *Journal of Cell Biochemistry, 81*(1), 102–113. https://doi.org/10.1002/1097-4644(20010401)81:1<102::aid-jcb1027>3.0.co;2-y

Hudson, B. I., Kalea, A. Z., Del Mar Arriero, M., Harja, E., Boulanger, E., D'Agati, V., & Schmidt, A. M. (2008). Interaction of the RAGE cytoplasmic domain with diaphanous-1 is required for ligand-stimulated cellular migration through activation of Rac1 and Cdc42. *Journal of Biological Chemistry, 283*(49), 34457–34468. https://doi.org/10.1074/jbc.M801465200

Huebschmann, A. G., Regensteiner, J. G., Vlassara, H., & Reusch, J. E. B. (2006). Diabetes and advanced glycoxidation end products. *Diabetes Care, 29*(6), 1420–1432. doi.org/10.2337/dc057-2096

Ishihara, K., Tsutsumi, K., Kawane, S., Nakajima, M., & Kasaoka, T. (2003). The receptor for advanced glycation end-products (RAGE) directly binds to ERK by a D-domain-like docking site. *FEBS Letters, 550*((1–3)), 107–113. https://doi.org/10.1016/s0014-5793(03)00846-9

Iwamoto, K., Kanno, K., Hyogo, H., Yamagishi, S., Takeuchi, M., Tazuma, S., & Chayama, K. (2008). Advanced glycation end products enhance the proliferation and activation of hepatic stellate cells. *Journal of Gastroenterology, 43*(4), 298–304. https://doi.org/10.1007/s00535-007-2152-7

Jakus, V., & Rietbrock, N. (2004). Advanced glycation end-products and the progress of diabetic vascular complications. *Physiological Research, 53*(2), 131–142. PMID: 15046548

Jiménez-Aspee, F., Theoduloz, C., Vieira, M. N., Rodríguez-Werner, M. A., Schmalfuss, E., Winterhalter, P., & Schmeda-Hirschmann, G. (2016). Phenolics from the Patagonian currants *Ribes* spp.: Isolation, characterization and cytoprotective effect in human AGS cells. *Journal of Functional Foods, 26*, 11–26. https://doi.org/10.1016/j.jff.2016.06.036

Kalousova, M., Hodkova, M., Kazderova, M., Fialova, J., Tesar, V., Dusilova-Sulkova, S., & Zima, T. (2006). Soluble receptor for advanced glycation end products in patients with decreased renal function. *American Journal of Kidney Diseases, 47*(3), 406–411. https://doi.org/10.1053/j.ajkd.2005.12.028

Khangholi, S., Majid, F. A. A., Berwary, N. J. A., Ahmad, F., & Aziz, R. B. A. (2016). The mechanisms of inhibition of advanced glycation end products formation through polyphenols in hyperglycemic conditions. *Planta Medica, 82*(1-2), 32–45.

Kheradia, N. & Cropper, S. (2018). Skin-gut axis: The potential of pomegranate. Retrieved from https://www.nutraingredients.com/.

Kierdorf, K., & Fritz, G. (2013). RAGE regulation and signaling in inflammation and beyond. *Journal of Leukocyte Biology*, *94*(1), 55–68. https://doi.org/10.1189/jlb.1012519

Kim, J., Jo, K., Lee, I.-S., Kim, C.-S., & Kim, J.-S. (2016). The extract of Aster Koraiensis prevents retinal pericyte apoptosis in diabetic rats and its active compound, chlorogenic acid inhibits AGE formation and AGE/RAGE interaction. *Nutrients*, *8*(9). https://doi.org/10.3390/nu8090585

Koch, M., Chitayat, S., Dattilo, B. M., Schiefner, A., Diez, J., Chazin, W. J., & Fritz, G. (2010). Structural basis for ligand recognition and activation of RAGE. *Structure*, *18*(10), 1342–1352. https://doi.org/10.1016/j.str.2010.05.017

Koschinsky, T., He, C.-J., Mitsuhashi, T., Bucala, R., Liu, C., Buenting, C., . . . Vlassara, H. (1997). Orally absorbed reactive glycation products (glycotoxins): An environmental risk factor in diabetic nephropathy. *Proceedings of the National Academy of Sciences of the United States of America*, *94*(12), 6474–6479. https://doi.org/10.1073/pnas.94.12.6474

Kuda, T., Eda, M., Kataoka, M., Nemoto, M., Kawahara, M., Oshio, S., . . . Kimura, B. (2016). Anti-glycation properties of the aqueous extract solutions of dried algae products and effect of lactic acid fermentation on the properties. *Food Chemistry*, *192*, 1109–1115. https://doi.org/10.1016/j.foodchem.2015.07.073

Kumagai, Y., Nakatani, S., Onoder, H., Nagatomo, A., Nishida, N., Matsuura, Y., . . . Wada, M. (2015). Anti-glycation effects of pomegranate (Punica granatum L.) fruit extract and its components *in vivo* and *in vitro*. *Journal of Agricultural and Food Chemistry*, *63*(35), 7760–7764. https://doi.org/10.1021/acs.jafc.5b02766

Kuo, C. T., Liu, T.-H., Hsu, T.-H., Lin, F.-Y., & Chen, H.-Y. (2015). Antioxidant and antiglycation properties of different solvent extracts from Chinese olive (Canarium album L.) fruit. *Asian Pacific Journal of Tropical Medicine*, *8*(12), 1013–1021. https://doi.org/10.1016/j.apjtm.2015.11.013

Lander, H. M., Tauras, J. M., Ogiste, J. S., Hori, O., Moss, R. A., & Schmidt, A. M. (1997). Activation of the receptor for advanced glycation end products triggers a p21(ras)-dependent mitogen-activated protein kinase pathway regulated by oxidant stress. *Journal of Biological Chemistry*, *272*(28), 17810–17814. https://doi.org/10.1074/jbc.272.28.17810

Leclerc, E., Fritz, G., Vetter, S. W., & Heizmann, C. W. (2009). Binding of S100 proteins to RAGE: An update. *Biochimica Biophysica Acta*, *1793*(6), 993–1007. https://doi.org/10.1016/j.bbamcr.2008.11.016

Lekshmi, P. C., Arimboor, R., Nisha, V. M., Menon, A. N., & Raghu, K. G. (2014). In vitro antidiabetic and inhibitory potential of turmeric (Curcuma longa L) rhizome against cellular and LDL oxidation and angiotensin converting enzyme. *Journal of Food Science and Technology*, *51*(12), 3910–3917. https://doi.org/10.1007/s13197-013-0953-7

Lertittikul, W., Benjakul, S., & Tanaka, M. (2007). Characteristics and antioxidative activity of Maillard reaction products from a porcine plasma protein-glucose model system as influenced by pH. *Food Chemistry*, *100*(2), 669–677. https://doi.org/10.1016/j.foodchem.2005.09.085

Li, J. L., Liu, D., Lu, Y., & Zhang, Z. (2012). Advanced glycation end products and neurodegenerative diseases: Mechanisms and perspective. *Journal of the Neurological Sciences*, *317*(1-2), 1–5. https://doi.org/10.1016/j.jns.2012.02.018

Lin, Y-H., Yu, C-H., & Chung, Y-M. (2021). Methods for skin whitening, ultraviolet radiation protection and suppression of glycation using green mango extracts and compounds obtained therefrom. *US Patent US2021/0069272 A1*; issued March 11, 2021.
Liu, G., Xia, Q., Lu, Y., Zheng, T., Sang, S., & Lv, L. (2017). Influence of quercetin and its methylglyoxal adducts on the formation of α-dicarbonyl compounds in a lysine/glucose model system. *Journal of Agricultural and Food Chemistry*, *65*(10), 2233–2239. https://doi.org/10.1021/acs.jafc.6b05811
Liu, J. H., Ru, Q. M., & Ding, Y. T. (2012). Glycation a promising method for food protein modification: Physicochemical properties and structure, a review. *Food Research International*, *49*(1), 170–183. https://doi.org/10.1016/j.foodres.2012.07.034
Lue, L. F., Walker, D. G., Brachova, L., Beach, T. G., Rogers, J., Schmidt, A. M., . . . Yan, S. D. (2001). Involvement of microglial receptor for advanced glycation endproducts (RAGE) in Alzheimer's disease: Identification of a cellular activation mechanism. *Experimental Neurology*, *171*(1), 29–45. https://doi.org/10.1006/exnr.2001.7732
Lue, L. F., Walker, D. G., Jacobson, S., & Sabbagh, M. (2009). Receptor for advanced glycation end products: its role in Alzheimer's disease and other neurological diseases. *Future Neurology*, *4*(2), 167–177. https://doi.org/10.2217/14796708.4.2.167
Lunceford, N., & Gugliucci, A. (2005). Ilex paraguariensis extracts inhibit AGE formation more efficiently than green tea. *Fitoterapia*, *76*, 419–427.
Mana, P., Das, J., Ghosh, J., & Sil, P. C. (2010). Contribution of type 1 diabetes to rat liver dysfunction and cellular damage via activation of NOS, PARP, IκBα/NF-κB, MAPKs, and mitochondria-dependent pathways: Prophylactic role of arjunolic acid. *Free Radical Biology and Medicine*, *48*(11), 1465–1484. https://doi.org/10.1016/j.freeradbiomed.2010.02.025
Marinakis, E., Bagkos, G., Piperi, C., Roussou, P., & Diamanti-Kandarakis, E. (2014). Critical role of RAGE in lung physiology and tumorigenesis: A potential target of therapeutic intervention? *Clinical Chemistry Laboratory Medicine*, *52*(2), 189–200. https://doi.org/10.1515/cclm-2013-0578
Melo J. B. (2018). *Correlation of the antiglycant effect and antioxidant capacity of phenolic compounds in vitro* (Dissertation of master degree, Food and Nutrition Department, Faculty of Food Engineering, State University of Campinas). Retrieved from https://doi.org/10.47749/T/UNICAMP.2018.1096631.
Monnier, V. M. (2003). Intervention against the Maillard reaction in vivo. *Archives of Biochemistry and Biophysics*, *419*(1), 1–15. https://doi.org/10.1016/j.abb.2003.08.014
Moreno, J., & Peinado, R. (2012). *Enologycal Chemistry*, Chapter 5 (pp. 53–76). Elsevier. eBook ISBN: 9780123884398. Hardcover ISBN: 9780123884381R
Najjar, F. M. (2017). Destructive effect of non-enzymatic glycation on catalase and remediation via curcumin. *Archives of Biochemistry and Biophysics*, *630*, 81–90. https://doi.org/10.1016/j.abb.2017.06.018
Neeper, M., Schmidt, A. M., Brett, J., Yan, S. D., Wang, F., Pan, Y. C., . . . Shaw, A. (1992). Cloning and expression of a cell surface receptor for advanced glycosylation end products of proteins. *Journal of Biological Chemistry*, *267*(21), 14998–15004. PMID: 1378843
Ostendorp, T., Leclerc, E., Galichet, A., Koch, M., Demling, N., Weigle, B., . . . Fritz, G. (2007). Structural and functional insights into RAGE activation by multimeric S100B. *EMBO Journal*, *26*(16), 3868–3878. https://doi.org/10.1038/sj.emboj.7601805

Park, H., Adsit, F. G., & Boyington, J. C. (2010). The 1.5 A crystal structure of human receptor for advanced glycation endproducts (RAGE) ectodomains reveals unique features determining ligand binding. *Journal of Biological Chemistry, 285*(52), 40762–40770. https://doi.org/10.1074/jbc.M110.169276

Paul, R. G., & Bailey, A. J. (1999). The effect of advanced glycation end-product formation upon cell-matrix interactions. *The International Journal of Biochemistry and Cell Biology, 31*(6), 653–660.

Peng, X. F., Ma, J., Chen, F., & Wang, M. (2011). Naturally occurring inhibitors against the formation of advanced glycation end-products. *Food & Function, 2*(6), 289–301. https://doi.org/10.1039/c1fo10034c

Prada, A. L., Achod, L. D. R., Keita, H., Carvalho, J. C. T., deSouza, T. P., & Amado, J. R. R. (2020). Development, pharmacological and toxicological evaluation of a new tablet formulation based on *Cassia grandis* fruit extract. *Sustainable Chemistry and Pharmacy, 16*, 100244. https://doi.org/10.1016/j.scp.2020.100244

Pullerits, R., Bokarewa, M., Dahlberg, L., & Tarkowski, A. (2005). Decreased levels of soluble receptor for advanced glycation end products in patients with rheumatoid arthritis indicating deficient inflammatory control. *Arthritis Research and Therapy, 7*(4), R817–R824. https://doi.org/10.1186/ar1749

Qin, Y. H., Dai, S. M., Tang, G. S., Zhang, J., Ren, D., Wang, Z. W., & Shen, Q. (2009). HMGB1 enhances the proinflammatory activity of lipopolysaccharide by promoting the phosphorylation of MAPK p38 through receptor for advanced glycation end products. *Journal of Immunology, 183*(10), 6244–6250. https://doi.org/10.4049/jimmunol.0900390

Ramos, A. S., Mar, J. M., daSilva, L. S., Acho, L. D., Silva, B. J. P., Lima, E. S., . . . Machado, M. B. (2019). Pedra-ume caá fruit: An Amazon cherry rich in phenolic compounds with antiglycant and antioxidant properties. *Food Research International, 123*, 674–683.

Rani, M. P., Krishna, M. S., Padmakumari, K. P., Raghu, K. G., & Sundaresan, A. (2012). Zingiber officinale extract exhibits antidiabetic potential via modulating glucose uptake, protein glycation and inhibiting adipocyte differentiation: And *in vitro* study. *Journal of the Science of Food and Agriculture, 92*(9), 1948–1955. https://doi.org/10.1002/jsfa.5567

Raucci, A., Cugusi, S., Antonelli, A., Barabino, S. M., Monti, L., Bierhaus, A., . . . Bianchi, M. E. (2008). A soluble form of the receptor for advanced glycation endproducts (RAGE) is produced by proteolytic cleavage of the membrane-bound form by the sheddase a disintegrin and metalloprotease 10 (ADAM10). *FASEB Journal, 22*(10), 3716–3727. https://doi.org/10.1096/fj.08-109033

Ruan, B. H., Li, X., Winkler, A. R., Cunningham, K. M., Kuai, J., Greco, R. M., . . . Winkler, D. G. (2010). Complement C3a, CpG oligos, and DNA/C3a complex stimulate IFN-alpha production in a receptor for advanced glycation end product-dependent manner. *Journal of Immunology, 185*(7), 4213–4222. https://doi.org/10.4049/jimmunol.1000863

Sakaguchi, M., Murata, H., Yamamoto, K., Ono, T., Sakaguchi, Y., Motoyama, A., . . . Huh, N. H. (2011). TIRAP, an adaptor protein for TLR2/4, transduces a signal from RAGE phosphorylated upon ligand binding. *PLoS One, 6*(8), e23132. https://doi.org/10.1371/journal.pone.0023132

Sampath, C., Rashid, M. R., Sang, S., & Ahmedna, M. (2017). Green tea epigallocatechin 3-gallate alleviates hyperglycemia and reduces advanced glycation end products via nrf2 pathway in mice with high fat diet-induced obesity. *Biomedicine and Pharmacotherapy, 87*, 73–81. https://doi.org/10.1016/j.biopha.2016.12.082

Sang, S., Shao, X., Bai, N., Lo, C. Y., Yang, C. S., & Ho, C. T. (2007). Tea polyphenol (-)-epigallocatechin-3-gallate: A new trapping agent of reactive dicarbonyl species. *Chemical Research in Toxicology, 20*(12), 1862–1870. https://doi.org/10.1021/tx700190s

Sanmartin, E., Arboleya, J.-C., Villamiel, M., & Moreno, J. F. (2009). Recent advances in the recovery and improvement of functional proteins from fish processing by-products: Use of protein glycation as an alternative method. *Comprehensive Reviews in Food Science and Food Safety, 8*(4), 332–344. https://doi.org/10.1111/j.1541-4337.2009.00083.x

Sato, R., Sawabe, T., & Saeki, H. (2005). Characterization of fish myofibrillar protein by conjugation with alginate oligosaccharide prepared using genetic recombinant alginate lyase. *Journal of Food Science, 70*(1), C58–C62. https://doi.org/10.1111/j.1365-2621.2005.tb09021.x

Schmidt, A. M., Vianna, M., Gerlach, M., Brett, J., Ryan, J., Kao, J., . . . Clauss, M. (1992). Isolation and characterization of two binding proteins for advanced glycosylation end products from bovine lung which are present on the endothelial cell surface. *Journal of Biological Chemistry, 267*(21), 14987–14997. PMID: 1321822

Sebekova, K., Kupcová, V., Schinzel, R., & Heiland, A. (2002). Markedly elevated levels of plasma advanced glycation endproducts in patients with liver cirrhosis—Amelioration by liver transplantation. *Journal of Hepatology, 36*(1), 66–71.

Shangari, N., & O'brien, P. J. (2004). The cytotoxic mechanism of glyoxal involves oxidative stress. *Biochemical Pharmacology, 68*(7), 1433–1442. https://doi.org/10.1016/j.bcp.2004.06.013

Shen, Y., Xu, Z., & Sheng, Z. (2017). Ability of resveratrol to inhibit advanced glycation end product formation and carbohydrate-hydrolyzing enzyme activity, and to conjugate methylglyoxal. *Food Chemistry, 216*, 153–160. https://doi.org/10.1016/j.foodchem.2016.08.034

Shinohara, M., Thornalley, P. J., Giardino, I., Beisswenger, P., Thorpe, S. R., Onorato, J., & Brownlee, M. (1998). Overexpression of glyoxalase-I in bovine endothelial cells inhibits intracellular advanced glycation endproduct formation and prevents hyperglycemia-induced increases in macromolecular endocytosis. *Journal of Clinical Investigation, 101*(5), 1142–1147. https://doi.org/10.1172/JCI119885

Smith, D. J., Yerkovich, S. T., Towers, M. A., Carroll, M. L., Thomas, R., & Upham, J. W. (2011). Reduced soluble receptor for advanced glycation end-products in COPD. *European Respiratory Journal, 37*(3), 516–522. https://doi.org/10.1183/09031936.00029310

Somensi, N., Brum, P. O., de Miranda Ramos, V., Gasparotto, J., Zanotto-Filho, A., Rostirolla, D. C., . . . Pens Gelain, D. (2017). Extracellular HSP70 activates ERK1/2, NF-kB and pro-inflammatory gene transcription through binding with RAGE in A549 human lung cancer cells. *Cell Physiology and Biochemistry, 42*(6), 2507–2522. https://doi.org/10.1159/000480213

Spínola V. A. R. (2018). *Nutraceuticals and functional foods for diabetes and obesity control* (Doctoral thesis, Universidade da Madeira). Retrieved from https://digituma.uma.pt/bitstream/10400.13/2218/1/DoutoramentoV%C3%ADtor%20Sp%C3%ADnola.pdf.

Sturchler, E., Galichet, A., Weibel, M., Leclerc, E., & Heizmann, C. W. (2008). Site-specific blockade of RAGE-Vd prevents amyloid-beta oligomer neurotoxicity. *Journal of Neuroscience, 28*(20), 5149–5158. https://doi.org/10.1523/JNEUROSCI.4878-07.2008

Sun, C., Zhao, C., Guven, E. C., Paoli, P., Simal-Gandara, J., Ramkumar, K. M., . . . Xiao, J. (2020). Dietary polyphenols as antidiabetic agents: Advances and opportunities. *Food Frontiers*, *1*, 18–44. https://doi.org/10.1002/fft2.15

Takeuchi, M., & Yamagishi, S. (2008). Possible involvement of advanced glycation end-products (AGEs) in the pathogenesis of Alzheimer's disease. *Current Pharmaceutical Design*, *14*(10), 973–978. https://doi.org/10.2174/138161208784139693

Thornalley, P. J., Battah, S., Ahmed, N., Karachalias, N., Agalou, S., Babaei-Jadidi, R., & Dawnay, A. (2003). Quantitative screening of advanced glycation endproducts in cellular and extracellular proteins by tandem mass spectrometry. *Biochemical Journal*, *375*, 581–592. https://doi.org/10.1042/BJ20030763

Uribarri, J., Cai, W., Peppa, M., Goodman, S., Ferrucci, L., Striker, G., & Vlassara, H. (2007). Circulating glycotoxins and dietary advanced glycation endproducts: Two links to inflammatory response, oxidative stress, and aging. *Journals of Gerontology Series a-Biological Sciences and Medical Sciences*, *62*(4), 427–433. https://doi.org/10.1093/gerona/62.4.427

Uribarri, J., Castillo, M. D. C., Maza, M. P. M., Filip, R., Gugliucci, A., Luevano-Contreras, C., . . . Garay-Sevilla, M. E. (2015). Dietary advanced glycation end products and their role in health and disease. *American Society for Nutrition*, *6*(4), 461–473. https://doi.org/10.3945/an.115.008433

Uribarri, J., Woodruff, S., Goodman, S., Cai, W., Chen, X., Pizyk, R., . . . Vlassara, H. (2010). Advanced glycation end products in foods and a practical guide to their reduction in the diet. *Journal of the American Dietetic Association*, *110*(6), 911–916. https://doi.org/10.1016/j.jada.2010.03.018

Vázquez-Sánchez, K., Martinez-Saez, N., Rebollo-Hernanz, M., DelCastillo, M. D., Gaytán-Martínez, M., & Campos-Vega, R. (2018). *In vitro* health promoting properties of antioxidant dietary fiber extracted from spent coffee (*Coffee arabica* L.) grounds. *Food Chemistry*, *261*, 253–259.

Vlassara, H., Cai, W., Candrall, J., Goldberg, T., Oberstein, R., Dardaine, V., . . . Rayfield, E. J. (2002). Inflammatory mediators are induced by dietary glycotoxins, a major risk factor for diabetic angiopathy. *Proceedings of the National Academy of Sciences of the United States of America*, *99*(24), 15596–15601. https://doi.org/10.1073/pnas.242407999

Wang, W., Liu, H., Wang, Z., Qi, J., Yuan, S., Zhang, W., . . . Jia, A.-Q. (2016). Phytochemicals from Camellia nitidissima Chi inhibited the formation of advanced glycation end-products by scavenging methylglyoxal. *Food Chemistry*, *205*, 204–211. https://doi.org/10.1016/j.foodchem.2016.03.019

Wang, X., Zhu, H., Luo, Z., & Tang, J. (2021). Composition having anti-glycation effect and application thereof. *US Patent 2021/0106642 A1*; issued April 15, 2021.

Wang, Z., Jiang, Y., Liu, N., Ren, L., Zhu, Y., An, Y., & Chen, D. (2012). Advanced glycation end-product N ε-carboxymethyl-Lysine accelerates progression of atherosclerotic calcification in diabetes. *Atherosclerosis*, *221*(2), 387–396. https://doi.org/10.1016/j.atherosclerosis.2012.01.019

Wu, C. H., Huang, S.-M., Lin, J.-A., & Yen, G.-C. (2011). Inhibition of advanced glycation end product formation by foodstuffs. *Food and Function*, *2*(5), 224–234. https://doi.org/10.1039/c1fo10026b

Xie, Y., & Chen, X. (2013). Structures required of polyphenols for inhibiting advanced glycation end products formation. *Current Drug Metabolism*, *4*(14). https://doi.org/10.2174/1389200211314040005

Xue, J., Rai, V., Singer, D., Chabierski, S., Xie, J., Reverdatto, S., . . . Shekhtman, A. (2011). Advanced glycation end product recognition by the receptor for AGEs. *Structure*, *19*(5), 722–732. https://doi.org/10.1016/j.str.2011.02.013

Yamamoto, Y., Harashima, A., Saito, H., Tsuneyama, K., Munesue, S., Motoyoshi, S., . . . Yamamoto, H. (2011). Septic shock is associated with receptor for advanced glycation end products ligation of LPS. *Journal of Immunology*, *186*(5), 3248–3257. https://doi.org/10.4049/jimmunol.1002253

Yan, S. D., Chen, X., Fu, J., Chen, M., Zhu, H., Roher, A., . . . Schmidt, A. M. (1996). RAGE and amyloid-beta peptide neurotoxicity in Alzheimer's disease. *Nature*, *382*(6593), 685–691. https://doi.org/10.1038/382685a0

Yonekura, H., Yamamoto, Y., Sakurai, S., Petrova, R. G., Abedin, M. J., Li, H., . . . Yamamoto, H. (2003). Novel splice variants of the receptor for advanced glycation end-products expressed in human vascular endothelial cells and pericytes, and their putative roles in diabetes-induced vascular injury. *Biochem J*, *370*(Pt 3), 1097–1109. https://doi.org/10.1042/BJ20021371

Zeng, J. M., Dunlop, R. A., Rodgers, K. J., & Davies, M. J. (2006). Evidence for inactivation of cysteine proteases by reactive carbonyls via glycation of active site thiols. *Biochemical Journal*, *398*, 197–206. https://doi.org/10.1042/BJ20060019

Zhang, D., Wang, Y., & Liu, H. (2020). Corn silk extract inhibit the formation of Nε-carboxymethyllysine by scavenging glyoxal/methyl glyoxal in a casein glucose-fatty acid model system. *Food Chemistry*, *309*, 125708. https://doi.org/10.1016/j.foodchem.2019.125708

Zhang, L., Bukulin, M., Kojro, E., Roth, A., Metz, V. V., Fahrenholz, F., . . . Postina, R. (2008). Receptor for advanced glycation end products is subjected to protein ectodomain shedding by metalloproteinases. *Journal of Biological Chemistry*, *283*(51), 35507–35516. https://doi.org/10.1074/jbc.M806948200

Zhou, Q., Cheng, K.-W., Gong, J., Li, E. T. S., & Wang, M. (2019). Apigenin and its methylglyoxal-adduct inhibit advanced glycation end products-induced oxidative stress and inflammation in endothelial cells. *Biochemical Pharmacology*, *166*, 231–241. https://doi.org/10.1016/j.bcp.2019.05.027

Zhou, Q., Cheng, K.-W., Xiao, J., & Wang, M. (2020). The multifunctional roles of flavonoids against the formation of advanced glycation end products (AGEs) and AGEs-induced harmful effects. *Trends in Food Science & Technology*, *103*, 333–347. https://doi.org/10.1016/j.tifs.2020.06.002

Zhu, H., Poojary, M. M., Andersen, M. L., & Lund, M. N. (2019). Effect of pH on the reaction between naringenin and methylglyoxal: A kinetic study. *Food Chemistry*, *298*, 125086.

2

Psychobiotics, a Special Type of Probiotics, and Their Potential Molecular Mechanisms to Ameliorate Symptoms of Stress and Anxiety

Tana Hernandez-Barrueta

Department of Food Science and Technology, University Of California, Davis, USA

2.1 Stress, Anxiety, and Depression, a Public Health Concern

Depressive and anxiety disorders represent the first and sixth leading cause of nonfatal health loss worldwide, according to the latest report of the World Health Organization (WHO) addressing mental health (WHO, 2017). In 2019, the Global Burden of Disease estimated that 3.3–4.1 and 3.4–4.7% of global population suffer from depressive and anxiety disorders, respectively, which translates to roughly 300 million people (GBD, 2019). These statistics, however, might vary between and within countries. For instance, in the United States, 11.2% of adults over 18 suffered regular feelings of worry, nervousness, and anxiety, while 4.7% of the population was diagnosed with depression (CDC, 2020). Although many circumstances can lead to these ailments, psychosocial stress is recognized as one of the major risk factors for both (Slattery et al., 2012; Yang et al., 2015).

Chronic stress, anxiety, and depression affect humans both personally and collectively. At the individual level, those who suffer from it might experience physical and psychological distress, along with the possible economic expenses that result from their pharmacological treatment and hospitalization (Simon, Ormel, VonKorff, & Barlow, 1995). At the community level, the best documented burden arises from the economic loss due to absence and disability from work (Wang, Simon, & Kessler, 2003).

Depression and anxiety disorders worldwide are estimated to cost US$1 trillion every year (WHO, 2019). In this regard, Chisholm et al. (2016) highlighted the need to respond to the public health and economic burden of these disorders. They proposed a scaled-up global treatment that would cost US$147 billion for 2016–2030, but with approximately US$4 return for each US$1 invested. Sharing the same spirit, WHO published its "Special Initiative for Mental Health (2019–2023)", a document that explained the plan to raise US$ 60 million over 5 years, in an effort to ensure universal mental health coverage in 12 priority countries accounting for more than 100 million people (WHO, 2019). Even though the target sum of funds is far from what Chisholm et al. (2016) recommended, WHO's

Molecular Mechanisms of Functional Food, First Edition. Edited by Rocio Campos-Vega and B. Dave Oomah.

initiative is perhaps one of the most ambitious projects addressing mental health across the globe. However, the results of such initiative and its impact can only be evaluated in the future in the prevalence of anxiety and depression disorders worldwide. Yet, there is a possibility that the COVID-19 epidemic may weaken such effort. Altogether, the statistics can only lead us to conclude what Kleinman et al. (2016) perfectly put into words in the title of their comment in The Lancet: it is "*time for mental health to come out of the shadows*."

In the past decade, scientists revealed that besides the ability of our brains to keep close communication with our guts and the microorganisms that inhabit it, this communication is in fact bidirectional in such a way that our commensal micoorganisms are capable to "talk back." Later, it was observed that ingestion of certain microorganism in adequate amounts could lead to health benefits in patients suffering from psychiatric illnesses; these microorganisms were named "psychobiotics" (Dinan, Stanton, & Cryan, 2013). Altogether, such findings lead to the hypothesis that food targeting the microbiota, such as the so-called psychobiotics, could potentially modulate behavior and thereby ameliorate the symptoms of stress, anxiety, and depression. This chapter reviews the scientific evidence suggesting that psychobiotics actually accomplish these functions. For this, we will first revise the physiological response to chronic stress and how the microbiota is involved, followed by a discussion on double-blind, randomized control trials where probiotics were tested for their effects in stressed, anxious, or depressed populations. Finally, we will disentangle some of the underlying mechanisms that advocate for probiotics as co-adjuvants in the treatment of such ailments.

2.2 Physiological Response to Stress: The HPA Axis

Several definitions of stress might exist in the literature, as largely debated in the past. Here, we define it as "*a process in which environmental demands tax or exceed the adaptive capacity of an organism, resulting in psychological and biological changes that may place persons at risk for disease*" (Cohen, Kessler, & Gordon, 1997). This definition according to Contrada and Baum (2010) accounts for important elements such as the entanglement of environmental, psychological, and biological phenomena, the acknowledgment of stress as a process, and its potential risk in the development of disease.

When the stressor originates from social interactions, we refer to it as psychosocial stress (Kaplan, 2013). Recognized chronic psychosocial stressors include, but are not limited to: early-life adversity (abuse, bullying, and family instability), financial strain, loss of a loved one, diagnose with a serious illness, divorce, work-related stress, loneliness, unpaid caregiving of a family member or friend (Rentscher, Carroll, & Mitchell, 2020), ethnic minority status, and some urban environments (McCutcheon, Bloomfield, Dahoun, Mehta, & Howes, 2019). Despite the broad nature of the stressors, those who suffer from chronic stress might experience a sustained threat perception that leads to a hyperactive hypothalamic–pituitary–adrenal (HPA) axis, characterized by increased cortisol levels. When prolonged, it can culminate in a generalized state of inflammation manifested as an increased amount of pro-inflammatory cytokines and the dysregulation of the HPA axis (McEwen, 2015). Some individuals can have a higher propensity to anxiety than others, if their responses to the stressors are exaggerated (O'Donovan, Slavich, Epel, & Neylan, 2013).

At a fundamental level (Figure 2.1), the amygdala responds to the perception of danger or threat by activating the hypothalamus so that it releases corticotropin-releasing

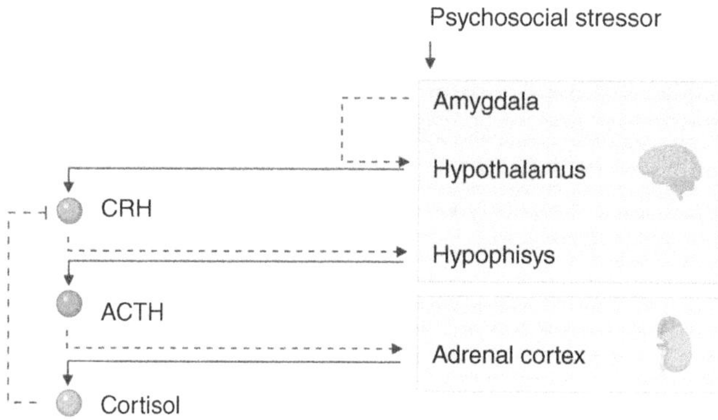

Figure 2.1 Fundamental mechanisms of the hypothalamic–pituitary–adrenal (HPA) axis in response to psychological stress. CRH: corticotropin-releasing hormone; ACTH: adrenocorticotropic hormone. Production →; Stimulation – – →; Inhibition – – –|. Created with BioRender.

hormone (CRH). Then, CRH signals the hypophysis to secrete adrenocorticotropin hormone (ACTH) to the bloodstream, which travels to the adrenal cortex where it induces cortisol and other glucocorticoids release (Chrousos & Gold, 1992; Hänsel, Hong, Cámara, & von Känel, 2010). Regardless of its reputation, cortisol operates via a negative feedback and fulfills various homeostasis-restoring functions. Among them, cortisol inhibits CRH and ACTH release while also terminating the secretion of pro-inflammatory cytokines upon its detection by the glucocorticoid receptors in immune cells. Prolonged exposure to cortisol occurring under chronic stressful events can push immune cells to a state of glucocorticoid receptors insensitivity, which accounts for the elevated levels of cortisol and pro-inflammatory cytokines in the bloodstream (Cohen et al., 2012).

This physiological process (Figure 2.1) remains incomplete until we include the gut. Almost anyone that has faced a stressful situation, either an acute or chronic one (for example, a highly demanding job, the loss of a loved one) can appreciate the interconnectedness of the mind and gut. This knowledge has prevailed not only among those who study the mind and body but through several generations. These historical associations and their evolution through history have been reported (Miller, 2011, 2018). The concept of the HPA axis was accordingly updated to incorporate the gut, from which the gut–brain axis rose as a widely accepted term in stress research. Yet, new information from the last decade required further updates in terminology with increasing understanding of human physiology. However, as will be discussed later, the vast community of microorganisms that inhabit the gut must be included as a responsive/responsible organ part of the coping mechanisms against stress.

2.3 Psychobiotics for the Treatment of Stress-Related Conditions

In anxiety and depressive disorders, one of the first-line treatments is a pharmacological family known as selective serotonin reuptake inhibitors, to which fluoxetine belongs. The effects of this type of treatment are perceived after 2 weeks of intake but might take up to 6 weeks to fully feel beneficial. Moreover, 30–50% of patients experience side effects that include nausea, diarrhea, headache, insomnia, jitteriness, restlessness, and a diminished sexual interest (Farach et al., 2012). The second-line intervention includes the prescription of benzodiazepines, such as diazepam, that specifically binds to the type A receptors of the neurotransmitter gamma-aminobutyric acid (GABA). These were previously considered as standard prescription, but their use was reduced as they tend to cause physiological dependence (Farach et al., 2012; Manthey et al., 2011).

These pharmacological approaches have existed for a long time, without significantly affecting the prevalence of chronic stress, anxiety, and depression disorders. Whether this gap is a consequence of underdiagnosis within the population, truncated treatments resulting from the undesired side effects of the medicine, hesitation to undergo medication, or intensification of socioeconomic conditions that perpetuate these ailments, complementary methods keep emerging as necessary tools to reduce their rate. Many have explored the stress and anxiety relief effect of exercise such as yoga (Shohani et al., 2018; Ströhle, 2009), outdoors activities (Kondo, Jacoby, & South, 2018), or pet ownership (Ein, Li, & Vickers, 2018), to mention some.

Despite these tactics, a dietary approach represents an encouraging alternative for populations unwilling or unable to radically change their habits (Kelly, Clarke, Cryan, & Dinan, 2016). Inspired by the large body of evidence pointing to certain probiotics as potential allies in fighting mental distress and disorders, the research group of John F. Cryan, pioneer in the field, proposed to define psychobiotics as "*live organism that, when ingested in adequate amounts, produces a health benefit in patients suffering from psychiatric illness*" (Dinan et al., 2013). Some argue this definition is based on a psychiatric criterion and therefore is far from representing an independent type of probiotic (Zendeboodi, Khorshidian, Mortazavian, & da Cruz, 2020). Table 2.1 summarizes some of the most recent double-blind, randomized, placebo-controlled trials for putative psychobiotics.

Rudzki et al. (2019) distributed 60 patients diagnosed with major depressive syndrome and that were prescribed with a selective serotonin reuptake inhibitor (escitaloptram, sertraline, paroxetine, or fluoxetine) into two groups, to complementary receive either 10×10^9 CFU day^{-1} of *Lactobacillus plantarum 299v* or placebo. After 8 weeks intervention, cognitive functions improved in patients that ingested the probiotics, in a time-dependent manner. However, no changes were detected in depression and anxiety symptoms, or in the cortisol or inflammatory cytokines. Another study of the same duration and dose allocated 81 patients equally diagnosed with major depressive disorder to receive either *Lactobacillus helveticus* R0052 and *Bifidobacterium longum* R0175, or placebo (Kazemi, Noorbala, Azam, & Djafarian, 2019). All subjects were, before and during the intervention, under one of the following medications: citalopram, fluoxetine, sertraline, or amitriptyline. After 8 weeks, patients consuming probiotics reported lower levels of depressive symptoms (Beck Depression Inventory score) and 20% less serum cortisol, but no changes in serum inflammatory cytokines. A final example was provided by Akkasheh et al. (2016), who distributed 40 patients with the same disorder but under no medication, into two groups to receive either probiotics or placebo during 8 weeks. The probiotic group ingested a mixture of *Lactobacillus acidophilus*, *Lactobacillus casei*, and *Bifidobacterium bifidum* (2×10^9 CFU of each strain). Patients ingesting the probiotics displayed reduced symptoms of depression (Beck Depression Inventory) and improved insulin resistance.

Table 2.1 Psychobiotic double-blind randomized placebo-controlled trials for stress-related conditions in human.

Condition	Strains	Dose and duration	Subjects	Results	Reference
Major depressive syndrome	*L. acidophilus, L. casei, B. bifidum*	2×10^9 CFU day^{-1} each strain 8 weeks	40 patients 20–55 years old	*↓ Beck Depression Inventory score *↓ Serum insulin *↓ Insulin resistance	Akkasheh et al. (2016)
	*L. plantarum*299v	10×10^9 CFU day^{-1} 8 weeks	40 patients 26–51 years old Under one of the following medications: escitalopram, sertraline, paroxetine, or fluoxetine	*↓ Cognitive functions *↓Kynurenine *↑ 3-hydroxynurenine *No changes in anxiety, depression symptoms, or cytokines *No change in cortisol *	Rudzki et al. (2019)
	L. helveticus, B. longum	10×10^9 CFU day^{-1} each strain 8 weeks	81 patients 28–44 years old Under one of the following medications: citalopram, fluoxetine, sertraline, or amitriptyline	*↓ Beck Depression Inventory score on probiotic but not prebiotic group *↓ Cortisol *No change in cytokines	Kazemi, Noorbala, Azam, Eskandari, and Djafarian (2019)
Generalized anxiety disorder	*B. longum, B. bifidum, B. lactis, L. acidophilus*	18×10^9 CFU day^{-1} total 8 weeks	39 patients 8–40 years old Under sertraline treatment	*↓ Score of Hamilton Rating Scale for anxiety *No changes in the score of quality of life	Eskandarzadeh et al. (2021)
Stress under educational examination	*Lactobacillus gasseri CP2305*	1×10^{10} CFU day^{-1} 24 weeks	60 healthy college medical students 25 years old	*↓ Anxiety *↓ Sleep disturbance *↑ Sleep quality *Attenuated reduction of *Bibidobacterium* and elevation of *Streptococcus*	Nishida, Sawada, Kuwano, Tanaka, and Rokutan (2019)
	Multi-strain formulation with 10–20 different strains (unspecified)	$1–5\times10^9$ CFU day^{-1} 4 weeks	86 healthy college students 18–31 years old	*Improved certain forms of anxiety depending on the formulation and dose	Tran, Zhebrak, Yacoub, Pelletier, and Hawley (2019)

	B. coagulans Unique IS2, *L. rhamnosus* UBLR58, *B. lactis* UBBLa70, *L. plantarum* UBLP40, *B. breve* UBBr01, *B. infantis* UBBI01	1–2 × 10^9 CFU day^{-1} each 4 weeks	74 students 18–24 years old	*↓Perceived stress *↓ Depression and anxiety scores ↓ Serum cortisol	Venkataraman et al. (2020)
	L. casei strain *Shitora*	100-mL fermented milk with 2 × 10^{10} CFU mL^{-1} 8 weeks	149 Japanese medical students 35–32 years old on average	*No changes in perceived stress levels *↓Cortisol in saliva *Rate of physical discomfort	Takada et al. (2016)
Moderate stress	*L. plantarum* p8	1 × 10^{10} CFU day^{-1} 12 weeks	81 subjects 20–42 years old Moderately stressed	After 4 weeks: *↓ Stress *↓ Anxiety *No change in cortisol *↓ Pro-inflammatory cytokines (TNF-α and INF-γ) *↑ Memory *Sex gender differences identified	Lew et al. (2019)
Healthy	*L. rhamnosus* HN001	6 × 10^9 CFU day^{-1} Until 6 months after labor	423 pregnant women in week 14–16 gestation	*↓ Depression score (Edinburgh Postnatal Depression Scale) *↓ Anxiety score	Slykerman et al. (2017)
	B. bifidum BGN4, *B. longum* BORI	1 × 10^{10} CFU day^{-1} 12 weeks	63 healthy elders ≥65 years old	*Improved in mental flexibility test and stress score *↑ Blood–brain-derived neurotrophic factor	Lew et al. (2019)

For generalized anxiety disorder, Eskandarzadeh et al. (2021) evaluated the effect of co-administering probiotic capsules (18×10^9 CFU of *B. longum, B. bifidum, Bifidobacterium lactis, and L. acidophilus*) or placebo, to 39 patients under sertraline treatment. The probiotic group reported greater reduction in anxiety symptoms. Clearly, the probiotic approach is incomparable to a pharmacological intervention alone. It should rather be envisioned, due to the encouraging results, as a complementary strategy to improve the life quality of those diagnosed with major depressive or anxiety disorders. The outcomes of the interventions depend on the probiotic strains, dose, and duration. Probiotics may also be beneficial in populations with pharmacological restrictions, such as pregnant or lactating women. Slykerman et al. (2017) administered 6×10^9 CFU of *Lactobacillus rhamnosus* HN001 cells to 14–16 weeks pregnant women. The treatment occurred from recruitment and until 6 months post-labor. Depression and anxiety scores were lower in women receiving the probiotic strain, thus proposing it as an intervention to prevent or treat postpartum depression. The probiotic NVP-1704 (a mixture of *Lactobacillus reuteri* NK33 [2×10^9 CFU] and *Bifidobacterium adolescentis* NK98 [0.5×10^9 CFU $500\,mg^{-1}$ NVP-1704 capsule]) was evaluated for its efficacy to improve stress, depression, anxiety, and sleep disturbances in healthy adults ($n = 63$, 18M; average age 39 years, $24\,kg\,m^{-2}$ BMI) with subclinical symptoms of depression, anxiety, and insomnia (Lee et al., 2021). NVP-1704 intake (two capsules daily for 8 weeks) significantly reduced serum pro-inflammatory cytokine IL-6 levels, insomnia symptoms (improved Pittsburg Sleep Quality Index [PSQI] and Insomnia Severity Index [ISI] scores), *Enterobacteriaceae*/*Bifidobacteriaceae* and *Enterobacteriaceae*/*Lactobacillaceae* ratios (altered gut microbiota), alleviated depression/anxiety, and improved sleep quality, particularly sleep induction. Furthermore, NVP-1704 treatment increased *Bifidobacterium* and *Lactobacillus* strains and reduced *Proteobacteria* population at the phylum level, restoring balanced gut microbiota thereby alleviating psychiatric symptoms.

Beyond populations clinically diagnosed with anxiety or depressive disorders, psychobiotics also represent a promising alternative to improve mood or lower the effects of chronic stress and anxiety in otherwise healthy subjects.

Lew et al. (2019) reported reduced self-perceived anxiety (DASS-42 questionnaire) and pro-inflammatory cytokines (tumor necrosis factor alpha [TNF-α] and interferon-gamma [INF-γ]) in moderately stressed adults ingesting *L. plantarum* P8 at 1×10^{10} CFU day^{-1} during at least 4 weeks. Notably, consumption of the probiotic exerted different psychological effects according to the subject's sex. Women showed enhanced socio-emotional cognition features and basic attention, whereas men improved their verbal memory and mental focus. Acknowledging and exploring these sex differences can help to reduce the gap in the prevalence of anxiety and depression that affect women in higher proportion (Baxter et al., 2014). Moreover, subjects who ingested *L. plantarum* P8 during the trial displayed an enhanced diversity of species-level genome bins of *B. adolescentis*, *B. longum*, and *Fecalibacterium prausnitzii*, while the abundances of *Roseburia faecis* and *Fusicatenibacter saccharivorans* decreased (Ma et al., 2021). *L. plantarum* P8, originally isolated from Mongolian fermented sour milk, is a probiotic with a remarkable capacity to colonize the host. It was detected in fecal samples for up to 4 weeks upon a single ingestion (Song et al., 2018). Its beneficial effects were significantly different from placebo only after 4 weeks of administration (Lew et al., 2019). However, not all probiotic strains exert such great colonizing capacity, which highlights the importance of dose and length of administration. For example, ingesting 8×10^9 CFU cells of *B. longum* every day during a single week had no effect on the self-perception of stress, anxiety, or depression of 79 undergraduate students (Siegel & Conklin, 2020). In contrast, the dose-dependent effect of probiotics in relieving anxiety symptoms was confirmed after 28 days of ingesting a multi-strain probiotic supplement, administrated among healthy students (Tran et al., 2019).

It has been proposed that as students feel stressed during exams they might benefit from regular consumption of probiotics. For instance, daily consumption of fermented milk containing *L. casei* strain *Shitora* resulted in lower cortisol levels in saliva and rate of negative physical symptoms of 149 medical students prior to examination (Takada et al. 2016). The students recruited at Tokushima University (Japan) were preparing for a 1-day national exam required to advance to their fifth year,arguably the most stressful situation that the Japanese medical students have to undergo. The intervention consisted in daily administration of 100 mL of fermented milk containing 2×10^{10} CFU mL^{-1} of *L. casei* strain *Shirota* during 8 weeks prior to the exam. The study was conducted 3 years in a row during 2012–2014. Others showed that students under stress can benefit from probiotic consumption by feeling less stress and anxious (Tran et al., 2019; Venkataraman et al., 2020) or increasing the quantity and quality of their sleep (Nishida et al., 2019).

Of note, psychobiotics have also been studied for their effects on mental health beyond stress. For example, regarding their effects in neurodegenerative diseases such as Alzheimer's and Parkinson's disease; and neurodevelopmental disorders such as Tourette syndrome, attention-deficit hyperactivity disorder, and autism spectrum disorder (Cheng, Liu, Wu, Wang, & Tsai, 2019).

2.4 Potential Mechanisms of Psychobiotic Activity

The studies presented in the previous section (Table 2.1) were designed to understand the therapeutic effects of potential psychobiotics consumption, rather than the underlying mechanisms of action. Therefore, only biomarkers of chronic stress, anxiety, or depression were assessed. Probiotics can act as psychobiotics via multiple pathways. The most studied mechanisms originate from their ability to produce or regulate the production of compounds that, in our central nervous system, function as neurotransmitters: GABA and serotonin. The metabolism of these neurotransmitters is targeted in the two pharmacological approaches most broadly used to treat anxiety and depression.

2.4.1 Production of Gamma-Aminobutyric Acid

Compounds other than cortisol play a role in regulating the HPA axis. GABA is one of the main inhibitory neurotransmitters in the central nervous system of humans. The paraventricular nerve located in the hypothalamus contains hypophysiotropic CRH neurons, the cells responsible for releasing CRH after a stressful stimuli (Cullinan, 2000). The GABA receptors within these neurons stop CRH secretion upon sensing GABA released from tissues in the vicinity. In fact, a third of the inhibitory inputs received by these cells apparently originate from GABA (Cullinan, Ziegler, & Herman, 2008). People exposed to chronic stress or diagnosed with anxiety and depressive disorders showed reduced GABA levels in the brain, as well as alteration in the levels of GABA receptors. Because of this, GABAergic systems are possible therapeutic targets (Fogaça & Duman, 2019). Beyond the nervous system, GABA and GABA receptors are also widely distributed along the intestinal tract, in enteric nerves and endocrine cells. There, GABA is involved in modulating intestinal motility, gastric content evacuation, gastric acid secretion, and even in the visceral sense of painful colonic stimuli (Hyland & Cryan, 2010). Remarkably, some commensal bacteria from the gut microbiota, including some probiotic strains, are known to produce GABA, and furthermore host and commensal bacteria share the metabolism of GABA synthesis (Fenalti et al., 2007), although they produce it for different reasons.

Bacteria's ability to produce GABA depends on the strain, and its widely distributed among certain lactobacilli and bifidobacterium species (Cui, Miao, Niyaphorn, & Qu, 2020; Sarasa et al., 2020). In these GABA-producing microorganisms, its synthesis signifies one of the main strategies against intra- and extracellular acidification (Wu, Tun, Law, Khafipour, & Shah, 2017). Production of GABA in probiotics arises from the presence of the *gad* genes within their genomes, which contains the instructions to synthetize the enzyme glutamate decarboxylase and the GLU/GABA antiporter (Yunes et al., 2016). Briefly, bacteria introduce glutamate to its cytosol through the GLU/GABA antiporter, which is then transformed into GABA in an irreversible reaction catalyzed by the glutamate decarboxylase. This reaction utilizes one cytosolic proton and releases CO_2, accounting for its pH-protective mechanism. Newly synthetized GABA can then evacuate the cell through the same antiporter or undergo further metabolization to a succinate molecule able to enter the Krebs cycle (Sarasa et al., 2020).

Oral administration of GABA in placebo-controlled human trials have shown some beneficial effects on relieving sleep quality and stress symptoms (Hepsomali, Groeger, Nishihira, & Scholey, 2020), although the correct doses have yet to be established. Thus, justified by its potential anxiolytic and antidepressive effect, probiotics have been used to produce GABA-rich products as one would expect such foods to exert the same benefits. Table 2.2 summarizes reports of GABA-producing strains, some of which were used to develop fermented food. Many of these strains were originally isolated from diverse traditionally fermented foods such as cheese, kimchi (Kwon et al., 2016; Wu & Shah, 2015), marcha from the Indian state of Sikkim (Das & Goyal, 2015), Thai fermented sausage (Kanklai et al., 2020) among others. Thus, it is important to recognize the ageless practices of food fermentation not only as a source of many key bioactive compounds and probiotic strains, but as an inspiration for the current development of more nutritious food. In the specific case of isolating microbes capable of producing GABA, one can only agree with the title of Selhub, Logan, and Bested (2014) paper: *ancient practice meets nutritional psychiatry*.

Many of the fermented food containing GABA encompass dairy products. However, it is important to develop alternative products in order to meet the increasing demand for non-dairy ones. For instance, Das and Goyal (2015) qualitatively demonstrated the ability of *L. plantarum* DM5 to produce GABA and reported that such probiotic at 1×10^{10} CFU mL^{-1} can inhibit ascorbate autooxidation by 20% when present in the media at 0.05 mM L^{-1}. Their results advocate for the potential to use *L. plantarum* DM5 to develop GABA-containing orange juice, where maintaining the concentration of ascorbic acid during shelflife is vital to achieve real vitamin C increase in the plasma of consumers (Johnston & Hale, 2005).

Table 2.2 GABA-producing bacteria and some fermented food containing gamma-aminobutyric acid. MSG: monosodium glutamate.

Strain	Source of isolation	Fermented food developed	GABA yield	Fermentation conditions	Reference
Lactobacillus brevis DSM 32386	Italian Alpine cheese	—	—	—	Mancini et al. (2019)
Lactobacillus plantarum DM5	Indian Marcha	—	—	—	Das and Goyal (2015)
Lactobacillus brevis	Korean Kimchi	—	—	—	Wu and Shah (2015)
Leuconostoc mesenteroides SMl, *Lactobacillus plantarum* K154	Korean Kimchi	Fermented water dropwort	10 g mL^{-1}	3% MSG 7 days	Kwon, Garcia, Song, and Lee (2016)
Lactobacillus plantarum NDC75017	Chinese fermented products	Yogurt	314 mg 100 g^{-1}	80 mM MSG 36 °C	Shan et al. (2015)
Lactobacillus helveticus ND01	Fermented horse milk	Fermented skim milk	165 mg mL^{-1}	37 °C	Sun et al. (2009)
Streptococcus thermophilus fmb5	-	Yogurt	9.66 mg mL^{-1}	20 g L^{-1} MSG 38 °C, 120 h	Chen, Alcazar, Yang, Lu, and Lu (2018)
Levilactobacillus brevis F064A	Thai fermented sausage	Fermented mulberry	3.31 mg mL^{-1}	20 mg mL^{-1} MSG 37 °C, 48 h	Kanklai, Somwong, Rungsirivanich, and Thongwai (2020)

Dietary sources of GABA rely not only on fermented food, as this amino acid is found in certain beans, spinach, oats, and mushrooms (Briguglio et al., 2018). In this regard, the rationale behind the design of probiotic food products is not limited to GABA production within the food matrix, but rather the promise of the *in situ* GABA production upon host colonization, a power that might not be achieved in other way. Hence, what ultimately matters is not only the ability of microorganism to produce GABA, but whether or not, upon ingestion, it can actually reduce the psychophysiological symptoms of stress, anxiety, and depression. Some studies (Table 2.3) respond to this question, while trying to elucidate the mechanisms of action.

Lactobacillus brevis FPA3709 isolated from fish intestines was used to ferment black soybean milk containing (w/v) 1% brown sugar, 1% monosodium glutamate, and 0.1% peptone (Ko et al., 2013). The maximum GABA concentration was 5.42 ± 0.39 mg mL^{-1} obtained after 48 h at 37 °C. The beverage successfully reduced the depressive-like behavior of mice subjected to a forced swimming test, when administered 5 mg of GABA per kg of body weight daily for 28 days. In a more comprehensive effort, a multi-strain fermented adzuki bean sprouts (*Vigna angularis*) milk containing 0.24 mg mL^{-1} of GABA ameliorated behavioral and physiological symptoms of mild depression in mice (Wu, Wang, et al., 2021). For this, rats were subjected to a social frustration stress for 10 days, while food was administered with 0.2–0.4 mg of GABA daily. The behavior was evaluated by performing a sucrose preference, social interaction, forced swimming, and tail suspension test; the physiological markers were serotonin, norepinephrine, and dopamine. The regulatory effect was attributed to the signaling pathway of GABAb-cyclic AMP-protein kinase A-cAPM-response element binding protein.

In recognition of the bidirectional association that insomnia displays with anxiety and depression (Jansson-Fröjmark & Lindblom, 2008), milk fermented by GABA-producing strain *L. brevis* DL1-11 was evaluated for sleep quantity and quality of mice challenged with the open field and elevated plus-maze test (Yu et al., 2020). The anxiety-like behavior was reduced, and the sleep quality and quantity improved in mice fed the fermented milk. The results were comparable to the effect of diazepam at 3 mg kg^{-1} of body weight. The antidepressive and anxiolytic effect of the fermented GABA-containing beverages were statistically comparable to pharmacological approaches (fluoxetine at 20 mg kg^{-1} of body weight, or diazepam, accordingly), but without exerting the undesired side effects (Ko et al., 2013; Wu, Wang, et al., 2021; Yu et al., 2020). However, these studies were conducted only with male rats which further narrows the prospective translational power of their results (issue that will be discussed in more detail later in this chapter).

The stress and anxiety relief effect of some probiotic strains may partly be explained by a vagus nerve-dependent increase in GABA concentration in the brain, as inferred after observing the magnetic resonance imaging of mice following ingestion of *L. rhamnosus* JB-1 (Janik et al., 2016). Beyond food development, other technological advances have occurred related to GABA-producing microbes. Recently, Strandwitz et al. (2019) grew in the lab a bacterial isolate KLE1783, which required the presence of *Bacteriodes fragilis* in order to thrive. Further analysis revealed that GABA was the growth factor desperately needed by the KLE1783 isolate produced by *B. fragilis*. This discovery draws attention to GABA not only as a bacterial mechanism to cope with acidity, but also as an indispensable growth factor for certain microbial communities within the gut. Moreover, it justifies further studies of microorganisms that were previously impossible to culture in the laboratory.

Table 2.3 Effect of administering fermented food containing GABA or GABA-producing probiotics to stressed male mice.

Strain	Food administered	GABA dose	Stressor	Results	Reference
Lactobacillus brevis FPA 3709	Fermented black soy bean milk	5 mg kg^{-1} of body weight 28 days	Forced swimming test	*↓Depressive-like behavior *Comparable to fluoxetine *Results dose-dependent on GABA	Ko, Lin, and Tsai (2013)
Lactobacillus plantaum 90sk, *Bifidobacterium adolescentis*150	Probiotic supplement	1×10^8 CFU 14 days		*↓Depressive-like behavior *Comparable to fluoxetine	Yunes et al. (2020)
Lactobacillus. brevis J1	Multi-strain fermented adzuki bean sprout milk	3.3 mg kg^{-1} 10 days	Chronic depression model	*↓Depressive-like behavior *Comparable to fluoxetine	Wu, Wang, et al. (2021)
Lactobacillus brevis DL1-11	Fermented milk	33.3 mg kg^{-1} 30 days	Open field test and elevatedplus maze	*↓ Anxiety behavior *↑Sleep time *Comparable to diazepam * Results dose-dependent on GABA	Yu et al. (2020)

2.4.2 Modulation of Host Serotonin Production

Serotonin (5-hydroxytryptamine) is a monoamine produced from tryptophan in two steps, catalyzed first by the tryptophan hydroxylase (1 or 2) enzyme, and then by an aromatic amino acid decarboxylase (Bader, 2020). In the brain, serotonin acts as a neurotransmitter involved in modulating mood, cognition, reward, and memory (Maffei, 2020), and in motility and gastrointestinal inflammatory response in the gut (Gershon, 2013). Since this molecule is unable to cross the blood–brain barrier, its synthesis occurs separately in the brain and the gut (El-Merahbi, Löffler, Mayer, & Sumara, 2015). Yet, as it will be discussed later, its role in the gut can equally affect behavior.

Dysregulations in the metabolism of serotonin as well as their transporters and receptors are associated with the pathogenesis of stress-related disorders such as irritable bowel syndrome (Jin et al., 2016), social anxiety (Frick et al., 2015; Lanzenberger et al., 2007), migraine attacks (Izzati-Zade, 2008) and depression (Drevets et al., 2007). Despite its role as a neurotransmitter, about 90% of the serotonin (if what observed in mice is true for humans) is actually synthetized in the gut by enterochromaffin cells (Hata et al., 2017). Apparently, serotonin production by these cells is promoted by indigenous microorganisms. Yano et al. (2015) reported colonic and fecal serotonin deficit in germ-free mice, but no difference in enterochromaffin cells abundance, suppression of tryptophan hydroxylase 1 (rate-limiting enzyme), and increased tryptophan concentration in the feces and serum. However, oral supplementation of serotonin but not tryptophan ameliorated the serotonin deficit in the colon and serum. Altogether, their results demonstrated the role of gut microbiota in promoting serotonin biosynthesis by upregulating tryptophan hydroxylase 1 expression in colonic enterochromaffin cells. Moreover, spore-forming microbes (particularly *Clostridium* genre) restored the serum and colonic levels of serotonin and alleviated the associated deficient intestinal motility through the production of signaling molecules, as inferred by the use of cell-free extracts (Yano et al., 2015). Further illustrating the bidirectional association between the microbiota and the host, Fung et al. (2019) later determined that oral serotonin supplementation to mice elevated the serotonin levels in feces that correlated with increased relative abundance of certain spore-forming gut microorganisms.

The follow-up question is which microbial metabolites [referred to as "signaling molecules" according to Yano et al., 2015] can trigger serotonin production by enterochromaffin cells. Ye et al. (2020) attributed such serotonin-enhancing capability to two bacterial metabolites derived from tryptophan: indole and indole-3-carboxaldehyde. Their results were based on *in vivo* (zebra-fish model) experiments, as well as *in vitro* studies using human serotonin receptors expressed in human embryonic kidney (HEK) cells. Another candidate was recently proposed by Engevik et al. (2021) using a combination of gnotobiotic mice as well as mice and human enteroid cultures. They demonstrated that the short chain fatty acid acetate, produced by *Bifidobacterium dentium*, stimulates serotonin production by enterochromaffin cells. Furthermore, they demonstrated that *B. dentium* upon colonization can modulate the expression of serotonin receptors both in the gut and brain, and thereby change the anxiety-like behavior of mice, previously linked to such receptors. Interestingly, Engevik et al. (2021) also reported that other acetate-producing bacteria such as the human commensal *Escherichia coli* MG1655 failed to stimulate serotonin secretion, suggesting that other microbial metabolites from *B. dentium* might be working coordinately. Apparently, GABA was identified as the component of interest since it can be produced by *B. dentium* (Pokusaeva et al., 2017), and enterochromaffin cells are also responsive to it.

Previously in this chapter, the inability of gut-derived serotonin to cross the blood–brain barrier was mentioned. Accordingly, the reader should question how its production in the gut, to some extent due to the influence of the microbiota, can then affect one's behavior. It is well documented that the communication between the gut and the brain relies on the vagus nerve (Bonaz, Bazin, & Pellissier, 2018; Fülling, Dinan, & Cryan, 2019). Further supporting these findings, Ye et al. (2020) also revealed that certain microbial communities through tryptophan catabolism enhance serotonin secretion from a specific line of enterochromaffin cells ultimately resulting in vagal sensory ganglia activation.

Although some commensal bacteria can produce serotonin, the major role of microbiota appears to be its ability to regulate its production in the gut. In this regard, new probiotic strains emerge as potential psychobiotics that can ameliorate the psychological and physiological effects of stress, anxiety, and depression (Table 2.4). For example, administration of *Lactobacillus reuteri* decreased the acetate concentration and depressive-like behavior in mice and increased the serotonin in blood and colon (Xie et al., 2020). These results initially appear counterintuitive to the previous discussion of *B. deuterium* ability to stimulate serotonin production in enterochromaffin cells via acetate production (Engevik et al., 2021). However, in-depth evaluation of Xie et al. (2020) data reveals that although *L. reuteri* administration suppressed the apparently higher concentration of acetate in the depressed mice, its proportion relative to the total short chain fatty acid remained constant within all groups of mice. These results illustrate that different probiotic strains can utilize strain-specific mechanisms to modulate serotonin metabolism.

In human trials, Chong et al. (2019) administered *L. plantarum* DR7, previously isolated from Malaysian fresh cow's milk, to 111 stressed but otherwise healthy women and men between 18 and 60 years old, in a randomized, double-blind, placebo-controlled trial. Subjects ingested a daily dose of 1×10^9 log CFU for 12 weeks. After just 8 weeks, those receiving probiotics had reduced levels of stress, anxiety, and total psychological scores evaluated with the Depression, Anxiety and Stress Scale (DASS-42) questionnaire. Moreover, pro-inflammatory cytokines (interferon-γ and transforming growth factor-α) levels decreased and anti-inflammatory cytokines (interleukin 10) levels increased compared to the control group. These effects were attributed to the modulation of serotonin pathways, as inferred by an increase in tryptophan hydroxylase-2 and 5-hydroxytryptamine receptor-6 observed in the probiotic group. Later, the same research group analyzed the microbiota changes associated with *L. plantarum* DR7 administration in their previous study. The administration of the probiotic strain modulated the bowel movements induced by stress and prevented the reduction of gut microbiota diversity while maintaining their relative abundance (Liu, Chong, Chung, Li, & Liong, 2020).

Serotonin also serves as a precursor of melatonin (N-acetyl-5-methoxytryptamine), hormone produced via enzymatic acetylation and methylation of serotonin. Melatonin is best known for its role in modulating the circadian rhythm but is also involved in gut permeability (Gao et al., 2018). Beck et al. (2019) reported the ability of *L. reuteri* ATG-F4 (isolated from newborn humans) to modulate gene expression related to circadian rhythm.

Table 2.4 Probiotic interventions in stress-related conditions, where the potential psychobiotic mechanisms were partially attributed to serotonin pathways.

Strain	Dose	Subject	Stressor	Results	Reference
Lactobacillus reuteri 3	1×10^{10} CFU day^{-1} 4 weeks	Male mice	Chronic social defeat	*↓ Depression-like behavior *Prevented the ↑*Clostridiales* ↓*Lactobacillus*↓ in blood and colonic serotonin *↑*Bifidobacterium* *↓ Acetate *↑ Serotonin receptors	Xie et al. (2020)
Lactobacillus rhamnosus JB-1	1×10^{9} CFU day^{-1} 4 weeks			*↑ Aggressor avoidance *↓ Sociability	Liu, Steinhausen, et al. (2020)
	1.67×10^{9} CFU day^{-1} 4 weeks			*↓Anxiety behavior *Prevented decrease in social interactions *↑ Anti-inflammatory cytokines	Bharwani, Mian, Surette, Bienenstock, and Forsythe (2017)
Lactobacillus plantarum IS-10506	2.67×10^{9} CFU day^{-1} 1 week		No stressor	*↑ Serotonin in the intestine *↑Expression of serotonin transporter in the brain	Ranuh et al. (2019)
Bifidobacterium longum subsp. *infantis* CCFM687	1×10^{9} CFU mL^{-1} 6 weeks	Mice	Forced swimming, restrain, water deprivation, isolation, food depravation, wet bedding, no bedding, tilt cages, clip tail, crowded space, and continuous illumination	*↑ Serotonin in the brain *Alleviated depressive-like behavior comparable to fluoxetine *Prevented the loss of alpha diversity and abundance of butyrate-producing bacteria	Tian et al. (2019)

Therefore, the chronoregulation ability of probiotics was suggested as an equally important criterion to determine their psychobiotic capacity. This, in accordance with the disruption in the circadian rhythm, is known to be part of the physiopathology of stress-related disorders. However, the role of melatonin in human health is out of the scope of this chapter, as it will be further discussed later in this book in the section concerning chronobiotics.

2.5 Limitations of Psychobiotic Research

Complementary treatment of mental disorders with probiotic interventions requires further research to correctly establish adequate dosage and duration of the treatments. Moreover, it should not be assumed that all probiotic strains are beneficial across all mental disorders as each illness has a specific associated physiopathology. For instance, the independent administration of sertraline or *L. rhamnosus* JB-1 was detrimental in male rats during early stages of a model of post-traumatic stress disorder (Liu, Steinhausen, et al., 2020). If a positive outcome could be achieved in mice under advanced stage of post-traumatic stress disorder remains unclear, these results establish the need to include a disorder's stage as a factor to consider.

Other unexpected outcomes have been associated with *L. rhamnosus* JB-1. Kelly et al. (2017) attempted to translate their findings in stressed mice to healthy human. The probiotic strain administered to 29 healthy male volunteers for 8 weeks did not change their mood, perceived levels of stress and anxiety, as well as sleep quality. When these men were subjected to a stress test, the inflammatory cytokines concentration showed no significant difference compared to placebo group. Two possible explanations were proposed: that the beneficial effects of this probiotic strain is limited to subjects with mental disorders thus excluding healthy individuals, or that *L. rhamnosus* JB-1 does not exert the same effects in humans. As Kelly et al. (2017) reflected over the possibility of host-specific probiotic effects contributing to failure in translational studies, they opened the discussion on whether mice, largely utilized as a model in stress-related research, accurately translates to the human context. Pigs were suggested as more appropriate model for psychosocial stress and the underling mechanisms behind brain-to-immune and immune-to-brain communications (Gimsa, Tuchscherer, & Kanitz, 2018). This recommendation is based on cortisol as the main corticosteroid hormone in humans and pigs, whereas corticosterone is the equivalent in rats; and that the circadian rhythm of glucocorticoids in mice and rat is opposite to that of human and pigs. Pigs already form part of the studies addressing cardiovascular disease, gastrointestinal and pharmacological research. In this context, Menneson et al. (2019) validated a porcine model of psychosocial stress (based on isolation, environmental impoverishment, and unpredictability); they proposed it as a tool to study the gut–brain and behavior, in order to shorten the bridge for translational research. However, we acknowledge that using pigs as a model represents a bigger challenge in terms of cost, time, and space.

Epidemiological data show that the prevalence of anxiety and depression is higher among women (which accounts for half the human population worldwide). However, several preclinical and clinical trials continue to use only male subjects to understand the mechanisms from which potential psychobiotics exert their therapeutic effects. This limitation is mostly neglected or never addressed in many publications. When it is mentioned, two reasons are more often provided: the effect of hormones that might mask the phenomenon, and that the lack of difference in the evaluated phenomenon across the two sexes make males representative enough. For example, in the methodology section of their paper, Kelly et al. (2017) wrote "*Male participants were selected to avoid the influence of the menstrual cycle, which can impact upon cortisol output and other read-outs and all preclinical studies with this bacteria to date have focused on male animals.*". Still, one can argue that both rationale more vigorously account for the use of female subjects. The third possible reason would be that incorporating females increases the cost of the study. Upon this last argument, one can only wonder: is there no experimental design powerful enough to include them without exponentially increasing the number of subjects needed?

2.6 Current and Future Perspectives of Psychobiotics Market

The concept of psychobiotics is relatively novel. The research about this topic remains in its infancy, although it continues to gather attention from the scientific community as depicted by the increasing number of publications released (Figure 2.2).

The studies discussed in Section 2.3. suggest the beneficial effects of complementing medication with the consumption of certain probiotics in patients suffering from depressive and anxiety disorders, especifically certain strains of *L. acidophilus, L. casei, B. longum, B. bifidum*, or *L. plantarum* (299v). In this case, psychobiotic interventions can be considered as an adjuvant in the treatment of such illness.

Beyond these clinical ailments, the concept of psychobiotics has already been exploited both by the supplement and food industry regarding the treatment of chronic stress (Table 2.5). For instance, a patent claims that a product containing *L. plantarum* spp. can prevent and treat mental illness as well as symptoms associated with chronic stress (Stenman et al., 2020). The patent includes *L. paracasei* Lpc-37, which is comercialized

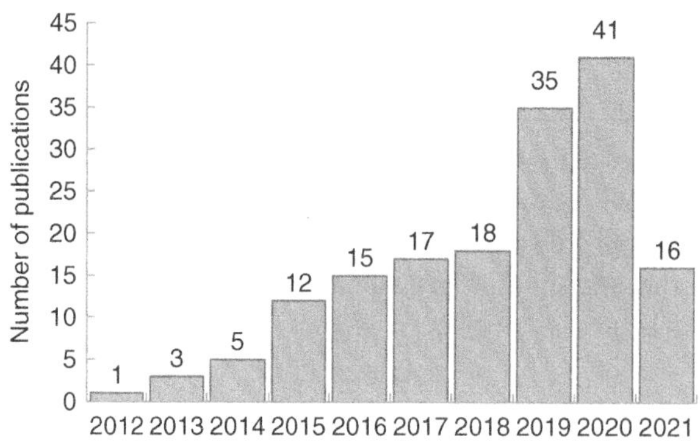

Figure 2.2 Publications about psychobiotics according to Pub Med data.

Table 2.5 Patented psychobiotics and their claimed effects on mental health.

Psychobiotic	Mental health benefit claim	Patent number	Reference
L. plantarum spp.	Prevent or treat symptoms associated with chronic stress	US Patent 2020/0306327	Stenman, Lehtinen, and Mao (2020)
Bifidobacterium longum strain NCIMB 41676	Improve sleep quality	US Patent 2021/0127695	Murphy, Groeger, Dinan, and Cryan (2021)
Lactobacillus plantarum subsp. *plantarum* PS128	Prevent or treat movement disorder	US Patent 2019/0201461	Tsai et al. (2019)
Heat-killed *Lactobacillus fermentum* and *Lactobacillus delbrueckii*; 9 : 1 ratio	Treat anxiety, depression, mood disturbances, or stress-related conditions	US Patent 2020/0405787	Hill, Dinan, Warda, and Perrett (2020)

under the name of Huwaru® Calm. Recently, a clinical trial (The Sisu study) reported that *L. paracasei* LPc-37 can reduce perceived stress, as well as other benefits such as ameliorating hearth rate during accute stress and normalizing cortisol level dependending on sex and stress levels (Patterson, Griffin, Ibarra, Ellsiepen, & Hellhammer, 2020).

Bened Biomedical Co., Ltd. (Taipei, Taiwan) commercializes a patented psychobiotic: *Lactobacillus plantarum* PS129™, strain originally isolated from spontaneously fermented mustard greens in Taiwan (Tsai, Hsu, Liao, Cheng, & You, 2019). Eight weeks of 1×10^{10} CFU day^{-1} of *L. plantarum* PS129™ administered to Taiwanese Information Technology workers improved their self-perceived stress, overall job stress, job burden, cortisol level, and sleep disturbances (Wu, Wu, et al. 2021). They concluded that the product can be beneficial to relieve stress and improve mental health of people with high-stress jobs. These results advocate for this type of products to be available at work cafeterias, particularly in places such as hospitals where doctors, nurses, and caregivers converge, all of them previously reported as subgroups with high levels of chronic stress and burnout syndrome. *L. plantarum* PS129 ™ patent declares that it effectively treats or prevents numerous and diverse neurological conditions associated with basal ganglia dysfunction by increasing dopamine turnover rate in the prefrontal cortex (Tsai et al., 2019); the company recently announced they will move to randomized controlled trials for the same strain in patients suffering from Parkinson's disease, and Tourette and Rett syndrome (Koe, 2021).

A patent for *B. longum* 1714™ strain NCIMB 41676 (10^9 CFU day^{-1}) claims its administration for 5 weeks improves sleep quality (measured as increased sleep duration, reduced sleep disturbance, amount of time to fall asleep, and daytime dysfunction due to sleep disturbance) in humans. Additionally, 8 weeks intake of the same strain prior to examination improved sleep duration and efficiency and prevented the negative effects of stressful period on sleep quality (according to the Pittsburgh Sleep Quality Index) of healthy individuals (20 subjects of both sex, 18–24 years old) in a double-blind, randomized, placebo-controlled, cross-over study. Moreover, in an open label study, the combination of *B. longum* 1714™ and *B. longum* 35624® strain improved sleep quality in adults (40 female, 18–55 years old) with irritable bowel syndrome and/or mild to moderate anxiety or depression for 8 weeks followed by an 8-week washout (Murphy et al. 2021). Notably, inventors of the patent for *B. longum* 1714™ include pioneers in the field of psychobiotics and co-authors of the paper who first proposed such a definition (Dinan et al., 2013).

Lacteol® is derived from a culture solution containing heat-killed cells of *Lactobacillus* LB strain (a combination of *Lactobacillus fermentum* and *Lactobacillus delbrueckii*; 9 to 1 ratio) and fermented culture medium. Lacteol® produces psychobiotic effect (treating anxiety, depression, mood disturbances, or stress-related conditions) in animal and humans by modifying (reducing) the amount of *Alistipes* and/or *Odoribacter* species (Hill et al., 2020).

Regarding psychobiotics in the food industry, the company Yakult Honsha Co., Ltd., which produces arguably the world's leading probiotic beverage Yakult™, released a product in the Japanese market called Yakult 1000™. This product is advertised as a beverage capable of relieving stress and improving the quality of sleep under temporary situations that cause mental stress (Yakult Honsha Co., Ltd, 2021). This product perhaps derives from the work of Takada et al. (2016), regarding the reduction of stress in medical students (discussed in Section 2.3). Interestingly, Takada et al. (2016) used a stress model in rats to elucidate the mechanism behind the results observed in humans. Here, they showed that the ingestion of *L. casei* strain *Shirota* suppressed elevated glucocorticoids caused by stressful events by inhibiting the activation of corticotropin releasing factor-producing cells in the paraventricular nucleus of the hypothalamus. According to their data, these events were most likely mediated by the vagus nerve. From this and the results discussed in Section 2.3 (Table 2.1), one could envision a future where probiotic beverages are available at school cafeterias as part of the daily menu to promote mental health among students.

2.7 Conclusion

This chapter reviewed some results supporting the promise of a future where our decisions in the supermarket become a stronger ally as we seek a holistic health that includes our minds. As discussed, the ability of probiotics to produce or modulate GABA and serotonin production account for one of the well-documented mechanisms of action to exert their psychobiotic activity (Figure 2.3). Yet, current knowledge still requires further insights into which stress-related ailments will benefit from the consumption of specific strains along with the correct doses and durations. Once these relationships are established, the food industry will have the responsibility to provide supplements or food with the adequate amounts of psychobiotics (or psychobiotic supplements) and to advertise them in an evidence-based manner. On the other hand, the consumer will have the responsibility to continually incorporate them in their diet as their effects are regularly not observed after a single dose. Finally, it should be unambiguously stated: the information provided in this chapter is under no circumstance a suggestion that depression, anxiety, and chronic stress can or should be treated entirely through our diets; it is no substitute for professional advice. Instead, psychobiotics should be considered as a complementary strategy upon proof that they can ameliorate the physiological or psychological symptoms of very specific stress-related ailments.

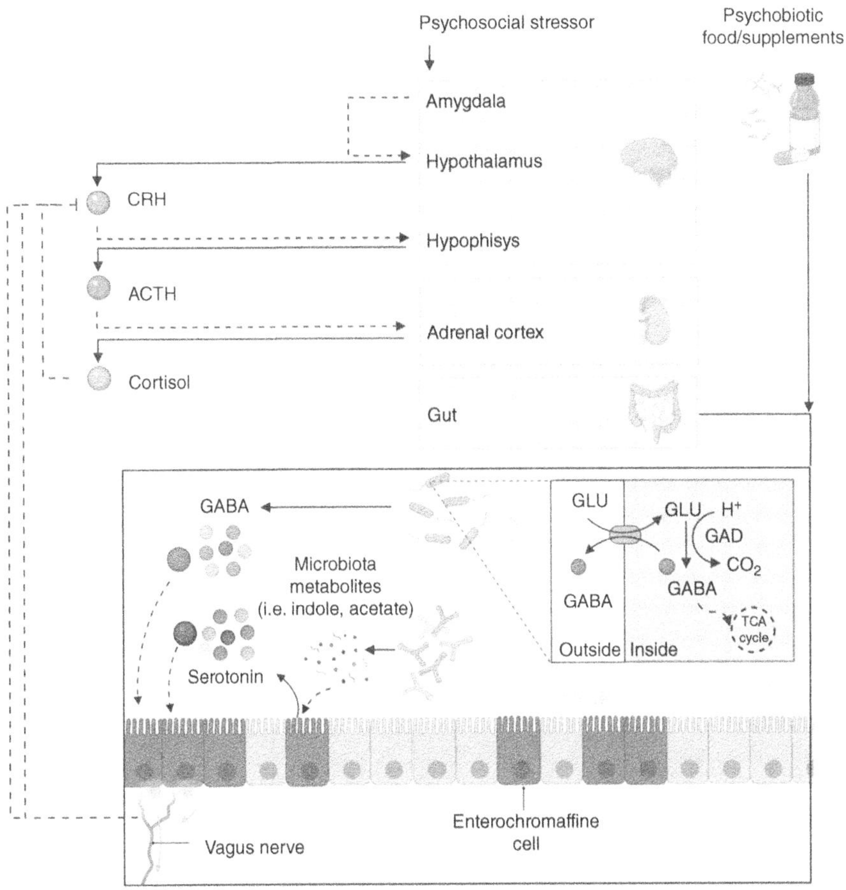

Figure 2.3 Mechanisms of psychobiotic activity that affect the hypothalamic–pituitary–adrenal (HPA) axis in response to psychological stress. CRH: corticotropin-releasing hormone; ACTH: adrenocorticotropic hormone. Production →; Stimulation — — →; Inhibition — — —|. Created with BioRender.

References

Akkasheh, G., Kashani-Poor, Z., Tajabadi-Ebrahimi, M., Jafari, P., Akbari, H., Taghizadeh, M., . . . Esmaillzadeh, A. (2016). Clinical and metabolic response to probiotic administration in patients with major depressive disorder: A randomized, double-blind, placebo-controlled trial. *Nutrition*, *32*(3), 315–320. https://doi.org/10.1016/j.nut.2015.09.003

Bader, M. (2020). Inhibitison of serotonin synthesis: A novel therapeutic paradigm. *Pharmacology and Therapeutics*, *205*, 107423. https://doi.org/10.1016/j.pharmthera.2019.107423

Baxter, A. J., Scott, K. M., Ferrari, A. J., Norman, R. E., Vos, T., & Whiteford, H. A. (2014). Challenging the myth of an "epidemic" of common mental disorders: trends in the global prevalence of anxiety and depression between 1990 and 2010. *Depression and Anxiety*, *31*(6), 506–516. https://doi.org/10.1002/da.22230

Beck, B. R., Park, G.-S., Jeong, D. Y., Lee, Y. H., Im, S., Song, W. H., & Kang, J. (2019). Multidisciplinary and comparative investigations of potential psychobiotic effects of *lactobacillus* strains isolated from newborns and their impact on gut microbiota and ilealtranscriptome in a healthy murine model. *Frontiers in Cellular and Infection Microbiology*, *9*, 269. https://doi.org/10.3389/fcimb.2019.00269

Bharwani, A., Mian, M. F., Surette, M. G., Bienenstock, J., & Forsythe, P. (2017). Oral treatment with *Lactobacillus rhamnosus* attenuates behavioural deficits and immune changes in chronic social stress. *BMC Medicine*, *15*(1), 7. https://doi.org/10.1186/s12916-016-0771-7

Bonaz, B., Bazin, T., & Pellissier, S. (2018). The vagus nerve at the interface of the microbiota-gut-brain axis. *Frontiers in Neuroscience*, *12*, 49. https://doi.org/10.3389/fnins.2018.00049

Briguglio, M., Dell'Osso, B., Panzica, G., Malgaroli, A., Banfi, G., Zanaboni Dina, C., . . . Porta, M. (2018). Dietary neurotransmitters: A narrative review on current knowledge. *Nutrients*, *10*(5). https://doi.org/10.3390/nu10050591

CDC. (2020). National Center for Health Statistics. Mental Health.Retrieved from https://www.cdc.gov/nchs/fastats/mental-health.htm.

Chen, L., Alcazar, J., Yang, T., Lu, Z., & Lu, Y. (2018). Optimized cultural conditions of functional yogurt for γ-aminobutyric acid augmentation using response surface methodology. *Journal of Dairy Science*, *101*(12), 10685–10693. https://doi.org/10.3168/jds.2018-15391

Cheng, L. H., Liu, Y. W., Wu, C. C., Wang, S., & Tsai, Y. C. (2019). Psychobiotics in mental health, neurodegenerative and neurodevelopmental disorders. *Journal of Food and Drug Analysis*, *27*(3), 632–648. https://doi.org/10.1016/j.jfda.2019.01.002 Epub 2019 Feb 10. PMID: 31324280

Chisholm, D., Sweeny, K., Sheehan, P., Rasmussen, B., Smit, F., Cuijpers, P., & Saxena, S. (2016). Scaling-up treatment of depression and anxiety: A global return on investment analysis. *The Lancet, Psychiatry*, *3*(5), 415–424. https://doi.org/10.1016/S2215-0366(16)30024-4

Chong, H. X., Yusoff, N. A. A., Hor, Y. Y., Lew, L. C., Jaafar, M. H., Choi, S. B., . . . Liong, M. T. (2019). *Lactobacillus plantarum* DR7 alleviates stress and anxiety in adults: a randomised, double-blind, placebo-controlled study. *Beneficial Microbes*, *10*(4), 355–373. https://doi.org/10.3920/BM2018.0135

Chrousos, G. P., & Gold, P. W. (1992). The concepts of stress and stress system disorders. Overview of physical and behavioral homeostasis. *The Journal of the American Medical Association*, *267*(9), 1244–1252.

Cohen, S., Janicki-Deverts, D., Doyle, W. J., Miller, G. E., Frank, E., Rabin, B. S., & Turner, R. B. (2012). Chronic stress, glucocorticoid receptor resistance, inflammation, and disease risk. *Proceedings of the National Academy of Sciences of the United States of America*, *109*(16), 5995–5999. https://doi.org/10.1073/pnas.1118355109

Cohen, S., Kessler, R. C., & Gordon, L. U. (1997). *Measuring stress: A guide for health and social scientists* (pp. 3). Oxford University Press.

Contrada, R., & Baum, A. (2010). *The handbook of stress science: Biology, psychology, and health (Illustrated.)*. Springer Publishing Company.

Cui, Y., Miao, K., Niyaphorn, S., & Qu, X. (2020). Production of gamma-aminobutyric acid from lactic acid bacteria: A systematic review. *International Journal of Molecular Sciences*, *21*(3). https://doi.org/10.3390/ijms21030995

Cullinan, W. E. (2000). GABA(A) receptor subunit expression within hypophysiotropic CRH neurons: A dual hybridization histochemical study. *The Journal of Comparative Neurology*, *419*(3), 344–351. https://doi.org/10.1002/(sici)1096-9861(20000410)419:3<344::aid-cne6>3.0.co;2-z

Cullinan, W. E., Ziegler, D. R., & Herman, J. P. (2008). Functional role of local GABAergic influences on the HPA axis. *Brain Structure and Function*, *213*(1–2), 63–72. https://doi.org/10.1007/s00429-008-0192-2

Das, D., & Goyal, A. (2015). Antioxidant activity and γ-aminobutyric acid (GABA) producing ability of probiotic *Lactobacillus plantarum* DM5 isolated from Marcha of Sikkim. *LWT - Food Science and Technology*, *61*(1), 263–268. https://doi.org/10.1016/j.lwt.2014.11.013

Dinan, T. G., Stanton, C., & Cryan, J. F. (2013). Psychobiotics: A novel class of psychotropic. *Biological Psychiatry*, *74*(10), 720–726. https://doi.org/10.1016/j.biopsych.2013.05.001

Drevets, W. C., Thase, M. E., Moses-Kolko, E. L., Price, J., Frank, E., Kupfer, D. J., & Mathis, C. (2007). Serotonin-1A receptor imaging in recurrent depression: replication and literature review. *Nuclear Medicine and Biology*, *34*(7), 865–877. https://doi.org/10.1016/j.nucmedbio.2007.06.008

Ein, N., Li, L., & Vickers, K. (2018). The effect of pet therapy on the physiological and subjective stress response: A meta-analysis. *Stress and Health: Journal of the International Society for the Investigation of Stress*, *34*(4), 477–489. https://doi.org/10.1002/smi.2812

El-Merahbi, R., Löffler, M., Mayer, A., & Sumara, G. (2015). The roles of peripheral serotonin in metabolic homeostasis. *FEBS Letters*, *589*(15), 1728–1734. https://doi.org/10.1016/j.febslet.2015.05.054

Engevik, M. A., Luck, B., Visuthranukul, C., Ihekweazu, F. D., Engevik, A. C., Shi, Z., . . . Versalovic, J. (2021). Human-derived bifidobacterium dentium modulates the mammalian serotonergic system and gut-brain axis. *Cellular and Molecular Gastroenterology and Hepatology*, *11*(1), 221–248. https://doi.org/10.1016/j.jcmgh.2020.08.002

Eskandarzadeh, S., Effatpanah, M., Khosravi-Darani, K., Askari, R., Hosseini, A. F., Reisian, M., & Jazayeri, S. (2021). Efficacy of a multispecies probiotic as adjunctive therapy in generalized anxiety disorder: A double blind, randomized, placebo-controlled trial. *Nutritional Neuroscience*, *24*(2), 102–108. https://doi.org/10.1080/1028415X.2019.1598669

Farach, F. J., Pruitt, L. D., Jun, J. J., Jerud, A. B., Zoellner, L. A., & Roy-Byrne, P. P. (2012). Pharmacological treatment of anxiety disorders: Current treatments and future directions. *Journal of Anxiety Disorders*, *26*(8), 833–843. https://doi.org/10.1016/j.janxdis.2012.07.009

Fenalti, G., Law, R. H., Buckle, A. M., Langendorf, C., Tuck, K., Rosado, C. J., . . . Whisstock, J. C. (2007). GABA production by glutamic acid decarboxylase is regulated by a dynamic catalytic loop. *Nature Structural and Molecular Biology*, *14*(4), 280–286. https://doi.org/10.1038/nsmb1228

Fogaça, M. V., & Duman, R. S. (2019). Cortical gabaergic dysfunction in stress and depression: New insights for therapeutic interventions. *Frontiers in Cellular Neuroscience*, *13*, 87. https://doi.org/10.3389/fncel.2019.00087

Frick, A., Åhs, F., Engman, J., Jonasson, M., Alaie, I., Björkstrand, J., . . . Furmark, T. (2015). Serotonin synthesis and reuptake in social anxiety disorder: A positron emission tomography study. *JAMA Psychiatry*, *72*(8), 794–802. https://doi.org/10.1001/jamapsychiatry.2015.0125

Fülling, C., Dinan, T. G., & Cryan, J. F. (2019). Gut microbe to brain signaling: What happens in vagus. . .. *Neuron, 101*(6), 998–1002. https://doi.org/10.1016/j.neuron.2019.02.008

Fung, T. C., Vuong, H. E., Luna, C. D. G., Pronovost, G. N., Aleksandrova, A. A., Riley, N. G., . . . Hsiao, E. Y. (2019). Intestinal serotonin and fluoxetine exposure modulate bacterial colonization in the gut. *Nature Microbiology, 4*(12), 2064–2073. https://doi.org/10.1038/s41564-019-0540-4

Gao, J., Xu, K., Liu, H., Liu, G., Bai, M., Peng, C., . . . Yin, Y. (2018). Impact of the gut microbiota on intestinal immunity mediated by tryptophan metabolism. *Frontiers in Cellular and Infection Microbiology, 8*, 13. https://doi.org/10.3389/fcimb.2018.00013

GBD.(2019). Global Burden of Disease Study 2019. Retrieved from http://ghdx.healthdata.org/gbd-results-tool

Gershon, M. D. (2013). 5-Hydroxytryptamine (serotonin) in the gastrointestinal tract. *Current Opinion in Endocrinology, Diabetes, and Obesity, 20*(1), 14–21. https://doi.org/10.1097/MED.0b013e32835bc703

Gimsa, U., Tuchscherer, M., & Kanitz, E. (2018). Psychosocial stress and immunity-what can we learn from pig studies? *Frontiers in Behavioral Neuroscience, 12*, 64. https://doi.org/10.3389/fnbeh.2018.00064

Hänsel, A., Hong, S., Cámara, R. J. A., & von Känel, R. (2010). Inflammation as a psychophysiological biomarker in chronic psychosocial stress. *Neuroscience and Biobehavioral Reviews, 35*(1), 115–121. https://doi.org/10.1016/j.neubiorev.2009.12.012

Hata, T., Asano, Y., Yoshihara, K., Kimura-Todani, T., Miyata, N., Zhang, X.-T., . . . Sudo, N. (2017). Regulation of gut luminal serotonin by commensal microbiota in mice. *PLoS One, 12*(7), e0180745. https://doi.org/10.1371/journal.pone.0180745

Hepsomali, P., Groeger, J. A., Nishihira, J., & Scholey, A. (2020). Effects of oral gamma-aminobutyric acid (GABA) administration on stress and sleep in humans: A systematic review. *Frontiers in Neuroscience, 14*, 923. https://doi.org/10.3389/fnins.2020.00923

Hill, C., Dinan, T., Warda, A. & Perrett, S. (2020). New use of microbiological composition.*U.S. Patent* 2020/0405787 A1. Issued December 31, 2020.

Hyland, N. P., & Cryan, J. F. (2010). A gut feeling about GABA: Focus on GABA(B) receptors. *Frontiers in Pharmacology, 1*, 124. https://doi.org/10.3389/fphar.2010.00124

Izzati-Zade, K. F. (2008). The role of serotonin in the pathogenesis and clinical presentations of migraine attacks. *Neuroscience and Behavioral Physiology, 38*(5), 501–505. https://doi.org/10.1007/s11055-008-9008-3

Janik, R., Thomason, L. A. M., Stanisz, A. M., Forsythe, P., Bienenstock, J., & Stanisz, G. J. (2016). Magnetic resonance spectroscopy reveals oral Lactobacillus promotion of increases in brain GABA, N-acetyl aspartate and glutamate. *Neuroimage, 15*(125), 988–995. https://doi.org/10.1016/j.neuroimage.2015.11.018

Jansson-Fröjmark, M., & Lindblom, K. (2008). A bidirectional relationship between anxiety and depression, and insomnia? A prospective study in the general population. *Journal of Psychosomatic Research, 64*(4), 443–449. https://doi.org/10.1016/j.jpsychores.2007.10.016

Jin, D.-C., Cao, H.-L., Xu, M.-Q., Wang, S.-N., Wang, Y.-M., Yan, F., & Wang, B.-M. (2016). Regulation of the serotonin transporter in the pathogenesis of irritable bowel syndrome. *World Journal of Gastroenterology, 22*(36), 8137–8148. https://doi.org/10.3748/wjg.v22.i36.8137

Johnston, C. S., & Hale, J. C. (2005). Oxidation of ascorbic acid in stored orange juice is associated with reduced plasma vitamin C concentrations and elevated lipid peroxides. *Journal of the American Dietetic Association, 105*(1), 106–109. https://doi.org/10.1016/j.jada.2004.10.026

Kanklai, J., Somwong, T. C., Rungsirivanich, P., & Thongwai, N. (2020). Screening of GABA-producing lactic acid bacteria from Thai fermented foods and probiotic potential of *Levilactobacillusbrevis* F064A for GABA-fermented mulberry juice production. *Microorganisms, 9*(1). https://doi.org/10.3390/microorganisms9010033

Kaplan, H. B. (2013). *Psychosocial stress: Trends in theory and research*. Academic Press.

Kazemi, A., Noorbala, A. A., Azam, K., & Djafarian, K. (2019). Effect of prebiotic and probiotic supplementation on circulating pro-inflammatory cytokines and urinary cortisol levels in patients with major depressive disorder: A double-blind, placebo-controlled randomized clinical trial. *Journal of Functional Foods, 52*, 596–602. https://doi.org/10.1016/j.jff.2018.11.041

Kazemi, A., Noorbala, A. A., Azam, K., Eskandari, M. H., & Djafarian, K. (2019). Effect of probiotic and prebiotic vs placebo on psychological outcomes in patients with major depressive disorder: A randomized clinical trial. *Clinical Nutrition, 38*(2), 522–528. https://doi.org/10.1016/j.clnu.2018.04.010

Kelly, J. R., Allen, A. P., Temko, A., Hutch, W., Kennedy, P. J., Farid, N., . . . Dinan, T. G. (2017). Lost in translation? The potential psychobiotic*Lactobacillus rhamnosus* (JB-1) fails to modulate stress or cognitive performance in healthy male subjects. *Brain, Behavior, and Immunity, 61*, 50–59. https://doi.org/10.1016/j.bbi.2016.11.018

Kelly, J. R., Clarke, G., Cryan, J. F., & Dinan, T. G. (2016). Brain-gut-microbiota axis: Challenges for translation in psychiatry. *Annals of Epidemiology, 26*(5), 366–372. https://doi.org/10.1016/j.annepidem.2016.02.008

Kim, C.-S., Cha, L., Sim, M., Jung, S., Chun, W. Y., Baik, H. W., & Shin, D.-M. (2021). Probiotic supplementation improves cognitive function and mood with changes in gut microbiota in community-dwelling older adults: A randomized, double-blind, placebo-controlled, multicenter trial. *The Journals of Gerontology. Series A, Biological Sciences and Medical Sciences, 76*(1), 32–40. https://doi.org/10.1093/gerona/glaa090

Kleinman, A., Estrin, G. L., Usmani, S., Chisholm, D., Marquez, P. V., Evans, T. G., & Saxena, S. (2016). Time for mental health to come out of the shadows. *The Lancet, 387*(10035), 2274–2275. https://doi.org/10.1016/S0140-6736(16)30655-9

Ko, C. Y., Lin, H.-T. V., & Tsai, G. J. (2013). Gamma-aminobutyric acid production in black soybean milk by *Lactobacillus brevis* FPA 3709 and the antidepressant effect of the fermented product on a forced swimming rat model. *Process Biochemistry, 48*(4), 559–568. https://doi.org/10.1016/j.procbio.2013.02.021

Koe, T. (2021, March 16). *Psychobiotic RCTs: Bened to trial PS128 on Tourette and Rett syndrome patients following new findings on Parkinson's [Online]*. Retrieved from https://www.nutraingredients-asia.com.

Kondo, M. C., Jacoby, S. F., & South, E. C. (2018). Does spending time outdoors reduce stress? A review of real-time stress response to outdoor environments. *Health and Place, 51*, 136–150. https://doi.org/10.1016/j.healthplace.2018.03.001

Kwon, S.-Y., Garcia, C. V., Song, Y.-C., & Lee, S.-P. (2016). GABA-enriched water dropwort produced by co-fermentation with *Leuconostocmesenteroides* SM and Lactobacillus plantarum K154. *LWT - Food Science and Technology*, *73*, 233–238. https://doi.org/10.1016/j.lwt.2016.06.002

Lanzenberger, R. R., Mitterhauser, M., Spindelegger, C., Wadsak, W., Klein, N., Mien, L.-K., . . . Tauscher, J. (2007). Reduced serotonin-1A receptor binding in social anxiety disorder. *Biological Psychiatry*, *61*(9), 1081–1089. https://doi.org/10.1016/j.biopsych.2006.05.022

Lee, H. J., Hong, J. K., Kim, J. K., Kim, D. H., Jang, S. W., Han, S. W., & Yoon, I. Y. (2021). Effects of probiotic NVP-1704 on mental health and sleep in healthy adults: An 8-week randomized, double-blind, placebo-controlled trial. *Nutrients*, *13*(8), 2660. https://doi.org/10.3390/nu13082660

Lew, L.-C., Hor, Y.-Y., Yusoff, N. A. A., Choi, S.-B., Yusoff, M. S. B., Roslan, N. S., . . . Liong, M.-T. (2019). Probiotic *Lactobacillus plantarum* P8 alleviated stress and anxiety while enhancing memory and cognition in stressed adults: A randomised, double-blind, placebo-controlled study. *Clinical Nutrition*, *38*(5), 2053–2064. https://doi.org/10.1016/j.clnu.2018.09.010

Liu, G., Chong, H.-X., Chung, F. Y.-L., Li, Y., & Liong, M.-T. (2020). *Lactobacillus plantarum* DR7 modulated bowel movement and gut microbiota associated with dopamine and serotonin pathways in stressed adults. *International Journal of Molecular Sciences*, *21*(13). https://doi.org/10.3390/ijms21134608

Liu, Y., Steinhausen, K., Bharwani, A., Mian, M. F., McVey Neufeld, K.-A., & Forsythe, P. (2020). Increased persistence of avoidance behaviour and social deficits with L.rhamnosus JB-1 or selective serotonin reuptake inhibitor treatment following social defeat. *Scientific Reports*, *10*(1), 13485. https://doi.org/10.1038/s41598-020-69968-y

Ma, T., Jin, H., Kwok, L.-Y., Sun, Z., Liong, M.-T., & Zhang, H. (2021). Probiotic consumption relieved human stress and anxiety symptoms possibly via modulating the neuroactive potential of the gut microbiota. *Neurobiology of Stress*, *14*, 100294. https://doi.org/10.1016/j.ynstr.2021.100294

Maffei, M. E. (2020). 5-Hydroxytryptophan (5-HTP): Natural occurrence, analysis, biosynthesis, biotechnology, physiology and toxicology. *International Journal of Molecular Sciences*, *22*(1). https://doi.org/10.3390/ijms22010181

Mancini, A., Carafa, I., Franciosi, E., Nardin, T., Bottari, B., Larcher, R., & Tuohy, K. M. (2019). *In vitro* probiotic characterization of high GABA producing strain *Lactobacilluasbrevis* DSM 32386 isolated from traditional "wild" Alpine cheese. *Annals of Microbiology*, *69*(13), 1435–1443. https://doi.org/10.1007/s13213-019-01527-x

Manthey, L., van Veen, T., Giltay, E. J., Stoop, J. E., Neven, A. K., Penninx, B. W. J. H., & Zitman, F. G. (2011). Correlates of (inappropriate) benzodiazepine use: The Netherlands study of depression and anxiety (NESDA). *British Journal of Clinical Pharmacology*, *71*(2), 263–272. https://doi.org/10.1111/j.1365-2125.2010.03818.x

McCutcheon, R. A., Bloomfield, M. A. P., Dahoun, T., Mehta, M., & Howes, O. D. (2019). Chronic psychosocial stressors are associated with alterations in salience processing and corticostriatal connectivity. *Schizophrenia Research*, *213*, 56–64. https://doi.org/10.1016/j.schres.2018.12.011

McEwen, B. S. (2015). Biomarkers for assessing population and individual health and disease related to stress and adaptation. *Metabolism: Clinical and Experimental*, *64*(3 Suppl 1), S2–S10. https://doi.org/10.1016/j.metabol.2014.10.029

Menneson, S., Ménicot, S., Ferret-Bernard, S., Guérin, S., Romé, V., Le Normand, L., . . . Val-Laillet, D. (2019). Validation of a psychosocial chronic stress model in the pig using a multidisciplinary approach at the gut-brain and behavior levels. *Frontiers in Behavioral Neuroscience*, *13*, 161. https://doi.org/10.3389/fnbeh.2019.00161
Miller, I. (2011). *A modern history of the stomach: Gastric illness, medicine and British society* (pp. 1800–1950). Pickering and Chatto.
Miller, I. (2018). The gut-brain axis: Historical reflections. *Microbial Ecology in Health and Disease*, *29*(1), 1542921. https://doi.org/10.1080/16512235.2018.1542921
Murphy, E. F., Groeger, D., Dinan, T. & Cryan, J. (2021). *Bifidobacterium longum* NCIMB 41676. *US Patent 2021/0127695* A1; issuedMay 6, 2021.
Nishida, K., Sawada, D., Kuwano, Y., Tanaka, H., & Rokutan, K. (2019). Health benefits of *Lactobacillus gasseri* CP2305 tablets in young adults exposed to chronic stress: A randomized, double-blind, placebo-controlled study. *Nutrients*, *11*(8). https://doi.org/10.3390/nu11081859
O'Donovan, A., Slavich, G. M., Epel, E. S., & Neylan, T. C. (2013). Exaggerated neurobiological sensitivity to threat as a mechanism linking anxiety with increased risk for diseases of aging. *Neuroscience and Biobehavioral Reviews*, *37*(1), 96–108. https://doi.org/10.1016/j.neubiorev.2012.10.013
Patterson, E., Griffin, S. M., Ibarra, A., Ellsiepen, E., & Hellhammer, J. (2020). *Lacticaseibacillusparacasei* Lpc-37® improves psychological and physiological markers of stress and anxiety in healthy adults: A randomized, double-blind, placebo-controlled and parallel clinical trial (the Sisu study). *Neurobiology of Stress*, *24*(13), 100277. https://doi.org/10.1016/j.ynstr.2020.100277 PMID: 33385020; PMCID: PMC7770962
Pokusaeva, K., Johnson, C., Luk, B., Uribe, G., Fu, Y., Oezguen, N., . . . Versalovic, J. (2017). GABA-producing *Bifidobacterium dentium* modulates visceral sensitivity in the intestine. *Neurogastroenterology and Motility*, *29*(1). https://doi.org/10.1111/nmo.12904
Ranuh, R., Athiyyah, A. F., Darma, A., Risky, V. P., Riawan, W., Surono, I. S., & Sudarmo, S. M. (2019). Effect of the probiotic *Lactobacillus plantarum* IS-10506 on BDNF and 5HT stimulation: Role of intestinal microbiota on the gut-brain axis. *Iranian Journal of Microbiology*, *11*(2), 145–150.
Rentscher, K. E., Carroll, J. E., & Mitchell, C. (2020). Psychosocial stressors and telomere length: A current review of the science. *Annual Review of Public Health*, *41*, 223–245. https://doi.org/10.1146/annurev-publhealth-040119-094239
Rudzki, L., Ostrowska, L., Pawlak, D., Małus, A., Pawlak, K., Waszkiewicz, N., & Szulc, A. (2019). Probiotic *Lactobacillus Plantarum*299v decreases kynurenine concentration and improves cognitive functions in patients with major depression: A double-blind, randomized, placebo controlled study. *Psychoneuroendocrinology*, *100*, 213–222. https://doi.org/10.1016/j.psyneuen.2018.10.010
Sarasa, S. B., Mahendran, R., Muthusamy, G., Thankappan, B., Selta, D. R. F., & Angayarkanni, J. (2020). A brief review on the non-protein amino acid, gamma-amino butyric acid (GABA): Its production and role in microbes. *Current Microbiology*, *77*(4), 534–544. https://doi.org/10.1007/s00284-019-01839-w
Selhub, E. M., Logan, A. C., & Bested, A. C. (2014). Fermented foods, microbiota, and mental health: Ancient practice meets nutritional psychiatry. *Journal of Physiological Anthropology*, *33*, 2. https://doi.org/10.1186/1880-6805-33-2

Shan, Y., Man, C. X., Han, X., Li, L., Guo, Y., Deng, Y., . . . Jiang, Y. J. (2015). Evaluation of improved γ-aminobutyric acid production in yogurt using *Lactobacillus plantarum* NDC75017. *Journal of Dairy Science*, *98*(4), 2138–2149. https://doi.org/10.3168/jds.2014-8698

Shohani, M., Badfar, G., Nasirkandy, M. P., Kaikhavani, S., Rahmati, S., Modmeli, Y., . . . Azami, M. (2018). The effect of yoga on stress, anxiety, and depression in women. *International Journal of Preventive Medicine*, *9*, 21. https://doi.org/10.4103/ijpvm.IJPVM_242_16

Siegel, M. P., & Conklin, S. M. (2020). Acute intake of B. longum probiotic does not reduce stress, anxiety, or depression in young adults: A pilot study. *Brain, Behavior, and Immunity -Health*, *2*, 100029. https://doi.org/10.1016/j.bbih.2019.100029

Simon, G., Ormel, J., VonKorff, M., & Barlow, W. (1995). Health care costs associated with depressive and anxiety disorders in primary care. *The American Journal of Psychiatry*, *152*(3), 352–357. https://doi.org/10.1176/ajp.152.3.352

Slattery, D. A., Uschold, N., Magoni, M., Bär, J., Popoli, M., Neumann, I. D., & Reber, S. O. (2012). Behavioural consequences of two chronic psychosocial stress paradigms: anxiety without depression. *Psychoneuroendocrinology*, *37*(5), 702–714. https://doi.org/10.1016/j.psyneuen.2011.09.002

Slykerman, R. F., Hood, F., Wickens, K., Thompson, J. M. D., Barthow, C., Murphy, R., . . . Probiotic in Pregnancy Study Group. (2017). Effect of *Lactobacillus rhamnosus* HN001 in pregnancy on postpartum symptoms of depression and anxiety: A randomised double-blind placebo-controlled trial. *EBioMedicine*, *24*, 159–165. https://doi.org/10.1016/j.ebiom.2017.09.013

Song, Y., He, Q., Zhang, J., Qiao, J., Xu, H., Zhong, Z., . . . Zhang, H. (2018). Genomic variations in probiotic *Lactobacillus plantarum* P-8 in the human and rat gut. *Frontiers in Microbiology*, *9*, 893. https://doi.org/10.3389/fmicb.2018.00893

Stenman, L., Lehtinen, M., & Mao, Y. (2020). Probiotics for cognitive and mental health. *US Patent 2020/0306327* A1, issued Oct. 1. 2020.

Strandwitz, P., Kim, K. H., Terekhova, D., Liu, J. K., Sharma, A., Levering, J., . . . Lewis, K. (2019). GABA-modulating bacteria of the human gut microbiota. *Nature Microbiology*, *4*(3), 396–403. https://doi.org/10.1038/s41564-018-0307-3

Ströhle, A. (2009). Physical activity, exercise, depression and anxiety disorders. *Journal of Neural Transmission*, *116*(6), 777–784. https://doi.org/10.1007/s00702-008-0092-x

Sun, T., Zhao, S., Wang, H., Cai, C., Chen, Y., & Zhang, H. (2009). ACE-inhibitory activity and gamma-aminobutyric acid content of fermented skim milk by Lactobacillus helveticus isolated from Xinjiang koumiss in China. *European Food Research and Technology*, *228*(4), 607–612. https://doi.org/10.1007/s00217-008-0969-9

Takada, M., Nishida, K., Kataoka-Kato, A., Gondo, Y., Ishikawa, H., Suda, K., . . . Rokutan, K. (2016 Jul). Probiotic *Lactobacillus casei* strain Shirota relieves stress-associated symptoms by modulating the gut-brain interaction in human and animal models. *Neurogastroenterology and Motility*, *28*(7), 1027–1036. https://doi.org/10.1111/nmo.12804

Tian, P., Zou, R., Song, L., Zhang, X., Jiang, B., Wang, G., . . . Chen, W. (2019). Ingestion *of Bifidobacterium longum* subspecies infantis strain CCFM687 regulated emotional behavior and the central BDNF pathway in chronic stress-induced depressive mice through reshaping the gut microbiota. *Food and Function*, *10*(11), 7588–7598. https://doi.org/10.1039/c9fo01630a

Tran, N., Zhebrak, M., Yacoub, C., Pelletier, J., & Hawley, D. (2019). The gut-brain relationship: Investigating the effect of multispecies probiotics on anxiety in a randomized placebo-controlled trial of healthy young adults. *Journal of Affective Disorders, 252*, 271–277. https://doi.org/10.1016/j.jad.2019.04.043

Tsai, Y-C.,Hsu, C-C., Liao, J-F., Cheng, Y-F. & You, S-T. (2019). Method for preventing or treating movement disorder with lactic acid bacterium. *World Patent 2018/014225* A1; issuedJanuary 25, 2018.

Venkataraman, R., Madempudi, R. S., Neelamraju, J., Ahire, J. J., Vinay, H. R., Lal, A., . . . Stephen, S. (2020). *Effect of Multi-strain probiotic formulation on students facing examination stress: A double-blind*. Placebo-Controlled Study: *Probiotics and Antimicrobial Proteins*. https://doi.org/10.1007/s12602-020-09681-4

Wang, P. S., Simon, G., & Kessler, R. C. (2003). The economic burden of depression and the cost-effectiveness of treatment. *International Journal of Methods in Psychiatric Research, 12*(1), 22–33. https://doi.org/10.1002/mpr.139

WHO. (2017). *Depression and other common mental disorders. Global Health Estimates*. World Health Organization.

WHO. (2019). The WHO special initiative for mental health (2019–2023): Universal health coverage for mental health (No. WHO/MSD/19.1). World Health Organization. Retrieved from https://apps.who.int/iris/handle/10665/310981

Wu, Q., & Shah, N. P. (2015). Gas release-based prescreening combined with reversed-phase HPLC quantitation for efficient selection of high-γ-aminobutyric acid (GABA)-producing lactic acid bacteria. *Journal of Dairy Science, 98*(2), 790–797. https://doi.org/10.3168/jds.2014-8808

Wu, Q., Tun, H. M., Law, Y. S., Khafipour, E., & Shah, N. P. (2017). Common distribution of gad Operon in *Lactobacillus brevis* and its GadA contributes to efficient GABA synthesis toward cytosolic near-neutral pH. *Frontiers in Microbiology, 14*(8), 206. https://doi.org/10.3389/fmicb.2017.00206

Wu, S.-I., Wu, C.-C., Tsai, P.-J., Cheng, L.-H., Hsu, C.-C., Shan, I.-K., . . . Tsai, Y.-C. (2021). Psychobiotic supplementation of PS128^{TM} improves stress, anxiety, and insomnia in highly stressed information technology specialists: A pilot study. *Frontiers in Nutrition, 8*, 614105.

Wu, Z., Wang, P., Pan, D., Zeng, X., Guo, Y., & Zhao, G. (2021). Effect of adzuki bean sprout fermented milk enriched in γ-aminobutyric acid on mild depression in a mouse model. *Journal of Dairy Science, 104*(1), 78–91. https://doi.org/10.3168/jds.2020-19154

Xie, R., Jiang, P., Lin, L., Jiang, J., Yu, B., Rao, J., . . . Qiao, Y. (2020). Oral treatment with *Lactobacillus reuteri* attenuates depressive-like behaviors and serotonin metabolism alterations induced by chronic social defeat stress. *Journal of Psychiatric Research, 122*, 70–78. https://doi.org/10.1016/j.jpsychires.2019.12.013

Yakult Honsha Co., Ltd. (2021, April 16). President's new year's message [Online]. Retrieved from https://www.yakult.co.jp/english/ir/message/.

Yang, L., Zhao, Y., Wang, Y., Liu, L., Zhang, X., Li, B., & Cui, R. (2015). The effects of psychological stress on depression. *Current Neuropharmacology, 13*(4), 494–504.

Yano, J. M., Yu, K., Donaldson, G. P., Shastri, G. G., Ann, P., Ma, L., . . . Hsiao, E. Y. (2015). Indigenous bacteria from the gut microbiota regulate host serotonin biosynthesis. *Cell, 161*(2), 264–276. https://doi.org/10.1016/j.cell.2015.02.047

Ye, L., Bae, M., Cassilly, C. D., Jabba, S. V., Thorpe, D. W., Martin, A. M., . . . Rawls, J. F. (2020). Enteroendocrine cells sense bacterial tryptophan catabolites to activate enteric and vagal neuronal pathways. *Cell Host and Microbe*. https://doi.org/10.1016/j.chom.2020.11.011

Yu, L., Han, X., Cen, S., Duan, H., Feng, S., Xue, Y., . . . Chen, W. (2020). Beneficial effect of GABA-rich fermented milk on insomnia involving regulation of gut microbiota. *Microbiological Research*, *233*, 126409. https://doi.org/10.1016/j.micres.2020.126409

Yunes, R. A., Poluektova, E. U., Dyachkova, M. S., Klimina, K. M., Kovtun, A. S., Averina, O. V., . . . Danilenko, V. N. (2016). GABA production and structure of gadB/gadC genes in *Lactobacillus* and *Bifidobacterium*strains from human microbiota. *Anaerobe*, *42*, 197–204. https://doi.org/10.1016/j.anaerobe.2016.10.011

Yunes, R. A., Poluektova, E. U., Vasileva, E. V., Odorskaya, M. V., Marsova, M. V., Kovalev, G. I., & Danilenko, V. N. (2020). A multi-strain potential probiotic formulation of GABA-Producing *Lactobacillus plantarum* 90sk and Bifidobacteriumadolescentis 150 with antidepressant effects. *Probiotics and Antimicrobial Proteins*, *12*, 973–979. https://doi.org/10.1007/s12602-019-09601-1

Zendeboodi, F., Khorshidian, N., Mortazavian, A. M., & da Cruz, A. G. (2020). Probiotic: Conceptualization from a new approach. *Current Opinion in Food Science*, *32*, 103–123. https://doi.org/10.1016/j.cofs.2020.03.009

3

Molecular Mechanisms of Chronobiotics as Functional Foods

Elisa Dufoo-Hurtado[1], Abraham Wall-Medrano[2], and Rocio Campos-Vega[1]

[1] *Research and Graduate Studies in Food Science, School of Chemistry, Universidad Autónoma de Querétaro (UAQ), Querétaro, Mexico*
[2] *Instituto de Ciencias Biomédicas, Universidad Autónoma de Ciudad Juárez, Chihuahua, Mexico*

3.1 Introduction

The establishment of *chrononutrition* as an emerging topic within *chronobiology* has a peculiar history. Empirical observations on rhythmic "sunrise to sunset" patterns in living organisms, the object of study in *chronobiology* (a.k.a. *circadian biology*), comes from the Hippocratic era, the term desynchronosis and its relationship of environmental light and temperature with sleep patterns date from the fifties (Folk Jr, 1957). The first review on chronobiology indexed in Scopus appeared in 1920 (Halberg, 1969). Jürgen Aschoff and Collin Pittendrigh are considered the fathers of modern chronobiology with their pioneering work on the natural entrainment (synchronization) of circadian (circa "around" or "approximately", diem "day") systems, derived from their so-called "resonance" and "bunker" experiments (Daan, 2000). Konopka and Benzer (1971) were perhaps the first to report that certain genes located in the X chromosome controls the timekeeping system in *D. melanogaster*, while the CLOCK (*circadian locomotor output cycles protein kaput*) gene reported by Takahashi in 1997 is one of the circadian genes controlling a plethora of physiological process (including the sleep–wake cycle) in mammals (Semenova, Madaeva, & Kolesnikova, 2021). However, the 2017 Nobel Prize in Physiology or Medicine was jointly awarded to Jeffrey Hall, Michael Rosbash, and Michael Young for their work on the mechanism underlying the circadian rhythm. Concurrently to modern chronobiology, chrononutrition was also leaving its mark. The scientific basis of modern chrononutrition (Dufoo-Hurtado, Wall-Medrano, & Campos-Vega, 2020; Halberg, 1989) is based on the biological periodicity of liver glycogen deposition and/or changes in body temperature in birds, humans, and other mammals (Ågren, Wilander, & Jorpes, 1931; Higgins & Mann, 1926), the body core temperature altered by starvation and *MESOR* (*Midline Estimating Statistic of Rhythm*) (Nelson & Halberg, 1986) and the role of dietary tryptophan (Trp) as a synchronizer of the circadian rhythm with external zeitgebers (environmental cues) (Wurtman, Shoemaker, Larin and Rosenthal, 1968). Although the use of the term

Molecular Mechanisms of Functional Food, First Edition. Edited by Rocio Campos-Vega and B. Dave Oomah.

Table 3.1 Advances in scientific publication on chronobiology, chrononutrition, and chronobiotics.

Year	Chronobiology	Chronobiotic	Chrononutrition
1900–1959	20	0	0
1960–1969	37	0	0
1970–1979	840	26	0
1980–1989	2630	41	3
1990–1999	5100	271	6
2000–2009	11400	1080	76
2010–2019	18300	2030	887
2020–2021*	5380	411	455

* Publications as of June 2021.
Source: Google scholar.

chronobiotic began in the seventies for drugs with chronotypic effect such as pentobarbital and theophylline–a green coffee bioactive (Ehret, Potter, & Dobra, 1975), the term chrononutrition appeared in scientific publications from 1980 onward (Table 3.1) and formal Japanese textbooks between 2005 and 2009 (Oda, 2015).

The body's circadian system consists of the central (SCN) and peripheral (organs) clocks that allow metabolic, physiological, and behavioral functions to oscillate between daylight and darkness (Ba-Ali et al., 2019). Such rhythmicity avoids the concurrence of conflicting behaviors (e.g. hunger vs. sleep) and cellular homeostasis (e.g. anabolic vs. catabolic processes), reducing the odds for metabolic disturbances (Challet, 2015; Plano et al., 2017). Chronodisruption should actually be considered bidirectional since wake–sleep cycles are severely disrupted by diseases affecting cytoplasmic homeostasis (Beesley et al., 2020). Also, the food anticipatory activity (FAA) attributed to the food-entrainable oscillator located outside the SCN precede circadian responses such as gene expression, enzyme secretion, hormone release, and afferent signals necessary for gastrointestinal digestion, nutrients bioconversion, and absorption before feeding (Shibata, Hirao, & Tahara, 2010). Chrononutrition has a remarkable translational, clinical, and therapeutic potential in regulating the circadian rhythm by modifying dietary patterns (e.g. timing and diet quality) and promoting the intake of chronobiotics such as phytomelatonin, prebiotics, glucose, amino acids, alcohol, retinoic acid, and short-chain fatty acids (SCFAs) (Dufoo-Hurtado, Olvera-Bautista, Wall-Medrano, Loarca-Piña, & Campos-Vega, 2021; Froy, 2010). The dietary sources, biological mechanisms, and health benefits of natural chronobiotics are discussed in depth in the following sections.

3.2 Circadian Clock and Diseases

Chronobiology is responsible for the study of circadian rhythms, defined as physical, biochemical, and cognitive changes that follow a 24-hr cycle. The central clock (brain) coordinates the biological clock of each cell causing convenient rhythmicity to sustain homeostatic

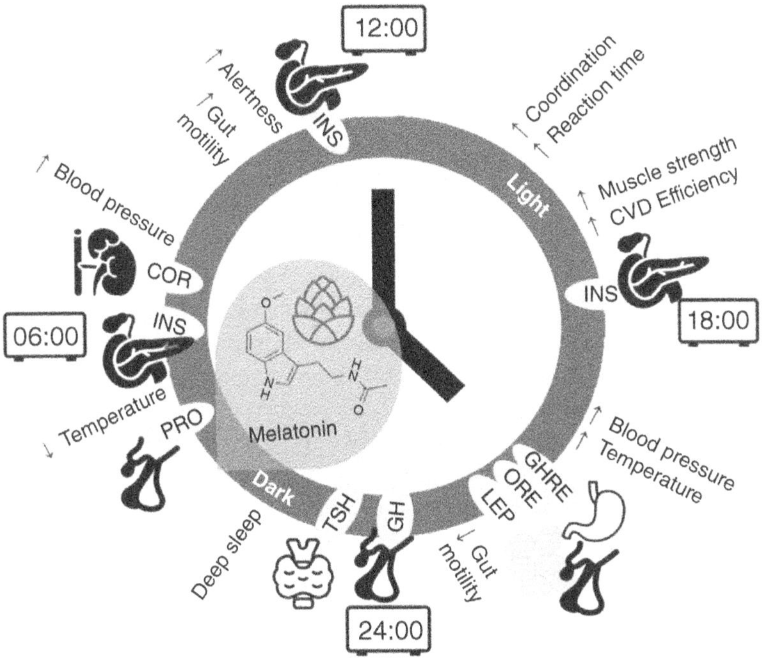

Figure 3.1 Circadian rythm, endocrine response, and body functions are controlled by melatonin. Cortisol (COR), Cardiovascular (CV), ghrelein (GHRE), growth hormone (GH), insulin (INS), leptin (LEP), maximal (↑) and minimal (↓) level, orexins (ORE), metabolite *peak max* (yellow ellypsoid), prolactin (PRO), thyroid-stimulating hormone (TSH). Circadian synthesis of melatonin and peak max ~ 3AM (gray ballon). *Source:* The authors.

processes and survival. Pineal melatonin (N-acetyl-5-hydroxytryptamine) is synthesized from tryptophan in the dark phase of the day and is the main endocrine output signal of the body's timekeeping system (Aulinas, 2020). The rhythmic secretion in mammals is driven by the central (master) circadian clock (oscillator) located in the suprachiasmatic nucleus, a small region in the hypothalamus, situated directly above the optic chiasm (Challet, 2015). Nocturnal secretion of melatonin (peak max at ~ 3 AM; Figure 3.1) is the kick-off signal that delivers the "circadian message" to peripheral circadian clocks during sleep time while decreased daytime melatonin promotes alertness, and this daily melatonin rhythmicity controls the timing of many body functions (Ba-Ali et al., 2019). Binding sites [G protein coupled melatonin receptors (MT_1 and MT_2) and non-receptors (amphipathic sites)] for melatonin have been detected in many areas of the brain (e.g. pars tuberalis and hypothalamus), the immune system, gonads, kidney, liver, and the cardiovascular (CV) system (Aulinas, 2020; Cecon, Liu, & Jockers, 2019).

Peripheral clocks (a.k.a. oscillators) occur in many organs, all synchronized to the master oscillator via direct autonomic and/or endocrine outputs (e.g. leptin, ghrelin, and cortisol; Figure 3.1) and indirect blood-borne metabolites, temporally gated by the CNS (Morris, Aeschbach, & Scheer, 2012; Tsujino & Sakurai, 2012; Challet, 2015; Plano et al., 2017; Chellappa, Vujovic, Williams, & Scheer, 2019). This interconnectivity results in a tightly controlled molecular transcription-translation feedback loops (TTFL) in a 24-hr temporal framework, and in mammals this TTFL include several genes including CLOCK, *Bmal*1 (aryl hydrocarbon receptor nuclear translocator-like 1; *Arntl/Bmal*1), and clock-controlled genes (*Ccgs*) such as *Periodcircadian protein homologue* 1–3 (*Per*1, *Per*2 and *Per*3) and *Cryptochrome* 1 and 2 (*Cry*1 and *Cry*2) (Beesley et al., 2020; Huang, Lu, & Ho, 2021). Very detailed reviews on the TTFL process and melatonin circadian production have been published (Aulinas, 2019; Crnko, Du Pré, Sluijter, & Van Laake, 2019; Semenova et al., 2021). Any genetic, physiological, or environmental factor leading to chronodisruption and impairment of melatonin synthesis will alter all peripheral clocks, initiating many pathological processes. The intensity of nocturnal light, wakefulness, use of antihypertensives, benzodiazepines, or nonsteroidal anti-inflammatory drugs or alcohol consumption, cause phase-shifting or hypomelatoninemia, while posture at night, exercise, antidepressants, and caffeine increase blood melatonin(Aulinas, 2019). Circadian misalignments further increase the risk for chronic diseases, including obesity (Froy, 2010; Laermans & Depoortere, 2016), cardiovascular disease (CVD) (Chellappa et al., 2019; Crnko et al., 2019), diabetes and insulin resistance (Khandelwal, Dutta, Chittawar, & Kalra, 2017; McMullan, Schernhammer, Rimm, Hu, & Forman, 2013; Stenvers, Scheer, Schrauwen, La Fleur, & Kalsbeek, 2019), osteoporosis (Winter et al., 2021), and cancer (Song et al., 2018), to name a few.

Obesity. The "energetic centers" located in the hypothalamus and brain stem control food intake, sleep/wakefulness, locomotor activity, and energy expenditure (Laermans & Depoortere, 2016). Several cells connected to the network of storage, distribution, and metabolism of energy substrates (e.g. adipose tissue, liver, skeletal muscle, and intestine) have their own circadian machinery (Froy, 2010), and several detailed experiments have revealed the relationship between clock proteins and the expression of enzymes related to basal (during sleep), postprandial (during daylight), and physical activity energy expenditure (Orozco-Solis et al., 2016; Zitting et al., 2018). Clock gene mutant and knockout animals (e.g. ClockΔ19, *Per*1/*Per*2, *Bmal*1, *Cry*1, *and Rev-erbα* mice) often exhibit morbid/early-onset obesity, hyperphagia, hyperleptinemia, hepatic steatosis, and insulin resistance (Laermans & Depoortere, 2016). Conversely, time-imposed physical exercise commonly recommended for obese patients entrains behavioral rhythms (Tahara, Aoyama, & Shibata, 2017), while bariatric surgery modifies the expression patterns of clock proteins in rodents (Kim, Son, Kang, Ha, & Ha, 2015) and night eating syndrome in humans (De Zwaan, Marschollek, & Allison, 2015).

Cardiovascular diseases. Epidemiological observations indicate that individual chronotypes, a measure of preferred timing of sleep and activity, defines its mortality risk from CVD; evening types (e.g. second-shift workers) being at higher risk (Chellappa et al., 2019). Virtually all CV cells (endothelial and vascular smooth muscle cells, myocardial fibroblasts, cardiomyocytes, and cardiac progenitor-like cells) have clock oscillators and the rhythmic expression of clock genes begins during cardiac differentiation. CV cells collectively regulate endothelial function, thrombus formation, blood pressure, and heart rate, while chronodisruption leads to heart failure, cardiomyopathy, myocardial infarction, bradycardia, and TTFL-related proteome/transcriptome abnormalities (Crnko et al., 2019). Nowadays, chronotherapy for CVD considers precise-timing pharmacology to maximize efficacy, novel circadian "omic" chronobiomarkers to define their clinical course, and prognosis and novel pharmacologic compounds (chronobiology drugs) targeting circadian mechanism (Tsimakouridze, Alibhai, & Martino, 2015).

Diabetes and insulin resistance. While the central clock regulates food intake, energy expenditure and systemic insulin sensitivity, peripheral clocks located in enterocytes, hepatocytes, adipocytes, and miocytes regulate the timely absorption, transport, and cellular entry of glucose collectively mediated with the pancreatic oscillator responsible of insulin/glucagon fine-tuning (Stenvers et al., 2019). The strict switching from gluconeogenesis to glycolysis and vice versa, often interrupted in diabetes, is controlled by the hepatic circadian apparatus, while the pancreatic clock is synchronized to the light–dark cycle and the oscillatory expression of its clock genes is glucose-dependent (Qian, Block, Colwell, & Matveyenko, 2013). However, sleep disturbances are very common in diabetic patients (Khandelwal et al., 2017) and functional mutations in the melatonin receptor and lower melatonin secretion linked to insulin resistance (McMullan et al., 2013). Melatonin can be restored by supplementation since it promotes the growth and differentiation of pancreatic cells by stimulating insulin growth factor receptor (IGF-R) and insulin receptor (IR) tyrosine phosphorylation (Sharma, Singh, Ahmad, Mishra, & Tiwari, 2015).

Osteoporosis and cancer. Studies in rodents have demonstrated that bone tissue also has an intrinsic circadian system and disruption of clock genes expression lead to disturbed bone structure and quality. Pronounced rhythms in the bone resorption (e.g. CTX: carboxy-terminal collagen crosslinks) and formation (e.g. osteocalcin) markers and rhythmic production of glucocorticoids indicate that osteoclasts and osteoblasts possess robust diurnal activity patterns in humans with a peak during the rest phase (Winter et al., 2021). Moreover, agonist of the circadian regulators REV-ERBs inhibits osteoclast differentiation, and therefore clock synchronizing therapy may help to correct abnormal bone metabolism (Song et al., 2018). On the other hand, the circadian machinery in many organs modifies the 24-hr rhythmicity of key epigenetic events such as DNA methylation, miRNA expression, and histone modification (Reszka & Zienolddiny, 2018); genetic variation in circadian rhythm and melatonin pathways are common in prostate cancer (Gu et al., 2017) and chemotherapy worsens the rest–activity rhythm and quality of life of breast cancer patients (Sultan, Choudhary, & Parganiha, 2017).

3.3 Regulation of Circadian Clocks

Circadian clocks are internal timing systems that enable organisms to adjust their behavioral and physiological rhythms to their daily environmental changes. These clocks generate self-sustained oscillations at the cellular, tissue, and behavioral level. The rhythm-generating mechanism is based on a gene expression network with delayed negative feedback loop that causes the transcript to oscillate with a period of approximately 24 hr (Srikanta & Cermakian, 2020 and references therein). The circadian clock can be divided into two parts: the master clock, residing in the suprachiasmatic nucleus of the hypothalamus (SCN), and the peripheral clocks that reside in various tissues throughout the body.

Many clock-controlled genes (CCGs) are transcripted by the heterodimerization of two transcription factors: CLOCK, and BMAL1 that binds to E-box sites within their promoters. Those genes generate a tightly self-regulated feedback loop by controlling the transcription of their repressors, period (PER) and cryptochrome (CRY) family members. In response to light, CLOCK interacts with BMAL1 to activate *Per* and *Cry* transcription; the increase in

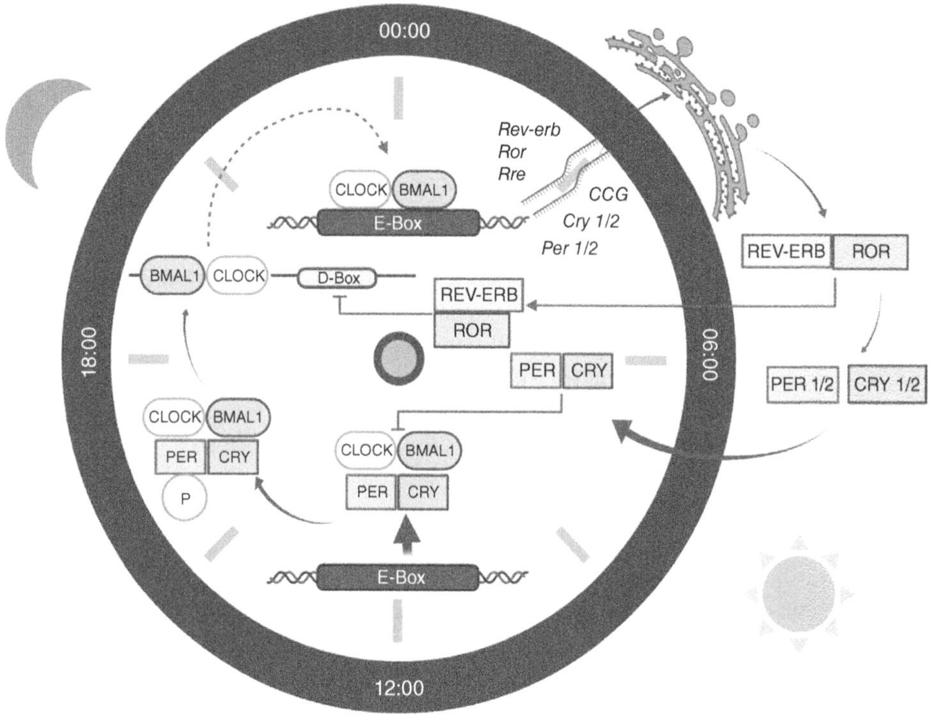

Figure 3.2 The molecular clock. Figure created with BioRender.

these genes transcription results in the accumulation of circadian repressors PER and CRY which heterodimerize, translocate to the nucleus, and inhibit CLOCK/BMAL1-mediated transcription. In response to the decrease/absence of light, the PER/CRY repressor complex degrades, and the cycle begins again. In a secondary feedback cycle, REV-ERB (nuclear receptor) and ROR (retinoic acid receptor-related orphan receptors) participate in regulating BMAL1 expression (Figure 3.2) (Dufoo-Hurtado et al., 2020 and references therein).

The master clock is synchronized or entrained by the light–dark cycle and, in turn, synchronizes clocks present in peripheral tissues and organs. The primary time cue for the molecular clock in the SCN is light detected by the eye. The photoreceptors involved in this process include the rods and cones that mediate vision, as well as the recently identified melanopsin-expressing photosensitive retinal ganglion cells (pRGCs). Light information is conveyed to the SCN via the retinohypothalamic tract, resulting in an intracellular signaling cascade which converges on cAMP response elements in the promoters of several key clock genes (Dannerfjord, Brown, Foster, & Peirson, 2021 and references therein).

Most cells have molecular circadian clocks, and most peripheral tissues are synchronized locally at least at the tissue level showing circadian oscillation of clock genes (peripheral clocks); the phases of peripheral clocks are similar among many tissues. The evidence supports the requirement of the SCN for temporal orchestration of the periphery in living animals. However, a recent study found that even in the absence of SCN or extra-SCN clocks, peripheral clocks remain rhythmic, indicating previously controversial circadian oscillator coupling within peripheral tissues (Finger & Kramer, 2021). Systemic cues constantly entrain peripheral cell clocks that are easily desynchronized, among them the rhythms of body temperature, the physiological levels of oxygen rhythms, feeding cycles, hormones as well as glucocorticoid and blood concentrations (Oike, 2017 and references therein).

The connections of the mammalian clock system have a conceptual model based on knowledge from the turn of the century, suggesting that the SCN regulates peripheral clocks mainly through feeding-fasting cycles, body temperature cycles, hormones, and blood concentrations. However, a new model of the system has been proposed where in addition to the SCN, other distinct hypothalamic clocks control daily energy homeostasis. The autonomic nervous system (ANS) differentially innervates peripheral tissues and modulates their clocks by adjusting sympathetic tone. All clocks are connected via the circulatory system by an ever-growing list of synchronizing factors, many of which are under local circadian control in peripheral tissues (Koronowski & Sassone-Corsi, 2021 and references therein).

3.3.1 Food as Synchronizer

A general model of circadian entrainment is recognized in which the SCN regulates the phase of peripheral clocks via its control of feeding behavior. Accordingly, the timing of food intake is highlighted as a powerful environmental cue with the potential to destroy or restore the synchrony of circadian rhythms in metabolism (Pickel & Sung, 2020). Restricted feeding has no impact on the central pacemaker in the SCN but alters the secondary oscillators from peripheral tissues such as liver, kidney, heart, and pancreas, indicating that diet regulates peripheral tissue clock oscillators (Dragoi et al., 2019 and references therein). Yamamuro et al. (2020) suggested that constant light exposure induced disruption of both central and peripheral circadian rhythms; time-restricted feeding in mice restored the circadian rhythms of most genes, including *Clock* and *Cry* except *Per*1, to various degrees in both liver and white adipose tissue (WAT), as well as genes involved in lipid metabolism (*Hmgcr, Dgat*1*, Dgat*2*, Atgl, Lipe, Fasn, Lxrα, Pparα, Pparγ, Srebp*1*c, and Srebp*2).

Feeding at the wrong time of the day and diet composition (free choice high-fat-high-sugar diet; fcHFHS) disturb the peripheral clocks in skeletal muscle (SM) and brown adipose tissue (BAT), but to different degrees and thereby result in a further desynchronization between metabolically active tissues such as SM, BAT, WAT, and liver. For example, *Pdk*4 and *Ucp*3 were phase-shifted, while the rhythmic expression was lost for *Ucp*3 in SM on the fcHFHS diet during the light phase (De Goede, Hellings, Coopmans, Ritsema, & Kalsbeek, 2020).

A carbohydrate-containing diet causes strong entrainment through insulin secretion. However, it is unknown whether human diets entrain peripheral circadian clocks. Per gene expression rhythm is altered in monocytes and delays the amplitude of cortisol in humans by changing a higher carbohydrate and lower fat to isoenergetic diet (Pivovarova et al., 2015). The adherence of low carbohydrate diet can improve the circadian rhythm by reducing levels of inflammatory markers in intermediate- and morning-type groups of women (n=304) affected by overweight and obesity, IL-1β and Galectin-3 being the mediating markers (Tavakoli, Mirzababaei, Sajadi, & Mirzaei, 2021). In PER2::LUC knock-in mice, feeding the control mouse diet (AIN-93M) with a protein: fat: carbohydrate ratio of 14.7 : 9.5 : 75.8 for 2 days during inactive periods, produced significant phase advance (7.3 hr) in the liver clock compared with that in 24-hr fasted mice, whereas human diets caused significant but smaller phase advances (4.7–6.2 hr). Phase advance decreased in adenine-induced chronic kidney disease mice with a normal diet compared with healthy and high-fat/sucrose-induced type 2 diabetes mice. These results support the food-induced entrainment of peripheral clocks in human clinical trials (Yasuda et al., 2019).

The phase *Per*2 mRNA rhythms in WAT was significantly delayed (by 0.97 ± 0.29 hr) with delayed meals (meals at 12:00, 17:00, and 22:00) compared to regular meal pattern (meals at 07:00, 12:00, and 17:00) in 10 healthy young men (Wehrens et al., 2017). However, skipping breakfast for 6 days does not alter clock gene expression in leukocytes albeit the delay (42.0 ± 16.2 min) on the circadian rhythm of the body temperature (Ogata et al., 2020). Breakfast consumption acutely affects clock and CCG expression leading to normal oscillation in healthy individuals and individuals with diabetes (Jakubowicz et al., 2017). Moreover, the level of *Bmal*1, *Rorα*, and *Sirt*1 expression was higher ($P < 0.05$) after lunch on breakfast and lunch day, whereas the other clock genes remained unchanged in healthy individuals. In individuals with type 2 diabetes, *Bmal*1, *Per*1, *Per*2, *Rev-erbα*, and *Ampk* increased ($P < 0.05$) after lunch on the breakfast and lunch day. Interestingly, the omission of breakfast altered clock and metabolic gene expression in both healthy and with type 2 diabetic individuals.

Only a single human study demonstrated that meal timing sufficiently fulfills at least one of the proposed zeitgeber criteria (Aschoff, 1954) according to a recent review (Lewis, Oster, Korf, Foster, & Erren, 2020). The review also suggests preference for nonparametric analyses instead of modeling, and that such studies should be undertaken under real-life conditions aiming to devise practical recommendations for chronodisruptive work or environmental conditions.

Recent studies also suggest that natural food intake has little or null synchronizing effect on peripheral circadian clocks. Xie et al. (2020a) evaluated the effects of normal feeding patterns on phase shift or entrain peripheral tissues by measuring circadian rhythms of the liver, kidney, and submandibular gland in *mPer2*^{Luc} mice under different food schedules. In experiment 1, a 12-hr shift of the simulated natural food provision schedule resulted in small shifts in the liver and kidney and no shift in the submandibular gland. In experiment 2, the light–dark cycle was removed to determine entrain peripheral clocks on continued scheduled food. Peripheral organ phase was nevertheless best predicted by the rest–activity cycle; food significantly entrained only one of 12 kidneys and one of 12 livers. As a positive control, time-restricted feeding was an effective zeitgeber in both experiments, unlike the simulated natural feeding. Thus, they infer that under normal conditions of food intake, the SCN remains the dominant synchronizer of peripheral clocks without requiring the intermediate step of feeding behavior. The same group (Xie et al., 2020b) evaluated the efficacy of naturalistic feeding patterns on phase shift and peripheral tissues entrainment, measuring circadian rhythms of the liver, kidney, and submandibular gland in *mPer2*^{Luc}mice under different feeding schedules. In *ad lib* feeding, as well as in a schedule designed to mimic the *ad lib* pattern, PER2::LUC bioluminescence peaked during the night as expected. Surprisingly, shifting the scheduled feeding by 12 hr caused only small advances (< 3 hr). To isolate the effects of feeding from the light–dark cycle, clock phase was then measured in mice acclimated to scheduled feeding and housed in constant darkness. In these conditions, peripheral clock phases were better predicted by the rest–activity cycle than the food schedule. Under natural feeding patterns, the master pacemaker in the brain sets the phase of peripheral organs independent of feeding behavior.

3.4 Chronobiotics as Functional Foods

Circadian rhythm disorders negatively influence health and have no current medical treatment. However, chrononutrition is becoming a tool to enhance nutritional health-related status; particularly, some natural compounds have proved their potential to synchronize and restore the clocks, identified as chronobiotics (Dufoo-Hurtado et al., 2020 and references therein). Chronobiotics are defined as drugs able to synchronize or increase the amplitude of the circadian rhythms. Melatonin, produced mainly by the pineal gland, an unusual phylogenetically conserved compound present in all known aerobic phyla is the most relevant prototype (Cardinali, 2019 and references therein).

Dietary compounds entrain the circadian rhythm in many peripheral tissues. For example, the consumption of nobiletin, L-theanine, L-ornithine, and resveratrol can benefit people in maintaining proper circadian rhythms and keeping good health condition; however, the mechanism is still unknown (Cheng et al., 2021 and references therein). CLOCK, BMAL1, and PER respond to metabolites and xenobiotics due to their Per-Arnt-Sim (PAS) domains containing structures, and hormonal ligands and metabolites such as sterols can activate the nuclear receptors REV-ERB and ROR (Koronowski & Sassone-Corsi, 2021 and references therein).

A comparative circadian metabolomic analysis on serum and liver in mice under high-fat diet revealed that the nutritional challenge reduces serum metabolite rhythmicity compared with liver, indicating a circadian misalignment between the analyzed tissues (Abbondante, Eckel-Mahan, Ceglia, Baldi, & Sassone-Corsi, 2016). A high-fat diet, which induces the disruption of the normal circadian cycle enables large-scale genesis of de novo oscillating transcripts, resulting in reorganization of the coordinated oscillations between coherent transcripts and metabolites. This reprogramming involves both the impairment of CLOCK: BMAL1 chromatin recruitment and a pronounced cyclic activation of surrogate pathways through the transcriptional regulator PPARγ (Eckel-Mahan et al., 2013).

3.4.1 Nutrients

Several circadian clock proteins as well as their accessory proteins (such as nuclear receptors) are highly sensitive to nutrient metabolism. Macro and micronutrients can function as zeitgebers for the clock in a tissue-specific way and can thus impair synchrony between clocks across the body, or potentially restore synchrony in circadian misalignment. Circadian nuclear receptors are particularly sensitive to nutrient metabolism and can alter tissue-specific rhythms in response to changes in the diet. The SNPs in human clock genes is apparently correlated with diet-specific responses and along with chronotype eventually may provide valuable information from a clinical perspective using diet and nutrition to treat metabolic disorders (Ribas-Latre & Eckel-Mahan, 2016 and references therein).

Many circulating nutrients and hormones, such as glucose, amino acids, insulin, glucagon, glucocorticoids, and others, possess zeitgeber capacity (Reinke & Asher, 2019 and references therein).

Timing is important for nutrition along with quality and quantity. Dietary components, such as fiber, unsaturated fatty acids, and polyphenols, at suitable times can promote health (Oike, Oishi, & Kobori, 2014). Interestingly, the daily access to a palatable meal can entrain the SCN; several stimuli can be implicated in this process including motivation and arousal (Mendoza, Angeles-Castellanos, & Escobar, 2005). In the SCN, *c-Fos* expression peaked at palatable meal time and PER-1 showed a peak during the onset of subjective night, as predicted according to the behavioral entrained pattern. In Wistar rats, a daily piece of chocolate coupled to the onset of the active phase (breakfast) accelerated re-entrainment in a jet lag model by setting the activity of the SCN to the new cycle. Furthermore, in a rat model of shift work, a piece of chocolate for breakfast prevented circadian desynchrony by increasing the amplitude of the day–night *c-Fos* activation in the SCN. Contrastingly, chocolate for dinner prevented re-entrainment in the jet lag condition and favored circadian desynchrony in the shift work models. Moreover, chocolate for breakfast resulted in low body weight gain, while chocolate for dinner increased body weight (Escobar et al., 2020).

3.4.1.1 Glucose

Carbohydrates, including glucose, have been proposed as zeitgeber in humans for the food-entrainable peripheral oscillators. The SCN is the principal driver of circadian blood glucose fluctuations by scheduling food ingestion to the activity phases. Low glucose levels during fasting in the rest phase are replenished via circadian glucose excretion from the liver, which is mediated by clock dependent expression of GLUT2, which peaks during the rest (fasting) phase (Gachon, Loizides-Mangold, Petrenko, & Dibner, 2017 and references therein). Two of the most abundant circulating nutrients for cellular energy generation besides glucose, namely, glutamine and lactate, have been linked to feedback signaling pathways for circadian clock control. Glucose and glutamine are converted through the hexosamine pathway into uridine diphosphate *N*-acetylglucosamine (UDP-GlcNAc), the substrate for the covalent modification of proteins by *O*–β-d-*N*-acetylglucosamine (*O*-GlcNAc). *O*-glcNAcylation of BMAL1, CLOCK, and PER2 regulates their transcriptional activity and stability affecting entrainment and period length of the circadian clock (Reinke & Asher, 2019 and references therein).

Evening carbohydrate-rich meal, approximately 8 hr after the consumption, increases the core body temperature. While, morning carbohydrate-rich meals significantly advance circadian phase position in the same parameter (+59 ± 12 min) and heart rate (+43 ± 18 min) compared to evening carbohydrate-rich meals (Kräuchi, Cajochen, Werth, & Wirz-Justice, 2002). Hirao, Tahara, Kimura, and Shibata (2009) showed that balanced diets containing carbohydrates/sugars and proteins are good for restricted feeding-induced entrainment of the peripheral circadian clock, and that a balanced diet that increases blood glucose, but not by sugar alone, is suitable for entertainment. However, early investigation suggested that glucose induced clock resetting by downregulating *Per*1 and *Per*2 in Rat-1 fibroblasts (Hirota et al., 2002), while other group showed that glucose has zeitgeber properties for the feeding entrainable oscillator in rats (Stephan & Davidson, 1998). The circadian behavior and peripheral circadian clocks in mice are altered when animals are fed during 2 weeks with a low-carbohydrate, high-protein diet. This diet induced hypoglycemia without affecting body weight, although the mice consumed more calories. Furthermore, the circadian mRNA expression of clock genes, such as *BMAL*1, *Cry*1, *NPAS*2, and *Rev-erbα*, was significantly phase-advanced, and mean expression levels of *BMAL*1 and *Cry*1mRNAs were significantly elevated in the liver and kidneys of the animals (Oishi, Uchida, & Itoh, 2012).

Rapidly digested starch is appropriate for peripheral clock entrainment. An α-potato starch-substituted diet in PER2::LUCIFERASE knock-in mice induced larger phase delays of the liver clock than did β-potato starch. *Per*2 and *Ror-γ* expression increased in response to α-potato starch administration (Itokawa et al., 2013).

3.4.1.2 Amino Acids

Some of the first studies in the sixties reported the daily rhythms of amino acids concentrations in human plasma (Feigin, Klainer, & Beisel, 1967; Rapoport, Feigin, Bruton, & Beisel, 1966; Wurtman, Rose, Chou, & Larin, 1968). Krüppel-like factor 15 (KLF15) is a critical mediator of branched-chain amino acids (BCAAs) leucine, isoleucine, and valine catabolism and rhythmicity, influencing processes ranging from gluconeogenesis during fasting to nitrogen homeostasis (Fan, Hsieh, Sweet, & Jain, 2018 and references therein). The promoter region of KLF15 has four E-box binding sites for the core clock genes CLOCK and BMAL1, which are positive KLF15 regulators. Therefore, KLF15 is crucial for imparting diurnal rhythmicity to the release, uptake, and utilization of several substrates, notably BCAAs.

This has been supported by the lack of a direct relation between dietary protein and blood amino acid values; for example, normal periodicity or amino acids are maintained in adults despite long starvation periods or complete restriction of dietary protein intake (Feigin, Beisel, & Wannemacher Jr, 1971 and references therein). Also, amino acids availability is reduced by high-protein diets (Moundras, Remesy, & Demigne, 1993). However, amino acids rhythmicity is altered by the onset of an acute illness, including obesity. Valine, leucine, isoleucine, tyrosine, and phenylalanine increased and glycine decreased in obese subjects compared with age- and sex-matched normal-weight subjects, phenomena linked to insulin ineffectiveness (Felig, Marliss, & Cahill Jr, 1969).

Similar results have recently been published. Diets of 100% glucose, 100% casein, 100% starch, and 100% soybean oil fail to significantly advance phase in *Per*2::luciferase knock-in mice. However, 86% glucose with 14% casein significantly advance phase compared to 100% sucrose alone. Nevertheless, a balanced diet properly entertains signals of the mouse liver clock. Nutrients containing glucose and amino acids rapidly change the expression of clock genes, especially *Per*2 and *Rev-erbα*, resulting in a phase shift in mice (Tahara, Otsuka, Fuse, Hirao, & Shibata, 2011).

Conversely, tryptophan and some products modulate light-dependent regulation of circadian rhythm by triggering aryl hydrocarbon receptor (AhR). Both tryptophan and its product 6-formylindolo (3,2-b) carbazole (FICZ) increase cytochrome P4501A1 (*CYP*1*A*1) and alter the circadian expression of *Per*1, *Cry*1, and *Cry*2 genes in SCN 2.2 cells (Mukai & Tischkau, 2007). Methionine feeding significantly influences the fluctuation of relative mRNA expression of *Clock, Bmal*1, *Cry*1, *Cry*2, *Per*2, and *Per*3 during a day, affecting the rhythmic fluctuation of amino acid transporter in hen's jejunum (Yi-lin et al., 2018).

Elevated supply of methionine and arginine influences α-s1-casein protein synthesis of primary bovine mammary epithelial cells through changes in the mTORC1, circadian clock (*Per*1 *and Clock*), and AMPK pathways (Hu et al., 2020). Yamaguchi (2018) proposes that the neuronal circuit mediated by arginine vasopressin (AVP)/V1 receptor signaling in the SCN plays a crucial role in the resilience of the circadian clock to jet lag. Spengler, Kuropatwinski, Schumer, and Antoch (2009) identified a conserved cluster of serine that include Ser431 which is a prerequisite phosphorylation site for generating BMAL-dependent phospho-primed CLOCK and for the potential GSK-3 phosphorylation at Ser427. Serine loop underlies differential affinity of cryptochromes for *Clock: Bmal*1 to control circadian timing (Fribourgh et al., 2020).

3.4.1.3 Dietary Fiber

Microbiota-derived SCFAs and organic acids produced by the fermentation of nondigestible fiber can communicate from the microbiome to host tissues and modulate homeostasis in mammals. The microbiome has circadian rhythmicity and helps the host circadian clock function (Tahara et al., 2018). Due to this, dietary fibers and prebiotics have been proposed as novel dietary approach to attenuate circadian misalignment, modulating the expression and phase of circadian clock genes.

Tahara et al. (2018) provided the first evidence that high-fiber diets (cellulose) enhanced refeeding-induced peripheral clock entrainment. They found that microbiota-derived SCFA or lactate modulates the phase of host peripheral clocks *in vivo*. Thereafter, our research group demonstrated that antioxidant dietary fiber isolated from spent coffee grounds improved chronotype and circadian locomotor activity in young adults (Oseguera-Castro et al., 2019). In C57BL/6J mice subjected to weekly shifted light–dark cycle under a high-fat diet, beta-glucan (0.2 g day^{-1}) and insulin (0.2 g day^{-1}) reverses the phase delay of *Period* 1 and *Period* 3 in the hypothalamus and reverses the phase delay of *Period* 2 in the liver (Cheng, Lam, Kong, & Cheung, 2020). The observed effects may be due to small chemical compounds exclusively produced by the gut microbiota from undigested carbohydrates. Among them, 3-(4-hydroxyphenyl)propionic acid (4-OH-PPA) and 3-phenylpropionic acid (PPA), the major metabolites exclusively produced by *Clostridium sporogenes*, increase the amplitude of both PER2 and *Bmal*1 oscillation of peripheral and central clock machineries of mice, in a dose-dependent manner following their administration immediately after the nadir or the peak of their rhythm (Ku et al., 2020).

Butyrate, one of the SCFAs produced from dietary fiber by gut microbiota, regulates B cell genes in human submandibular gland cells by increasing IL-10-producing B (B10) cells and decreasing IL-17-producing B cells, through the circadian clock genes RAR-related orphan receptor alpha and nuclear receptor subfamily 1 group D member 1. This acid may ameliorate Sjögren's syndrome, an autoimmune disease caused by inflammation of the exocrine gland (Woo et al., 2021).

The direct addition of butyrate to an *in vitro* hepatic model significantly affects clock component expression. Butyrate heightened the amplitude and shifted the phase of the expression of clock components *Bmal*1 and *Per*2, suggesting that bacterial metabolites can potentiate clock oscillations (Leone et al., 2015). Conversely, the circadian clock regulates the diurnal levels of microbial SCFAs and their rhythmic effects on colon contractility in mice. Diurnal microbial SCFA levels regulate the rhythm of the SCFA receptor *Ffar*3 expression in the colonic myenteric plexus, which causes rhythmicity in SCFA-induced colonic motility. The deletion of *Bmal1* abolishes rhythmicity of SCFA levels and their downstream effects (Segers et al., 2019).

3.4.2 Phenolic Compounds

Phenolic compounds are secondary metabolites widely distributed in plants. They are synthesized by different pathways such as pentose phosphate, shikimate, and phenylpropanoid; they are classified according to their chemical structure and biosynthesis route (Vuolo, Lima, & Junior, 2019). Phenolic acids, flavonoids, and tannins are the major phenolic compound classes found in foods and are associated with beneficial effects on obesity, diabetes, cancer, and CVDs (Heleno, Martins, Queiroz, & Ferreira, 2015). Furthermore, the potential of phenolic compounds on the adjustment of circadian clocks in peripheral tissues has recently been evaluated.

3.4.2.1 Phenolic Acids

Phenolic acids are a group of phenolic compounds that contain carboxylic acid and are present in various foods. They are commonly linked to amides, esters, or glycosides and are very rarely found free. In its classification, two subgroups are distinguished: hydroxybenzoic acids and hydroxycinnamic acids, and the most representative of both groups are ferulic, coumaric, gallic, vanillic, and protocatechuic acids, among others. This group is characteristic for its small size, high *in vitro* antioxidant capacity, its effects as an antidiabetic and antimicrobial agent, and cancer treatment (Kumar & Goel, 2019). However, its effect on circadian clocks has recently been studied.

Cinnamic acids and their derivatives are bioactive compounds of natural origin present in plants. They are synthesized from precursors such as phenylalanine and tyrosine via the shikimic acid route. The best known and most abundant are cinnamic, ferulic, and caffeic acids and their derivatives. These compounds have received great attention recently due to their numerous health benefits, including neuroprotective, analgesic, anti-inflammatory, antioxidant, and antidepressant effects, among others (Adisakwattana, 2017). The administration of this group of compounds evaluated recently *in vivo* models showed that cinnamic acid shortens the circadian period of circadian rhythms directly and indirectly through the central clock in the SCN, and the locomotor rhythm in animals. All the above suggests that dietary intake of cinnamic acids and their derivatives can help prevent and treat circadian disorders. However, so far, the molecular mechanisms through which the effect is expressed have not been explored (Oishi, Yamamoto, Oike, Ohkura, & Taniguchi, 2017).

Chicoric acid is a hydroxycinnamic acid, an ester of caffeic acid, a member of the phenylpropanoid family. It is found mainly in the roots of plants such as chicory, purple coneflower, and basil; it is primarily associated with antiviral, anti-inflammatory, glucose, and lipid homeostasis effects (Peng, Sun, & Park, 2019). Its regulatory effects on lipid metabolism have recently been associated with the circadian clock. Chicoric acid improves the morphology and levels of hepatic lipids in a dose-dependent manner. Also, this phenolic acid regulates the expression of the circadian clock genes and their daily oscillations. The beneficial effects of this acid in the regulation of lipid metabolism were associated with the

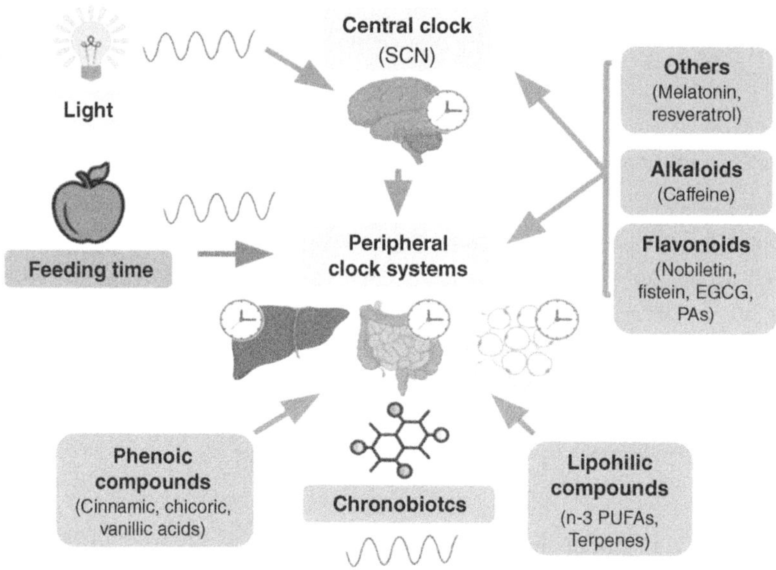

Figure 3.3 Food bioactive compounds with chronobiotic potential. *Source:* The authors.

regulation of circadian clocks through BMAL-1, a key component of the circadian clock that plays an essential role in adipocyte maturation and differentiation. It is suggested that the mechanism may involve BMAL1 regulating the expression of essential components of lipid synthesis and degradation such as fatty acid synthase (FAS), acetyl CoA carboxylase (ACC), peroxisome proliferator-activated receptor α (PPARα), and liver X receptor α (Lxrα) through the Akt/GSK3β signaling pathway in a BMAL-1-dependent manner (Guo et al., 2018; Figure 3.3). Therefore, chicoric acid is a promising chronobiotic that modulates circadian rhythms to improve metabolic syndrome, nonalcoholic fatty liver disease, among other conditions.

Vanillic acid is a hydroxybenzoic acid with several potential health benefits, including its antioxidant capacity and anticancer properties. The administration of vanillic acid in rats with induced endometrial carcinoma slows the development of tumors, improves the amplitude, and reverses the phase delay of the circadian rhythm of enzymes of the antioxidant system such as superoxide dismutase (SOD) (Bhavani, Subramanian, & Shanmugapriya, 2016; Figure 3.3).

3.4.2.2 Flavonoids

Flavonoids constitute the group of phenolic compounds with the greatest distribution in plants. They are characterized by their low molecular weight and their structure consisting of two aromatic rings joined by a three-carbon bridge, often in the form of a heterocyclic ring. Variations in substitution patterns result in various flavonoid subgroups such as flavonols, flavones, flavanones, flavanols (catechins), isoflavones, flavanonols, and anthocyanidins (Vuolo et al., 2019). Flavonoids have been reported to affect the circadian rhythm. This group of compounds includes nobiletin, fisetin, myricetin, and hesperidin, which mainly affect the amplitude of the circadian rhythm (Figure 3.4). Fisetin is a flavonoid

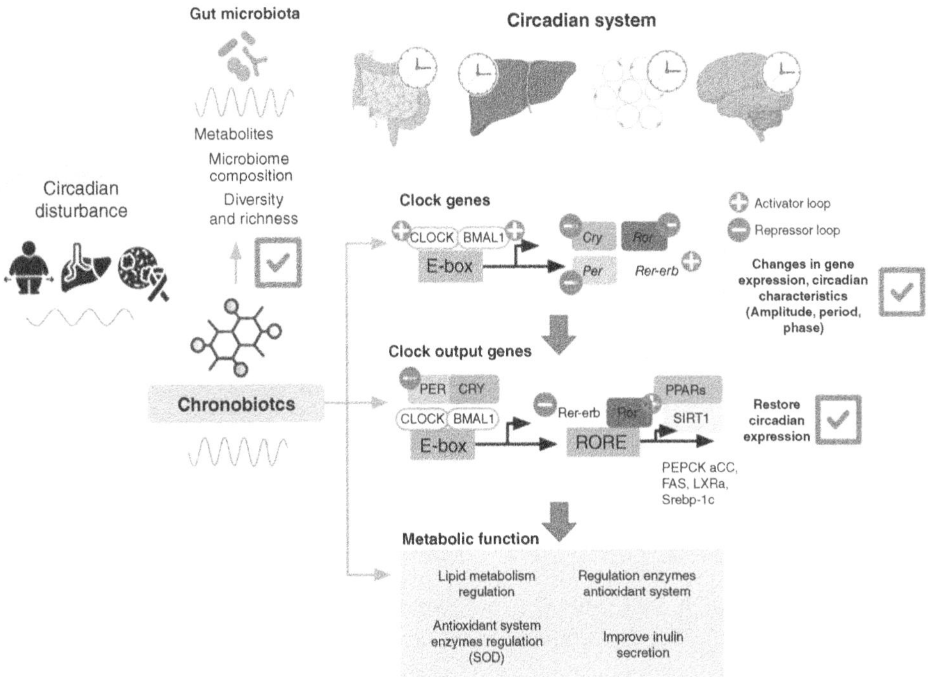

Figure 3.4 Food bioactive compounds with chronobiotic potential: molecular mechanisms and metabolic effects. *Source:* The authors.

distributed mainly in fruits. The effect of fisetin consumption evaluated in an *in vivo* model reduced fragmentation and restoration of the circadian pattern of rest–activity in animals with hyperammonemia (Subramanian et al., 2015). However, there are no molecular studies of the mechanism through which this is carried out.

Nobiletin is a polymethoxy flavonoid widely distributed in the peel of citrus fruits and the pulp of fruits. It is recognized for its effects in attenuating inflammatory, metabolic, Alzheimer's processes, among others. Its effects have also been evaluated on circadian rhythms. In the first investigation with an *in vivo* model of diet-induced obesity, nobiletin improved circadian rhythms in the liver through the activation of the orphan receptor tyrosine kinase (ROR), increasing circadian genes expression, resulting in the prevention of metabolic syndrome in a circadian clock-dependent manner (He et al., 2016). One year later, Shinozaki et al. (2017) evaluated *in vitro* the chronic or transient exposure of 18 different flavonoids (flavones, flavonols, isoflavones, catechins, and polymethoxy flavonoids) on the amplitude, period, and phase of *Per2*. The results showed that although the compounds share a large part of the chemical structure, *Per2* expression varied, even when the compounds were from the same subfamily. However, it is emphasized that only polymethoxy flavones such as tangeretin and nobiletin improve amplitude, period, and phase delay.

Epigallocatechin gallate (EGCG), better known as epigallocatechin-3-gallate, is the most abundant bioactive compound in various teas, including green, black, and white, in fruits, nuts, among others. Structurally, this molecule is an ester of epigallocatechin and gallic acid and has been studied for its beneficial effects, including its antioxidant, anti-inflammatory, and antiproliferative capacity, among others (Aggarwal et al., 2020). Besides, there has been particular interest in the research of this compound as a regulator of the circadian system. EGCG modifies clock gene oscillations (*Clock* and *Bmal*1) in the hypothalamus, liver, and adipose tissue in *in vivo* studies. It also reduces circadian desynchronization and energy imbalance in BAT, the metabolism controlled by the clock maintains lipid homeostasis in WAT, and improves insulin sensitivity in an animal model of circadian misalignment caused by a Western diet. These processes occur through changes in genes expression of the clock system such as *Clock, Bmal*1 and *Cry*1, *Sirt*1 and PGC1α, which in turn modify the expression of lipid metabolism genes such as *PPARγ*, and negatively regulate lipids accumulation (Li, Kek, Lim, Gelling, & Han, 2016; Mi et al., 2017; Figure 3.3). *In vivo* and *in vitro* models of neuroblastoma cells and mouse primary hepatocytes exposure to EGCG modifies the expression of circadian clock genes and prevents oxidative stress and mitochondrial dysfunction in a BMAL1-dependent manner (Qi et al., 2017; Qi et al., 2018). In an *in vivo* model of constant darkness, insulin resistance and glucose and lipid metabolism improve in the liver, suggesting that green tea polyphenols such as EGCG can also synchronize the center clock in the suprachiasmatic nucleus and positively affect cognitive impairment (Qi et al., 2017). Previous evidence suggests the potential for this chronobiotic to maintain the relationship between the circadian clock and oxidative stress and thus mitigate age-related neurodegenerative and metabolic diseases.

EGCG also has proven beneficial effects in different eye diseases. Retinitis pigmentosa (RP) is a disease that involves visual function and patients with this condition present dysfunction of circadian rhythmicity. Furthermore, the retina plays an essential role in synchronizing the circadian clocks. RP exhibits characteristic age-related degenerative effects on the melanopsin system and is associated with weaker circadian patterns (Lax et al., 2016). The effect of EGCG consumption was evaluated on the loss of photoreceptors and its association on biological rhythms of circadian clock output (temperature and circadian activity). However, the results did not show a clear relationship with the effect of EGCG on circadian rhythmicity in the evaluated output variables, and more studies are necessary (Perdices, Fuentes-Broto, Segura, Cuenca, & Pinilla, 2020).

The gut microbiome is of vital importance in maintaining the host's circadian rhythm and affecting its health. Most bacterial communities of the microbiota have circadian characteristics, which help maintain homeostasis through metabolic processes. As mentioned above, catechins can modify the rhythmicity of the circadian clocks of peripheral organs (Leone et al., 2015). Furthermore, EGCG and derivatives of these compounds such as epicatechin gallate (ECG), epigallocatechin (EGC), and epicatechin, and methylated forms of EGCG exert prebiotic potential (Zhang et al., 2013). An emerging area is the study of chronobiotics on the intestinal microbiota and their relationship with the circadian rhythm. The modification of the intestinal microbiota is believed to be the molecular mechanism of action of this chronobiotic. Oolong tea is a fermented beverage that contains numerous catechins such as EGCG. Guo et al. (2019) evaluating oolong tea on the intestinal microbiota in an *in vivo* model with the circadian disorder showed improved richness and diversity of the colonic microbiota after treatment. The molecular mechanism through which this beneficial effect occurs includes interactions with the development of beneficial bacteria and producers of SCFAs, metabolic aspects of the bacteria, and interference with the function of the cell membrane and bacterial energy metabolism. All the above suggest the chronobiotic potential of catechins such as EGCG to reduce conditions such as metabolic syndrome associated with circadian disorders. Zhang, Yan, and Wu (2020) recently evaluated the main catechins in green tea on the intestinal microbiota composition. Their results agree with previous report (Guo et al., 2019). In addition, an analysis of the associated metabolic pathways suggests that the molecular mechanism that may be involved in modifying the microbiota includes the activation of genes relevant to ubiquinone, another biosynthesis of terpenoids-quinones, pentose, and tautomeric glucuronate, and the biosynthesis of lipopolysaccharides which were found to be differentially expressed.

Proanthocyanidins (PAs) are polymers of flavonoids known as condensed tannins. These compounds are found abundantly in flowers, nuts, fruits, and seeds. They are formed by catechin and epicatechin units. They are recognized for their impact on astringency and their biological potential as antioxidant, neuroprotective, cardioprotective, antidiabetic agent, among others (Rauf et al., 2019). However, their effect as a chronobiotic has also been evaluated.

Grape seeds are a good source of PA. The effect of extracts from this natural source on the expression of the circadian clock system genes has been evaluated *in vivo* and *in vitro* models of different peripheral tissues. An acute dose of PA acts as a synchronizer to maintain nocturnal melatonin levels in the first 3 hr of the light phase. Also, PA administration modulates the molecular clock in the liver. This effect may occur due to regulation in the expression of BMAL1, RORα, *Nampt*, and NAD, which suggests the transactivation of RORα, overexpressing BMAL1. Its administration also affects the expression of genes of the central clock system of the hypothalamus (Ribas-Latre, Baselga-Escudero et al., 2015; Ribas-Latre, Del Bas, Baselga-Escudero, Casanova, Arola-Arnal, Salvadó, Bladé et al., 2015; Ribas-Latre, Ribas-Latre, Del Bas, Baselga-Escudero, Casanova, Arola-Arnal, Salvadó, Arola et al., 2015; Figure 3.3).

3.4.2.3 Others

Resveratrol is a stilbenoid polyphenol with several recognized therapeutic effects. A good source of this compound is grapes and wine (Singh et al., 2019). It has been associated with the prevention and progression of various diseases through various mechanisms, including a chronobiotic mechanism (Figure 3.4). Its chronobiotic potential has been evaluated *in vivo* with mice fed a high-fat diet. The results demonstrate that resveratrol intake affects the circadian expression of *Sirt*1 and *Rev-Erbα* in the liver to restore clock genes and the circadian expression of PPARα, which in turn regulates the expression of genes related to lipid metabolism like Srebp-1c, Fas, and Acc1 (Sun et al., 2015, Figure 3.3).

3.4.3 Lipophilic Compounds

Hydrophobicity is the main characteristic of lipophilic compounds. These compounds exert positive effects on different physiological, nutritional, and structural processes in the human body. Examples of these compounds are polyunsaturated lipids, oil-soluble vitamins, phytosterols, curcuminoids, carotenoids, terpenes, among others (Raikos & Ranawana, 2017).

3.4.3.1 Unsaturated Fatty Acids

Polyunsaturated fatty acids (PUFAs) contain two or more double bonds and are classified as *n*–3 or *n*–6, according to the position of the last double bond in the carbon chain concerning the terminal methyl end of the molecule. The main dietary sources of these compounds are vegetable oils (*n*–3) and foods of animal origin such as fish, salmon, tuna, and trout (*n*–6). The three main n-3 fatty acids are alpha-linolenic acid (ALA), eicosapentaenoic acid (EPA), and docosahexaenoic acid (DHA) (Schmitz & Ecker, 2008). The importance of adequate consumption of PUFA *n*–3 derived from fatty marine fish, specifically EPA and DHA, in development, CV health, and maintaining health has been highlighted (Calder, 2015). Furthermore, the investigation of the effect of PUFAs as chronobiotic potentials has also had a significant impact (Figure 3.4). In an *in vivo* model, fish oil EPA/DHA consumption significantly induced phase changes in the expression of circadian clock genes (*Per2*) in the liver. It has a positive physiological effect through increase in insulin secretion due to GPR120 (a polyunsaturated fatty acids receptor) interaction (Furutani et al., 2015; Figure 3.3). Recently, Serhiyenko, Segin, and Serhiyenko (2018) evaluated the effect of *n*–3 PUFA consumption on circadian rhythmicity measured through heart rate variability in patients with diabetes mellitus 2 and autonomic CV neuropathy in a clinical study. Patients consuming one capsule mainly containing EPA/DHA per day for 3 months showed positive changes in parameters of heart rate variability during periods of activity and rest.

Finally, the effect of *n*–3 PUFAs in fish and algae oils has been evaluated *in vivo* in improving hyperlipidemia by modifying the circadian rhythm of the intestinal microbiota; changes of bacteria were significantly different between ZT0 (active period) and ZT12 (resting period). The gut microbiota in mice fed high-fat and fructose diets can be improved by supplementing *n*–3 PUFAs-rich oils by restoring the circadian rhythms of the microbiota, modulating some intestinal bacteria associated with the metabolism of serum lipids, and the production of SCFAs (Gui et al., 2019).

3.4.3.2 Terpenoids

Terpenes are organic compounds that are characterized by giving plants smell and color. They consist of isoprene units and are named for every five carbon atoms as hemiterpenes (C5), monoterpenes (C10), diterpenes (C20), and triterpenes (C30) (Mukherjee, 2019). This last group of compounds has been widely evaluated for its pharmacological activities, including anti-inflammatory, antidiabetic, anticancer, and antioxidant activities. Some triterpenes include corosolic acid, celastrol, and cucurbitacins present in Banaba tea leaves (*Lagerstroemia speciosa* L.), *Tripterygium wilfordii*, and cucurbits such as pumpkin, cucumber, melon, among others. The chronobiotic potential of these compounds has recently been evaluated in an *in vitro* model. Suzuki, Fukumitsu, and Oike (2021) evaluating the effect of 15 commercial triterpenes, including analogs of 3 main groups: corosolic acid, celastrol, and cucurbitacins, in osteosarcoma and fibroblast cell lines on PER2 expression showed a dose-dependent effect for each group of compounds with only some of the compounds modifying the circadian clock presumably due to the structure of each compound.

Corosolic acid, celastrol, and cucurbitacin B can reset and synchronize the circadian clock by increasing the expression of clock genes such as *Per2* (Figure 3.3). In particular, celastrol shortens the period length in fibroblasts. This result illustrates the potential of this group of compounds as future chronobiotic ingredients that can prevent various diseases through circadian clocks adjustment (Figure 3.4).

3.4.4 Alkaloids

Alkaloids are an important group of secondary metabolites widely distributed in nature. Their structure contains one or more nitrogen atoms, particularly as part of a heterocyclic ring, and most have toxic effects on organisms. Among the most important are caffeine, capsaicin, piperine, theobromine, theophylline, among others. They are characteristic for their effects as activators or depressants of the central nervous system (Debnath et al., 2018).

3.4.4.1 Caffeine

Caffeine is the most consumed stimulator of the nervous system, which works by blocking adenosine binding to its receptor. Dietary sources of caffeine include coffee, tea, yerba mate, caffeinated soft drinks (cola-type), and energy drinks (Reyes & Cornelis, 2018); which are consumed to relieve drowsiness for a long time, which is why its effect as a chronobiotic has also been of particular interest. *In vivo*, *in vitro*, and *ex vivo* models indicate that this psychostimulant affects the circadian rhythm by lengthening the circadian period in mouse liver explants. Simultaneously, in the suprachiasmatic nucleus, it delays circadian rhythm and lengthens the circadian locomotor rhythm of animals (Oike, Kobori, Suzuki, & Ishida, 2011; Figure 3.4). Human studies suggest that caffeine also influences circadian physiology increasing daytime exposure to sunlight and reducing light at night. Consuming caffeine causes a phase delay; nonetheless, this depends on its consumption time. Its chronobiotic effects are explained by affecting the cyclic adenosine monophosphate (cAMP)-dependent receptor, suggesting that caffeine modifies the circadian rhythm through a blockade of this receptor (Burke et al., 2015). Caffeine also improves alertness without affecting circadian clock synchronization and light responsiveness at the level of the suprachiasmatic nucleus (Hilaire & Lockley, 2015; McHill, Smith, & Wright Jr, 2014; Van Diepen et al., 2014).

3.4.5 Others

3.4.5.1 Melatonin

Melatonin is a ubiquitous hormone found in various organisms. According to its chemical structure, it belongs to the family of indoleamine. In humans, this hormone is produced mainly in the pineal gland and is closely associated with regulating the circadian clock and peripheral oscillators, allowing the temporary organization of biological functions through circadian rhythms (Tordjman et al., 2017). This hormone has been linked to the circadian clock through oxidative stress due to its antioxidant properties. Due to its lipophilic characteristics, melatonin can cross biological barriers, such as the blood–brain and the placental barriers. Several studies suggest its potential to regulate endogenous antioxidant systems (Liu et al., 2019). Melatonin is produced from the essential amino acid tryptophan in the pineal gland of humans, but it is also found in microorganisms and plants, where it is known as "phytomelatonin." The concentration of melatonin in foods varies and has been reported in various food groups, including nuts, cereals, fruits and vegetables, coffee, green tea, products of animal origin, among others (Meng et al., 2017)

This hormone is essential in the clock system (Figure 3.4); however, the amount of physiologically produced melatonin decreases with age and may be related to the development of pathological conditions related to the interruption of the clock. This effect can be enhanced by administering melatonin-rich foods and increasing its circulating levels and total plasma antioxidant potential. However, evidence is lacking on the use of phytomelatonin, plant origin, or foods that contain it as a chronobiotic (Dufoo-Hurtado et al., 2020 and references therein). In this regard, evidence in humans suggests a beneficial effect of consuming spent coffee fiber (SCF) on markers of the circadian clock (improvement of sleep, physical activity, and chronotype) of people, suggesting that this effect could be due to the amount of phytomelatonin present in coffee; nevertheless, more studies are needed to demonstrate this hypothesis (Oseguera-Castro et al., 2019). It has recently been reported in an *in vitro* model that pistachios phytomelatonin is bioaccessible and permeable during its passage in the gastrointestinal tract, which suggests its potential bioavailability to exert its chronobiotic effect and alleviate health problems related to circadian alterations (Dufoo-Hurtado et al., 2021).

3.5 Current Trends and Perspectives

Chrononutrition as a subdiscipline of chronobiology has been consolidating for two decades. During this time, the empirical knowledge that certain rhythmic biological processes are linked to eating patterns and that some modifiers of the biological clock (e.g. herbal teas and coffee) reduce the probability of suffering from certain diseases has been supported by solid scientific evidence outlining its mechanisms at the cellular and systemic level. We have also witnessed progress in the understanding of the chemical biotransformation processes of some food bioactives of animal and plant origin, along with the discovery of the genetic and epigenetic mechanisms that govern them. Despite all this, studies in chrononutrition are still scarce and its recognition as an applied scientific discipline will be consolidated in the coming years. Nevertheless, studies published in chronobiotics nutrition provide guidelines to offer recommendations on alternative and/or complementary therapy to the already recognized chronobiotics.

Chronobiotics is slowly setting a foothold in mainstream product development. For example, two companies (Unilever and Microba) are collecting information on gut bacteria's role in regulating circadian rhythm, promoting sleep quality and the production of substances that aid and encourage sleep (Chu, 2021). Their initiative is based on emerging evidence on the link between gut microbiome and inflammation and sleep loss, circadian misalignment, affective disorders, and metabolic disease. Moreover, the gut phyla (Bacteroidetes and Firmicutes) positively associated with sleep efficiency influences both circadian rhythm and food intake thereby impacting sleep quality. InnoBev, a US-based start-up has introduced BioLift, a beverage featuring natural ingredients (guarana, ginkgo biloba, and elderberry) to synchronize with circadian rhythm by elevating focus, alertness, and performance. The energizing beverage apparently increases vigilance, improves cognitive function, delivers mental stimulation, and boosts the immune system. Givaudan Active Beauty released Synchronolight (gardenia fruit extract), an active cosmetic ingredient for skincare, designed to protect the skin from digital stress or artificial blue lights that impact circadian rhythm. In chrononutrition, supplements with energetic properties such as cordyceps and fermented ginseng are promoted for morning consumption to help the body focus throughout the day (Nutrition Insight, 2020). Mantra Labs developed comprehensive biorhythm-matched products "Rise"([vitamins, hydration minerals, electrolytes nootropics), "Go" (fermented tea leaf caffeine), and "Rest" to address chrononutrition requirements for morning, afternoon, and evening, respectively (Masterson, 2020).

References

Abbondante, S., Eckel-Mahan, K. L., Ceglia, N. J., Baldi, P., & Sassone-Corsi, P. (2016). Comparative circadian metabolomics reveal differential effects of nutritional challenge in the serum and liver. *Journal of Biological Chemistry*, *291*(6), 2812–2828.

Adisakwattana, S. (2017). Cinnamic acid and its derivatives: Mechanisms for prevention and management of diabetes and its complications. *Nutrients*, *9*(2), 163–190.

Aggarwal, V., Tuli, H. S., Tania, M., Srivastava, S., Ritzer, E. E., Pandey, A., . . . Bishayee, A. (2020). Molecular mechanisms of action of epigallocatechin gallate in cancer: Recent trends and advancement. *Seminars in Cancer Biology.*; https://doi.org/10.1016/j.semcancer.2020.05.011

Ågren, G., Wilander, O., & Jorpes, E. (1931). Cyclic changes in the glycogen content of the liver and the muscles of rats and mice: Their bearing upon the sensitivity of the animals to insulin, and their influence on the urinary output of nitrogen. *Biochemical Journal*, *25*(3), 777.

Aschoff, J. (1954). Zeitgeber der tierischen Tagesperiodik. *Naturwissenschaften*, *41*(3), 49–56.

Aulinas, A. (2019). Physiology of the pineal gland and melatonin. In K. R. Feingold, B. Anawalt, A. Boyce, et al. (Eds.), *Endotext [Internet]*. South Dartmouth.

Aulinas, A. (2020). Physiology of the pineal gland and melatonin. In K. R. Feingold, B. Anawalt, A. Boyce, M. Levy, M. Korbonits, K. Dungan, et al. (Eds.), *Endotext [Internet]*. South Dartmouth (MA). MDText.com, Inc.; 2000-. Available from:https://www.ncbi.nlm.nih.gov/sites/books/NBK550972/

Ba-Ali, S., Brøndsted, A. E., Andersen, H. U., Sander, B., Jennum, P. J., & Lund-Andersen, H. (2019). Assessment of diurnal melatonin, cortisol, activity, and sleep—wake cycle in patients with and without diabetic retinopathy. *Sleep Medicine*, *54*, 35–42.

Beesley, S., Kim, D. W., D'Alessandro, M., Jin, Y., Lee, K., Joo, H., . . . Lee, C. (2020). Wake-sleep cycles are severely disrupted by diseases affecting cytoplasmic homeostasis. *Proceedings of the National Academy of Sciences*, *117*(45), 28402–28411.

Bhavani, P., Subramanian, P., & Shanmugapriya, S. (2016). Modulating effects of vanillic acid on circadian pattern of indices of redox homeostasis in N-Methly-N′-Nitro-N-Nitrosoguanidine induced endometrial carcinoma in rats. *Biological Rhythm Research*, *47*(4), 609–619.

Burke, T. M., Markwald, R. R., McHill, A. W., Chinoy, E. D., Snider, J. A., Bessman, S. C., . . . Wright, K. P. (2015). Effects of caffeine on the human circadian clock *in vivo* and *in vitro*. *Science Translational Medicine*, *7*(305), 305ra146-305ra146. https://doi.org/10.1126/scitranslmed.aac5125

Calder, P. C. (2015). Functional roles of fatty acids and their effects on human health. *Journal of Parenteral and Enteral Nutrition*, *39*, 18S–32S.

Cardinali, D. P. (2019). Melatonin: Clinical perspectives in neurodegeneration. *Frontiers in Endocrinology*, *10*(480), 1–22.

Cecon, E., Liu, L., & Jockers, R. (2019). Melatonin receptor structures shed new light on melatonin research. *Journal of Pineal Research*, *67*(4), e12606.

Challet, E. (2015). Keeping circadian time with hormones. *Diabetes, Obesity and Metabolism*, *17*, 76–83.

Chellappa, S. L., Vujovic, N., Williams, J. S., & Scheer, F. A. (2019). Impact of circadian disruption on cardiovascular function and disease. *Trends in Endocrinology and Metabolism*, *30*(10), 767–779.

Cheng, H., Liu, Z., Wu, G., Ho, C. T., Li, D., & Xie, Z. (2021). Dietary compounds regulating the mammal peripheral circadian rhythms and modulating metabolic outcomes. *Journal of Functional Foods*, *78*, 104370.

Cheng, W. Y., Lam, K. L., Kong, A. P. S., & Cheung, P. C. K. (2020). Prebiotic supplementation (beta-glucan and inulin) attenuates circadian misalignment induced by shifted light-dark cycle in mice by modulating circadian gene expression. *Food Research International*, *137*, 109437.

Chu, W. (2021). Unilever inks deal with Microba to explore the sleep-gut microbiome link. https://nutraingredients.com; 18-mar-21

Crnko, S., Du Pré, B. C., Sluijter, J. P., & Van Laake, L. W. (2019). Circadian rhythms and the molecular clock in cardiovascular biology and disease. *Nature Reviews Cardiology*, *16*(7), 437–447.

Daan, S. (2000). Colin Pittendrigh, Jürgen Aschoff, and the natural entrainment of circadian systems. *Journal of Biological Rhythms*, *15*(3), 195–207.

Dannerfjord, A. A., Brown, L. A., Foster, R. G., & Peirson, S. N. (2021). Light input to the mammalian circadian clock. In S. A. Brown (Ed.), *Circadian Clocks* (pp. 233–247). New York, NY: Humana.

De Goede, P., Hellings, T. P., Coopmans, T. V., Ritsema, W. I., & Kalsbeek, A. (2020). After-effects of time-restricted Feeding on whole-body metabolism and gene expression in four different peripheral tissues. *Obesity*, *28*, S68–S80.

De Zwaan, M., Marschollek, M., & Allison, K. C. (2015). The night eating syndrome (NES) in bariatric surgery patients. *European Eating Disorders Review*, *23*(6), 426–434.

Debnath, B., Singh, W. S., Das, M., Goswami, S., Singh, M. K., Maiti, D., & Manna, K. (2018). Role of plant alkaloids on human health: A review of biological activities. *Materials Today Chemistry*, *9*, 56–72.

Dragoi, C. M., Morosan, E., Dumitrescu, I. B., Nicolae, A. C., Arsene, A. L., Draganescu, D., . . . Nicolae, C. (2019). Insights into chrononutrition: The innermost interplay amongst nutrition, metabolism and the circadian clock, in the context of epigenetic reprogramming. *Farmacia*, *67*(4), 557–571.

Dufoo-Hurtado, E., Olvera-Bautista, R., Wall-Medrano, A., Loarca-Piña, G., & Campos-Vega, R. (2021). *In vitro* gastrointestinal digestion and simulated colonic fermentation of pistachio nuts determine the bioaccessibility and biosynthesis of chronobiotics. *Food and Function*, *12*(11), 4931–4934.

Dufoo-Hurtado, E., Wall-Medrano, A., & Campos-Vega, R. (2020). Naturally-derived chronobiotics in chrononutrition. *Trends in Food Science and Technology*, *95*, 173–182.

Eckel-Mahan, K. L., Patel, V. R., De Mateo, S., Orozco-Solis, R., Ceglia, N. J., Sahar, S., . . . Sassone-Corsi, P. (2013). Reprogramming of the circadian clock by nutritional challenge. *Cell, 155*(7), 1464–1478.

Ehret, C. F., Potter, V. R., & Dobra, K. W. (1975). Chronotypic action of theophylline and of pentobarbital as circadian zeitgebers in the rat. *Science, 188*(4194), 1212–1215.

Escobar, C., Espitia-Bautista, E., Guzmán-Ruiz, M. A., Guerrero-Vargas, N. N., Hernández-Navarrete, M. Á., Ángeles-Castellanos, M., . . . Buijs, R. M. (2020). Chocolate for breakfast prevents circadian desynchrony in experimental models of jet-lag and shift-work. *Scientific Reports, 10*(1), 1–14.

Fan, L., Hsieh, P. N., Sweet, D. R., & Jain, M. K. (2018). Krüppel-like factor 15: Regulator of BCAA metabolism and circadian protein rhythmicity. *Pharmacological Research, 130*, 123–126.

Feigin, R. D., Beisel, W. R., & Wannemacher, R. W., Jr. (1971). Rhythmicity of plasma amino acids and relation to dietary intake. *The American Journal of Clinical Nutrition, 24*(3), 329–341.

Feigin, R. D., Klainer, A. S., & Beisel, W. R. (1967). Circadian periodicity of blood amino-acids in adult men. *Nature, 215*(5100), 512–514.

Felig, P., Marliss, E., & Cahill, G. F., Jr. (1969). Plasma amino acid levels and insulin secretion in obesity. *New England Journal of Medicine, 281*(15), 811–816.

Finger, A. M., & Kramer, A. (2021). Peripheral clocks tick independently of their master. *Genes and Development, 35*(5-6), 304–306.

Folk, G. E., Jr. (1957). Twenty-four hour rhythms of mammals in a cold environment. *The American Naturalist, 91*(858), 153–166.

Fribourgh, J. L., Srivastava, A., Sandate, C. R., Michael, A. K., Hsu, P. L., Rakers, C., . . . Partch, C. L. (2020). Dynamics at the serine loop underlie differential affinity of cryptochromes for CLOCK: BMAL1 to control circadian timing. *Elife, 9*, e55275.

Froy, O. (2010). Metabolism and circadian rhythms—implications for obesity. *Endocrine Reviews, 31*(1), 1–24.

Furutani, A., Ikeda, Y., Itokawa, M., Nagahama, H., Ohtsu, T., Furutani, N., . . . Shibata, S. (2015). Fish oil accelerates diet-induced entrainment of the mouse peripheral clock via GPR120. *PLoS One, 10*(7), e0132472.

Gachon, F., Loizides-Mangold, U., Petrenko, V., & Dibner, C. (2017). Glucose homeostasis: regulation by peripheral circadian clocks in rodents and humans. *Endocrinology, 158*(5), 1074–1084.

Gu, F., Zhang, H., Hyland, P. L., Berndt, S., Gapstur, S. M., Wheeler, W., . . . Caporaso, N. E. (2017). Inherited variation in circadian rhythm genes and risks of prostate cancer and three other cancer sites in combined cancer consortia. *International Journal of Cancer, 141*(9), 1794–1802.

Gui, L., Chen, S., Wang, H., Ruan, M., Liu, Y., Li, N., . . . Liu, Z. (2019). ω-3 PUFAs Alleviate High-Fat Diet–Induced Circadian Intestinal Microbes Dysbiosis. *Molecular Nutrition and Food Research, 63*(22), 1900492. https://doi.org/10.1002/mnfr.201900492

Guo, R., Zhao, B., Wang, Y., Wu, D., Wang, Y., Yu, Y., . . . Liu, X. (2018). Cichoric acid prevents free-fatty-acid-induced lipid metabolism disorders via regulating Bmal1 in HepG2 cells. *Journal of Agricultural and Food Chemistry, 66*(37), 9667–9678.

Guo, T., Song, D., Ho, C. T., Zhang, X., Zhang, C., Cao, J., & Wu, Z. (2019). Omics analyses of gut microbiota in a circadian rhythm disorder mouse model fed with oolong tea polyphenols. *Journal of Agricultural and Food Chemistry*, *67*(32), 8847–8854.

Halberg, F. (1969). Chronobiology. *Annual Review of Physiology*, *31*(1), 675–726.

Halberg, F. (1989). Some aspects of the chronobiology of nutrition: more work is needed on "when to eat". *The Journal of Nutrition*, *119*(3), 333–343.

He, B., Nohara, K., Park, N., Park, Y. S., Guillory, B., Zhao, Z., . . . Chen, Z. (2016). The small molecule nobiletin targets the molecular oscillator to enhance circadian rhythms and protect against metabolic syndrome. *Cell Metabolism*, *23*(4), 610–621.

Heleno, S. A., Martins, A., Queiroz, M. J. R. P., & Ferreira, I. C. F. R. (2015). Bioactivity of phenolic acids: Metabolites versus parent compounds: A review. *Food Chemistry.*, *173*, 501–513. https://doi.org/10.1016/j.foodchem.2014.10.057

Higgins, G. M., & Mann, F. C. (1926). Observations on the emptying of the gall bladder. *American Journal of Physiology-Legacy Content*, *78*(2), 339–348.

Hilaire, M. A. S., & Lockley, S. W. (2015). Caffeine does not entrain the circadian clock but improves daytime alertness in blind patients with non-24-hour rhythms. *Sleep Medicine*, *16*(6), 800–804.

Hirao, A., Tahara, Y., Kimura, I., & Shibata, S. (2009). A balanced diet is necessary for proper entrainment signals of the mouse liver clock. *PLoS One*, *4*(9), e6909.

Hirota, T., Okano, T., Kokame, K., Shirotani-Ikejima, H., Miyata, T., & Fukada, Y. (2002). Glucose down-regulates Per1 and Per2mRNA levels and induces circadian gene expression in cultured rat-1 fibroblasts. *Journal of Biological Chemistry*, *277*(46), 44244–44251.

Hu, L., Chen, Y., Cortes, I. M., Coleman, D. N., Dai, H., Liang, Y., . . . Loor, J. J. (2020). Supply of methionine and arginine alters phosphorylation of mechanistic target of rapamycin (mTOR), circadian clock proteins, and α-s1-casein abundance in bovine mammary epithelial cells. *Food and Function*, *11*(1), 883–894.

Huang, J., Lu, M., & Ho, C. T. (2021). Health benefits of dietary chronobiotics: Beyond resynchronizing internal clocks. *Food and Function*, *12*, 6136–6156.

Itokawa, M., Hirao, A., Nagahama, H., Otsuka, M., Ohtsu, T., Furutani, N., . . . Shibata, S. (2013). Time-restricted feeding of rapidly digested starches causes stronger entrainment of the liver clock in PER2:: LUCIFERASE knock-in mice. *Nutrition Research*, *33*(2), 109–119.

Jakubowicz, D., Wainstein, J., Landau, Z., Raz, I., Ahren, B., Chapnik, N., . . . Froy, O. (2017). Influences of breakfast on clock gene expression and postprandial glycemia in healthy individuals and individuals with diabetes: A randomized clinical trial. *Diabetes Care*, *40*(11), 1573–1579.

Khandelwal, D., Dutta, D., Chittawar, S., & Kalra, S. (2017). Sleep disorders in type 2 diabetes. *Indian Journal of Endocrinology and Metabolism*, *21*(5), 758–761.

Kim, M., Son, Y. G., Kang, Y. N., Ha, T. K., & Ha, E. (2015). Changes in glucose transporters, gluconeogenesis, and circadian clock after duodenal–jejunal bypass surgery. *Obesity Surgery*, *25*(4), 635–641.

Konopka, R. J., & Benzer, S. (1971). Clock mutants of Drosophila melanogaster. *Proceedings of the National Academy of Sciences*, *68*(9), 2112–2116.

Koronowski, K. B., & Sassone-Corsi, P. (2021). Communicating clocks shape circadian homeostasis. *Science*, *371*(6530), eabd0951.

Kräuchi, K., Cajochen, C., Werth, E., & Wirz-Justice, A. (2002). Alteration of internal circadian phase relationships after morning versus evening carbohydrate-rich meals in humans. *Journal of Biological Rhythms*, *17*(4), 364–376.

Ku, K., Park, I., Kim, D., Kim, J., Jang, S., Choi, M., . . . Kim, K. (2020). Gut microbial metabolites induce changes in circadian oscillation of clock gene expression in the mouse embryonic fibroblasts. *Molecules and Cells*, *43*(3), 276–285.

Kumar, N., & Goel, N. (2019). Phenolic acids: Natural versatile molecules with promising therapeutic applications. *Biotechnology Reports*, *24*, e00370. https://doi.org/10.1016/j.btre.2019.e00370

Laermans, J., & Depoortere, I. (2016). Chronobesity: Role of the circadian system in the obesity epidemic. *Obesity Reviews*, *17*(2), 108–125.

Lax, P., Esquiva, G., Fuentes-Broto, L., Segura, F., Sánchez-Cano, A., Cuenca, N., & Pinilla, I. (2016). Age-related changes in photosensitive melanopsin-expressing retinal ganglion cells correlate with circadian rhythm impairments in sighted and blind rats. *Chronobiology International*, *33*(4), 374–391.

Leone, V., Gibbons, S. M., Martinez, K., Hutchison, A. L., Huang, E. Y., Cham, C. M., . . . Chang, E. B. (2015). Effects of diurnal variation of gut microbes and high-fat feeding on host circadian clock function and metabolism. *Cell Host and Microbe*, *17*(5), 681–689.

Lewis, P., Oster, H., Korf, H. W., Foster, R. G., & Erren, T. C. (2020). Food as a circadian time cue—evidence from human studies. *Nature Reviews Endocrinology*, *16*(4), 213–223.

Li, H., Kek, H. C., Lim, J., Gelling, R. W., & Han, W. (2016). Green tea (−)-epigallocatechin-3-gallate counteracts daytime overeating induced by high-fat diet in mice. *Molecular Nutrition and Food Research*, *60*(12), 2565–2575.

Liu, F., Zhang, X., Zhao, B., Tan, X., Wang, L., & Liu, X. (2019). Role of food phytochemicals in the modulation of circadian clocks. *Journal of Agricultural and Food Chemistry*, *67*(32), 8735–8739.

Masterson, D. (2020). NutraCast podcast: Mantra labs on chrononutrition. https://nutraingredients.com; 13-Nov-2020.

McHill, A. W., Smith, B. J., & Wright, K. P., Jr. (2014). Effects of caffeine on skin and core temperatures, alertness, and recovery sleep during circadian misalignment. *Journal of Biological Rhythms*, *29*(2), 131–143.

McMullan, C. J., Schernhammer, E. S., Rimm, E. B., Hu, F. B., & Forman, J. P. (2013). Melatonin secretion and the incidence of type 2 diabetes. *Jama*, *309*(13), 1388–1396.

Mendoza, J., Angeles-Castellanos, M., & Escobar, C. (2005). A daily palatable meal without food deprivation entrains the suprachiasmatic nucleus of rats. *European Journal of Neuroscience*, *22*(11), 2855–2862.

Meng, X., Li, Y., Li, S., Zhou, Y., Gan, R. Y., Xu, D. P., & Li, H. B. (2017). Dietary sources and bioactivities of melatonin. *Nutrients*, *9*(4), 367.

Mi, Y., Qi, G., Fan, R., Ji, X., Liu, Z., & Liu, X. (2017). EGCG ameliorates diet-induced metabolic syndrome associating with the circadian clock. *Biochimica et Biophysica Acta (BBA)-Molecular Basis of Disease*, *1863*(6), 1575–1589.

Morris, C. J., Aeschbach, D., & Scheer, F. A. (2012). Circadian system, sleep and endocrinology. *Molecular and Cellular Endocrinology*, *349*(1), 91–104.

Moundras, C., Remesy, C., & Demigne, C. (1993). Dietary protein paradox: Decrease of amino acid availability induced by high-protein diets. *American Journal of Physiology-Gastrointestinal and LiverPhysiology, 264*(6), G1057–G1065.

Mukai, M., & Tischkau, S. A. (2007). Effects of tryptophan photoproducts in the circadian timing system: searching for a physiological role for aryl hydrocarbon receptor. *Toxicological Sciences, 95*(1), 172–181.

Mukherjee, P. K. (2019). Bioactive phytocomponents and their analysis. In P. K. Mukherjee (Ed.), *Quality control and evaluation of herbal drugs*. Netherlands: Amsterdam.

Nutrition Insight. (2020). "Biohacking" the circadian rhythm with botanical beverage BioLift; https://nutritioninsight.com/

Oda, H. (2015). Chrononutrition. *Journal of Nutritional Science and Vitaminology, 61*(Supplement), S92–S94.

Ogata, H., Horie, M., Kayaba, M., Tanaka, Y., Ando, A., Park, I., . . . Tokuyama, K. (2020). Skipping breakfast for 6 days delayed the circadian rhythm of the body temperature without altering clock gene expression in human leukocytes. *Nutrients, 12*(9), 2797.

Oike, H. (2017). Modulation of circadian clocks by nutrients and food factors. *Bioscience, Biotechnology, and Biochemistry, 81*(5), 863–870.

Oike, H., Kobori, M., Suzuki, T., & Ishida, N. (2011). Caffeine lengthens circadian rhythms in mice. *Biochemical and Biophysical Research Communications, 410*(3), 654–658.

Oike, H., Oishi, K., & Kobori, M. (2014). Nutrients, clock genes, and chrononutrition. *Current Nutrition Reports, 3*(3), 204–212.

Oishi, K., Uchida, D., & Itoh, N. (2012). Low-carbohydrate, high-protein diet affects rhythmic expression of gluconeogenic regulatory and circadian clock genes in mouse peripheral tissues. *Chronobiology International, 29*(7), 799–809.

Oishi, K., Yamamoto, S., Oike, H., Ohkura, N., & Taniguchi, M. (2017). Cinnamic acid shortens the period of the circadian clock in mice. *Biochemistry and Biophysics Reports, 9*, 232–237.

Orozco-Solis, R., Aguilar-Arnal, L., Murakami, M., Peruquetti, R., Ramadori, G., Coppari, R., & Sassone-Corsi, P. (2016). The circadian clock in the ventromedial hypothalamus controls cyclic energy expenditure. *Cell Metabolism, 23*(3), 467–478.

Oseguera-Castro, K. Y., Madrid, J. A., Madrid, M. J. M., García, O. P., Del Castillo, M. D., & Campos-Vega, R. (2019). Antioxidant dietary fiber isolated from spent coffee (*Coffea arabica* L.) grounds improves chronotype and circadian locomotor activity in young adults. *Food and Function, 10*(8), 4546–4556.

Peng, Y., Sun, Q., & Park, Y. (2019). The bioactive effects of chicoric acid as a functional food ingredient. *Journal of Medicinal Food, 22*(7), 645–652.

Perdices, L., Fuentes-Broto, L., Segura, F., Cuenca, N., & Pinilla, I. (2020). Epigallocatechin gallate slows retinal degeneration, reduces oxidative damage, and modifies circadian rhythms in P23H Rats. *Antioxidants, 9*(8), 718–739.

Pickel, L., & Sung, H. K. (2020). Feeding rhythms and the circadian regulation of metabolism. *Frontiers in Nutrition, 7*(39). https://doi.org/10.3389/fnut.2020.00039

Pivovarova, O., Jürchott, K., Rudovich, N., Hornemann, S., Ye, L., Möckel, S., . . . Pfeiffer, A. F. (2015). Changes of dietary fat and carbohydrate content alter central and peripheral clock in humans. *The Journal of Clinical Endocrinology and Metabolism, 100*(6), 2291–2302.

Plano, S. A., Casiraghi, L. P., García Moro, P., Paladino, N., Golombek, D. A., & Chiesa, J. J. (2017). Circadian and metabolic effects of light: Implications in weight homeostasis and health. *Frontiers in Neurology, 8*(558). https://doi.org/10.3389/fneur.2017.00558

Qi, G., Mi, Y., Fan, R., Zhao, B., Ren, B., & Liu, X. (2017). Tea polyphenols ameliorates neural redox imbalance and mitochondrial dysfunction via mechanisms linking the key circadian regular Bmal1. *Food and Chemical Toxicology, 110*, 189–199.

Qi, G., Wu, W., Mi, Y., Shi, R., Sun, K., Li, R., . . . Liu, X. (2018). Tea polyphenols direct Bmal1-driven ameliorating of the redox imbalance and mitochondrial dysfunction in hepatocytes. *Food and Chemical Toxicology, 122*, 181–193.

Qian, J., Block, G. D., Colwell, C. S., & Matveyenko, A. V. (2013). Consequences of exposure to light at night on the pancreatic islet circadian clock and function in rats. *Diabetes, 62*(10), 3469–3478.

Raikos, V., & Ranawana, V. (2017). Designing emulsion droplets of foods and beverages to enhance delivery of lipophilic bioactive components–a review of recent advances. *International Journal of Food Science and Technology, 52*(1), 68–80.

Rapoport, M. I., Feigin, R. D., Bruton, J., & Beisel, W. R. (1966). Circadian rhythm for tryptophan pyrrolase activit and its circulating substrate. *Science, 153*(3744), 1642–1644.

Rauf, A., Imran, M., Abu-Izneid, T., Patel, S., Pan, X., Naz, S., . . . Suleria, H. A. R. (2019). Proanthocyanidins: A comprehensive review. *Biomedicine and Pharmacotherapy, 116*, 108999. https://doi.org/10.1016/j.biopha.2019.108999

Reinke, H., & Asher, G. (2019). Crosstalk between metabolism and circadian clocks. *Nature Reviews Molecular Cell Biology, 20*(4), 227–241.

Reszka, E., & Zienolddiny, S. (2018). Epigenetic basis of circadian rhythm disruption in cancer. *Cancer Epigenetics for Precision Medicine, 1856*, 173–201.

Reyes, C. M., & Cornelis, M. C. (2018). Caffeine in the diet: Country-level consumption and guidelines. *Nutrients, 10*(11), 1772.

Ribas-Latre, A., Baselga-Escudero, L., Casanova, E., Arola-Arnal, A., Salvadó, M. J., Bladé, C., & Arola, L. (2015). Dietary proanthocyanidins modulate BMAL1 acetylation, Nampt expression and NAD levels in rat liver. *Scientific Reports, 5*(1), 1–13.

Ribas-Latre, A., Del Bas, J. M., Baselga-Escudero, L., Casanova, E., Arola-Arnal, A., Salvadó, M. J., . . . Arola, L. (2015). Dietary proanthocyanidins modulate the rhythm of BMAL1 expression and induce RORα transactivation in HepG2 cells. *Journal of Functional Foods, 13*, 336–344.

Ribas-Latre, A., Del Bas, J. M., Baselga-Escudero, L., Casanova, E., Arola-Arnal, A., Salvadó, M. J., . . . Bladé, C. (2015). Dietary proanthocyanidins modulate melatonin levels in plasma and the expression pattern of clock genes in the hypothalamus of rats. *Molecular Nutrition and Food Research, 59*(5), 865–878.

Ribas-Latre, A., & Eckel-Mahan, K. (2016). Interdependence of nutrient metabolism and the circadian clock system: Importance for metabolic health. *Molecular Metabolism, 5*(3), 133–152.

Schmitz, G., & Ecker, J. (2008). The opposing effects of $n-3$ and $n-6$ fatty acids. *Progress in Lipid Research, 47*(2), 147–155.

Segers, A., Desmet, L., Thijs, T., Verbeke, K., Tack, J., & Depoortere, I. (2019). The circadian clock regulates the diurnal levels of microbial short-chain fatty acids and their rhythmic effects on colon contractility in mice. *Acta Physiologica, 225*(3), e13193.

Semenova, N. V., Madaeva, I. M., & Kolesnikova, L. I. (2021). Clock Gene, Melatonin, and the Sleep–Wake Cycle. *Russian Journal of Genetics, 57*(3), 251–257.

Serhiyenko, V. A., Segin, V. B., & Serhiyenko, A. A. (2018). Effects of omega-3 polyunsaturated fatty acids on the circadian rhythm of heart rate variability parameters in patients with type 2 diabetes mellitus and cardiovascular autonomic neuropathy. *Russian Journal of Cardiology, 23*(5), 56–60.

Sharma, S., Singh, H., Ahmad, N., Mishra, P., & Tiwari, A. (2015). The role of melatonin in diabetes: Therapeutic implications. *Archives of Endocrinology and Metabolism, 59*(5), 391–399.

Shibata, S., Hirao, A., & Tahara, Y. (2010). Restricted feeding-induced entrainment of activity rhythm and peripheral clock rhythm. *Sleep and Biological Rhythms, 8*(1), 18–27.

Shinozaki, A., Misawa, K., Ikeda, Y., Haraguchi, A., Kamagata, M., Tahara, Y., & Shibata, S. (2017). Potent effects of flavonoid nobiletin on amplitude, period, and phase of the circadian clock rhythm in PER2:: LUCIFERASE mouse embryonic fibroblasts. *PLoS One, 12*(2), e0170904. https://doi.org/10.1371/journal.pone.0170904

Singh, A. P., Singh, R., Verma, S. S., Rai, V., Kaschula, C. H., Maiti, P., & Gupta, S. C. (2019). Health benefits of resveratrol: Evidence from clinical studies. *Medicinal Research Reviews, 39*(5), 1851–1891.

Song, C., Wang, J., Kim, B., Lu, C., Zhang, Z., Liu, H., . . . Li, F. (2018). Insights into the role of circadian rhythms in bone metabolism: a promising intervention target? *BioMed Research International, 2018*. https://doi.org/10.1155/2018/9156478

Spengler, M. L., Kuropatwinski, K. K., Schumer, M., & Antoch, M. (2009). A serine cluster mediates BMAL1-dependent CLOCK phosphorylation and degradation. *Cell Cycle, 8*(24), 4138–4146.

Srikanta, S. B., & Cermakian, N. (2020). To Ub or not to Ub: Regulation of circadian clocks by ubiquitination and deubiquitination. *Journal of Neurochemistry, 157*(1), 11–30.

Stenvers, D. J., Scheer, F. A., Schrauwen, P., La Fleur, S. E., & Kalsbeek, A. (2019). Circadian clocks and insulin resistance. *Nature Reviews Endocrinology, 15*(2), 75–89.

Stephan, F. K., & Davidson, A. J. (1998). Glucose, but not fat, phase shifts the feeding-entrained circadian clock. *Physiology and Behavior, 65*(2), 277–288.

Subramanian, P., Jayakumar, M., Singaravel, M., Kumar, D., Basu, P., Jayapalan, J. J., & Hashim, O. H. (2015). Fisetin, a dietary flavonoid, attenuates hyperammonemia and improves circadian locomotor deficits, redox balance, and astrocytic markers in rats. *Journal of Functional Foods, 12*, 409–419.

Sultan, A., Choudhary, V., & Parganiha, A. (2017). Worsening of rest-activity circadian rhythm and quality of life in female breast cancer patients along progression of chemotherapy cycles. *Chronobiology International, 34*(5), 609–623.

Sun, L., Wang, Y., Song, Y., Cheng, X. R., Xia, S., Rahman, M. R. T., . . . Le, G. (2015). Resveratrol restores the circadian rhythmic disorder of lipid metabolism induced by high-fat diet in mice. *Biochemical and Biophysical Research Communications, 458*(1), 86–91.

Suzuki, C., Fukumitsu, S., & Oike, H. (2021). Modulation of cellular circadian clocks by triterpenoids. *Phytochemistry, 181*, 112539. https://doi.org/10.1016/j.phytochem.2020.112539

Tahara, Y., Aoyama, S., & Shibata, S. (2017). The mammalian circadian clock and its entrainment by stress and exercise. *The Journal of Physiological Sciences, 67*(1), 1–10.

Tahara, Y., Otsuka, M., Fuse, Y., Hirao, A., & Shibata, S. (2011). Refeeding after fasting elicits insulin-dependent regulation of Per2 and Rev-erbαwith shifts in the liver clock. *Journal of Biological Rhythms*, *26*(3), 230–240.

Tahara, Y., Yamazaki, M., Sukigara, H., Motohashi, H., Sasaki, H., Miyakawa, H., . . . Shibata, S. (2018). Gut microbiota-derived short chain fatty acids induce circadian clock entrainment in mouse peripheral tissue. *Scientific Reports*, *8*(1), 1–12.

Tavakoli, A., Mirzababaei, A., Sajadi, F., & Mirzaei, K. (2021). Circulating inflammatory markers may mediate the relationship between low carbohydrate diet and circadian rhythm in overweight and obese women. *BMC Women's Health*, *21*(1), 1–10.

Tordjman, S., Chokron, S., Delorme, R., Charrier, A., Bellissant, E., Jaafari, N., & Fougerou, C. (2017). Melatonin: Pharmacology, functions and therapeutic benefits. *Current Neuropharmacology*, *15*(3), 434–443.

Tsimakouridze, E. V., Alibhai, F. J., & Martino, T. A. (2015). Therapeutic applications of circadian rhythms for the cardiovascular system. *Frontiers in Pharmacology*, *6*, 77.

Tsujino, N., & Sakurai, T. (2012). Circadian rhythm of leptin, orexin and ghrelin. *Nihon rinsho. Japanese Journal of. Clinical Medicine*, *70*(7), 1121–1125.

Van Diepen, H. C., Lucassen, E. A., Yasenkov, R., Groenen, I., Ijzerman, A. P., Meijer, J. H., & Deboer, T. (2014). Caffeine increases light responsiveness of the mouse circadian pacemaker. *European Journal of Neuroscience*, *40*(10), 3504–3511.

Vuolo, M. M., Lima, V. S., & Junior, M. R. M. (2019). Phenolic compounds: Structure, classification, and antioxidant power. In M. R. S. Campos (Ed.), *Bioactive Compounds* (pp. 33–50). Woodhead Publishing.

Wehrens, S. M., Christou, S., Isherwood, C., Middleton, B., Gibbs, M. A., Archer, S. N., . . . Johnston, J. D. (2017). Meal timing regulates the human circadian system. *Current Biology*, *27*(12), 1768–1775.

Winter, E. M., Kooijman, S., Appelman-Dijkstra, N. M., Meijer, O. C., Rensen, P. C., & Schilperoort, M. (2021). Chronobiology and chronotherapy of osteoporosis. *JBMR Plus*, *5*(10), e10504.

Woo, J. S., Min, H. K., Choi, J. W., Moon, J. H., Park, M. J., Kwok, S. K., . . . Cho, M. L. (2021). Short-chain fatty acid butyrate induces IL-10-producing B cells by regulating circadian-clock-related genes to ameliorate Sjögren's syndrome. *Journal of Autoimmunity*, *119*, 102611.

Wurtman, R. J., Rose, C. M., Chou, C., & Larin, F. F. (1968). Daily rhythms in the concentrations of various amino acids in human plasma. *New England Journal of Medicine*, *279*(4), 171–175.

Xie, X., Kukino, A., Calcagno, H. E., Berman, A., Garner, J. P., & Butler, M. P. (2020a). Natural food intake patterns do not synchronize peripheral clocks. *BMC Biology*, *18*(1), 1–11.

Xie, X., Kukino, A., Calcagno, H. E., Berman, A., Garner, J. P., & Butler, M. P. (2020b). Natural food intake patterns do not synchronize peripheral clocks. *BioRxiv*. CC-BY-ND 4.0 International license

Yamaguchi, Y. (2018). Arginine vasopressin signaling in the suprachiasmatic nucleus on the resilience of circadian clock to jet lag. *Neuroscience Research*, *129*, 57–61.

Yamamuro, D., Takahashi, M., Nagashima, S., Wakabayashi, T., Yamazaki, H., Takei, A., . . . Ishibashi, S. (2020). Peripheral circadian rhythms in the liver and white adipose tissue of mice are attenuated by constant light and restored by time-restricted feeding. *PLoS One*, *15*(6), e0234439.

Yasuda, S., Iwami, S., Tamura, K., Ikeda, Y., Kamagata, M., Sasaki, H., . . . Shibata, S. (2019). Phase resetting of circadian peripheral clocks using human and rodent diets in mouse models of type 2 diabetes and chronic kidney disease. *Chronobiology International*, *36*(6), 851–869.

Yi-lin, L., Ke, Z., Dan, W., Xi-hong, Z., Zheng, R., Xin, W., & Yu-long, Y. (2018). Dynamic feeding low and high methionine diets affect the diurnal rhythm of amino acid transporters and clock related genes in jejunum of laying hens. *Biological Rhythm Research*, *49*(5), 671–679.

Zhang, L., Yan, R., & Wu, Z. (2020). Metagenomics analysis of intestinal flora modulatory effect of green tea polyphenols by a circadian rhythm dysfunction mouse model. *Journal of Food Biochemistry*, *44*(10), e13430. https://doi.org/10.1111/jfbc.13430

Zhang, X., Zhu, X., Sun, Y., Hu, B., Sun, Y., Jabbar, S., & Zeng, X. (2013). Fermentation *in vitro* of EGCG, "GCG and EGCG3" Me isolated from Oolong tea by human intestinal microbiota. *Food Research International*, *54*(2), 1589–1595.

Zitting, K. M., Vujovic, N., Yuan, R. K., Isherwood, C. M., Medina, J. E., Wang, W., . . . Duffy, J. F. (2018). Human resting energy expenditure varies with circadian phase. *Current Biology*, *28*(22), 3685–3690.

Nelson, W., & Halberg, F. (1986). Schedule-shifts, circadian rhythms and lifespan of freely-feeding and meal-fed mice. *Physiology & Behavior*, *38*(6), 781–788.

Part II

Functional Food

4

Common Beans Bioactive Components and Their Potential to Modulate Molecular Markers of Obesity and Type 2 Diabetes

Montserrat Alcazar-Valle, Eugenia Lugo-Cervantes, David Fonseca-Hernandez, and Luis Mojica

Tecnología Alimentaria, Centro de Investigación y Asistencia en Tecnología y Diseño del Estado de Jalisco, A. C., CIATEJ, Zapopan, Jalisco, Mexico

4.1 Introduction

Around 20 legumes species are used as dry grains for human consumption and are considered as an essential source of nutrients (Hayat, Ahmad, Masud, Ahmed, & Bashir, 2014). Legumes are part of the human diet; they are the second most consumed food worldwide, only after cereals (Singh, Singh, & Kaur, 2020). Common bean (*Phaseolus vulgaris* L.) is the most consumed legume around the world. It has been cultivated for several hundred years and plays an important role in the traditional diet of various regions from America, Africa, and Asia. The origin and domestication of common beans (*P. vulgaris*) took place in the Mesoamerican region. The genus *Phaseolus* of the *Fabaceae* family has around 50 plant species, all native of America. Thus, variations are extensive in the relationship of the cultivation patterns, the characteristics of the maturation, and adaptation of bean seeds, bringing about 40,000 varieties (Chávez-Mendoza & Sánchez, 2017; Hernández-López, Vargas-Vázquez, Muruaga-Martínez, Hernández-Delgado, & Mayek-Pérez, 2013; Romero-Arenas et al., 2013; Suárez-Martínez et al., 2016).

An increase in scientific evidence supports the importance of bioactive compounds in disease prevention due to their beneficial effects on health. Besides their high protein content, carbohydrates, dietary fiber, minerals, and vitamins, common beans also contain several bioactive compounds with biological potential, such as flavonoids, anthocyanins, condensed tannins, phytosterols, and saponins. Some clinical studies have shown that common bean consumption reduces the risk of cardiovascular diseases, stomach, and prostate cancer, promotes weight control, reduces obesity and stress, and controls anxiety and depression, among others (Chávez-Mendoza & Sánchez, 2017; Ganesan & Xu, 2017).

Molecular Mechanisms of Functional Food, First Edition. Edited by Rocio Campos-Vega and B. Dave Oomah.

4.1.1 Common Beans and Their Importance

Common beans are one of the most important legumes worldwide; about 300 million people worldwide consume them. The main bean-consuming countries are Brazil, India, Mexico, United States, Tanzania, and Uganda. Moreover, common beans are considered the second source of protein in East and Southeast Africa and the fourth in America (Romero-Arenas et al., 2013; Suárez-Martínez et al., 2016). According to Food and Agriculture Organization (FAO) data, common bean production was 30,434,280 tons in 2018, of which 47.4% was produced in Asia, 26.1% in America, 23.1% in Africa, 3.3% in Europe, and 0.1% in Oceania. Besides, according to the FAO, from 2015 to 2018, the main bean-producing countries were India, Myanmar, Brazil, United States, China, Tanzania, Mexico, Uganda, Kenya, and Ethiopia.

Epidemiological evidence shows that the consumption of common beans has many health benefits, including weight control, glucose and insulin regulation, and lower blood lipid and LDL-cholesterol levels, due to their antioxidant and other properties (Lin, O'Keefe, Duncan, & Fernández-Fraguas, 2020; Suárez-Martínez et al., 2016). Some companies have commercialized food products formulated using functional ingredients derived from beans. Thanks to the number of bioactive components, they can be used to formulate nutraceutical products such as pasta, snack bars, chips, among others (Moreno-Valdespino, Luna-Vital, Camacho-Ruiz, & Mojica, 2020).

4.1.1.1 *Common Beans (Phaseolus vulgaris* L.)

Common beans are annual plants grown in temperate and semitropical regions. They were first cultivated in Peru and Mexico 8,000 years ago and are now produced around the world. Common bean varieties do not vary considerably in nutritional composition in most cases, but they differ slightly in taste and texture (Table 4.1). Mexico is recognized as the center of origin of common beans since 47 of the 52 varieties identified are endemic to this country (Chávez-Mendoza, Hernández-Figueroa, & Sánchez, 2018). Common beans consist of seed coat, cotyledon, and the embryonic axis (9, 90, and 1% of the dry seed weight, respectively). Common beans are a good source of protein and bioactive compounds with high antioxidant potential (Chávez-Mendoza et al., 2018), mainly related to the phenolic compounds in the seed coat. Dark-colored seeds present higher concentration of phenolic compounds and flavonoids compared to light-colored ones (Figure 4.1).

4.1.1.2 Nutritional Composition of Common Bean

The genus *Phaseolus* ssp is considered a significant source of proteins, carbohydrates, vitamins, and minerals and a rich source of unsaturated fatty acids, such as linoleic and oleic acids (Celmeli et al., 2018). Common beans (*P. vulgaris*) represent an excellent source of protein (22–33%), where albumin and globulins account for 80–90% of the total bean protein and essential amino acids (lysine). Moreover, common bean has high dietary fiber content (29–47%), which includes soluble and insoluble fiber and resistant starch (9.2–18.1%), lipids (1.1–2.8%), and oligosaccharides (1.6–9.1%) such as raffinose (Chávez-Mendoza et al., 2018; Suárez-Martínez et al., 2016). Beans are also rich in minerals such as copper, iron, manganese, zinc, and sulfur (18, 60, 23, 29, and 234 mg kg^{-1}, respectively), vitamins, and other important secondary metabolites such as phenolic compounds (Carbas et al., 2020; Kan et al., 2017).

Table 4.1 Main commercialized common beans and their health benefits.

Common bean	Physical characteristics	Health benefits
Kidney beans	Small-sized, white color, and oval-shaped	Modulate anti-inflammatory biomarkers both locally and systematically in diseases such as colitis Regulate cholesterol levels and lipid profiles
Kidney beans	Large-sized, firm texture, pink/red (shiny) color, and kidney-shaped	Reduce the glycemic index in diabetic mice
Red beans	Soft texture, small-sized, and oval-shaped	Exhibit anti-inflammatory effect, antifungal potential, regulate the immune system, and promote apoptosis in tumor cells
Black beans (turtle beans)	Sweet taste, smooth texture, medium size, and oval-shaped	Excellent source of anthocyanins and other phenolic compounds with the potential use as a natural food coloring, with an exceptional antidiabetes approach
Pinto beans	Medium-sized, brown color, and oval-shaped	Antifungal, antidiabetes, and antitumor activities
Cranberry beans (Roman beans)	Creamy red texture, medium-sized, and oval-shaped	They are rich in phenolic compounds and nondigestible fermentable components, which can alleviate colitis and mitigate the severity of pathologies associated with the intestinal mucosa.

Source: Adapted from Ganesan and Xu (2017).

Figure 4.1 Native *Phaseolus vulgaris* L. from Oaxaca, Mexico.

According to the United States Department of Agriculture (USDA), 125 g of cooked common beans provides a good amount of recommended macro- and micronutrients in the daily diet of an adult (Table 4.2). However, raw beans contain several antinutritional components limiting their consumption, presenting adverse effects, including decreased

Table 4.2 Recommend daily percentage of macro and micronutrients vs. percentage provided by common beans (*Phaseolus vulgaris* L.).

Nutrient	Common bean (%) of recommended intake	Recommended intake (%)
Protein	10.6	15.0
Dietary fiber	37.5	40.0
Phytosterols	16.4	26.0
Linolenic acid	21.5	35.0
Polyphenols	4.0	48.0

Source: Adapted from Arnoldi, Zanoni, Lammi, and Boschin (2015).

intestinal absorption of sugars and amino acids and alterations in the immune response. Among the antinutritional components are trypsin inhibitors, lectin, phytates, tannins, and indigestible oligosaccharides; these compounds tend to degrade with temperature during cooking, and paradoxically low concentrations of some antinutritional factors exhibit nutraceutical effects (Suárez-Martínez et al., 2016).

4.1.2 Bioactive Components in Common Beans

The importance of legumes, especially common beans, relies in their diverse bioactive components. They can be classified into phenolic compounds, phytosterols, oligosaccharides, fiber, resistant starch, saponins, carotenoids, tocopherols, proteins, and peptides (Moreno-Valdespino et al., 2020; Singh et al., 2020).

4.1.2.1 Phenolic Compounds

Phenolic compounds or polyphenols are secondary metabolites, which are decisive in the sensory and nutritional quality of fruits, vegetables, and other plants. Recently, phenolic compounds have received more attention due to their functions in the plant, such as protection against insects, seed pigmentation, and the associated health benefits such as anti-inflammatory potential, protection against UV light, among others (Ganesan & Xu, 2017).

Phenolic compounds have at least one phenolic unit with one or more hydroxyl group subunits. These compounds can be present in free, solubilized conjugate or insoluble form. They have been identified mainly in legumes such as peas, chickpeas, lentils, and common beans. The total content of phenolic compounds (TPC) in beans varies significantly from 0.6 to 7.0 mg gallic acid equivalents (GAE) g^{-1} (Singh et al., 2020). Phenolic compounds are found in both bean seed coats and cotyledon.

However, the seed coat contains higher concentration of phenolic compounds compared to cotyledons (Table 4.3).

Further, some studies in nutrition, including epidemiological, animal, and cell line studies, have shown that adequate consumption of polyphenols can have beneficial effects on metabolic syndrome, some types of cancer, and cognitive functions (Sandoval et al., 2020). Phenolic acids, flavonoids, and pro-anthocyanidins (condensed tannins) are the main polyphenols in common beans, where the most colorful beans are the richest in these compounds

Table 4.3 Main bioactive components in common beans.

Phenolic compounds		Subclassification of phenolic compounds	Phenolic compounds in common beans (*P. vulgaris* L.)
Phenolic acids		Hydroxycinnamic acids	Caffeic, *p*-coumaric, sinapic
			Ferulic
		Hydroxybenzoic acids	Gallic, *p*-hydroxybenzoic
			Protocatechuic, vanillin
			Syringic
Flavonoids		Flavonols	Kaempferol, quercetin
			Myricetin, all its derivatives
		Flavones	Apigenin, luteolin
		Isoflavonoid	Genistein, daidzein, glycitein
		Flavanols	Catechin, epicatechin
			Epigallocatechin, epicatechin gallate
		Flavanones	Naringenin,
			Hesperetin (glucosides)
		Anthocyanins	Delphinidin, petunidin
			Malvidin, cyanidin
	Pro-anthocyanidins		Procyanidin
			Procyanidin dimer, trimer
	Coumarins		Coumestrol
	Carotenoids		Lutein
	Tocopherols		δ-Tocopherol
	Phytosterols		β-Sitosterol, β-campesterol Δ-7-Avenasterol
	Saponins		Phaseoside I Deacetyl soyasaponin Af Soyasaponin Af, αg, βa Soyasaponin βg, γg

Source: Adapted from Yang, Gan, Ge, Zhang, and Corke (2018); and Chávez-Santoscoy, Gutiérrez-Uribe, and Serna-Saldivar (2013).

(Ganesan & Xu, 2017; Yang et al., 2018). Genotype, environmental conditions, storage, and thermal processes influence the composition of phenolic compounds in common beans (García-Díaz, Aquino-Bolaños, Chávez-Servia, Vera-Guzmán, & Carrillo-Rodríguez, 2018). Numerous studies have shown that common bean polyphenols exhibit antimutagenic, antimicrobial, antidiabetic, and anti-inflammatory potential, which is strongly related to the antioxidant characteristics of polyphenols (Chávez-Mendoza et al., 2018; Ganesan & Xu, 2017; Yang et al., 2018). Moreover, the total polyphenol content in common beans is higher compared to lentils, chickpeas, or soybeans (García-Díaz et al., 2018).

4.1.2.2 Carbohydrates

Legumes contain important amounts of carbohydrates, including dietary fiber, resistant starch, and oligosaccharides that can exert bioactive properties. These bioactive carbohydrates improve microorganisms activity in the colon and increase the production of short-chain fatty acids (SCFAs), among other benefits (Singh et al., 2020). Furthermore, regular consumption of common beans provides the amount of fiber necessary to reduce the glycemic index in humans. Some studies show that diets that include common beans reduce low-density lipoprotein (LDL) and increase high-density lipoprotein (HDL), which reduces risk factors of developing cardiovascular diseases, obesity, and type 2 diabetes (T2D) (Arnoldi et al., 2015; Ganesan & Xu, 2017). The consumption of a standardized white kidney bean extract induced the development of *Bifidobacterium*, *Lactobacillus*, and *Akkermansi* and ameliorated complications related to obesity and T2D in a high-fat diet animal model (Song et al., 2016). This effect may be associated with the presence of undigested carbohydrates acting as resistant starch and the presence of glycoproteins. In other studies, two varieties of white and cream beans with similar nutritional content showed comparable effects by reducing blood glucose and triglycerides in T2D patients (Olmedilla-Alonso et al., 2013). Fiber content varies in cooked and raw beans. Cooked beans Negro 8025, Pinto Durango, Bayo Madero, and Azufrado Higuera fiber content ranges from 36.5 to 55% and 25.6 to 37.6% in raw beans. For soluble fiber, the content ranged from 5.5 to 14% and 0.3 to 1.6% in cooked and raw beans, respectively, while the insoluble fiber content was 31–41% in cooked beans and 25.2–36.1% in raw beans, respectively. In a clinical trial, pinto beans (120–162 g day^{-1}) supplementation reduced serum cholesterol and LDL in hyperlipidemic men (Vergara-Castañeda, Oomah, & Campos-Vega, 2013).

The use of dietary supplements rich in α-amylase inhibitors from common beans has demonstrated the potential to cause weight loss in clinical trials. A meta-analysis evaluated a white bean extract supplementation, resulting in weight loss with daily doses ranging from 400 to 3000 mg per day (Udani, Tan, & Molina, 2018). Red kidney bean supplementation reduced the cholesterol levels in rats fed a diet of whole beans rich in dietary fiber and resistant starch. These results suggest that the consumption of dietary fiber and resistant starch from common beans can prevent chronic disease development and provide substantial health benefits (Kilua, Chihiro, Han, Homma, & Fukuma, 2020).

4.1.2.3 Phytosterols

Phytosterols present in all plants' cells are important structural components that stabilize plants' biological membranes. Phytosterols are part of the triterpene family found in high concentrations in oilseeds and nuts (Moreau et al., 2018). In legume seeds, phytosterols are present in free and glycosylated form, where the main compounds are β-sitosterol, campesterol, and Δ-7-avenasterol (Singh et al., 2020). Phytosterols comprise β-sitosterol (85 and 86.5 mg 100 g^{-1}), campesterol (15.2 and 6.5 mg 100 g^{-1}), and stigmasterol (86.2 and 41.4 mg 100 g^{-1}) in butter beans and kidney beans, respectively (Campos-Vega, Vergara-Castañeda, & Oomah, 2012). Other phytosterols have been identified in common bean sprouts, including fucosterol, β-sitosterol, β-campesterol, Δ7-avenasterol, campesteryl 3-β-glucopyranoside, and stigmasteryl 3-β-glucopyranoside (Table 4.3) (Mendoza-Sánchez et al., 2019). In the last decade, phytosterols have been studied for their cholesterol-lowering effects, anti-inflammatory and antioxidant potential, in addition to their benefits to the immune system (Arnoldi et al., 2015; Vezza et al., 2020). Moreover, β-sitosterol inhibits vascular smooth muscle cell proliferation by increasing p21cip1 protein levels, inhibiting the CDK2 activity, interrupting cell cycle progress. It also induces G2/M arrest and apoptosis through the Bcl-2 and P13K/Akt signaling pathways in human leukemia cells, U937 and HL60 (Vergara-Castañeda et al., 2013).

4.1.2.4 Saponins

Saponins are bioactive compounds widely distributed in plants, made up of a nonpolar part called sapogenin, which, based on its structure, could be steroidal, triterpenoid, or steroidal amine, linked to a chain of carbohydrates. These compounds can capture free radicals and stimulate antioxidant enzymes that inhibit the formation of complexes between radicals and metal ions (Moreno-Valdespino et al., 2020). In legumes, saponins are categorized into three groups: A, B, and E, where group B of saponins has been reported to be the main compounds (Table 4.3). Saponin content is high in kidney and haricot beans (35 and 41 mg/g, respectively). *Phaseolus vulgaris* contain a mixture of both soyasaponin I and V. Moreover, navy bean starch contains sapogenin with significant hypocholesterolemic effect (Campos-Vega et al., 2012).

4.1.2.5 Carotenoids and Tocopherols.

Carotenoids and tocopherols are lipophilic antioxidants. Besides, carotenoids provide plants with red, orange, and yellow colors. These molecules are responsible for some organoleptic properties; lutein is the main carotenoid identified in red beans, and δ-tocopherol is the major compound in black and pinto beans (Moreno-Valdespino et al., 2020). Nowadays, interest has increased in the extraction of these bioactive compounds to determine their biological potential to develop functional ingredients for the food industry.

The best method to extract bioactive compounds in legumes must be selected according to their chemistry and distribution in the seed (Moreno-Valdespino et al., 2020). Therefore, the use of different solvents such as methanol, ethanol, hexane, or solvent matrices such as 50% acetone to extract bioactive compounds must be carefully selected according to the chemical affinity of the compound of interest for the solvent and the final application of the extracts, since the majority of solvents are considered toxic (Figure 4.2).

4.1.2.6 Proteins and Peptides

Common beans contain between 16 and 33% proteins, depending on the variety. Some important common bean proteins are phaseolins, lectins, Bowman-Birk inhibitors, and α-amylase inhibitors. Bioactive peptides are short sequences of amino acids (2–20) obtained from proteins through fermentation, processing, and hydrolysis using gastrointestinal enzymes and other commercial enzymes obtained from microorganisms. These molecules can promote health benefits in humans such as antioxidant, antidiabetes, antiobesity, antimicrobial, anti-inflammatory, and antihypertensive activities (De Fátima Garcia, de Barros, & de Souza Rocha, 2020; Luna-Vital, Mojica, González de Mejía, Mendoza, & Loarca-Piña, 2015).

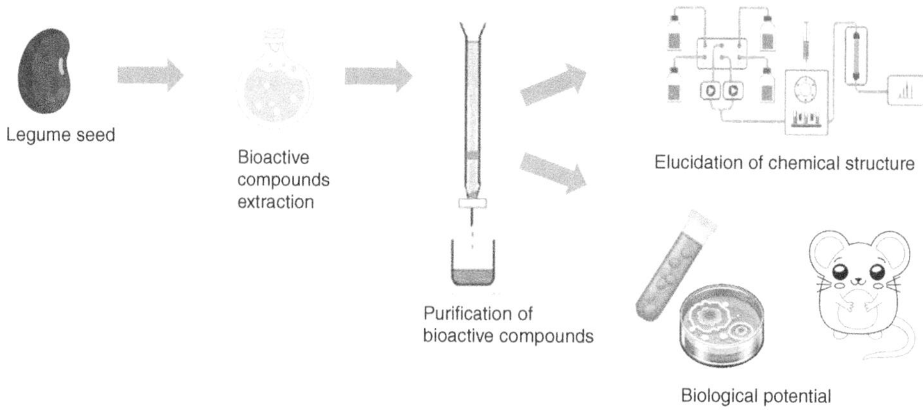

Figure 4.2 Diagram for the extraction, purification, and identification of bioactive compounds. *Source:* Based on Singh et al. (2020).

4.1.3 Obesity and Diabetes in the World

Obesity is defined as excess body weight for a given height, with the body mass index (BMI) to define overweight (BMI 25–29.9) and obesity (BMI ≥ 30). An analysis of data from 195 countries reveals that the prevalence of obesity has doubled in more than 70 countries since 1980. Nearly, 600 million adults were considered obese in 2015; obesity continues to be a disturbing disease throughout the world (Gadde, Martin, Berthoud, & Heymsfield, 2018). Moreover, overweight and obesity were responsible for approximately 4 million deaths globally in 2015, representing 7.2% of total deaths that year (Nolan et al., 2020). Furthermore, childhood obesity has also increased, with 41 million children being overweight or obese in 2016 (Nolan et al., 2020). Therefore, obesity is currently considered a global pandemic. It is widely recognized that in addition to the metabolic disorders associated with obesity, such as nonalcoholic fatty liver disease, obesity also increases the risk for cardiovascular disease, cancer, and T2D (Thompson, McGinley, Neil, & Brick, 2017). Diabetes is one of the major causes of mortality globally; it is expected that by 2040, 10.4% of the world population will have this condition (Di Stefano, Oliviero, & Udenigwe, 2018).

Therefore, it is a complex chronic disease that requires ongoing medical care; it represents a high financial cost for individuals and society. The annual cost of diagnosed diabetes in 2017 was $327 billion approximately, an increase of 26% compared to 2012 (ADA, 2020). Diabetes can be classified into the following categories:

1) Type 1 diabetes (absolute insulin deficiency, due to β-cells destruction).
2) T2D (frequently related to insulin resistance and the progressive loss of β-cells).
3) Gestational diabetes mellitus (diagnosed in the second or third trimester of pregnancy).
4) Specific cases of diabetes, due to other cases, for example, monogenic diabetes syndrome, diseases of the exocrine pancreas, drug- or chemical-induced diabetes.

T2D is the most common condition, affecting about 90% of people diagnosed with diabetes, where increasing age in population, unhealthy diets, limited physical activity, and obesity are important factors that promote the development of this disease (Di Stefano et al., 2018).

4.1.3.1 Pathophysiology of Obesity and Diabetes

Studies implicate the hypothalamus and the pituitary gland in the pathophysiology of various complex diseases, such as obesity (Figure 4.3). A multitude of neuropeptides of the hypothalamus are affected modulating the parameters of homeostasis energy. Thus, changes in body weight can occur due to alteration of several variables (Mendoza-Sánchez et al., 2019):

1) Amount and type of food eaten.
2) Central control of satiety.
3) Hormonal control of assimilation or storage.
4) Physical activity or metabolic rate.

Obesity appears to be a consequence of multiple mechanisms; several studies establish an imbalance in the hypothalamic and brain–intestine neuroendocrine systems. Hence, obesity can be both a cause and a consequence of diseases, depending on the disorder. For example, T2D sometimes presents initial clinical manifestations through sudden weight gain, reflecting the insulin-resistant character of the obese state. Adipose tissue secretes several molecules, which regulate energy intake, inflammation processes, among others. The most studied molecules are leptin and adiponectin. Thus, leptin exerts pro-inflammatory actions and adiponectin anti-inflammatory properties (Gadde et al., 2018; Graβmann et al., 2017). Besides, fat accumulation induces the production of free fatty acids and activates cascades of pro-inflammatory serine kinases, thus initiating the secretion of cytokines and promoting an inflammatory environment that leads to obesity disorders (Graβmann et al., 2017). Besides, some data supported a positive correlation between intake of refined

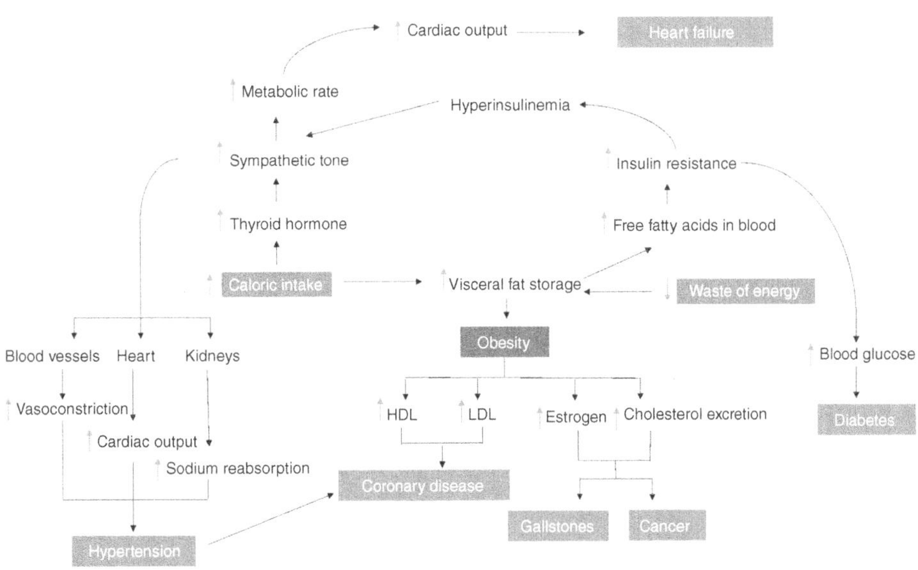

Figure 4.3 Role of obesity in the pathophysiology of the disease (Gadde et al., 2018; Graβmann, Wirsching, Eichelmann, & Aleksandrowa, 2017).

carbohydrates and obesity, since carbohydrates are the macronutrients that provide energy and the excess energy intake subsequent contribution to weight gain (Wang et al., 2020).

Obesity is also associated with the accumulation of free fatty acids and cholesterol that causes white tissue dysfunction and lipotoxicity. The accumulation of lipids causes adipocytes to expand, promoting the inflammation process and increasing the pro-inflammatory cytokines such as tumor necrosis factor-alpha (TNF-α), interleukins, and C-reactive protein. Besides insulin resistance increase, lipolysis increases the free fatty acids in plasma (Coronado-Cáceres et al., 2019).

T2D development is associated with multiple genetic and environmental factors. Some relevant factors are aging, >25 BMI, sedentary lifestyle, smoking, and family history of T2D (De Fronzo et al., 2015). The pathophysiology is characterized by insulin resistance, the decline of insulin production, and finally, pancreatic β-cells failure. Insulin resistance is the initial defect in patients that develop T2D. Normally, insulin binds to its receptors by activating the tyrosine kinase insulin receptor and activating the insulin receptor substrates (IRS) through phosphorylation. These IRS activate intracellular signaling molecules such as phosphatidylinositol 3-kinase (PI3K). PI3K promotes glucose transporter type 4 (GLUT4) translocation to the cell membrane, allowing glucose uptake into skeletal muscle and adipose tissue. T2D insulin resistance is associated with a defect in the PI3K signaling pathway that reduces the translocation of glucose transporter protein GLUT4 (De Fronzo et al., 2015; Oseguera, Gonzalez, & Mejia, 2016). On the other hand, the low-rate secretion of insulin by β-cells is due to diverse factors, including the resistance or deficiency of incretin hormones glucagon-like peptide 1 (GLP-1), which is a glucose-insulin stimulator and glucose-dependent insulinotropic polypeptide (GIP) (Ferrannini & Mari, 2014). These hormones have a short half-life in the plasma because they are inactivated by dipeptidyl peptidase-4 (DPP-IV) (Gilbert & Pratley, 2020). DPP-IV is a cell surface protease mainly expressed in endothelial and epithelial cells, and it has a soluble form present in plasma. DPP-IV has an important role in glucose homeostasis by inactivating incretins hormones through proteolysis (Lambeir, Durinx, Scharpé, & De Meester, 2003). Additionally, in advanced states of T2D, β-cells mass is reduced through cell apoptosis in patients with this pathology, leading to pancreatic β-cells failure (Tomita, 2016).

4.1.3.2 Obesity and Type 2 Diabetes Treatments

Several professional guidelines developed in the United States and Europe highlight that weight loss and control in patients diagnosed with obesity require a change in their lifestyle, including changes in diet, increased physical activity, and in severe cases, drug therapy or bariatric surgery. Therefore, pharmacotherapy is recommended to be used for patients whose BMI is greater than 30 kg m^{-2} and who may have comorbidities such as T2D; some of the most widely used drugs are orlistat, lorcaserin, liraglutide, phentermine/topiramate, and naltrexone/bupropion, and these five medications are currently approved in the United States for the management of obesity (Gadde et al., 2018).

Pharmacological approaches to T2D use specific drugs such as metformin, thiazolidinediones, sulfonylureas, DPP-IV inhibitors, GLP-1 receptor agonists, sodium-glucose transport protein (SGLT)-2, α-amylase and α-glucosidase inhibitors, or their combinations (Chatterjee, Khunti, & Davies, 2017). Metformin is the most common drug used to treat diabetes; it helps to reduce hepatic glucose production and stimulates GLP-1 secretion, consequently lowering the glucose concentration in the blood (Moreno-Valdespino et al., 2020). Thiazolidinediones act as insulin sensitizers and aid in lowering postprandial glucose and free fatty acid concentrations (Chatterjee et al., 2017). DPP-IV inhibitors such as sitagliptin and saxagliptin enhance the disponibility and effects of physiological GLP-1 (Yki-Jarvinen, 2004). Also, sulfonylureas act directly on pancreatic β-cells by stimulating insulin secretion (Ashcroft, 1996). However, the use of drugs for T2D treatment can cause side effects in patients. In this sense, it is important to search for alternative treatments or adjuvants in the treatment of T2D, such as bioactive compounds. Several studies have reported the potential of plant-based protein hydrolysates and peptides to modulate molecular markers associated with the prevention or treatment of T2D. These include the inhibition of starch-degrading (α-amylase and α-glucosidase) enzymes, DPP-IV inhibition, reduction of reactive oxygen species (ROS) levels, reduction of glucose uptake by modulating intestinal glucose transporters, among other mechanisms.

4.1.4 Potential Molecular Mechanisms of Common Bean Bioactive Components to Treat Obesity

Lifestyle modification is recommended as a cornerstone of obesity treatment (Gadde et al., 2018). The connection between common beans intake and obesity is not entirely clear. However, consuming common beans in moderate to high portions can reduce body fat and waist circumference. Therefore, common beans may have dietary components that can be beneficial in the treatment of obesity (Tucker, 2020).

Scientific evidence shows that α-amylase inhibitors in white kidney beans can reduce body weight in obese human subjects (Wang et al., 2020). White kidney beans have three isoforms of α-amylase inhibitor, α-AI, α-AI2, and α-AIL, which can effectively block starch degradation (Wang et al., 2020). The total visceral adipose cells in rats fed a diet containing common beans differed from those receiving a control diet (Thompson et al., 2017). The study showed that common beans consumption increases the expression of four proteins involved in fatty acids oxidation: acyl-CoA synthase long-chain fatty acid isoform 4 (ACSL4), acyl-CoA dehydrogenase for long-chain fatty acids (ACADL), fatty acid translocase (CD36), and carnitine palmitoyltransferase 1 (CPT1) suggesting these proteins may be responsible for reducing obesity in rats.

Moreover, consumption of common bean extracts containing flavonoids and saponins stimulates the expression of reverse cholesterol transporters in the liver (ABCG5/ABCG8 and CYP7A1), decreasing liver lipid deposits and promoting bile acid secretion (Calderón-Guzmán et al., 2020). Flavonoids are known as the bioactive compounds most associated with the prevention of cardiovascular diseases, T2D, and obesity. Flavonoids activate the protein AMP kinase (AMPK), which controls lipid metabolism and adipogenesis (Sandoval et al., 2020). Besides, some polyphenols and saponins in common bean sprouts were reported to be inhibitors of pancreatic lipase activity and interfere with intestinal absorption of triglycerides (Mendoza-Sánchez et al., 2019). Also, phenolic compounds and protein hydrolysates modulated inflammation homeostasis by suppressing some pro-inflammatory markers such as TNF-α and interleukins-6 and -1β (IL-6 and IL-1β) (Flores-Medellín et al., 2021).

Black bean varieties are rich in polyphenols such as anthocyanins. An ethanolic extract obtained from black bean flour contained two main anthocyanins, delphinidin 3-glucoside

(D3G) and petunidin 3-glucoside (P3G). In silico studies demonstrated that these anthocyanins potentially inhibit protein tyrosine phosphatase (PTP) which is related to T2D development. The increase of PTP expression in the pancreas leads to a dysfunction of insulin secretion by β-cells. In muscle and adipose tissues, the expression of specific PTPs is associated with the T2D development. These suggest that polyphenols related to PTP inhibition can prevent T2D development and its complications (Damián-Medina et al., 2020). Also, these phenolic compounds showed potential to modulate molecular markers associated with inflammation, insulin resistance, oxidative stress, glucose, and lipid metabolism (Damián-Medina et al., 2020).

Other phenolic compounds present in common beans such as flavonoids and their glucosides of catechin, myricetin, epicatechin, vanillic acid, syringic acid, cyanidin, malvidin, delphinidin, and petunidin present antihyperglycemic potential. *In vitro* studies showed interesting theoretical inhibition of α-glucosidase, α-amylase, and DPP-IV with these compounds (Ganesan & Xu, 2017). Furthermore, beans containing phenolic compounds reduce blood glucose, glycosylated hemoglobin, and high insulin levels in *in vivo* studies. A regular intake of common beans decrease total serum cholesterol, LDL, and elevated HDL in healthy and obese patients (Ganesan & Xu, 2017). Also, bean consumption has been related to reduced blood glucose. In a clinical trial, patients with T2D consumed a bean-based diet for 3 days to analyze the effects on the blood's postprandial glucose and insulin levels. Compared to a white-rice-based diet, the bean-based diet significantly increased insulin levels and reduced postprandial glucose levels. Common beans can reduce blood glucose due to their low glycemic index (Xiong, Li, Nie, Meng, & Yang, 2021). Cooked navy bean consumption improved intestinal health in lean mice and improved intestinal dysbiosis and inflammation in obese mice (Monk et al., 2021).

Other studies showed that the consumption of resistant starch (RS) present in common beans can prevent obesity, since bacteria in the intestine ferment the RS and produce SCFAs that upregulate expression of anti-inflammatory, antiobesity, and antidiabetic biological markers (Luzardo-Ocampo et al., 2018; Tanaka, Honda, Miwa, Akahori, & Matsumoto, 2019). Studies referring to the potential effect of white kidney beans to promote weight loss in both animal and human studies may be attributed to the three isoforms of α-amylase inhibitors (α-AI1, α-AI2, and α-AIL) that interact with the active site of pancreatic α-amylase, reducing free glucose concentration from starch (Nolan et al., 2020). Regarding phytosterols, *in vitro* 3T3-L1 cells studies revealed beneficial effects of these compounds on the metabolism of adipose tissue (Vezza et al., 2020). (Table 4.4).

4.1.5 Potential Molecular Mechanisms of Common Bean Bioactive Components to Treat Type 2 Diabetes

Similar to obesity, the primary mechanism of action to prevent or control T2D is lifestyle modification, including changes in diet and physical activity (Wang et al., 2020). The most common strategy to control T2D is to reduce sugar consumption; the enzymes α-amylase and α-glucosidase are key in carbohydrate metabolism. The enzyme α-glucosidase is the last enzyme to control glucose release from polysaccharides in the intestine, and the inhibition of its activity is one of the most sought in T2D treatment; some studies report that

Table 4.4 Antidiabetes and antiobesity potential of common beans.

Common bean	Compound	Model/subject	Doses	Activity	Molecular marker
White and cream bean	Protein, dietary fiber, and starch	Human	275 g day^{-1}	Antidiabetes	↓ Blood glucose ↓ Triglycerides
	White bean extract (powder)	Human (meta-analysis)	0.4–3 g day^{-1}	Antiobesity Antidiabetes	Glycemic control and weight loss
	Aqueous extract	Human	2.4 g day^{-1} (35 days)	Antiobesity	α-Amylase inhibition
Kidney bean	Total phenols, anthocyanin	Overweight subjects, randomized	50 g/bw/p.o (60 days)	Antidiabetes, antiobesity	α-Amylase inhibition
	Phenolic acids and bioactive peptide fractions	*In vitro* (24 h)	<1, 1–3.5, 3.5–5, 5–10, and 10 kDa	Antidiabetes	Suppress NF-κB pathway
	Total phenols	Rats	50 mg kg^{-1} bw/p.o (7 days)	Antidiabetes	α-Amylase inhibition
	Phenolic acids (chlorogenic, gallic, p-hydroxybenzoic, and caffeic)	Rats	0.4, 0.8, and 1.2 g kg^{-1} bw po (21 days)	Antidiabetes	↑ Insulin levels
	Quercetin, kaempferol, *p*-coumaric acid, ferulic acid, *p*-hydroxybenzoic acid, and vanillic acid	Rats and mice	7.5%, 15%, 30%, and 60% w/w (7 days)	Antiobesity	↓ Cholesterol levels and LDL-C, regulatory effects on enzymes involved in fatty acid or cholesterol biosynthesis
Red kidney bean	Dietary fiber and resistance starch	Rats	Whole bean	Antiobesity	↓ Cholesterol levels
Navy bean	Total phenols, anthocyanin	3T3-L1 adipocytes	50 g/bw/p.o (60 days)	Antidiabetes, antiobesity	↓ Expression of TNF-α

(Continued)

Table 4.4 (Continued)

Common bean	Compound	Model/subject	Doses	Activity	Molecular marker
	Protein hydrolysates	*In vitro*	IC_{50}=0.095 mg/Ml	Antidiabetes	DPP-IV inhibition
	Total phenols	Diabetic patients	50 g/bw/p.o (24 h)	Antidiabetes	AMPK regulation
Pinto bean	Delphinidin glucoside, petunidin glucoside, malvidin glucoside, catechin, myricetin 3-O-arabinoside, epicatechin, vanillic acid, syringic acid, and O-coumaric acid	*In vitro*	50 ml (24 h)	Antidiabetes	α-Amylase and α-glucosidase inhibition
	Total phenols	Human (3h)	100 g/bw/p.o	Antidiabetes	α-Amylase inhibition
	Flavonoids	Rats	50, 200, and 500 mg kg^{-1} bw	Antiobesity and hypoglycemic effect	α-Amylase inhibition
Black bean	Delphinidin 3-O-glucoside, petunidin-3-O-glucoside, malvidin-3-O-glucoside	Caco-2 cells	1 mg mL^{-1} (24–36 h)	Antidiabetes	α-Glucosidase, α-amylase, DPP-IV inhibition
	Total phenols, tannins, anthocyanin	Rats	200 mg kg^{-1} bw (45 days)	Antidiabetes	↓ Lipogenic and hexokinase enzyme activities ↑ Gluconeogenic enzyme activities
	Total phenols	*In vitro* (24 h)	100 μg	Antidiabetes	α-Amylase inhibition
	Phenolic acids and bioactive peptide fractions	*In vitro*	10.2–0.34 mg (24 h)	Antidiabetes	α-Glucosidase, α-amylase inhibition
	Quercetin, kaempferol, *p*-coumaric acid, ferulic acid, *p*-hydroxybenzoic acid, vanillic acid	Rats and mice	7.5%, 15%, 30%,and 60% w/w (7 days)	Antiobesity	↓ Cholesterol levels and LDL-C, regulatory effects on enzymes involved in fatty acid or cholesterol biosynthesis

	Flavonoid	Rats	50 and 500 mg kg^{-1} bw	Antiobesity and hypoglycemic effect	α-Amylase inhibition
	Flavonoids, total phenols, tannins, and anthocyanins	Mice	200, 400 mg kg^{-1} bw	Antiobesity and hypoglycemic effect	Protection of pancreatic β-cells
	Protein hydrolysates	*In vitro* (3T3-L1 cells)	IC_{50}=0.57 mg mL^{-1} IC_{50}=5.55 mg mL^{-1} 0.1 mg mL^{-1}	Antidiabetes Antiobesity	α-Amylase inhibition α-Glucosidase inhibition ↓ Lipid accumulation
	Mixture of peptides	*In vitro*	IC_{50}=1.2 mg SP mL^{-1}	Antidiabetes	DPP-IV inhibition
Black Otomi bean	Peptides	*In vitro*	IC_{50}=0.14 mg DW mL^{-1}	Antidiabetes	DPP-IV inhibition
	Protein hydrolysates	Hyperglycemic rat model Caco-2 cells	100–200 mg HPI kg^{-1} bw/day	Antidiabetes	↓ Postprandial glucose ↓ Glucose uptake ↓ ROS
Pinto Durango and Black bean 8025	Protein hydrolysates	INS-1E cells 3T3-L1 cells	100 µg mL^{-1}	Antidiabetes Antiobesity	↑ GLP-1R ↓ Lipid accumulation ↓ Expression of DPP-IV
BRS pontal	Protein hydrolysates	*In vitro*	10 mg mL^{-1}	Antidiabetes Antiobesity	α-Amylase inhibition α-Glucosidase inhibition
Eureka bean seeds	Peptides	*In vitro*	IC_{50}=1.94 mg mL^{-1}	Antidiabetes	α-Amylase inhibition

*bw = body weight; p.o = per oral. DW = dried weight, SP = soluble protein, HPI = hydrolyzed protein isolate, GLP-1R = GLP-1 receptor, INS-1E = rat insulinoma, DPP-IV= dipeptidyl peptidase-IV, ↑ increase, ↓ decrease.

Source: adapted from Ganesan and Xu (2017); Wang et al. (2020); Oseguera et al. (2016); Hutchins and Winham (2020); Jakubczyk, Karaś, Złotek, and Szymanowska (2017); Mojica, Gonzalez de Mejia, Granados-Silvestre, and Menjivar (2017); Valencia-Mejía, Batista, Fernández, and Fernandes (2019).

phenolic compounds, bioactive peptides, terpenoids, and non-starch polysaccharides inhibit α-glucosidase activity (Di Stefano et al., 2018).

Insulin resistance is an important risk factor for T2D; some clinical studies have shown that the consumption of legumes such as common beans can reduce this insulin resistance, besides other parameters related to T2D due to their antioxidant potential (Calderón-Guzmán et al., 2020). Moreover, glucose control is essential to delay the progression of T2D in patients; therefore, some studies demonstrate that pinto beans (*P. vulgaris*) help reduce blood glucose, postprandial glucose, and hemoglobin A1c (Hutchins & Winham, 2020).

Dry legumes can prevent T2D development by lowering insulin resistance. Common beans improve glycemic control; therefore, consuming beans will reduce the glycemic index and increase dietary fiber intake to control blood glucose concentration. T2D patients lowered glucose concentration by ingesting kidney beans compared to glucose (50 g) ingestion (Vergara-Castañeda et al., 2013). Moreover, the protein hydrolysates and phenolic compounds of fermented black beans inhibited α-amylase and α-glucosidase enzymes (Wang et al., 2020). In silico studies revealed the molecular interactions between common bean peptides such as KTYGL, KKSSG, CPGNK, GGGLHK, AKSPLF, QTPF, FEELN, and LSKSVL with the active site of α-glucosidase, and flavonols such as kaempferol that interacted outside the catalytic site of the enzyme (Di Stefano et al., 2018) (Table 4.3).

Furthermore, DPP-IV inhibition is a common approach to raise the insulin levels in plasma, since this enzyme degrades incretin hormones GLP-1 and GIP. Peptides from a black bean cultivar hydrolyzed with alcalase can inhibit DPP-IV activity (Mojica, Luna-Vital, & González de Mejía, 2017). Also, a black bean protein hydrolysate showed an IC_{50} of 1.2 mg of soluble protein/mL to block DPP-IV. Common bean protein hydrolysates can be used as functional ingredients in T2D treatment (De Souza Rocha et al., 2015). Navy bean protein hydrolysates also presented high potential to block DPP-IV, with an IC_{50} of 0.095 mg mL^{-1} (Mojica, Chen, & González, 2015). α-amylase and α-glucosidase are enzymes involved in the degradation of carbohydrates. α-amylase breaks down dietary carbohydrates such as starch into monosaccharides, then α-glucosidase degrades them to glucose, which enters the bloodstream by absorption in the gut enterocytes.

The inhibition of both enzymes can delay carbohydrate digestion and glucose uptake, helping to regulate blood sugar levels. Protein hydrolysates obtained from fermented black bean inhibited both α-amylase and α-glucosidase (IC_{50} = 0.57 and 5.55 mg soluble protein mL^{-1}, respectively) (Flores-Medellín et al., 2021). Microwave treatment significantly increases α-amylase inhibition in red, black, and pinto beans (Oomah, Kotzeva, Allen, & Bassinello, 2014). In another study, bioactive peptides were obtained by fermenting eureka bean seeds; the peptide fraction with a molecular mass of 3.5–7 kDa inhibited α-amylase (IC_{50} = 1.94 mg mL^{-1}) (Jakubczyk et al., 2017). Protein hydrolysates obtained by the fermentation process can be used as ingredients to inhibit these enzymes and reduce blood sugar levels in T2D patients. Another approach to lowering glucose uptake is through blockage of the gastrointestinal glucose transporters. Hydrolyzed protein from black bean (10 mg mL^{-1}) reduced glucose uptake by 21.5% in Caco-2 cells under 24 hr of glucose stimulus. Also, the black bean protein hydrolysates reduced intracellular ROS. ROS are species that contain one or more unpaired electrons and, when found in high concentrations can cause oxidative damage in tissues and cells. High glucose levels induce high ROS concentration through several pathways, developing oxidative stress that has been related to insulin resistance. In *in vivo* studies, total intake of 100–200 mg of HPI kg^{-1} BW/day reduced postprandial glucose levels (23–48%) in a hyperglycemic rat model, suggesting the potential of the protein isolates in reducing glucose absorption (Mojica, Gonzalez de Mejia, et al., 2017). The peptides fraction (<1kDa; 100 µg mL^{-1}) obtained from black and Pinto Durango beans increased the GLP-1R expression in INS-1E cells. GLP-1R improves glucose homeostasis by increasing insulin levels. Additionally, the increase of adipose tissue is associated with the development of insulin resistance. In 3T3-L1 adipocytes, the same concentration of peptides fraction (<1kDa) was able to inhibit lipid accumulation by 28% (Oseguera et al., 2016). Common beans lectins are associated with the reduction of food intake and weight control. They can act as α-amylase inhibitors and the phytohemagglutinin can stimulate the cholecystokinin (CCK) and GLP-1 release, which play a role in food intake modulation. The inhibition of starch degradation and weight control can prevent obesity and consequently, the development of T2D (Astell, Mathai, & Su, 2013). Bioactive peptides and proteins from common beans can be used as adjuvants to prevent obesity and T2D.

Common beans could also modulate molecular markers associated with satiety. This effect related to food intake can indirectly affect obesity and T2D, which are highly related to food consumption habits. For instance, common beans nondigestible fraction (NDF) is fermented by microbiota in the colon (Campos-Vega et al., 2009). As a result, it generates SCFAs that can activate the free fatty acid receptor 2 and 3 (FFAR2, FFAR3) that promote the release of anorexigenic hormones such as CCK, peptide tyrosine-tyrosine (PYY), GLP-1 and leptin and block the anorectic peptides such as ghrelin (GH) (Loarca-Piña, Vital, Mojica, Ramírez-Jiménez, & Ocampo, 2018). A clinical trial showed that brown beans consumption increased SCFAs generation, increased PYY and reduced GH (Nilsson, Johansson, Ekström, & Björck, 2013). Furthermore, some proteins such as phytohemagglutinin and α-amylase inhibitor decreased food intake, probably mediated by CCK and reduced GH secretion (Fantini et al., 2009; Spadafranca et al., 2013). In addition, other bioactive components in common beans can promote satiety, including slowly digestible carbohydrates, fiber, and proteins (Clark & Duncan, 2017). The use of common bean flour (56%) in cookie formulation increased their nutritional composition and reduced glycemic response and satiety perception after ingestion in a clinical trial (Mecha et al., 2021).

4.2 Conclusion

Common beans are rich in bioactive components, including phenolic compounds, phytosterols, oligosaccharides, fiber, resistant starch, saponins, carotenoids, tocopherols, proteins, and peptides. Several studies demonstrate the potential of common beans bioactive components to modulate molecular markers related to obesity and T2D, including inflammation markers, decreasing DPP-IV expression and activity, blocking carbohydrate-degrading enzymes, inhibiting enzymes involved in fatty acid or cholesterol biosynthesis, regulating satiety hormones, among other mechanisms. Common bean bioactive components exert great potential to modulate molecular markers of noncommunicable diseases and can be used as adjuvants preventing and treating obesity and T2D.

References

American Diabetes Association. (2020). ADA Standards of Medical Care in Diabetes. *The Journal of Clinical and Applied Research and Education, 49*(1), S1–S183.

Arnoldi, A., Zanoni, C., Lammi, C., & Boschin, G. (2015). The role of grain legumes in the prevention of hypercholesterolemia and hypertension. *Critical Reviews in Plant Sciences, 34*, 144–168. https://doi.org/10.1080/07352689.2014.897908

Ashcroft, F. M. (1996). Mechanisms of the clycaemic effects of sulfonylureas. *Hormone and Metabolic Research, 28*(9), 456–463. https://doi.org/10.1055/s-2007-979837

Astell, K. J., Mathai, M. L., & Su, X. Q. (2013). A review on botanical species and chemical compounds with appetite suppressing properties for body weight control. *Plant Foods Human Nutrition, 68*, 1–9. https://doi.org/10.1007/s11130-013-0361-1

Calderón-Guzmán, D., Juárez-Olguín, H., Veloz-Corona, Q., Ortiz-Herrera, M., Osnaya-Brizuela, N., & Barragán-Mejía, G. (2020). Consumption of cooked common beans or saponins could reduce the risk of diabetic complications. *Diabetes, Metabolic Syndrome and Obesity: Targets and Therapy, 13*, 3481–3486. https://doi.org/10.2147/DMSO.S270564

Campos-Vega, R., Reynoso-Camacho, R., Pedraza-Aboytes, G., Acosta-Gallegos, J. A., Guzman-Maldonado, S. H., Paredes-Lopez, O., . . . Loarca-Piña, G. (2009). Chemical composition and *in vitro* polysaccharide fermentation of different beans (Phaseolus vulgaris L.). *Journal of Food Science, 74*, 59–65. https://doi.org/10.1111/j.1750-3841.2009.01292.x

Campos-Vega, R., Vergara-Castañeda, H. A., & Oomah, B. D. (2012). Functional food sources: Beans. In E. Popescu & I. Golubev (Eds.), *Beans: Nutrition, consumption and health* (pp. 53). New York: Nova Science Publishers.

Carbas, B., Machado, N., Oppolzer, D., Ferreira, L., Brites, C., Rosa, E. A. S., & Barros, A. I. R. N. A. (2020). Comparison of near-infrared (NIR) and mid-infrared (MIR) spectroscopy for the determination of nutritional and antinutritional parameters in common beans. *Food Chemistry, 306*, 125509. https://doi.org/10.1016/j.foodchem.2019.125509

Celmeli, T., Sari, H., Canci, H., Sari, D., Adak, A., Eker, T., & Toker, C. (2018). The nutritional content of common bean (*Phaseolus vulgaris* L.) landraces in comparison to modern varieties. *Agronomy, 8*, 166. https://doi.org/10.3390/agronomy8090166

Chatterjee, S., Khunti, K., & Davies, M. J. (2017). Type 2 diabetes. *The Lancet, 6736*(17), 1–13. https://doi.org/10.1016/S0140-6736(17)30058-2

Chávez-Mendoza, C., Hernández-Figueroa, K. I., & Sánchez, E. (2018). Antioxidant capacity and phytonutrient content in the seed coat and cotyledon of common beans (*Phaseolus vulgaris* L.) from various regions in Mexico. *Antioxidants, 8*, 5. https://doi.org/10.3390/antiox8010005

Chávez-Mendoza, C., & Sánchez, E. (2017). Bioactive compounds from Mexican varieties of common bean (*Phaseolus vulgaris*): Implications for health. *Molecules, 22*, 1360. https://doi.org/10.3390/molecules22081360

Chávez-Santoscoy, R. A., Gutiérrez-Uribe, J. A., & Serna-Saldivar, S. O. (2013). Effect of flavonoids and saponins extracted from black bean (*Phaseolus vulgaris* L.) seed coats as cholesterol micelle disruptors. *Plant Foods for Human Nutrition, 68*, 416–423. https://doi.org/10.1007/s11130-013-0384-7

Clark, S., & Duncan, A. M. (2017). The role of pulses in satiety, food intake and body weight management. *Journal of Functional Foods, 38*, 612–623. https://doi.org/10.1016/j.jff.2017.03.044

Coronado-Cáceres, L. J., Rabadán-Chávez, G., Quevedo-Corona, B., Hernández-Ledesma, B., Garcia, A. M., Mojica, L., & Lugo-Cervantes, E. (2019). Antiobesity effect of cocoa proteins (*Theobroma cacao* L.) variety "Criollo" and the expression of genes related to the dysfunction of white adipose tissue in high-fat diet-induced obese rats. *Journal of Functional Foods, 62*, 103519. https://doi.org/10.1016/j.jff.2019103519

Damián-Medina, K., Salinas-Moreno, Y., Milenkovic, D., Figueroa-Yáñez, L., Marino-Marmolejo, E., Higuera-Ciapara, I., . . . Lugo-Cervantes, E. (2020). In silico analysis of antidiabetic potential of phenolic compounds from blue corn (Zea mays L.) and black bean (Phaseolus vulgaris L.). *Heliyon, 6*(3), e03632. https://doi.org/10.1016/j.heliyon.2020.e03632

De Fátima Garcia, B., de Barros, M., & de Souza Rocha, T. (2020). Bioactive peptides from beans with the potential to decrease the risk of developing non-communicable chronic diseases. *Critical Reviews in Food Science and Nutrition, 0*(0), 1–19. https://doi.org/10.1080/10408398.2020.1768047

De Fronzo, R. A., Ferrannini, E., Groop, L., Henry, R. R., Herman, W. H., Holst, J. J., . . . Weiss, R. (2015). Type 2 diabetes mellitus. *Nature Reviews Disease Primers, 1*, 1–23. https://doi.org/10.1038/nrdp.2015.19

De Souza Rocha, T., Real-Hernandez, L. M., Mojica, L., Johnson, M. H., Chang, Y. K., & González de Mejía, E. (2015). Germination of Phaseolus vulgaris and alcalase hydrolysis of its proteins produced bioactive peptides capable of improving markers related to Type-2 diabetes in vitro. *Food Research International, 76*, 150–159. https://doi.org/10.1016/j.foodres.2015.04.041

Di Stefano, E., Oliviero, T., & Udenigwe, C. (2018). Functional significance and structure-activity relationship of food-derived α-glucosidase inhibitors. *Current Opinion in Food Science, 20*, 7–12.

Fantini, N., Cabras, C., Lobina, C., Colombo, G., Gessa, G. L., Riva, A., . . . Carai, M. A. M. (2009). Reducing effect of a *Phaseolus vulgaris* dry extract on food intake, body weight, and glycemia in rats. *Journal of Agricultural and Food Chemistry, 57*(19), 9316–9323.

Ferrannini, E., & Mari, A. (2014). β-Cell function in type 2 diabetes. *Metabolism: Clinical and Experimental, 63*(10), 1217–1227. https://doi.org/10.1016/j.metabol.2014.05.012

Flores-Medellín, S. A., Camacho-Ruíz, R. M., Guízar-González, C., Rivera-Leon, E. A., Llamas-Covarrubias, I. M., & Mojica, L. (2021). Protein hydrolysates and phenolic compounds from fermented black beans inhibit markers related to obesity and Type-2 diabetes. *Legume Science, 3*(1), 1–17. https://doi.org/10.1002/leg3.64

Gadde, K. M., Martin, C. K., Berthoud, H. R., & Heymsfield, S. B. (2018). Obesity: Pathophysiology and management. *Journal of the American College of Cardiology, 71*(1), 69–84. https://doi.org/10.1016/j.jacc.2017.11.011

Ganesan, K., & Xu, B. (2017). Polyphenol-rich dry common beans (*Phaseolus vulgaris* L.) and their health benefits. *International Journal of Molecular Sciences, 18*, 2331. https://doi.org/10.3390/ijms18112331

García-Díaz, Y. D., Aquino-Bolaños, E. N., Chávez-Servia, J. L., Vera-Guzmán, A. M., & Carrillo-Rodríguez, J. C. (2018). Bioactive compounds and antioxidant activity in the common bean are influenced by cropping season and genotype. *Chilean Journal of Agricultural Research, 78*(2), 255–265. https://doi.org/10.4067/S0718-58392018000200255

Gilbert, M. P., & Pratley, R. E. (2020). GLP-1 analogs and DPP-4 inhibitors in Type 2 diabetes therapy: Review of head-to-head clinical trials. *Frontiers in Endocrinology*, *11*, 1–13. https://doi.org/10.3389/fendo.2020.00178

Graβmann, S., Wirsching, J., Eichelmann, F., & Aleksandrowa, K. (2017). Association between peripheral adipokines and inflammation markers: A systematic review and meta-analysis. *Obesity*, *25*(10), 1776–1785. https://doi.org/10.1002/oby.21945

Hayat, I., Ahmad, A., Masud, T., Ahmed, A., & Bashir, S. (2014). Nutritional and health perspectives of beans (*Phaseolus vulgaris* L.): An overview. *Critical Reviews in Food Science and Nutrition*, *54*, 580–592. https://doi.org/10.1080/10408398.2011.596639

Hernández-López, V. M., Vargas-Vázquez, M. L., Muruaga-Martínez, J. S., Hernández-Delgado, S., & Mayek-Pérez, N. (2013). Origin, domestication and diversification of common beans. Advances and Perspectives. *Revista Fitotecnica Mexicana*, *36*(2), 95–104.

Hutchins, A., & Winham, D. M. (2020). Pinto beans and green beans result in comparable glycemic control in adults with Type 2 diabetes: A randomized pilot trial. *Food Science amd Nutrition Technology*, *5*(1), 000211. https://doi.org/10.23880/fsnt-16000211

Jakubczyk, A., Karaś, M., Złotek, U., & Szymanowska, U. (2017). Identification of potential inhibitory peptides of enzymes involved in the metabolic syndrome obtained by simulated gastrointestinal digestion of fermented bean (Phaseolus vulgaris L.) seeds. *Food Research International*, *100*, 489–496. https://doi.org/10.1016/j.foodres.2017.07.046

Kan, L., Nie, S., Hu, J., Wang, S., Cui, S. W., Li, Y., . . . Xie, M. (2017). Nutrients, phytochemicals and antioxidant activities of 26 kidney bean cultivars. *Food and Chemical Toxicology*, *108*, 467–477. https://doi.org/10.1016/j.fct.2016.09.007

Kilua, A., Chihiro, H., Han, K., Homma, K., & Fukuma, N. (2020). Whole kidney bean (Phaseolus vulgaris) and bean hull reduce the total serum cholesterol, modulate the gut microbiota and affect the caecal fermentation in rats. *Bioactive Carbohydrates and Dietary Fibre*, *24*, 100232. https://doi.org/10.1016/j.bcdf.2020.100232

Lambeir, A. M., Durinx, C., Scharpé, S., & De Meester, I. (2003). Dipeptidyl-peptidase IV from bench to bedside: An update on structural properties, functions, and clinical aspects of the enzyme DPP IV. *Critical Reviews in Clinical Laboratory Sciences*, *40*(3), 209–294. https://doi.org/10.1080/713609354

Lin, T., O'Keefe, S., Duncan, S., & Fernández-Fraguas, C. (2020). Manipulation of the dry bean (*Phaseolus vulgaris* L.) matrix by hydrothermal and high-pressure treatments: Impact on *in vitro* bile salt-binding ability. *Food Chemistry*, *310*, 125699. https://doi.org/10.1016/j.foodchem.2019.125699

Loarca-Piña, G., Vital, D. L., Mojica, L., Ramírez-Jiménez, A. K., & Ocampo, I. L. (2018). Health benefits of beans. In R. Campos-Vega, P. Z. Bassinello, & B. D. Oomah (Eds.), *Phaseolus vulgaris: Cultivars, Production and Uses* (pp. 201–227). Nova Science Publishers.

Luna-Vital, D. A., Mojica, L., González de Mejía, E., Mendoza, S., & Loarca-Piña, G. (2015). Biological potential of protein hydrolysates and peptides from common bean (Phaseolus vulgaris L.): A review. *Food Research International*, *76*, 39–50. https://doi.org/10.1016/j.foodres.2014.11.024

Luzardo-Ocampo, I., Campos-Vega, R., Cuellar-Nuñez, M. L., Vázquez-Landaverde, P. A., Mojica, L., Acosta-Gallegos, J. A., & Loarca-Piña, G. (2018). Fermented non-digestible fraction from combined nixtamalized corn (*Zea mays* L.)/cooked common bean (*Phaseolus vulgaris* L.) chips modulate antiinflammatory markers on RAW 264.7 macrophages. *Food Chemistry*, *259*, 7–17. https://doi.org/10.1016/j.foodchem.2018.03.096

Mecha, E., Correia, V., da Silva, A. B., Ferreira, A., Sepodes, B., Figueira, M. E., . . . Bronze, M. R. (2021). Improvement of wheat cookies' nutritional quality, by partial substitution with common bean and maize flours, sustained human glycaemia and enhanced satiety perception. *Cereal Chemistry*, Accepted for publication. https://doi.org/10.1002/cche.10460

Mendoza-Sánchez, M., Pérez-Ramírez, I. F., Wall-Medrano, A., Martínez-Gonzalez, A. I., Gallegos-Corona, M. A., & Reynoso-Camacho, R. (2019). Chemically induced common bean (*Phaseolus vulgaris* L.) sprouts ameliorate dyslipidemia by lipid intestinal absorption inhibition. *Journal of Functional Foods*, *52*, 54–62. https://doi.org/10.1016/j.jff.2018/10.032

Mojica, L., Chen, K., & González, E. (2015). Impact of commercial precooking of common bean (*Phaseolus vulgaris*) on the generation of peptides, after pepsin – pancreatin hydrolysis, capable to inhibit dipeptidyl peptidase-IV. *Journal of Food Science*, *80*(1), H188–H198. https://doi.org/10.1111/1750-3841.12726

Mojica, L., Gonzalez de Mejia, E., Granados-Silvestre, M. Á., & Menjivar, M. (2017). Evaluation of the hypoglycemic potential of a black bean hydrolyzed protein isolate and its pure peptides using *in silico*, *in vitro* and *in vivo* approaches. *Journal of Functional Foods*, *31*, 274–286. https://doi.org/10.1016/j.jff.2017.02.006

Mojica, L., Luna-Vital, D. A., & González de Mejía, E. (2017). Characterization of peptides from common bean protein isolates and their potential to inhibit markers of Type-2 diabetes, hypertension, and oxidative stress. *Journal of the Science of Food and Agriculture*, *97*(8), 2401–2410. https://doi.org/10.1002/jsfa.8053

Monk, J. M., Wu, W., Lepp, D., Pauls, K. P., Robinson, L. E., & Power, K. A. (2021). Navy bean supplementation in established high-fat diet-induced obesity attenuates the severity of the obese inflammatory phenotype. *Nutrients*, *13*(3), 757. https://doi.org/10.3390/nu13030757

Moreau, R. A., Nyström, L., Whitaker, B. D., Winkler-Moser, J. K., Baer, D. J., Gebauer, S. K., & Hicks, K. B. (2018). Phytosterols and their derivatives: Structural diversity, distribution, metabolism, analysis, and health-promoting uses. *Progress in Lipid Research*, *70*, 35–61. https://doi.org/10.1016/j.plipres.2018.04.001

Moreno-Valdespino, C. A., Luna-Vital, D., Camacho-Ruiz, R. M., & Mojica, L. (2020). Bioactive proteins and phytochemicals from legumes: Mechanisms of action preventing obesity and type-2 diabetes. *Food Research International*, *130*, 108905. https://doi.org/10.1016/j.foodres.2019.108905

Nilsson, A., Johansson, E., Ekström, L., & Björck, I. (2013). Effects of a brown beans evening meal on metabolic risk markers and appetite regulating hormones at a subsequent standardized breakfast: A randomized cross-over study. *PloS one*, *8*(4), e59985. https://doi.org/10.1371/journal.pone.0059985

Nolan, R., Shannon, O. M., Robinson, N., Joel, A., Houghton, D., & Malcomson, F. C. (2020). It's no has bean: a review of the effects of white kidney bean extract on body composition and metabolic health. *Nutrients*, *12*(5), 1398. https://doi.org/10.3390/nu12051398

Olmedilla-Alonso, B., Pedrosa, M. M., Cuadrado, C., Brito, M., Asensio-S-Manzanera, C., & Asensio-Vegas, C. (2013). Composition of two Spanish common dry beans (Phaseolus vulgaris), "Almonga" and "Curruquilla", and their postprandial effect in type 2 diabetics. *Journal of the Science of Food and Agriculture*, *93*(5), 1076–1082. https://doi.org/10.1002/jsfa.5852

Oomah, B. D., Kotzeva, L., Allen, M., & Bassinello, P. Z. (2014). Microwave and micronization treatments affect dehulling characteristics and bioactive contents of dry beans (*Phaseolus vulgaris* L.). *Journal of the Science of Food and Agriculture, 94*, 1349–1358. https://doi.org/10.1002/jsfa.6418

Oseguera, M. E., Gonzalez, E., & Mejia, D. (2016). Common bean (*Phaseolus vulgaris* L.) protein- derived peptides increased insulin secretion, inhibited lipid accumulation, increased glucose uptake and reduced the phosphatase and tensin homologue activation in vitro. *Journal of Functional Foods, 27*, 160–177. https://doi.org/10.1016/j.jff.2016.09.001

Romero-Arenas, O., Damián-Huato, M. A., Rivera-Tapia, J. A., Báez-Simón, A., Huerta-Lara, M., & Cabrera-Huerta, E. (2013). The nutritional value of beans (*Phaseolus vulgaris* L.) and its importance for feeding of rural communities in Puebla-Mexico. *International Research Journal of Biological Sciences, 2*(8), 59–65.

Sandoval, V., Sanz-Lamora, H., Arias, G., Marrero, P. F., Haro, D., & Relat, J. (2020). Metabolic impact of flavonoids consumption in obesity: From central to peripheral. *Nutrients, 12*(8), 2393. https://doi.org/10.3390/nu12082393

Singh, J. P., Singh, B., & Kaur, A. (2020). Bioactive compounds of legume seeds. In H. N. Murthy & K. Y. Paek (Eds.), *Bioactive compounds in underutilized vegetables and legumes. Reference series in phytochemistry*. Springer: Cham. https://doi.org/10.1007/978-3-030-44578-2_33-1

Song, H., Han, W., Yan, F., Xu, D., Chu, Q., & Zheng, X. (2016). Dietary *Phaseolus vulgaris* extract alleviated diet-induced obesity, insulin resistance and hepatic steatosis and alters gut microbiota composition in mice. *Journal of Functional Foods, 20*, 236–244. https://doi.org/10.1016/j.jff.2015.10.022

Spadafranca, A., Rinelli, S., Riva, A., Morazzoni, P., Magni, P., Bertoli, S., & Battezzati, A. (2013). *Phaseolus vulgaris* extract affects glycometabolic and appetite control in healthy human subjects. *British Journal of Nutrition., 109*(10), 1789–1795.

Suárez-Martínez, S. E., Ferriz-Martínez, R. A., Campos-Vega, R., Elton-Puente, J. E., de la Torre Carbot, K., & García-Gasca, T. (2016). Bean seeds: Leading nutraceutical source for human health. *CyTA-Journal of Food, 14*(1), 131–137. https://doi.org/10.1080/19476337.2015.1063548

Tanaka, M., Honda, Y., Miwa, S., Akahori, R., & Matsumoto, K. (2019). Comparison of the effects of roasted and boiled red kidney beans (*Phaseolus vulgaris* L.) on glucose/lipid metabolism and intestinal immunity in a high-fat diet-induced murine obesity model. *Journal of Food Science, 84*(5), 1180–1187. https://doi.org/10.1111/1750-3841.14583

Thompson, H. J., McGinley, J. N., Neil, E. S., & Brick, M. A. (2017). Beneficial effects of common bean on adiposity and lipid metabolism. *Nutrients, 9*(9), 998. https://doi.org/10.3390/nu9090998

Tomita, T. (2016). Apoptosis in pancreatic β-islet cells in Type 2 diabetes. *Bosnian Journal of Basic Medical Sciences, 16*(3), 162–179. https://doi.org/10.17305/bjbms.2016.919

Tucker, L. A. (2020). Bean consumption accounts for differences in body fat and waist circumference: A cross-sectional study of 246 women. *Journal of Nutrition and Metabolism, 2020*. https://doi.org/10.1155/2020/9140907

Udani, J., Tan, O., & Molina, J. (2018). Systematic review and meta-analysis of a proprietary alpha-amylase inhibitor from white bean (Phaseolus vulgaris L.) on weight and fat loss in humans. *Foods, 7*(4), 1–10. https://doi.org/10.3390/foods7040063

Valencia-Mejía, E., Batista, K. A., Fernández, J. J. A., & Fernandes, K. F. (2019). Antihyperglycemic and hypoglycemic activity of naturally occurring peptides and protein hydrolysates from easy-to-cook and hard-to-cook beans (Phaseolus vulgaris L.). *Food Research International, 121*(February), 238–246. https://doi.org/10.1016/j.foodres.2019.03.043

Vergara-Castañeda, H. A., Oomah, B. D., & Campos-Vega, R. (2013). Pulses facing the new age: Functional compounds on gene expression and health connection. In A. Robinson & D. Emerson (Eds.), *Functional foods: Sources, biotechnology applications and health challenges* (pp. 1–64). New York: Nova Science Publishers, Inc.

Vezza, T., Canet, F., de Marañón, A. M., Bañuls, C., Rocha, M., & Víctor, V. M. (2020). Phytosterols: Nutritional health players in the management of obesity and its related disorders. *Antioxidants, 9*(12), 1266. https://doi.org/10.3390/antiox9121266

Wang, S., Chen, L., Yang, H., Gu, J., Wang, J., & Ren, F. (2020). Regular intake of white kidney beans extract (*Phaseolus vulgaris* L.) induces weight loss compared to placebo in obese human subjects. *Food Science and Nutrition., 8*, 1315–1324. https://doi.org/10.1002/fsn3.1299

Xiong, Q., Li, Z., Nie, R., Meng, X., & Yang, X. J. (2021). Comparison of the effects of a bean-based and a white rice-based breakfast diet on postprandial glucose and insulin levels in Chinese patients with Type 2 diabetes. *Medical Science Monitor, 27*, 1–8. https://doi.org/10.12659/MSM.930349

Yang, Q. Q., Gan, R. Y., Ge, Y. Y., Zhang, D., & Corke, H. (2018). Polyphenols in common beans (*Phaseolus vulgaris* L.): Chemistry, analysis, and factors affecting composition. *Comprehensive Reviews in Food Science and Food Safety, 17*, 1518–1539. https://doi.org/10.1111/1541-4337.12391

Yki-Jarvinen, H. (2004). Drug therapy: thiazolidinediones. *The New England Journal of Medicine, 351*, 1106–1118.

5

Benefits of Carioca Beans (*Phaseolus vulgaris*): Molecular Mechanisms to Human Health and Nutrition

Juliana Aparecida Correia Bento[1], Mariana Juste Contin Gomes[2], Priscila Zaczuk Bassinello[3], Hércia Stampini Duarte Martino[2], Menandes Alves de Souza Neto[1], and B. Dave Oomah[4]

[1] School of Agronomy, Federal University of Goiás - UFG, Goiânia, Brazil
[2] Department of Nutrition and Health, Federal University of Viçosa - UFV, Viçosa, Brazil
[3] Embrapa Rice and Beans, Santo Antônio de Goiás, Goiás, Brazil
[4] (Retired) Formerly with Summerland Research and Development Centre, Agriculture and Agri-Food Canada, Summerland, BC, Canada

5.1 Introduction

Carioca beans, as common dry beans, play an important role in human nutrition because of their high amount of protein (with elevated lysine content), complex carbohydrates, dietary fiber (both soluble and insoluble) as well as several vitamins and minerals. This pulse is used as a source of protein in vegan diets. Additionally, beans are low in fat and glycemic index. They contribute significantly to satiety and improve the gut microbiome (Tucker, 2020).

Among the various commercial types of Mesoamerican common beans, carioca is preferred by Brazilian consumers, corresponding to more than 70% of beans consumed in Brazil (Bento et al., 2020). This type of grain is characterized by a light cream color, with the presence of light brown streaks (Figure 5.1). They look slightly like pinto beans, except for the smaller size of the grains and the stripes instead of spots in the tegument. After grain harvest, the carioca type quickly loses its commercial value due to chemical changes that alter the light hue, probably related to proanthocyanidin oxidation, causing the depreciation of grains by the processing and packaging units. This is because dark grains are associated with old grains and long cooking time, being rejected by consumers. High levels of polyphenols (tannins) are also related to the darkening process in beans. There is genetic variation for grain darkening (Alvares et al., 2019; Islam, Bett, Pauls, Marsolais, & Dhaubhadel, 2020; Spitti et al., 2019), which makes it possible to obtain cultivars with slow browning or darkening (Figure 6.1). Carioca beans yield increased by 0.72–6.74% in Brazil due to genetic improvement over the past decades, although the main Brazilian bean cultivars possess relatively low genetic diversity. Moreover, carioca beans among 215 common bean accessions (including 85 carioca beans) from IDR-Paraná gene bank had the highest

Molecular Mechanisms of Functional Food, First Edition. Edited by Rocio Campos-Vega and B. Dave Oomah.

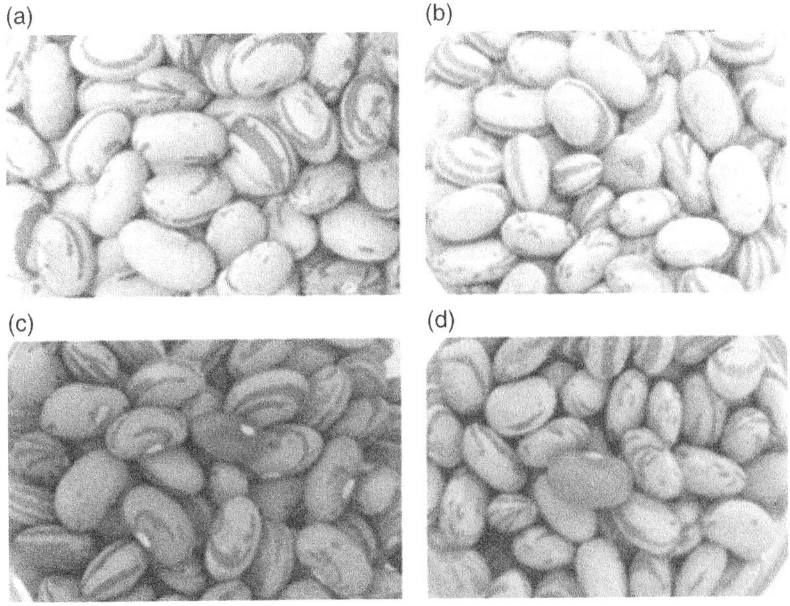

Figure 5.1 Carioca beans: a and b – freshly harvested; c and d – grains stored up to 180 days. Cultivars: Pérola (a, c) and Madreperola (b, d). *Source:* Personal Archive.

average value for total phenolic content and DPPH antioxidant scavenging activity (Nogueira et al., 2021). In addition to the carioca, other bean types ("mulatinho", brindle, cranberry, sugar beans, and pinto beans) also vary in grain darkening.

Beans are good sources of bioactive compounds and recent evidence, mainly for a nondigestible fraction (NDF) of beans, have provided information of their impact and mechanism of action on colon cancer (Campos-Vega, Oomah, Loarca-Piña, & Vergara-Castañeda, 2013). Bioactive compounds are food components capable of regulating metabolic processes in humans and animals and improving health. These compounds can modulate molecular mechanisms involved in the antioxidant, anti-hyperlipidemic, and anti-inflammatory systems. The bioactive compounds of carioca beans, such as flavones, anthocyanins, and condensed tannins, give color to the grains and have potent antioxidant activity, but other compounds such as dietary fibers, proteins, and bioactive peptides also exert anti-hyperlipidemic and anti-inflammatory effects and protect the vascular endothelium (de Lima et al., 2019; Gomes et al., 2020).

Phenolic acids, flavonoids, and proanthocyanidins can be found in the cotyledons of grains but are most concentrated in the seed coats (Yang, Gan, Ge, Zhang, & Corke, 2018). They are notable compounds responsible for the common bean grains diverse colors and for their potent antioxidant properties (Bento, Bassinello, Morais, et al., 2021) as well as for their presumed prebiotic effect in the colon (Cires, Wong, Carrasco-Pozo, & Gotteland, 2017). Colorectal cancer is one of the most common neoplasms in industrialized societies and can be prevented, for the most part, by adequate diets including *pulses* (beans, lentils, peas, and peanuts) and regular physical activities. It has been demonstrated that anthocyanin-rich extracts from berries and grapes and several pure anthocyanins and anthocyanidins exhibit proapoptotic effects in multiple cell types such as colon, breast, prostate, and leukemia cancer cells (Campos-Vega et al., 2013). The main anthocyanins in beans seed coats have been identified as delphinidin 3-glucoside (65.7%), petunidin 3-glucoside (24.3%), and malvidin 3-glucoside (8.7%) (Alves, de Mejía, Vasconcelos, Bassinello, & Martino, 2016).

By increasing the daily consumption of a wide variety of common beans in the diet, we can improve consumers' nutrition and health and reduce the risk of cancer and other health problems. Further research is warranted regarding the implications and the molecular mechanisms in which common beans and their bioactive compounds modulate the development of different types of chronic diseases. Most of the nutritional and functional properties of common bean grains are also displayed by broken and aged grains for different cultivars (Alves, de Mejía, et al., 2016; Alves, Vasconcelos, Bassinello, de Mejia, & Martino, 2016; de Lima et al., 2019), which can improve their commercial value as well as their potential use as an ingredient for the development of new products, gluten-free food, and as a protein source for plant-based food, as we will highlight at the end of this chapter.

5.2 Health Benefits of Compounds in Carioca Beans

5.2.1 Protein

The carioca beans protein content ranges between 20 and 30% depending on cultivars, with higher digestibility (80%), two to three times more than cereal proteins (Alves et al., 2021; Luna-Vital, Mojica, González de Mejía, Mendoza, & Loarca-Piña, 2015). Although deficient in the sulfur amino acids (methionine and cysteine, ~ 2.5–3.0 g/100 g proteins and 0.6–0.7 mg/100 g proteins, respectively), carioca beans have high lysine content (6.4–7.6 g/100 g protein), phenylalanine, and tyrosine (5.3–8.2 g/100 g protein, combined) (Chávez-Mendoza & Sánchez, 2017; Ribeiro et al., 2007). These amino acids are essential for human health. The main protein fractions in beans are globulin (50–70%) and albumin (10%). Beans more specifically common beans (as carioca beans) contain 38.5% G1 globulins (phaseolin), 31.5% albumins, 22.4% glutelins, 13.8% G2 globulins (phytohemagglutinin), and 1.7% prolamines (Alves, Vasconcelos, et al., 2016; Hayat, Ahmad, Masud, Ahmed, & Bashir, 2014). The phaseolin, an active glycoprotein, consists of about 50% of the total bean nitrogen, whereas the globulin fraction is only 10% (Ganesan & Xu, 2017). Moreover, carioca beans have been described as a potential source of bioactive peptides with potential health benefits (Alves, Gomes, et al., 2021; Alves, Vasconcelos, et al., 2016; de Lima et al., 2019; Gomes et al., 2020; Luna-Vital et al., 2015).

Phaseolin, a glycoprotein containing mannose consists of three polypeptide subunits with molecular weights between 43 and 53 kDa (Lawrence, Izard, Beuchat, Blagrove, & Colman, 1994) (Figure 5.2). Its nutritive value is limited by its low content in sulfur amino acids (methionine and cysteine) and its high resistance to enzymatic hydrolysis (Luna-Vital et al., 2015; Mojica & de Mejía, 2015). Phaseolin is considered an allergenic protein and has demonstrated the ability to bind with IgE in allergic human subjects and enhance the levels of pro-inflammatory factors such as IgG1, mouse mast cell protease-1, IL-4, IL-5, IL- mRNA expressions 13, and GATA-3 in the lungs, spleen, and intestine (Kumar et al., 2014).

Figure 5.2 Phaseolin subunits. *Source:* Adapted from Lawrence et al. (1994).

On the other hand, the phaseolin fractions, isolated from adzuki beans, at low dosages are effective in reducing serum cholesterol levels in hypercholesterolaemic rats, thus confirming that this class of proteins, regardless of the species, may exert similar biological activities in animal models (Ferreira et al., 2015). Moreover, the albumin and globulin fractions obtained from black bean seeds demonstrated the ability to stabilize the erythrocyte membrane in patients with sickle cell anemia and promoted reduction of vascular hemolysis in a study conducted in the West Region of Cameroon (Kenmoe et al., 2020).

A study with digested protein isolates from fresh and stored carioca beans (cultivar *BRSMG Madrepérola*) showed that parental proteins such as phytohemagglutinin, α- and β- phaseolin, α-amylase inhibitor 1, and α-amylase inhibitor 2 reduced markers of atherosclerosis in oxidized low-density lipoprotein (LDL)-induced THP-1 macrophages (Alves, Vasconcelos, et al., 2016). The author also suggested that these proteins may exhibit protective activity on weight gain and food consumption. The mechanism involves stimulating the secretion of cholecystokinin (CCK) by enteroendocrine cells, stimulating the secretion of anorexic hormones, and reducing glucose absorption, which leads to increased satiety and reduced food intake (Alves, Vasconcelos, et al., 2016; de Lima et al., 2019). Some phytohemagglutinin or plant lectin binds ACE2 of SARS-CoV-1 to inhibit virus entry into host cells thereby conferring protection against viral infection (Palit et al., 2021). Although lectins and α-amylase inhibitor have long been considered as antinutrients, they have been proposed to be used in formulating therapeutic products for treatment and prevention of diseases such as cardiovascular diseases (CVDs), cancer, diabetes, osteoporosis, hypertension, and obesity (Carbonaro, 2021). Purified white kidney bean α-amylase inhibitor

(10–50 kDa) [PVE] diet (180 mg kg^{-1}/daily for 8 weeks) ameliorated serum lipid profiles (TG, TC, LDL-C, and HDL-C) and improved inflammatory response (by reducing inflammatory cytokines IL-1β and TNF-α [mediator of insulin resistance] and serum concentrations) of hyperglycemic mice (Wang et al., 2021). Moreover, mixture of PVE with plain yogurt ([YPVE]; 180 mg kg^{-1} PVE + 30 ml kg^{-1} yogurt) regulated intestinal flora with significant changes in *Clostridia* and *Bacteroidetes* (*Coprococcus*, *Allobaculum*, *Oscillospira*, and *AF*12), increased *S*24–7 (that correlates inversely with type 2 diabetes) with abundant *Ruminococcaceae* (reducing lipid metabolism). YPVE increased functions of several metabolisms (amino acid [alanine, aspartate, and glutamate; higher alanine metabolism improves triglyceride metabolism beneficially affecting the cardiometabolic status of type 2 diabetic men], energy [sulfur and carbon fixation], and nucleotide) and reduced metabolism of glycan biosynthesis (Wang et al., 2021). Therefore, α-amylase inhibitor can exert prebiotic effects and reshape the gut microbiota benefiting the host.

Finally, we have previously shown that whole carioca bean flour and carioca bean protein hydrolysate (700 mg kg^{-1} day^{-1}) decreased glycemia, total cholesterol, lipid peroxidation, and inflammation in mice. Additionally, carioca bean protein hydrolysate decreased weight gain and food intake (de Lima et al., 2019). It also avoids endothelial dysfunction and, as a result, the atherogenic risk, by reducing inflammation, tumor necrosis factor-α (TNF-α), and angiotensin II, and increasing endothelial nitric oxide synthase (Gomes et al., 2020), which enables common beans to enhance human health and avoid chronic diseases (Alves, Gomes, et al., 2021; Mullins & Arjmandi, 2021).

5.2.2 Carbohydrates, Fiber, Resistant Starch, and Oligosaccharides

The carbohydrate (~60–65%) in common beans is composed of starch, with small amounts of monosaccharides and disaccharides. As the main form of carbohydrate, starch content in pulses includes high amount of resistant starch (RS), as reflected in 30 to 40% amylose content resistant to digestion. The RS consists of insoluble and soluble dietary fiber as well as nondigestible oligosaccharides, which resist hydrolysis in the small intestine but are fermented by the colonic microflora in the large intestine at varying rates and metabolites (Hayat et al., 2014; Mullins & Arjmandi, 2021). Dietary fiber plays a role in the body that makes it a functional food, due to its resistance to digestion and absorption (Barber, Kabisch, Pfeiffer, & Weickert, 2020; Feregrino-Perez et al., 2014; Vergara-Castañeda et al., 2010). The European Food Safety Authority (EFSA) provides a long list of dietary fiber that includes non-starch polysaccharides, fructo-oligosaccharides, cellulose, pectin, hydrocolloids, and "resistant starch." These compounds are retained in the NDF, the portion of food that resists enzymatic and pH changes throughout the gastrointestinal tract (Vergara-Castañeda et al., 2010). The main components of NDF found in common beans are nondigestible carbohydrates, for example, the raffinose family oligosaccharides (RFO) (raffinose, stachyose, and verbascose), cellulose, lignin, pectin, galactose, arabinose, and xylose; some polyphenols (phenolic acids, condensed tannins, and flavonoids, among others) and associated compounds; minerals strongly bound to the cell wall and resistant peptides (Feregrino-Perez et al., 2014; Luna-Vital et al., 2015; Vergara-Castañeda et al., 2010).

Studies evaluating the specific effects of NDF from carioca beans are scarce. It is known that NDF fermentation from common beans results in the production of short-chain fatty acids (SCFAs) whose concentration and distribution depend on microflora as well as carbohydrate content in the intestine (Barber et al., 2020; Feregrino-Perez et al., 2014; Feregrino-Pérez et al., 2008; Vergara-Castañeda et al., 2010). Some studies showed protective effect of NDF from raw and cooked common beans against colon cancer. The mechanism involves the NDF that is fermented in the large intestine to produce SCFAs such as acetate, propionate, and butyrate. The butyrate was related to the reduction in aberrant crypt foci (ACF), which inhibited colon cancer through overexpression of genes (*β-catenin*, *p*53, *p*21, *rb*, *bax*, and *caspase*-3) involved in proliferation, cellular arrest, and apoptosis (Feregrino-Perez et al., 2014; Feregrino-Pérez et al., 2008; Moreno-Jiménez et al., 2019; Vergara-Castañeda et al., 2010). Moreover, the results reported by Reverri et al. (2015) support the potential induction of DNA repair by the NDF from common bean due to synergism among different genes (*Tp*53, *Ccmb*2, *Gadd*45*a*, *Pcna*, *Msb*2, and *Xrcc*5) as an alternative protective mechanism that contributes to the chemoprotective effect of common bean on early-stage colon cancer. Meanwhile, a Korean kidney bean husk extract exhibited a series of antitumor effects such as cell death and apoptotic body appearance. These antitumor effects were accompanied by increases in p-AMPK (AMP-activated protein kinase, target molecule of tumor control) and phospho-acetyl-CoA carboxylase (p-ACC), as well as antitumor proteins p53 and p21 (Campos-Vega et al., 2013).

Maybe the most widely known and appreciated beneficial effects of dietary fiber is its effect on gut motility and inhibition of constipation. The literature provides ample evidence that common bean's soluble and insoluble fiber, including RS, improve biological mechanisms of cardiometabolic risk factors including improved glycemic index, postprandial lipemia modulation, and blood pressure control (Mullins & Arjmandi, 2021; Reverri et al., 2015) and support a healthy gut bacterial environment (Barber et al., 2020; Mullins & Arjmandi, 2021; Vergara-Castañeda et al., 2010). The limited dietary fiber intake stymies the establishment and maintenance of a healthy, viable, and diverse colonic microbiota that, in turn, limit the local production of SCFAs, including butyrate. Signaling pathways that implicate nuclear factor kappa-B (NF-kB) and deacetylase inhibition influence inflammatory processes both in the neighborhood and systemically. These inflammatory processes are influenced by butyrate levels within the colon (Bach Knudsen et al., 2018; Wong, Harris, & Ferguson, 2016). The associations between dietary fiber with favorable insulin sensitivity, body composition, appetite regulation, and diversity and viability of the gut microflora can presumably explain its effect on global rates of CVD (Barber et al., 2020). The mechanism with the relationship between fiber intake and reduced CVD risk involves the association of reduced levels of plasma trimethylamine-N-oxide (TMAO, a by-product of certain gut microbiota, derived from choline) with reduced CVD risk (Mullins & Arjmandi, 2021; Reverri et al., 2015).

Other important compounds are the oligosaccharides, especially the RFO present in beans, with an influence on human health. RFO are considered antinutrients because of their involvement in gastrointestinal discomfort (formation of gases in people that consume this product infrequently, flatulence). Even so, the current judgment of RFO in main food crops has changed (Liu, Ragaee, Marcone, & Abdel-Aal, 2020). Regular consumption of RFO may be a crucial dietary tool in the prevention of chronic diseases (Parnell, Raman, Rioux, & Reimer, 2012) and provide other health benefits such as stimulation of mineral uptake and deposition (Yeung, Glahn, Welch, & Miller, 2005). Raffinose and stachyose have beneficial physiological effects, such as those of dietary fiber. The NDF from carioca beans contains considerable amounts of RFO (Alves, Gomes, et al., 2021; Moreno-Jiménez et al., 2019). These compounds exert prebiotic function essential for the maintenance of a healthy gastrointestinal microbiome and potentially reduce carcinogenic N-nitroso compounds levels in the gut (Johnson et al., 2015). Finally, oligosaccharides induce cell cycle arrest in different human colon cancer cell lines through the regulation of *p*21, cyclins, and some kinase expression. Thus, these compounds influence cell cycle arrest by modulating these genes (Feregrino-Perez et al., 2014; Moreno-Jiménez et al., 2019).

5.2.3 Micronutrients (Vitamins and Minerals)

Beans are considered an important source of minerals (Fe, Zn, Mg, Cu, Ca, Al, and others) (Hayat et al., 2014). Zinc and iron are present in various beans in high quantities. These elements are essential for human health as they are cofactors for many physiological and metabolic functions (Campos-Vega, Loarca-Piña, & Oomah, 2010; de Oliveira, Mateó, Fioroto, de Oliveira, & Naozuka, 2018).

Iron is present in several biomolecules such as heme (e.g., hemoglobin, myoglobin, cytochrome proteins, myeloperoxidase, nitric oxide synthetases), iron–sulfur clusters (e.g., respiratory complexes I-III, coenzyme Q10, mitochondrial aconitase, DNA primase), or other functional groups (e.g., hypoxia-inducible factor prolyl hydroxylases). In this way, it participates in several biological processes such as DNA replication and repair, oxidative phosphorylation, and oxygen transport (Dev & Babitt, 2017; Lal, 2020). Due to the high iron content in carioca beans (Table 5.1), they are commonly consumed to prevent the emergence of deficiency anemia, including through the consumption of iron-biofortified beans (Junqueira-Franco et al., 2018; Tako et al., 2015).

Iron homeostasis depends on a complex system where iron absorbed via duodenal enterocytes is transported in the bloodstream linked to transferrin. Most are mobilized to the bone marrow where the heme groups that make up red blood cells (RBCs) will be produced, with smaller quantities going to other tissues for fundamental cellular processes and the excess transported to the liver for storage linked to ferritin. This homeostasis still

Table 5.1 Mineral content (mg kg^{-1}) of carioca beans.

Cultivars	P	Fe	Zn	Mn
Fresh				
Estilo	2140	53	43	18
Madrepérola	2470	104	41	21
Pontal	3030	106	38	18
Stored up to 180 days				
Estilo	3070–4030	69–145	42–50	19–26
Madrepérola	2750–2970	65–132	44–53	17–33
Pontal	2980–3230	48–121	36–45	19–32

Source: Adapted from Silochi et al. (2016).

depends on the recycling of iron from erythrocytes by means of reticuloendothelial macrophages and by the unregulated loss of iron due to flaking or bleeding and can also be affected by inflammatory processes, hypoxia, genetic changes, liver problems, and others (Knutson, 2017). On the other hand, in healthy individuals, iron supplementation can lead to poisoning and is related to risks of development of certain chronic disorders such as glucose intolerance and consequently diabetes, nonalcoholic fatty liver, and CVDs. Iron excess induces cellular injury and functional abnormalities in hepatocytes by the process of lipid peroxidation and DNA damage through the formation of free radicals by Fenton's reaction (Deugnier, Bardou-Jacquet, & Lainé, 2017; Ma et al., 2019; Silva, Silva, de Paula, Carneiro, & Pedrosa, 2008). Studies on common beans have shown that after cooking, more than 70% of the Fe species became insoluble due to protein denaturation and association with polyphenols and phytate (de Oliveira et al., 2018). Moreover, the Fe-biofortified carioca bean provided similar concentration of phytate and polyphenols. Nevertheless, higher protein content, in parallel to the Fe standard bean, can potentially enhance intestinal functionality and Fe bioavailability. Additionally, the current research proposes that increased Fe content may not certainly result in an enhanced absorbable Fe, and a key factor is the measurement of dietary Fe bioavailability in Fe-biofortified crop varieties-based diets, and as part of the breeding process (Dias et al., 2018).

Zinc (Table 5.1) is an essential micronutrient in many biological processes, including enzyme cofactor, the development and function of the immune system, reproductive system and quality of sperm antioxidant, an anti-inflammatory agent, and modulates antiviral and antibacterial immunity and regulates inflammatory response (Fallah, Mohammad-Hasani, & Colagar, 2018; Lee, 2018; Marreiro et al., 2017). Therefore, zinc can play a protective role against chronic diseases that are associated with oxidative stress and chronic inflammation, including cancer, CVD, and diabetes, and was cited as a drug in COVID-19 treatment (Maret, 2013; Skalny et al., 2020). The storage capacity of zinc in the human body is limited, with food as the main source.

Calcium is the most abundant mineral in the body, present in three different forms: ionized Ca (~ 50%), Ca bound to protein (~ 40%), and 10% is complexed with other molecules, such as citrate and phosphate. Plasma Ca balance is controlled by the actions of several hormones, including parathyroid hormone (PTH), calcitonin, and 1,25 $(OH)_2$-vitamin D (Correa-Rodríguez et al., 2020). Adequate calcium consumption is related to the prevention of hypertensive disorders of pregnancy and blood pressure control, reduction of LDL cholesterol levels, and prevention of osteoporosis and colorectal adenomas (Cormick & Belizán, 2019).

Magnesium is the fourth most abundant mineral in the body and the second most abundant intracellular cation. In the plasma, about 1/3 of the Mg is bound to plasma proteins and the rest is filtered at the renal level (Mihatsch et al., 2018). This mineral acts as a cofactor for more than 300 enzymes, regulating muscle contraction, neuromuscular conduction, glycemic control, myocardial contraction, and blood pressure. Besides, magnesium participates in metabolic pathways related to energy production, active transmembrane transport, and bone development (Al Alawi, Majoni, & Falhammar, 2018). When evaluating aging indicators for carioca beans, Bento et al. (2020) showed that calcium and magnesium are components of the cell walls with higher concentration in the integument (about 30–50 times that of the cotyledon); *BRSMG Madrepérola* and *CNFC* 10467 contain the highest calcium levels and *BRS Estilo* the lowest (about 3 to 4 times those in cotyledons) with pinto and *BRS Pontal* presenting the highest and lowest values, respectively.

In addition to minerals, beans are a source of vitamin B complex: folate, tocopherols, thiamine, riboflavin, niacin, biotin, and pyridoxamine (Campos-Vega et al., 2010). Vitamin B complex is essential for the healthy balance of the immune and nervous systems, metabolism of lipids, carbohydrates, and proteins, as a coenzyme in several oxidation and reduction reactions in the body and in bicarbonate-dependent carboxylation reactions, hormone synthesis, synthesis of the neurotransmitters serotonin and noradrenaline, for myelin formation, and the promotion of RBC formation and maturation, the synthesis of purines and pyrimidines (Harirari, Schellack, & Schellack, 2019). Riboflavin deficiency in blood is presumed to be related to a disturbance in riboflavin absorption. Disturbances in the steps in intermediary metabolism may occur if riboflavin intake is inadequate (Campos-Vega et al., 2013). Villavicencio, Mancini-Filho, Delincée, and Bognár (2000) evaluated the effect of household cooking and storage after irradiation on vitamin contents in carioca beans and concluded that thiamine loss does not occur in *Macaçar* variety. Riboflavin concentration increased with radiation dose, and at doses up to 1 kGy, B6 vitamin was not affected.

5.3 Bioactive Compounds in Carioca Beans

Carioca beans (*Phaseolus vulgaris*) have been increasingly studied due to their promising nutritional composition and the high content of soluble and insoluble dietary fibers, phytates, and polyphenols, imparting numerous physiological benefits that lead to reduce risk of developing noncommunicable diseases (NCDs) responsible for the death of 41 million people annually. CVDs account for most NCD deaths (~17.9 million people annually), followed by cancers (9 million), respiratory diseases (3.9 million), and diabetes (1.6 million). For many years, this group of diseases has affected the global population at an increasing rate. Altogether, about 80% of all premature NCD deaths are caused by these four diseases (WHO, 2018), but the inclusion of foods rich in bioactive compounds in the diet can be a strategy to improve health conditions.

Bioactive compounds act by many mechanisms of action in the body, which will be described in detail in the following sections of this chapter. It is known that in type 2 diabetes, bioactive compounds inhibit the α-amylase and α-glucosidase (α-GLA) hydrolytic enzymes, reduce inflammation induced by cytokines and oxidative stress, leading to the beneficial maintenance of β-cells activity (Li, Yuan, Pan, Wang, & Chen, 2017). In Alzheimer's disease, they reduce the hyperphosphorylation of the tau protein, present in the neurons of the central nervous system and responsible for stabilizing the microtubules, and the β-Site APP-cleaving enzyme 1 (BACE1), a protease responsible for β-amyloid peptides production in neurons (Li, Zhang, Rasool, Geetha, & Babu, 2019). In CVD, the mechanism of action of bioactive compounds includes hypolipidic, antioxidant, anti-inflammatory activities, and reduction of endothelial dysfunction, among others (Cao et al., 2019). In obesity, a state of chronic low-grade inflammation, bioactive compounds inhibit the production of pro-inflammatory cytokines, lipids accumulation in adipocytes, and LDL cholesterol oxidation (Moreno-Valdespino, Luna-Vital, Camacho-Ruiz, & Mojica, 2020), and modulate hunger and satiety center in the hypothalamus (Gomes et al., 2020).

5.3.1 Bioactive Peptides

Bioactive peptides are formed by amino acids capable of conducting a diverse set of actions in the body. The biological activity of bioactive peptides found in carioca beans (*Phaseolus vulgaris* L.) is used to classify their functions, which include anti-inflammatory responses, antioxidant, nitrogen metabolism, storage protein, carbohydrate metabolism, and immune system modulation (Luna-Vital et al., 2015). Besides, a recent study showed that the large amount of amino acids that form the protein molecules of carioca beans generate several peptides with bioactive potential, which have been described as inhibition of dipeptidyl peptidase IV (DPP-IV) and dipeptidyl peptidase III (DPP-III) improving glycemic control; inhibition of angiotensin-converting enzyme I (ACE-I) improving blood pressure control; peptides with antioxidative effects, related to carbohydrate metabolism and modulation of the immune system (Alves, Gomes, et al., 2021).

Several studies demonstrate the beneficial effects of bioactive peptides from carioca beans *in vitro* and *in vivo* (Table 5.2). *BRS Pontal* and *BRSMG Madrepérola* carioca beans digested protein isolates used to treat THP-1 cells decreased prostaglandin E-2 (PGE-2) and TNF-α. In addition, the beans reduced the lectin-type oxidized LDL receptor (LOX-1), the matrix metalloproteinase-9 (MMP-9) expression, and the expression of several atherosclerosis-related cytokines, with anti-atherosclerotic potential equivalent to simvastatin (Alves, Vasconcelos, et al., 2016). Carioca beans protein hydrolysate also exerts anti-inflammatory potential *in vitro*. For example, protein hydrolysates from *BRS Pontal* reduced TNF-α secretion, whereas those from *BRSMG Madrepérola* reduced interleukin 1β (IL-1β) and PGE-2 in human THP-1 macrophage-like cells (Alves, de Mejía, et al., 2016).

The protective *in vitro* effects of the cardiovascular system were confirmed in an *in vivo* study demonstrating reduction in weight gain and food intake in animals fed with *BRSMG Madrepérola* protein hydrolysate; the consumption of both bean flour and its protein hydrolysate reduced total cholesterol, TNF-α concentration, and NF-κB gene expression. Moreover, SOD concentration increased and peroxisomal proliferator-activated receptors alpha (PPAR-α) decreased with the consumption of carioca bean flour compared with the high-fat diet. These data indicate anti-inflammatory activity, a reduction of risk factors for CVDs and glycemic control due to the action of bioactive peptides in inhibiting DPP-IV and DPP-III, and reduction in blood glucose in animals that consumed bean-derived treatments (de Lima et al., 2019). In relation to the cardiovascular system, the consumption of carioca bean protein hydrolysate rich in bioactive peptides modulated lipid metabolism and prevented vascular endothelium dysfunction *in vivo*. The mechanism involved reduction of TNF-α and angiotensin II gene expressions, angiotensin II concentration in cardiac tissue, and increased expression of endothelial nitric oxide synthase (e-NOS) and nitric oxide concentration within physiological levels (acting as a vasodilating agent) in male adult BALB/mice fed an atherogenic diet supplemented with bean protein hydrolysate (700 mg kg^{-1} day^{-1} for 9 weeks) compared to atherogenic diet group (Gomes et al., 2020).

Carioca bean peptides are multifunctional (Table 5.2) most frequently exhibiting ACE and DPP-IV inhibitory activities along with antioxidant and glucose uptake stimulating abilities. Angiotensin receptor-1 (AR1) blocker such as angiotensin 1 inhibitor peptides can attenuate the angiotensin-II (ACE2)-mediated vasoconstriction, inflammation, cell damage, lung injury, and fibrosis in COVID-19-infected hypertensive patients. Moreover,

Table 5.2 Bioactive peptides of carioca beans and their *in vivo* and *in vitro* effects.

Cultivars	Peptide sequence	Biological activity
Fresh		
Pontal	VELVGPK	ACE inhibitor, anti-amnesic, antioxidative, antithrombotic, DPP-IV inhibitor, GUSP, PRSM
	LAMPV	ACE inhibitor, DPP IV inhibitor, UMPA
	ALVPVGSKPK	ACE inhibitor, DPP-IV inhibitor, antioxidative, GUSP
Madrepérola	LSSLEM	DPP-IV inhibitor
	LVSCLVDL	DPP-IV inhibitor, GUSP
	VELVGPK	ACE inhibitor, antioxidative, DPP-IV, GUSP, PRSM, PEI
Stored up to 90 days		
Pontal	VELVGPK	ACE inhibitor, antioxidative, DPP-IV inhibitor, GUSP, PRSM, PEI
	SLPVLV	GUSP; DPP-IV inhibitor
Madrepérola	VELVGPK	ACE inhibitor, antioxidative, DPP-IV, GUSP, PRSM, PEI
	LSSLEM	DPP-IV inhibitor
	VELVGPK	ACE inhibitor, antioxidative, DPP-IV, GUSP, PRSM, PEI
	KLLALQKMRK	ACE inhibitor, UMPA, DPPIV, GUSP, PRSM, PEI
Stored up to 180 days		
Pontal	TRGVLV	ACE inhibitor; DPP-IV inhibitor; GUSP
Madrepérola	LVTTTVDL	GUSP, DPP-IV inhibitor
	QTSTPLFS	ACE inhibitor, DPP-IV inhibitor
	VELVGPK	ACE inhibitor, antioxidative, DPP-IV, GUSP, PRSM, PEI
	TRGVLV	ACE inhibitor, DPP-IV inhibitor GUSP

ACE inhibitor, angiotensin-converting-enzyme inhibitor; DPP-IV inhibitor, dipeptidyl peptidase IV inhibitor; GUSP, glucose uptake stimulating peptide; PRSM, peptide regulating the stomach mucosal membrane activity; PEI, prolyl endopeptidase inhibitor; UMPA, ubiquitin-mediated proteolysis activating peptide.
Source: adapted from Alves, de Mejía, et al. (2016) and Gomes et al. (2020).

peptides exhibiting ACE2 and/or DPP-IV inhibitory activities can block SARS-CoV-2 invasion in host cells of COVID-19 patients (Palit et al., 2021). Peptide ALVPVGSKPK has two proline residues 2P enabling it to maintain its biological activity during simulated gastrointestinal tract (pepsin and pancreatin) digestion (Parthasarathy et al., 2018). Peptide LVSCLVDL contains high amount (63 % of the peptides) of the neurotransmitter amino acids leucine and valine (Suzuki et al., 2020), two branched chain amino acids that can also influence gut microbiota. Moreover, proline and leucine of the SLPVLP peptide promote synergy with non-peptide antioxidants (Etemadian et al., 2021). Peptides LVTTTVDL and KLLALQKMRK contain leucine, lysine, and threonine (71 and 60% of the total peptides, respectively, that are associated with increased satiety (Volk et al., 2020). Moreover, the presence of threonine (LVTTTVDL) particularly is presumed to account for this increased satiety. Peptide QTSTPLFS has the hydrophobic hypoglycemic amino acids proline-leucine

Table 5.3 Hydrophobicity of bioactive peptides present in carioca bean.

Peptides	Length	Hydrophobicity		H-Index
		HPB%	kcal mol^{-1}	
ALVPVGSKPK	10	60	13.7	0.2
KLLALQKMRK	10	50	14.9	−0.4
LAMPV	5	100	6.2	2.0
LSSLEM	6	50	9.2	0.7
LVSCLVDL	8	63	7.3	2.2
LVTTTVDL	8	50	8.8	1.3
QTSTPLFS	8	38	7.2	−0.1
SLPVLP	6	83	5.1	2.2
TRGVLV	6	50	8.9	1.1
VELVGPK	7	57	13.5	0.4

HPB%, percent hydrophobicity; H-Index, Hydropathicity Index.
Source: Adapted from Alves, de Mejía, et al. (2016).

(Ren et al., 2016) and contains the memory improving threonine and proline [TP] amino acid motif (Tanaka et al., 2020).

Carioca bean peptides are highly hydrophobic (≥ 50%) except for QTSTPLFS (Table 5.3) containing hydrophilic amino acids as indicated by the negative hydropathicity value (Alves, de Mejía, et al., 2016). Peptide LAMPV is extremely hydrophobic (100% HPB), and peptide VELVGPK has the hydrophobic amino acids proline and leucine that contribute to α-GLA inhibitory activity (Ren et al., 2016). VELVGPK contains the memory improving amino acids glycine and proline [GP] motif (Tanaka et al., 2020). Studies evaluating the effects of consumption of this specific commercial group of beans on human health are scarce despite those *in vivo* and *in vitro* investigations in animals regarding the health effects of bioactive peptides from carioca beans. However, some studies in the literature have evaluated the effects of several common bean cultivars (*Phaseolus vulgaris*) on human health related to weight loss in obese individuals.

Peptide fraction (<3 kDa) [PV3] from dehulled (hand-peeled extracted with acetonitrile/water/formic acid 25/24/1 ratio) carioca bean (cultivar *BRS Pontal*) hardened (greenhouse storage 120 days, 40 °C, 75% RH) protects human umbilical vein endothelial cells from oxidative stress by increasing nitric oxide (NO) release without cell death (apoptosis), reducing H_2O_2-induced reactive oxygen species (ROS) production and vasodilatory effects due to endothelium-dependent NO release (Graziani et al., 2021). Thirty-five peptides with an average length of 10.2 amino acids (5–23 amino acids range) and mass of 1.14 kDa were identified in the PV3 extract corresponding to 20 *Phaseolus vulgaris* proteins. All peptides had ACE and DPP-IV inhibitory activities and many displayed variegated and critical function. For example, 59% of the peptides exhibited DPP-III inhibition, 41% antioxidant activity, 29% stimulated glucose uptake or vasoactive substance release, 26% renin inhibition, 15% ion flow regulation, 15% ubiquitin proteolysis activation, 15% calmodulin-dependent cyclic nucleotide phosphodiesterase (caMPDE) inhibition [Ca channel blocker] and 9% each anti-amnesia, anti-inflammatory antithrombic and hypolipidemic activities. ACE inhibition is important in regulating blood pressure and therefore for the management of hypertension and other CVDs. PV3 peptides exert cardiovascular effects by releasing NO (confirmed *ex vivo* in isolated aortic rings), a potent endothelial-derived relaxing factor that counteracts angiotensin II constricting effects. The DPP-IV inhibition improves glycemic regulation in type 2 diabetics. Moreover, PV3 can potentially improve endothelial function in diseases such as arterial hypertension, since PV3 actions rely on endothelial-dependent oxidonitrergic paths and can modify platelet aggregation. This peptide extract is presumed to increase NO bioavailability concomitantly with antioxidant effect (Graziani et al., 2021).

5.3.2 Phenolic Compounds

The group of phenolic compounds consists of various molecules, formed both by compounds with simple chemical structures, such as phenolic acids, and by complex compounds, such as tannins and lignin. Carioca beans are rich in phenolic compounds, especially polyphenols, with high antioxidant potential (Table 5.4). The polyphenolic profile of two carioca bean cultivars, *BRS Pérola* and *BRS Cometa*, shows the presence of kaempferol 3-glucoside (17.3 and 16.2 μM, respectively), catechin (26.1 and 25.9 μM, respectively), epicatechin (12.8 and 11 μM, respectively), procyanidin B1 (1.4 and 1.2 μM, respectively), and quercetin 3-glucoside (0.2 μM for *BRS Pérola*) as the five most prevalent polyphenolic compounds in the bean seed coats (Dias et al., 2018). In addition to the systemic effects in the body, phenolic compounds can improve the composition of the intestinal microbiota *in vivo* by protecting outgrowth of opportunistic and pathogenic bacteria (Dias et al., 2018; Reed, Neuman, Glahn, Koren, & Tako, 2017).

In addition, it is important to clarify the relationship between phytochemicals content and the storage time of carioca beans, since beans are generally stored for some time on

Table 5.4 Phenolic compounds and antioxidant activity of carioca beans.

Cultivars	Phenolic compounds (mg GAEq g^{-1})	Antioxidant activity (ORAC-mmol TE g^{-1})
Fresh		
Pontal	1.32	262
Madrepérola	1.25	239
Stored up to 90 days		
Pontal	1.68	264
Madrepérola	1.18	243
Stored up to 180 days		
Pontal	1.29	259
Madrepérola	1.09	243

GAEq–gallic acid equivalents; TE –trolox equivalent
Source: Adapted from Alves, Gomes, et al. (2021).

supermarket shelves or even at home. A recent study showed that the storage time of two carioca bean cultivars (*BRS Pontal* and *BRSMG Madrepérola*) for up to 6 months did not affect the concentration of total phenolic in the grains. Although the storage time increases the oxidation of the compounds, and the reduction of the total phenolic content is expected, the study showed that there may be a modification only in the types of phytochemicals, mainly due to the presence of primary metabolites and specific ions that participate in oxidation reactions, which is even associated with darkening and hardening during storage. Regarding the concentration of tannins, storage at 0, 3, and 6 months did not trigger significant changes, but the study showed that this compound can vary for different cultivars, being higher in the *BRS Pontal* than in *BRSMG Madrepérola* bean (Alves, Gomes, et al., 2021).

5.3.2.1 Antioxidant Activity

The antioxidant potential of carioca beans demonstrated in *in vitro* and *in vivo* studies is directly related to the concentration of phenolic compounds, such as phenolic acids, flavonoids, and tannins, and the presence of bioactive peptides with high antioxidant activity. Polyphenols stand out concerning antioxidant activity due to their ability to scavenge free radicals and bind to free molecules forming chelates, which prevent the oxidative process and the production of ROS. In addition, *BRS Pontal* and *BRSMG Madrepérola* carioca beans, after simulated gastrointestinal digestion of its protein isolates, generated bioactive peptides profile with high antioxidant and anti-inflammatory potential (Table 5.2). The antioxidant capacity of the digested protein isolates did not change after commercial storage (6 months), and the peptides profile or their biological potential remained unaffected during storage (Alves, Vasconcelos, et al., 2016). The mechanism involved with the antioxidant protection of bioactive peptides is related to the presence of very small sequences, composed of aromatic amino acids, such as Trp, Tyr, and Phe, and basic amino acids such as His and Arg, which act as protons donors for potentially reactive molecules (Luna-Vital et al., 2015).

The antioxidant capacity of carioca beans was unaffected when stored for 0, 3, or 6 months (Alves, Gomes, et al., 2021). Stored and freshly cooked carioca beans presented about 252 mmol trolox equivalent/g of dried flour (Alves, Gomes, et al., 2021), higher than for raw and fresh *BRS Pontal* (210 mmol trolox equivalent/g for seed coats) (Mojica, Meyer, Berhow, & de Mejía, 2015). Among the phenolic compounds present in carioca beans, catechin and kaempferol 3-glucoside act directly against oxidative damage, reduce DNA damage and LDL oxidation, and contribute to the prevention of CVDs (de Lima et al., 2019). Consumption of 346.6 g of bean flour kg^{-1} of diet/day increased the activity of the antioxidant enzyme superoxide dismutase (SOD) in the liver of mice fed a high-fat diet and indicated a promising antioxidant effect (de Lima et al., 2019). Regarding the consumption of carioca bean protein hydrolysate, a study showed that 700 mg kg^{-1} body weight/day reduced lipid peroxidation assessed by the malondialdehyde concentration in serum of adult mice compared to mice fed a high-fat diet, but it did not change the total antioxidant capacity (Gomes et al., 2020).

In addition to the promising phytochemical composition of conventional carioca beans, the study of biofortified foods has been gaining pace in the scientific community. Iron (Fe)-biofortified carioca beans can be an effective alternative to increase the intake of this mineral in the population at risk of deficiency. In addition, the antioxidant capacity of these new cultivars has been evaluated.

Table 5.5 Polyphenolic compounds of Fe-biofortified and Fe standard carioca beans

Compound (μmol g^{-1})	Fe-biofortified carioca beans	Fe standard carioca beans
3,4–dihydroxybenzoic acid	0.21	0.20
Catechin	0.18	0.17
Quercetin 3–glucoside *	0.08	0.00
Kaempferol 3–glucoside *	0.30	0.21
Kaempferol	0.01	0.01

* Significant difference between the cultivars.
Source: Adapted from Tako et al. (2015).

Among the five most prevalent polyphenolic compounds (Table 5.5) found in the cream seeded carioca bean, the concentration of kaempferol 3-glycoside and quercetin 3-glycoside increased significantly in the biofortified cultivar compared to the conventional one, showing antioxidant potential and promising capacity to increase iron intake bioavailable in populations that consume these grains, but that have low access to nutritious foods (Tako et al., 2015).

5.3.2.2 Phytate

Phytates are molecules present mainly in cereals, seeds, and legumes such as beans and soybeans and are responsible for storing and providing essential phosphorus for energy production and for the development of plant cell membranes. Despite its well-recognized antioxidant and anti-inflammatory potential (Chen et al., 2017), phytates act as antinutritional factors due to their metal binding ability to ions such as Fe, zinc (Zn), and calcium (Ca) and reduce their bioavailability (Carbas et al., 2020). Common beans naturally have high phytate content; however, cultivars with more intense coloring have higher phytate content and consequently greater antioxidant activity (Carbas et al., 2020). Phytate concentration in carioca beans is about 120 mg/100 g and does not change with commercial storage for up to 6 months (Alves, Gomes, et al., 2021).

The antioxidant and anti-inflammatory mechanism of action of phenolic compounds and phytate is regulated by chelates formation with free metal ions or activation of antioxidant enzymes such as SOD, catalase (CAT), and glutathione peroxidase (GPX). SOD acts on the superoxide radical, converting it to hydrogen peroxide and thus reducing its reactivity. Consumption of *BRSMG Madrepérola* bean flour, containing 122 mg of phytate/100 g, for 9 weeks, increased SOD activity in animals fed a high-fat diet, reduced inflammation and the risk factors for CVDs *in vivo* (de Lima et al., 2019).

Because beans undergo heat processing to be consumed in cooked form, the concentration of phytates, phenolic compounds, saponins, and protease inhibitors decreases, representing a positive point in terms of anti-nutritional compounds, but negative in terms of antioxidant, anti-inflammatory, and for the treatment of chronic diseases such as CVD, cancer, type 2 diabetes, and obesity. In addition, anti-nutritional activity should not be related to just one food or chemical compound but account for the diversity of the diet, the support of micronutrients and proteins of high biological value, and also the health status of the individual (Díaz-Batalla, Widholm, Fahey Jr., Castaño-Tostado, & Paredes-López, 2006), especially concerning the intestinal microbiota, which has a primary role in the digestion and absorption processes and will be addressed in this chapter.

5.3.2.3 Saponins

Saponins are compounds present in low amounts in cooked beans, but with already established antioxidant activity, allowing its designation as a bioactive compound. Like phytate, saponin is considered an anti-nutritional factor for its ability to bind proteins, carbohydrates, vitamins, and minerals, and reduce their bioavailability. However, they are also able to bind free radicals and stimulate antioxidant enzymes, thus establishing an antioxidant potential. They are found as glycosides and their structure may have a four-ringed steroid nucleus (a steroid aglycone), or a five-ringed structure (triterpene aglycone) (Chávez-Mendoza & Sánchez, 2017).

Studies that have assessed the specific effects of saponins from carioca beans are limited. The bean seed coat or hull concentrates the largest amount of phytochemicals and responsible for the antioxidant, anti-inflammatory, and antidiabetic properties observed for these compounds. In mung bean (*Vigna radiate* L.) and adzuki bean (*Vigna angularis* L.) without thermal processing or soaking, the saponin content was high in the husk (24.3 and 73.6 mg of saponin equivalent/g, respectively), while in the whole grains the concentration was much lower, 2.2 mg of saponin equivalent/g of mung beans and 10.8 mg of saponin equivalent/g of adzuki beans (Luo, Cai, Wu, & Xu, 2016). However, saponins are sensitive to heat and because they are tensioactive compounds, soaking and cooking reduce their concentration in the grains (Shi et al., 2004). Additionally, a study of phenolic and saponin profile in carioca beans during storage confirms its reduction during storage (3 and 6 months) in the cultivar *Notável* (Bento, Ribeiro, Bassinello, de Brito et al., 2021).

Several health-promoting activities of legumes saponins have recently been reviewed, including antioxidants, anti-inflammatory, hypocholesterolemic, and hypoglycemic *in vivo* and *in vitro* models, as well as hypolipidemic, hypoglycemic, and strong anticarcinogenic activity, as shown in clinical studies. Mechanisms of action include the ability to bind and form complexes with lipids, preventing the action of bile acid and thus reducing cholesterol absorption or bile acid reabsorption. In addition, saponins reduce the cytotoxicity of cancer cells and normalize carcinogen-induced cell proliferation due to their immunomodulatory activity (Singh, Singh, Singh, & Kaur, 2017).

5.3.3 Anthocyanins

Anthocyanins, flavonoids belonging to the group of phenolic compounds, are widely distributed in nature and because they are natural pigments, they give color to foods such as fruits, vegetables, and legumes such as beans. In general, the color intensity of some food is related to their anthocyanin concentration, which provides high antioxidant activity. The concentration of anthocyanins in common beans varies according to genetics, but anthocyanins of seed coats expressed as delphinidin-3-glucoside equivalents ranged from 0.013 to 0.21%. The genetic characteristic of the common bean cultivar to provide high anthocyanin content is regulated by only a few alleles, that is, a small number of genes that express this characteristic. In addition, beans with a more intense red tegument color had genetic characteristics to express higher levels of anthocyanins (Díaz, Caldas, & Blair, 2010).

Delphinidin 3-glucoside, petunidin 3-glucoside, and malvidin 3-glucoside are the main anthocyanins present in the common bean seed coat and exhibit high antioxidant activity. Generally, the intake of flavonoids is directly related to the quality of the diet and with the reference dietary guidelines, indicating that individuals with higher flavonoids intake, including anthocyanins, have a healthier diet (Sebastian et al., 2015). The antioxidant properties of anthocyanins are based on their potential to carry free radicals and prevent CVDs, by reducing LDL oxidation and damage to the vascular endothelium, modulation of inflammation, platelet aggregation, and anticancer action. Several flavonoids have been inversely related to the risk of cancers. A study showed that the high intake of anthocyanidins reduced the risk of colorectal cancer (OR = 0.67) (Rossi, Bosetti, Negri, Lagiou, & La Vecchia, 2010) and the intake of anthocyanins (6 mg kg^{-1} and 24 mg kg^{-1}) for 40 days reduced body weight, food intake, and the weight of adipose tissue compared to the placebo group. The mechanism of action is due to suppression of neuropeptide Y and increase of the γ-amino butyric acid receptor (GABAB1R) in the hypothalamus, modulating the center of hunger and satiety (Badshah et al., 2013).

Both bioactive peptides and phenolic compounds modulate dysfunction of vascular endothelium by attenuating markers of inflammation and altering the lipid metabolism at the early stages of atherogenic process based on the proposed mechanism of action (Gomes et al., 2020) (Figure 5.3). The action of carioca bean bioactive peptides from protein hydrolysate (simulated gastrointestinal digest) is associated with protein expression of LOX-1 signaling pathway (Figure 6.3). These peptides can inhibit oxidized LOX-1 receptor, suppress adhesion molecules (ICAM-1) and chemoattractant proteins (MCP-1) expressions, and decrease foam cell formation and matrix-degrading proteases (MMP-9) and inflammatory cytokines (Alves, Vasconcelos, et al., 2016).

5.4 Potential Benefits of Compounds in Carioca Beans in The Modulation of The Gut Microbiota

Recently, the study of foods that potentially modulate the gut microbiota has increased due to the role of the microbiota in activating metabolic pathways involved in health and disease. The role of dietary fibers in the activation of intestinal motility, regulation of intestinal transit, satiety, and fermentation (with the production of SCFAs) is well established. However, gene expression related to health and disease has only recently been revealed explaining the mechanisms of action of dietary fiber-rich foods and bioactive compounds in modulating the gut microbiota composition and also its functionality.

Carioca beans are a source of soluble and insoluble dietary fibers, phenolic compounds, bioactive peptides, and countless other bioactive compounds that come directly into contact with the bacteria that colonize the gastrointestinal tract of humans and animals. In addition, new cultivars of Fe-biofortified beans are studied as a new nutritional approach to increase the intake of specific nutrients for vulnerable populations, and these foods have also been evaluated for their effects on the gut microbiota (Table 5.6). Carioca bean

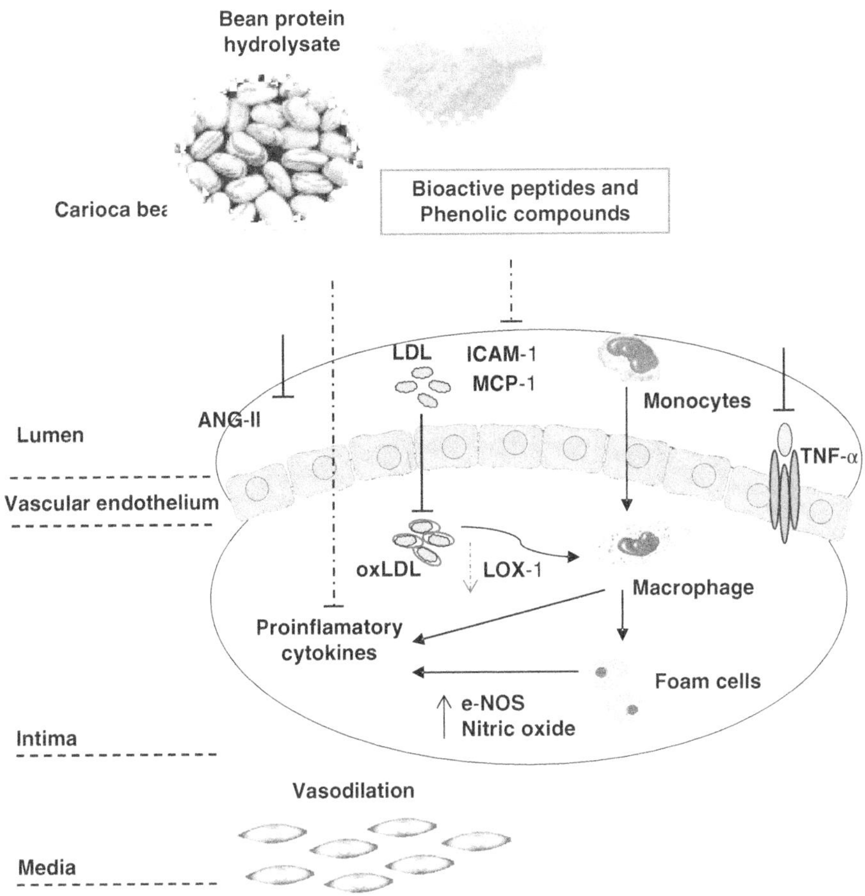

Figure 5.3 Potential mechanism of action of bioactive peptides and phenolic compounds from carioca bean and bean protein hydrolysate after simulated gastrointestinal digestion in the dysfunction of vascular endothelium *in vivo* and related to protein expression of LOX-1 signaling pathway *in vitro*. *In vitro* mechanisms are indicated by dashed lines. Ang II: angiotensin II; LDL: low-density lipoprotein; ICAM-1: intracellular adhesion molecule-1; MCP-1: monocyte chemoattractant protein-1; TNF-α: tumor necrosis factor-α; oxLDL: oxidized LDL; LOX-1: lectin-type oxLDL receptor; e-NOS: endothelial nitric oxide synthase. *Source:* Gomes et al. (2020)/ELSEVIER.

cultivars *BRS Cometa* (Fe-biofortified) and *BRS Pérola* (Fe standard) were evaluated in association with a Brazilian-based diet to investigate Fe bioavailability and gut microbiota modulation. Despite the similar phytate concentration and polyphenol profile among cultivars, *BRS Cometa* (Fe-biofortified) increased the total body Hb-Fe values in the experimental broiler model (*Gallus gallus*), the abundance of beneficial SCFA-producing bacteria, bacteria linked to phenolic catabolism, and general changes in phylogenetic diversity among the groups. In addition, intestinal villi height was higher in the Fe-biofortified group compared to the Fe standard, indicating beneficial effects on specific bacterial populations (Dias et al., 2018).

Table 5.6 Phylum and genus abundance (%) in the gut microbiota with the Fe-biofortified and Fe standard carioca beans diet.

Level	Fe-biofortified carioca bean diet	Fe standard carioca bean diet
Phylum		
Verrucomicrobia	—	0.7
Tenericutes	0.7	4.0
Proteobacteria	7.5	4.5
Firmicutes	91.5	89.0
Order		
Clostridiales	15.0	11.5
Family		
Ruminococcaceae	10.0	16.5
Lachnospiraceae	20.0	15.0
Genera		
Streptococcus	41.5	26.0
Ruminococcus	12.0	20.5
Oscillospira	—	9.7

Source: Adapted from Dias et al. (2018).

Recently, a review showed that the consumption of Fe-biofortified beans improves the composition of the gut microbiota due to the greater supply of minerals, which are essential for the growth of several families of beneficial Gram + bacteria to the microbiome. In addition to the presence of soluble and insoluble dietary fibers, proteins, and carbohydrates, secondary metabolites such as phytochemicals, phenolic compounds, terpenoids, and alkaloids have antimicrobial activity, naturally selecting specific bacterial groups that will modify the composition and function of the intestinal microbiota. For example, these compounds modulate the abundance of bacteria involved in phenolic catabolism, increase the predominance of SCFA-producing bacteria, and beneficially modulate the abundance of species that convert organic compounds, such as catechin and epicatechin into their metabolic derivatives, increasing their bioavailability and improving gut health (Gomes, Martino, & Tako, 2021)

The mechanisms of action of the compounds present in carioca beans in modulating the gut microbiota are based on the contribution of essential substrates for the maintenance of beneficial bacterial taxa, leading to the growth of SCFA-producing bacteria, which from the fermentation of soluble fibers, increase the production of SCFAs.

The SCFAs serve as a substrate for the nutrition of the cells that make up the brush border membrane of the enterocytes, improving the absorptive capacity of the intestine, reducing intestinal permeability, which is regulated by tight junctions, reducing inflammation and gut-associated lymphoid tissue (GALT) activation. These effects are also proven by the increase in goblet cell number, which synthesize and secrete the mucus, and the increase in villi height and diameter (Dias et al., 2018; Gomes et al., 2021). Thus, the beneficial modulation of bacterial taxa regulates functional pathways and modulates the expression of genes of the host, improving general health and reducing the risk of developing chronic diseases.

5.5 Potential Functional Foods/Industry Products

The nutritional benefits of beans make them suitable to be considered as ingredients that can be incorporated into food products. Therefore, there is an increasing interest in the development of novel functional foods with common beans. A new way of including beans in food products is the use of bean flours. The bean flour constitutes an alternative for using materials with low added value (as aged grains and broken grains), which can contribute to the sustainability of the food industries and still be aligned with new trends and consumption habits, based on sensory quality, practicality, diversity, and health. Moreover, flour from carioca beans is an option for improving the nutritional quality of foods, enriching them with protein, insoluble dietary fiber, and RS (Bassinello, Bento, Gomes, Caliari, & Oomah, 2020; Ramírez-Jiménez, Gaytán-Martínez, Morales-Sánchez, & Loarca-Piña, 2018). Another important aspect that must be considered in the development of bean-based foods is the reduction of carbohydrates, in relation to wheat flour traditionally used in the preparation of convenience products. Moreover, bean flour can be used in the development of gluten-free products, which in turn is one of the most prosperous markets in the field of the food industry (Bascunan, Vespa, & Araya, 2017). Therefore, bean flours have been used as a base ingredient in various foods, such as spaghetti and ravioli (Gallegos-Infante, Bello-Perez, Rocha-Guzman, Gonzalez-Laredo, & Avila-Ontiveros, 2010; Ringuette, Finley, Prinyawiwatkul, & King, 2018), penne, baked snacks, and extruded snacks (Bassinello, Carvalho, Rios, Maciel, & Berrios, 2015), snack bars (Ramírez-Jiménez et al., 2018), cookies (Pérez-Ramírez et al., 2018), cakes (Bassinello et al., 2020), and bread and chips (Hooper, Glahn, & Cichy, 2019).

A study verified the nutritional and bioactive components of a carioca bean (*Phaseolus vulgaris* L) tempeh. The tempeh of carioca beans is a relevant source of protein, minerals, and carbohydrates and presents a satisfactory content of dietary fibers, often superior to traditional soybean tempeh. Additionally, the final product presented a relevant antioxidant potential. Finally, it was concluded that given its nutritional profile, functional potential, and adequate sensory quality (color and texture), carioca bean tempeh can be another alternate use of stored grains and to become a component of healthy diets (Bento, Bassinello, Colombo, Vital, & Carvalho, 2020). Carioca bean tempeh (Bento, Bassinello, Colombo, Vital, & Carvalho, 2021) was used for vegan burger development (Figure 5.4). The sensorial results indicated a low overall acceptance index (58%); thus, the authors recommended that the hamburger made with bean tempeh needs to be improved further. Moreover, the common bean tempeh is an innovative product that may contribute to the use of depreciated grains, as well as an alternative for the consumption of healthy foods like vegan hamburgers.

Another study demonstrated that the incorporation of extruded carioca bean flour in gluten-free cake mix improves their nutritional value. The protein digestibility and total dietary fiber were improved, besides increasing ash, iron, and zinc contents. In

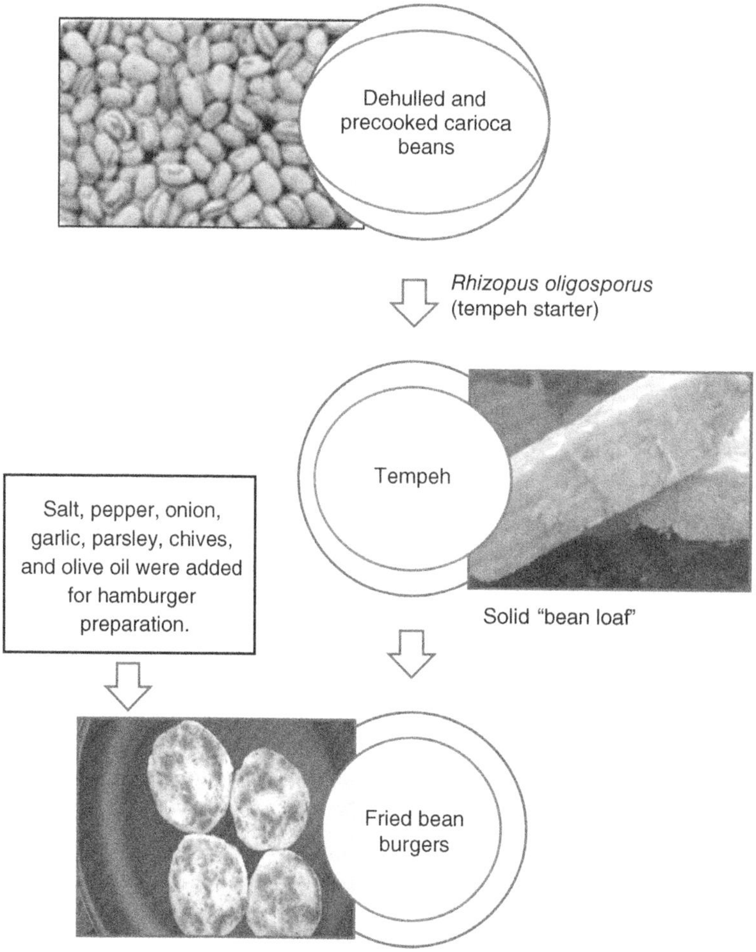

Figure 5.4 Flowchart of carioca bean tempeh preparation and its hamburger. *Source:* Adapted from Bento et al. (Bento, Bassinello, et al., 2020; Bento, Bassinello, Colombo, et al., 2021). *Photos*: Personal Archive.

combination with rice grit flour, it enables the preparation of a cake mix with gluten-free protein, resulting in higher nutritional quality, without limiting the content of the amino acids. The cakes, containing up to 75% of extruded bean flour, presented high nutritional value since they had high levels of protein, fiber, minerals, and good protein digestibility (Bassinello et al., 2020). In the latest study, instant pasta and baked snacks made from pre-gelatinized flours of broken carioca and black beans had high protein and dietary fiber (19 and 18 g/100 g, respectively) and excelled in sensorial test (Bento, Ribeiro, Alexandree Silva et al., 2021). Moreover, the gluten-free penne-type pasta from the pre-gelatinized flours had short cooking time (3–4 min). The development of reduced glycemic load/high-fiber ravioli using bean flour was evaluated. The nutritional value of frozen pasta product increased by substituting navy bean flour for a portion of the standard pasta flour due to a large increase in fiber content, a moderate decrease in glycemic load, and an increase in protein content (Ringuette et al., 2018). The combination of pinto beans and oat was used for developing cookies. The administration of 1.2 g kg^{-1} of oat-beans cookies slightly decreased serum glucose levels in rats, which was associated with an increased insulin production and/or secretion, as well as decreased glucose intestinal absorption. Furthermore, supplementation with 1.2 g kg^{-1} of oat-beans cookies exerted the greatest beneficial effect on the atherogenic index and C-reactive protein, suggesting their cardio-protective potential. The beneficial effect of oat-beans cookies was associated with their content of dietary fiber, RS, and galactooligosaccharides, as well as their high content of chlorogenic acid, rutin, protocatechuic acid, caffeic acid, β-campesterol, β-sitosterol, and soyasaponins. Finally, these results suggest that pinto bean flour could be used as a functional ingredient to develop bakery products, such as cookies, with an antidiabetic effect (Pérez-Ramírez et al., 2018). Carioca bean flour can replace egg yolk up to 50% in mayonnaise, improving the nutritional characteristics (increased protein and reduced cholesterol and lipid contents) without affecting the behavior, functionality and quality of the mayonnaise (Alves et al., 2021). The latest study (Álvares-da-Silva et al., 2021) demonstrates that the higher consumption of bean in the Brazilian diet is associated with beneficial microbial diversity compared with US-based patients with cirrhosis. This higher microbial diversity enables metabolism of complex carbohydrates to SCFAs, thereby stimulating the immune system and better gut health.

References

Al Alawi, A. M., Majoni, S. W., & Falhammar, H. (2018). Magnesium and human health: Perspectives and research directions. *International Journal of Endocrinology, 2018*, 9041694. https://doi.org/10.1155/2018/9041694

Alvares, R. C., Stonehouse, R., Souza, T. L. P. O., Melo, P. G. S., Miklas, P. N., Bett, K. E., . . . Pereira, H. S. (2019). Generation and validation of genetic markers for the selection of carioca dry bean genotypes with the slow-darkening seed coat trait. *Euphytica, 215*(8), 141. https://doi.org/10.1007/s10681-019-2461-y

Álvares-da-Silva, M. R., Oliveira, C. P., Fagan, A., Longo, L., Thoen, R. U., Yoshimura Zitelli, P. M., . . . Bajaj, J. S. (2021). Interaction of microbiome, diet, and hospitalizations between Brazilian and American patients with cirrhosis. *Clinical Gastroenterology and Hepatology, S1542–3565*(21), 00355–00354. https://doi.org/10.1016/j.cgh.2021.03.045

Alves, J. D. S., Rodrigues, A. S., Moro, K. I. B., Boeira, C. P., Holkem, A. T., Roggia, I., . . . Rosa, C. S. D. (2021). Elaboration of mayonnaises containing landrace common bean flour as a partial egg yolk substitute. *Ciência Rural, 51*(9), e20200543. https://doi.org/10.1590/0103-8478cr20200543

Alves, N. E. G., de Mejía, E. G., Vasconcelos, C. M., Bassinello, P. Z., & Martino, H. S. D. (2016). Postharvest storage of Carioca bean (Phaseolus vulgaris L.) did not impair inhibition of inflammation in lipopolysaccharide-induced human THP-1 macrophage-like cells. *Journal of Functional Foods, 23*, 154–166. https://doi.org/10.1016/j.jff.2016.02.029

Alves, N. E. G., Gomes, M. J. C., Vasconcelos, C. M., Lima, A. C., de Lima, S. L. S., Brito, E. S., . . . Martino, H. S. D. (2021). Six months under uncontrolled relative humidity and room temperature changes technological characteristics and maintains the physicochemical and functional properties of carioca beans (Phaseolus vulgaris L.). *Food Chemistry*, *342*, 128390. https://doi.org/10.1016/j.foodchem.2020.128390

Alves, N. E. G., Vasconcelos, C. M., Bassinello, P. Z., de Mejia, E. G., & Martino, H. S. D. (2016). Digested protein isolate from fresh and stored Carioca beans reduced markers of atherosclerosis in oxidized LDL-induced THP-1 macrophages. *Journal of Functional Foods*, *24*, 97–111. https://doi.org/10.1016/j.jff.2016.03.027

Bach Knudsen, K. E., Lærke, H. N., Hedemann, M. S., Nielsen, T. S., Ingerslev, A. K., Gundelund Nielsen, D. S., . . . Hermansen, K. (2018). Impact of diet-modulated butyrate production on intestinal barrier function and inflammation. *Nutrients*, *10*(10). https://doi.org/10.3390/nu10101499

Badshah, H., Ullah, I., Kim, S. E., Kim, T. H., Lee, H. Y., & Kim, M. O. (2013). Anthocyanins attenuate body weight gain via modulating neuropeptide Y and GABAB1 receptor in rats hypothalamus. *Neuropeptides*, *47*(5), 347–353. https://doi.org/10.1016/j.npep.2013.06.001

Barber, T. M., Kabisch, S., Pfeiffer, A. F. H., & Weickert, M. O. (2020). The health benefits of dietary fibre. *Nutrients*, *12*(10). https://doi.org/10.3390/nu12103209

Bascunan, K. A., Vespa, M. C., & Araya, M. (2017). Celiac disease: Understanding the gluten-free diet. *Europena Journal of Nutrition*, *56*(2), 449–459. https://doi.org/10.1007/s00394-016-1238-5

Bassinello, P. Z., Bento, J. A. C., Gomes, L. D. O. F., Caliari, M., & Oomah, B. D. (2020). Nutritional value of gluten-free rice and bean based cake mix. *Ciência Rural*, *50*. https://doi.org/10.1590/0103-8478cr20190653

Bassinello, P. Z., Carvalho, A. V., Rios, A. D. O., Maciel, R. D. A., & Berrios, J. D. J. (2015). Expanded gluten-free extrudates made from rice grits and bandinha (Bean) flour mixes: Main quality properties. *Journal of Food Processing and Preservation*, *39*(6), 2267–2275. https://doi.org/10.1111/jfpp.12472

Bento, J. A. C., Bassinello, P. Z., Colombo, A. O., Vital, R. J., & Carvalho, R. N. (2020). Nutritional and bioactive components of carioca common cean (*Phaseolus vulgaris* L.) Tempeh and yellow soybean (*Glycine max* L.) Tempeh. *Current Nutrition and Food Science*, *16*, 1–8. https://doi.org/10.2174/1573401316666200121111854

Bento, J. A. C., Bassinello, P. Z., Colombo, A. O., Vital, R. J., & Carvalho, R. N. (2021). Vegan tempeh burger: Prepared with aged bean grains fermented by Rhizopus oligosporus inoculum. *Research, Society and Development*, *10*(2), e38110212503. https://doi.org/10.33448/rsd-v10i2.12503

Bento, J. A. C., Bassinello, P. Z., Morais, D. K., de Souza Neto, M. A., Bataus, L. A. M., Carvalho, R. N., . . . Soares Júnior, M. S. (2021). Pre-gelatinized flours of black and carioca bean by-products: Development of gluten-free instant pasta and baked snacks. *International Journal of Gastronomy and FoodScience*, *25*, 100383. https://doi.org/10.1016/j.ijgfs.2021.100383

Bento, J. A. C., Lanna, A. C., Bassinello, P. Z., Oomah, B. D., Pimenta, M. E. B., Carvalho, R. N., & Moreira, A. S. (2020). Aging indicators for stored carioca beans. *Food Research International*, *134*(109249), 1–11. doi: 10.1016/j.foodres.2020.109249

Bento, J. A. C., Ribeiro, P. R. V., Alexandree Silva, L. M., Alves Filho, E. G., Bassinello, P. Z., de Brito, E. S., . . . Soares Júnior, M. S. (2021). Chemical profile of colorful bean (Phaseolus vulgaris L) flours: Changes influenced by the cooking method. *Food Chemistry, 356*, 129718. https://doi.org/10.1016/j.foodchem.2021.129718

Bento, J. A. C., Ribeiro, P. R. V., Bassinello, P. Z., de Brito, E. S., Zocollo, G. J., Caliari, M., & Soares Júnior, M. S. (2021). Phenolic and saponin profile in grains of carioca beans during storage. *LWT-Food Science Technology, 139*, 110599. https://doi.org/10.1016/j.lwt.2020.110599

Campos-Vega, R., Loarca-Piña, G., & Oomah, B. D. (2010). Minor components of pulses and their potential impact on human health. *Food Research International, 43*(2), 461–482. https://doi.org/10.1016/j.foodres.2009.09.004

Campos-Vega, R., Oomah, B. D., Loarca-Piña, G., & Vergara-Castañeda, H. A. (2013). Common beans and their non-digestible fraction: Cancer inhibitory activity-An overview. *Foods, 2*(3), 374–392. https://doi.org/10.3390/foods2030374

Cao, S. Y., Zhao, C. N., Gan, R. Y., Xu, X. Y., Wei, X. L., Corke, H., . . . Li, H. B. (2019). Effects and mechanisms of tea and Iis bioactive compounds for the prevention and treatment of cardiovascular diseases: An updated review. *Antioxidants (Basel), 8*(6). https://doi.org/10.3390/antiox8060166

Carbas, B., Machado, N., Oppolzer, D., Ferreira, L., Queiroz, M., Brites, C., . . . Barros, A. I. (2020). Nutrients, antinutrients, phenolic composition, and antioxidant activity of aommon aean cultivars and their potential for food applications. *Antioxidants (Basel), 9*(2). https://doi.org/10.3390/antiox9020186

Carbonaro, M. (2021). Nutraceutical perspectives of pulses. In B. K. Tiwari, A. Gowen, & B. McKenna (Eds.), *Pulse foods* (2nd ed., pp. 423–460). Oxford, UK: Academic Press.

Chávez-Mendoza, C., & Sánchez, E. (2017). Bioactive compounds from Mexican varieties of the common bean (Phaseolus vulgaris): Implications for health. *Molecules, 22*(8), 1360.

Chen, P. X., Zhang, H., Marcone, M. F., Pauls, K. P., Liu, R., Tang, Y., . . . Tsao, R. (2017). Anti-inflammatory effects of phenolic-rich cranberry bean (*Phaseolus vulgaris* L.) extracts and enhanced cellular antioxidant enzyme activities in Caco-2 cells. *Journal of Functional Foods, 38*, 675–685. https://doi.org/10.1016/j.jff.2016.12.027

Cires, M. J., Wong, X., Carrasco-Pozo, C., & Gotteland, M. (2017). The gastrointestinal tract as a key target organ for the health-promoting effects of dietary proanthocyanidins. *Frontiers in nutrition, 3*, 57.

Cormick, G., & Belizán, J. M. (2019). Calcium intake and health. *Nutrients, 11*(7), 1606. https://doi.org/10.3390/nu11071606

Correa-Rodríguez, M., Schmidt-RioValle, J., Ramírez-Vélez, R., Correa-Bautista, J. E., González-Jiménez, E., & Rueda-Medina, B. (2020). Influence of calcium and vitamin D intakes on body composition in children and adolescents. *Clinical Nursing Research, 29*(4), 243–248. https://doi.org/10.1177/1054773818797878

de Lima, S. L. S., Gomes, M. J. C., da Silva, B. P., Alves, N. E. G., Toledo, R. C. L., Theodoro, J. M. V., . . . Martino, H. S. D. (2019). Whole flour and protein hydrolysate from common beans reduce the inflammation in BALB/c mice fed with high fat high cholesterol diet. *Food Research International, 122*, 330–339. https://doi.org/10.1016/j.foodres.2019.04.013

de Oliveira, A. P., Mateó, B. D. S. O., Fioroto, A. M., de Oliveira, P. V., & Naozuka, J. (2018). Effect of cooking on the bioaccessibility of essential elements in different varieties of beans (Phaseolus vulgaris L.). *Journal of Food Composition and Analysis, 67*, 135–140. https://doi.org/10.1016/j.jfca.2018.01.012

Deugnier, Y., Bardou-Jacquet, É., & Lainé, F. (2017). Dysmetabolic iron overload syndrome (DIOS). *La Presse Médicale, 46*(12, Part 2), e306–e311. https://doi.org/10.1016/j.lpm.2017.05.036

Dev, S., & Babitt, J. L. (2017). Overview of iron metabolism in health and disease. *Hemodialysis International, 21*(S1), S6–S20. https://doi.org/10.1111/hdi.12542

Dias, D. M., Kolba, N., Binyamin, D., Ziv, O., Regini Nutti, M., Martino, H. S. D., . . . Tako, E. (2018). Iron biofortified carioca bean (*Phaseolus vulgaris* L.)-based brazilian diet delivers more absorbable iron and affects the gut microbiota *in vivo* (*Gallus gallus*). *Nutrients, 10*(12). https://doi.org/10.3390/nu10121970

Díaz, A. M., Caldas, G. V., & Blair, M. W. (2010). Concentrations of condensed tannins and anthocyanins in common bean seed coats. *Food Research International, 43*(2), 595–601. https://doi.org/10.1016/j.foodres.2009.07.014

Díaz-Batalla, L., Widholm, J. M., Fahey, G. C., Jr., Castaño-Tostado, E., & Paredes-López, O. (2006). Chemical components with health implications in wild and cultivated Mexican common bean seeds (Phaseolus vulgaris L.). *Journal of Agricultural Food Chemistry, 54*(6), 2045–2052. https://doi.org/10.1021/jf051706l

Etemadian, Y., Ghaemi, V., Shaviklo, A. R., Pourashouri, P., Sadeghi Mahoonak, A. R., & Rafipour, F. (2021). Development of animal/plant-based protein hydrolysate and its application in food, feed and nutraceutical industries: State of the art. *Journal of Cleaner Production, 278*, 123219. https://doi.org/10.1016/j.jclepro.2020.123219

Fallah, A., Mohammad-Hasani, A., & Colagar, A. H. (2018). Zinc is an essential element for male fertility: A review of Zn roles in men's health, germination, sperm quality, and fertilization. *Journal of Reproduction and Infertility, 19*(2), 69–81.

Feregrino-Pérez, A. A., Berumen, L. C., García-Alcocer, G., Guevara-Gonzalez, R. G., Ramos-Gomez, M., Reynoso-Camacho, R., . . . Loarca-Piña, G. (2008). Composition and chemopreventive effect of polysaccharides from common beans (Phaseolus vulgaris L.) on azoxymethane-induced colon cancer. *Journal of Agric Food Chem, 56*(18), 8737–8744. https://doi.org/10.1021/jf8007162

Feregrino-Perez, A. A., Piñol-Felis, C., Gomez-Arbones, X., Guevara-González, R. G., Campos-Vega, R., Acosta-Gallegos, J., & Loarca-Piña, G. (2014). A non-digestible fraction of the common bean (Phaseolus vulgaris L.) induces cell cycle arrest and apoptosis during early carcinogenesis. *Plant Foods forHuman Nutrition, 69*(3), 248–254. https://doi.org/10.1007/s11130-014-0428-7

Ferreira, E. S., Amaral, A. L. S., Demonte, A., Zanelli, C. F., Capraro, J., Duranti, M., & Neves, V. A. (2015). Hypocholesterolaemic effect of rat-administered oral doses of the isolated 7S globulins from cowpeas and adzuki beans. *Journal of Nutritional Science, 4*, e7. https://doi.org/10.1017/jns.2014.70

Gallegos-Infante, J.-A., Bello-Perez, L. A., Rocha-Guzman, N. E., Gonzalez-Laredo, R. F., & Avila-Ontiveros, M. (2010). Effect of the addition of sommon bean (*Phaseolus vulgaris* L.) flour on the *in vitro* Ddgestibility of starch and undigestible carbohydrates in spaghetti. *Journal of Food Science, 75*(5), H151–H156. https://doi.org/10.1111/j.1750-3841.2010.01621.x

Ganesan, K., & Xu, B. (2017). Polyphenol-rich dry common beans (*Phaseolus vulgaris* L.) and their health benefits. *International Journal of Molecular Sciences, 18*(11), 2331. https://doi.org/10.3390/ijms18112331

Gomes, M. J. C., Lima, S. L. S., Alves, N. E. G., Assis, A., Moreira, M. E. C., Toledo, R. C. L., . . . Martino, H. S. D. (2020). Common bean protein hydrolysate modulates lipid metabolism and prevents endothelial dysfunction in BALB/c mice fed an atherogenic diet. *Nutrition, Metabolism and Cardiovascular Diseases*, *30*(1), 141–150. https://doi.org/10.1016/j.numecd.2019.07.020

Gomes, M. J. C., Martino, H. S. D., & Tako, E. (2021). Effects of iron and zinc biofortified foods on gut microbiota *in vivo* (*Gallus gallus*): A systematic review. *Nutrients*, *13*(1). https://doi.org/10.3390/nu13010189

Graziani, D., Ribeiro, J. V. V., Cruz, V. S., Gomes, R. M., Araújo, E. G., Santos Júnior, A. C. M., . . . Xavier, C. H. (2021). Oxidonitrergic and antioxidant effects of a low molecular weight peptide fraction from hardened bean (*Phaseolus vulgaris*) on endothelium. *Brazilian Journal of Medical and Biological Research*, *54*(6), e10423.

Harirari, P., Schellack, G., & Schellack, N. (2019). Vitamin B-complex deficiency, supplementation and management. *Pharmaceutical Journal*, *86*(3), 23–29. https://doi.org/10.10520/EJC-16fb37f64d

Hayat, I., Ahmad, A., Masud, T., Ahmed, A., & Bashir, S. (2014). Nutritional and health perspectives of beans (*Phaseolus vulgaris* L.): An overview. *Critical Reviews in Food Science and Nutrition*, *54*(5), 580–592. https://doi.org/10.1080/10408398.2011.596639

Hooper, S. D., Glahn, R. P., & Cichy, K. A. (2019). Single varietal dry bean (*Phaseolus vulgaris* L.) pastas: Nutritional profile and consumer acceptability. *Plant Foods for Human Nutrition*, *74*(3), 342–349. https://doi.org/10.1007/s11130-019-00732-y

Islam, N. S., Bett, K. E., Pauls, K. P., Marsolais, F., & Dhaubhadel, S. (2020). Postharvest seed coat darkening in pinto bean (Phaseolus vulgaris) is regulated by Psd, an allele of the basic helix-loop-helix transcription factor P. *Plants, People, Planet*, *2*(6), 663–677. https://doi.org/10.1002/ppp3.10132

Johnson, C. R., Thavarajah, D., Thavarajah, P., Payne, S., Moore, J., & Ohm, J.-B. (2015). Processing, cooking, and cooling affect prebiotic concentrations in lentil (*Lens culinaris* Medikus). *Journal of Food Composition and Analysis*, *38*, 106–111. https://doi.org/10.1016/j.jfca.2014.10.008

Junqueira-Franco, M. V. M., de Oliveira, J. E. D., Nutti, M. R., Pereira, H. S., de Carvalho, J. L. V., Abrams, S. A., . . . Marchini, J. (2018). Iron absorption from beans with different contents of iron, evaluated by stable isotopes. *Clinical Nutrition ESPEN*, *25*, 121–125. https://doi.org/10.1016/j.clnesp.2018.03.120

Kenmoe, L., Kotue, T. C., Chandra, K., Fm, D., Pieme, A., Kansci, G., . . . Arumugam, N. (2020). Albumin and globulin fractions from black bean seeds (*Phaseolus vulgaris* L.) used in the management of sickle cell disease (SCD) in the West Region of Cameroon have antisickling and antioxidant properties. *Journal Biotechnology Biomedical*, *3*, 078–092. https://doi.org/10.26502/jbb.2642-91280029

Knutson, M. D. (2017). Iron transport proteins: Gateways of cellular and systemic iron homeostasis. *Journal of Biological Chemistry*, *292*(31), 12735–12743. https://doi.org/10.1074/jbc.R117.786632

Kumar, S., Verma, A. K., Sharma, A., Roy, R., Kumar, D., Bh, G., . . . Dwivedi, P. D. (2014). Phaseolin: A 47.5kDa protein of red kidney bean (Phaseolus vulgaris L.) plays a pivotal role in hypersensitivity induction. *International Immunopharmacology*, *19*(1), 178–190. https://doi.org/10.1016/j.intimp.2014.01.014

Lal, A. (2020). Iron in health and disease: An update. *The Indian Journal of Pediatrics, 87*(1), 58–65. https://doi.org/10.1007/s12098-019-03054-8

Lawrence, M. C., Izard, T., Beuchat, M., Blagrove, R. J., & Colman, P. M. (1994). Structure of phaseolin at 2·2 Å resolution: Implications for a common vicilin/legumin structure and the genetic engineering of seed storage proteins. *Journal of Molecular Biology, 238*(5), 748–776. https://doi.org/10.1006/jmbi.1994.1333

Lee, S. R. (2018). Critical role of zinc as either an antioxidant or a prooxidant in cellular systems. *Oxidative Medicine and Cellular Longevity, 2018*, 9156285. https://doi.org/10.1155/2018/9156285

Li, R., Zhang, Y., Rasool, S., Geetha, T., & Babu, J. R. (2019). Effects and underlying mechanisms of bioactive compounds on Type 2 diabetes mellitus and alzheimer's disease. *Oxidative Medicine and Cellular Longevity, 2019*, 8165707. https://doi.org/10.1155/2019/8165707

Li, W., Yuan, G., Pan, Y., Wang, C., & Chen, H. (2017). Network pharmacology studies on the bioactive compounds and action mechanisms of natural products for the treatment of diabetes Mmellitus: A review. *Frontiers in Pharmacology, 8*, 74. https://doi.org/10.3389/fphar.2017.00074

Liu, Y., Ragaee, S., Marcone, M. F., & Abdel-Aal, E. S. M. (2020). Effect of different cooking methods and heating solutions on nutritionally-important starch fractions and flatus oligosaccharides in selected pulses. *Cereal Chemistry, 97*, 1216–1226. https://doi.org/10.1002/cche.10344

Luna-Vital, D. A., Mojica, L., González de Mejía, E., Mendoza, S., & Loarca-Piña, G. (2015). Biological potential of protein hydrolysates and peptides from common bean (*Phaseolus vulgaris* L.): A review. *Food Research International, 76*, 39–50. https://doi.org/10.1016/j.foodres.2014.11.024

Luo, J., Cai, W., Wu, T., & Xu, B. (2016). Phytochemical distribution in hull and cotyledon of adzuki bean (*Vigna angularis* L.) and mung bean (*Vigna radiate* L.), and their contribution to antioxidant, anti-inflammatory and anti-diabetic activities. *Food Chemistry, 201*, 350–360. https://doi.org/10.1016/j.foodchem.2016.01.101

Ma, W., Feng, Y., Jia, L., Li, S., Li, J., Wang, Z., . . . Du, H. (2019). Dietary iron modulates glucose and lipid homeostasis in diabetic Mice. *Biological Trace Element Research, 189*(1), 194–200. https://doi.org/10.1007/s12011-018-1446-3

Maret, W. (2013). Zinc Biochemistry: from a single zinc enzyme to a key element of life. *Advances in Nutrition, 4*(1), 82–91. https://doi.org/10.3945/an.112.003038 %J Advances in Nutrition

Marreiro, D. D. N., Cruz, K. J. C., Morais, J. B. S., Beserra, J. B., Severo, J. S., & De Oliveira, A. R. S. (2017). Zinc and oxidative stress: Current mechanisms. *Antioxidants, 6*(2), 24. https://doi.org/10.3390/antiox6020024

Mihatsch, W., Fewtrell, M., Goulet, O., Molgaard, C., Picaud, J. C., Senterre, T., . . . Yan, W. (2018). ESPGHAN/ESPEN/ESPR/CSPEN guidelines on pediatric parenteral nutrition: Calcium, phosphorus and magnesium. *Clinical Nutrition, 37*(6), 2360–2365. https://doi.org/10.1016/j.clnu.2018.06.950

Mojica, L., & de Mejía, E. G. (2015). Characterization and comparison of protein and peptide profiles and their biological activities of improved common bean cultivars (*Phaseolus vulgaris* L.) from Mexico and Brazil. *Plant Foods for Human Nutrition, 70*(2), 105–112. https://doi.org/10.1007/s11130-015-0477-6

Mojica, L., Meyer, A., Berhow, M. A., & de Mejía, E. G. (2015). Bean cultivars (*Phaseolus vulgaris* L.) have similar high antioxidant capacity, in vitro inhibition of α-amylase and α-glucosidase while diverse phenolic composition and concentration. *Food Research International, 69*, 38–48. https://doi.org/10.1016/j.foodres.2014.12.007

Moreno-Jiménez, M. R., López-Barraza, R., Cervantes-Cardoza, V., Pérez-Ramírez, I. F., Reyna-Rojas, J. A., Gallegos-Infante, J. A., . . . Rocha-Guzmán, N. E. (2019). Mechanisms associated to apoptosis of cancer cells by phenolic extracts from two canned common beans varieties (Phaseolus vulgaris L.). *Journal of Food Biochemistry, 43*(6), e12680. https://doi.org/10.1111/jfbc.12680

Moreno-Valdespino, C. A., Luna-Vital, D., Camacho-Ruiz, R. M., & Mojica, L. (2020). Bioactive proteins and phytochemicals from legumes: Mechanisms of action preventing obesity and type-2 diabetes. *Food Research International, 130*, 108905. https://doi.org/10.1016/j.foodres.2019.108905

Mullins, A. P., & Arjmandi, B. H. (2021). Health benefits of plant-based nutrition: Focus on beans in cardiometabolic diseases. *Nutrients, 13*(2). https://doi.org/10.3390/nu13020519

Nogueira, A. F., Moda-Cirino, V., Delfini, J., Brandão, L. A., Mian, S., Constantino, L. V., . . . Gonçalves, L. S. A. (2021). Morpho-agronomic, biochemical and molecular analysis of genetic diversity in the Mesoamerican common bean panel. *PLoS One, 16*(4), e0249858. https://doi.org/10.1371/journal.pone.0249858

Palit, P., Chattopadhyay, D., Thomas, S., Kundu, A., Kim, H. S., & Rezaei, N. (2021). Phytopharmaceuticals mediated Furin and TMPRSS2 receptor blocking: can it be a potential therapeutic option for Covid-19? *Phytomedicine, 85*, 153396. https://doi.org/10.1016/j.phymed.2020.153396

Parnell, J. A., Raman, M., Rioux, K. P., & Reimer, R. A. (2012). The potential role of prebiotic fibre for treatment and management of non-alcoholic fatty liver disease and associated obesity and insulin resistance. *Liver International, 32*(5), 701–711. https://doi.org/10.1111/j.1478-3231.2011.02730.x

Parthasarathy, A., Cross, P. J., Dobson, R. C. J., Adams, L. E., Savka, M. A., & Hudson, A. O. (2018). A three-ring circus: Metabolism of the three proteogenic aromatic amino acids and their role in the health of plants and animals. *Frontiers in Molecular Biosciences, 5*, 29.

Pérez-Ramírez, I. F., Becerril-Ocampo, L. J., Reynoso-Camacho, R., Herrera, M. D., Guzmán-Maldonado, S. H., & Cruz-Bravo, R. K. (2018). Cookies elaborated with oat and common bean flours improved serum markers in diabetic rats. *Journal of the Science of Food and Agriculture, 98*(3), 998–1007. https://doi.org/10.1002/jsfa.8548

Ramírez-Jiménez, A. K., Gaytán-Martínez, M., Morales-Sánchez, E., & Loarca-Piña, G. (2018). Functional properties and sensory value of snack bars added with common bean flour as a source of bioactive compounds. *LWT-Food Science and Technology, 89*, 674–680. https://doi.org/10.1016/j.lwt.2017.11.043

Reed, S., Neuman, H., Glahn, R. P., Koren, O., & Tako, E. (2017). Characterizing the gut (Gallus gallus) microbiota following the consumption of an iron biofortified Rwandan cream seeded carioca (Phaseolus Vulgaris L.) bean-based diet. *PLoS One, 12*(8), e0182431. https://doi.org/10.1371/journal.pone.0182431

Ren, Y., Liang, K., Jin, Y., Zhang, M., Chen, Y., Wu, H., & Lai, F. (2016). Identification and characterization of two novel α-glucosidase inhibitory oligopeptides from hemp (Cannabis sativa L.) seed protein. *Journal of Functional Foods, 26*, 439–450. https://doi.org/10.1016/j.jff.2016.07.024

Reverri, E. J., Randolph, J. M., Steinberg, F. M., Kappagoda, C. T., Edirisinghe, I., & Burton-Freeman, B. M. (2015). Black beans, fiber, and antioxidant capacity pilot study: Examination of whole foods *vs.* functional components on postprandial metabolic, oxidative stress, and inflammation in adults with metabolic syndrome. *Nutrients, 7*(8), 6139–6154. https://doi.org/10.3390/nu7085273

Ribeiro, N. D., Londero, P. M. G., Cargnelutti Filho, A., Jost, E., Poersch, N. L., & Mallmann, C. A. (2007). Composição de aminoácidos de cultivares de feijão e aplicações para o melhoramento genético. *Pesquisa Agropecuária Brasileira, 42*(10), 1393–1399.

Ringuette, C., Finley, J., Prinyawiwatkul, W., & King, J. M. (2018). Development of reduced glycemic load/high Ffber ravioli using bean flour. *Journal of Culinary Science and Technology, 16*(4), 336–356. https://doi.org/10.1080/15428052.2017.1391147

Rossi, M., Bosetti, C., Negri, E., Lagiou, P., & La Vecchia, C. (2010). Flavonoids, proanthocyanidins, and cancer risk: A network of case-control studies from Italy. *Nutrition and Cancer, 62*(7), 871–877. https://doi.org/10.1080/01635581.2010.509534

Sebastian, R. S., Wilkinson Enns, C., Goldman, J. D., Martin, C. L., Steinfeldt, L. C., Murayi, T., & Moshfegh, A. J. (2015). A new database facilitates characterization of flavonoid intake, sources, and positive associations with diet auality among US adults. *Journal of Nutrition, 145*(6), 1239–1248. https://doi.org/10.3945/jn.115.213025

Shi, J., Arunasalam, K., Yeung, D., Kakuda, Y., Mittal, G., & Jiang, Y. (2004). Saponins from edible legumes: Chemistry, processing, and health benefits. *Journal Medicinal Food, 7*(1), 67–78. https://doi.org/10.1089/109662004322984734

Silochi, R. M. H. Q., Coelho, S. R. M., Bischoff, T. Z., Cassol, F. A. D. R., do Prado, N. V., & Bassinello, P. Z. (2016). Nutritional technological characterization and secondary metabolites in stored carioca bean cultivars. *African Journal of Agricultural Research, 11*(24), 2102–2111. https://doi.org/10.5897/AJAR2016.11013

Silva, M., Silva, M. E., de Paula, H., Carneiro, C. M., & Pedrosa, M. L. (2008). Iron overload alters glucose homeostasis, causes liver steatosis, and increases serum triacylglycerols in rats. *Nutrition Research, 28*(6), 391–398. https://doi.org/10.1016/j.nutres.2008.02.009

Singh, B., Singh, J. P., Singh, N., & Kaur, A. (2017). Saponins in pulses and their health promoting activities: A review. *Food Chemistry, 233*, 540–549. https://doi.org/10.1016/j.foodchem.2017.04.161

Skalny, A. V., Rink, L., Ajsuvakova, O. P., Aschner, M., Gritsenko, V. A., Alekseenko, S. I., . . . Tinkov, A. A. (2020). Zinc and respiratory tract infections: Perspectives for COVID-19 (Review). *International Journal of Molecular Medicine, 46*(1), 17–26. https://doi.org/10.3892/ijmm.2020.4575

Spitti, A. M. D. S., Carbonell, S. A. M., Dias, C. T. D. S., Sabino, L. G., Carvalho, C. R. L., & Chiorato, A. F. (2019). Carioca bean genotypes for tolerance to grain darkening by natural and accelerated methods. *Ciência e Agrotecnologia., 43*, e012519. doi: https://doi.org/10.1590/1413-7054201943012519

Suzuki, H., Yamashiro, D., Ogawa, S., Kobayashi, M., Cho, D., Iizuka, A., . . . Fujiwara, Y. (2020). Intake of seven essential amino acids improves cognitive function and psychological and social function in middle-aged and older adults: A double-blind, randomized, placebo-controlled trial. *Frontiers in Nutrition*, *7*, 240.

Tako, E., Reed, S., Anandaraman, A., Beebe, S. E., Hart, J. J., & Glahn, R. P. (2015). Studies of cream seeded carioca beans (*Phaseolus vulgaris* L.) from a rwandan efficacy trial: *in vitro* and in vivo screening tools reflect human studies and predict beneficial results from iron biofortified beans. *PLoS One*, *10*(9), e0138479. https://doi.org/10.1371/journal.pone.0138479

Tanaka, M., Kiyohara, H., Yoshino, A., Nakano, A., Takata, F., Dohgu, S., . . . Matsui, T. (2020). Brain-transportable soy dipeptide, Tyr-Pro, attenuates amyloid β peptide25-35-induced memory impairment in mice. *npj Science of Food*, *4*(1), 7. https://doi.org/10.1038/s41538-020-0067-3

Tucker, L. A. (2020). Bean consumption accounts for differences in body fat and waist circumference: A cross-sectional study of 246 women. *Journal of Nutritional Metabolism*, *2020*, 9140907. https://doi.org/10.1155/2020/9140907

Vergara-Castañeda, H. A., Guevara-González, R. G., Ramos-Gómez, M., Reynoso-Camacho, R., Guzmán-Maldonado, H., Feregrino-Pérez, A. A., . . . Loarca-Piña, G. (2010). Non-digestible fraction of cooked bean (Phaseolus vulgaris L.) cultivar Bayo Madero suppresses colonic aberrant crypt foci in azoxymethane-induced rats. *Food Function*, *1*(3), 294–300. https://doi.org/10.1039/c0fo00130a

Villavicencio, A. L. C. H., Mancini-Filho, J., Delincée, H., & Bognár, A. (2000). Effect of gamma irradiation on the thiamine, riboflavin and vitamin B6 content in two varieties of Brazilian beans. *Radiation Physics and Chemistry*, *57*(3), 299–303. https://doi.org/10.1016/S0969-806X(99)00395-3

Volk, C., Brandsch, C., Schlegelmilch, U., Wensch-Dorendorf, M., Hirche, F., Simm, A., . . . Stangl, G. I. (2020). Postprandial metabolic response to rapeseed protein in healthy subjects. *Nutrients*, *12*(8). https://doi.org/10.3390/nu12082270

Wang, S., Guo, C., Xing, Z., Li, M., Yang, H., Zhang, Y., . . . Mi, S. (2021). Dietary intervention with α-amylase inhibitor in white kidney beans added yogurt modulated gut microbiota to adjust blood glucose in mice. *Frontiers in Nutrition*, *8*, 664976. https://doi.org/10.3389/fnut.2021.664976

WHO. (2018). World Health Organization. Noncommunicable diseases. Retrieved from https://www.who.int/news-room/fact-sheets/detail/noncommunicable-diseases

Wong, C., Harris, P. J., & Ferguson, L. R. (2016). Potential benefits of dietary Ffbre intervention in inflammatory bowel disease. *International Journal of MolecularScience*, *17*(6). https://doi.org/10.3390/ijms17060919

Yang, Q.-Q., Gan, R.-Y., Ge, Y.-Y., Zhang, D., & Corke, H. (2018). Polyphenols in common beans (*Phaseolus vulgaris* L.): Chemistry, analysis, and factors affecting composition. *Comprehensive Reviews in Food Science and Food Safety*, *17*(6), 1518–1539. https://doi.org/10.1111/1541-4337.12391

Yeung, C. K., Glahn, R. E., Welch, R. M., & Miller, D. D. (2005). Prebiotics and iron bioavailability-Is there a connection? *Journal of Food Science*, *70*(5), R88–R92. https://doi.org/10.1111/j.1365-2621.2005.tb09984.x

6

Molecular Effects of Bioactive Compounds from Semi-Desert Plants and Their Uses as Potential Ingredient in Food Products

Lissethe Palomo-Ligas, Sendar Daniel Nery Flores, Jesus David García-Ortiz, Claudia Mariana Pérez-Juárez, Adriana Carolina Flores-Gallegos, Raúl Rodríguez-Herrera, and Marisol Cruz-Requena

School of Chemistry, Autonomous University of Coahuila, Saltillo, Coahuila, Mexico

6.1 Introduction

The desert and semi-desert regions in Northern Mexico are important due to the great variety of plants that grow under extreme climate conditions with many chemical compounds that are used as defense mechanisms (Wong Paz, Muñiz Márquez, Martínez Ávila, Belmares Cerda, & Aguilar, 2015). In addition, interesting data on a phylogeographic area suggest the existence of latitudinal gradients in genetic diversity in Northern Mexican plants, but very few studies have examined this fact (Sánchez-Del Pino, Alfaro, & Andueza-Noh, 2020). Some of these plants have been used as traditional remedies and others are a very important economic resource for the people living in these areas, due to the extreme climatic conditions which restrict agricultural activities (Rojas, Tafolla-Arellano, & Martínez-Ávila, 2020). The physicochemical properties and the composition of semi-desert plants (minerals, carotenoids, phenols, flavonoids, vitamins, lignans, triterpenes, tannins, naphthoquinones, among others) are essential for their wide range of biological effects (Abou-Gazar, Bedir, Takamatsu, Ferreira, & Khan, 2004; Pinedo-Espinoza et al., 2020). The utilization of different tissues (flowers, seeds, leaves, green stems, woody stems, and roots), the methodology of extraction, solvent selection, sample treatment, etc., are also important for extraction of phytochemicals and their activities (Hyder, Fredrickson, Estell, Tellez, & Gibbens, 2002).

Recently, only few of these semi-desert plants have been evaluated, with promising effects on human health. Through the analysis and characterization of the components and the search of their mechanisms of action, there is a large field of unexplored plants and compounds for study. The plants discussed in this chapter include *Jatropha dioica*, *Flourensia cernua*, *Turnera diffusa*, *Eucalyptus camaldulensis* (Wong Paz et al., 2015) *Euphorbia antisyphilitica* Zucc (Rojas et al., 2020), *Larrea tridentata* (Hyder et al., 2002),

Molecular Mechanisms of Functional Food, First Edition. Edited by Rocio Campos-Vega and B. Dave Oomah.

Lippia graveolens (González Güereca, Soto Hernández, Kite, & Martínez Vázquez, 2007), *Opuntia, Agave,* and *Prosopis* genus (Harlev, Nevo, Solowey, & Bishayee, 2013; Pinedo-Espinoza et al., 2020; Sanchez et al., 2016). Among the main activities reported are antioxidant effects which are important to module oxidative stress, a phenomenon involved in several human diseases like inflammatory diseases, cardiovascular diseases, cancer, diabetes, neurodegenerative diseases, aging, and skin diseases (Baek & Lee, 2016; Lü et al., 2010). Total extracts, fractions, and isolated compounds from semi-desert plants have demonstrated oxidative capacity *in vitro* and *in vivo* through their free radical scavenging activity, modulation of antioxidant enzymes, attenuation of radical product, among others (Barriada-Bernal et al., 2013; Bassino, Gasparri, & Munaron, 2019; González Güereca et al., 2007; Matias et al., 2014). However, components such as carotenoids, polyphenols, vitamins, and flavonoids contribute to its wide spectrum of activities. These heterogeneous class of molecules in the diversity of the vegetal sources are related with the numerous activities: anticancer, hepatoprotective, vasodilator, antibacterial, antifungal, antiparasitic effect, reduction in glucose, triglycerides, cholesterol, to mention a few (Ahumada-Santos et al., 2013; Anguiano-Sevilla, Lugo-Cervantes, Ordaz-Pichardo, Rosas-Trigueros, & Jaramillo-Flores, 2018; Avelino-Flores, Cruz-Lopez, Jimenez-Montejo, & Reyes-Leyva, 2015; Budinsky et al., 2001; Cárdenas et al., 2019; Gutiérrez-Tlahque et al., 2019; Kim et al., 2017; Luna-Vazquez et al., 2018; Nuñez-López, Paredes-López, & Reynoso-Camacho, 2013; Quintanilla-Licea, Vargas-Villarreal, Verde-Star, Rivas-Galindo, & Torres-Hernandez, 2020; Rufino-Gonzalez et al., 2012). Besides, the multiple beneficial effects on health provide interesting information about their potential use as ingredients in functional food and other promising applications. Finally, semi-desert plants are plant sources with wide potential for research.

6.2 Traditional Uses of Semi-Desert Plants

6.2.1 Some Characteristics of the Semi-Desert and its Plants

A region is considered desert if the annual rainfall is less than 250 mm, while in the semi-desert areas, the annual rainfall is between 250 and 350 mm, and these areas are generally located near the Tropics of Cancer and Capricorn (Pimienta-Barrios, Méndez-Morán, Ramírez-Hernández, García de Alba-García, & Domínguez-Arias, 2008). In the semi-desert, attention is drawn to vegetation scarcity, high temperatures during the day, low temperatures at night, long drought periods, low relative humidity, high osmotic pressures, and the different morphological forms that the plants have acquired to adapt and survive under extreme environmental conditions in these areas (Lira-Saldívar,2003). However, these conditions have guided the evolution of plants to display greater phytochemicals diversity for their defense, deeper and more dispersed roots for greater water absorption, morphological leaf modifications, to reduce water evaporation, stem modification for water storage, some with different photosynthesis types, known as plants with crassulacean acid metabolism (CAM) (Torres-Ponce, Morales-Corral, Ballinas-Casarrubias, & Nevarez-Morillon, 2015). In addition, many semi-desert plants have been adapting to adverse conditions such as soil type (calcaceous, saline, etc.), water scarcity and low fertility (Flores-Hernández et al., 2011).

6.2.2 Traditional Uses

6.2.2.1 Food

Semi-desert plants have been used for thousands of years as food, nonalcoholic and alcoholic beverages, feed, fuel source, construction, medicine, a source of textile fibers, wax, and for making ornaments, among others. Among the semi-desert plants used as food are the nopal (*Opuntia* spp.), of which the tender leaves called cladodes are used, after thorn removal. Nopal can be prepared in pickles, salads, fried, roasted, smoothies, weathered, stewed, etc., and its fruits called tunas (prickly pear) are consumed fresh, used for cheese production, and used in jams, smoothies, liquor etc., including its pests such as cochineal are used to obtain pigments (Torres-Ponceet al., 2015). Xoconostle (*Opuntia joconostle*) has a sour fruit that after the removal of its peels, is used in sauces, juices, broths, and smoothies (Pimienta-Barrios et al., 2008). The creosote bush (*Larrea tridentata*) fruits are edible (Lira-Saldívar, 2003). The tender flowers of maguey (*Agave* spp.) are used as food, and the cooked inflorescence (quiote) is chewed and the juice is consumed (VázquezGarcía et al., 2016). The tender flowers of the palm (*Yucca carnerosana*), produced in March and April, are cooked and fried with different vegetables, and they can also be used to make patties (Flores-Hernández et al., 2011). Mezquite (*Prosopis* spp.) pods once ripe are chewed and the juice is consumed, once dry the pods can be ground to obtain a flour for gluten-free bakery products for celiac, and its gum can be used as a gelling agent or natural thickener, with properties similar to Arabic gum (Rodríguez-Sauceda et al., 2014). The immature flowers (known as cabuches) of the biznaga (*Ferocaptus pilosus*) are consumed in salads, pickled, cooked, or fried (Alanís-Flores & Velasco-Macias, 2008). While, to flavor different soups or broths, dried oregano leaves (*Lippia* spp.) (Arcila-Lozano, Loarca-Piña, Lecona-Uribe, & González de Mejía, 2004) and creosote bush flowers are used (Lira-Saldívar, 2003).

6.2.2.2 Nonalcoholic Beverages

The fruits of nopal and xoconostle can be used to prepare juices or smoothies (Pimienta-Barrios et al., 2008; Torres-Ponce et al., 2015), although the unfermented maguey sap (known as aguamiel) is the best-known nonalcoholic drink obtained from semi-desert plants. To obtain aguamiel, the central part of the maguey is removed when it is between 8 and 10 years old, and a cavity is made in the center part of the plant and the sap emanating from this cavity is collected, which can be consumed, generally after cooking, as a form of sterilization (VázquezGarcíaet al., 2016).

6.2.2.3 Alcoholic Beverages

Different plants of the semi-desert are used to obtain alcoholic beverages, although the best known is tequila produced using *Agave tequilana*. Other maguey species are used to obtain liquors similar to tequila; however, by using other species of *Agave* different to *tequilana*, they cannot be called tequila and they acquire names like mezcal, bacanora, raicilla, and tuxca, among other names. For both tequila and mezcal, there is a designation of origin (VázquezGarcíaet al., 2016). Other alcoholic beverages that are obtained from semi-desert plants are the liqueur of prickly pear cactus and the liqueur obtained from sotol (*Dasylirion* spp.), which is known with the same name (sotol) (Flores-Gallegoset al., 2019). One alcoholic beverage with great popularity and tradition is pulque, which is obtained after fermenting the sap (aguamiel) of the maguey (VázquezGarcíaet al., 2016).

6.2.2.4 Forage

The maguey stalks are cut into small pieces and used to feed cattle and goats (VázquezGarcíaet al., 2016), and the sotol stems have also been used to feed cattle, especially in times of forage scarcity (Flores-Gallegoset al., 2019). The mature nopal stalks are used to feed cattle, for this, first, the spines of the stalks are burned (Torres-Ponce et al., 2015). Leaves of plants such as mesquite are also used for animal feed (Rodríguez-Saucedaet al., 2014).

6.2.2.5 Source of Fuel

For food preparation or heat source, people from semi-desert areas have used mesquite wood (Rodríguez-Saucedaet al., 2014), branches of creosote bush (Lira-Saldívar, 2003), stems of different palms (Flores-Hernández et al., 2011), and dried maguey stems (VázquezGarcíaet al., 2016).

6.2.2.6 Building

Maguey, palms, and nopal are used as living fences for the delimitation of areas (Flores-Gallegoset al., 2019; Torres-Ponceet al., 2015; VázquezGarcíaet al., 2016), the dried inflorescences of sotol, lechugilla (*Agave lechuguilla*) and magueys, and the stems of palms are used for fences and roofs (Flores-Gallegos et al., 2019; Flores-Hernández et al., 2011; VázquezGarcíaet al., 2016). Mesquite wood is classified as semi-precious and is used in flooring and furniture (Rodríguez-Saucedaet al., 2014). Meanwhile, the creosote brush branches are used on roofs (Lira-Saldívar, 2003).

6.2.2.7 Medicine

The creosote bush is used to treat bad odors and foot infections, stomach pain, gynecological problems such as female sterility, inflammation, kidney stones, pain due to arthritis, gout and chickenpox, itchy skin rashes as well as to heal blisters, bruises, and wounds naturally, and the leaves are used for stomach pain and infections, gastrointestinal problems, mainly constipation, because it has strong laxative and purgative properties (Lira-Saldívar, 2003). Hojasen (*Flourensia cernua*) has been used for digestion problems, flatulence, and parasites (Wong-Paz et al., 2015). In addition, it is employed as purgative and antimicrobial (Martínez, 1989). Nopal is used for the treatment of diarrhea and gastric ulcers, has regenerative and moisturizing properties, reduces body temperature, cholesterol, and inflammation, improves digestion, regulates blood sugar and body weight, and strengthens bones (Torres-Ponce et al., 2015).

6.2.2.8 Textile Fibers

Plants such as maguey (Flores-Hernández et al., 2011), palms (VázquezGarcíaet al., 2016), and lechuguilla (*Agave lechugilla*) have been used for a long time to obtain natural fibers, which are used to make brushes, hammocks, cribs, chairs, rugs, toys, scouring pads, ropes, clothing, and sacks (Carmona, Morales-Martínez, Mussatto, Castillo-Quiroz, & Ríos-González, 2017).

6.2.2.9 Wax and Ornaments

One of the major sources of wax is candelilla (*Euphorbia antisyphilitica*), and it is used to make wax candles (Cabello-Alvaradoet al., 2013). The basal part of the sotol leaves, called spoons, are used to make religious ornaments or for civil ceremonies (weddings, birthdays, etc.) (Flores-Gallegoset al., 2019).

6.2.3 Industrial Uses

The main industrial use of semi-desert plants (maguey and sotol) is the production of alcoholic beverages (tequila, mezcal, pulque, and sotol) (Flores-Gallegoset al., 2019; VázquezGarcíaet al., 2016) and as natural fibers (lechuguilla and palm) (Carmona et al., 2017). Creosote bush is used to dye leather, manufacture paints, plastics, and insecticides, wax (shoes), oils, lubricants, and varnishes as descaler of saline materials in boilers (Lira-Saldívar, 2003). Candelilla wax is used as a plasticizer, which has many applications in the production of lipsticks, hair preparations, and body creams. In addition, wax can be used to coat paper and cardboard, to make adhesives, crayons, inks, paints, lubricants, anticorrosives, and drugs (Cabello-Alvaradoet al., 2013). While, palms have been used to obtain contraceptive molecules (Flores-Hernández et al., 2011) and maguey is used to obtain inulin, a prebiotic, and production of fructan-rich syrup (Vázquez-Garcíaet al., 2015).

6.3 Semi-Desert Plants and Bioactive Compounds

6.3.1 Chihuahua Desert

The Chihuahuan desert has a great diversity of plants possessing various bioactive compounds such as antioxidants. The extension of the Chihuahuan desert is around 70 million hectares and includes the states of Coahuila, Chihuahua, Durango, Zacatecas, large portions of San Luis Potosi, Nuevo Leon, and significant areas of Texas and New Mexico in the United States (Pérez-Solano, Gallina-Tessaro, & Sánchez-Rojas, 2016). Some of the characteristics of this desert are the hot summers, and cool and dry winters, with intermittent rainfall mostly of monsoonal origins during the summer months. This extreme climate has generated over time a great variety of plant species throughout this territory (Briggs, Lozano-Cavazos, Mills Poulos, Ochoa-Espinoza, & Rodríguez-Pineda, 2020).

6.3.2 Semi-Desert Plants

The characteristic vegetation of the Chihuahuan semi-desert is made up of desert scrub and grassland, with areas of woodland and chaparral in the mountains and narrow ribbons of riparian forest and scrub along stream channels and springs (Granados-Sánchez, Sánchez-González, Granados Victorino, & Borja de la Rosa, 2011). There is an extensive number of plants; however, some of them stand out, such is the case of creosote bush (*Larrea tridentata*) that dominates in the driest sites and frequently found close to another plant species such as lechuguilla (*Agave lechuguilla*) and tarbush (*Flourensia cernua*). Likewise, this diversity comprises highly drought-tolerant plants such as yuccas (*Yucca* spp.) and sotol (*Dasylirion wheeleri*). On the other hand, grasslands integrate grass and shrub mosaics, with species such as bush muhly (*Muhlenbergia porteri*), purple three-awn (*Aristida purpurea*), and bluegrass (*Bouteloua gracilis*) (Zavala-Hurtado & Jiménez, 2020). Some other characteristic plants of this desert region are guayule (*Parthenium argentatum*), sweet mesquite (*Prosopis glandulosa*), mariola (*Parthenium incanum*), ocotillo (*Fouquieria spendens*), sotol (*Dasylirion sp.*), fishhook barrel cactus (*Ferocactus wislizenii*), and peyote (*Lophophora williamsii*) (Montaño-Arias, García-Sánchez, Ochoa-de la Rosa, & Monroy-Ata, 2006).

6.3.3 Bioactive Compounds

The first step to define a bioactive compound is to determine its etymological Greek root: bios' means life and Latin "activus" means dynamic or full of energy. Taking this as a basis, a bioactive compound is a molecule or substance that has a biological activity on living organism, positive or negative, depending on the nature of the substance, its dose, and its bioavailability (Walia, Gupta, & Sharma, 2019). These compounds are amply distributed in nature, principally plants, and vary in chemical structure and function (Altemimi, Lakhssassi, Baharlouei, Watson, & Lightfoot, 2017).

Bioactive compounds are present in plants, fruits, flowers, and by-product from arid zones such as Chihuahuan desert. Among these compounds can be founded antioxidants as polyphenols, carotenoids, anthocyanin, vitamins, terpenes, flavonoids, tannins, saponins, ellagic acid, tannic acid, gallic acid, catechin, quercetin, methyl gallate, cinnamic acid, 2-hydroxycinamic acid, and resorcinol, and others (Ruíz-Martínez, Ascacio, Rodríguez-Herrera, Morales, & Aguilar, 2011). There is an extensive list of plants and molecules and their repercussions on health, for example prickly pears, the fruit from nopal (*Opuntia* spp.), rich in carotenoids or another substance with greatest antioxidant properties as phenolic acids and flavonoids, which can be founded in their seeds (Alqurashi, Al-Mssallem, & Al-Khayri, 2020). Creosote bush (*Larrea tridentata*), Dragon's blood (*Jatropha dioica*), far bush (*Flourensia cernua*), Eucalyptus (*Eucalyptus camaldulensis*), and Damiana (*Turnera diffusa*) are par excellence a representative of the semi-desert plants most widely studied and with various active compounds such as flavonoids (phytoestrogens), polyphenols, phenolic compounds, among other (Martins, Aguilar, Teixeira, & Mussatto, 2012; Wong-Paz et al., 2015).

6.4 Antioxidants from Semi-Desert Plants and Health Effects

Oxidative stress is defined as an imbalance between oxidants and antioxidants in favor of the oxidants, leading to a disruption of redox signaling, and control and/or molecular damage (Sies, 2015). Reactive oxygen species (ROS) are highly reactive molecules, which are produced normally in organisms as a result of normal cellular metabolism and environmental factors. However, ROS can damage lipids, proteins, and nucleic acids, altering their functions (Gupta et al., 2014). Therefore, oxidative stress contributes to the pathogenesis of inflammatory diseases, cardiovascular diseases, cancer, diabetes, neurodegenerative diseases, and aging (Lü et al., 2010). The human body has the ability to counteract these oxidizing agents through the action of antioxidants, which can be produced in the body (endogenous) or obtained through food or supplements (exogenous). Therefore, an antioxidant is a molecule capable of slowing or preventing the oxidation of a substrate (Pisoschi & Pop, 2015). Exogenous antioxidants have been shown to participate in ROS detoxification and play an important role in helping endogenous antioxidants to neutralize oxidative stress. Epidemiologic studies have indicated that the dietary intake of antioxidant substances is inversely associated with mortality from diseases related to oxidative stress (Salazar et al., 2008).

Mexico has around 55,000 species of plants. The desert Northern Mexico region is important due to diverse plant that can grow under extreme climate conditions with various chemical compounds that are used as defense mechanisms, and these secondary metabolites have antioxidants properties (Wong Paz et al., 2015). Some of the semi-desert plants that have potential antioxidant activity are *Larrea tridentata*, *Flourensia cernua*, *Jatropha dioica*, *Turnera diffusa*, *Agave spp.*, *Opuntia ficus*, *Opuntia joconostle*, *Euphorbia antisyphilitica*, *Lippia graveolens*, and *Prosopis spp* among others.

Larrea tridentata (creosote bush) extracts from its stems and leaves are useful to treat tuberculosis, menstrual pains, diabetes, and various types of cancers. Various studies indicate that the main constituents of its extracts are lignans, flavonoids, triterpenes, tannins, and naphthoquinones (Abou-Gazar et al., 2004; Ortega-Ramirez et al., 2014). *L. tridentata* is a notable source of natural products with approximately 50% of the leaves dry weight as extractable matter. The resin that covers the leaves is a rich source of lignans, notably, the nordihydroguaiaretic acid (NDGA), which corresponds to 80% of the polyphenols in the resin (Arteaga, Andrade-Cetto, & Cárdenas, 2005). NDGA is a phenolic lignan that has four phenolic hydroxyls; the hydrogen atoms of these hydroxyl groups have the ability to react with ROS. Thus, the extracts of *L. tridentata* are used as powerful antioxidants (Lü et al., 2010). The antioxidant effect of NDGA is due to several mechanisms of action: (a) ROS scavenger; (b) inhibitory effect on lipoxygenases; and (c) the activation of the endogenous antioxidant response by means of the transcription factor Nrf2 (Manda, Rojo, Martínez-Klimova, Pedraza-Chaverri, & Cuadrado, 2020). Floriano-Sánchez et al. (2009) reported that NDGA is a potent *in vitro* scavenger of peroxynitrite ($ONOO^-$), singlet oxygen ($^1O_2$), hydroxyl radical (•OH), superoxide anion ($O_2^-\bullet$), hydrogen peroxide (H_2O_2) and hypochlorous acid (HOCl). Moreover, NDGA at 20 mg $kg^{-1}day^{-1}$ for 7 days prevented ozone (0.25 ppm, 4 $hday^{-1}$ by 7 days) induced tyrosine nitration in rat lungs. Morán-Santibañez et al. (2019) found that extracts from *L. tridentata* leaves using ethanol:water (60:40% v/v) at 15 μgmL^{-1} decrease H_2O_2 oxidative stress and cytotoxicity in human neuroblastoma SH-SY5Y cells. The potential mechanism implicated in improving cell viability is presumably attenuation of H_2O_2-triggered ROS accumulation, mitochondrial depolarization, phosphatidylserine externalization, and caspase-3/7 activation which are hallmarks of apoptosis. Other studies have shown that NDGA from *L. tridentata* mitigates 12-*O*-tetradecanoylphorbol-13-acetate (TPA)-induced inflammatory and oxidative stress responses in mouse skin. Pretreatment and treatment with NDGA (15 and 25 µmol) reduced glutathione level and activities of antioxidant enzymes and also lowered activities of myeloperoxidase, xanthine oxidase, and skin edema formation (Rahman, Ansari, Rehman,

Parvez, & Raisuddin, 2011). However, despite its extensive benefits, excessive consumption of NDGA can damage various organs such as kidneys and liver (Manda et al., 2020).

Flourensia cernua has been used as an infusion to treat various gastrointestinal diseases. In addition, *F. cernua* exhibits antioxidant activity. Alvárez-Pérez et al. (2020) reported that purified phenolic compounds exhibit a radical inhibition of 85% for (2,2-azino-bis (3-ethyl benzothiazoline-6-sulfonic acid) ABTS, 58% for 1,1-dipheny-2-picrylhydrazyl (DDPH), and 58% for lipid oxidation inhibition (LOI). Another study that used *F. cernua* stem and leaves extracts showed a radical inhibition of 62% for DPPH radical scavenging and 23% for LOI (Wong Paz et al., 2015). However, more studies on its antioxidant activity are necessary.

The infusions of *Jatropha dioica* branches and roots have been traditionally used against hair loss, tooth decay, and cancer (Ramírez-Moreno et al., 2020). Martínez et al. (2014) reported the presence of terpenoid compounds, polyphenols, reducing sugars and alkaloids as well as its cytoprotective effect in the root. The administration of *J. dioica* decoction at 3.7, 10.7, and 21.4 $mLkg^{-1}$ effectively prevents genotoxic damage in liver, kidney, and bone marrow of mice and against three mutagens (cyclophosphamide, daunorubicin, and metyl metanesulfonate) with different mechanisms of action. The chemoprotective effect was greater over those that involve an increase in oxidative stress as part of its toxic effect; thus, the antioxidant effect of the extract is an alternative to reduce oxidative damage in different pathologies. Another study showed that the highest concentration of polyphenols were found in the root of the plant and 89% of the antioxidant activity of *J. dioica* extracts root are due to the polyphenols (Ramírez Moreno et al., 2016). Gutiérrez-Tlahque et al.(2019) found that the root extracts contained alkyl esters of acetic and ferulic acids, as well as the known diterpenes citlalitrione, jatrophatrione, jatropholone A, and jatropholone B. These metabolites may be responsible for the extract antioxidant activities. Furthermore, Gutiérrez-Tlahque et al. (2018) showed that ethanol (70% v/v) can extract more of these antioxidant compounds from the plant compared to water.

Turnera diffusa is a medicinal plant traditionally used as stimulant, diuretic, and aphrodisiac; it apparently possesses large amount of essential oils and antioxidants, particularly flavonoids. Urbizu-González, Castillo-Ruiz, Martínez-Ávila, and Torres-Castillo (2017) reported that the essential oils in the stems and leaves of *T. diffusa* are eucalyptol, 1,4-methanocycloocta [d] pyridazine, ethanone, 1-(1,3-dimethyl-3-cyclohexen-1-yl), among 21 other compounds. These essential oils inhibit ABTS radical (72%) in addition to exerting antioxidant activity against FRAP equivalent to 21 mg of gallic acid (GAE)/g of crude matter. Bezerra, Negri, Duarte-Almeida, Smaili, and Carlini (2016) showed that flavonoids were the main antioxidant component of a hydroalcoholic extract from the leaves and stem of *T. diffusa*. In addition, the toxic dose for astrocytes (support cells in the nervous system) was 1 $mgmL^{-1}$ of the extract. Likewise, Esquivel-Gutiérrez, Alcaraz-Meléndez, Hernández-Herrera, Torres, and Rodríguez-Jaramillo (2018) demonstrated the antioxidant effect of a hydroalcoholic extract of *T. diffusa* in a model of streptozotocin-induced diabetes mellitus. The extract decreased levels of malondialdehyde and nitric oxide in the kidney of diabetic mice. Therefore, the extract of *T. diffusa* can potentially be used to prevent oxidative damage observed in type 2 diabetes mellitus.

Agave spp. has been used as a source of food and beverages; it also contains phytochemicals with antioxidant activity. Ahumada-Santos et al. (2013) evaluated the antioxidant activity of six different *Agave* species. The hexane extract of *Agave rzedowskiana* showed the highest antioxidant effect associated with nonpolar compounds, specifically α-tocopherol.

Other antioxidant compounds described are phenolic compounds that include flavonols such as kaempferol and quercetin in various glycosylated forms, which were obtained from the edible flowers of *Agave durangensis* (Barriada-Bernal et al., 2013). Likewise, Puente-Garza, Meza-Miranda, Ochoa-Martínez, and García-Lara (2017) described that the antioxidant activity in *Agave* extracts is not related to the total content of polyphenols, and this is due to the presence of non-phenolic antioxidants such as saponins, and the content of saponins is directly related to its antioxidant activity, where the main saponins in *Agave salmiana* were tigogenin and chlorogenin. This highlights the use of *Agave* extracts not only for nutritional purposes but also as an important source of antioxidant compounds.

Opuntia (Nopal) is considered a functional food because it is a proven source of dietary fiber and bioactive compounds with antioxidant activity. It is rich in antioxidant compounds such as vitamin C, polyphenols, flavonoids, betaxanthin, betacyanin, and carotenoids. Santos-Zea, Gutierrez-Uribe, and Serna-Saldivar (2011) reported the antioxidant profile of cladode flours from different varieties of *Opuntia spp. Opuntia robusta* contained the highest antioxidant capacity (739 μmol Trolox equivalents (TE)/g), and the most abundant flavonol was isorhamnetin, which was found in six glycosidic forms. Another *in vitro* study (Matias et al., 2014) described the antioxidant and anti-inflammatory activity of flavonoid-rich extract from *Opuntia ficus* juice on human colon carcinoma cells. The main constituents identified in the juice are isohamnetins and phenolic acids such as ferulic, piscidic, and eucomic acids. These compounds inhibited the formation of H_2O_2-induced radicals and reduced protein oxidation. Simultaneously, the juice prevented the activation of NF-κB, a transcription factor involved in inflammatory response; therefore, the expression of inflammatory markers such as iNOS and TNF-α is reduced. Likewise, the antioxidant activity of nopal consumption has been evaluated in healthy volunteers and patients with type 2 diabetes mellitus (Avila-Nava et al., 2014; López-Romero et al., 2014). In both cases, plasma antioxidant activity increased, in addition to preventing postprandial blood glucose peaks. So, the consumption of nopal is not a complete replacement for prescription, but it can be an adjuvant in the treatment of diabetes. Another plant of the *Opuntia* genus with high levels of antioxidants is *Opuntia joconostle*. The pulp and seeds are a source of fiber, phenolics, flavonoids, and tocopherols which provide a good antioxidant capacity (Morales, Ramírez-Moreno, Sanchez-Mata, Carvalho, & Ferreira, 2012). Xoconostle peel showed higher antioxidant activity compared to that of skin (twofold) and pulp (sixfold). Polyphenolic compounds are mainly located in the peel. Seven phenolics were identified as protocatechuic, 4-hydroxybenzoic, caffeic, vanillic and syringic acids, rutin, and quercetin (Guzmán-Maldonado et al., 2010). Other compounds with antioxidant potential like betacyanins (betanin, isobetanin, betanidin, isobetanidin, and phyllocactin) are mainly concentrated in the endocarp (Osorio-Esquivel, Alicia Ortiz, Álvarez, Dorantes-Álvarez, & Giusti, 2011). Therefore, the consumption of the pulp as well as the peel of the fruit of *O. joconostle* can be a viable alternative to improve human health.

Euphorbia antisyphilitica (Candelilla) has proven to be a good source of useful phytochemicals such as fiber, wax, and polyphenolic compounds (catechin and ellagic acid). Burboa, Ascacio-Valdés, Zugasti-Cruz, Rodríguez-Herrera, and Aguilar (2014) reported the free radical scavenging activity of Candelilla by-products (stalks without wax). The 50% ethanol extracts showed an inhibition of more than 88% of the DPPH radicals. Stem residues were an important source of ellagic acid (20 mg per g). Ellagic acid is used as a potent antioxidant; therefore, Candelilla waste can be a viable alternative for ellagic acid extraction and its application in the food, pharmaceutical, and cosmetic industries (Ascacio-Valdés, Aguilera-Carbó, Martínez-Hernández, Rodríguez-Herrera, & Aguilar, 2010). Other antioxidant compounds reported in *E. antisyphilitica* are tannins, gallic acid, and catechins (Rojas et al., 2020). However, despite the high content of antioxidant compounds in *E. antisyphilitica*, studies on their biological activities are lacking.

Lippia graveolens know as Mexican oregano is an important source of essential oil as well as antioxidant compounds and traditionally is used as an antiseptic, antipyretic, analgesic, and anti-inflammatory agent. The essential oil contains hydrocarbonated and oxygenated monoterpenes, carvacrol, thymol, and *p*-cymene. Mexican oregano has higher essential oil and carvacrol contents than European oregano. Also the plant is rich in polyphenolic compounds like naringenin, pinocembrin, lapachenole, kaempferol, and rosmarinic acid which are powerful antioxidants (Martínez-Rocha, Puga, Hernández-Sandoval, Loarca-Piña, & Mendoza, 2008). In addition, the antioxidant activity of extracts from the stem and leaves of *L. graveolens* has been reported. González Güereca et al. (2007) compared the antioxidant activity of extracts of *L. graveolens* with different solvents, such as water, methanol, ethyl acetate, and hexane. The antioxidant activity in the extracts was evaluated by the DPPH radical method. The percentage of radical inhibition was 87.6% for the aqueous extract, 91.5% with methanol, 94% with ethyl acetate, and 3.4% with hexane. Also, Soto-Domínguez, García-Garza, Ramírez-Casas, Morán-Martínez, and Serrano-Gallardo (2012) reported that the aqueous extract presented antioxidant activity at 160 $mg mL^{-1}$, in addition, the liver and kidney toxicity of the aqueous extract was evaluated in mice. Doses of 10, 100, 1000, and 5000 $mg kg^{-1}$ of body weight were used orally, which did not show toxicity, so the use of the aqueous extract of oregano can be an alternative for the treatment of pathologies that exhibit oxidative stress.

Prosopis spp. (mesquite) is a plant traditionally used by humans as food in certain regions of Mexico. In addition, the plant has high levels of phytochemicals such as alkaloids, tannins, phenols, flavonoids, terpenes, and steroids (Henciya et al., 2017). Ramírez-Rojo et al. (2019) determined the total content of phenols and flavonoids of *Prosopis velutina* extracts using three different extraction solvents: water, ethanol:water (1 : 1), and ethanol, as well as their antioxidant activity against the DPPH radical. The ethanolic extract of the mesquite leaves showed the highest number of phenolic compounds and flavonoids as well as a higher antioxidant activity compared to the other extracts. In another study, García-Andrade et al. (2013) demonstrated the antioxidant activity and the potential cardioprotective activity of a *Prosopis leavigata* acetone extract. The extract displayed phenolic compounds such as gallic acid, catechin, gallocatechin, epicatechin, rutin and luteonin, and DPPH scavenging capacity at 2986 $\mu g mL^{-1}$. However, the antioxidant capacity was increased by fractionating the total extract, obtaining values of 312–93 $\mu g mL^{-1}$. In addition, the extract showed antihypertensive effects inhibiting angiotensin-converting enzyme. Thus, mesquite leaves can be a source of bioactive phenolic compounds.

6.5 Carotenoids from Semi-Desert Plants and Health Effects

Carotenoids are known as natural fat-soluble terpenoid pigments, mostly attached to a protein, and that colors in yellow, red, and orange shades are mainly due to its physicochemical characteristics. These pigments are naturally synthesized by plants, algae and even, photosynthetic bacteria or fungi, and its main function is to capture the light energy for its use during photosynthesis. Therefore, they can be found in photosynthetic tissue, along with chlorophylls (Alós, Rodrigo, & Zacarias, 2016; Llansola-Portoles, Pascal, & Robert, 2017; Meléndez-Martínez, Vicario, & Heredia, 2004; Quintana-López, Hurtado-Oliva, Hernández, & Palacios-Mechetnov, 2018; Rao & Rao, 2007; Solovchenko & Neverov, 2017). Besides, these carotenoids act as photoprotective molecule during the first steps of photosynthesis preventing the photobleaching of the organism (Llansola-Portoles et al., 2017).

Carotenoids are always found in a 1:3 relation alongside with chlorophyll. Mature fruit or vegetable contains high concentration of the carotenoids (Meléndez-Martínez et al., 2004). Carotenoids are generally divided into two groups: carotenes and xanthophylls, based on their molecular composition (Quintana-López et al., 2018) (Table 6.1).

Carotenoids cannot be produced by animals and can be produced only by plants and some microorganisms. Therefore, animals must consume fruits and vegetables that are rich in carotenoids (like carrots, some edible algae, oranges, tomatoes, etc.) to maintain certain metabolic activities. Besides, because carotenoids are the main source of provitamin A, which is transformed by the organism into vitamin A, making its consumption vital for animals to fulfill their nutritional requirements (Llansola-Portoles et al., 2017; Meléndez-Martínez et al., 2004; Rao & Rao, 2007).

The semi-desert is an ecosystem with lack of rain and high evaporation rates through the year. Under these conditions, the structure and characteristics of the semi-desert plants evolved to survive with small amounts of water and to retain it. This ecosystem occupies 41% of land surface on the planet (Barbosa-Briones, Cardona-Benavides, Reyes-Hernández, & Muñoz-Robles, 2019; Delgado et al., 2014; Grubb, Marañón, Pugnaire, & Sack, 2015; INEGI. 2005; Seidl et al., 2020). Another characteristic developed in the evolution of the typical vegetation of the semi-desert for their survival is the production of many

Table 6.1 Differences between carotenes and xanthophylls including examples and sources.

	Carotenes	**Xanthophylls**
Definition	Carotenoid molecule contains carbon and hydrogen	Carotenoid molecule containing carbon and hydrogen as well as, at least, one oxygen atom.
Examples	• β-Carotene • α-Carotene • Lycopene • Cryptoxanthin	• Lutein • Zeaxanthin • Capsanthin • Astaxanthin
Semi-desert plant sources	• Fresh nopal (245 $\mu g g^{-1}$) • Chiltepin pepper mature (6 mg βCE g − 1 DM)	• Fresh nopal (0.38 $mg g^{-1}$)
Other sources	Algae and in some ground growth vegetables like carrots, oranges, and cantaloupes	Algae, bacteria, and superior plants like peppers and cranberries.

Source: Betancourt-Domínguez, Hernández-Pérez, García-Saucedo, Cruz-Hernández, and Paredes-López (2006), Rao and Rao (2007), Llansola-Portoles et al. (2017), Quintana-López et al. (2018), González-Cruz, Filardo-Kerstupp, Bello-Pérez, Güemes-Vera, and Bernardino-Nicanor (2012), Vázquez-Flores et al. (2020).

phytochemical substances to better exploit nutrients and water available in the soil or survival of extreme environmental conditions. A very good example of those phytochemical substances is the already described carotenoids (Gebhardt, Fehmi, Rasmussen, & Gallery, 2017; Grubb et al., 2015; Hirschberg, 2001; Solovchenko & Neverov, 2017).

Carotenoids are usually formed as a photo stabilizer in the plant, and they are synthesized within the plastids by nuclear-encoded enzymes. Carotenoids and their derivative are formed to participate in biological function of the plant such as phytohormones biosynthesis and protection of the plant's structure against stress factors such as soil salinity, extreme temperatures, and intense sunlight (Alvarado, Cardador-Martinez, Garcia-Trejo, & Magallan-Hernandez, 2018; Hirschberg, 2001; Solovchenko & Neverov, 2017; Yang et al., 2019). Carotenoids also protect the plant against oxidative damage; they absorb light and accumulate it in the plant cell and can remain there for long time. Carotenoid's protection acts by quenching the excess excitation provoked by the sunlight in a non-radiative manner, and this is known as non-photochemical quenching. As a plant pigment, carotenoids act along the chloroplast for energy catching and transportation to chlorophylls (Dreuw, Fleming, & Head-Gordon, 2005; Franco, Matsubara, & Orthen, 2007; Hirschberg, 2001; Kirilovsky, 2007; Solovchenko & Neverov, 2017; Vítek, Ascaso, Artieda, Casero, & Wierzchos, 2017). The quenching step is extremely necessary in plants that grow in semi-arid zones with long intense sunlight, since it functions as a mechanism to protect the chlorophyll from bleaching and oxidizing by dispersing the excess photo excitation acting as an antioxidant for the plant (Hirschberg, 2001; Solovchenko & Neverov, 2017; Vítek et al., 2017).

6.5.1 Examples of Semi-Desert Plants Rich in Carotenoids

Opuntia ficus-indica better known as "edible nopal" is a very common edible cactus that grows in semi-arid zones. This plant is native to the American continent, and it is recognized for growing in regions with lack of vegetal production. It is a very nutritious plant with high water and fiber contents, plus high quantity of iron (0.027%) and calcium (0.22%). This cactus is mostly known and eaten in North and Central America. (Griffith, 2004; Guzmán Loayza & Chávez, 2007; Medina, Rodríguez, & Romero, 2007). This cactus grows in segments and forms different limbs and branches with stems capable of growing a fruit at the top. This fruit is called cactus pear that turns red at maturity indicating high carotenoids content. The nopal itself also has a considerable number of carotenoids, highlighting β-carotene content in the pulp (Guzmán Loayza & Chávez, 2007; Ramadan & Mörsel, 2003; Saenz, Tapia, Chavez, & Robert, 2009).

Capsicum annuum L. var. Glabriusculum, better known as the chiltepin pepper, is another species native from America, from the semi-desert zones of the continent (Brito-Vega, Rodríguez-Maturino, González-Mendoza, Grimaldo-Juarez, & Ceceña-Duran, 2014; Campos, Gómez, Ordo, & Ancona, 2013). It is the most common hot pepper in Mexico and used to prepare traditional Mexican dishes and soups (Campos et al., 2013; Rodriguez-Maturino et al., 2015). This pepper has the green and the red stages, signs of different fruit maturity phases, and different substances produced and metabolized by the pepper like antioxidants, polyphenolic compounds, and carotenoids. The presence of the carotenoids is most evident and high at the red stage. Its nutritional value is very important in Mexican food, since it is the second most consumed source of carotenoids in the country

(Campos et al., 2013; Rodriguez-Maturino et al., 2015; Vazquez-Flores et al., 2020). Carotenoids extracted from the chiltepin are not only important and useful in culinary environment but also have been tested in fruits and vegetables storage as an antifungal substance preventing the growth of fungus (mainly fungus but can also prevent pathogen bacteria growth). Some of these carotenoid extracts from the chiltepin strongly inhibit (99%) conidia of *Botrytis cinerea* (Rodriguez-Maturino et al., 2015).

Other important plant in North America with a significant content of carotenoids is the short shrub *Larrea tridentata*. It proliferates and commonly grows in arid zones of Mexico and generally used in Coahuila and Chihuahua States as an herbal remedy for certain sexual-transmitted diseases, tuberculosis, chickenpox, dysmenorrhea, and snake bite, because of its high polyphenolic substances and carotenoids contents (Medina-Guillén, Cantú-Silva, Estrada-Castillón, González-Rodríguez, & Delgadillo-Villalobos, 2017; Park, Lee, Lim, & Kim, 2019; Ríos et al., 2016). Its homemade extract is not recommended for frequent use because it can be cytotoxic for liver and kidney cells (Liu & Guan, 2012; Park et al., 2019).

6.5.2 Properties and Applications

Carotenoid-rich foods are one of the main sources of provitamin A which allows the formation of vitamin A. This vitamin is important in animals' diet, because it mainly helps to maintain the proper functioning of the immunological system and vision; it also ensures good and healthy reproduction, embryonic development, postnatal growth, and tissue and epithelia differentiation (Meléndez-Martínez et al., 2004; Alós et al., 2016; Khalid, Saeed-ur-Rahman, Iqbal, & Huang, 2019).The transformation of provitamin A into vitamin A is very important for vision acuity. It generally promotes and protects good nocturnal vision. The compounds chromophores that form vitamin A enable better light absorption for the eye, allowing better vision and protection of the retinal tissue. (Gammone, Riccioni, & D'Orazio, 2015; Khalid et al., 2019).

These compounds have antioxidant potential because of their nature and structure. Moreover, carotenoids consumption can prevent the formation of cancerous cells and tissue, by limiting their abnormal growth and enhancing the gap junctional communication. Carotenoids are also linked to heart disease prevention by avoiding the formation and oxidation of some low-density lipoproteins and scavenging already formed free radicals (Di Pietro, Di Tomo, & Pandolfi, 2016; Eggersdorfer & Wyss, 2018; Gammone et al., 2015). In many studies, carotenoids potentiate in a positive way the performance of the immune response, because of their association with CD4 and CD8 lymphocytes amplification. This information may be useful in the search for the treatment of diseases in which the immune system is compromised such as cancer, leukemia, or HIV (Eggersdorfer & Wyss, 2018; Meléndez-Martínez et al., 2004).

6.6 Polyphenols from Semi-Desert Plants and Health Effects

Global deaths during 2019 were mainly due to cardiovascular diseases, respiratory diseases, cancer, and diabetes, creating a great concern in recent years to be able to treat those diseases (WHO, 2021). Based on this, the incorporation of new ways to treat these diseases has been sought; an alternative is the use of natural compounds such as polyphenols, which have been investigated during the last 25 years (Tomás-Barberán & Andrés-Lacuevas, 2012). These compounds incorporated in diets can provide large benefits in prevention and treatment of chronic degenerative diseases such as obesity, type 2 diabetes, cardiovascular and neurodegenerative diseases, and cancer (Ávila-Gálvez, González-Sarrías, & Espín, 2018; Cory, Passarelli, Szeto, Tamez, & Mattei, 2018).

Polyphenols are present mainly in plants, as well as in foods derived from fruits and vegetables, and these belong to a group of substances characterized by having one or more aromatic rings with two or more hydroxyl groups (Skrovankova, Sumczynski, Mlcek, Jurikova, & Sochor, 2015). These compounds can be classified in different ways as phenolic acids, stilbenes, flavonoids, and lignans (Fernandes De Araújo, De Paulo Farias, & Neri-Numa, 2021; Rodr & Cristina, 2019; Tanase, Cosarca, & Muntean, 2019; Vuolo, Lima, Roberto, & Junior, 2019).

6.6.1 Phenolic Acids

Phenolic acids are one of the main subclasses of polyphenols found in plants such as esters and glycosides, which have been associated with various functions related to the absorption of nutrients, intervention in protein synthesis and photosynthesis. These are composed of a phenolic ring and at least one carboxylic acid, the most important being hydroxycinnamic and hydroxybenzoic acids (Goleniowski, Bonfil,Cusido, & Palazón, 2013; Tanase et al., 2019).

The chemical compounds from semi-desert plants have been documented and associated with numerous applications in the pharmaceutical area to treat diseases of interest. Protocatechuic acid has anti-inflammatory, antihyperglycemic, and neuroprotective activities, inhibits *in vitro* chemical carcinogenesis, exerts proapoptotic and antiproliferative effects, alleviates the decrease in the quality of sperm and reproductive hormones (animal model), improves cognitive deterioration due to Alzheimer's disease (animal model), and attenuates neuronal oxidative stress and neuroinflammation (Kho et al., 2018; Lin, Huang, Huang, & Yin, 2009; Ma, Chen, Yang, Chen, & Shi, 2018; Masella, Superiore, Volti, & Superiore, 2015; Ormazabal et al., 2018; Owumi, Adedara, Farombi, & Oyelere, 2020; Ran, Hyun, Lee, Ju, & Young, 2020; Semaming Pannengpetch, Chattipakorn, & Chattipakorn, 2015; Tanaka, Tanaka, & Tanaka, 2011). The 4-hydroxybenzoic acid is valuable for the synthesis of bioproducts with potential applications in food, cosmetics, pharmacy, and fungicides, due to their potential as an antioxidant agent; it also has neuroprotective capacity avoiding neurodegeneration, completely restores the biosynthesis of coenzyme Q10, induces apoptosis in cancer cell lines, and can inhibit abnormal neuronal activation (Herebian et al., 2017; Wang, Wang, Zhang, Yang, & Li, 2018; Winter et al., 2017). Syringic acid has antioxidant, antimicrobial, anti-inflammatory, anti-endotoxic, anti-nitrosing, and antidiabetic activities, attenuating the blockage of insulin signaling and modulating glucose consumption, chemopreventive in carcinogenesis, anti-peroxidative of lipids, and neuroprotective effect in depression and may have potential in orthopedic applications (Arumugam, Balagangadharan, & Selvamurugan, 2018; Cheemanapalli, Mopuri, Golla, & Anuradha, 2018; Dalmagro, Camargo, & Zeni, 2017; Fang, Zhu, Niu, & Yin, 2019; Gheena & Ezhilarasan, 2019; Kim, Song, Chang, Jeong, & Keun, 2018; Periyannan, 2020; Shahzad, Mateen, Naeem, & Rizvi, 2019; Vo et al., 2020). Chlorogenic acid greatly exhibits antioxidant, antidiabetic, anticarcinogenic, anti-inflammatory, and antiobesity activities. Vanillic acid has neuroprotective, antidiabetic, and antioxidant activities, and it fights oxidative stress by activating tissue antioxidants, thereby alleviating the progression of osteoarthritis (Ali et al., 2017; Chaves-Ulate & Esquivel-Rodríguez, 2019; Vega, León, & Reyes, 2018). NDGA is a phenolic lignin and is commonly found in the creosote bush (*Larrea tridentate*). This bioactive compound has been used in traditional medicine for treating various diseases such as cancer, renal, cardiovascular, neurological, and immunological disorders. On the other hand, there are preclinical studies in cell culture and rodents which suggest that NDGA is a potential drug for the treatment of many chronic diseases and cancer, and this effect is due to their antioxidant activity (scavenging of ROS), but also indirectly by activating endogenous antioxidant responses mediated by transcription factor Nrf2 (Hano, Dinkova-Kostova, Davin, Cort, & Lewis, 2021).

6.6.2 Flavonoids

Flavonoids are one of the compounds found mostly in vegetables where they play an important role in color, texture, enzymatic reactions, and vitamins, and they are made up of two aromatic rings joined by an oxygenated heterocyclic ring; in this category, other subcategories can be found such as chalcones, flavones, flavanols, anthocyanins, and proanthocyanidins. (Vuolo et al., 2019; Yonekura-sakakibara, 2019; Zugasti-Cruz, Rodríguez-Herrera, & Sierra-Rivera, 2020).

Green technology such as supercritical CO_2 extraction yields higher flavonoids and phenolic acid contents from *L. graveolans* aerial parts (leaves, flowers, and stems) than methanol (80% v/v) extraction (Picos-Salas et al., 2021). The supercritical CO_2 extract exhibited superior antioxidant capacity and α-amylase and α-glucosidase inhibitory activities than methanol extract due to differences in phenolic compounds. Naringenin and kaempferide were the main constituents of both extracts, whereas cirsimaritin and kaemferol-glucoside were predominant in the supercritical CO_2 and methanol extracts, respectively. Moreover, both extracts had higher α-glucosidase inhibitory activity than acarbose indicating strong antidiabetic potential of their phenolics (Picos-Salas et al., 2021). Semi-desert plants are rich in flavonoids, which in turn have different positive effects for human health (Figure 6.1).

6.6.3 Stilbenes

They are phenolic compounds with a lower molecular weight and are made up of two connected aromatic rings (Roupe, Remsberg, Yáñez, & Davies, 2006) and are mainly related to the stress produced by pathogens and abiotic stress (Tanase et al., 2019).

6.6.4 Lignans

They are the result of the union of two units of coniferyl or sinapyl alcohol forming phenylpropanoid dimers, and these have been of great interest due to their chemotherapeutic potential (Rodr & Cristina, 2019).

Semi-desert regions provide a wide range of variables in climate and soil characteristics, which encourages a varied flora and a richness of species which stand out for their tolerance to environmental stress (Encina-Dominguez, Arévalo-Sierra, Villarreal-Quintanilla,

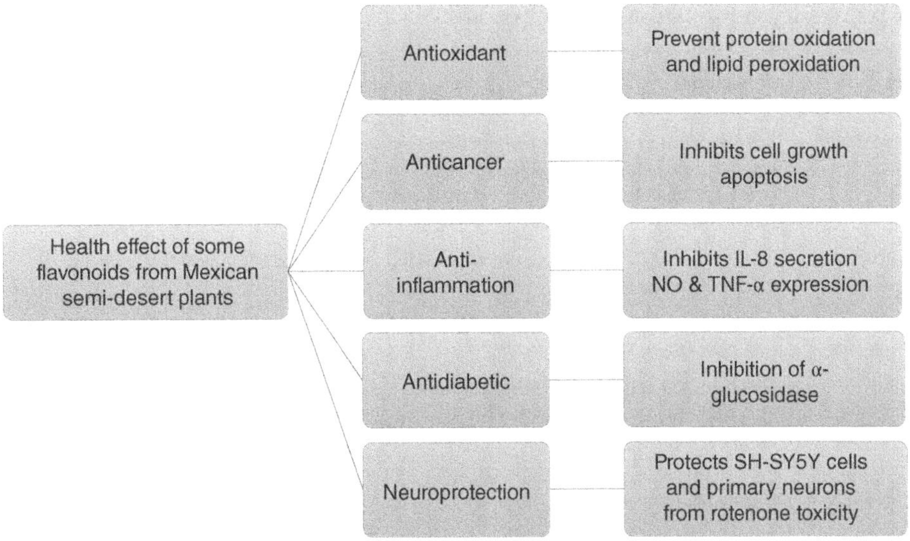

Figure 6.1 Health effects from Mexican semi-desert plants flavonoids and mechanisms.

& Estrada Castillon, 2020). Some of these plants have been used as traditional medicine in the region due to their different functional properties (Torres-Ponce et al., 2015). In recent years, the plants of semi-desert regions have gained interest due to the species with high polyphenols content as well as high antioxidant activity (Table 6.2), and these compounds can help development, as well as the preservation of the plant under stress conditions (Rodríguez & An, 2020; Wong et al., 2015).

Catechin has beneficial effects for human health, such as anticancer, antiobesity, antidiabetic, anticardiovascular, anti-infective, and neuroprotective hepatoprotective effects, and increases immunological activity (Cheng, Tan, Sun, Gu, & Liu, 2019; Ganeshpurkar & Saluja, 2018; Isemura, 2019; Jiang et al., 2019; Samarghandian et al., 2017). Rutin has anti-hyperglycemic effects, decreasing the absorption of carbohydrates in the small intestine, stimulates insulin secretion, reduces sorbitol formation, and has antioxidant, anti-inflammatory, neuroprotective, cardioprotective, and hepatoprotective activities (Enogieru, Haylett, Hiss, Bardien, & Ekpo, 2018; Ganeshpurkar & Saluja, 2016; Ghorbani, 2017; Iriti et al., 2017; Lú-chau, Moreira, Lema, & Eibes, 2017; Nordin, Jalil, Yusof, & Kamisah, 2019). Isohamnetin has been documented to have antioxidant, anti-inflammatory, antiproliferative, antitumor, and hepatoprotective activities (Gao et al., 2017; Lee, 2018; Lu et al., 2018; Wang, Quan, 2018; Zheng et al., 2019). Kaemferol has antimicrobial, anti-inflammatory, antioxidant, antitumor, cardioprotective, neuroprotective, hepatoprotective, and antidiabetic activities (Du, An, He, Zhang, & He, 2018; Kashyap et al., 2017; Marra, 2019; Mran et al., 2019; Wang, Wang, et al., 2018). Ferulic acid has an anti-inflammatory, antimicrobial, anticancer, and antidiabetic antioxidant effect, and it also can protect the main structures of the skin, inhibits melanogenesis, and improves angiogenesis (Zduńska, Dana, Kolodziejczak, & Rotsztejn, 2018; Zheng et al., 2019). Sinapic acid has antioxidant and anti-inflammatory activities and attenuates nephrotoxicity, apoptosis, and inflammation; it is a chemopreventive compound to control hypertension and cardiovascular diseases.

Table 6.2 Semi-desert plants and their polyphenols.

Plant	Compounds	References
Opuntia spp.	Protocatechuic acid, 4-hydroxybenzoic acid, chlorogenic acids, syringic acid, caffeic acid, vanilic acid, catechin, rutin, isorhamnetin, kaempferol.	Astello-García et al. (2015) Santos-Zea et al. (2011)
Lippia graveolens	Naringenin, kaempferol, isokaempferide, pilosin, cirsimaritin	González Güereca et al. (2007),
Agave lechuguilla	Kaempferol, quercetin, biflavonoid (afzelequin 4β-8 quercetin)	Anguiano-Sevilla et al. (2018)
Larrea tridentata	Sesamin, galocatechin, methyl galangin, epigallocatechin 7-O-glucuronide, epigalocatechin	García et al. (2018)
Leucophyllum ambiguum	(+) – yangambin, sesartemin, veratric acid, β-sitosterol	Rios, Ocampo-Acuña,Ramírez-Cisneros and Salazar-Rios, (2019)
Fouquieria splendens	Phenolic acids, elagic acid, quercetin-3-O-glycoside, rutin, kaempferol derivative	Monreal-García et al. (2019)
Prosopis spp.	Mesquitol, juliflorine, juliprosine	Ibrahim, Nadir, Ali, Ahmad, and Rasheed (2013)
Coryphantha macromeris	Ferulic acid, synapic acid, phenolic glycosides	Cabañas-García, Arache, Jáuregui-Rincón, Cruz-Sosa, and Pérez-Molphe Balch (2019)

The consumption of this type of polyphenols can modulate the intestinal microbiota (Aldubayan, Ahmed, Emara, Ahmed, & Elgharabawy, 2020; Ansari, 2017; Eroğlu, Avc, Vural, & Kurar, 2018; Yang et al., 2018).

Elagic acid has antioxidant, anticancer, antitumor, antimetastatic, and antiangiogenic activities (Ceci et al., 2018; Chen, Chen, & Zhou, 2018; Gupta, Mohammad, Khan, Alajmi, & Hussain, 2019; Rios, Giner, Marin, & Frecio, 2018). Other compound found in semi-desert plants is sesamin, which has antioxidant and anti-apoptotic, photoresist, anti-inflammatory, antiproliferative, proapoptotic, and antimetastatic activities (Lin, Wu, Hou, Chien, & Chang, 2019; Liu et al., 2017; Majdalawieh, Massri, & Gheyath, 2017; Majdalawieh et al., 2020; Zhao et al., 2018).

The plants of the semi-desert region present a great option for obtaining polyphenols, which mostly provide an antioxidant, anti-inflammatory, antidiabetic, or anticancer effect, and these become potential source of compounds as promising agents in the treatment of multiple diseases, some of which have a high mortality.

6.7 Vitamins from Semi-Desert Plants and Health Effects

Vitamins are organic molecules that the organism requires in small quantities for the correct functioning of metabolism; they are classified as water-soluble and fat-soluble. The organism cannot synthesize them, so they must be obtained through the diet.

Some semi-desert plants have been studied as an important source of nutrients with beneficial effects on health. Among these plants are some cacti, such as those belonging to the *Opuntia* genus. *Opuntia ficus-indica* is a cactus that produces edible fruits known as tunas. According to several studies on different *Opuntia* species, the fruits are an important source of vitamin C, flavonoids, phenolic acids, carotenoids, and betalains (El Kharrassi et al., 2016), while the cladodes are rich in sterols, minerals, amino acids, and vitamins (Aragona, Lauriano, Pergolizzi, & Faggio, 2018). Another product derived from this cactus and commercially exploited is oil from its seed, which is rich in tocopherols, as well as phytosterols, phenols, and polyunsaturated fatty acids (PUFAs). Bakar, Çakmak, Ibrahim, Özer, and Saydam (2020) studied the effect of different pretreatments on the content of different nutrients in Opuntia fruits, including fat-soluble vitamins (vitamins A and E) and water-soluble vitamins (ascorbic acid and vitamins B_1, B_2, B_3, B_6, B_9, and B_{12}). These fruits were harvested in Southern Turkey and showed that pretreatment significantly influences the content of these vitamins. In the case of vitamin A, expressed as the sum of retinol and retinoic acid, the highest content was reported in fresh fruit ($2.89 \pm 0.20\,\mu g dw^{-1}$), followed by frozen ($2.74 \pm 0.16\,\mu g dw^{-1}$), microwave-dried ($1.07 \pm 0.06\,\mu g g^{-1}$ dw), and sun-dried ($0.95 \pm 0.06\,\mu g dw^{-1}$). Vitamin A plays an essential role in normal vision, cell growth and development, epithelial cell integrity, immune response, and reproduction (Amidzic, Brboric, Cudina, & Vladimirov, 2005).

On the other hand, Vitamin E plays a primordial role in the protection of molecules and tissues against free radicals by participating in the stabilization of biological membranes and is also related to immune and neurological functions (Ognjanovic et al., 2003). The different evaluated pretreatments showed the same trend as for vitamin A, with fresh fruit again showing the highest vitamin E content and sun-dried fruit the lowest ($63.2 \pm 5.2\,\mu g g^{-1}$ dw and $33.8 \pm 2.12\,\mu g g^{-1}$ dw, respectively). Vitamin E is abundant in prickly pear seed oil, where the main components are α- and γ-tocopherol (80% of vitamin E (El Kharrassi et al., 2018). In studies conducted by El Kharrassi et al. (2020), *O. ficus-indica* and *O. megacantha* species recorded higher total tocopherol content (815 and $792\,mg kg^{-1}$, respectively), with γ-tocopherol being the major component (98%) in the various evaluated species. The γ-tocopherol has a positive impact on human nutrition and health; it is involved in the activity of PUFAs and contributes to the quality of oils, protecting them from oxidation and prolonging their shelf life (Chahdoura et al., 2017).

Another vitamin with a protective effect against oxidative stress is ascorbic acid (vitamin C), as it can neutralize superoxide and hydroxyl radicals due to its ability to donate electrons (Muhammad et al., 2014). Vitamin C is important in collagen formation, reduction of plasma cholesterol, iron absorption, inhibition of nitrosamine formation, strengthening the immune system, and maintaining healthy gums, skin, and blood vessels. Some reports indicate that it reduces the risk of cardiovascular diseases, arteriosclerosis, and some types of cancer (Lee & Kader, 2000). To date, it is well known that nopal fruits are rich in vitamin C (20–40 mg/100 g) (El Kossori, Villaume, El Boustani, Sauvaire, & Méjean, 1998). However, this content can be influenced by environmental conditions, plant genotype, processing, or storage. For example, El Kharrassi et al. (2020) studied the vitamin C content in the juice of fruits of different *Opuntia* species, reporting the highest values (24.6 mgL^{-1}) for *O. megacantha*, while those of *O. ficus-indica* and *O. leucotricha* showed no significant difference (around 5.5 mgL^{-1}); the lowest value (2.40 mgL^{-1}) was reported for *O. aequatorialis*. Other vitamin C values have been reported for *Opuntia* species in Mexico: *O. robusta* (4 mg/100 g fresh weight), *O. lindheimeri* (121 μgg^{-1} fresh weight), *O. stricta*, and *O. streptacantha* (815 μgg^{-1} fresh weight). Regarding the effect of storage, Tarabih and EL Eryan (2020) reported a gradual decrease in vitamin C content from 100 mg/100 g fresh weight at harvest to 70 after 30 days of storage at 2 °C and 90–95% relative humidity. In the study by Bakar et al. (2020), the highest vitamin C content was reported in fresh and frozen fruit (479 and 475 μg/g dw), while in sun-dried or microwaved fruit, the values decreased by more than 50%. Despite their decrease, these values represent twice the vitamin C reported in fresh tomatoes (106 μgg^{-1}); other fruits such as mandarin, orange, or apple contain 40, 750, and 120 μgg^{-1}, respectively (Lee & Kader, 2000; Rasanu, Magearu, Nicoleta Matei, & Soceanu, 2005).

Complex B vitamins participate as coenzymes in biochemical reactions, significantly influencing the nervous and digestive systems, and wound healing. Their deficiency can lead to various health problems (Kennedy, 2016). Concentrations of around 8 μg/g dry weight (dw) were reported in fresh and frozen prickly pears for vitamin B_1, while sun-dried and microwaved prickly pears presented 4.53 ± 0.22 and 4.85 ± 0.30 μgg^{-1} dw, respectively. On the other hand, Isaac (2016) reported 0.12 μgg^{-1} thiamine in this same fruit. This vitamin is indispensable for glucose metabolism (Ramsey & Muskin, 2013). Riboflavin (vitamin B2) is essential for converting tryptophan to niacin and obtaining energy from food (Ashoori & Saedisomeolia, 2014). Bakar et al. (2020) reported values close to 6.5 μg/g dw of this vitamin in fresh and frozen fruits; its value decreased three times in the sun-drying or microwave process. Isaac (2016) also reported values of 0.41 μgg^{-1} of *Opuntia ficus-indica*. The behavior of vitamin B_3 was similar to that of other vitamins under different evaluated pretreatments; the highest content was reported in fresh fruits of *Opuntia ficus-indica* (307 ± 23.6 μg/g dw). This vitamin is a coenzyme of dehydrogenases for hydrogen transfer. Pyridoxine (vitamin B_6) is required for the synthesis of serotonin, noradrenaline, myelin, and neurotransmitters (Schellack, 2015). Bakar et al. (2020) reported that drying *Opuntia* fruits decreased its vitamin's content by more than 50% (from 21 to 10.5 μg/g dw).

Folate (vitamin B_9) is a vitamin involved in protein metabolism and red blood cell formation. Values close to 700 μg/g dw were reported in fresh and frozen prickly pears; as with other vitamins in this study, the drying process significantly decreased its content. In conjunction with folic acid, vitamin B_{12} (cyanocobalamin) is important during the formation of red blood cells and promotes the proper functioning of the nervous system (Bellows & Moore, 2012). Bakar et al. (2020) reported contents of 4.00 ± 0.40, 3.80 ± 0.32, 1.96 ± 0.11, and 1.84 ± 0.14 μg vit B_{12}/g dw in fresh, frozen, sun-dried, and microwaved prickly pears, respectively. Values 0.585 μgg^{-1} of this vitamin have been reported in peach (Płonka, Toczek, & Tomczyk, 2012).

The vitamin content varies depending on the condition of the plant material, that is, whether it has undergone the drying treatment or not. These data have been evaluated in different studies that have been carried out on the nopal or its fruit (Table 6.3).

Table 6.3 Vitamins content in fresh and dry Opuntia fruit.

Vitamin	Fresh (μgdw^{-1})	Dry (μgdw^{-1})
A	2.89 ± 0.20	0.95 ± 0.06
B_1	8 μgg^{-1}	4.53 μgg^{-1}
B_9	4 μgg^{-1}	1.96 μgg^{-1}
C	20–40 mg/100 g	Decreased by more than 50%.

6.8 Flavonoids from Semi-Desert Plants and Health Effects

Flavonoids are phenolic compounds ubiquitously occurring in practically all parts of plants. Many have several biological properties with medical implications (Barriada-Bernal et al., 2013). They can be classified based on the number and arrangement of their carbon atoms into flavonols, flavones, flavan-3-ols, anthocyanidins, flavanones, isoflavones, and others, they are commonly found conjugated to sugars and organic acids (Gutierrez-Grijalva et al., 2017). Several plants that are rich in polyphenolic compounds and exhibit biological properties grow in the semi-desert region of Mexico under extreme climate conditions (Wong Paz et al., 2015). In this section, we discuss studies using plant from semi-desert and the beneficial effects of its flavonoids on human health.

6.8.1 Anticancer Activity

Despite major advances in therapy, cancer continues to be a leading cause of mortality (Mun, Babiker, Weinberg, Kirson, & Von Hoff, 2018). Recently, flavonoids have been studied due to their antioxidant capacity and effects against different types of cancer, reducing cellular viability, migration, and invasion. A study evaluated the cytotoxic activity against lung cells using a fraction enriched in flavonoids from *Agave lechuguilla* extract. The ethanolic extract at 50 μgmL^{-1} showed growth inhibition of 75.7 ± 1.3% in SK-LU-1 cell line. Three flavonoids were identified by chromatographic analysis: kaempferol, quercetin, and a flavonoid dimer formed by afzelechin-4-8-quercetin. The induction of apoptosis was evaluated by phosphatidylserine, and the results demonstrated that the extract induces cell death by apoptosis with 39.8% at 24 h. Finally, through molecular docking, it was found that flavonoids bind with death receptors. The bioflavonoid and quercetin bind to the receptor TNF-R1 and DR4, activating death cell signaling by apoptosis, while kaempferol binds to DR5 and Fas/CD95 which activate the intrinsic pathway, and the kaempferol interaction with caspase-3 that helps the same apoptotic amplification. The authors conclude that a mixture of this flavonoid can be used as therapeutic targets due to the interactions with death receptors that causes apoptosis by extrinsic and intrinsic pathways (Anguiano-Sevilla et al., 2018).

Other flavonoids of interest are isorhamnetin and its different glycosides, which have been identified in *Opuntia* genus and possess different biological effects depending on their structures. The compound isorhamnetin-3-O-glucosyl pentoside (IGP) isolated from *Opuntia ficus-indica* (L.) Mill has been investigated as an inductor of apoptosis in human colon cancer cells (HT-29). The exposition of IGP reduces cells confluence and morphological changes like rounded cells and the presence of apoptotic bodies. Besides, an increase in the exposure of phosphatidylserine (48% in IGP-treated cells) and membrane permeabilization (30% in IGP-treated cells) was observed. Other effects by IGP exposure were increased chromatin condensation (54%) and induction of cell cycle arrest in G2/M phase. To investigate the mechanism of IGP action, the expression of Bax/Bcl-2 was evaluated by western blot. Bax/Bcl2 ratio increased (threefold) compared to Bcl2/actin ratio, suggesting mitochondrial damage, which was confirmed by the loss of mitochondrial membrane potential in IGP-treated cells. However, the cell death induction of IGP was caspase-dependent confirmed using a pan-caspase inhibitor (Antunes-Ricardo et al., 2019).

Another plant with cytotoxic activity is *Turnera diffusa* Willd. The methanolic extract of this plant showed important cytotoxic activity on MDAMB-231 breast cancer cells (IC_{50} 31 $\mu g mL^{-1}$) compared with other cancer cell lines. This difference in activity can be attributed to the distinct type of cells and cellular receptors. Four organic fractions of this extract exhibited activity on this cancer cell line. In the most active fraction (F4), the trihydroxyflavone apigenin was identified. This compound can induce apoptosis in various tumor cell lines. The authors conclude that the methanolic extract of *T. diffusa* is a potential candidate in cancer therapy (Avelino-Flores et al., 2015). Other cells affected by *T. difffusa* are multiple myeloma (MM) cell lines. The ethyl acetate fraction of a crude extract decreased the viability of NCI-H929, U266, and MM1S cell lines. By chromatography purification steps, the isolated compounds of this fraction were the following: naringenin, apigenin 7-O-(6"-O-p-E-coumaroyl)-glucoside, apigenin 7-O-(6"-O-p-Z-cou maroyl)-glucoside, apigenin 7-O-(4"-O-p-E-coumaroyl)-glucoside, acacetin, genkwanin, velutin, gonzalitosin I, and acacetin 7-O-methyl ether. Also in this study, the evaluation of the flavonoid naringenin showed high cytotoxicity and decreased viability of NCI-H929 and U266 cell lines even at low concentrations (Willer, Johrer, Greil, Zidorn, & Cicek, 2019).

6.8.2 Antioxidant Activity

The antioxidant activity is one of the main biological effects of flavonoids and influenced by their chemical structure. The presence of double bonds, hydroxyl substitutions, among others is related to an increase in this activity. In 2007, in a study using crude extracts from *Lippia graveolens*, several flavonoids were detected (pilosin, cirsimartin, narigenin, kaempferol, isokaemferide, a derivative of catechin, and a non-identified hexoside of quercetin). The evaluation of the antioxidant activity of the extracts by the DPPH method showed a reduction of the radical, especially in the ethyl acetate extract. The identified compounds showed hydroxyl groups in its structure and the authors observed that the substitution of –OH groups by –OMe reduces the free radical scavenging activity (González Güereca et al., 2007).

Another source of polyphenolic compounds is *Agave durangensis*. In 2012, a study of several extracts from this plant identified different flavonoids: five quercetin glycosides and three kaempferol glycosides by liquid chromatography with photodiode array detection (HPLC–DAD). The total extracts of the entire flowers showed the highest flavonoid content (1210.4 $\mu g g^{-1}$ dry extract) and the most complex flavonoid profile (eight compounds).

All the extracts showed important antioxidant activity (total antioxidant capacity, free radical scavenging activity, and iron-reducing power). Some of the identified compounds from the entire edible flowers of this plant were kaempferol-3,7-O-diglucoside, quercetin-3-O-[rhamnosyl-(1–6)-galactoside], and kaempferol-3-O-[rhamnosyl-(1–6)-glucoside] (Barriada-Bernal et al., 2013).

A flavonoid-rich concentrate from *Opuntia ficus indica* juice was used to obtain bioactive compounds by different methods. The effect on induced oxidative stress and inflammation using Caco-2 cells was evaluated. The flavonoids detected were isorhamnetins and its derivatives such as isorhamnetin 3-O-rhamnose, rutinoside and isorhamnetin 3-O-rutinoside. The results showed attenuation in the oxidation products by H_2O_2-induced radicals, and the authors demonstrated reduced carbonyl proteins content, especially when cells are co-incubated with the flavonoids concentrate (than pre-incubation). The flavonols present in concentrate (isorhamnetins glucosides) could be responsible for this action, by preventing protein oxidation. Besides, the impact on inflammation response was determined by quantification of different inflammatory mediators. IL-8 secretion decreased significantly, and degradation of IκBα was blocked for the co-incubation method. The flavonoid concentrate also reduces NO and TNF-α expression and modulates Caco-2 cells permeability. These findings indicate that the compounds in *Opuntia ficus indica* juice are promising alternatives in the treatment of intestinal inflammation diseases (Matias et al., 2014). The extraction methodology and the solvent also are important elements for the isolation of polyphenols. In this sense, ultrasound-assisted extraction was used for the extraction of phenolic compounds from *Jatropha dioica, Flourensia cernua, Turnera diffusa*, and *Eucalyptus camaldulensis*. The highest antioxidant activity was obtained in the *E. camaldulensis* extracts: DPPH radical scavenging (80%) and LOI (70%). The results indicate the potential of these extracts for use in scavenging free radicals for general health and lipid oxidation prevention in foods (Wong Paz et al., 2015).

Regarding the role of phytochemicals in diseases associated with oxidative stress, several plants have been reported as effective on dermal wounds. *Opuntia ficus-indica* (L.) Mill. modified the metabolism (increased MMP-9 release from damaged cells) and the migratory properties of Human Adult Keratinocyte cell line (HaCaT). In addition, HaCaT modulated response to oxidative stress. Low concentrations of *Opuntia* reduced lipid peroxidation in HaCaT exposed to oxidative stress, while the highest dose ($5\,mgml^{-1}$) enhanced lipid peroxidation. Finally, all tested doses of *Opuntia* increased HaCaT migration when added to poor medium (DMEM 0%), and lower concentrations of *Opuntia* ($12.5\,\mu gml^{-1}$ and $1.25\,\mu gml^{-1}$) strongly promoted wound healing. The characterization of compounds that modulated HaCaT response to oxidative stress is an interesting option to study (Bassino et al., 2019).

Recently, the cactus berry (*Myrtillocactus geometrizans*) was studied. Their phenolic composition and the *in vitro* antioxidant, anti-inflammatory, and antihyperglycemic activity of their extracts were determined. Thirteen flavonoids were identified in cactus berry such as isorhamnetin glycosides (the most abundant) and other flavonoid glycosides such as rutin rhamnosyl-glucoside and kaempferol-7-O-neohesperidoside. The pulp extract showed the highest antioxidant activity (ORAC assay) correlated ($p \leq 0.05$) with ascorbic acid and flavonoid content due to its high isorhamnetin glycoside content. Isorhamnetin glycosides produced high anti-inflammatory activity, and the peel and pulp extracts had the highest hyaluronidase-inhibiting activity (antihyperglycemic activity). The authors concluded that the bioactive compounds in cactus berry are promising options for the management of some degenerative chronic diseases (Montiel-Sanchez, Garcia-Cayuela, Gomez-Maqueo, Garcia, & Cano, 2020).

Other semi-desert plants are *Agave salmiana, Aloe vera, Erythrina americana*, and *Myrtillocactus geometrizans,* whose edible flowers have been characterized recently. *M. geometrizans* showed the highest antioxidant activity, measured by DPPH and FRAP, and this activity correlates with the phenolic compounds content. The main phenolic compounds detected in flowers were rutin and phloridzin, belonging to the flavonoids family (Pinedo-Espinoza et al., 2020).

6.8.3 Hepatoprotective Activity

Threatening cases of liver disorder such as fatty liver, fibrosis, and cirrhosis were associated with alcohol consumption. Several phytochemicals have been proposed as hepatoprotective agents. A *n*-butanol fraction of *O. ficus-indica* fruit extract and the isolated flavonoids (+)-taxifolin, quercetin 3-O-methyl ether, isorhamnetin 3-O-β-D-glucoside, and narcissin exhibited protective activities on hepatocytes against alcoholic oxidative stress; (+)-taxifolin showed the most significant protective activity (55% at 50 μM). This effect may be associated with powerful antioxidant activity because of GSH level maintained by *in vitro* cellular glutathione. Further studies are needed to elucidate the mechanism of action, but these flavonoids are interesting options in the search of molecules to reduce hepatic damage (Kim et al., 2017).

6.8.4 Vasodilator Activity

Arterial hypertension remains the most important risk factor for cardiovascular and renal diseases (Handschin, Henny-Fullin, Buess, Leuppi, & Dieterle, 2015). In a screening of 19 compounds as promising candidates for the hypertension treatment, the molecule 5,8,4′-trihydroxy-3,7-dimethoxyflavone isolated from *Larrea tridentata* showed vasorelaxant activity on isolated rat aorta model. Also, the vasodilator effect involves stimulation of the NO/cGMP pathway. Studies to clarify the mechanism of action are needed, but this compound is an interesting and potent vasodilator agent (Luna-Vazquez et al., 2018).

6.8.5 Neurological Effect

In the search for new drugs to treat neurodegenerative disorders, the acacetin 7-methyl ether (5-hydroxy-4′, 7-dimethoxyflavone) is a natural flavone isolated from *Turnera diffusa* (damiana). Studies *in vitro* indicate that it is a potent selective inhibitor of MAO-B (IC_{50} = 198 nM). By molecular docking assays, the selectivity of binding of this molecule to MAO-B over MAO-A was confirmed, and the molecular dynamics revealed a stable and strong complex ligand protein. Recently, the search of MAO-B inhibitors are oriented for the treatment of Parkinson's and Alzheimer's diseases; in this sense, further investigations are needed to evaluate if acacetin 7-methyl as a potential new drug (Chaurasiya et al., 2019).

6.8.6 Antibacterial Activity

A bacterial infection is a common clinical disease that can affect various organs and tissues (Xiao & Cai, 2020). The prevalence of infectious diseases of the respiratory tract caused by opportunistic bacteria and the resistance to antimicrobial agents necessitates the identification of new alternative treatment. Flavonoids isolated from *Larrea tridentata* were tested against bacteria. The results showed that 5,4′-dihydroxy-3,7,8,3′-tetramethoxyflavone and 5,4′-dihydroxy-3,7,8-trimethoxyflavone were active against one and three resistant strains of *Mycobacterium tuberculosis*, respectively, and these compounds contributed to the antimycobacterial activity found in *L. tridentata* (Favela-Hernandez, Garcia, Garza-Gonzalez, Rivas-Galindo, & Camacho-Corona, 2012).

Microbial biofilms are communities of aggregated microbial cells embedded in a self-produced matrix of extracellular polymeric substances (Yin, Wang, Liu, & He, 2019). Several bacterial and fungal infections are associated with biofilms, also the resistance to antibiotic complicates treatment options. In the search for new strategies for combating microbial biofilm, medicinal plants have been proposed. Methanolic extracts of *Prosopis laevigata* and *Opuntia ficus-indica* were tested against five clinical bacterial isolates. Both extracts showed growth inhibition against *Klebsiella pneumoniae*, *Enterococcus faecalis*, *Escherichia coli*, and *Staphylococcus aureus*. *Opuntia ficus-indica* extract significantly reduced biofilm formation dose-dependently. This effect could be due to the presence of flavonoids (quercetin, kaempferol, naringenin, and apigenin), because these molecules can reduce biofilm synthesis, by suppressing the activity of the autoinducer-2, responsible for cell-to-cell communication (Sanchez et al., 2016).

Another interesting source of flavonoids is xoconostle (*Opuntia oligacantha*). The antibacterial activity against *Salmonella typhimurium* was tested in a study of active films alone and with the addition of bioextract (BE) or microencapsulated bioextract (MBE) from xoconostle. The films with 0.8 and 1.2 g of BE and MBE had antibacterial properties against *S. typhimurium*, and the best result was for the film containing 1.2 g of MBE, due to the micro-encapsulates slowly liberating the antibacterial compounds. The presence of the flavonoids was determined, but the identification of the compounds was not evaluated. These findings are a valuable option for the development of films with functional properties (Cenobio-Galindo et al., 2019).

Recently, several medicinal plants have been tested in the search of new treatments against *Streptococcus mutans*, one of the major etiological agents of human dental caries. Extracts from the semi-desert plant *Jatropha dioica* effectively inhibit bacterial growth (IC_{50} 250 ppm). The chromatographic characterization revealed the presence of sinensetin and kaempferol 3,7,4′-otriglucoside, molecules that could be responsible for their antibacterial effect; these estimations need to be confirmed to propose new molecules as alternatives for cariogenic bacteria control (Vargas-Segura et al., 2020).

6.8.7 Antiparasitic Effect

The drug resistance in parasites is considered to be one of the foremost concerns in health and disease management (Pramanik, Alam, Roy Chowdhury, & Chakraborti, 2019). *Giardia lamblia* is a ubiquitous parasitic pathogen that causes significant morbidity and mortality worldwide (Abraham et al., 2019), and several phytochemicals have been tested as new treatment alternatives. A fraction of *Lippia graveolens* extract showed potent antigiardiasic effect (2.87 μgmL^{-1}) at 2 hr. This fraction altered the morphology and disrupted the nuclear envelope and injured the ventral suckling disk. The flavonoids identified in fraction 6 of the extract were naringenin and pinocembrin; the authors suggest that the effect could be due to a synergic activity of all compounds because the LD_{50} for individual compounds were higher than those observed for the fraction of oregano extract (Rufino-Gonzalez et al., 2012).

Amebiasis caused by *Entamoeba histolytica* is another serious worldwide public health problem. Recently, the flavonoids pinocembrin, sakuranetin, cirsimaritin, and naringenin were detected by several chromatographic techniques in a methanolic *L. graveolens* extract, showing a 50% inhibitory concentration (IC_{50}) ranging from 28 to 154 μgmL^{-1}. With these results, the authors confirmed the use of this plant in traditional Mexican medicine for the treatment of dyspepsia (Quintanilla-Licea et al., 2020). Through the identification and characterization of compounds in semi-desert plants, it is possible to find various biological activities (Table 6.4). Also, studies about the possible mechanisms of action are necessary to propose these molecules in the future as new pharmacological treatments.

6.9 Bioactive Compounds as Potential Ingredient in Food Products

A food ingredient is any substance that is added to achieve a desired effect, and they increase the attractiveness of food for consumers, such as visual appearance, product preservation, and flavor enhancement (Song & Hepp, 2005). The food industry has evolved since the prehistoric times where man used to preserve fresh food in caves, after that the use of salt was popular to preserve food products till the apparition of refrigeration in the twentieth century and several other methods as fermentation and drying (Hamad, Han, Kim, & Rather, 2018). During the past century, one of the most significant changes in food industry was the appearance and/or increase in the use of different additives as artificial sweeteners, nitrites, nitrates, high fructose corn syrup, among others, which brought with them an endless number of health problems related to the consumption of foods that cause them such as obesity, hypertension, diabetes, and cancer (WHO, 2018).

In the last decades, the interest in functional food ingredients has increased due to their use and application providing nutritional food products to health-conscious consumers and because it has been proven a variety of functional effects such as antibacterial, anticoagulant, anti-inflammatory, antifungal, anthelmintic, antiplatelet, antiprotozoal, antiviral activities and actions on the cardiovascular, endocrine, immune and nervous systems, as well as other miscellaneous mechanisms of action (Shahidi & Ambigaipalan, 2015). The activity of several plant bioactive compounds has been proved through different research studies and their effectiveness in the prevention of various chronic health diseases, in addition to the normal nutritive function. In this same sense, plants-derived compounds exhibiting strong antimicrobial and antioxidant activity could be used as an eco-friendly food preservative (Prakash, Kujur, Singh, Kumar, & Yadav, 2017). As previously mentioned, plants from semi-desert are rich in several bioactive components as antioxidants, polyphenols, catechins, anthocyanins, among others. In this same way,

Table 6.4 Beneficial effects of flavonoids from semi-desert plant in human health.

Flavonoids from semi-desert plant		Type of study	Health effect	Main findings	Reference
Agave lechuguilla	Ethanolic extract from dried powder. Kaempferol, quercetin, afzelechin-4-8-quercetin	*In vitro/ in silico*	Anticancer activity	-Cytotoxic activity against SK-LU-1 cell line. -Induction of apoptosis. -Interaction of flavonoids with death receptors: TNF-R1, DR4, Fas/CD95, DR5, among others.	Anguiano-Sevilla et al. (2018)
Opuntia ficus-indica (L.) Mill	Methanolic extract. Isorhamnetin-3-O-glucosyl pentoside	*In vitro*	Anticancer activity	-Cytotoxic effect in metastatic human colon cancer cells (HT-29). -Identification of mechanism involved in apoptosis. -Caspase-dependent mitochondrial damage.	Antunes-Ricardo et al. (2019)
Turnera diffusa Willd.	Methanolic extract from leaves and stems. Apigenin.	*In vitro*	Anticancer activity	-Cytotoxic effect in MDA-MB-231 cells. -Detection of later apoptosis and necrosis in cells exposed at 24 h.	Avelino-Flores et al. (2015)
Turnera diffusa	Ethyl acetate fraction. Naringenin, apigenin 7-O-(6"-O-p-E-coumaroyl)-glucoside, apigenin 7-O-(6"-O-p-Z-coumaroyl)-glucoside, apigenin 7-O-(4"-O-p-E-coumaroyl)-glucoside, acacetin, genkwanin, velutin, gonzalitosin I, and acacetin 7-O-methyl ether.	*In vitro*	Anticancer activity	-Cytotoxic effect in NCI-H929, U266, and MM1S cell lines -Naringenin decreased viability of NCI-H929 and U266	Willer et al. (2019)

(Continued)

Table 6.4 (Continued)

Flavonoids from semi-desert plant		Type of study	Health effect	Main findings	Reference
Lippia graveolens	Ethyl acetate extract from the stem. Pilosin, cirsimartin, narigenin, kaempferol, isokaemferide, catechin derivative, and a hexoxide of quercetin	Chemical analysis	Antioxidant activity	-Free radical scavenging activity by DPPH method	González Güereca et al. (2007)
Agave durangensis	Extracts of different hydrophobicity from ground flowers. Kaempferol-3,7-O-diglucoside, quercetin-3-O-[rhamnosyl-(1–6)-galactoside], and kaempferol-3-O-[rhamnosyl-(1–6)-glucoside]	Chemical analysis	Antioxidant activity	-Total antioxidant capacity by DPPH, free radical scavenging activity, and iron-reducing power	Barriada-Bernal et al. (2013)
Opuntia ficus indica	Flavonoids concentrate from the juice. Isorhamnetin 3-O-rhamnose8 rutinoside and isorhamnetin 3-O-rutinoside	*In vitro*	Antioxidant activity	-Attenuates radical production, reduction in protein oxidation -Decrease IL-8 secretion, blocked degradation of IκBα, reduce NO and TNF-α expression	Matias et al. (2014)
Jatropha dioica, Flourensia cernua, Turnera diffusa, and *Eucalyptus camaldulensis*	Ultrasound-assisted extraction of powder of dried plants	Chemical analysis	Antioxidant activity	-Antioxidant activity (DPPH and lipid oxidation inhibition): *E. camaldulensis* > *T. diffusa* > *F. cernua* > *J. dioica* extracts	Wong Paz et al. (2015)
Opuntia ficus-indica (L.) Mill.	Extract of cladodes	*In vitro*	Antioxidant activity	-Reduction of lipid peroxidation in HaCaT exposed to oxidative stress -Increased MMP-9 release from damaged cells -Increased HaCaT migration in poor medium	Bassino et al. (2019)

Myrtillocactus geometrizans	Ultrasonic extraction of peels, pulps, and whole fruits	*In vitro*	Antioxidant activity	-Antioxidant capacity by ORAC assay -High anti-inflammatory activity -Hyaluronidase-inhibiting activity	Montiel-Sanchez et al. (2020)
Agave salmiana, Aloe vera, Erythrina americana, and *Myrtillocactus geometrizans*	Edible flowers. Phloridzin, Phloretin, Naringenin, Myricetin, Apigenin, and Galangin	*In vitro*	Antioxidant activity	-DPPH and FRAP antioxidant activity -Highest activity found in *Myrtillocactus geometrizans*	Pinedo-Espinoza et al. (2020)
Opuntia ficus-indica	Fruit extracts and isolated compounds. Quercetin 3-O-methyl ester, (+)-taxifolin, isorhamnetin 3-O-β-d-glucoside, and narcissin	*In vitro*	Hepatoprotective activity	-Protection of rat primary hepatocytes against ethanol exposure	Kim et al. (2017)
Larrea tridentata	5,8,4′-trihydroxy-3,7-dimethoxyflavone	*In vitro*	Vasodilator activity	-Maximum vasodilator effect, higher than the positive control -Involvement in NO/cGMP pathway	Luna-Vazquez et al. (2018)
Turnera diffusa	Acacetin 7-methyl ether	*In vitro/ in silico*	Neurological effect	-Selective inhibition of MAO-B -Residues Leu171, Cys172, Tyr188, Ile199, Gln206, Tyr326, Tyr398, and Tyr435 are involved in protein–ligand interaction.	Chaurasiya et al. (2019)
Larrea tridentata	5,4′-dihydroxy-3,7,8,3′-tetramethoxyflavone and 5,4′-dihydroxy-3,7,8-trimethoxyflavone	*In vitro*	Antibacterial activity	-Growth inhibition in resistant strains of *M. tuberculosis*	Favela-Hernandez et al. (2012)

(*Continued*)

Table 6.4 (Continued)

Flavonoids from semi-desert plant		Type of study	Health effect	Main findings	Reference
Prosopis laevigata and Opuntia ficus-indica	Methanolic extract. Quercetin, kaempferol, naringenin, and apigenin	*In vitro*	Antibacterial activity	-Growth inhibition against nosocomial pathogens -Inhibition of biofilm formation, especially for the *Opuntia ficus-indica* extract	Sanchez et al. (2016)
Opuntia oligacantha	Active films with bioextract of the fruit	*In vitro*	Antibacterial activity	-Antioxidant activity (ABTS, DPPH) -Inhibition of *S. typhimurium*	Cenobio-Galindo et al. (2019)
Jatropha dioica	Aqueous extract. Kaempferol 3,7,4′-otriglucoside and Sinensetin	*In vitro*	Antibacterial activity	-Growth inhibition in *S. mutans* (IC_{50} 250 ppm)	Vargas-Segura et al. (2020)
Lippia graveolens	Ethanolic extract of leaves. Naringenin, pinocembrin	*In vitro*	Antiparasitic effect	-Antigiardiasic effect of fraction 6 -Perforations in the nuclear membrane -Protein pattern destruction in ventral disk	Rufino-Gonzalez et al. (2012)
Lippia graveolens Kunth	Methanolic extract of leaves. Pinocembrin, sakuranetin, cirsimaritin, and naringenin	*in vitro*	Antiparasitic effect	-Cytotoxicity against *Entamoeba histolytica* trophozoites	Quintanilla-Licea et al. (2020)

each component has its particular functionality, and each of them can contribute a characteristic in a food matrix as technological or functional compound.

6.9.1 Antioxidant as Food Additive

One of the features of natural plant antioxidants is that they can protect food components from oxidation under stress of processing and storage, delaying autoxidation through the formation of free radicals or by interrupting propagation of the free radicals (Lavelli, Kerr, García-Lomillo, & González-San José, 2017). Between these mechanisms are singlet oxygen quenching, breaking the autoxidative chain reaction, scavenging radical species that initiate peroxidation, chelating metal ions such that they are unable to generate reactive species or decompose lipid peroxides, and reducing localized O_2 concentrations (Decker, Chen, Panya,& Elias, 2010; Lü, Lin, Yao, & Chen, 2010). Synthetic antioxidant additives that have been traditionally used are butylated hydroxyanisole (BHA), butylated hydroxytoluene (BHT), tert-butylhydroquinone (TBHQ), and propyl gallate (PG), which have shown great effectiveness in preserving food-retarding lipid oxidation and the development of undesirable rancidity, related with free radical formation and off-flavors (Nikoo, Regenstein, & Gavlighi, 2018). Besides their use as alternative to synthetic preservers, natural plant extracts rich in antioxidants provide additional value to the final products, and these have been previously used in bakery, dairy, and meat products, where natural extract from aromatic plants, species, and fruits were added for antioxidant purposes (Caleja, Barros, Antonio, Oliveira, & Ferreira, 2017). There are several examples and different studies of some plant extracts and antioxidants which have been proved as food additives. One of the most popular bioactive compounds are anthocyanins, which are water-soluble natural pigments with strong activity and these are commonly used in food to inhibit lipid oxidation (Santos-Sánchez, Salas-Coronado, Valadez-Blanco, Hernández-Carlos, & Guadarrama-Mendoza, 2017).

Meat is of the food products which are exposed to lipid and protein oxidation causing its deterioration. Studies suggest that phenolic compounds found in extracts and in the by-products of berries inhibit the oxidation on meat, making it an alternative to synthetic antioxidants and actually present potential functional properties for consumers (Lorenzo et al., 2017). Another natural plant extract is rosemary, rich in antioxidants, which has been added to animal products and oils reducing the color loss of carotenoids and delaying lipid oxidation in oils and meat products (Nieto, Ros, & Castillo, 2018). Other antioxidant compounds than have demonstrated antioxidant protection of meat are carotenoids, hydroxycinnamic acids, flavonoids, terpenes, and antioxidant vitamins (Nimse & Pal, 2015), compounds that are present in semi-desert plants.

In this same way, different plant extracts have been used in dairy products demonstrating retardation of oxidation in these products, increasing the shelf life and with fewer side effects to humans than synthetic antioxidants (Alenisan, Alqattan, Tolbah, & Shori, 2017). Different antioxidant compounds have been added to dairy products, and an example is grape seed polyphenols that suppress flavor characteristics (Axten, Wohlers, & Wegrzyn, 2008). Another study carried out on cheese supplemented with polyphenolic compounds showed high retention coefficient values. The retention coefficient values were highest in flavone, followed by hesperetin, and lowest in catechin added cheese compared to control; these compounds enhanced the antioxidant activity and storage stability of cheese (Han et al., 2011).

6.9.2 Antioxidant Compounds as Antimicrobial Additive

One of the principal functions of food additive is preservation, and a way to preserve is by protecting against microorganisms that can affect product stability. Generally, antimicrobial compounds are present in food, food packaging, food processing, and food contact surfaces in order to prevent pathogenic growth and/or spoilage microorganisms (Bensid, El Abed, Houicher, Regenstein, & Özogul, 2020). The list of antimicrobials is extensive for food applications including inorganic acids and their sodium salts like nitrites and sulfate, and weak organic acids, such as benzoic, sorbic, propionic, acetic, and lactic acids, potassium sorbate, sodium benzoate, and sodium propionate (Davidson, Taylor, & Schmidt, 2013), which have proven effectiveness. However, bioactive compounds such as phenolic compounds possess antimicrobial activity and can inactivate essential enzymes, react with cell membrane, or alter the function of material genetic, making them ideal candidates for use as an antimicrobial agent in the food industry (Martillanes, Rocha-Pimienta, Cabrera-Bañegil, Martín-Vertedor, & Delgado-Adámez, 2017).

There are several examples of natural antioxidants that can act as antimicrobial agent in food. Carvacrol and thymol phenols have the highest levels of activity against Gram-negative microorganisms, such as *Salmonella typhimurium, E. coli, Klebsiella pneumoniae, Enterobacter cloacae*, and *Yersinia enterocolitica*; thymol being the most active. Such activity is also against Gram-positive bacteria such as *Staphylococcus aureus, Staphylococcus epidermidis, L. monocytogenes*, and *Bacillus subtilis* (Aligiannis, Kalpoutzakis, Mitaku, & Chinou, 2001). It is clear that bioactive components such as polyphenols have antimicrobial activity against food pathogens, and its activity has been tested *in vitro*; however, the effectiveness of these antioxidants within a food matrix remains to be proven.

6.10 Future Trends

Bioactive compounds from semi-desert plants have the potential to be functional food additive in several products. Polyphenols, carotenoids, vitamins, and flavonoids are compounds that are present in these types of plants, and their antioxidant power has been demonstrated in different studies (Jasso de Rodríguez et al., 2017; Salazar et al., 2008). In addition to the antioxidant potential, these bioactive compounds benefit health characteristics that are optimal for any ingredient or food additive, and these substances also have the potential as antimicrobial and as coloring agents. However, there is still much to investigate regarding the application of these compounds and their interaction with food, as well as the organoleptic properties and stability in the food matrix.

6.11 Conclusions

The extreme conditions of semi-desert regions are associated with biota diversity. Plants that grow in the arid and semi-arid areas varied according to climate, species, soil composition, georography, and topography. Recently, some of these semi-desert plants have been the focus of many studies because of their desirable properties. These capacities are related to the evolution and the ability to adapt and survive under environmental conditions in these areas. As a result, plants have modified their morphology, metabolism, and composition to survive. Molecules like minerals, carotenoids, phenols, flavonoids, vitamins, lignans, triterpenes, and tannins are responsible for their larger and diverse biological effects. The uses of semi-desert plants vary depending on the species: as medicinal or edible plants, as a source of products like fiber, pigments, fuel, as important economic resources for the people living in these areas, as an ornament, and recently, as components of functional foods due to their nutraceutical properties; however, other semi-desert plants remain unexplored. Through *in silico*, *in vitro*, or *in vivo* studies, numerous activities related to benefits on human health have been described, primarially due to the high antioxidant capacity exhibited for these vegetal sources. Recently, efforts are being made to verify their medicinal uses, and among the activities reported are effects on metabolic and hemodynamic parameters (lipid and carbohydrate metabolism, maintenance of glucose, cholesterol levels, and blood pressure), anti-inflammatory, cardioprotective, anticancer, hepatoprotective, neuroprotective, and skin health activities. The characterization of semi-desert plants (botanic, phytochemistry, composition, and isolation of bioactive molecules), the discovery of nutritional and pharmacological effects, and elucidation of molecular mechanisms that are implicated in these biological activities are important strategies to propose and develop new and valuable products potentially useful in pharmaceutical, food, and cosmetic industries, placing greater interest on their use as food additive. The potential of these bioactive components as functional ingredient in food matrices would generate functional foods, which are and will be in high demand by consumers who are increasingly looking for healthier foods that provide nutraceutical characteristics.

References

Abou-Gazar, H., Bedir, E., Takamatsu, S., Ferreira, D., & Khan, I. A. (2004). Antioxidant lignans from *Larrea tridentata*. *Phytochemistry*, *65*(17), 2499–2505. https://doi.org/10.1016/j.phytochem.2004.07.009

Abraham, R. J., Abraham, S., Stevens, A. J., Page, S. W., McCluskey, A., Trott, D. J., & O'Handley, R. M. (2019). Aminoguanidines: New leads for treatment of Giardia duodenalis infection. *International Journal for Parasitology: Drugs Drug Resistance*, *10*, 38–44. https://doi.org/10.1016/j.ijpddr.2019.04.003

Ahumada-Santos, Y. P., Montes-Avila, J., de Uribe-Beltrán, M., Díaz-Camacho, S. P., López-Angulo, G., Vega-Aviña, R., . . . Delgado-Vargas, F. (2013). Chemical characterization, antioxidant and antibacterial activities of six Agave species from Sinaloa, Mexico. *Industrial Crops and Products*, *49*, 143–149. https://doi.org/10.1016/j.indcrop.2013.04.050

Alanís-Flores, G. J., & Velasco-Macias, C. G. (2008). Importancia de las cactáceas como recurso natural en el Noreste de México. *Ciencia UANL*, *9*(1), 5–11.

Aldubayan, M. A., Ahmed, A. S., Emara, A. M., Ahmed, A. A., & Elgharabawy, R. M. (2020). Sinapic acid attenuates cardiovascular disorders in rats by modulating reactive oxygen species and angiotensin receptor expression. *Hindawi Oxidative Medicine and Cellular Longevity*, *2020*(Cvd), 1–14. https://doi.org/10.1155/2020/1436858

Alenisan, M. A., Alqattan, H. H., Tolbah, L. S., & Shori, A. B. (2017). Antioxidant properties of dairy products fortified with natural additives: A review. *Journal of the Association of Arab Universities for Basic and Applied Sciences, 24*, 101–106. https://doi.org/10.1016/j.jaubas.2017.05.001

Ali, N., Rashid, S., Nafees, S., Hasan, S. K., Shahid, A., Majed, F., & Sultana, S. (2017). Protective effect of Chlorogenic acid against Methotrexate induced oxidative stress inflammation and apoptosis in rat liver: An experimental approach. *Chemico-Biological Interactions, 1–32*. https://doi.org/10.1016/j.cbi.2017.05.002

Aligiannis, N., Kalpoutzakis, E., Mitaku, S., & Chinou, I. B. (2001). Composition and antimicrobial activity of the essential oils of two Origanum species. *Journal of Agricultural and Food Chemistry, 49*, 4168–4170. https://doi.org/10.1021/jf001494m

Alós, E., Rodrigo, M. J., & Zacarias, L. (2016). Manipulation of carotenoid content in plants to improve human health. *Carotenoids in Nature, 79*, 311–343. https://doi.org/10.1007/978-3-319-39126-7_12

Alqurashi, R. M., Al-Mssallem, M. Q., & Al-Khayri, J. M. (2020). Bioactive compounds of arid and semiarid fruits: Impact on human health. In H. Murthy & V. Bapat (Eds.), *Bioactive Compounds in Underutilized Fruits and Nuts, Reference Series in Phytochemistry* (pp. 21–34). Springer, Cham.

Altemimi, A., Lakhssassi, N., Baharlouei, A., Watson, D. G., & Lightfoot, D. A. (2017). Phytochemicals: Extraction, isolation and identification of bioactive compounds from plants extract. *Plants, 6*(4), 42. https://doi.org/10.3390/plants6040042

Alvarado, M. A., Cardador-Martinez, A. M., Garcia-Trejo, F., & Magallan-Hernandez, F. (2018). Determination of edaphoclimatic conditions and total flavonoids in populations of "Poleo" (*Clinopodium mexicanum*), from the semi-desert of Queretaro, Mexico. In *2018 XIV international engineering congress (CONIIN)*. IEEE. https://doi.org/10.1109/CONIIN.2018.8489797

Alvárez-Pérez, O. B., Ventura-Sobrevilla, J. M., Ascacio-Valdés, J. A., Rojas, R., Verma, D. K., & Aguilar, C. N. (2020). Valorization of *Florensia cernua DC* as source of antioxidants and antifungal bioactives. *Industrial Crops and Products, 152*(15), 112422. https://doi.org/10.1016/j.indcrop.2020.112422

Amidzic, R., Brboric, J., Cudina, O., & Vladimirov, S. (2005). RP-HPLC determination of vitamins B1, B3, B6, folic acid and B12 in multi-vitamin tablets. *Journal of the Serbian Chemical Society, 70*(10), 1229–1235.

Anguiano-Sevilla, L. A., Lugo-Cervantes, E., Ordaz-Pichardo, C., Rosas-Trigueros, J. L., & Jaramillo-Flores, M. E. (2018). Apoptosis induction of agave lechuguilla Torrey extract on human lung adenocarcinoma cells (SK-LU-1). *International Journal of MolecularScience, 19*(12). https://doi.org/10.3390/ijms19123765

Ansari, M. A. (2017). ScienceDirect Sinapic acid modulates Nrf2 / HO-1 signaling pathway in cisplatin-induced nephrotoxicity in rats. *Biomedicine et Pharmacotherapy, 93*, 646–653. https://doi.org/10.1016/j.biopha.2017.06.085

Antunes-Ricardo, M., Hernandez-Reyes, A., Uscanga-Palomeque, A. C., Rodriguez-Padilla, C., Martinez-Torres, A. C., & Gutierrez-Uribe, J. A. (2019). Isorhamnetin glycoside isolated from *Opuntia ficus-indica* (L.) MilI induces apoptosis in human colon cancer cells through mitochondrial damage. *Chemico-Biological Interactions, 310*, 108734. https://doi.org/10.1016/j.cbi.2019.108734

Aragona, M., Lauriano, E. R., Pergolizzi, S., & Faggio, C. (2018). *Opuntia ficus indica* (L.) miller as source of bioactivity compounds for health and nutrition. *Natural Product Research, 32*, 2037–2049. https://doi.org/10.1080/14786419.2017.1365073

Arcila-Lozano, C. C., Loarca-Piña, G., Lecona-Uribe, S., & González de Mejía, E. (2004). El orégano: Propiedades, composición y actividad biológica de sus componentes. *Archivos Latinoamericanos de Nutrición*, *54*(1), 100–111.

Arteaga, S., Andrade-Cetto, A., & Cárdenas, R. (2005). *Larrea tridentata* (creosote bush), an abundant plant of Mexican and US-American deserts and its metabolite nordihydroguaiaretic acid. *Journal of Ethnopharmacology*, *98*(3), 231–239. https://doi.org/10.1016/j.jep.2005.02.002

Arumugam, B., Balagangadharan, K., & Selvamurugan, N. (2018). Syringic acid, a phenolic acid, promotes osteoblast differentiation by stimulation of Runx2 expression and targeting of Smad7 by miR-21 in mouse mesenchymal stem cells. *Journal of Cell Communication and Signaling*, *12*(3), 561–573. https://doi.org/10.1007/s12079-018-0449-3

Ascacio-Valdés, J., Aguilera-Carbó, A., Martínez-Hernández, J., Rodríguez-Herrera, R., & Aguilar, C. (2010). Euphorbia antisyphilitica residues as a new source of ellagic acid. *Chemical Papers*, *64*(4). https://doi.org/10.2478/s11696-010-0034-6

Ashoori, M., & Saedisomeolia, A. (2014). Riboflavin (vitamin B2) and oxidative stress: A review. *British Journal Nutrition*, *111*, 1985–1991. https://doi.org/10.1017/S0007114514000178

Astello-García, M. G., Cervantes, I., Nair, V., Santos-Díaz, M. S., Reyes-Agüero, A., Guéraud, F., . . . Barba de la Rosa, A. P. (2015). Chemical composition and phenolic compounds profile of cladodes from *Opuntia* spp. cultivars with different domestication gradient. *Journal of Food Composition and Analysis*, *43*, 119–130. https://doi.org/10.1016/j.jfca.2015.04.016

Avelino-Flores, M. C., Cruz-Lopez, M. C., Jimenez-Montejo, F. E., & Reyes-Leyva, J. (2015). Cytotoxic activity of the methanolic extract of *Turnera diffusa* Willd on breast cancer cells. *Journal of Medicinal Food*, *18*(3), 299–305. https://doi.org/10.1089/jmf.2013.0055

Ávila-Gálvez, M. Á., González-Sarrías, A., & Espín, J. C. (2018). In vitro research on dietary polyphenols and health: A call of caution and a guide on how to proceed. *Journal of Agricultural and Food Chemistry*, *66*(30), 7857–7858. doi: https://doi.org/10.1021/acs.jafc.8b03377

Avila-Nava, A., Calderón-Oliver, M., Medina-Campos, O. N., Zou, T., Gu, L., Torres, N., . . . Pedraza-Chaverri, J. (2014). Extract of cactus (*Opuntia ficus indica)* cladodes scavenges reactive oxygen species *in vitro* and enhances plasma antioxidant capacity in humans. *Journal of Functional Foods*, *10*, 13–24. https://doi.org/10.1016/j.jff.2014.05.009

Axten, L., Wohlers, M., & Wegrzyn, T. (2008). Using phytochemicals to enhance health benefits of milk: Impact of polyphenols on flavor profile. *Journal of Food Science*, *73*(6), H122–H126. https://doi.org/10.1111/j.1750-3841.2008.00808.x

Baek, J., & Lee, M. G. (2016). Oxidative stress and antioxidant strategies in dermatology. *Redox Report*, *21*(4), 164–169. https://doi.org/10.1179/1351000215Y.0000000015

Bakar, B., Çakmak, M., Ibrahim, M. S., Özer, D., & Saydam, S. (2020). Investigation of amounts of vitamins, lycopene, and elements in the fruits of *Opuntia ficus-indica* subjected to different pretreatments. *Biological Trace Element Research*, *198*(1), 315–323. https://doi.org/10.1007/s12011-020-02050-w

Barbosa-Briones, E., Cardona-Benavides, A., Reyes-Hernández, H., & Muñoz-Robles, C. (2019). Ecohydrological function of vegetation patches in semi-arid shrublands of Central Mexico. *Journal of Arid Environments*, *168*, 36–45. https://doi.org/10.1016/J.JARIDENV.2019.05.005

Barriada-Bernal, L. G., Almaraz-Abarca, N., Delgado-Alvarado, E. A., Gallardo-Velázquez, T., Ávila-Reyes, J. A., Torres-Morán, M. I., . . . Herrera-Arrieta, Y. (2013). Flavonoid composition and antioxidant capacity of the edible flowers ofAgave durangensis(Agavaceae). *CyTA - Journal of Food, 12*(2), 105–114. https://doi.org/10.1080/19476337.2013.801037

Bassino, E., Gasparri, F., & Munaron, L. (2019). Natural dietary antioxidants containing flavonoids modulate keratinocytes physiology: *In vitro* tri-culture models. *Journal of Ethnopharmacology, 238*, 111844. https://doi.org/10.1016/j.jep.2019.111844

Bellows, L., & Moore, R. (2012). *Water-soluble vitamins: B-complex and vitamin C. Colorado*. Fort Collins: State University.

Bensid, A., El Abed, N., Houicher, A., Regenstein, J. M., & Özogul, F. (2020). Antioxidant and antimicrobial preservatives: Properties, mechanism of action and applications in food – A review. *Critical Reviews in Food Science and Nutrition, 18*, 1–17. https://doi.org/10.1080/10408398.2020.1862046

Betancourt-Domínguez, M. A., Hernández-Pérez, T., García-Saucedo, P., Cruz-Hernández, A., & Paredes-López, O. (2006). Physico-chemical changes in cladodes (nopalitos) from cultivated and wild cacti (*Opuntia* spp). *Plant Human for Human Nutrition, 61*, 115–119. https://doi.org/10.1007/s11130-006-0008-6

Bezerra, A. G., Negri, G., Duarte-Almeida, J. M., Smaili, S. S., & Carlini, E. A. (2016). Phytochemical analysis of hydroethanolic extract of Turnera diffusa Willd and evaluation of its effects on astrocyte cell death. *Einstein (São Paulo), 14*(1), 56–63. https://doi.org/10.1590/S1679-45082016AO3386

Briggs, M. K., Lozano-Cavazos, E. A., Mills Poulos, H., Ochoa-Espinoza, J., & Rodríguez-Pineda, J. A. (2020). The Chihuahuan Desert: A binational conservation response to protect a global treasure. In M. I. Goldstein & D. A. DellaSala (Eds.), *Encyclopedia of the World's biomes* (pp. 126–138). *Elsevier*.

Brito-Vega, H., Rodríguez-Maturino, A., González-Mendoza, D., Grimaldo-Juarez, O., & Ceceña-Duran, C. (2014). Antioxidant activity and total carotenoids of selected capsicum species from TabascoMexico. *Research on Crops, 15*(2), 432–436. https://doi.org/10.5958/2348-7542.2014.00133.8

Budinsky, A., Wolfram, R., Oguogho, A., Efthimiou, Y., Stamatopoulos, Y., & Sinzinger, H. (2001). Regular ingestion of *Opuntia robusta* lowers oxidation injury. *Prostaglandins, Leukotrienes, and Essential Fatty Acids, 65*(1), 45–50. https://doi.org/10.1054/plef.2001.0287

Burboa, E. A., Ascacio-Valdés, J. A., Zugasti-Cruz, A., Rodríguez-Herrera, R., & Aguilar, C. N. (2014). Capacidad antioxidante y antibacteriana de extractos de residuos de candelilla. *Revista Mexicana de Ciencias Farmacéuticas, 45*, 51–56.

Cabañas-García, E., Arache, C., Jáuregui-Rincón, J., Cruz-Sosa, F., & Pérez-Molphe Balch, E. (2019). Phytochemical profiling of *Coryphantha macromeris* (Cactaceae) growing in greenhouse conditions using ultra-high-performance liquid chromatography–tandem mass spectrometry. *Molecules, 24*(4), 705. https://doi.org/10.3390/molecules24040705

Cabello-Alvarado, C. J., Sáenz-Galindo, A., Barajas-Bermúdez, L., Pérez-Berumen, C., Ávila-Orta, C., & Valdés-Garza, J. A. (2013). Cera de candelilla y sus aplicaciones. *Avances en Química, 8*(2), 105–110.

Caleja, C., Barros, L., Antonio, A. L., Oliveira, M. B. P. P., & Ferreira, I. S. F. R. (2017). A comparative study between natural and synthetic antioxidants: Evaluation of their performance after incorporation into biscuits. *Food Chemistry*, *216*(1), 342–346. https://doi.org/10.1016/j.foodchem.2016.08.075

Campos, M. R. S., Gómez, K. R., Ordo, Y. M., & Ancona, D. B. (2013). Polyphenols, ascorbic acid and carotenoids contents and antioxidant properties of habanero pepper (*Capsicum chinense*) fruit. *Food and Nutrition Sciences*, *4*, 47–54. https://doi.org/10.4236/FNS.2013.48A006

Cárdenas, Y., Ríos-Silva, M., Huerta, M., López, M., Bricio-Barrios, J., Ortiz-Mesina, M., . . . Trujillo, X. (2019). The comparative effect of nopal and mucilage in metabolic parameters in rats with a high-fructose diet. *Journal of Medicinal Food*, *22*(5), 538–541. https://doi.org/10.1089/jmf.2018.0124

Carmona, J. E., Morales-Martínez, T. K., Mussatto, S. I., Castillo-Quiroz, D., & Ríos-González, L. J. (2017). Propiedades químicas, estructurales y funcionales de la lechuguilla (*Agave lechuguilla* Torr.). *Revista Mexicana de Ciencias Forestales*, *8*(42), 100–122.

Ceci, C., Lacal, P.M., Tentori, L., Martino, M.G. De, Miano,R., & Graziani, G. (2018). Experimental Evidence of the Antitumor, Antimetastatic and Antiangiogenic Activity of Ellagic Acid. *Nutrients*, *10*(1756), 1–23. doi: https://doi.org/10.3390/nu10111756

Cenobio-Galindo, A. J., Pimentel-Gonzalez, D. J., Del Razo-Rodriguez, O. E., Medina-Perez, G., Carrillo-Inungaray, M. L., Reyes-Munguia, A., & Campos-Montiel, R. G. (2019). Antioxidant and antibacterial activities of a starch film with bioextracts microencapsulated from cactus fruits (*Opuntia oligacantha*). *Food Science and Biotechnology*, *28*(5), 1553–1561. https://doi.org/10.1007/s10068-019-00586-9

Chahdoura, H., Barreira, J. C., Adouni, K., Mhadhebi, L., Calhelha, R. C., Snoussi, M., . . . Ferreira, I. C. F. R. (2017). Bioactivity and chemical characterization of *Opuntia macrorhiza* Engelm. Seed oil: Potential food and pharmaceutical applications. *Food and Function*, *8*, 2739–2747. https://doi.org/10.1039/c7fo00731k

Chaurasiya, N. D., Zhao, J., Pandey, P., Doerksen, R. J., Muhammad, I., & Tekwani, B. L. (2019). Selective inhibition of human monoamine oxidase B by acacetin 7-methyl ether isolated from *Turnera diffusa* (Damiana). *Molecules*, *24*(4). https://doi.org/10.3390/molecules24040810

Chaves-Ulate, E. C., & Esquivel-Rodríguez, P. (2019). Ácidos clorogénicos presentes en el café: capacidad antimicrobiana y antioxidante. *Agronomia Mesoamericana*, *30*(1), 299–311. https://doi.org/10.15517/am.v30i1.32974

Cheemanapalli, S., Mopuri, R., Golla, R., & Anuradha, C. M. (2018). *Biomedicine & pharmacotherapy syringic acid (SA) – A review of its occurrence. Biosynthesis, Pharmacological and Industrial Importance*, *108*(September), 547–557. https://doi.org/10.1016/j.biopha.2018.09.069

Chen, P., Chen, F., & Zhou, B. (2018). Antioxidative, anti-inflammatory and and anti-apoptotic effects of ellagic acid in liver and brain of rats treated by D-galactose. *Scientific Reports*, *8*(1465), 2–11. https://doi.org/10.1038/s41598-018-19732-0

Cheng, A. W., Tan, X., Sun, J. Y., Gu, C. M., & Liu, C. (2019). Catechin attenuates TNF- α induced inflammatory response via AMPK-SIRT1 pathway in 3T3-L1 adipocytes. *PLoS ONE*, *14*(5), 1–15. https://doi.org/10.1371/journal.pone.0217090.pone.0217090

Cory, H., Passarelli, S., Szeto, J., Tamez, M., & Mattei, J. (2018). The role of polyphenols in human health and food systems : A mini-review. *Frontiers in Nutrition*, *5*(87), 1–9. https://doi.org/10.3389/fnut.2018.00087

Dalmagro, A. P., Camargo, A., & Zeni, A. L. B. (2017). Morus nigra and its major phenolic, syringic acid, have antidepressant-like and neuroprotective effects in mice. *Metabolic Brain Disease*, *32*(6), 1963–1973. https://doi.org/10.1007/s11011-017-0089-y

Davidson, P. M., Taylor, T. M., & Schmidt, S. E. (2013). Chemical preservatives and natural antimicrobial compounds. In M. P. Doyle & E. L. Buchanan (Eds.), *Food microbiology: Fundamentals and frontiers* (pp. 765–802). Washington, DC: American Society for Microbiology Press.

De Rodríguez, D. J., & An, U. A. A. (2020). *Extractos de plantas del semidesierto en la inducción del crecimiento de tomate (Lycopersicon esculentum Mill) Extracts of plants from the semidesert in the induction of tomato (Lycopersicon esculentum Mill)* growth. *Ecosystems and Agricultural Resources*, *7*(1), 1–10. https://doi.org/10.19136/era.a7n1.2342

Decker, E. A., Chen, B., Panya, A., & Elias, R. J. (2010). Understanding antioxidant mechanisms in preventing oxidation in foods. In E. A. Decker (Ed.), *Oxidation in Foods and Beverages and Antioxidant Applications* (pp. 225–248). Woodhead Publishing.

Delgado, M., Mendez, J., Rodríguez-Herrera, R., Aguilar, C. N., Cruz-Hernández, M., & Balagurusamy, N. (2014). Characterization of phosphate-solubilizing bacteria isolated from the arid soils of a semi-desert region of north-East Mexico. *Biological Agriculture and Horticulture*, *30*(3), 211–217. https://doi.org/10.1080/01448765.2014.909742

Di Pietro, N., Di Tomo, P., & Pandolfi, A. (2016). Carotenoids in cardiovascular disease prevention. *JSM Atherosclerosis*, *1*(1), 1002.

Dreuw, A., Fleming, G. R., & Head-Gordon, M. (2005). Role of electron-transfer quenching of chlorophyll fluorescence by carotenoids in non-photochemical quenching of green plants. *Biochemical Society Transactions*, *33*(4), 858–862. https://doi.org/10.1042/BST0330858

Du, W., An, Y., He, X., Zhang, D., & He, W. (2018). Protection of kaempferol on oxidative stress-induced retinal pigment epithelial cell damage. *Hindawi Oxidative Medicine and Cellular Longevity*, *1610751*, 14. https://doi.org/10.1155/2018/1610751

Eggersdorfer, M., & Wyss, A. (2018). Carotenoids in human nutrition and health. *Archives of Biochemistry and Biophysics*, *652*, 18–26. https://doi.org/10.1016/j.abb.2018.06.001

El Kharrassi, Y., El Maaidan, E., Chakhchar, A., Lamaoui, M., Bourhim, T., Mabrouk, A., . . . Nasser, B. (2020). Physicochemical, phytochemical and antioxidant properties of juice and seed oil of cactus pear *Opuntia aequatorialis* and *Opuntia leucotricha*. *Vegetos*, *33*(4), 682–689. https://doi.org/10.1007/s42535-020-00158-x

El Kharrassi, Y., Maata, N., Mazri, M. A., El Kamouni, S., Talbi, M., El Keb-baj, R., . . . Nasser, B. (2018). Chemical and phytochemical characterizations of argan oil (*Argania spinosa* L. skeels), olive oil (*Olea europaea* L. cv. *Moroccan picholine*), cactus pear (*Opuntia megacantha* salm-dyck) seed oil and cactus cladode essential oil. *Journal of Food Measurement and Characterization*, *12*, 747–754. https://doi.org/10.1007/s11694-017-9688-x

El Kharrassi, Y., Mazri, M. A., Benyahia, H., Benaouda, H., Mzouri, E. L., & Nasser, E. (2016). Fruit and juice characteristics of two cactus pear species (*Opuntia ficus indica* and *Opuntia megacantha*) from different regions of Morocco. *LWT Food Science and Technology*, *65*, 610–617. https://doi.org/10.1016/j.lwt.2015.08.044

El Kossori, R. L., Villaume, C., El Boustani, E., Sauvaire, Y., & Méjean, L. (1998). Composition of pulp, skin and seeds of prickly pears fruit *(Opuntia ficus indica)*. *Plant Foods for Human Nutrition*, *52*, 263–270. https://doi.org/10.1023/a:1008000232406

Encina-Dominguez, J. A., Arévalo-Sierra, J. R., Villarreal-Quintanilla, J. A., & Estrada Castillon, E. (2020). Composition, structure and richness of vascular plants of the desert scrub in the nirth of coahuila. *Méxicio. Botanical Sciences*, *98*(1), 1–15. https://doi.org/10.17129/botsci.2251

Enogieru, A. B., Haylett, W., Hiss, D. C., Bardien, S., & Ekpo, O. E. (2018). Review article rutin as a potent antioxidant : implications for neurodegenerative disorders. *Hindawi Oxidative Medicine and Cellular Longevity*, *6241017*, 17. https://doi.org/10.1155/2018/6241017

Eroğlu, C., Avcı, E., Vural, H., & Kurar, E. (2018). Anticancer mechanism of Sinapic acid in PC-3 and LNCaP human prostate cancer cell lines. *Gene*, *671*(2017). #pagerange#. https://doi.org/10.1016/j.gene.2018.05.049

Esquivel-Gutiérrez, E. R., Alcaraz-Meléndez, L., Hernández-Herrera, R., Torres, A., & Rodríguez-Jaramillo, C. (2018). Effects of damiana (*Turnera diffusa*; var. *diffusa* and var. *aphrodisiaca*) on diabetic rats. *Acta Universitaria*, *28*, 84–92.

Fang, W., Zhu, S., Niu, Z., & Yin, Y. (2019). The protective effect of syringic acid on dextran sulfatesodium-induced experimental colitis in BALB / c mice. *Wiley Periodicals*, January, 1–10.https://doi.org/10.1002/ddr.21524

Favela-Hernandez, J. M., Garcia, A., Garza-Gonzalez, E., Rivas-Galindo, V. M., & Camacho-Corona, M. R. (2012). Antibacterial and antimycobacterial lignans and flavonoids from *Larrea tridentata*. *Phytotherapy Research*, *26*(12), 1957–1960. https://doi.org/10.1002/ptr.4660

Fernandes De Araújo, F., De Paulo Farias, D., & Neri-numa, I. A. (2021). Polyphenols and their applications: An approach in food chemistry and innovation potential. *Food Chemistry*, *338*, 127535. https://doi.org/10.1016/j.foodchem.2020.127535

Flores-Gallegos, A. C., Cruz-Requena, M., Castillo-Reyes, F., Rutiaga-Quiñones, O. M., Sepulveda-Torre, L., Paredes-Ortíz, A., . . . Rodriguez-Herrera, R. (2019). Sotol, an alcoholic beverage with rising importance in the worldwide commerce. In A. M. Grumezescu & A. M. Holban (Eds.), *Alcoholic beverages* (pp. 141–160). Woodhead Publishing.

Flores-Hernández, A., Hernández-Herrera, J. A., Madinaveitia-Rios, H., Valenzuela-Nuñez, L. M., Murillo-Amador, B., Rueda-Puente, E. O., . . . Ortiz-Cano, H. G. (2011). Evaluación de la población natural y hábitat de palma azul (*Yucca rigida*) en Mapimi, DurangoMéxico. *Tropical and Subtropical Agroecosystems*, *14*(1), 315–321.

Floriano-Sánchez, E., Villanueva, C., Medina-Campos, O. N., Rocha, D., Sánchez-González, D. J., Cárdenas-Rodríguez, N., & Pedraza-Chaverrí, J. (2009). Nordihydroguaiaretic acid is a potent *in vitro* scavenger of peroxynitrite, singlet oxygen, hydroxyl radical, superoxide anion and hypochlorous acid and prevents in vivo ozone-induced tyrosine nitration in lungs. *Free Radical Research*, *40*(5), 523–533. https://doi.org/10.1080/10715760500419365

Franco, A. C., Matsubara, S., & Orthen, B. (2007). Photoinhibition, carotenoid composition and the co-regulation of photochemical and non-photochemical quenching in neotropical savanna trees. *Tree Physiology*, *27*(5), 717–725. https://doi.org/10.1093/treephys/27.5.717

Gammone, M. A., Riccioni, G., & D'Orazio, N. (2015). Marine carotenoids against oxidative stress: Effects on human health. *Marine Drugs*, *13*(10), 6226–6246. https://doi.org/10.3390/md13106226

Ganeshpurkar, A., & Saluja, A. K. (2016). The pharmacological potential of Rutin. *Saudi Pharmaceutical Journal*, *25*(2), 149–164. https://doi.org/10.1016/j.jsps.2016.04.025

Ganeshpurkar, A., & Saluja, A. K. (2018). International immunopharmacology protective effect of catechin on humoral and cell mediated immunity in rat model. *International Immunopharmacology*, *54*(September 2017), 261–266. https://doi.org/10.1016/j.intimp.2017.11.022

Gao, L., Yao, R., Liu, Y., Wang, Z., Huang, Z., & Du, B. (2017). Isorhamnetin protects against cardiac hypertrophy through blocking PI3K – AKT pathway. *Molecular and Cellular Biochemistry*, *429*(1), 167–177. https://doi.org/10.1007/s11010-017-2944-x

García, J. E., Gómez, L., Mendoza-de-Gives, P., Rivera-Corona, J. L., Millán-Orozco, J., Ascacio, J. A., . . . Mellado, M. (2018). Anthelmintic efficacy of hydro-methanolic extracts of *Larrea tridentata* against larvae of *Haemonchus contortus*. *Tropical Animal Health and Production*, *50*(5), 1099–1105. https://doi.org/10.1007/s11250-018-1535-5

García-Andrade, M., González-Laredo, R. F., Rocha-Guzmán, N. E., Gallegos-Infante, J. A., Rosales-Castro, M., & Medina-Torres, L. (2013). Mesquite leaves (*Prosopis laevigata*), a natural resource with antioxidant capacity and cardioprotection potential. *Industrial Crops and Products*, *44*, 336–342. https://doi.org/10.1016/j.indcrop.2012.11.030

Gebhardt, M., Fehmi, J. S., Rasmussen, C., & Gallery, R. E. (2017). Soil amendments alter plant biomass and soil microbial activity in a semi-desert grassland. *Plant and Soil*, *419*(1–2), 53–70. https://doi.org/10.1007/s11104-017-3327-5

Gheena, S., & Ezhilarasan, D. (2019). Syringic acid triggers reactive oxygen species – mediated cytotoxicity in HepG2 cells. *Human and Experimental Toxicology*, *1*(162), 1–9. https://doi.org/10.1177/0960327119839173

Ghorbani, A. (2017). Biomedicine & pharmacotherapy mechanisms of antidiabetic eff ects of flavonoid rutin. *Biomedicine & Pharmacotherapy*, *96*(August), 305–312. https://doi.org/10.1016/j.biopha.2017.10.001

Goleniowski, M., Bonfil, l. M., Cusido, R., & Palazón, J. (2013). In K. Ramawat & J. M. Mérillon (Eds.), *Ácidos fenólicos* (Vol. 3, pp. 22144). Berlin Heidelberg: Springer.

González Güereca, M. C., Soto Hernández, M., Kite, G., & Martínez Vázquez, M. (2007). Actividad antioxidante de flavonoides del tallo de orégano mexicano (*Lippia graveolens* HBK var. *berlandieri* Schauer). *Revista Fitotecnia Mexicana*, *30*(1), 43–49.

González-Cruz, L., Filardo-Kerstupp, S., Bello-Pérez, L. A., Güemes-Vera, N., & Bernardino-Nicanor, A. (2012). Carotenoid content, antioxidant activity and sensory evaluation of low-calorie nopal (*Opuntian ficus-indica*) marmalade. *Journal of Food Processing and Preservation*, *36*, 267–275. https://doi.org/10.1111/j.1745-4549.2011.00589.x

Granados-Sánchez, D., Sánchez-González, A., Granados Victorino, R. L., & Borja de la Rosa, A. (2011). Ecología de la vegetación del desierto chihuahuense. *Revista Chapingo Series Ciencias Forestales y del Ambiente*, *17*, 111–130. https://doi.org/10.5154/r.rchscfa.2010.10.102

Griffith, M. P. (2004). The origins of an important cactus crop, *Opuntia ficus-indica* (Cactaceae): New molecular evidence. *American Journal of Botany*, *91*(11), 1915–1921. https://doi.org/10.3732/ajb.91.11.1915

Grubb, P. J., Marañón, T., Pugnaire, F. I., & Sack, L. (2015). Relationships between specific leaf area and leaf composition in succulent and non-succulent species of contrasting semi-desert communities in South-Eastern Spain. *Journal of Arid Environments*, *118*, 69–83. https://doi.org/10.1016/j.jaridenv.2015.03.001

Gupta, P., Mohammad, T., Khan, P., Alajmi, M. F., & Hussain, A. (2019). Biomedicine & pharmacotherapy evaluation of ellagic acid as an inhibitor of sphingosine kinase 1: A targeted approach towards anticancer therapy. *Biomedicine & Pharmacotherapy, 118*, 109245. https://doi.org/10.1016/j.biopha.2019.109245

Gupta, R. K., Patel, A. K., Shah, N., Choudhary, A. K., Jha, U. K., Yadav, U. C., . . . Pakuwal, U. (2014). Oxidative stress and antioxidants in disease and cancer: A review. *Asian Pacific Journal of Cancer Prevention, 15*(11), 4405–4409. https://doi.org/10.7314/APJCP.2014.15.11.4405

Gutierrez-Grijalva, E. P., Picos-Salas, M. A., Leyva-Lopez, N., Criollo-Mendoza, M. S., Vazquez-Olivo, G., & Heredia, J. B. (2017). Flavonoids and phenolic acids from oregano: Occurrence, biological activity and health benefits. *Plants (Basel), 7*(1). https://doi.org/10.3390/plants7010002

Gutiérrez-Tlahque, J., Aguirre-Mancilla, C. L., López-Palestina, C., Sánchez-Fernández, R. E., Hernández-Fuentes, A. D., & Martín Torres-Valencia, J. (2019). Constituents, antioxidant and antifungal properties of *Jatropha dioica* var. *dioica*. *Natural Product Communications, 14*(5). https://doi.org/10.1177/1934578X19852433

Gutiérrez-Tlahque, J., Leobardo, C., Raya-Pérez, J., Ramírez-Pimentel, J., Jiménez-Alvarado, R., & Hernández-Fuentes, A. (2018). Effect of climate conditions on total phenolic content and antioxidant activity of *Jatropha dioica* Cerv. Var. *dioica*. *Ciencia e Investigación Agraria, 45*(1), 70–81. https://doi.org/10.7764/rcia.v45i1.1832

Guzmán Loayza, D., & Chávez, J. (2007). Estudio bromatológico del cladodio del nopal (*Opuntia ficus-indica*) para el consumo humano. *Revista de la Sociedad Química del Perú, 73*(1), 41–45.

Guzmán-Maldonado, S. H., Morales-Montelongo, A. L., Mondragón-Jacobo, C., Herrera-Hernández, G., Guevara-Lara, F., & Reynoso-Camacho, R. (2010). Physicochemical, nutritional, and functional characterization of fruits Xoconostle (*Opuntia matudae*) pears from Central-México region. *Journal of Food Science, 75*(6), C485–C492. https://doi.org/10.1111/j.1750-3841.2010.01679.x

Hamad, A. F., Han, J. H., Kim, B. C., & Rather, I. A. (2018). The intertwine of nanotechnology with the food industry. *Saudi Journal of Biological Sciences, 25*, 27–30. https://doi.org/10.1016/j.sjbs.2017.09.004

Han, J., Britten, M., St-Gelais, D., Champagne, C. P., Fustier, P., Salmieri, S., & Lacroix, M. (2011). Polyphenolic compounds as functional ingredients in cheese. *Food Chemistry, 124*(4), 1589–1594. https://doi.org/10.1016/j.foodchem.2010.08.021

Handschin, A., Henny-Fullin, K., Buess, D., Leuppi, J., & Dieterle, T. (2015). Hypertension in the elderly. *Therapeutische Umschau, 72*(6), 397–403. https://doi.org/10.1024/0040-5930/a000692

Hano, C. F., Dinkova-Kostova, A. T., Davin, L. B., Cort, J. R., & Lewis, N. G. (2021). Lignans: Insights in to their biosynthesis, metabolic engineering, analytical methods and health benefits. *Frontiers in Plant Science, 11*. https://doi.org/10.3389/fpls.2020.630327

Harlev, E., Nevo, E., Solowey, E., & Bishayee, A. (2013). Cancer preventive and curative attributes of plants of the Cactaceae family: A review. *Planta Medica, 79*(9), 713–722. https://doi.org/10.1055/s-0032-1328632

Henciya, S., Seturaman, P., James, A. R., Tsai, Y.-H., Nikam, R., Wu, Y.-C., . . . Chang, F. R. (2017). Biopharmaceutical potentials of Prosopis spp. (Mimosaceae, Leguminosa). *Journal of Food and Drug Analysis, 25*(1), 187–196. https://doi.org/10.1016/j.jfda.2016.11.001

Herebian, D., Seibt, A., Smits, S. H. J., Rodenburg, R. J., Mayatepek, E., & Distelmaier, F. (2017). 4-Hydroxybenzoic acid restores CoQ 10 biosynthesis in human COQ2 deficiency. *Annals of Clinical and Translational Neurology*, *4*, 902–908. https://doi.org/10.1002/acn3.486

Hirschberg, J. (2001). Carotenoid biosynthesis in flowering plants. *Current Opinion in Plant Biology*, *4*(3), 210–218. https://doi.org/10.1016/s1369-5266(00)00163-1

Hyder, P. W., Fredrickson, E. L., Estell, R. E., Tellez, M., & Gibbens, R. P. (2002). Distribution and concentration of total phenolics, condensed tannins, and nordihydroguaiaretic acid (NDGA) in creosotebush (Larrea tridentata). *Biochemical Systematics and Ecology*, *30*(10), 905–912. https://doi.org/10.1016/S0305-1978(02)00050-9

Ibrahim, M., Nadir, M., Ali, A., Ahmad, V. U., & Rasheed, M. (2013). Phytochemical analyses of *Prosopis juliflora*Swartz dc. *Pakistan Journal of Botany*, *45*(6), 2101–2104.

Iriti, M., Kubina, R., Cochis, A., Sorrentino, R., Varoni, E. M., Kaba, A., . . . Wojtyczka, R. D. (2017). *Rutin, a quercetin glycoside. Restores Chemosensitivity in Human Breast Cancer Cells.*, *1538*(July), 1529–1538. https://doi.org/10.1002/ptr.5878

Isaac, A. A. (2016). Overview of cactus *Opuntia Ficus-Indica* (L): A myriad of alternatives. *Studies on Ethno Medicine*, *10*(2), 195–205. https://doi.org/10.1080/09735070.2016.11905488

Isemura, M. (2019). In human health and disease catechin in human health and disease. *Molecules*, *24*, 528. https://doi.org/10.3390/molecules24030528

Jasso de Rodriguez, D., Carrillo-Lomeli, D. A., Rocha-Guzman, N. E., Moreno-Jimenez, M. R., Rodriguez-Garcia, R., Diaz-Jimenez, M. L. V., Flores-Lopez, M. L., Villarreal-Quintanilla, J. A.(2017). Antioxidant, anti-inflammatory and apoptotic effects of Flourensiamicrophylla on HT-29 colon cancer cells. *Industrial Crops and Products*, 107, 472–481. DOI: https://doi.org/10.1016/j.indcrop.2017.03.034

Jiang, Y., Ding, S., Li, F., Zhang, C., Sun-waterhouse, D., Chen, Y., & Li, D. (2019). Effects of (+) -catechin on the differentiation and lipid metabolism of 3T3- L1 adipocytes. *Journal of Functional Foods*, *62*(103558). https://doi.org/10.1016/j.jff.2019.103558

Kashyap, D., Sharma, A., Singh, H., Sak, K., Punia, S., & Mukherjee, T. K. (2017). Kaempferol – A dietary anticancer molecule with multiple mechanisms of action : Recent trends and advancements. *Journal of Functional Foods*, *30*, 203–219. https://doi.org/10.1016/j.jff.2017.01.022

Kennedy, D. O. (2016). B vitamins and the brain: Mechanisms, dose and efficacy-a review. *Nutrients*, *8*, 68–97. https://doi.org/10.3390/nu8020068

Khalid, M., Saeed-ur-Rahman, Bilal, M., Iqbal, H. M. N., & Huang, D. F. (2019). Biosynthesis and biomedical perspectives of carotenoids with special reference to human health-related applications. *Biocatalysis and Agricultural Biotechnology*, *17*, 399–407. https://doi.org/10.1016/j.bcab.2018.11.027

Kho, A. R., Choi, B. Y., Lee, S. H., Hong, D. K., Lee, S. H., Jeong, J. H., . . . Choi, H. C. (2018). Effects of protocatechuic acid (PCA) on global cerebral ischemia-induced hippocampal neuronal death. *Hindawi Oxidative Medicine and Cellular Longevity*, *19*(5), 1–19. https://doi.org/10.3390/ijms19051420

Kim, E., Song, K., Chang, P., Jeong, C., & Keun, S. (2018). Syringic acid prevents skin carcinogenesis via regulation of NoX and EGFR signaling. *Biochemical Pharmacology*, *18*(2952). https://doi.org/10.1016/j.bcp.2018.06.007

Kim, J. W., Kim, T. B., Kim, H. W., Park, S. W., Kim, H. P., & Sung, S. H. (2017). Hepatoprotective flavonoids in *Opuntia ficus-indica*fruits by reducing oxidative stress in primary rat hepatocytes. *Pharmacognosy Magazine*, *13*(51), 472–476. https://doi.org/10.4103/pm.pm_232_16

Kirilovsky, D. (2007). Photoprotection in cyanobacteria: The orange carotenoid protein (OCP)-related non-photochemical-quenching mechanism. *Photosynthesis Research*, *93*(1–3), 7–16. https://doi.org/10.1007/s11120-007-9168-y

Lavelli, V., Kerr, W. L., García-Lomillo, J., & González-San José, M. L. (2017). Applications of recovered bioactive compounds in food product. In *Handbook of grape processing by-products* (pp. 233–266). Academic Press.

Lee, M. (2018). Effects of isorhamnetin on adipocyte mitochondrial biogenesis and AMPK activation. *Molecules*, *23*(1853), 1–12. https://doi.org/10.3390/molecules23081853

Lee, S. K., & Kader, A. A. (2000). Preharvest and postharvest factors influencing vitamin C content of horticultural crop. *Postharvest Biology and Technology*, *20*(3), 207–220. https://doi.org/10.1016/S0925-5214(00)00133-2

Lin, C.Y., Huang,C.S., Huang,C.Y., & Yin, M.C. (2009). Anticoagulatory, Antiinflammatory, and Antioxidative Effects of Protocatechuic Acid in Diabetic Mice. *Journal of Lipid Research*, *57*, 6661–6667. https://doi.org/10.1021/jf9015202

Lin, T., Wu, P., Hou, C., Chien, T., & Chang, Q. (2019). Protective eff ects of sesamin against UVB-induced skin inflammation and photodamage in vitro and in vivo. *Biomolecules*, *9*(479). https://doi.org/10.3390/biom9090479

Lira-Saldívar, R. H. (2003). Estado actual del conocimiento sobre las propiedades biocidas de la gobernadora [*Larrea tridentata* (D.C.) Coville]. *Revista Mexicana de Fitopatología*, *21*(2), 214–222.

Liu, N., & Guan, L. (2012). Linkages between woody plant proliferation dynamics and plant physiological traits in southwestern North America. *Journal of Plant Ecology*, *5*(4), 407–416. https://doi.org/10.1093/jpe/rts002

Liu, Y., Xu, Z., Yang, G., Yang, D., Ding, J., Chen, H., . . . Tian, H. (2017). Sesamin alleviates blood-brain barrier disruption in mice with experimental traumatic brain injury. *Nature Publishing Group*, *38*(11), 1445–1455. https://doi.org/10.1038/aps.2017.103

Llansola-Portoles, M. J., Pascal, A. A., & Robert, B. (2017). Electronic and vibrational properties of carotenoids: From *in vitro* to *in vivo*. *Journal of the Real Society Interface.*, *14*(135), 20170504. https://doi.org/10.1098/rsif.2017.0504

López-Romero, P., Pichardo-Ontiveros, E., Avila-Nava, A., Vázquez-Manjarrez, N., Tovar, A. R., Pedraza-Chaverri, J., & Torres, N. (2014). The effect of nopal (*Opuntia Ficus Indica*) on postprandial blood glucose, incretins, and antioxidant activity in mexican patients with type 2 diabetes after consumption of two Ddfferent composition breakfasts. *Journal of the Academy of Nutrition and Dietetics*, *114*(11), 1811–1818. https://doi.org/10.1016/j.jand.2014.06.352

Lorenzo, J. M., Pateiro, M., Domínguez, R., Barba, F. J., Putnik, P., Kovačević, D. B., . . . Franco, D. (2017). Berries extracts as natural antioxidants in meat products: A review. *Food Research International*, *106*, 1095–1104. https://doi.org/10.1016/j.foodres.2017.12.005

Lü, J. M., Lin, P. H., Yao, Q., & Chen, C. (2010). Chemical and molecular mechanisms of antioxidants: Experimental approaches and model systems. *Journal of Cellular and Molecular Medicine*, *14*(4), 840–860. https://doi.org/10.1111/j.1582-4934.2009.00897.x

Lü, J.-M., Nurko, J., Weakley, S. M., Jiang, J., Kougias, P., Lin, P. H., . . . Chen, C. (2010). Molecular mechanisms and clinical applications of nordihydroguaiaretic acid (NDGA) and its derivatives: An update. *Medical Science Monitor: International Medical journal of Experimental and Clinical Research*, *16*(5), RA93–RA100.

Lu, X., Liu, T., Chen, K., Xia, Y., Dai, W., Xu, S., . . . Fan, X. (2018). Biomedicine & pharmacotherapy isorhamnetin : A hepatoprotective flavonoid inhibits apoptosis and autophagy via P38 / PPAR- α pathway in mice. *Biomedicine & Pharmacotherapy*, *103*(2017), 800–811. https://doi.org/10.1016/j.biopha.2018.04.016

Lú-chau, T. A., Moreira, M. T., Lema, J. M., & Eibes, G. (2017). A review on extraction, identification and purification methods, biological activities and approaches to enhance its bioavailability. *Trends in Food Science & Technology Rutin*, *67*, 220–235. doi: https://doi.org/10.1016/j.tifs.2017.07.008

Luna-Vazquez, F. J., Ibarra-Alvarado, C., Camacho-Corona, M. D. R., Rojas-Molina, A., Rojas-Molina, J. I., Garcia, A., & Bah, M. (2018). Vasodilator activity of compounds isolated from plants used in mexican traditional medicine. *Molecules*, *23*(6). https://doi.org/10.3390/molecules23061474

Ma, Y., Chen, F., Yang, S., Chen, B., & Shi, J. (2018). Biomedicine & pharmacotherapy protocatechuic acid ameliorates high glucose-induced extracellular matrix accumulation in diabetic nephropathy. *Biomedicine & Pharmacotherapy*, *98*(8), 18–22. https://doi.org/10.1016/j.biopha.2017.12.032

Majdalawieh, A. F., Dalibalta, S., & Yousef, S. M. (2020). Effects of sesamin on fatty acid and cholesterol metabolism, macrophage cholesterol homeostasis and serum lipid profile: A comprehensive review. *European Journal of Pharmacology*, *885*(July), 173417. https://doi.org/10.1016/j.ejphar.2020.173417

Majdalawieh, A. F., Massri, M., & Gheyath, K. (2017). A comprehensive review on the anti-cancer properties and mechanisms of action of sesamin, alignan in sesame seeds (Sesamum indicum). *European Journal of Pharmacology*, *17*, 20670. https://doi.org/10.1016/j.ejphar.2017.10.020

Manda, G., Rojo, A. I., Martínez-Klimova, E., Pedraza-Chaverri, J., & Cuadrado, A. (2020). Nordihydroguaiaretic acid: From herbal medicine to clinical development for cancer and chronic diseases. *Frontiers in Pharmacology*, *11*. https://doi.org/10.3389/fphar.2020.00151

Marra, M. V. (2019). Dietary Quercetin and Kaempferol: Bioavailability in humans. *Nutrients*, *11*(2288). https://doi.org/i:10.3390/nu11102288

Martillanes, S., Rocha-Pimienta, J., Cabrera-Bañegil, M., Martín-Vertedor, D., & Delgado-Adámez, J. (2017). Application of phenolic compounds for food preservation: Food additive and active packaging. In M. Soto-Hernandez, M. Palma-Tenango, & M. del Rosario Garcia-Mateos (Eds.), *Phenolic Compounds - Biological Activity* (pp. 39–58). IntechOpen.

Martínez, M. (1989). *Plantas medicinales de México* (pp. 187). México, D.F: Editorial Botas.

Martínez, N., Almaguer, G., Vázquez-Alvarado, P., Figueroa, A., Zuñiga, C., & Hernández-Ceruelos, A. (2014). Análisis fitoquímico de *Jatropha dioica* y determinación de su efecto antioxidante y quimioprotector sobre el potencial genotóxico de ciclofosfamida, daunorrubicina y metilmetanosulfonato evaluado mediante el ensayo cometa. *Boletín Latinoamericano y del Caribe de Plantas Medicinales y Aromáticas*, *13*(5), 437–457.

Martínez-Rocha, A., Puga, R., Hernández-Sandoval, L., Loarca-Piña, G., & Mendoza, S. (2008). Antioxidant and antimutagenic activities of mexican oregano (*Lippia graveolens Kunth*). *Plant Foods for Human Nutrition*, *63*(1), 1–5. https://doi.org/10.1007/s11130-007-0061-9

Martins, S., Aguilar, C. N., Teixeira, J. A., & Mussatto, S. I. (2012). Bioactive compounds (phytoestrogens) recovery from *Larrea tridentata* leaves by solvents extraction. *Separation and Purification Technology*, *88*, 163–167. https://doi.org/10.1016/j.seppur.2011.12.020

Masella, R., Superiore, I., Volti, G. L., & Superiore, I. (2015). Protocatechuic acid and human disease prevention: biological activities and molecular mechanisms. *Current Medicinal Chemistry*, *19*(June). https://doi.org/10.2174/092986712800672102

Matias, A., Nunes, S. L., Poejo, J., Mecha, E., Serra, A. T., Madeira, P. J. A., . . . Duarte, C. M. M. (2014). Antioxidant and anti-inflammatory activity of a flavonoid-rich concentrate recovered from *Opuntia ficus-indica* juice. *Food and Function*, *5*(12), 3269–3280. https://doi.org/10.1039/c4fo00071d

Medina, E. M. D., Rodríguez, E. M. R., & Romero, C. D. (2007). Chemical characterization of *Opuntia dillenii* and *Opuntia ficus indica* fruits. *Food Chemistry*, *103*(1), 38–45. https://doi.org/10.1016/j.foodchem.2006.06.064

Medina-Guillén, R., Cantú-Silva, I., Estrada-Castillón, E., González-Rodríguez, H., & Delgadillo-Villalobos, J. A. (2017). Estructura y diversidad del matorral desértico rosetófilo rehabilitado con rodillo aireador, Coahuila, México. *Polibotánica*, *44*, 95–107. https://doi.org/10.18387/polibotanica.44.7

Meléndez-Martínez, A. J., Vicario, I. M., & Heredia, F. J. (2004). Importancia nutricional de los pigmentos carotenoides. *Archivos Latinoamericanos de Nutrición*, *54*(2), 149–155. http://ve.scielo.org/scielo.php?script=sci_arttext&pid=S0004-06222004000200003&lng=es&tlng=es

Monreal-García, H. M., Almaraz-Abarca, N., Ávila-Reyes, J. A., Torres-Ricario, R., González-Elizondo, M. S., Herrera-Arrieta, Y., & Gutierrez-Velázquez, M. V. (2019). Phytochemical variation among populations of *Fouquieria splendens* Engelm. (Fouquieriaceae). *Botanical Sciences*, *97*(3), 398–412. https://doi.org/10.17129/botsci.2191

Montaño-Arias, N. M., García-Sánchez, R., Ochoa-de la Rosa, G., & Monroy-Ata, A. (2006). Relación entre la vegetación arbustiva, el mezquite y el suelo de un ecosistema semiárido en México. *Terra Latinoamericana*, *24*(2), 193–205. https://www.redalyc.org/articulo.oa?id=57311108006

Montiel-Sanchez, M., Garcia-Cayuela, T., Gomez-Maqueo, A., Garcia, H. S., & Cano, M. P. (2020). *In vitro* gastrointestinal stability, bioaccessibility and potential biological activities of betalains and phenolic compounds in cactus berry fruits (*Myrtillocactus geometrizans*). *Food Chemistry*, *128087*. https://doi.org/10.1016/j.foodchem.2020.128087

Morales, P., Ramírez-Moreno, E., Sanchez-Mata, M. C., Carvalho, A. M., & Ferreira, I. C. F. R. (2012). Nutritional and antioxidant properties of pulp and seeds of two xoconostle cultivars (*Opuntia joconostle* F.A.C. Weber ex Diguet and *Opuntia matudae Scheinvar*) of high consumption in Mexico. *Food Research International*, *46*(1), 279–285. https://doi.org/10.1016/j.foodres.2011.12.031

Morán-Santibañez, K., Vasquez, A. H., Varela-Ramirez, A., Henderson, V., Sweeney, J., Odero-Marah, V., . . . Skouta, R. (2019). *Larrea tridentata* extract mitigates oxidative stress-induced cytotoxicity in human neuroblastoma SH-SY5Y cells. *Antioxidants*, *8*(10), 427. https://doi.org/10.3390/antiox8100427

Mran, M., Salehi, B., Sharifi-rad, J., Gondal, T. A., Arshad, M. U., Khan, H., & Guerreiro, S. G. (2019). Kaempferol: A key emphasis to its anticancer potential. *Molecules*, *24*(12), 277. https://doi.org/10.3390/molecules24122277

Muhammad, I., Ashiru, S., Ibrahim, D. I., Kanoma, A. I., Sani, I., & Garba, S. (2014). Effect of ripening stage on vitamin C content in selected fruits. *International Journal of Agriculture, Forestry and Fisheries*, *2*(3), 60–65. https://doi.org/10.22161/ijeab/2.2.19

Mun, E. J., Babiker, H. M., Weinberg, U., Kirson, E. D., & Von Hoff, D. D. (2018). Tumor-treating fields: A fourth modality in cancer treatment. *Clinical Cancer Research*, *24*(2), 266–275. https://doi.org/10.1158/1078-0432.CCR-17-1117

Nieto, G., Ros, G., & Castillo, J. (2018). Antioxidant and antimicrobial properties of rosemary (*Rosmarinus officinalis*,L.): A review. *Medicines*, *5*(3), 98. https://doi.org/10.3390/medicines5030098

Nikoo, M., Regenstein, J. M., & Gavlighi, H. A. (2018). Antioxidant and antimicrobial activities of (−) -epigallocatechin-3-gallate (EGCG) and its potential to preserve the quality and safety of foods. *Comprehensive Reviews in Food Science and Food Safety*, *17*(3), 732–753. https://doi.org/10.1111/1541-4337.12346

Nimse, S. B., & Pal, D. (2015). Free radicals, natural antioxidants, and their reaction mechanisms. *RSC Advances*, *5*, 27986–28006. 10.1039/C4RA13315C

Nordin, H., Jalil, J., Yusof, A., & Kamisah, Y. (2019). Roles of rutin in cardiac remodeling. *Journal of Functional Foods*, *May*, 103606. https://doi.org/10.1016/j.jff.2019.103606

Nuñez-López, M. A., Paredes-López, O., & Reynoso-Camacho, R. (2013). Functional and hypoglycemic properties of nopal cladodes (*O. ficus-indica*) at different maturity stages using *in vitro* and *in vivo* tests. *Journal of Agricultural and Food Chemistry*, *61*(46), 10981–10986. https://doi.org/10.1021/jf403834x

Ognjanovic, B. I., Pavlovic, S. Z., Maletic, S. D., Zikic, R. V., Stajn, A. S., Radojicic, R. M., . . . Petrović, V. M. (2003). Protective influence of vitamin E on antioxidant defense system in the blood of rats treated with cadmium. *Physiological Research*, *52*, 563–570.

Ormazabal, P., Scazzocchio, B., Varì, R., Santangelo, C., Archivio, M. D., Silecchia, G., . . . Masella, R. (2018). Effect of protocatechuic acid on insulin responsiveness and in fl ammation in visceral adipose tissue from obese individuals : possible role for PTP1B. *International Journal of Obesity*, *42*(2012). https://doi.org/10.1038/s41366-018-0075-4

Ortega-Ramirez, L. A., Rodriguez-Garcia, I., Leyva, J. M., Cruz-Valenzuela, M. R., Silva-Espinoza, B. A., Gonzalez-Aguilar, G. A., . . . Ayala-Zavala, J. F. (2014). Potential of medicinal plants as antimicrobial and antioxidant agents in food industry: A hypothesis. *Journal of Food Science*, *79*(2), R129–R137. https://doi.org/10.1111/1750-3841.12341

Osorio-Esquivel, O., Alicia Ortiz, M., Álvarez, V. B., Dorantes-Álvarez, L., & Giusti, M. M. (2011). Phenolics, betacyanins and antioxidant activity in *Opuntia joconostle* fruits. *Food Research International*, *44*(7), 2160–2168. https://doi.org/10.1016/j.foodres.2011.02.011

Owumi, S.E., Adedara,I.A., Farombi,E.O., & Oyelere, A.K.(2020). Protocatechuic acid modulates reproductive dysfunction linked to furan exposure in rats. *Toxicology*, *442*, 152556. https://doi.org/10.1016/j.tox.2020.152556

Park, Y., Lee, H., Lim, J. W., & Kim, H. (2019). Inhibitory effect of β-carotene on helicobacter pylori-induced TRAF expression and hyper-proliferation in gastric epithelial cells. *Antioxidants*, *8*(12), 637. https://doi.org/10.3390/antiox8120637

Pérez-Solano, L. A., Gallina-Tessaro, S., & Sánchez-Rojas, G. (2016). Individual variation in mule deer (*Odocoileus hemionus*) habitat and home range in the Chihuahuan Desert, Mexico. *Journal of Mammalogy*, *97*(4), 1228–1237. https://doi.org/10.1093/jmammal/gyw075

Periyannan, V. (2020). Syringic acid modulates molecular marker – involved cell proliferation, survival, apoptosis, inflammation, and angiogenesis in DMBA – induced oral squamous cell carcinoma in Syrian hamsters. *Journal of Biochemical Molecular Tocicology*, *April*, 1–9. https://doi.org/10.1002/jbt.22574

Picos-Salas, M. A., Gutiérrez-Grijalva, E. P., Valdez-Torres, B., Angulo-Escalante, M. A., López-Martínez, L. X., Delgado-Vargas, F., & Basilio Heredia, J. (2021). Supercritical CO_2 extraction of oregano (*Lippia graveolens*) phenolic compounds with antioxidant, α-amylase and α-glucosidase inhibitory capacity. *Journal of Food Measurement and Characterization.*. https://doi.org/10.1007/s11694-021-00928-4

Pimienta-Barrios, E., Méndez-Morán, L., Ramírez-Hernández, B. C., García de Alba-García, J. E., & Domínguez-Arias, R. M. (2008). Efecto de la ingestión del fruto de Xoconostle (*Opuntia joconostle* Web.) sobre la glucosa y lípidos séricos. *Agrociencia*, *42*(6), 645–653.

Pinedo-Espinoza, J. M., Gutierrez-Tlahque, J., Santiago-Saenz, Y. O., Aguirre-Mancilla, C. L., Reyes-Fuentes, M., & Lopez-Palestina, C. U. (2020). Nutritional composition, bioactive compounds and antioxidant activity of wild edible flowers consumed in semiarid regions of Mexico. *Plants Foods for Human Nutrition*, *75*(3), 413–419. https://doi.org/10.1007/s11130-020-00822-2

Pisoschi, A. M., & Pop, A. (2015). The role of antioxidants in the chemistry of oxidative stress: A review. *European Journal of Medicinal Chemistry*, *97*, 55–74. https://doi.org/10.1016/j.ejmech.2015.04.040

Płonka, J., Toczek, A., & Tomczyk, V. (2012). Multivitamin analysis of fruits, fruit-vegetable juices, and diet supplements. *Food Analytical Methods*, *5*(5), 1167–1176. https://doi.org/10.1007/s12161-011-9349-3

Prakash, B., Kujur, A., Singh, P. P., Kumar, A., & Yadav, A. (2017). Plants-derived bioactive compounds as functional food ingredients and food preservative. *Journal of Nutrition and Food Science*, *1*(1), 004.

Pramanik, P. K., Alam, M. N., Roy Chowdhury, D., & Chakraborti, T. (2019). Drug resistance in protozoan parasites: An incessant wrestle for survival. *Journal of Global Antimicrobial Resistance*, *18*, 1–11. https://doi.org/10.1016/j.jgar.2019.01.023

Puente-Garza, C. A., Meza-Miranda, C., Ochoa-Martínez, D., & García-Lara, S. (2017). Effect of *in vitro* drought stress on phenolic acids, flavonols, saponins, and antioxidant activity in *Agave salmiana*. *Plant Physiology and Biochemistry*, *115*, 400–407. https://doi.org/10.1016/j.plaphy.2017.04.012

Quintana-LópezA., Hurtado-Oliva, M. A., Hernández, C., &Palacios-Mechetnov, E. (2018). Carotenoides. ¿Qué son y para qué se usan?, de Facultad de Ciencias del Mar de la Universidad Autónoma de Sinaloa:https://www.revistaciencia.amc.edu.mx/images/revista/69_4/PDF/10_69_4_1106_Carotenoides_L.pdf

Quintanilla-Licea, R., Vargas-Villarreal, J., Verde-Star, M. J., Rivas-Galindo, V. M., & Torres-Hernandez, A. D. (2020). Antiprotozoal activity against entamoeba histolytica of flavonoids isolated from *Lippia graveolens* Kunth. *Molecules*, *25*(11). https://doi.org/10.3390/molecules25112464

Rahman, S., Ansari, R. A., Rehman, H., Parvez, S., & Raisuddin, S. (2011). Nordihydroguaiaretic acid from creosote bush (Larrea tridentata) mitigates 12-O-tetradecanoylphorbol-13-acetate-induced inflammatory and oxidative stress responses of tumor promotion cascade in mouse skin. *Evidence-Based Complementary and Alternative Medicine*, *2011*, 1–10. https://doi.org/10.1093/ecam/nep076

Ramadan, M. F., & Mörsel, J.-T. (2003). Oil cactus pear (*Opuntia ficus-indica* L.). *Food Chemistry*, *82*(3), 339–345. https://doi.org/10.1016/S0308-8146(02)00550-2

Ramírez Moreno, A., Serrano Gallardo, L. B., Barragán Ledezma, E., Quintanar Escorza, M. A., Arrellano Pérez-Vertti, R. D., & Delgadillo Guzmán, D. (2016). Determinación de los compuestos polifenólicos en extractos de Jatropha dioica y su capacidad antioxidante. *Revista Mexicana de Ciencias Farmacéuticas*, *47*(4), 42–48.

Ramírez-Moreno, A., Delgadillo-Guzmán, D., Bautista-Robles, V., Marszalek, J. E., Keita, H., Kourouma, A., . . . Tavares-Carvalho, J. C. (2020). Jatropha dioica, an Aztec plant with promising pharmacological properties: A systematic review. *African Journal of Pharmacy and Pharmacology*, *14*(6), 169–178. https://doi.org/10.5897/AJPP2020.5147

Ramírez-Rojo, M. I., Vargas-Sánchez, R. D., Hernández-Martínez, J., Martínez-Benavidez, E., Sánchez-Escalante, J. J., Torrescano-Urrutia, G. R., & Sánchez-Escalante, A. (2019). Actividad antioxidante de extractos de hoja de mezquite (*Prosopis velutina*). *Biotecnia*, *XXI*(1), 113–119.

Ramsey, D., & Muskin, P. R. (2013). Vitamin deficiencies and mental health: How are they linked? *Current Psychiatry*, *12*(1), 37–43.

Ran, J., Hyun, J., Lee, S., Ju, E., & Young, H. (2020). Protective effects of protocatechuic acid against cognitive impairment in an amyloid beta-induced Alzheimer's disease mouse model. *Food and Chemical Toxicology*, *144*(July), 111571. https://doi.org/10.1016/j.fct.2020.111571

Rao, A., & Rao, L. (2007). Carotenoids and human health. *Pharmacological Research*, *55*(3), 207–216. https://doi.org/10.1016/j.phrs.2007.01.012

Rasanu, N., Magearu, V., Nicoleta Matei, N., &Soceanu, A. (2005). Determination of vitamin C in different stages of fruits growth. In Analele UniversităŃii din Bucureşti - Chimie, Anul XIV (serie nouă), vol. I-II, pg. 167–172. Analele UniversităŃii din Bucureşti.

Rios, J. L., Giner, R. M., Marin, M., & Frecio, M. C. (2018). A pharmacological update of ellagic acid. *Planta Medica*, *84*(15), 1068–1093. https://doi.org/10.1055/a-0633-9492

Ríos, M. D., Muñoz, K. G., Tabizón, E. F., Márquez, D., Avitia, A. Y. C., Contreras, L. E. E. S., & González, J. R. (2016). Caracterización y capacidad de adsorción de la especie vegetal *Larrea tridentata* como adsorbente de color. *Cultura Científica y Tecnológica*, *54*(1), 208–224.

Rodr, C., & Cristina, S. (2019). Naturally lignan-rich foods: a dietary tool for health promotion? *Molecules*, *24*(917). https://doi.org/10.3390/molecules24050917

Rodriguez-Maturino, A., Troncoso-Rojas, R., Sánchez-Estrada, A., González-Mendoza, D., Ruiz-Sanchez, E., Zamora-Bustillos, R., . . . Aviles-Marin, M. (2015). Efecto antifúngico de extractos fenólicos y de carotenoides de chiltepín (*Capsicum annum* var. glabriusculum) en *Alternaria alternata* y *Fusarium oxysporum*. *Revista Argentina de Microbiología*, *47*(1), 72–77. https://doi.org/10.1016/j.ram.2014.12.005

Rodríguez-Sauceda, E. N., Rojo-Martínez, G. E., Ramírez-Valverde, B., Martínez-Ruiz, R., Cong- Hermida, M. C., Medina-Torres, S. M., & Piña-Ruiz, H. H. (2014). Análisis técnico del árbol del mezquite (*Prosopis laevigata* Humb. & Bonpl. ex Willd.) en México. *Ra Ximhai*, *10*(3), 173–193.

Rojas, R., Tafolla-Arellano, J. C., & Martínez-Ávila, G. C. G. (2020). *Euphorbia antisyphilitica Zucc*: A source of phytochemicals with potential applications in industry. *Plants (Basel)*, *10*(1), 8. https://doi.org/10.3390/plants10010008

Roupe, K. A., Remsberg, C. M., Yáñez, J. A., & Davies, N. M. (2006). Pharmacometrics of stilbenes: seguing towards the clinic. *Current Clinical Pharmacology*, *1*, 81–101.

Rufino-Gonzalez, Y., Ponce-Macotela, M., Gonzalez-Maciel, A., Reynoso-Robles, R., Jimenez-Estrada, M., Sanchez-Contreras, A., & Martinez-Gordillo, M. N. (2012). *In vitro* activity of the F-6 fraction of oregano against Giardia intestinalis. *Parasitology*, *139*(4), 434–440. https://doi.org/10.1017/S0031182011002162

Ruíz-Martínez, J., Ascacio, J. A., Rodríguez-Herrera, R., Morales, D., & Aguilar, C. N. (2011). Phytochemical screening of extracts from some Mexican plants used in traditional medicine. *Journal of Medicinal Plants Research*, *5*(13), 2791–2797. https://doi.org/10.5897/JMPR.9001010

Saenz, C., Tapia, S., Chavez, J., & Robert, P. (2009). Microencapsulation by spray drying of bioactive compounds from cactus pear (*Opuntia ficus-indica*). *Food Chemistry*, *114*(2), 616–622. https://doi.org/10.1016/j.foodchem.2008.09.095

Salazar, R., Pozos, M. E., Cordero, P., Perez, J., Salinas, M. C., & Waksman, N. (2008). Determination of the antioxidant activity of plants from Northeast Mexico. *Pharmaceutical Biology*, *46*(3), 166–170. https://doi.org/10.1080/13880200701498952

Samarghandian, S., Azimi-nezhad, M., & Farkhondeh, T. (2017). *Catechin treatment ameliorates diabetes and its complications in streptozotocin-induced diabetic rats* (pp. 1–7). *SAGE*. https://doi.org/10.1177/1559325817691158

Sanchez, E., Rivas Morales, C., Castillo, S., Leos-Rivas, C., Garcia-Becerra, L., & Ortiz Martinez, D. M. (2016). Antibacterial and antibiofilm activity of methanolic plant extracts against nosocomial microorganisms. *Evidence-Based Complementary and Alternative Medicine*, *2016*, 1572697. https://doi.org/10.1155/2016/1572697

Sánchez-Del Pino, I., Alfaro, A., & Andueza-Noh, R. H. (2020). High phylogeographic and genetic diversity of Tidestromia lanuginosa supports full-glacial refugia for arid-adapted plants in southern and Central Coahuila, Mexico. *American Journal of Botany*, *107*(9), 1296–1308. https://doi.org/10.1002/ajb2.1536

Santos-Sánchez, N. F., Salas-Coronado, R., Valadez-Blanco, R., Hernández-Carlos, B., & Guadarrama-Mendoza, P. C. (2017). Natural antioxidant extracts as food preservatives. *Acta Scientiarum Polonorum, Technologia Alimentaria*, *16*(4), 361–370. https://doi.org/10.17306/J.AFS.0530

Santos-Zea, L., Gutierrez-Uribe, J. A., & Serna-Saldivar, S. O. (2011). Comparative analyses of total phenols, antioxidant activity, and flavonol glycoside profile of cladode flours from different varieties of *Opuntia spp*. *Journal of Agricultural and Food Chemistry*, *59*(13), 7054–7061. https://doi.org/10.1021/jf200944y

Schellack, G. (2015). B-complex vitamin deficiency and supplementation. *South African Pharmaceutical Journal*, *82*(4), 28–32.

Seidl, A., Pérez-Collazos, E., Tremetsberger, K., Carine, M., Catalán, P., & Bernhardt, K. G. (2020). Phylogeny and biogeography of the Pleistocene Holarctic steppe and semi-desert goosefoot plant *Krascheninnikovia ceratoides*. *Flora*, *262*, 151504. https://doi.org/10.1016/j.flora.2019.151504

Semaming, Y., Pannengpetch, P., Chattipakorn, S. C., & Chattipakorn, N. (2015). Pharmacological properties of protocatechuic acid and its potential roles as complementary medicine. *Hindawi Oxidative Medicine and Cellular Longevity*, *2015*. https://doi.org/10.1155/2015/593902

Shahidi, F., & Ambigaipalan, P. (2015). Novel functional food ingredients from marine sources. *Current Opinion in Food Science*, *2*, 123–129. https://doi.org/10.1016/j.cofs.2014.12.009

Shahzad, S., Mateen, S., Naeem, S. S., & Rizvi, W. (2019). Syringic acid protects from isoproterenol induced cardiotoxicity in rats. *European Journal of Pharmacolog*, *849*, 135–145. https://doi.org/10.1016/j.ejphar.2019.01.056

Sies, H. (2015). Oxidative stress: A concept in redox biology and medicine. *Redox Biology*, *4*, 180–183. https://doi.org/10.1016/j.redox.2015.01.002

Skrovankova, S., Sumczynski, D., Mlcek, J., Jurikova, T., & Sochor, J. (2015). Bioactive compounds and antioxidant activity in different types of berries. *International Journal of Molecular Sciences, 16*, 24673–24706. https://doi.org/10.3390/ijms161024673

Solovchenko, A., & Neverov, K. (2017). Carotenogenic response in photosynthetic organisms: A colorful story. *Photosynthesis Research, 133*(1–3), 31–47. https://doi.org/10.1007/s11120-017-0358-y

Song, Y. S., & Hepp, M. A. (2005). US Food and Drug Administration approach to regulating intelligent and active packaging components. In J. H. Han (Ed.), *Innovations in Food Packaging* (pp. 475–481). Academic Press.

Soto-Domínguez, A., García-Garza, R., Ramírez-Casas, Y., Morán-Martínez, J., & Serrano-Gallardo, L. B. (2012). El extracto acuoso de orégano (Lippia graveolens HBK) del Norte de México tiene actividad antioxidante sin mostrar un efecto tóxico *in vitro* e *in vivo*. *International Journal of Morphology, 30*, 937–944.

Tanaka, T., Tanaka, T., & Tanaka, M. (2011). Potential cancer chemopreventive activity of protocatechuic acid. *Journal of Experimental and Clinical Medicine, 3*(1), 27–33. doi: https://doi.org/10.1016/j.jecm.2010.12.005

Tanase, C., Cosarca, S., & Muntean, D. L. (2019). A critical review of phenolic compounds extracted from the bark of woody vascular plants and their potential biological activity. *Moleculaes, 24*(1182), 1182. doi: 10.3390/molecules24061182

Tarabih, M., & EL Eryan, E. (2020). Glycine betaine and proline with thinning technique for resistance abiotic stress of Cristalina cactus pear. *Pakistan Journal of Biological Sciences, 23*, 68–80. https://doi.org/10.3923/pjbs.2020.68.80

Tomás-Barberán, F. A., & Andrés-Lacuevas, C. (2012). Polyphenols and health: current state and progress. *Journal of Agricultural and Food Chemistry, 60*(36), 8773–8775. https://doi.org/10.1021/jf300671j

Torres-Ponce, R. L., Morales-Corral, D., Ballinas-Casarrubias, M. L., & Nevarez-Morillon, G. V. (2015). El nopal: planta del semidesierto con aplicaciones en farmacia, alimentos y nutrición animal. *Revista Mexicana de Ciencias Agricolas, 6*(5), 1129–1142.

Urbizu-González, A. L., Castillo-Ruiz, O., Martínez-Ávila, G. C. G., & Torres-Castillo, J. A. (2017). Natural variability of essential oil and antioxidants in the medicinal plant *Turnera diffusa*. *Asian Pacific Journal of Tropical Medicine, 10*(2), 121–125. https://doi.org/10.1016/j.apjtm.2017.01.013

Vargas-Segura, A. I., Silva-Belmares, S. Y., Segura-Ceniceros, E. P., Ascacio-Valdes, J. A., Mendez-Gonzalez, L., & Ilyina, A. (2020). Screening and characterization of medicinal plants extracts with bactericidal activity against Streptococcus mutans. *Natural Product Researcch, 34*(18), 2672–2676. https://doi.org/10.1080/14786419.2018.1550757

Vázquez García, A., Fernández, A., Manuel, M., Chulim, E., Gabriel, N., Ortiz-Torres, E., . . . Ramírez, A. M. (2016). El maguey pulquero: Una planta multifuncional y polifacética: los usos desde una visión mestiza e indígena. *Scripta Ethnologica, 38*, 65–87.

Vazquez-Flores, A. A., Góngora-Pérez, O., Olivas-Orduña, I., Muñoz-Bernal, O. A., Osuna-Avila, P., Rodrigo-García, J., . . . Alvarez-Parilla, E. (2020). Phytochemical profile and antioxidant activity of chiltepin chili (*Capsicum annuum* var. *glabriusculum*), Sonora, Mexico. *Journal of Food Bioactives, 11*. http://cathi.uacj.mx/20.500.11961/15823

Vázquez-Flores, A. A., Góngora-Pérez, O., Olivas-Orduña, I., Muñoz-Bernal, O. A., Osuno-Ávila, P., Rodrigo-García, J., . . . Álvarez-Parilla, E. (2020). Phytochemical profile and antioxidant activity of chilpetin chili (*Capsicum annuum* var. *glabriusculum*), Sonora, México. *Journal of Food Bioactives*, *11*, 57–65. https://doi.org/10.31665/JFB.2020.11237

Vega, A., De León, J. A., & Reyes, S. M. (2018). Bioactive components of 34 commercial brands of coffee from panama: relationship between chlorogenic acids and caffeine. *Información Tecnológica*, *29*(4), 43–54. doi: http://dx.doi.org/10.4067/S0718-07642018000400043

Vítek, P., Ascaso, C., Artieda, O., Casero, M. C., & Wierzchos, J. (2017). Discovery of carotenoid red-shift in endolithic cyanobacteria from the Atacama Desert. *Scientific Reports*, *7*(1). https://doi.org/10.1038/s41598-017-11581-7

Vo, Q. V., Van Bay, M., Nam, P. C., Quang, D. T., Flavel, M., Hoa, N. T., & Mechler, A. (2020). Theoretical and Experimental Studies of the Antioxidant and Antinitrosant Activity of Syringic Acid. *Journal of Organic Chemistry*, *85*(23), 15514–15520. https://doi.org/10.1021/acs.joc.0c02258

Vuolo, M. M., Lima, V. S., Roberto, M., & Junior, M. (2019). Classification, and antioxidant power. In *Bioactive compounds*. Elsevier Inc. https://doi.org/10.1016/B978-0-12-814774-0.00002-5

Walia, A., Gupta, A. K., & Sharma, V. (2019). Role of bioactive compounds in human health. *Acta Scientific Medical Sciences*, *3*(9), 25–33.

Wang, J., Quan, Q., Ji, R., Guo, X., Zhang, J., & Li, X. (2018). Biomedicine & pharmacotherapy isorhamnetin suppresses PANC-1 pancreatic cancer cell proliferation through S phase arrest. *Biomedicine & Pharmacotherapy*, *108*(September), 925–933. https://doi.org/10.1016/j.biopha.2018.09.105

Wang, X., Wang, K., Zhang, X., Yang, C., & Li, X. (2018). Biochemical and Biophysical Research Communications 4-hydroxybenzoic acid (4-HBA) enhances the sensitivity of human breast cancer cells to adriamycin as a specific HDAC6 inhibitor by promoting HIPK2/p53 pathway. *Biochemical and Biophysical Research Communications*, *90*, 4–11. https://doi.org/10.1016/j.bbrc.2018.08.043

Willer, J., Johrer, K., Greil, R., Zidorn, C., & Cicek, S. S. (2019). Cytotoxic properties of Damiana (Turnera diffusa) extracts and constituents and a validated quantitative UHPLC-DAD assay. *Molecules*, *24*(5). https://doi.org/10.3390/molecules24050855

Winter, A. N., Brenner, M. C., Punessen, N., Snodgrass, M., Byars, C., Arora, Y., & Linseman, D. A. (2017). Comparison of the neuroprotective and anti-inflammatory effects of the anthocyanin metabolites protocatechuic acid and 4-hydroxybenzoic acid. *Hindawi Oxidative Medicine and Cellular Longevity*, *2017*. https://doi.org/10.1155/2017/6297080 Research

Wong, J. E., Muñiz, D. B., Martínez, G. C. G., Belmares, R. E., & Aguilar, C. N. (2015). Ultrasonics Sonochemistry Ultrasound-assisted extraction of polyphenols from native plants in the Mexican desert. *Ultrasonics - Sonochemistry*, *22*, 474–481. https://doi.org/10.1016/j.ultsonch.2014.06.001

Wong Paz, J. E., Muñiz Márquez, D. B., Martínez Ávila, G. C. G., Belmares Cerda, R. E., & Aguilar, C. N. (2015). Ultrasound-assisted extraction of polyphenols from native plants in the Mexican desert. *Ultrasonics Sonochemistry*, *22*, 474–481. https://doi.org/10.1016/j.ultsonch.2014.06.001

Wong-Paz, J. E., Contreras Esquivel, J. C., Rodríguez-Herrera, R., Carrillo Inungaray, M. L., Lopez, L. I., Nevarez-Moorillon, G. V., & Aguilar, C. N. (2015). Total phenolic content, *in vitro* antioxidant activity and chemical composition of plant extracts from semiarid Mexican region. *Asian Pacific Journal of Tropical Medicine, 2015*, 104–111. https://doi.org/10.1016/S1995-7645(14)60299-6

World Health Organization. (2018). Food Additives. https://www.who.int/news-room/fact-sheets/detail/food-additives

World Health Organization (WHO). (2021). The top 10 causes of death. https://www.who.int/es/news-room/fact-sheets/detail/the-top-10-causes-of-death

Xiao, Y., & Cai, W. (2020). Autophagy and bacterial infection. *Advances in Experimental Medicine Biology, 1207*, 413–423. https://doi.org/10.1007/978-981-15-4272-5_29

Yang, C., Deng, Q., Xu, J., Wang, X., Hu, C., Tang, H., & Huang, F. (2018). Sinapic acid and resveratrol alleviate oxidative stress with modulation of gut microbiota in high-fat diet-fed rats. *Food Research International, 116*, 1202–1211. https://doi.org/10.1016/j.foodres.2018.10.003

Yang, Y.-W., Yin, Y.-C., Li, Z.-K., Huang, D., Shang, J.-L., Chen, M., & Qiu, B.-S. (2019). Orange and red carotenoid proteins are involved in the adaptation of the terrestrial cyanobacterium Nostoc flagelliforme to desiccation. *Photosynthesis Research, 140*(1), 103–113. https://doi.org/10.1007/s11120-019-00629-6

Yin, W., Wang, Y., Liu, L., & He, J. (2019). Biofilms: The microbial "protective clothing" in extreme environments. *International Journal of Molecular Sciences, 20*(14). https://doi.org/10.3390/ijms20143423

Yonekura-sakakibara, K. (2019). The origin and evolution of plant flavonoid metabolism. *Frontiers in Plant Science, 10*(August), 1–16. https://doi.org/10.3389/fpls.2019.00943

Zavala-Hurtado, J. A., & Jiménez, M. (2020). Diversity and uniqueness at its best: Vegetation of the Chihuahuan desert. In M. Mandujano, I. Pisanty, & L. Eguiarte (Eds.), *Plant diversity and ecology in the Chihuahuan Desert* (pp. 1–17). Cuatro Ciénegas Basin: An Endangered Hyperdiverse Oasis.

Zduńska, K., Dana, A., Kolodziejczak, A., & Rotsztejn, H. (2018). Antioxidant properties of ferulic acid and its possible application. *Skin Pharmacol Physiol, 31*, 332–336. https://doi.org/10.1159/000491755

Zhao, Y., Wang, Q., Jia, M., Fu, S., Pan, J., Chu, C., . . . Liu, Z. (2018). (+)-Sesamin attenuates chronic unpredictable mild stress-induced depressive-like behaviors and memory deficits via suppression of neuroinflammation. *The Journal of Nutritional Biochemistry, 64*, 61–71. https://doi.org/10.1016/j.jnutbio.2018.10.006

Zheng, Q., Tong, M., Ou, B., Liu, C., Hu, C., & Yang, Y. U. (2019). Isorhamnetin protects against bleomycin-induced pulmonary fibrosis by inhibiting endoplasmic reticulum stress and epithelial-mesenchymal transition. *International Journal of Molecular Medicine, 43*, 117–126. https://doi.org/10.3892/ijmm.2018.3965

Zugasti-Cruz, A., Rodríguez-Herrera, R., & Sierra-Rivera, C. A. (2020). Polyphenols: An overview of food sources and associated bioactivities. In S. Thacheril, S. Shiburaj, & S. Sabu (Eds.), *Plant Metabolites: Methods, Applications and Prospects*, (pp. 465–496). Singapore: Springer.

7

Opuntia, ficus-indica [L.] Mill. and Other Species: Source of Bioactives and Their Molecular Mechanisms of Action to Promote Human Health

Federica Blando[1], Clara Albano[1], Cristian Jiménez-Martínez[2], and Anaberta Cardador-Martínez[3]

[1] *Institute of Sciences of Food Production (ISPA), CNR, Lecce, Italy*
[2] *Escuela Nacional de Ciencias Biológicas, Instituto Politécnico Nacional, Mexico City, Mexico*
[3] *Escuela de Ingeniería y Ciencias, Departamento de Bioingenierías, Tecnologico de Monterrey, Queretaro, Mexico*

7.1 Introduction

The different species of *Opuntia* (Cactaceae family) originated in the tropical and sub-tropical area of the American continent (domesticated in Mexico), spreading to the rest of the world after XVI century. Today, *Opuntia* spp., cultivated or wild, can be found in many agro-climatic conditions, although often associated with warm climate.

Opuntia spp. shares a common characteristic with other succulent plants that is spines. Their anatomy and morphology reveal a physiological adaptation to extreme habitats, where other plants cannot survive. Roots adapt to drought in various ways, with a morphology that reduces water flow to the shoots. The morphology and anatomy of the shoots have evolved especially to store considerable amounts of water (Nobel, 1995). The shoots have modified into "areoles," bearing hairs (glochids) and spines (morphological equivalent of leaves), the stem has become succulent with green, photosynthetic cortex, and the absence of foliage leaves.

In addition to these morphological characteristics, cacti have a distinctive "Crassulacean Acid Metabolism" (CAM), characterized by a daily pattern of carbon uptake and water loss, both occurring during the night. The stomata open at night to fix CO_2 that is incorporated into oxaloacetate and then converted to malate. The organic acid is stored inside vacuoles of the chlorenchyma cells until the next day. In the daylight, CAM plants do not open their stomata, but they can still photosynthesize. In fact, organic acids are transported out of the vacuole and broken down to release CO_2, which enters the photosynthetic Calvin cycle. The transpiration (water loss) in CAM plants is 3–5 times lower than that of other plants due to this special metabolism, enabling these plants very efficient water use in semi-arid

Molecular Mechanisms of Functional Food, First Edition. Edited by Rocio Campos-Vega and B. Dave Oomah.

habitats. Some cacti like *Opuntia ficus-indica* in these challenging environments show elevated annual biomass productivity exceeded only by those of a few cultivated C_3 and C_4 species (Nobel & Bobich, 2002). In regards to progressive CO_2 increasing concentrations, plant productivity can be enhanced, up to a certain level. In *O. ficus-indica* grown at elevated CO_2 levels, both growth enhancement and photorespiration decrease as an acclimation of plant respiration to growth conditions (Gomez-Casanovas, Blanc-Betes, Gonzalez-Meler, & Azcon-Bieto, 2007).

The potential of biomass production and carbon sequestering of some cacti must be taken into account, particularly because these plants grow where no other crops can. The capacity of these plants to resist high temperatures and long drought periods makes *Opuntia* highly suitable for production in arid and semi-arid zones. In particular, *O. ficus-indica* is a multipurpose dryland crop destined to become more important in view of an ever increasing world population and water and land scarcity (Nefzaoui, 2009; Nefzaoui, Louhaichi, & Ben Salem, 2014). However, cacti can also become an invading species in some part of the world. Careful agronomic management and innovative utilization of these cacti, mainly from the genus *Opuntia*, may be one way to partly alleviate the threat (Zimmermann, 2017).

Almost 300 species of the genus *Opuntia* are known, but so far, only a dozen species are cultivated. Some species are used for fruit production (*O. ficus-indica, O. amyclae, O. xoconostle, O. megacantha,* and *O. streptacantha*). *O. ficus-indica* [L.] Mill. fruit is also known as prickly pear, cactus fig, Indian fig, cactus pear, and Barbary fig and is available across the five continents from early summer until late autumn. Mexico is the main cactus pear world producer and accounts for over 45% of global production; however, only 1.5% of this production is exported (Madrigal-Santillán et al., 2013). Italy, with a 6.6% global cactus pear production, is instead the second largest producer and the world largest exporter of this fruit to the European Union (Melgar, Dias, et al., 2017; Timpanaro & Foti, 2014). *O. ficus-indica* is the most widely cultivated *Opuntia* species throughout the world and the only one cultivated in the Mediterranean basin, where it is used mainly for fruit production, but also as forage, in arid and semi-arid lands, during drought period when there is shortage of herbaceous plants. *Opuntia* spp. (*O. ficus-indica* particularly) is also cultivated for the production of tender pads, called "nopalito," limited to Mexico and Southern US, where it is traditionally consumed both as fresh or processed vegetable (Corrales-García & Sáenz, 2013).

Due to the scientific interest and economic potentials, *Opuntia* spp. (and particularly *O. ficus-indica*) has been the object of a huge literature in the last decade. For example, numerous publications (Figure 7.1) have appeared in the last 4 years, especially focusing on biological and functional properties (Angulo-Bejarano, del Rocío Gómez-García, Valverde, & Paredes-López, 2019; Aruwa, Amoo, & Kudanga, 2018; Barba et al., 2020; Das et al., 2021; Hernández García et al., 2020; Silva et al., 2021 and references therein). Moreover, the Food and Agriculture Organization of the United Nations (FAO) and the International Center for Agricultural Research in the Dry Areas (ICARDA) have released several publications on this species (Inglese, Mondragon, Nefzaoui, & Sáenz, 2017; Sáenz et al., 2013). The importance of the current topic is demonstrated by a recently published book on the chemistry, bioactivity, and industrial applications of *Opuntia* spp. (Ramadan, Moussa Ayoub, & Rohn, 2021).

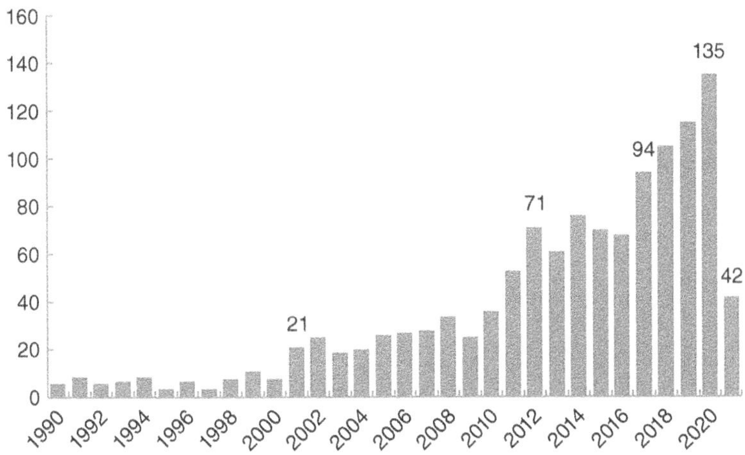

Figure 7.1 Number of publications on *Opuntia* according to PubMed.

7.2 Bioactive Compounds in the *Opuntia* Genus

Opuntia, a genus of the Cactaceae family, has been used for centuries as a medicinal plant, with different parts of the plant, including whole fruit, pulp, flowers, seeds, peels, stems, and cladodes. The genus *Opuntia* includes several species that contain bioactive compounds (Table 7.1). The most studied species is *O. ficus-indica* (Alimi et al., 2011; Berrabah et al., 2019; De Santiago et al., 2019; Graça Miguel et al., 2018; Ramírez-Moreno et al., 2017; Stintzing et al., 2005). Other species have also been studied, for instance *O. macrorhiza* (Amamou et al., 2020), *O. dillenii* (Betancourt et al., 2017; Liu et al., 2009), *O. robusta* (González-Ponce et al., 2020; Santos-Zea et al., 2011), *O. microdasys* (Chahdoura et al., 2014), among others.

7.2.1 Phenolic Compounds in *Opuntia* Genus

Phenolic compounds, a family of compounds that includes phenolic acids, flavonoids, and tannins (Figure 7.2), have been identified in *O. ficus-indica* (Albano et al., 2015; Alimi et al., 2011; Blando et al., 2019; Cejudo-Bastante et al., 2014; El-Beltagi et al., 2019; Mabrouki et al., 2015), *O. dillenii* (Betancourt et al., 2017; Liu et al., 2009), *O. stricta* (Koubaa et al., 2017), *O. joconostle* (Morales et al., 2012), *O. matudae* (Morales et al., 2012), *O. lindheimeri* (Santos-Zea et al., 2011), *O. undulata* (Santos-Zea et al., 2011), and *O. littoralis* (Abd El-Moaty et al., 2020). Moreover, anthocyanins have been identified in *O. ficus-indica* cladodes (Rocchetti, Pellizzoni, Montesano, & Lucini, 2018).

The content of phenolic compounds varies depending on *Opuntia* species. For instance, Abd El-Moaty et al. (2020) reported 504 and 429 mg/g of total flavonoids content in cladodes and fruits of *O. littoralis*, respectively. Lower concentration of total phenolics has been reported in *O. ficus-indica* cladodes (1.1–10.5 mg GAE/g) (Aruwa et al., 2019a, 2019b). On a dry weight (DW) basis, Blando et al. (2019) found a range of 18.6 to 54 mg GAE/g, depending

Table 7.1 Bioactive compounds present in different *Opuntia* species.

Species	Phytochemical	References
O. ficus-indica	Phenolics, betalains, vitamin C, carotenoids, polysaccharides, PUFAs	Butera et al. (2002); Stintzing and Carle (2005); Stintzing et al. (2005); Alimi, Hfaiedh, Bouoni, Sakly, and Ben Rhouma (2011); Cejudo-Bastante, Chaalal, Louaileche, Parrado, and Heredia (2014); Mabrouki, Zougari, Bendhifi, and Borgi (2015); Albano et al. (2015); Ramírez-Moreno et al. (2017); Cano, Gómez-Maqueo, García-Cayuela, and Welti-Chanes (2017); Graça Miguel et al. (2018); Aruwa, Amoo & Kudanga (2019a, 2019b); Berrabah, Taïbi, Ait Abderrahim, and Boussaid (2019); De Santiago et al. (2019); El-Beltagi, Mohamed, Elmelegy, Eldesoky, and Safwat (2019); Blando, Russo, Negro, De Bellis, and Frassinetti (2019); García-Cayuela, Gómez-Maqueo, Guajardo-Flores, Welti-Chanes, and Cano (2019); Farag, Sallam, Fekry, Zaghloul, and El-Dine (2020); Kolniak-Ostek et al. (2020)
O. macrorhiza	Polysaccharides	Amamou et al. (2020)
O. monacantha	Polysaccharides	Yang et al., (2008)
O. dillenii	Polyphenols, betalains, PUFAs	Liu et al. (2009); Betancourt, Cejudo-Bastante, Heredia, and Hurtado (2017)
O. microdasys	Sugars, organic acids, PUFAs	Chahdoura et al. (2014)
O. robusta	Betacyanins	Santos-Zea, Gutiérrez-Uribe, and Serna-Saldivar (2011); González-Ponce et al. (2020)
O. streptacantha	Betacyanins	Mabrouki et al. (2015); González-Ponce et al. (2020)
O. humifusa	Megastigmane	Jo et al. (2020)
O. ficus-barbaric	Phenolics	Juhaimi et al. (2020)
O. stricta	Polyphenols, vitamins, PUFAs	Koubaa et al. (2017); Gharby et al. (2020)
O. engelmannii	Phenolics	Melgar, Dias, et al. (2017)
O. joconostle	Phenolics, PUFAs	Morales, Ramírez-Moreno, Sanchez-Mata, Carvalho, and Ferreira, (2012)
O. matudae	Phenolics, PUFAs	Morales et al. (2012)
O. lindheimeri	Flavonoids	Santos-Zea et al. (2011)
O. albicarpa	Fatty acids	Ramírez-Moreno et al. (2017)
O. undulata	Flavonoids	Santos-Zea et al. (2011)
O. littoralis	Flavonoids	Abd El-Moaty, Sorour, Youssef, and Gouda (2020)

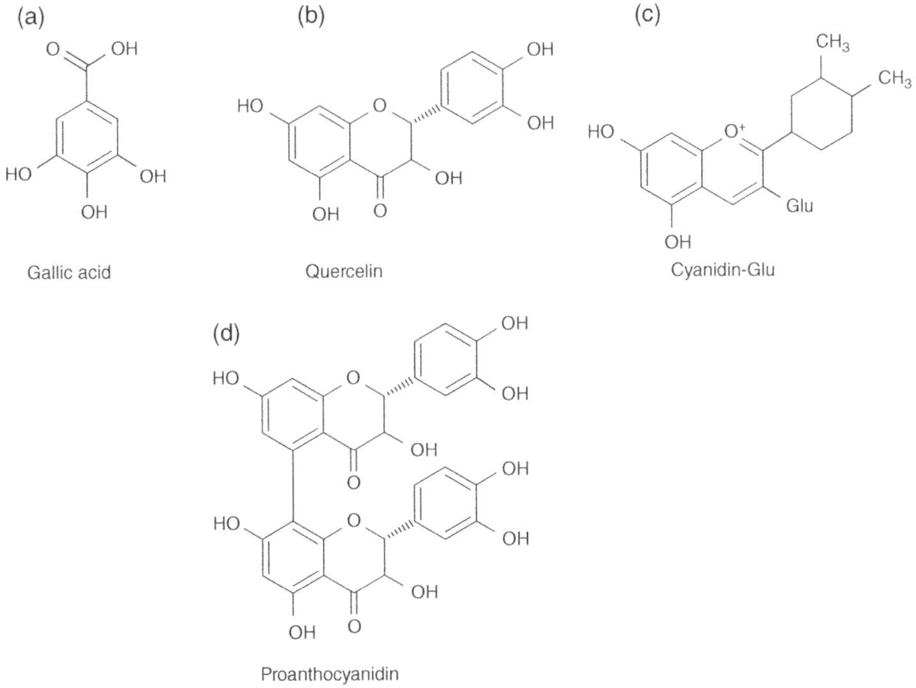

Figure 7.2 Phenolic compounds identified in *Opuntia* genus. (a) Phenolic acid, (b) flavonol, (c) anthocyanin, and (d) proanthocyanidin. (Structures were done with ACD ChemSketch freeware).

on the developmental stage of cladode. Instead, in fruits of different color of *O. ficus-indica*, the content was around 70–90 mg GAE/100 g fresh weight (FW), depending on the variety (Albano et al., 2015).

Not only fruits and cladodes contain phenolics but also *Opuntia* flowers (Ammar, Ennouri, & Attia, 2015; Berrabah et al., 2019). Moroccan *O. ficus-indica* flowers contained on average 3 mg GAE/g phenolics (Benayad, Martinez-Villaluenga, Frias, Gomez-Cordoves, & Es-Safi, 2014). In contrast, Ammar, Ennouri, and Attia (2015) reported higher content, up to 271 mg GAE/g in a methanolic extract of *O. ficus-indica* flowers. Similar results (2.9 mg/g extract) were reported in the flowers of *O. microdasys* (Chahdoura et al., 2014).

7.2.2 Betalains (Betacyanins and Betaxanthins) in *Opuntia* Genus

Betalains are water-soluble compounds classified according to their structure and color, and are found in *Opuntia* genus, as well as other few species of the order Caryophyllales. Betacyanins (betanidin and betanin) are red to purple color, while betaxanthins (indicaxanthin) are yellow (Figure 7.3). The mixture of these compounds results in various colors in *Opuntia* fruit (Castellanos-Santiago & Yahia, 2008; Stintzing et al., 2005). Betalains are promising protective agents against inflammation, oxidative stress and liver damage, and dyslipidemia-related diseases (González-Ponce et al., 2020; Rahimi, Abedimanesh, Mesbah-Namin, & Ostadrahimi, 2019). Indicaxanthins are stable in the gut, not metabolized by the

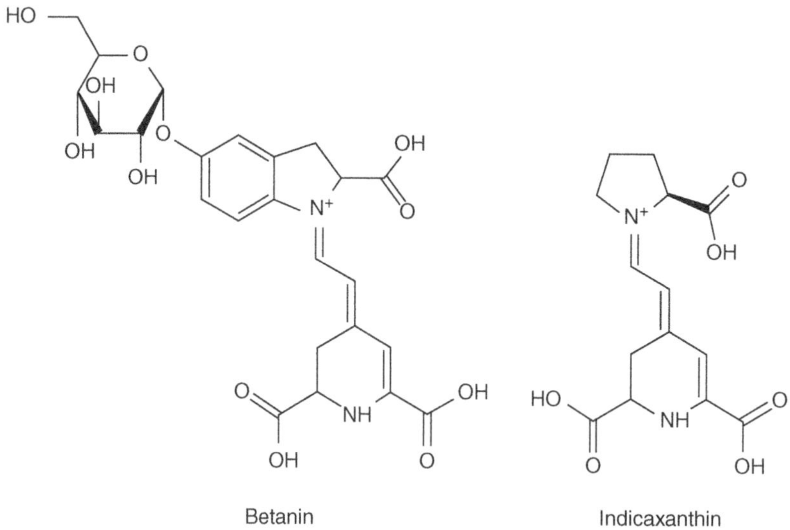

Figure 7.3 Structure of some betalains present in *Opuntia*. (Drawings were done with ACDLabs freeware).

enterocytes, and highly bioavailable in humans in its native form (Naselli et al., 2014; Tesoriere, Fazzari, Angileri, Gentile, & Livrea, 2008). Moreover, anti-inflammatory effects have been described based on the inhibition of NF-κB activation, a key transcription factor in the whole inflammatory cascade (Allegra et al., 2014). The anti-inflammatory activity of betalains was first demonstrated by betanin and indicaxanthin isolated from fresh *O.ficus-indica* pulp in TNF-α-stimulated human primary endothelial cells isolated from umbilical cords (Moreno-Ley, Osorio-Revilla, Hernández-Martínez, Ramos-Monroy, & Gallardo-Velázquez, 2021). Betanin inhibited cell adhesion molecule ICAM-1 cell expression better than indicaxanthin, even at low micromolar concentrations.

Betanin, isobetanin, neobetanin and indicaxanthin have been identified in *O. ficus-indica* purple fruit from South Italy, with a total betacyanin content of around 40 mg/100 g FW (Albano et al., 2015). Betanin and isobetanin have also been identified in *O. dillenii* as the main betacyanins, while gomphrenin (betanidin-6-*O*-glucoside) was in lower quantity. The total content of betacyanin was 16.6 mg betanin/100 g FW. Indicaxanthin, a yellow pigment, was also present (Betancourt et al., 2017). Much higher betanin content (~ 150 mg/100 g FW) was reported in the pulp of *O. stricta* var. *dillennii*, from Canary Island (Gómez-López, Lobo-Rodrigo, Portillo, & Cano, 2021). The total betacyanin contents in *O. robusta* and *O.streptacantha* fruits were 468 and 149 mg betacyanin equivalent/L of juice, respectively (González-Ponce et al., 2020). Similarly, betanin and isobetanin were identified as the main red pigments. Melgar, Pereira, et al. (2017) identified seven betalains in *O. ficus-indica* and *O. engelmannii* fruits; the latter contains more betacyanins. Castellanos-Santiago and Yahia (2008) identified a total of 18 betaxanthins and 6 betacyanins in *O. robusta* and *O. alba-carpa* fruits.

Opuntia fruit peels also contain betalains (Melgar, Dias, et al., 2017). For instance, peels of three Moroccan *Opuntia* varieties contain total betalain ranging from around 6.2 to 8.5 mg/100 g FW (Bourhia, Elmahdaoui, Ullah, Bari, & Benbacer, 2019). When Moroccan prickly pear was grown under climate stress, the total betalain content in peels ranged from 2 to 37 mg/100 g DW (Bourhia, Elmahdaoui, Moussa, Ullah, & Bari, 2020). Moreover, Abou-Elella and Ali (2014) determined 2.94 mg/L and 2.06 mg/L of betacyanins and betaxanthins, respectively, in *O. ficus-indica* peels. Bioaccessibility of betalains and phenolic compounds were relatively high in fruit peel (46–22% and 54–31%, respectively) in an *in vitro* simulated gastrointestinal [oral, gastric, and intestinal] digestion process (INFOGEST) (Gómez-López et al., 2021). Moreover, fruit pulp betalains bioacessibility was 47, 29, 20, and 7.6% for betanidin, isobetanin, 2'-O-apiosyl-4-O-phyllocactin, and neobetanin, respectively.

7.2.3 Fatty Acids

Unsaturated fatty acids have been detected in *O. ficus-indica* seed oil, comprising palmitic, stearic, oleic, and linoleic fatty acids, with predominant linoleic acid, ranging from 58 to 63% of the total fatty acids (Chougui et al., 2013; Ennouri, Fetoui, Bourret, Zeghal, & Attia, 2006; Kolniak-Ostek et al., 2020). Moreover, myristic and arachidonic acids have been found (Özcan & Al Juhaimi, 2011). The general fatty acid profile of *O. ficus-indica* seed oil resembles that of soybean oil with 15 and 22% of saturated and monounsaturated fatty acids, respectively, but higher amounts of linoleic acid (62 vs 54% of total fatty acids) (Gharby et al., 2020). Linoleic acid was also the major component in *O. joconostle* and *O. matudae* seed oils, reaching 72.5 and 79.2%, respectively (Morales et al., 2012). Surprisingly, in *O. dillenii,* linolenic acid was predominant (66.6%), followed by palmitic, stearic, and linoleic acids (19.8, 9, and 2.7%, respectively) (Liu et al., 2009).

O. stricta oil also contains phenolic compounds and vitamins such as vitamin E (88 mg/kg according to Gharby et al., 2020) prized as high value cosmetic ingredient. Generally, the total tocopherols content of cactus oil (788 mg/kg consisting of 18, 685, and 75 mg/kg of α-, γ-, and δ-tocopherols, respectively) is higher than that of extra virgin olive oil (Gharby et al., 2020). The complex mixture exhibited antimicrobial and antioxidant activities (Koubaa et al., 2017; Nazzaro et al., 2021; Ramírez-Moreno et al., 2017).

7.3 Biological Activities of *Opuntia*

Review on the biological activities of *Opuntia* spp. has been extensively published recently (Angulo-Bejarano et al., 2019; Aruwa et al., 2018; Das et al., 2021; Silva et al., 2021).

Several parts of the *Opuntia* spp. plant display diverse biological and health-promoting properties, both from *in vitro* and *in vivo* systems, including antiatherogenic, cholesterol-lowering, antidiabetic, and anticancer activities (Del Socorro Santos Díaz, Barba de la, Héliès-Toussaint, Guéraud, & Nègre-Salvayre, 2017) (Table 7.2). The beneficial effects arise from the complex mixture present, which can have additive and synergistic effects, as it has been proposed generally for herbal remedies, and cactus too (Osuna-Martínez, Reyes-Esparza, & Rodríguez-Fragoso, 2014).

Table 7.2 Biological activities in different parts of *Opuntia* spp. plant.

Part of plant		References
Fruit	Antioxidant	Tesoriere, Butera, Pintaudi, Allegra, and Livrea (2004); Stintzing et al. (2005); Galati et al. (2005); Abdel-Hameed, Nagaty, Salman, and Bazaid (2014); Abou-Elella and Ali (2014); Albano et al. (2015); Betancourt et al. (2017); Graça Miguel et al., (2018); Aruwaet al. (2019a, 2019b); El-Beltagi et al. (2019); Gómez-Maqueo, García-Cayuela, Welti-Chanes, and Cano (2019); Mahloko, Silungwe, Mashau, and Kgatla (2019); Zeghad, Ahmed, Belkhiri, Heyden, and Demeyer (2019); González-Ponce et al. (2020); Karabagias, Karabagias, Prodromiti, Gatzias, and Badeka (2020); Amamou et al. (2020); Farag et al. (2020)
	Anticancer	Abou-Elella and Ali (2014); Naselli et al. (2014); El-Beltagi et al. (2019)
	Antidiabetic	Gómez-Maqueo, García-Cayuela, Fernández-López, Welti-Chanes, and Cano (2019); Amamou et al. (2020)
	Antibacterial	Melgar, Dias, et al. (2017); Aruwa et al. (2019a, 2019b); Zourgui et al., (2020)
	Anti-inflammatory	Smeriglio et al., (2019); Gómez-Maqueo, García-Cayuela, Fernández-López, et al. (2019); Gómez-Maqueo, García-Cayuela, Welti-Chanes, and Cano (2019)
Cladode	Antioxidant	Santos-Zea et al. (2011); Avila-Nava et al. (2014); Pretti, Bazzu, Serra, and Nieddu (2014); Du Toit, De Wit, Osthoff, and Hugo (2018); Aruwa, C.E. et al. (2019a, 2019b); Blando et al. (2019)
	Antidiabetic	López-Romero et al. (2014); Harrat et al. (2019); Abd El-Moaty et al. (2020)
	Anticancer	De Santiago et al. (2019); Abd El-Moaty et al. (2020)
	Antimicrobial	Rasoulpour, Afsharifar, Izadpanah, and Aminlari (2017); Rasoulpour, Afsharifar, & Izadpanah, (2018); Aruwa et al. (2019a, 2019b); Blando et al. (2019)
	Anti–hypertensive	Harrat et al. (2019)
	Anti-inflammatory	Roome et al. (2019)
Flower	Antioxidant	Alimi et al. (2011); Ammar, Bardaa, et al. (2015); Ammar, Ennouri, and Attia (2015); Benayad et al. (2014); Berrabah et al. (2019)
	Antibacterial	Ammar, Bardaa, et al. (2015)
Seeds	Antioxidant, Antimicrobial, Antidiabetic	Liu et al. (2009); Koubaa et al. (2017); Ortega-Ortega et al. (2017); Ramírez-Moreno et al. (2017); Khémiri, Essghaier Hédi, Sadfi Zouaoui, Ben Gdara, and Bitri (2019); Nazzaro et al. (2021); Bouhrim et al. (2021)

Source: Based on Del Socorro Santos Diaz et al., 2017.

7.3.1 Antioxidant Capacity

The antioxidant capacity of prickly pear fruit relies on the presence of polyphenolic compounds, betalains, and vitamin C (Albano et al., 2015; Stintzing et al., 2005). Cactus fruits antioxidant activity is twice as high as pears, apples, tomatoes, bananas, and white grapes and is comparable to red grapes, pink grapefruit, and red orange (Butera et al., 2002).

Antioxidant capacity of an extract can be assessed by *in vitro* assays (including cells) and by *in vivo* trials with animals or humans.

In vitro antioxidant capacity can be determined by several methods; the most common is the α,α-diphenyl-β-picrilhydrazyl (DPPH) scavenging capacity. Abou-Elella and Ali (2014) reported DPPH values from 13 to 91.7% depending on the solvent used to extract *O. ficus-indica* fruit and peel, water being the one with the highest activity. Similar results were reported by Aruwa et al. (2019a, 2019b) for prickly pear pulp and peel extracts. These extracts also exhibited FRAP (ferric-reducing antioxidant power) 15.5–192 μmol Fe (II)/g. El-Beltagi et al. (2019) also investigating *O. ficus-indica* pulp and peels reported that the ethanolic extract of peel had higher antioxidant capacity attributed to the phenolic and flavonoids contents. Also, Graça Miguel et al. (2018) reported higher antioxidant capacity in peels than in fruits of several *Opuntia* cultivars. Albano et al. (2015) assessed the antioxidant capacity of prickly pear fruit (from South Italy) by ABTS and ORAC methodology and reported values up to 0.61 and 1.28 mmol/100 g FW, depending on the color. The same assays (and others) were used on both pulp and peel extracts (Smeriglio et al., 2019). Results showed that cactus pear peel possessed the best antioxidant activity in all assays in the following order: orange fruit > red fruit > yellow fruit. Moreover, the antioxidant capacity of peel extract was 5–20 times greater than the respective colors of the cactus pear pulp extracts (Smeriglio et al., 2019). Methanolic *Opuntia* flowers extract showed DPPH scavenging capacity lower than ascorbic acid (EC_{50} value 147 vs 7.2 mg/mL) (Alimi et al., 2011).

The protective effects of *O. ficus-indica* fruit juice were demonstrated against carbon tetrachloride (CCl4)-induced hepatotoxicity in animals (Galati et al., 2005). Fruit juice administration showed protective and curative effects against the CCl4-induced degenerative process in rat liver. This hepatoprotection was probably related to the polyphenolic components of the juice, but other compounds, such as vitamin C and betalains, could have synergistically counteracted many degenerative processes due to their antioxidant activity (Galati et al., 2005).

Cactus pear fruit consumption positively affected the body's redox balance, decreased lipids oxidative damage, and improved antioxidant status in healthy humans. Oxidative markers concentration such as malondialdehyde (MDA) decreased and reduced glutathione (GSH) increased after prickly pear fruit consumption, suggesting that indicaxanthin and betanin may be involved in protecting LDL against ex vivo-induced oxidative modifications (Tesoriere et al., 2004). Furthermore, the consumption of cladode (300 g/day for 3 days) increased blood and plasma antioxidant activity (by 5 and 20%, respectively) in healthy subjects partly due to the presence of polyphenols and vitamin C (Avila-Nava et al., 2014). Betalains bind to LDL protecting it against oxidative stress. Indicaxanthin was incorporated into LDL (98 pmol/mg LDL protein) after a single ingestion of *Opuntia* cactus pear fruit (500 g containing 28 and 16 mg indicaxanthin and betanin, respectively) by healthy volunteers ($n = 8$; average age 33 years; 21 kg/m^2 BMI). Serum LDL of the volunteers showed marked elongation of the lag time (58 vs 40 min) preceding Cu^{2+}-induced oxidation onset compared to LDL isolated before fruit indicating improved LDL resistance to oxidation. Moreover, 76 and 3.7% of the consumed indicaxanthin and betanin were excreted in urine during 12 hr (Sadowska-Bartosz & Bartosz, 2021 and references therein). In another study, broiled *O. robusta* JC Wendl. Ex Pfeiffer intake (250 g/day for 4 months) reduced oxidation injury by lowering total cholesterol and LDL-cholesterol and attenuating plasma and urinary 8-epi-PGF2a (oxidative stress marker) in 15 young patients with familial heterozygous hypercholesterolemia with stronger effects in females than in males (Sadowska-Bartosz & Bartosz, 2021 and references therein).

O. ficus-indica cladode extract tested by *in vitro* assays was able to scavenge H_2O_2, •OH, O_2•−, ONOO−, and HOCl (Avila-Nava et al., 2014; Blando et al., 2019). *Opuntia* sp. cladode powder inhibited *in vitro* extracellular ROS generation and LDL oxidation, while *in vivo* it significantly reduced atherosclerotic lesions formation and 4-hydroxynonenal adducts accumulation in the vascular wall of apoE-KO mice, indicating that *Opuntia* cladodes prevent lipid oxidation at the vascular level (Garoby-Salom et al., 2016). Trials on animals (obese rats) demonstrated that 4% nopal addition to the diet for 7 weeks reduced serum triglycerols by 50% and lowered liver MDA concentrations, a lipid peroxidation biomarker, that can be attributed to the lower ROS concentrations in the liver. Thus, attenuation of hepatic steatosis by nopal administration was achieved (Morán-Ramos et al., 2012). In mice treated with the antitumor Cis-Platinum, the cytotoxic action of this drug often associated with oxidative stress was restored by a pre–treatment with cactus cladode extract (Brahmi et al., 2012).

Several studies have demonstrated that the antioxidant capacity is preserved after processing *Opuntia* fruits and cladodes. Du Toit et al. (2018) evaluated the effect of drying on cladodes, resulting in no damage either in the antioxidant content or capacity. Moreover, the antioxidant capacity increased (36.8%) in purple prickly pear peels after treatment with high hydrostatic pressure (350 MPa) (Gómez-Maqueo, García-Cayuela, Welti-Chanes, & Cano, 2019).

7.3.2 Anticancer Activity

The Trypan Blue assay has been widely used to investigate the anticancer ability of plant extracts and fractions and/or their pure compounds. Chloroform and ethanol extracts from *Opuntia* fruit showed high anticancer activity (96% and 83%, respectively) in this assay against Ehrlich ascites carcinoma cells. The anticancer activity was related to two compounds, 17-hydroxy betanin, betanin and antioxidant activity (Abou-Elella & Ali, 2014). Betanin from *O. ficus-indica* fruit dose- and time-dependently reduced human chronic myeloid leukemia cells proliferation. It also induced K562 cells apoptosis by altering mitochondrial membrane integrity resulting in PARP cleavage, Bcl-2 suppression, and reduced membrane potential (Harlev, Nevo, Solowey, & Bishayee, 2013 and references therein).

An earlier review reported the cancer preventive effects of plants from the Cactaceae family, including *Opuntia* (Harlev et al., 2013). *Opuntia* juice reduced viability of prostate, colon, mammary and pancreatic cells without affecting those of normal fibroblast. Furthermore, differences in anticancer effects of the nine evaluated juices were due to variations in their phytochemical contents (primarily flavonoids and betalains). Aqueous

O. ficus-indica extract significantly increased apoptosis and growth inhibition in immortalized ovarian and cervical epithelial cells, as well as ovarian, cervical and bladder cancer cells dose- and time-dependently. The extract also modulated cancer cell cycle by increasing G1 and decreasing G2 and S phases (Harlev et al., 2013 and references therein). Another aqueous extract (5 and 10%) increased ROS, DNA fragmentation levels and perturbed apoptotic-related genes (Bax, Bad, caspase-3, Bcl-2, p53 and p21) and ROS-sensitive genes (NF-κB and c-jun/c-fos) leading to apoptosis in ovarian cancer cells (OVCA420 and SKOV3). *O. humifusa* extracts (water, hexane or ethyl acetate) inhibited proliferation and induced G1 arrest of MCF-7 human breast cancer cell; induced apoptosis and G1 arrest in U87MG human glioblastoma cells. Moreover, the numbers of papilloma and hyperplasia effectively decreased in chemically (DMBA and TPA) induced skin carcinogenesis in *O. humifusa* fed mice. These chemopreventive effects were presumably associated with reduced oxidative stress due to elevated total antioxidant capacity (particularly phase II detoxifying enzyme), modulation of cutaneous lipid peroxidation, and partial apoptotic influence (Harlev et al., 2013 and references therein).

An indicaxanthin-rich aqueous extract of *O. ficus-indica* showed dose-dependent apoptosis of proliferating (Caco-2) cells at nutritionally relevant amounts. Epigenomic activity of indicaxanthin was evident as de-methylation of the tumor suppressor p16INK4a gene promoter and other biochemical features, and suggests a chemopreventive potential of betaxanthin in intestinal cells (Naselli et al., 2014). The viability (%) decreased in different cell culture lines (liver [HepG2], colorectal adenocarcinoma [Caco-2] and breast carcinoma [MCF-7]) with increasing concentrations of *O. ficus-indica* aqueous ethanolic extracts from fruit pulp and peel. Phytochemical screening of bioactive plant extracts revealed the presence of alkaloids, tannins, and flavonoids (El-Beltagi et al., 2019). Melgar, Dias, et al. (2017) evaluated the cytotoxic activity of *Opuntia* sp. peel extracts against four cell lines (breast adenocarcinoma – MCF-7, non-small cell lung cancer – NCIH460, cervical carcinoma – HeLa, and hepatocellular carcinoma – HepG2); the extracts displayed a beneficial antitumor activity, although lower than the control, ellipticine. Phenolics and betalains can be responsible for cancer cell growth inhibition.

When MCF-7 cells were treated with different concentrations of hexane, ethyl acetate, and water extracts of *O. humifusa* fruit, the water extracts significantly decreased viable cell numbers in a concentration-dependent manner, indicating that water extracts can inhibit MCF-7 human breast cancer cell proliferation and induce G1 arrest (Yoon, Hahm, Park, & Son, 2009). Extracts of cladode and fruit from *O. littoralis* showed low cytotoxic activity against breast carcinoma cell line (MCF-7) with the ethyl acetate extract of cladodes expressing the best effect against MCF-7. Several flavonoids, such as 8-carbomethoxy-5-hydroxy-6-methyl isoflavone, were identified in the polar extracts of *O. littoralis* (Abd El-Moaty et al., 2020). It is known that flavonoids are among the most biologically active compounds.

7.3.3 Antimicrobial activity

The increased bacterial resistance to available antibiotics has pushed toward identifying novel compounds with antimicrobial activity. Novel antibiotic agents from *Opuntia* tissues to tackle antimicrobial resistance are expected to be identified as well as their mechanisms of action (Aruwa et al., 2018). *O. ficus-indica* fruit, peels, and cladodes extracts exert antibacterial activities against Gram-positive and Gram-negative bacteria (Aruwa et al., 2019a, 2019b). The antibacterial activity of *Opuntia* is generally associated with its antioxidant activity related primarily to the presence of phenolic compounds. These phenolics and other phytochemicals, including seed oil, inhibit the metabolic activity and biofilm of bacterial cells, conferring antibacterial activity (Blando et al., 2019; Nazzaro et al., 2021). Melgar, Dias, et al. (2017) evaluated the bacteriostatic and bactericidal effects of *O. ficus-indica* and *O. engelmannii* fruit peel extracts against several bacterial strains. The extracts were more effective against *Staphylococcus aureus* than ampicillin and streptomycin. Moreover, the same extracts showed antibacterial activity against *Trichoderma viride* and *Penicillum ochrochloron*. Ethanolic and aqueous extracts of *Opuntia* fruit peel were tested for antibacterial activity *in vitro*. The minimum inhibitory concentration for ethanolic extract was 4.75 mg/ml against *S. aureus*. Also, alcoholic extract was more effective than aqueous (Zourgui et al., 2020). Extracts of fruit peel and cladode showed antimycotoxigenic and antimycotic functions against *Fusarium* and *Aspergillus* fungi (Badr et al., 2020). *Opuntia* fruit juice also possesses *in vitro* antibacterial activity probably due to its betalain and phenolic content (Abd El-Moaty et al., 2020). Blando et al. (2019) showed that *O. ficus-indica* cladode extracts (mature and immature) inhibited *S. aureus* and *Escherichia coli* growth at 1 mg/mL concentration.

The phenolic compounds in supercritical CO_2 extracted *O. stricta* seed oil showed high antibacterial activity in the disc diffusion method (Koubaa et al., 2017). *Opuntia* spp. seed oil exhibited antibacterial activity against several bacteria strains such as *S. aureus*, *Pseudomonas aeruginosa,* and *Listeria monocytogenes*. Yeasts *Saccharomyces cerevisiae*, *Candida parapsilosis* and *Candida sake* were also sensitive to *Opuntia* seed oil. Also fungi like *Aspergillus niger* were sensitive to the *Opuntia* essential oil. The antibacterial activity was comparable to antimicrobial compounds such as ampicillin, streptomycin, sulfamethoxazole/trimethoprim and amphoterine B (Khémiri et al., 2019; Ramírez-Moreno et al., 2017). However, Brahmi et al. (2020) reported weak antibacterial activity against several bacteria and fungi, probably due to methodology issues.

Opuntia seed oil was evaluated against *Enterobacter cloacae,* resulting in bactericidal effect at 1/128 dilution. This oil was ineffective against Gram-positive bacteria (Khémiri et al., 2019). These authors attributed the antibacteria effect to the high content of free fatty acids (oleic and linoleic acids) which can act by inhibiting membrane enzymatic activities such as glucosyltransferase and on the other hand by activating autolytic enzymes in the pathogen cell wall. On the other hand, Ortega-Ortega et al. (2017) suggested that *Opuntia* seed oil is generally more effective against Gram-positive bacteria compared to Gram-negative bacteria, which are more resistant mainly because their outer membrane is less permeable due to the presence of lipopolysaccharides. For instance, *O. ficus-indica* seed oil inhibited Gram-negative bacteria growth (*E. coli* up to 38% and *P. aeruginosa* by almost twice, 71.8%) (Nazzaro et al., 2021).

7.3.4 Antidiabetic Activity (with Special Focus on *Opuntia* Cladode)

Diabetes is the most common endocrine disorder and one of the fastest growing noncommunicable diseases in the world, with an estimated prevalence of nearly 9% of the adult population and an expected further 50% increase over the next decades (Standl, Khunti,

Hansen, & Schnell, 2019). The incidence of Type 2 Diabetes Mellitus (T2DM) and the corresponding demand for its management strategies has stimulated increasing interest in the prevention and treatment based on alternative low-cost nutrition strategies, including plant-based supplements and functional foods. Plants (fruits and herbs) and their derivatives have been used for centuries in the traditional treatment of diabetes and are being studied intensively for the mechanism of action of their active principles. Several dietary components (soluble fibers, phenolic compounds, and peptides) are believed to exert similar physiological effects to those of synthetic antidiabetic drugs. Through *in vitro* and animal model studies, as well as clinical studies, dietary constituents have been shown to display various antihyperglycemic properties, possibly through the inhibition of digestive enzymes, the stimulation of insulin secretion, and activation of receptors involved in glucose metabolism (Lacroix & Li-Chan, 2014).

Opuntia spp. has been used in traditional Mexican medicine to control diabetes (Reynoso-Camacho & González de Mejía, 2008). In the past decade, several clinical trials reviewed the effect of *Opuntia* spp. supplementation to counteract hyperglycemia and related physiological disorders (Angulo-Bejarano et al., 2019; Gouws, Georgousopoulou, Mellor, McKune, & Naumovski, 2019; López-Romero et al., 2014; Nuñez-López, Paredes-López, & Reynoso-Camacho, 2013; Osuna-Martínez et al., 2014). The antidiabetic activity of *Opuntia* fruit extracts has been estimated by the α-amylase inhibition assay. Complex carbohydrates are digested by the pancreatic α-amylase; therefore, inhibition of this enzyme can decrease blood glucose level. Abd El-Moaty et al. (2020) evaluated the water extract of *O. littoralis* fruit using α-amylase *in vitro* assay, obtaining almost double the activity of acarbose, a compound used as control of α-amylase activity. Other *Opuntia* species, like *O. macrorhiza*, showed competitive α-amylase inhibition using *in vitro* assays. The compounds responsible for this inhibition were the polysaccharides, whose structure resembles the natural enzyme substrate (Amamou et al., 2020).

To manage hyperglycemia, α-glucosidase inhibition is also supposed to possess a pivotal role. Alpha-glucosidase is a membrane-bound enzyme located in the epithelium of the small intestine, which superintend to glucose cleavage from disaccharides and oligosaccharides. Inhibition of α-glucosidase activity consequently reduces glucose release and its subsequent uptake. Consumption of α-glucosidase inhibitors naturally occurring in food is an additional strategy in the management of postprandial hyperglycemia (Sakulnarmrat & Konczak, 2012). Several *O. ficus-indica* whole fruit varieties such as "Morada," "Vigor," and "Sanguinos" exhibited moderate α-amylase inhibition and higher α-glucosidase inhibition, resulting in low glucose absorption in postprandial hyperglycemia. The compound related to these activities was isorhamnetin glucosyl-rhamnosyl-pentoside (Gómez-Maqueo, García-Cayuela, Fernández-López et al., 2019).

Apart from the inhibitory activity against digestive enzymes, prickly pear fruit did not show other significant "anti-hyperglycemic" effects. Conversely, *in vitro* and *in vivo* trials showed cladode to exert significant effect on serum glucose reduction (Angulo-Bejarano et al., 2019; Gouws et al., 2019). Nuñez-López et al. (2013) reported that immature cladode flour (50 mg/kg body weight) treatment reduced postprandial blood glucose (46%) than control in streptozotocin-induced diabetic rats. Postprandial blood glucose reduction has been associated with glucose entrapment mechanism produced by the viscosity and swelling properties of soluble fiber inside the intestine. Thus, diet supplemented with viscous and no fermentable soluble fibers improved altered hepatic gene expression profiles associated with fatty liver, insulin resistance, and intestinal permeability (Kim, Bartley, Young, Seo, & Yokoyama, 2013).

Figure 7.4 shows the possible mechanisms of hypoglycemic action of *O. ficus-indica* fruit and cladode by enzyme inhibition and reduced glucose adsorption in the intestine. In streptozotocin-induced diabetic rats, a polysaccharide-rich purified extract of *O. monacantha* showed beneficial effects in improving the control of blood glucose and serum lipid level. The hypoglycemic action was hypothesized to be similar to that of dimethylbiguanide, a commonly used drug for diabetes treatment (Yang et al., 2008). In diabetic mouse strain model, *O. ficus-indica* extract (from cladode of the Korean variety *saboten*) treatment was used to evaluate the hypoglycemic effects. The extract inhibited glucose absorption from the intestine and enhanced glucose uptake from insulin-sensitive muscle cells through the AMPK/p38 MAPK signaling pathway (Leem, Kim, Hahm, & Kim, 2016).

The main flavonoid compound identified in *Opuntia* spp. cladode is differently glycosylated isorhamnetin. Isorhamnetin glycosides (0.3% and 0.6%; >90% purity) extracted from *O. ficus-indica* cladode improved glucose tolerance and hepatic insulin signaling, enhanced energy expenditure, reduced leptin concentration and hepatic steatosis, prevented adipose tissue hypertrophy, ameliorated endoplastic reticulum stress, and reduced body weight gain in high fat diet-induced obese male Wistar rats (Rodríguez-Rodríguez et al., 2015). Moreover, the cladode extract stimulated insulin secretion, in *in vitro* pancreatic islets, in association with increased glucose transporter 2 (GLUT2) and peroxisome proliferator-activated receptor gamma (PPARγ) mRNA content. These metabolic improvements were associated with reduced adipocyte size, increased hepatic IRS1 tyr-608 and S6 Kinase Thr-389 phosphorylation, and upregulated mitochondrial fatty acid oxidation genes (CPT1 and AOX) expressions (Gong et al., 2020; Rodríguez-Rodríguez et al., 2015) (Figure 7.5).

In a recent animal trial, freeze-dried *O. ficus-indica* nopalitos improved arterial blood pressure, glycemic control (reduced insulinemia and insulin resistance), metabolic pathway of cholesterol (low-density lipoprotein cholesterol reduction) and redox status (catalase [CAT], glutathione peroxidase [GPx], and superoxide dismutase [SOD]) in T2DM rats fed a high fat diet (Harrat et al., 2019). Recently, lyophilized *Opuntia* cladode powder was tested in an *in vitro* cellular model for adipocyte differentiation and an *in vivo* high-fat-diet (HFD)-induced obesity rat model (Héliès-Toussaint et al., 2020). Cladode flour induced antiadipogenic effect (reduced triglyceride storage in adipocytes and glucose uptake) in cells, while it improved some obesity-related metabolic parameters, increased triglyceride excretion in the feces, and reduced intestinal fat absorption in animals. *O. ficus-indica* cladodes supplementation (0.78 kg/animal/day for 34 days) activated energy (glucose and glycerolipids [primarily triacylglycerides]), tyrosine, and methane metabolism resulting in increased fertility and reproductive rate in adult female Rambouillet ewes without body weight increase (Rosales-Nieto et al., 2021).

Table 7.3 reports clinical studies on T2DM subjects. The first studies on the antidiabetic potential of *Opuntia* extracts (by Frati's group) focused on *O. streptacantha* (Frati et al., 1991; Frati-Munari et al., 1983, 1988). Butterweck et al. (2011) first experimented an aqueous *O. ficus-indica* cladode (and mixed cladode/fruit peel) extract in normal rats after oral dosing. Basal plasma insulin secretion was stimulated, while a direct action on ß-cells was hypothesized. The enhanced ß-cell function and attenuated insulin resistance are key

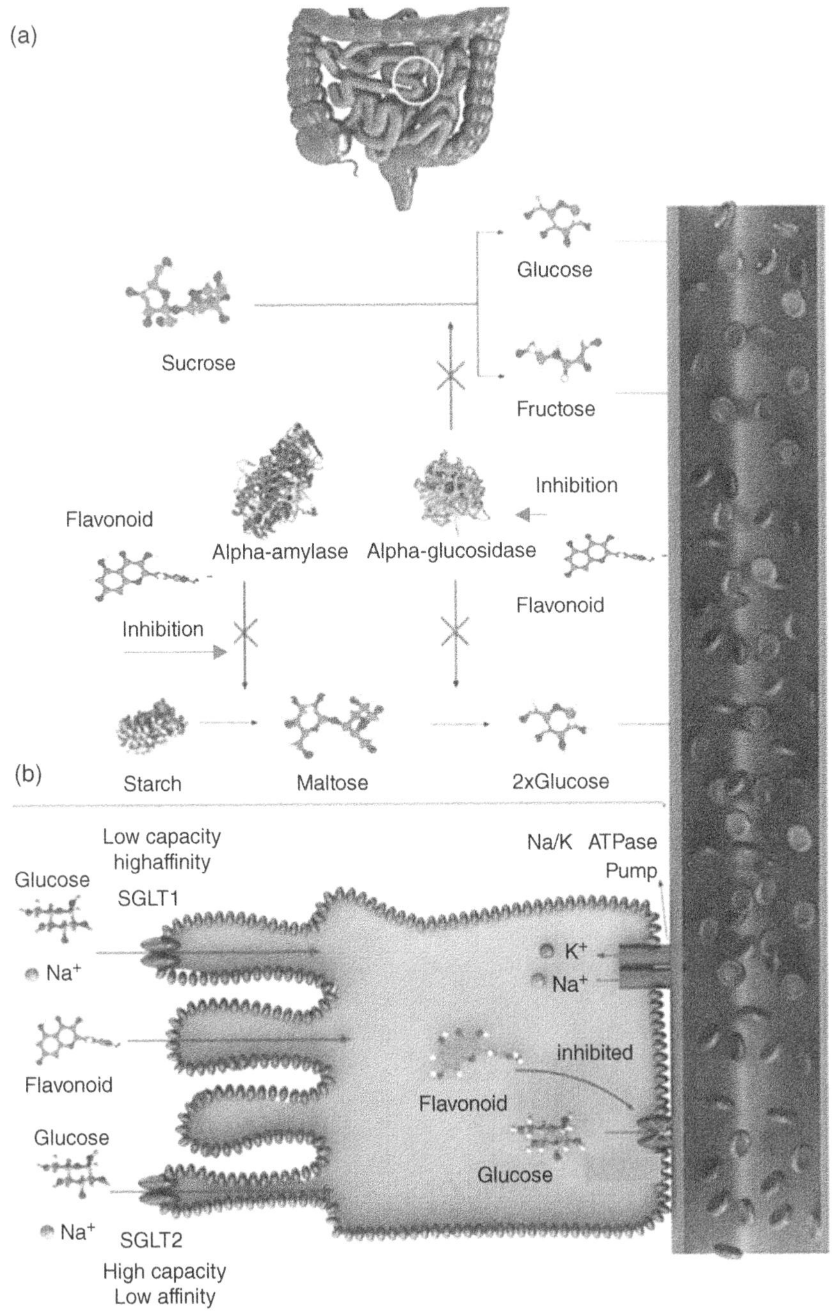

Figure 7.4 Mechanisms of hypoglycemic action of *O. ficus-indica* fruit and cladode by (a) α-amylase and α-glucosidase inhibition and (b) reduced glucose adsorption at intestine level. (From: Madadi et al., 2020 / MDPI / CC BY-4.0)

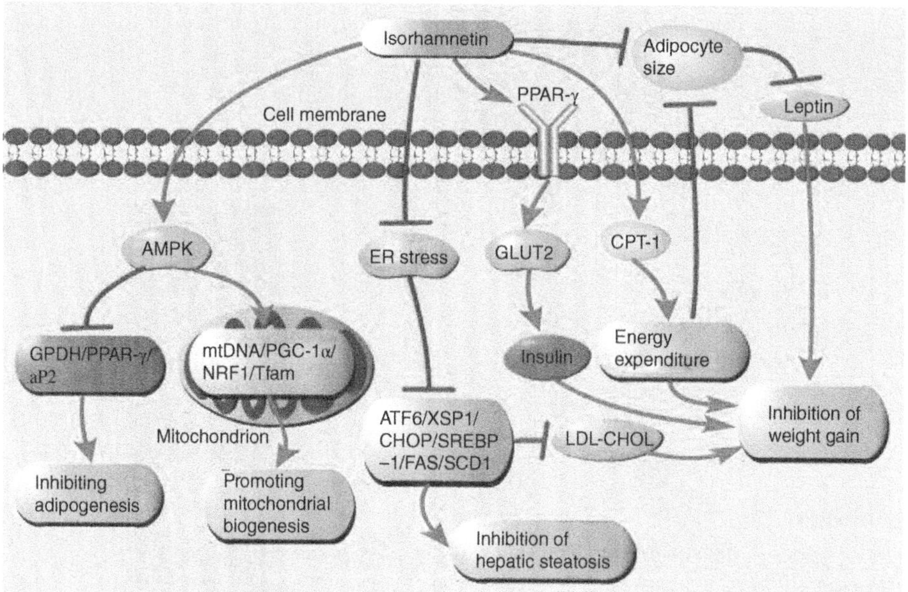

Figure 7.5 Obesity-preventing mechanism of isorhamnetin. Isorhamnetin can prevent obesity by inhibiting adipogenesis and promoting mitochondrial biogenesis. (From Gong et al., 2020, with permission)

features for improving metabolic control of diabetes, as it has also been described in alloxan-induced INS-1 cells. Bcl-2 expression increased, whereas Bax expression and caspase-3 and caspase-9 activities decreased (Li et al., 2020).

In a double blind randomized placebo-controlled crossover clinical trial with 20 healthy subjects, cactus fiber (a supplement made of dehydrated cladode of *O. ficus-indica* and fortified with soluble fiber from *Acacia* spp., 2 tablets thrice daily for 45 days) promoted fecal fat excretion (>3-folds), supporting the hypothesis that cactus fiber helps in reducing body weight (Uebelhack, Busch, Alt, Beah, & Chong, 2014). Those studies led to the patent of a product (Litramine IQP-G-002AS), an *O. ficus-indica*-derived fiber which promoted fat excretion and weight loss in four randomized, placebo-controlled clinical trials (Chong, Lau, Gruenwald, & Uebelhack, 2014). Litramine IQP-G-002AS (3 g/day) for 7 days increased fecal fat excretion (15.8 vs 4.6% for placebo) in healthy Caucasian subjects and significantly reduced weight, body fat composition, and waist circumference in overweight and obese subjects (25-35 kg/m^2 BMI) in a 12 week study. Furthermore, the weight loss attributed to fat binding, delayed nutrients absorption, satiating effects of the high viscosity fiber that attenuates insulin secretion by minimizing postprandial glucose spikes was maintained for another 12 weeks. Another *O. ficus-indica* standardized fiber cladode powder (NeOpuntia, Nexira Health) also binds dietary fat preventing intestinal fat digestion reducing thereby its absorption and increasing fat excretion (27% in healthy subjects receiving 4.8 g Litramine IQP-G-002AS/day). Dietary fiber (Litramine derived from *O. ficus-indica*) intake with a hypocaloric diet and moderate physical activity for 12 weeks reduced weight, body mass

Table 7.3 Main studies in human subjects with *Opuntia* spp. cladode consumption on blood glucose and insulin (adapted by Gouws et al., 2019).

References	Aim of the trial	Subjects	Design	Results
Frati-Munari, Fernández-Harp, de la Riva, Ariza-Andraca, and del Carmen Torres (1983)	The effect of *O. streptacantha* cladode consumption on serum lipids, glycemia and body weight	Healthy ($n = 8$) vs. T2DM ($n = 7$)	Length: 10 days Treatment: (1) Regular diet (2) Regular diet +100 g broiled cladode before meals, 3×/daily	GLU: ↓ (vs. fasting blood GLU) T2DM ($p < 0.001$), 'non-T2DM' ($p < 0.05$) ↑
Frati-Munari, Gordillo, Altamirano, and Ariza (1988)	The effect of *O. streptacantha* cladode consumption on T2DM	Three groups T2DM **Group 1**: (n = 16; 9 Males, 7 Females) Mean age: 48.3 ± 11.4 years (n = 12 received sulfonylureas) **Group 2**: (n = 10; 6 Males; 4 Females) Mean: 46.2 ± 10.8 y; (n = 10 receivedsulfonylureas) **Group 3**: (n = 6; 4 Males; 2 Females) Mean: 48 ± 11.7 years (n = 4 received sulfonylureas)	Length: Single consumption Hyperglycemic agents discontinued 72 hr before trial. Treatment: **Group 1**: 500 g of broiled cladode **Group 2**: 300 mL water **Group 3**: 500 g of broiled zucchini squash Randomized control-trial	GLU: ↓ (vs. basal; $p < 0.001$) **Group 1** vs. **2**: ↓60 min ($p < 0.025$); 120 min ($p < 0.001$); 180 min ($p < 0.001$) **Group 2**: No change **Group 3**: NS ($p > 0.05$) INS: ↓ Serum INS **Group 1** vs. **2**: ↓ 120 min, 180 min ($p < 0.001$) **Group 3**: NS ($p > 0.05$)
Frati, Jimenez, and Ariza (1990)	The effect of *O. ficus-indica* cladode on hyperglycemia (CLD) in T2DM	T2DM (n = 8; 2 Males; 6 Females) Age: 45–68 years (Mean: 55 years)	(1) Entire Broiled CLD (2) Blended and broiled CLD (3) Blended CLD (4) Blended and heated (60° C) CLD	↓ GLU at 120 min and 190 min (VS. control (baseline); $p < 0.01$) NS differences between heated and unheated ($p > 0.05$)

(Continued)

Table 7.3 (Continued)

References	Aim of the trial	Subjects	Design	Results
Frati et al. (1991)	The effect of *O. streptacantha* nopal consumption on blood and GLU of T2DM	**Group 1**: T2DM (n = 14; 9 Males, 5 Females) Age: 32–56 years (Mean: 43.4 years); **Group 2**: 'Healthy' (n = 14; 9 Males; 5 Females) Age: 23–51 years (Mean: 32.7 years)	Length: Single consumption (fasted) **Group 1**: 500 g grilled cladode **Group 2**: 400 mL water. Randomized control trial	GLU: ↓ in T2DM (vs. control) ($p < 0.005$, 60 min; $p < 0.005$; $p < 0.001$ min); 'Healthy'- NS ($p > 0.05$) INS: (n = 7, each Group) ↓- T2DM (vs. control) ($p < 0.001$; 120 min,180 min); 'Healthy': NS
Bacardi-Gascon, Dueñas-Mena, and Jimenez-Cruz (2007)	Post-prandial glycemic response to nopal cladode associated with typical Mexican meal	T2DM (n = 36) Age: 47–72 years GLU: 8.0 ± 2 mmol/L BMI: 24.86 ± 3.76 kg/m^2 **Group 1**: n = 11 **Group 2**: n = 9 **Group 3**: n = 9	Length: 3 weeks Treatment: 1 × week/3 weeks (fasted) (**1**) Chilaquiles (corn tortilla, vegetable oil, tomato sauce, cheese, pinto beans) (with vs without cladode) (**2**) Burrito (scrambled egg, tomato, onion, vegetable oil, flour tortilla, pinto beans) (with and without cladode) (**3**) Quesadillas (flour tortilla, low fat cheese, avocado, pinto beans) (with and without cladode) 7-day washout periods Control: Chilaquiles, Burrito and Quesadillas without cladode Placebo-controlled Cross-over trial	↓GLU (60 + min) both groups; ↓GLU (basal vs. 180 min) ($p < 0.001$) for both groups

López-Romero et al. (2014)	Determine glycemic index of cladode and effect of nopal consumption on blood GLU	**Study 1**: 'Healthy' (n = 7; 3 Males, 4 Females) Age: 26.3 ± 1.2 years; BMI: 23.5 ± 0.8 kg/m^2 **Study 2**: **Group 1**: T2DM (n = 14; 4 Males; 10 Females) Age: 40–60 years (Mean 48 ± 2.1 years); BMI: <30 kg/m^2 (Mean 28.9 ± 1 kg/m^2) **Group 2**: 'Healthy' (n = 7; 4 Males; 3 Females) Age: 25–54 years (Mean 21.1 ± 1.2 years) BMI: <25 kg/m^2 (Mean: 22.2 ± 0.6 kg/m^2)	Fasted (12 hr) Treatment: 150 g steamed Nopal **Study 1**: Length: Single consumption Dried nopal (CLD): 24.8% CHO; 32.2% insoluble fiber; 4.8% soluble fiber; <1.9% fat; 15.4% PRO. 50 g of Nopal vs. 50 g GLU Control trial **Study 2**: Length: Single consumption High CHO breakfast: 300 kcal; 89% CHO; 6% PRO; 5% fat (240 mL apple juice, 55.6 g white bread, 21 g strawberry jam) High Soy Protein Breakfast: 344 kcal; 42.2% CHO, 40.7% PRO; 16.9% fat (61.5 g soy hamburger, 230 mL soymilk beverage). Washout period: 1 week Cross-over trial	**Study 1**: Nopal ingestion (vs. GLU) ↓ AUC GLU ($p < 0.001$) Glycemic index of Nopal: 32.5 ± 4.0 Nopal ingestion (vs. GLU) ↓ AUC INS ($p < 0.05$) Insulinemic index of Nopal: 36.1 ± 6.1 **Study 2**: High CHO Breakfast + Nopal = ↓ GLU at 30 min ($p < 0.05$), 45 min ($p < 0.01$), 60 min ($p < 0.01$); ↓ AUC GLU ($p < 0.001$). ↓ AUC of INS ($p < 0.05$) in T2DM patients High soy protein breakfast: ↓ GLU ($p < 0.001$) regardless of group; Nopal GLU was NS ($p > 0.05$) Nopal INS was NS ($p > 0.05$)

index, body fat composition and waist circumference of 125 overweight and obese volunteers in a double blind, randomized, placebo controlled clinical trial (Grube, Chong, Lau, & Orzechowski, 2013).

Cladode flour consists of chemical components that are resistant to digestive enzymes such as cellulose, hemicelluloses, pectin, lignin, and gums. The viscosity and swelling properties are associated with soluble fibers which reduce carbohydrate absorption because of the gel forming properties of pectins, even if the mechanisms of the improved glycemic control associated with high fiber intake remain undefined (Chandalia et al., 2000). Generally, blood glucose level is lowered through improved insulin sensitivity by suppressing hepatic glucose production. In this mechanism, the glucagon-like peptide-1 (GLP-1) is involved; this incretin hormone increases insulin secretion in a glucose-dependent manner. Nondigestible fermentable dietary fibers increase the release of intestinal GLP-1, stimulating insulin secretion and attenuating hepatic glucose production (Klosterbuer, Thomas, & Slavin, 2012).

In a clinical study, nopal inclusion in a high-carbohydrate meal showed antihyperglycemic and antihyperinsulinemic effects (López-Romero et al., 2014). Nopal consumption increased the antioxidant activity in both healthy people and T2DM patients. The consumption of 300 g of steamed cladodes by diabetic patients reduced postprandial blood glucose, serum insulin, and plasma glucose-dependent insulinotropic peptide peaks, as well as increased the antioxidant activity. The glycemic, insulinemic, and glucose-dependent insulinotropic peptide (GIP) and glucagon-like peptide 1 (GLP-1) indexes of the nopal were 33, 36, 6.5, and 26, respectively. These antihyperglycemic and antihyperinsulinemic effects were attributed to the nopal polysaccharides which restored function of GLP-1 (which suppress hepatic glucose concentration) and of GIP (which stimulate insulin secretion) (López-Romero et al., 2014).

O. ficus-indica cladodes can inhibit early atherogenesis stage by NADPH oxidase and NFκB inhibition thereby exhibiting anti-inflammatory activity and reducing ROS involved in LDL oxidation. Nopal intake ameliorates glucose, insulin, GIP, and GLP-1 index postprandial response after high-carbohydrate or high soy protein breakfast in diabetic patients (Díaz, de la Rosa, Héliès-Toussaint, Guéraud, & Nègre-Salvayre, 2017 and references therein). Blood glucose level decreased in obese prediabetic patients treated with OpunDia™ (mixture of *O. ficus-indica* cladode and fruit extracts) for 16 weeks.

The *O. dillenii* seed oil has also shown antihyperglycemic effect in normal and streptozotocin (STZ)-diabetic rats (Bouhrim et al., 2021). The seed oil significantly decreased the intestinal D-glucose absorption *in situ*. The *ex vivo* test showed that the *O. dillenii* seed oil supplementation significantly blocked the intestinal SGLT1 (sodium-glucose cotransporter 1 (IC_{50} = 60.24 g/mL) and induced significant inhibition of intestinal α-glucosidase (IC_{50} = 278 g/mL) and pancreatic α-amylase (IC_{50} = 0.81 mg/mL) *in vitro*.

In conclusion, *Opuntia* spp. fruit, seed, and cladode can be recommended for T2DM patients. Although not as potent as synthetic drugs, the antidiabetic constituents from *Opuntia* can have the advantage of causing less undesirable side effects and thus can represent a suitable complementary approach in the starting stages of diabetes. Nopal, in particular, can represent not a complete replacement for antihyperglycemic and antihyperinsulinemic drugs, but the results of the studies till now support the traditional use of nopal as a complementary T2DM management without any side effects.

7.4 Effect of *Opuntia* spp. Extracts Against Several Diseases and Mechanisms of Action

Table 7.4, and Figures 7.6 and 7.7 describe the main findings in diverse *in vivo* and *in vitro* studies evaluating the possible mechanism of action of *Opuntia* fruit and cladode (and the associated biological effects). Adult male Wistar rats have been used as a model to evaluate anti-ulcerogenic, wound healing, hepatoprotective, neuroprotective, hypotensive, and diabetic nephropathy protection (Ammar et al., 2018; Bakour et al., 2017; El-Hawary et al., 2020; Sutariya & Saraf, 2017).

Flavonoids from *Opuntia* flowers showed preventive antiulcerogenic activity in Wistar rats previously treated with alcohol. The *Opuntia* methanolic flower extract prevented formation of the deep necrotic lesions of the gastric epithelium by preserving normal antioxidant enzymes (CAT, GPx, and SOD) activities, inhibiting lipid peroxidation biomarkers (malonaldialdehyde [MDA] and carbonyls protein [PC]), oxidation of protein, and DNA fragmentation in gastric mucosa (Alimi et al., 2011).

Table 7.4 Possible mechanisms of action of *Opuntia* spp. extracts tested in *in vivo* and *in vitro* models.

Model	Biological activity	Main findings	References
***in vivo* assays**			
Adult male Wistar albino rats with ethanol-induced gastric ulcers	Anti-ulcerogenic	Dose-dependently reduced ethanol-induced gastric damage; restituted antioxidant enzymes SOD, CAT, and GPx; decreased peroxidation biomarker MDA.	Alimi et al. (2011)
Adult male Wistar rats	Wound healing	Accelerated evolution of wound contraction rates; improved wound healing due to neovascularization, connective tissues, and collagen density.	Ammar, Bardaa, et al. (2015);
		Reduced reepithelialization phase.	Khémiri et al. (2019)
Adult male Wistar rats	Hepatoprotection	Decreased mRNA Sod2, Hmox1, Gclc, Gadd45b expression; decreased metabolic endotoxemia, glucose insulinotropic peptide, glucose intolerance, and lipogenesis.	González-Ponce et al. (2020); Sánchez-Tapia et al. (2017)
		Reduced oxidation markers: ASP, AST, ALT, LDH.	
Adult male Sprague Dawley rats	Neuroprotective	Increased total antioxidant capacity, SOD, GSH, and CAT levels and decreased the elevated MDA level; reduced norepinephrine (NE), dopamine (DA), and serotonin (5-HT) levels in brain	El-Hawary et al. (2020)

(*Continued*)

Table 7.4 (Continued)

Model	Biological activity	Main findings	References
Adult male Wistar rats	Anti-inflammatory	Reduced levels of proinflammatory markers PGE2, NO, IL-1β, and TNF-α as well as iNOS and COX2 mRNAs and proteins; reduced MDA level; increased CAT, SOD, and GSH activities.	Allegra et al. (2014)
Rat paw edema model Arthritic-induced adult female Wistar rats		Inhibition of arachidonic acid metabolites (PGE_2 and LTB_4), ROS, and cytokines (TNF-α, IL-1β, and IL-6) production	Ammar et al. (2018); Roome et al. (2019)
Adult male Wistar rats and Rabbits	Hypotensive	Decreased blood pressure due to urine volume increase	Bakour et al. (2017)
Rats	Diabetic nephropathy prevention	Inhibited protein and glucose presence in urine Reduced creatinine and restored antioxidant enzyme activities in the kidney tissue	Sutariya and Saraf (2017)
Non diabetic and streptozotocin diabetic- induced adult Wistar rats	Anti–hyperglycemic effect	Reduced postprandial hyperglycemia and intestinal D-glucose absorption due to inhibition of intestinal sodium-glucose cotransporter 1 (SGLT1), intestinal α-glucosidase, and pancreatic α-amylase activities.	Bouhrim et al., (2021)
Two-months old male mice	Anti–obesity and anti-T2DM	Improved glucose tolerance and hepatic insulin signaling, enhanced energy expenditure, reduced leptin concentration and hepatic steatosis, prevented adipose tissue hypertrophy, ameliorated endoplastic reticulum stress, and reduced body weight gain	Rodríguez-Rodríguez et al. (2015)
Four-months old male Wistar rats	Bone mineral density increase	Increased physical, mechanical, and microstructural bone strength, mineral content, and bone mineral density in the femurs of rats fed with cladodes.	Hernández-Becerra et al. (2017)
***In vitro* assays**			
HepG2 cells	Hypocholesterolemic	Decreased NPC1L1, ABCA1, ABCG5, and ABCG8 transporters Changes in the RNA amount	Ressaissi, Attia, Pacheco, Falé, and Serralheiro, (2020)

Table 7.4 (Continued)

Model	Biological activity	Main findings	References
HaCaT keratinocytes	Would healing	Scratch test on HaCaT cellular monolayer resulted in wound repairing process.	Catanzano et al. (2021); Moon, Ngoc, Chae, Tran, and Lee (2020)
	Reduced apoptosis and genotoxicity	Downregulation of the phosphorylating transcription factors (ERK, JNK, and p53), and scavenging ROS radicals. Prevention of aryl hydrocarbon receptor degradation.	
HT29 cells	Anti-proliferative	Induced cell cycle arrest at different checkpoints	Serra, Poejo, Matias, Bronze, and Duarte (2013)
Cortical neurons	Neuroprotective effects	Reduced intracellular ROS Decreased intracellular free Ca^{2+}	Chen et al. (2011)

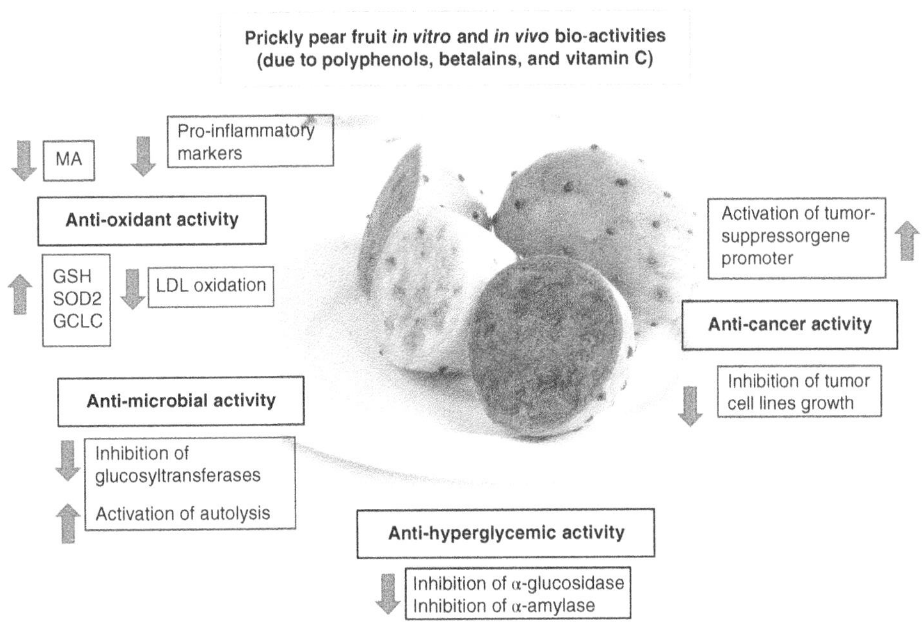

Figure 7.6 Prickly pear fruit biological activities and mechanisms of action. *Source:* salvatore / Adobe Stock.

Wound healing is a natural body reaction leading to the restoration of structural and functional integrity of injured tissues. New sources of healing compounds have been investigated due to antibacterial antibiotic resistance. The aqueous extract of *Opuntia* flowers containing mucilage, and the ethanolic extract induced cicatrization of wounds in Wistar rats, by inducing neovascularization, and formation of new connective tissue (Ammar, Bardaa, et al., 2015). Skin wound healing was investigated using an excisional

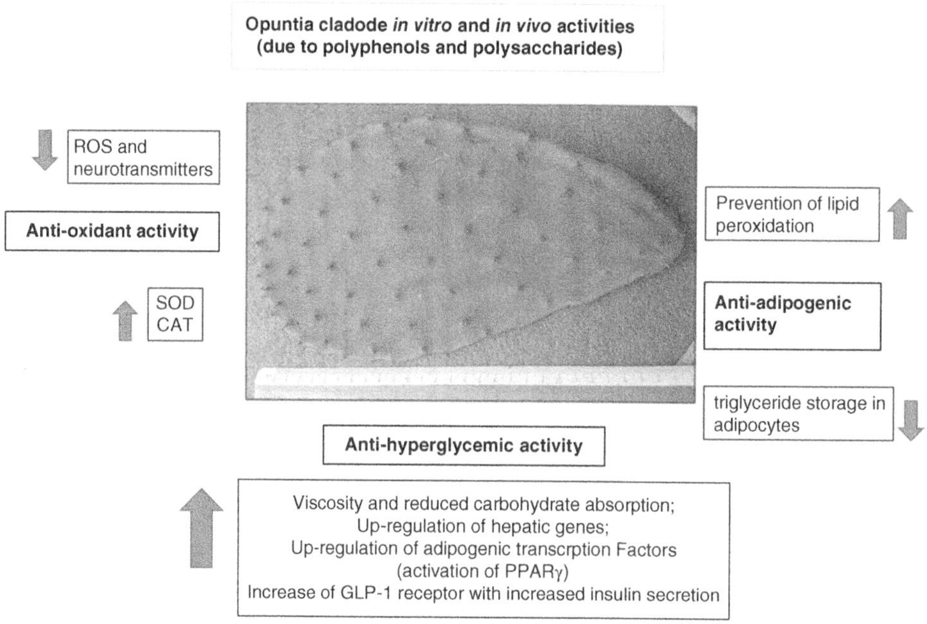

Figure 7.7 *Opuntia* cladode biological activities and mechanisms of action.

wound healing model in rat, treated with *O. ficus-indica* seed oil. The treatment (0.6 μL/mm^2 of OFI seed oil) prevented cutaneous infections and reduced the re-epithelialization phase (Khémiri et al., 2019). *O. ficus-indica* cladode extract was used in combination with chitosan as a film to accelerate skin injury repair. Swelling characteristics, water vapor transmission rate, and release kinetics demonstrated that these films are adequate for the proposed application. Also tests on keratinocytes confirmed *in vitro* the repairing activity of cladode extract (Catanzano et al., 2021).

Microwave-assisted *O. humifusa* leaf extract has been proposed as a cosmeceutical ingredient to prevent environmentally induced skin oxidative stress and inflammation (Moon et al., 2020). The extract exhibited higher antioxidant activity (98% DPPH radical scavenging efficiency), lower viscosity, and toxicity compared to conventionally prepared extract. Apoptosis and genotoxicity of human keratinocyte (HaCaT) cells were minimized during long-term treatment, thanks to the radical scavenging activity of the extract and downregulated phosphorylating transcription factors (ERK, JNK, and p53) (Moon et al., 2020). Betalains have been tested against inflammation, oxidative stress, liver damage, and dyslipidemia-related diseases (Rahimi et al., 2019). A systematic review of animal and human studies provides evidence that supplementation of betalain-rich cacti fruits (*Hylocereus* and *Opuntia*) and cladodes may have potential to influence endothelial and vascular function (Cheok, George, Rodriguez-Mateos, & Caton, 2020).

Betacyanins from *Opuntia* red fruits have good antioxidant capacity and have been used to prevent liver damage in rats treated with acetaminophen. Less histological damage was observed in betacyanin-supplemented rats, associated with higher expression of antioxidant enzymes. Moreover, mRNA for antioxidant enzymes (*Sod2* and *Gclc*) increased in acetaminophen treated hepatocytes after a single dose of betalains extract

(González-Ponce et al., 2020). Indicaxanthin possesses anti-inflammatory effects based on its inhibition of NF-κB activation, a key transcription factor in the whole inflammatory cascade. Oral indicaxanthin administration to rats rapidly decreased proinflammatory markers PGE2, NO, IL-1β, and TNF-α as well as iNOS and COX2 mRNAs and proteins in the leukocytes recruited in the pleural cavity at the early phase of the inflammatory response (Allegra et al., 2014). A methanolic extract of *O. ficus-indica* flowers significantly inhibited inflammation in a carrageenan induced rat paw edema model confirmed by hematological and the histological analysis. The anti-inflammatory effect of *Opuntia* flower extract was associated with the reduction of malonaldialdehyde level (MDA as a lipid peroxidation index) and increase in catalase (CAT), superoxide dismutase (SOD), and reduced glutathione (GSH) activities (Ammar et al., 2018).

Opuntiol (6-hydroxymethyl-4-methoxy-2H-pyran-2-one) and its glucoside, opuntioside (6-β-D-glucopyranosyloxymethyl-4-methoxy-2H-pyran-2-one) have been identified in *O. dillenii* cladode extract. Nano-formulations of opuntiol and opuntioside successfully restored the disease progression in Complete Freund's Adjuvant (CFA)-induced arthritic rats due to their immunomodulatory and anti-inflammatory potentials and can therefore be considered for rheumatoid arthritis therapy (Roome et al., 2019). This substantial antiarthritic activity is related to inhibition of arachidonic acid metabolites (PGE_2 and LTB_4), reactive oxygen species (ROS), and cytokines (TNF-α, IL-1β, and IL-6) production, which play predominant role in pathogenesis of rheumatoid arthritis (Roome et al., 2019 and references therein).

Diet cladode addition can modify gut microbiota and reduce metabolic endotoxemia and other obesity-related biochemical abnormalities. Nopal (5%) modified rat microbiota when administered with high-fat diet containing 5% sucrose (HFS) in drinking water for 1 month (Sánchez-Tapia et al., 2017). Moreover, hepatic steatosis and oxidative stress decreased in adipose tissue and brain, and cognitive function improved associated with *Bacteroides fragilis* increase. Furthermore, nopal-derived oligosaccharides, *O. ficus-indica* mucilage and pectin, also exert prebiotic effects by increasing bifidobacteria (lactobacilli) population and short-chain fatty acids (propionate and butyrate) in human colon microbiota (35 and 16%), respectively (Guevara-Arauza et al., 2012). Increased propionate and butyrate can suppress inflammation in the gastrointestinal tract and modulate immune function. These beneficial prebiotic effects occurred at 2.5 g/day *in vitro* equivalent to a human daily intake of 5 g, similar to or within the range of other *in vivo* probiotics studies.

Other antioxidant compounds, such as phenolics from *Opuntia*, have neuroprotective effect. The administration of *O. ficus-indica* var. *saboten* stem ethanolic extract combined with dextrin for 12 weeks significantly improved cognitive function in old (≤70 years) adults based on blood tests results, vital signs, and electrocardiograms in a randomized double-blind placebo-controlled study (Han et al., 2020). The beneficial effects may be attributed to the treatment of cholinergic blockade-induced cognitive dysfunction in mice (Kwon et al., 2018). The extract (100, 200, and 400 mg/kg) in that study ameliorated scopolamine, induced cognitive function by inhibiting acetylcholinesterase (AChE) activity in the brain, and increasing the levels of phosphorylated extracellular signal-regulated kinase and cAMP response element-binding protein (CREB) and upregulating brain-derived neurotrophic factor (BDNF) expression in the hippocampus (Kwon et al., 2018). Neurotransmitters norepinephrine, dopamine, and serotonin decreased after *Opuntia* extract treatment (El-Hawary et al., 2020). This neuroprotective effect was also observed in cortical neurons expressed as reduced ROS after *Opuntia* extracts application (Chen et al., 2011).

Opuntia phytochemicals have also been tested to treat hypertension. *In vivo* studies in male Wistar rats and rabbits demonstrated that consumption of cladodes increases urine excretion, resulting in reduced blood pressure (Bakour et al., 2017). The diuretic effect of *Opuntia* was demonstrated in a 7-day rat study where *Cacti-Nea*™ (a proprietary *O. ficus-indica* natural water extract [Bio Serae, France]) doubled the urine volume while preserving minerals compared to control (Bisson, Daubié, Hidalgo, Guillemet, & Linarès, 2010). The same *Cacti-Nea* (2 g/day for 28 days) positively benefited body composition (reduced waist [cellulite], ankle, and calf circumferences) of 49 women with normal BMI (20–25 kg/m^2) by limiting fat mass gain (www.bioserae.com). Dehydrated betalain-rich extract of *O. ficus indica* Cacti-Nea® had comparable diuretic effect to hydrochlorothiazide, a diuretic medication often used to treat high blood pressure and swelling due to fluid accumulation (Sadowska-Bartosz & Bartosz, 2021 and references therein). In a human MRI single-center, randomized, 3-treatment (wheat bran, nopal and psyllium) crossover study, nopal (dehydrated cactus leaf; OroVere Nopal Cactus Green Leaf Powder, 50.8% total dietary fiber with 13.2% soluble fiber) stimulated significantly more small bowel water in 11 young adults (average age 20 years, median 22.8 kg/m^2 BMI) than wheat bran due to the presence of mucilage and particulate fiber (Gunn et al., 2020). Moreover, nopal is rapidly fermented in the colon without colonic distension. Betanin can prevent diabetic kidney injury while inhibiting proteinuria, blood glucose, and creatinine excretion. In addition, betanin can increase the expression of antioxidant enzymes (Sutariya & Saraf, 2017).

Dehydrated cladodes of *O. ficus-indica* were added to rodent diet, providing 5 g/kg of calcium (as in the control diet). Results demonstrated that the calcium in cladodes, particularly the late maturity stage, was bioavailable and improved mineral density, mechanical and microstructural properties of the bones in rats. These findings suggest that the consumption of mature cladodes might have a beneficial impact on bone health, still needing to be confirmed in humans (Hernández-Becerra et al., 2017).

Opuntia prevents cardiovascular disease due to its high antioxidant polyphenols content that reduces lipid peroxidation and thereby confers atherogenic properties. The dietary fiber and proteins also exhibit lipid-lowering properties in humans and animals (Díaz et al., 2017). *Opuntia* fruit and juice consumption prevents oxidative stress and improves the redox status in healthy humans. Regular consumption of prickly pear significantly lowers plasma LDL cholesterol, 8-epi-prostaglandin F2α and F2 isoprostane in hypercholesterolemic patients. This hypocholesterolemic effect was confirmed with *O. ficus-indica* dried leaves intake as dietary supplement in women with metabolic syndrome and *O. robusta* prickly pear in hyperlipidemic nondiabetic patients. The lipid-lowering effect is attributed to dietary fiber, particularly *Opuntia* pectin and mucilage that can alter hepatic cholesterol metabolism and lower glucose absorption.

7.5 Current and Potential Applications for Functional Food

7.5.1 *O. ficus-indica* Fruits

Cactus pear fruit is composed of peel (40–50%) and pulp (50–60%). The pulp can be further subdivided into seeds and strained pulp (~45%) (Madrigal-Santillán et al., 2013). Therefore, fruit processing results in the accumulation of several quantities of by-products (Melgar, Dias, et al., 2017). The peel is generally considered a waste, despite the presence of valuable components such as mucilage and pectins (Majdoub, Roudesli, & Deratani, 2001), polyunsaturated fatty acids, and different antioxidants (Cardador-Martínez, Jiménez-Martínez, & Sandoval, 2011; Cejudo-Bastante et al., 2014; Ramadan & Mörsel, 2003). The pulp is the edible portion of prickly pear, and it is consumed mainly fresh; only a small portion is processed into dehydrated products, juices and nectars, jams and dehydrated sheets (Atef, Abou-Zaid, Ibrahim, Ramadan, & Nadir, 2013; El-Samahy, El-Mansy, Bahlol, El-Desouky, & Ahmed, 2008; Sáenz, 2000); sweeteners, ice cream, alcoholic and nonalcoholic beverages and wines (Karabagias et al., 2020; Sáenz, 2000); and for extraction of bioactive components (Del Socorro Santos Díaz & Camarena-Rangel, 2019).

Juices of prickly pear fruit are a water source enriched with sugars (fructose and glucose), organic acids, vitamins, and mineral salts (mainly potassium, calcium, and sodium), in addition to phenolic compounds and betalains (Hernández García et al., 2020). Alcoholic beverages from *O. ficus-indica* fruits have been formulated by inoculating their juice with *S. cerevisiae*, SO_2 (10 mg/L), and citric acid, to acidify up to pH 3.3 (Bustos, 1981). Prickly pear fruit pulp has also been added to other alcoholic beverages such as yakju, increasing the levels of alcohol, sugars, and antioxidant activity (Cho, Huh, & Kim, 2010). Jams are prepared from ripe fruits with high amount of added sugar. As prickly pear fruit already contains pectin, to define the right sugar/pulp ratio is important for the final thickening of the processed product (Sáenz, 2000).

O. ficus-indica fruits vary in color, with the red and purple fruits being particularly attractive. These colored fruits contain betalains, which are the same pigment present in red beet and are widely used in the food industry, under the EU code E162. Purple cactus pears are a promising source of betalains. The most interesting betalains feature for technological application is its pH stability in the 3–7 range, when processed at low pasteurization, in the presence of ascorbic acid, and stored at low temperature; these conditions make betalains suitable for application in many low acid and neutral foods (Moreno, García-Viguera, Gil, & Gil-Izquierdo, 2008). Due to its properties, juice concentrates from cactus pear are expected to be a suitable foodstuff coloring for low acid products such as ice cream or yogurt (Stintzing et al., 2005; Stintzing & Carle, 2007). *O. stricta* fruit extract was characterized and studied for stability. The high betalain content (800 mg/kg f. w.) (five times higher than in red fruits of *O. ficus-indica*) and the vibrant red-purple color, in addition to high storage stability, makes this fruit of particular interest as food colorant. The technological features were in the same range as those obtained for the commercial liquid concentrate from red beet, red carrot, and grape skin (Castellar, Obón, & Fernández-López, 2006)

7.5.2 Potential Applications of Fruit

The potential application of prickly pear fruit as functional food has been reviewed (Patel, 2013). In addition, excellent reviews have been recently published, presenting comprehensive reports on the different properties of *Opuntia* spp. fruits (as well as other part of the plant) (Aruwa et al., 2018; Silva et al., 2021).

Betalains are recognized as pigments with a potential to be applied in functional foods, not only for their action as colorants but also for their antioxidants, antimicrobial, antiproliferative, and hypolipidemic properties (Rahimi et al., 2019; Stintzing & Carle, 2007). Among the richest betalains sources, *O. stricta* var. *dillennii* has been thoroughly characterized for the phytochemical profile of this fruit and its products, revealing the potential of this species (Gómez-López et al., 2021). Betanin can affect and potentially be a new drug for acute myocardial infarction (AMI). It eliminates AMI-induced NF-κB protein expression, suppresses isoproterenol-induced iNOS protein, inhibits oxidative stress damage and ROS production, ameliorates isoproterenol-induced AMI through iNOS, inflammation, and oxidative stress-MPO/LDL in rats, elevates MPO activity, and reduces isoproterenol-induced LDL levels (Moreno-Ley et al., 2021 and references therein). Indicaxanthin offers an interesting approach to chronic inflammatory bowel disease (IBD) therapy. *O. ficus–indica* fruit indicaxanthin inhibited COX-2 and iNOS expression and cytokine (IL-6, IL-8, PGE2, and NO) secretion in an *in vitro* intestinal inflammation model (IL-1β [IBD initiator and amplifier] activated Caco-2 cell monolayers), (Moreno-Ley et al., 2021 and references therein).

Other components of *Opuntia* spp. fruit can be also exploited. Pintado et al. (2001) isolated a protein fraction from unripe cactus fruits with coagulant and caseinolytic activities, for use as plant-derived rennet, substitute for chymosin. Prickly pear peels agro-industrial by-product can be of great interest as an ingredient for the development of dietary supplements or improvement of functional foods (Amaya-Cruz et al., 2019).

Carboxymethyl cellulose (CMC) edible films have also been prepared using peel powder and/or its aqueous extract from *O. ficus-indica* peel fruit (Aparicio-Fernández et al., 2018). The study revealed the feasibility to use prickly pear peel, a vegetable waste, to obtain valuable antioxidant compounds to incorporate into biodegradable food packaging materials.

Recently, an alcoholic beverage by *S. cerevisiae*-fermented prickly pear juice and its pulp, in combination with sugar and blossom honey, was studied (Karabagias et al., 2020). A biofunctional alcoholic beverage (23% alcohol) was prepared, exhibiting high antioxidant activity and total phenolic content, and demonstrated the potential to obtain alcoholic beverages with biological activity using perishable fruits, such as the prickly pear. As prickly pear is a fruit rich in nutraceuticals, but also perishable, new preservation strategies are emerging, to enter the ready-to-eat market (Palma, Continella, La Malfa, & D'Aquino, 2018). The nutraceutical properties of prickly pear fruit have been considered in the development of supplements by a company (BIO SERAE Laboratoires SA, France). The company developed Cacti-Nea, a product from an aqueous extract of prickly pear fruit, rich in betalain, particularly indicaxanthin, which has been shown to be a highly bioavailable anti-inflammatory compound (Allegra et al., 2014; Tesoriere et al., 2008). A company trial evaluated the extract for its chronic diuretic and antioxidant effects in Wistar rats and showed a diuretic effect as well a reduction of weight gain (Bisson et al., 2010). The product is designed for unique weight management support.

7.5.3 *O. ficus-indica* Cladodes

Some species are used mostly for their tender cladodes (called "nopalitos"): *O. ficus-indica, O. streptacantha, O. amyclaea, O. robusta, O. leucotricha, O. robusta*, and *Nopalea cochenillifera*. In Mexico and the Southern US, where the consumption of *Opuntia* products is deeply embedded in local culture, young cactus pads are frequently consumed, representing an important part of the diet in those regions (Corrales-García & Sáenz, 2013; Mondragón Jacobo & De Jesus Méndez Gallegos, 2017; Ramírez-Moreno, Marques, Sánchez-Mata, & Goñi, 2011). Because of diurnal acidity changes, the cladodes should be harvested after 2 hr of sunshine for best use as vegetable (Stintzing & Carle, 2005). Then, spines and glochids are removed, together with the outer chlorenchyma. The peeled cladodes are sliced or cut into small cubes, cooked (as a soup or stew) or used as a fresh vegetable (salads, sauces) or as snacks, beverages, and desserts. While cladodes have been traditionally used as a meat substitute during fasting periods, they are nowadays served with meals similar to green beans.

Companies processing cladodes into various foods are exclusively found in Mexico. Consequently, the greatest product variety is found in Mexico and the USA. The small percentage of exported products constitutes mainly vegetable preserves (Sáenz, 2000; Sáenz, Corrales, & Aquino, 2002). Further products derived from cladodes are jam, chutney or pickles, candied nopales, liqueurs, etc. (Sáenz et al., 2013).

Cladodes contain about 90% water and have high mucilaginous substances that are a complex mixture of polysaccharides of L-arabinose (24.6–42%), D-galactose (21–40.1%), D-xylose (22–22.2%), L-rhamnose (7–13.1%), and D-galacturonic acid (8–12.7%) (Allegra et al., 2017). This dietary fiber is the most important carbohydrate fraction (40–50% on dry weight basis), which is a ß-polysaccharides with a ß-glucan equivalent higher than some Aloe species (Rocchetti et al., 2018). They also contain polyphenols, proteins, vitamins, and a good amount of minerals, especially potassium and calcium (Ayadi, Abdelmaksoud, Ennouri, & Attia, 2009; Ramírez-Moreno et al., 2011).

In Mexico, a natural concentrate of cactus pear fiber (which it called "cactus pear flour") is manufactured by cutting and drying 2- and 3-year-old cladodes. The cactus pear flour is used to develop breakfast products, commercially available under the name of "Cactu Fibra." Cactus pear flour addition was also tested in some foods such as breakfast cereals, biscuits, vegetable soup, and a gelled dessert (flan) (Sáenz et al., 2002; Stintzing & Carle, 2005). Another product available in Mexico is a powdered drink ("Polvo Rostenberg") prepared with water or milk. This drink is made with soy protein, maltodextrin, milk whey, cactus pear flour to 10%, a blend of fibers (polydextrose, soy fiber, wheat bran, barley, sesame, guar gum, carrageenan), malt extract, calcium phosphate, aspartame, ascorbic acid, citric acid, and flavors (Sáenz et al., 2002).

During severe drought, mature cactus pads serve as supplementary fodder for cows, sheep, and goats in Brazil, Chile, California, Morocco, Mexico, South Africa, Texas, and Tunisia (Dubeux, Ben Salem, & Nefzaoui, 2017). Low phenolic and tannin contents of cactus stems facilitate digestion and improve meat production. The laxative effect (due to the high oxalic acid content) must be taken into account, combining feeding with straw. Finally, cactus pads improve milk taste and color (Stintzing & Carle, 2005).

Several *Opuntia* species have been used for growing the cochineal (*Dactylopius coccus*) insect, from which the natural pigment carminic acid is extracted (Portillo & Vigueras, 2017). After a period of decline, today the use of natural pigments is increasing thanks to their health benefits. The cochineal production is being re-established with a range of breeding technologies in many tropical and subtropical areas of the world where this activity represents an income source (Corrales-García & Sáenz, 2013).

7.5.4 Potential Applications of Cladode

The mucilage extracted from the cladodes of *O. ficus-indica* is a promising matrix to be used in food industry thanks to its emulsifying properties, derived from amphiphilic protein fraction and the polyanionic pectin fraction (Guevara, Yahia, De La Fuente, & Biserka, 2003;

Majdoub et al., 2001; Medina-Torres, Brito-De La Fuente, Torrestiana-Sanchez, & Katthain, 2000). This material has been used as an edible coating to extend the shelf life of food products, such as fresh strawberries and breba fig (Allegra et al., 2017; Del Valle, Hernández-Muñoz, Guarda, & Galotto, 2005). Other potential utilization of the mucilage and pectins from the cladodes is as gelling or thickening agent but also as emulsifier in nonalcoholic beverages or dairy products.

O. ficus-indica cladodes can also be considered a rich source of dietary fibers and functional compounds, which make them an important candidate for the production of health-promoting and functional foods (Missaoui et al., 2020; Patel, 2014; Rocchetti et al., 2018). Sáenz, Yoong, Figuerola, Chiffelle, and Estevez (2012) obtained a purified natural dietary fiber with >80 g/100 g of total dietary fiber and ~20 g/100 g of soluble dietary fiber, with promising characteristics (rich in dietary fiber, low in flavor and color, and with high antioxidant capacity) for use as an ingredient in new food formulas.

Recently, several food products have been formulated with the incorporation of cladode flour or mucilage, in order to improve the functionality, as is the case of pasta (Attanzio et al., 2019; Giglio et al., 2020), or bread (Liguori et al., 2020; Msaddak et al., 2017; Sciacca et al., 2021), or biscuits and crackers (Dick, Limberger, Thys, de Oliveira Rios, & Flôres, 2020; Nabil, Ouaabou, Ouhammou, Essaadouni, & Mahrouz, 2020). In particular, cladode flour was used instead of wheat flour for gluten-free formulation, or to add bioactive compounds (phenolics and mucilage) to the formulation of functional food (Dick et al., 2020). Pasta (500 g/week) with 3% *O. ficus-indica* cladode extract intake for 30 days effectively prevented age-related metabolic disorders in 39 healthy adults (mean age 47 years) in a single-arm longitudinal intervention study (Aiello et al., 2018). Moreover, pasta intake reduced abdominal waist probably due to appetite reduction by soluble fiber, decreased total cholesterol (TC), increased paraoxonase [PON] (an endogenous antioxidant), and glutathione [GSH] (reduced oxidative stress) and exerted hypoglycemic and anti-inflammatory effects. The same treatment, pasta (500 g/week) with 3% *O. ficus-indica* soluble cladode extract (30% insoluble polysaccharides) intake for 30 days reduced cardiovascular risk (atherogenic lipoproteins) and metabolic parameters (waist circumference, plasma glucose, triglycerides, creatinine, urea, and aspartate transaminase) in 49 (13M + 36F) adults (mean age 56 ± 5 years) with metabolic syndrome (Giglio et al., 2020). Moreover, *O. ficus-indica* modulated lipid and glucose metabolism by reducing small dense low-density lipoproteins (45% sdLDL decrease) and significantly increasing large LDL-1 (31%) primarily due to the beneficial effects (hypoglycemic, hypolipidemic, and antioxidant activities) of its bioactives (phenolics, flavonoids, betaxanthins, and betacyanins).

Cladode has been considered by companies for the development of natural supplements. NeOpuntia®, from the French BIO SERAE, is made by cactus cladode powder, and it is a combination of soluble and insoluble fibers. It has also been clinically tested showing a positive effect on balancing blood lipid parameters, which are typically associated with metabolic syndrome (Linarès, Thimonier, & Degre, 2007). Also the supplement OpunDia showed the acute blood glucose-lowering effects and the long-term safety of the product, thus supporting the traditional use of *O. ficus-indica* for blood glucose management (Godard et al., 2010). In healthy subjects, the OpunDia oral intake increased serum insulin concentration while reducing blood glucose level for a given amount of glucose ingestion after an endurance exercise (Deldicque et al., 2013).

An Italian company (Bionap s.r.l., Belpasso, Catania, Italy) has recently produced a blend of two herbal extracts containing polysaccharides of prickle pear cladodes and biophenols of olive leaves (MUCOSAVE™FG). A randomized double-blind placebo-controlled trial to assess the effect of a standardized extract of *O. ficus-indica* cladodes and *Olea europaea* L. leaves on the symptoms of healthy adults with gastrointestinal discomfort was conducted. This standardized extract of *Opuntia* cladode and olive leaves (MUCOSAVEFG, 400 mg/day) supplementation for 8 weeks improved symptoms in healthy adults (mean age 41.6 years, $n = 60/40$ MUCOSAVEFG/placebo [maltodextrin]) with gastrointestinal discomfort (heartburn, abdominal pain, constipation, reflux, flatulence, and nausea) and gastroesophageal reflux disease (GERD) in a double-blind randomized controlled study (Malfa, Di Giacomo, Cardia, Sorbara, & Mannucci, & C., Calapai, G., 2021). These digestive benefits may be due to the antiulcerogenic and the gastroprotective activities of *O. ficus-indica* and the gastrointestinal mucosa regeneration of its polysaccharides. The beneficial effects were attributed to the active compounds present in the extract that protect gastric mucosa and oxidative damage and modulate inflammatory response (Malfa et al., 2021). Moreover, *O. ficus-indica* fiber supplementation (20 g/day) for 1 week reduced irritable bowel syndrome (> 50 vs 20% in placebo [dextrose]) in IBS (Rome IV) patients in a double-blind, parallel, randomized controlled trial (Remes-Troche et al., 2021).

7.6 Conclusions

Due to the high nutritional value and health-promoting properties of fruit and cladode of *Opuntia* spp. (and in particular *O. ficus-indica*), these species possess high potential for food and health, in addition to food security in arid and semi-arid area of the world. However, despite the abundance of published papers on the topic where the potential of these species is praised, they have not been fully exploited so far. This represents a profitless situation, since the multifunctional properties of cactus pear fruits, leaves, and flowers suggest a great prospective for this plant in the agroindustrial field. *Opuntia* spp. fit well inside the meaning of 'Circular Economy' objective promoted by the 'European Green Deal'. This chapter is expected to be useful for the wide recognition of *Opuntia* spp. potential and for evoking interest for innovative food research and sustainable agriculture.

References

Abd El-Moaty, H. I., Sorour, W. A., Youssef, A. K., & Gouda, H. M. (2020). Structural elucidation of phenolic compounds isolated from Opuntia littoralis and their anti-diabetic, antimicrobial and cytotoxic activity. *South African Journal of Botany*, *131*, 320–327.

Abdel-Hameed, E.-S. S., Nagaty, M. A., Salman, M. S., & Bazaid, S. A. (2014). Phytochemicals, nutritionals and antioxidant properties of two prickly pear cactus cultivars (*Opuntia ficus–indica* Mill.) growing in Taif, KSA. *Food Chemistry*, *160*, 31–38.

Abou-Elella, F. M., & Ali, R. F. M. (2014). Antioxidant and anticancer activities of different constituents extracted from Egyptian prickly pear cactus (Opuntia ficus-indica) Peel. *Biochemistry and Analytical Biochemistry*, *3*, 1–9.

Aiello, A., Di Bona, D., Candore, G., Carru, C., Zinellu, A., Di Miceli, G., . . . Accardi, G. (2018). Targeting aging with functional food: Pasta with *Opuntia* single-arm pilot study. *Rejuvenation Research, 21*(3), 249–256.

Albano, C., Negro, C., Tommasi, N., Gerardi, C., Mita, G., Miceli, A., ... Blando, F. (2015). Betalains, phenols and antioxidant capacity in cactus pear [Opuntia ficus-indica (L.) Mill.] fruits from Apulia (South Italy) genotypes. *Antioxidants*, 4, 269–280. https://doi.org/10.3390/antiox4020269.

Alimi, H., Hfaiedh, N., Bouoni, Z., Sakly, M., & Ben Rhouma, K. (2011). Evaluation of antioxidant and antiulcerogenic activities of *Opuntia ficus–indica* f. inermis flowers extract in rats. *Environmental Toxicology and Pharmacology*, 32(3), 406–416. https://doi.org/10.1016/j.etap.2011.08.007.

Allegra, A., Sortino, G., Inglese, P., Settanni, L., Todaro, A., & Gallotta, A. (2017). The effectiveness of Opuntia ficus-indica mucilage edible coating on post-harvest maintenance of Dottato'fig (Ficus carica L.) fruit. *Food Packaging and Shelf Life, 12*, 135–141.

Allegra, M., Ianaro, A., Tersigni, M., Panza, E., Tesoriere, L., & Livrea, M.A. (2014). Indicaxanthin from cactus pear fruit exerts anti-inflammatory effects in carrageenin-induced rat pleurisy. *The Journal of Nutrition*, 144(2), 185–192. https://doi.org/10.3945/jn.113.183657.

Amamou, S., Lazreg, H., Hafsa, J., Majdoub, H., Rihouey, C., Le Cerf, D., & Achour, L. (2020). Effect of extraction condition on the antioxidant, antiglycation and α-amylase inhibitory activities of Opuntia macrorhiza fruit peels polysaccharides. *LWT, 127*(109), 411.

Amaya-Cruz, D.M., Pérez-Ramírez, I.F., Delgado-García, J., Mondragón-Jacobo, C., Dector-Espinoza, A., & Reynoso-Camacho, R. (2019). An integral profile of bioactive compounds and functional properties of prickly pear (*Opuntia ficus -indica* L.) peel with different tonalities. *Food Chemistry*, 278, 568–578. https://doi.org/10.1016/j.foodchem.2018.11.031.

Ammar, I., Bardaa, S., Mzid, M., Sahnoun, Z., Rebaii, T., Attia, H., & Ennouri, M. (2015). Antioxidant, antibacterial and *in vivo* dermal wound healing effects of *Opuntia* flower extracts. *International Journal of Biological Macromolecules*, 81, 483–490. https://doi.org/10.1016/j.ijbiomac.2015.08.039.

Ammar, I., Ben Salem, M., Harrabi, B., Mzid, M., Bardaa, S., Sahnoun, Z., ... Ennouri, M. (2018). Anti-inflammatory activity and phenolic composition of prickly pear (Opuntia ficus-indica) flowers. *Industrial Crops and Products*, 112, 313–319. https://doi.org/10.1016/j.indcrop.2017.12.028.

Ammar, I., Ennouri, M., & Attia, H. (2015). Phenolic content and antioxidant activity of cactus (Opuntia ficus-indica L.) flowers are modified according to the extraction method. *Industrial Crops and Products, 64*, 97–104.

Angulo-Bejarano, P.I., del Rocío Gómez-García, M., Valverde, M.E., & Paredes-López, O. (2019). Nopal (*Opuntia* spp.) and its effects on metabolic syndrome: New insights for the use of a millenary plant. *Current Pharmaceutical Design, 25*, 3457–3477. https://doi.org/10.2174/1381612825666191010171819.

Aparicio-Fernández, X., Vega-Ahuatzin, A., Ochoa-Velasco, C. E., Cid-Pérez, S., Hernández-Carranza, P., & Ávila-Sosa, R. (2018). Physical and antioxidant characterization of edible films added with red prickly pear (Opuntia ficus-indica L.) cv. San Martín peel and/or its aqueous extracts. *Food and Bioprocess Technology, 11*(2), 368–379.

Aruwa, C. E., Amoo, S. O., & Kudanga, T. (2018). Opuntia (Cactaceae) plant compounds, biological activities and prospects – A comprehensive review. *Food Research International, 112*, 328–344.

Aruwa, C.E., Amoo, S.O., & Kudanga, T. (2019a). Extractable and macromolecular antioxidants of Opuntia ficus-indica cladodes: Phytochemical profiling, antioxidant and antibacterial activities. *South African Journal of Botany*, 125, 402–410. https://doi.org/10.1016/j.sajb.2019.08.007.

Aruwa, C.E., Amoo, S., & Kudanga, T. (2019b). Phenolic compound profile and biological activities of Southern African Opuntia ficus-indica fruit pulp and peels. *LWT*, 111, 337–344. https://doi.org/10.1016/j.lwt.2019.05.028.

Atef, A. M., Abou-Zaid, N., Ibrahim, I., Ramadan, M. T., & Nadir, A. (2013). Quality evaluation of sheets, jam and juice from prickly pear and melon blends. *Life Science Journal*, *10*(2), 200–208.

Attanzio, A., Diana, P., Barraja, P., Carbone, A., Spanò, V., Parrino, B., ... Montalbano, A. (2019). Quality, functional and sensory evaluation of pasta fortified with extracts from Opuntia ficus-indica cladodes. *Journal of the Science of Food and Agriculture*, *99*(9), 4242–4247.

Avila-Nava, A., Calderón-Oliver, M., Medina-Campos, O. N., Zou, T., Gu, L., Torres, N., ... Pedraza-Chaverria, J. (2014). Extract of cactus (*Opuntia ficus–indica*) cladodes scavenges reactive oxygen species *in vitro* and enhances plasma antioxidant capacity in humans. *Journal of Functional Foods*, *10*, 13–24.

Ayadi, M. A., Abdelmaksoud, W., Ennouri, M., & Attia, H. (2009). Cladodes from Opuntia ficus-indica as a source of dietary fiber: Effect on dough characteristics and cake making. *Industrial Crops and Products*, *30*(1), 40–47.

Bacardi-Gascon, M., Dueñas-Mena, D., & Jimenez-Cruz, A. (2007). Lowering effect on postprandial glycemic response of nopales added to Mexican breakfasts. *Diabetes Care*, *30*, 1264–1265.

Badr, A. N., Gramadzka, K., Shehata, M. G., Szablewska, K., Drzewiecka, K., & Abdel-Razek, A. G. (2020). Prospective antimycotoxigenic act of wild Opuntia ficus-indica by-products. *Czech Journal of Food Sciences*, *38*, 308–314. https://doi.org/10.17221/11/2020-CJFS

Bakour, M., Al-Waili, N., El-Haskoury, R., El-Menyiy, N., Al-Waili, T., Al-Waili, A., & Lyoussi, B. (2017). Comparison of hypotensive, diuretic and renal effects between cladodes of Opuntia ficus-indica and furosemide. *Asian Pacific Journal of Tropical Medicine*, 10(9), 900–906. https://doi.org/10.1016/j.apjtm.2017.08.016.

Barba, F. J., Garcia, C., Fessard, A., Munekata, P. E. S., Lorenzo, J. M., Aboudia, A., ... Remize, F. (2020). Opuntia ficus-indica edible parts: A food and nutritional security perspective. *Food Reviews International*. https://doi.org/10.1080/87559129.2020.1756844

Benayad, Z., Martinez-Villaluenga, C., Frias, J., Gomez-Cordoves, C., & Es-Safi, N. E. (2014). Phenolic composition, antioxidant and anti-inflammatory activities of extracts from Moroccan Opuntia ficus-indica flowers obtained by different extraction methods. *Industrial Crops and Products*, *62*, 412–420.

Berrabah, H., Taïbi, K., Ait Abderrahim, L., & Boussaid, M. (2019). Phytochemical composition and antioxidant properties of prickly pear (Opuntia ficus-indica L.) flowers from the Algerian germplasm. *Journal of Food Measurement and Characterization*, *13*, 1166–1174.

Betancourt, C., Cejudo-Bastante, M. J., Heredia, F. J., & Hurtado, N. (2017). Pigment composition and antioxidant capacity of betacyanins and betaxanthins fractions of Opuntia dillenii (Ker Gawl) Haw cactus fruit. *Food Research International*, *101*, 173–179.

Bisson, J.-F., Daubié, S., Hidalgo, S., Guillemet, D., & Linarés, E. (2010). Diuretic and antioxidant effects of Cacti-Nea®, a dehydrated water extract from prickly pear fruit, in rats. *Phytotherapy Research*, *24*, 587–594. https://doi.org/10.1002/ptr.2996

Blando, F., Russo, R., Negro, C., De Bellis, L., & Frassinetti, S. (2019). Antimicrobial and antibiofilm activity against Staphylococcus aureus of Opuntia ficus-indica (L.) Mill. cladode polyphenolic extracts. *Antioxidants*, 8, 117. https://doi.org/10.3390/antiox8050117.

Bouhrim, M., Ouassou, H., Boutahiri, S., Daoudi, N.E., Mechchate, H., Gressier, B., ... Bnouham, M. (2021). Opuntia dillenii (Ker Gawl.) Haw., seeds oil antidiabetic potential using *in vivo*, *in vitro*, in situ, and ex vivo approaches to reveal its underlying mechanism of action. *Molecules*, 26, 1677. https://doi.org/10.3390/molecules26061677.

Bourhia, M., Elmahdaoui, H., Moussa, S. I., Ullah, R., & Bari, A. (2020). Potential natural dyes food from the powder of prickly pear fruit peels (*Opuntia* spp.) growing in the mediterranean basin under climate stress. *Biomed Research International, 2020*, 7579430.

Bourhia, M., Elmahdaoui, H., Ullah, R., Bari, A., & Benbacer, L. (2019). Promising physical, physicochemical, and biochemical background contained in peels of prickly pear fruit growing under hard ecological conditions in the mediterranean countries. *Biomed Research International, 2019*, 9873146.

Brahmi, D., Ayed, Y., Hfaiedh, M., Bouaziz, C., Ben Mansour, H., Zourgui, L., & Bacha, H. (2012). Protective effect of cactus cladode extract against cisplatin induced oxidative stress, genotoxicity and apoptosis in balb/c mice: Combination with phytochemical composition. *BMC Complementary and Alternative Medicine, 12*, 111. https://doi.org/10.1186/1472-6,882-12-111

Brahmi, F., Haddad, S., Bouamara, K., Yalaoui-Guellal, D., Prost-Camus, E., de Barros, J.-P. P., ... Lizard, G. (2020). Comparison of chemical composition and biological activities of Algerian seed oils of Pistacia lentiscus L., Opuntia ficus indica (L.) Mill. and Argania spinosa L. Skeels. *Industrial Crops and Products, 151*, 112456. https://doi.org/https://doi.org/10.1016/j.indcrop.2020.112456

Bustos, O. E. (1981). Alcoholic beverage from Chilean Opuntia ficus-indica. *American Journal of Enology and Viticulture, 32*(3), 228–229.

Butera, D., Tesoriere, L., Di Gaudio, F., Bongiorno, A., Allegra, M., Pintaudi, A. M., ... Livrea, M. A. (2002). Antioxidant activities of Sicilian prickly pear (*Opuntia ficus–indica*) fruit extracts and reducing properties of its betalains: Betanin and indicaxanthin. *Journal of Agricultural and Food Chemistry, 50*(23), 6895–6901.

Butterweck, V., Semlin, L., Feistel, B., Pischel, I., Bauer, K., & Verspohl, E. J. (2011). Comparative evaluation of two different Opuntia ficus-indica extracts for blood sugar lowering effects in rats. *Phytotherapy Research, 25*, 370–375. https://doi.org/10.1002/ptr.3271

Cano, M.P., Gómez-Maqueo, A., García-Cayuela, T., & Welti-Chanes, J. (2017). Characterization of carotenoid profile of Spanish Sanguinos and Verdal prickly pear (Opuntia ficus-indica, spp.) tissues. *Food Chemistry*, 237, 612–622. http://dx.doi.org/10.1016/j.foodchem.2017.05.135.

Cardador-Martínez, A., Jiménez-Martínez, C., & Sandoval, G. (2011). Revalorization of cactus pear (*Opuntia* spp.) wastes as a source of antioxidants. *Food Science and Technology, 31*(3), 782–788.

Castellanos-Santiago, E., & Yahia, E. M. (2008). Identification and quantification of betalains from the fruits of 10 mexican prickly pear cultivars by high-performance liquid chromatography and electrospray ionization mass spectrometry. *Journal of Agricultural and Food Chemistry, 56*, 5758–5764.

Castellar, M. R., Obón, J. M., & Fernández-López, J. A. (2006). The isolation and properties of a concentrated red-purple betacyanin food colorant from Opuntia stricta fruits. *Journal of the Science of Food and Agriculture, 86*(1), 122–128.

Catanzano, O., Gomez d'Ayala, G., D'Agostino, A., Di Lorenzo, F., Schiraldi, C., Malinconico, M., ... Laurienzo, P. (2021). PEG-crosslinked-chitosan hydrogel films for in situ delivery of Opuntia ficus-indica extract. *Carbohydrate Polymers*, 264, 117,987. https://doi.org/10.1016/j.carbpol.2021.117987.

Cejudo-Bastante, M. J., Chaalal, M., Louaileche, H., Parrado, J., & Heredia, F. J. (2014). Betalain profile, phenolic content, and color characterization of different parts and varieties of Opuntia ficus-indica. *Journal of Agricultural and Food Chemistry*, *62*(33), 8491–8499.

Chahdoura, H., Barreira, J. C. M., Barros, L., Santos-Buelga, C., Ferreira, I. C. F. R., & Achour, L. (2014). Phytochemical characterization and antioxidant activity of Opuntia microdasys (Lehm.) Pfeiff flowers in different stages of maturity. *Journal of Functional Foods*, *9*, 27–37.

Chandalia, M., Garg, A., Lutjohann, D., Von Bergmann, K., Grundy, S. M., & Brinkley, L. J. (2000). Beneficial effects of high dietary fiber intake in patients with type 2 diabetes mellitus. *New England Journal of Medicine*, *342*(19), 1392–1398. https://doi.org/10.1056/NEJM200005113421903

Chen, Y., Zhao, B., Huang, X., Zhan, J., Zhao, Y., Zhou, M., & Guo, L. (2011). Purification and neuroprotective effects of polysaccharides from Opuntia Milpa Alta in cultured cortical neurons. *International Journal of Biological Macromolecules*, 49(4), 681–687. https://doi.org/10.1016/j.ijbiomac.2011.06.031.

Cheok, A., George, T. W., Rodriguez-Mateos, A., & Caton, P. W. (2020). The effects of betalain-rich cacti (dragon fruit and cactus pear) on endothelial and vascular function: A systematic review of animal and human studies. *Food and Function*, *11*(8), 6807–6817.

Cho, I. K., Huh, C. K., & Kim, Y. D. (2010). Quality characteristics of yakju (a traditional Korean beverage) after addition of different tissues of Opuntia ficus-indica from Shinan, Korea. *Korean Journal of Food Preservation*, *17*(1), 36–41.

Chong, P.-W., Lau, K.-Z., Gruenwald, J. & Uebelhack, R. (2014). A review of the efficacy and safety of Litramine IQP-G-002AS, an Opuntia ficus-indica derived fiber for weight management. *Evidence-Based Complementary and Alternative Medicine*, 2014, 943,713. http://dx.doi.org/10.1155/2014/943713.

Chougui, N., Tamendjari, A., Hamidj, W., Hallal, S., Barras, A., Richard, T., & Larbat, R. (2013). Oil composition and characterization of phenolic compounds of Opuntia ficus-indica seeds. *Food Chemistry*, *139*(1–4), 796–803.

Corrales-García, J., & Sáenz, C. (2013). Use of cladodes in food products. In C. Sáenz, H. Berger, A. Rodriguez-Felix, L. Galleti, J. Corrales-Garcia, E. Sepulveda, E., ... Rosell, C. (Eds.), *Agro-industrial utilization of cactus pear* (pp. 45–55). Rome: FAO.

Das, G., Lim, K. J., Tantengco, O. A. G., Carag, H. M., Gonçalves, S., Romano, A., ... Patra, J. K. (2021). Cactus: Chemical, nutraceutical composition and potential bio-pharmacological properties. *Phytotherapy Research*, *35*, 1248–1283. https://doi.org/10.1002/ptr.6889

De Santiago, E., Gill, C. I. R., Carafa, I., Tuohy, K. M., De Peña, M. P., & Cid, C. (2019). Digestion and colonic fermentation of raw and cooked Opuntia ficus-indica cladodes impacts bioaccessibility and bioactivity. *Journal of Agricultural and Food Chemistry*, *67*, 2490–2499.

Del Socorro Santos Díaz, M., Barba de la Rosa, A.P., Héliès-Toussaint, C., Guéraud, F., & Nègre-Salvayre, A. (2017). *Opuntia* spp.: Characterization and benefits in chronic diseases. *Oxidative Medicine and Cellular Longevity*, 2017, 1–17. https://doi.org/10.1155/2017/8634249.

Del Socorro Santos Díaz, M. & Camarena-Rangel, N.G. (2019). Cacti for production of metabolites: Current state and perspectives. *Applied Microbiology and Biotechnology*, 103: 8657–8667. https://doi.org/10.1007/s00253-019-10,125-5.

Del Valle, V., Hernández-Muñoz, P., Guarda, A., & Galotto, M. J. (2005). Development of a cactus-mucilage edible coating (Opuntia ficus-indica) and its application to extend strawberry (Fragaria x ananassa) shelf-life. *Food Chemistry*, *91*(4), 751–756.

Deldicque, L., Van Proeyen, K., Ramaekers, M., Pischel, I., Sievers, H. & Hespel, P. (2013). Additive insulinogenic action of Opuntia ficus-indica cladode and fruit skin extract and leucine after exercise in healthy males. *Journal of the International Society of Sports Nutrition*, 10, 45. https://doi.org/10.1186/1550-2,783-10-45.

Díaz, M. D. S. S., de la Rosa, A. P. B., Héliès-Toussaint, C., Guéraud, F., & Nègre-Salvayre, A. (2017). *Opuntia* spp.: Characterization and benefits in chronic diseases. *Oxidative Medicine and Cellular Longevity*. https://doi.org/10.1155/2017/8634249

Dick, M., Limberger, C., Thys, R.C.S., de Oliveira Rios, A., & Flôres, S.H. (2020). Mucilage and cladode flour from cactus (Opuntia monacantha) as alternative ingredients in gluten-free crackers. *Food Chemistry*, 314, 126,178. https://doi.org/10.1016/j.foodchem.2020.126178.

Du Toit, A., De Wit, M., Osthoff, G., & Hugo, A. (2018). Antioxidant properties of fresh and processed cactus pear cladodes from selected Opuntia ficus-indica and O. robusta cultivars. *South African Journal of Botany*, *118*, 44–51.

Dubeux, J. C. B., Ben Salem, H., & Nefzaoui, A. (2017). Forage production and supply for animal nutrition. In P. Inglese, C. Mondragon, A. Nefzaoui, & C. Saenz (Eds.), *Crop ecology, cultivation and uses of cactus pear* (pp. 73–91). Rome: FAO.

El-Beltagi, H. S., Mohamed, H. I., Elmelegy, A. A., Eldesoky, S. E., & Safwat, G. (2019). Phytochemical screening, antimicrobial, antioxidant, anticancer activities and nutritional values of cactus (Opuntia ficus-indica) pulp and peel. *Fresenius Environmental Bulletin*, *28*, 1545–1562.

El-Hawary, S. S., Sobeh, M., Badr, W. K., Abdelfattah, M. A. O., Ali, Z. Y., El-Tantawy, M. E., ... Wink, M. (2020). HPLC-PDA-MS/MS profiling of secondary metabolites from Opuntia ficus-indica cladode, peel and fruit pulp extracts and their antioxidant, neuroprotective effect in rats with aluminum chloride induced neurotoxicity. *Saudi Journal of Biological Sciences*, 27(10), 2829–2838. https://doi.org/10.1016/j.sjbs.2020.07.003.

El-Samahy, S. K., El-Mansy, H. A., Bahlol, H. E., El-Desouky, A. I., & Ahmed, A. E. (2008). Thermal process time and sensory evaluation for canned cactus pear nectar. *Journal of the Professional Association for Cactus Development*, *10*, 85–107.

Ennouri, M., Fetoui, H., Bourret, E., Zeghal, N., & Attia, H. (2006). Evaluation of some biological parameters of Opuntia ficus-indica. 1. Influence of a seed oil supplemented diet on rats. *Bioresource Technology*, *97*(12), 1382–1386.

Farag, M. A., Sallam, I. E., Fekry, M. I., Zaghloul, S. S., & El-Dine, R. S. (2020). Metabolite profiling of three Opuntia ficus-indica fruit cultivars using UPLC-QTOF-MS in relation to their antioxidant potential. *Food Bioscience*, *36*(100), 673.

Frati, A. C., Gordillo, B. E., Altamirano, P., Ariza, C. R., Cortes-Franco, R., Chavez-Negrete, A., & Islas-Andrade, S. (1991). Influence of nopal intake upon fasting glycemia in type II diabetics and healthy subjects. *Archivos de Investigación Médica (Mex)*, *22*, 51–56.

Frati, A. C., Jimenez, E., & Ariza, C. R. (1990). Hypoglycemic effect of *Opuntia-ficus-indica* in non-insulin dependent diabetes mellitus patients. *Phytotherapy Research*, *4*, 195–197.

Frati-Munari, A. C., Fernández-Harp, J. A., de la Riva, H., Ariza-Andraca, R., & del Carmen Torres, M. (1983). Effects of nopal (*Opuntia* sp.) on serum lipids, glycemia and body weight. *Archivos de Investigación Médica (Mex)*, *14*, 117–125.

Frati-Munari, A. C., Gordillo, B. E., Altamirano, P., & Ariza, C. R. (1988). Hypoglycemic effect of Opuntia streptacantha Lemaire in NIDDM. *Diabetes Care*, *11*, 63–66.

Galati, E. M., Mondello, M. R., Lauriano, E. R., Taviano, M. F., Galluzzo, M., & Miceli, N. (2005). Opuntia ficus-indica (L.) Mill. fruit juice protects liver from carbon tetrachloride-induced injury. *Phytotherapy Research, 19*(9), 796–800. https://doi.org/10.1002/ptr.1741

García-Cayuela, T., Gómez-Maqueo, A., Guajardo-Flores, D., Welti-Chanes, J., & Cano, M. P. (2019). Characterization and quantification of individual betalain and phenolic compounds in Mexican and Spanish prickly pear (Opuntia ficus-indica L. Mill) tissues: A comparative study. *Journal of Food Composition and Analysis, 76*, 1–13.

Garoby-Salom, S., Guéraud, F., Camaré, C., de la Rosa, A.P.B., Rossignol, M., Díaz, M. D.S.S., ... Negre-Salvayre, A. (2016). Dietary cladode powder from wild type and domesticated Opuntia species reduces atherogenesis in apoE knock-out mice. *Journal of Physiology and Biochemistry*, 72(1), 59–70. https://doi.org/10.1007/s13105-015-0461-5.

Gharby, S., Ravi, H. K., Guillaume, D., Vian, M. A., Chemat, F., & Charrouf, Z. (2020). 2-methyloxolane as alternative solvent for lipid extraction and its effect on the cactus (Opuntia ficus-indica L.) seed oil fractions. *OCL, 27*(2020), 27. https://doi.org/10.1051/ocl/2020021

Giglio, R. V., Carruba, G., Cicero, A. F. G., Banach, M., Patti, A. M., Nikolic, D., ... Rizzo, M. (2020). Pasta supplemented with Opuntia ficus-indica extract improves metabolic parameters and reduces atherogenic small dense low-density lipoproteins in patients with risk factors for the metabolic syndrome: A four-week intervention study. *Metabolites, 10*, 428. https://doi.org/10.3390/metabo10110428

Godard, M. P., Ewing, B. A., Pischel, I., Ziegler, A., Benedek, B., & Feistel, B. (2010). Acute blood glucose lowering effects and long-term safety of OpunDia™ supplementation in pre-diabetic males and females. *Journal of Ethnopharmacology, 130*(3), 631–634.

Gomez-Casanovas, N., Blanc-Betes, E., Gonzalez-Meler, M. A., & Azcon-Bieto, J. (2007). Changes in respiratory mitochondrial machinery and cytochrome and alternative pathway activities in response to energy demand underlie the acclimation of respiration to elevated CO_2 in the invasive Opuntia ficus-indica. *Plant Physiology, 145*(1), 49–61.

Gómez-López, I., Lobo-Rodrigo, G., Portillo, M.P., & Cano, M.P. (2021). Characterization, stability, and bioaccessibility of betalain and phenolic compounds from Opuntia stricta var. Dillenii fruits and products of their industrialization. *Foods*, 10, 1593. https://doi.org/10.3390/foods10071593.

Gómez-Maqueo, A., García-Cayuela, T., Fernández-López, R., Welti-Chanes, J., & Cano, M. P. (2019). Inhibitory potential of prickly pears and their isolated bioactives against digestive enzymes linked to type 2 diabetes and inflammatory response. *Journal of the Science of Food and Agriculture, 99*, 6380–6391.

Gómez-Maqueo, A., García-Cayuela, T., Welti-Chanes, J., & Cano, M. P. (2019). Enhancement of anti-inflammatory and antioxidant activities of prickly pear fruits by high hydrostatic pressure: A chemical and microstructural approach. *Innovative Food Science and Emerging Technologies, 54*, 132–142.

Gong, G., Guan, Y. Y., Zhang, Z. L., Rahman, K., Wang, S. J., Zhou, S., ... Zhang, H. (2020). Isorhamnetin: A review of pharmacological effects. *Biomedicine and Pharmacotherapy*, 128, 110301; https://doi.org/10.1015/j.biopha.2020.110301.

González-Ponce, H.A., Martínez-Saldaña, M.C., Tepper, P.G., Quax, W.J., Buist-Homan, M., Faber, K.N., & Moshage, H. (2020). Betacyanins, major components in *Opuntia* red-purple fruits, protect against acetaminophen-induced acute liver failure. *Food Research International*, 137, 109,461. https://doi.org/10.1016/j.foodres.2020.109461.

Gouws, C. A., Georgousopoulou, E. N., Mellor, D. D., McKune, A., & Naumovski, N. (2019). Effects of the consumption of prickly pear cacti (*Opuntia* spp.) and its products on blood glucose levels and insulin: A systematic review. *Medicina*, *55*(5), 138. https://doi.org/10.3390/medicina55050138

Graça Miguel, M., Gago, C., Valente, R., Guerreiro, A., Antunes, D., Manhita, A., & Barrocas-Dias, C. (2018). Qualitative evaluation of fruits from different Opuntia ficus-indica ecotypes/cultivars harvested in South Portugal. *Journal of Food Biochemistry*, *42*(6), e12652.

Grube, B., Chong, P. W., Lau, K. Z., & Orzechowski, H. D. (2013). A natural fiber complex reduces body weight in the overweight and obese: A double-blind, randomized, placebo-controlled study. *Obesity*, *21*(1), 58–64.

Guevara, J. C., Yahia, E. M., De La Fuente, E. B., & Biserka, S. P. (2003). Effects of elevated concentrations of CO_2 in modified atmosphere packaging on the quality of prickly pear cactus stems (*Opuntia* spp.). *Postharvest Biology and Technology*, *29*(2), 167–176.

Guevara-Arauza, J. C., de Jesús Ornelas-Paz, J., Pimentel-González, D. J., Mendoza, S. R., Guerra, R. E. S., & Maldonado, L. M. T. P. (2012). Prebiotic effect of mucilage and pectic-derived oligosaccharides from nopal (Opuntia ficus-indica). *Food Science and Biotechnology*, *21*(4), 997–1003.

Gunn, D., Murthy, R., Major, G., Wilkinson-Smith, V., Hoad, C., Marciani, L., ... Spiller, R. (2020). Contrasting effects of viscous and particulate fibers on colonic fermentation *in vitro* and *in vivo*, and their impact on intestinal water studied by MRI in a randomized trial. *The American Journal of Clinical Nutrition*, 112(3), 595–602; https://doi.org/10.1093/ajcn/nqaa173.

Han, E. H., Lim, M. K., Lee, S., Lee, S. H., Yun, S. M., Yu, H. J., ... Lim, Y.-H. (2020). Efficacy of ethanolic extract of Opuntia ficus-indica var. *saboten* stems for improving cognitive function in elderly subjects 55–85 years of age: A randomized, double-blind, placebo-controlled study. *Journal of Medicinal Food*, *23*(11), 1146–1154.

Harlev, E., Nevo, E., Solowey, E., & Bishayee, A. (2013). Cancer preventive and curative attributes of plants of the Cactaceae family: A review. *Planta Medica*, *79*(09), 713–722.

Harrat, N. El I., Louala, S., Bensalah, F., Affane, F., Chekkal, H., & Lamri-Senhadji, M. (2019). Anti-hypertensive, anti-diabetic, hypocholesterolemic and antioxidant properties of prickly pear nopalitos in type 2 diabetic rats fed a high-fat diet. *Nutrition and Food Science*, 49, 476–490. https://doi.org/10.1108/NFS-06-2018-0169.

Héliès-Toussaint, C., Fouché, E., Naud, N., Blas-Y-Estrada, F., del Socorro Santos-Diaz, M., Nègre-Salvayre, A., ... Guéraud, F. (2020). Opuntia cladode powders inhibit adipogenesis in 3 T3-F442A adipocytes and a high-fat-diet rat model by modifying metabolic parameters and favoring fecal fat excretion. *BMC Complementary Medicine and Therapies*, 20(1), 1–12. https://doi.org/10.1186/s12906-020-2,824-x.

Hernández García, F., Coll, L. A., Cano-Lamadrid, M., Lluch, D. L., Barrachina, Á. A. C., & Murcia, P. L. (2020). Valorization of Prickly Pear [Opuntia ficus-indica (L.) Mill]: Nutritional composition, functional properties and economic aspects. *In Invasive Species-Introduction Pathways, Economic Impact, and Possible Management Options*, IntechOpen, Hamadttu El-Shafie. https://doi.org/10.5772/intechopen.92009

Hernández-Becerra, E., Gutiérrez-Cortez, E., Del Real, A., Rojas-Molina, A., Rodríguez-García, M., Rubio, E., ... Rojas-Molina, I. (2017). Bone mineral density, mechanical, microstructural properties and mineral content of the femur in growing rats fed with cactus Opuntia ficus-indica (L.) Mill. (Cactaceae) cladodes as calcium source in diet. *Nutrients*, *9*, 108. https://doi.org/10.3390/nu9020108

Inglese, P., Mondragon, C., Nefzaoui, A., & Sáenz, C. (2017). *Crop ecology, cultivation and uses of cactus pear*. Rome: Food and Agriculture Organization of the United Nations (FAO).

Jo, M. S., Lee, S., Yu, J. S., Baek, S. C., Cho, Y. C., & Kim, K. H. (2020). Megastigmane derivatives from the cladodes of Opuntia humifusa and their nitric oxide inhibitory activities in macrophages. *Journal of Natural Products*, *83*, 684–692.

Juhaimi, F. A., Ghafoor, K., Uslu, N., Mohamed Ahmed, I. A., Babiker, E. E., Özcan, M. M., & Fadimua, G. J. (2020). The effect of harvest times on bioactive properties and fatty acid compositions of prickly pear (*Opuntia ficus-barbarica* A. Berger) fruits. *Food Chemistry*, *303*, 125387.

Karabagias, V. K., Karabagias, I. K., Prodromiti, M., Gatzias, I., & Badeka, A. (2020). Bio-functional alcoholic beverage preparation using prickly pear juice and its pulp in combination with sugar and blossom honey. *Food Bioscience*, *35*(100), 591.

Khémiri, I., Essghaier Hédi, B., Sadfi Zouaoui, N., Ben Gdara, N., & Bitri, L. (2019). The antimicrobial and wound healing potential of Opuntia ficus-indica L. inermis extracted oil from Tunisia. *Evidence-Based Complementary and Alternative Medicine*, *2019*, 9148782. https://doi.org/10.1155/2019/9148782

Kim, H., Bartley, G. E., Young, S. A., Seo, K. H., & Yokoyama, W. (2013). Altered hepatic gene expression profiles associated with improved fatty liver, insulin resistance, and intestinal permeability after hydroxypropyl methylcellulose (HPMC) supplementation in diet-induced obese mice. *Journal of Agricultural and Food Chemistry*, *61*(26), 6404–6411.

Klosterbuer, A. S., Thomas, W., & Slavin, J. L. (2012). Resistant starch and pullulan reduce postprandial glucose, insulin, and GLP-1, but have no effect on satiety in healthy humans. *Journal of Agricultural and Food Chemistry*, *60*(48), 11928–11934.

Kolniak-Ostek, J., Kita, A., Miedzianka, J., Andreu-Coll, L., Legua, P., & Hernandez, F. (2020). Characterization of bioactive compounds of Opuntia ficus-indica (L.) Mill. seeds from Spanish cultivars. *Molecules*, *25*(23), 5734. https://doi.org/10.3390/molecules25235734

Koubaa, M., Mhemdi, H., Barba, F. J., Angelotti, A., Bouaziz, F., Chaabouni, S. E., & Vorobiev, E. (2017). Seed oil extraction from red prickly pear using hexane and supercritical CO_2: Assessment of phenolic compound composition, antioxidant and antibacterial activities. *Journal of the Science of Food and Agriculture*, *97*, 613–620.

Kwon, Y., Liao, Y., Koo, B., Bae, H., Zhang, J., Han, E. H., ... Ryu, J. H. (2018). Ethanolic extract of Opuntia ficus-indica var. *saboten* ameliorates cognitive dysfunction induced by cholinergic blockade in mice. *Journal of Medicinal Food*, *21*(10), 971–978.

Lacroix, I. M. E., & Li-Chan, E. C. Y. (2014). Overview of food products and dietary constituents with antidiabetic properties and their putative mechanisms of action: A natural approach to complement pharmacotherapy in the management of diabetes. *Molecular Nutrition and Food Research*, *58*, 61–78. https://doi.org/10.1002/mnfr.201300223

Leem, K.-H., Kim, M.-G., Hahm, Y.-T., & Kim, H. (2016). Hypoglycemic effect of Opuntia ficus-indica var. saboten is due to enhanced peripheral glucose uptake through activation of AMPK/p38 MAPK pathway. *Nutrients*, *8*, 800.

Li, W., Lin, K., Zhou, M., Xiong, Q., Li, C., & Ru, Q. (2020). Polysaccharides from *Opuntia milpa alta* alleviate alloxan-induced INS-1 cells apoptosis via reducing oxidative stress and upregulating Nrf2 expression. *Nutrition Research*, 77, 108–118. https://doi.org/10.1016/j.nutres.2020.02.004.

Liguori, G., Gentile, C., Gaglio, R., Perrone, A., Guarcello, R., Francesca, N., ... Settanni, L. (2020). Effect of addition of Opuntia ficus-indica mucilage on the biological leavening, physical, nutritional, antioxidant and sensory aspects of bread. *Journal of Bioscience and Bioengineering*, 129(2), 184–191. https://doi.org/10. 1,016/j.jbiosc.2019.08.009.

Linarès, E., Thimonier, C., & Degre, M. (2007). The effect of neopuntia® on blood lipid parameters – Risk factors for the metabolic syndrome (Syndrome χ). *Advances in Therapy*, *24*(5), 1115–1125.

Liu, W., Fu, Y. J., Zu, Y. G., Tong, M. H., Wu, N., Liu, X. L., & Zhang, S. (2009). Supercritical carbon dioxide extraction of seed oil from Opuntia dillenii Haw. and its antioxidant activity. *Food Chemistry*, *114*, 334–339.

López-Romero, P., Pichardo-Ontiveros, E., Avila-Nava, A., Vázquez-Manjarrez, N., Tovar, A.R., Pedraza-Chaverri, J., & Torres, N. (2014). The effect of nopal (Opuntia ficus-indica) on postprandial blood glucose, incretins, and antioxidant activity in Mexican patients with type 2 diabetes after consumption of two different composition breakfasts. *Journal of the Academy of Nutrition and Dietetics*, 114(11), 1811–1818. http://dx.doi.org/10.1016/j.jand.2014.06.352.

Mabrouki, L., Zougari, B., Bendhifi, M., & Borgi, M. A. (2015). Evaluation of antioxidant capacity, phenol and flavonoid contents of Opuntia streptacantha and Opuntia ficus-indica fruits pulp. *Nature and Technology*, *13*, 2.

Madadi, E., Mazloum-Ravasan, S., Yu, J. S., Ha, J. W., Hamishehkar, H., & Kim, K. H. (2020). Therapeutic application of Betalains: A review. *Plants*, *9*(1), 219. https://doi.org/10.3390/plants9091219

Madrigal-Santillán, E., García-Melo, F., Morales-González, J. A., Vázquez-Alvarado, P., Muñoz-Juárez, S., Zuñiga-Pérez, C., ... Hernández-Ceruelos, A. (2013). Antioxidant and anticlastogenic capacity of prickly pear juice. *Nutrients*, *5*(10), 4145–4158.

Mahloko, L. M., Silungwe, H., Mashau, M. E., & Kgatla, T. E. (2019). Bioactive compounds, antioxidant activity and physical characteristics of wheat-prickly pear and banana biscuits. *Heliyon*, *5*, e02479.

Majdoub, H., Roudesli, S., & Deratani, A. (2001). Polysaccharides from prickly pear peel and nopals of Opuntia ficus-indica: Extraction, characterization and polyelectrolyte behavior. *Polymer International*, *50*(5), 552–560.

Malfa, G. A., Di Giacomo, C., Cardia, L., Sorbara, E. E., Mannucci, C., & Calapai, G. (2021). A standardized extract of Opuntia ficus-indica (L.) Mill and Olea europaea L. improves gastrointestinal discomfort: A double-blinded randomized-controlled study. *Phytotherapy Research*, *35*(7), 3756–3768. https://doi.org/10.1002/ptr.7074

Medina-Torres, L., Brito-De La Fuente, E., Torrestiana-Sanchez, B., & Katthain, R. (2000). Rheological properties of the mucilage gum (Opuntia ficus-indica). *Food Hydrocolloids*, *14*(5), 417–424.

Melgar, B., Dias, M. I., Ciric, A., Sokovic, M., Garcia-Castello, E. M., Rodriguez-Lopez, A. D., ... Ferreira, I. (2017). By-product recovery of *Opuntia* spp. peels: Betalainic and phenolic profiles and bioactive properties. *Industrial Crops and Products*, *107*, 353–359.

Melgar, B., Pereira, E., Oliveira, M. B. P., Garcia-Castello, E. M., Rodriguez-Lopez, A. D., Sokovic, M., ... Ferreira, I. (2017). Extensive profiling of three varieties of *Opuntia* spp. fruit for innovative food ingredients. *Food Research International*, *101*, 259–265.

Missaoui, M., D'Antuono, I., D'Imperio, M., Linsalata, V., Boukhchina, S., Logrieco, A. F., & Cardinali, A. (2020). Characterization of micronutrients, bioaccessibility and antioxidant activity of prickly pear cladodes as functional ingredient. *Molecules*, *25*(9), 2176. https://doi.org/10.3390/molecules25092176

Mondragón Jacobo, C., & De Jesus Méndez Gallegos, S. (2017). Nopalitos or vegetable cactus production and utilization. In P. Inglese, C. Mondragon, A. Nefzaoui, & C. Saenz (Eds.), *Crop ecology, cultivation and uses of cactus pear* (pp. 93–104). Rome: FAO.

Moon, J.-Y., Ngoc, L. T. N., Chae, M., Tran, V. V., & Lee, Y.-C. (2020). Effects of microwave-assisted Opuntia humifusa extract in inhibiting the impacts of particulate matter on human keratinocyte skin cell. *Antioxidants, 9*(4), 271. https://doi.org/10.3390/antiox9040271

Morales, P., Ramírez-Moreno, E., Sanchez-Mata, M. D. C., Carvalho, A. M., & Ferreira, I. C. F. R. (2012). Nutritional and antioxidant properties of pulp and seeds of two xoconostle cultivars (Opuntia joconostle F.A.C. Weber ex Diguet and *Opuntia matudae* Scheinvar) of high consumption in Mexico. *Food Research International, 46*, 279–285.

Morán-Ramos, S., Avila-Nava, A., Tovar, A. R., Pedraza-Chaverri, J., López-Romero, P., & Torres, N. (2012). Opuntia ficus-indica (Nopal) attenuates hepatic steatosis and oxidative stress in obese Zucker (fa/fa) rats. *The Journal of Nutrition, 142*, 1956–1963. https://doi.org/10.3945/jn.112.165563

Moreno, D.A., García-Viguera, C., Gil, J.I., & Gil-Izquierdo, A. (2008). Betalains in the era of global agri-food science, technology and nutritional health. *Phytochemistry Reviews*, 7(2), 261–280. https://doi.org/10.1007/s11101-007-9,084-y.

Moreno-Ley, C., Osorio-Revilla, G., Hernández-Martínez, D., Ramos-Monroy, O., & Gallardo-Velázquez, T. (2021). Anti-inflammatory activity of betalains: A comprehensive review. *Human Nutrition and Metabolism, 200*, 126. https://doi.org/10.1016/j.hnm.2021.200126

Msaddak, L., Abdelhedi, O., Kridene, A., Rateb, M., Belbahri, L., Ammar, E., ... Zouari, M. (2017). Opuntia ficus-indica cladodes as a functional ingredient: Bioactive compounds profile and their effect on antioxidant quality of bread. *Lipids in Health and Disease, 16*(1), 1–8. https://doi.org/10.1186/s12944-016-0397-y

Nabil, B., Ouaabou, R., Ouhammou, M., Essaadouni, L., & Mahrouz, M. (2020). Functional properties, antioxidant activity, and organoleptic quality of novel biscuit produced by Moroccan cladode flour "Opuntia ficus-indica". *Journal of Food Quality*, 2020, Article ID 3542398, 12 pages. https://doi.org/10.1155/2020/3542398.

Naselli, F., Tesoriere, L., Caradonna, F., Bellavia, D., Attanzio, A., Gentile, C., & Livrea, M. A. (2014). Anti-proliferative and pro-apoptotic activity of whole extract and isolated indicaxanthin from Opuntia ficus-indica associated with re-activation of the onco-suppressor p16INK4a gene in human colorectal carcinoma (Caco-2) cells. *Biochemical and Biophysical Research Communications, 450*(1), 652–658.

Nazzaro, F., Fratianni, F., D'Acierno, A., Caputo, L., De Feo, V., & Coppola, R. (2021). Antibiofilm properties exhibited by the prickly pear (Opuntia ficus-indica) seed oil. *Proceedings, 66*, 29. https://doi.org/10.3390/proceedings2020066029

Nefzaoui, A. (2009). Cactus: A crop to meet the challenges of climate change in dry areas. *Annals of Arid Zone, 48*(1), 1–18.

Nefzaoui, A., Louhaichi, M., & Ben Salem, H. (2014). Cactus as a tool to mitigate drought and to combat desertification. *Journal of Arid Land Studies, 24*(1), 121–124.

Nobel, P. S. (1995). Environmental biology. In G. Barbera, P. Inglese, & E. Pimienta-Barrios (Eds.), *Agro-ecology, cultivation and uses of cactus pear* (pp. 36–48). Rome: FAO.

Nobel, P. S., & Bobich, E. G. (2002). Environmental biology. In P. S. Nobel (Ed.), *Cacti, biology and uses* (pp. 57–74). Los Angeles, CA, USA: University of California Press.

Nuñez-López, M.A., Paredes-López, O., & Reynoso-Camacho, R. (2013). Functional and hypoglycemic properties of nopal cladodes (O. ficus-indica) at different maturity stages using *in vitro* and *in vivo* tests. *Journal of Agricultural and Food Chemistry*, 61(46), 10981–10986. https://doi.org/10.1021/jf403834x.

Ortega-Ortega, M. D. L. A., Cruz-Cansino, N. D. S., Alanís-García, E., Delgado-Olivares, L., Ariza-Ortega, J. A., Ramírez-Moreno, E., & de Jesús Manríquez-Torres, J. (2017). Optimization of ultrasound extraction of cactus pear (Opuntia ficus-indica) seed oil based on antioxidant activity and evaluation of its antimicrobial activity. *Journal of Food Quality, 2017*(9), 315360. https://doi.org/10.1155/2017/9315360

Osuna-Martínez, U., Reyes-Esparza, J., & Rodríguez-Fragoso, L. (2014). Cactus (Opuntia ficus-indica): A review on its antioxidants properties and potential pharmacological use in chronic diseases. *Natural Products Chemistry and Research*, *2*(6), 153. https://doi.org/10.4172/2329-6836.1000153

Özcan, M. M., & Al Juhaimi, F. Y. (2011). Nutritive value and chemical composition of prickly pear seeds (Opuntia ficus-indica L.) growing in Turkey. *International Journal of Food Sciences and Nutrition*, *62*, 533–536.

Palma, A., Continella, A., La Malfa, S., & D'Aquino, S. (2018). Changes in physiological and some nutritional, nutraceuticals, chemical–physical, microbiological and sensory quality of minimally processed cactus pears cvs 'Bianca', 'Gialla' and 'Rossa' stored under passive modified atmosphere. *Journal of the Science of Food and Agriculture*, 98(5), 1839–1849. https://doi.org/10.1002/jsfa.8660.

Patel, S. (2013). Reviewing the prospects of Opuntia pears as low cost functional foods. *Reviews in Environmental Science and Bio/Technology*, 12(3), 223–234. https://doi.org/10.1007/s11157-012-9,295-6.

Patel, S. (2014). Opuntia cladodes (nopal): Emerging functional food and dietary supplement. *Mediterranean Journal of Nutrition and Metabolism*, *7*(1), 11–19. https://doi.org/10.3233/MNM-140003

Pintado, A. I., Macedo, A. C., Teixeira, G., Pais, M. S., Clemente, A., & Malcata, F. X. (2001). Caseinolytic activity of fruit extract from Opuntia ficus-indica on Bovine, Caprine, and Ovine Sodium Caseinates. *Biotechnology Progress*, *17*(4), 643–646.

Portillo, L., & Vigueras, A. L. (2017). Cochineal breeding. In P. Inglese, C. Mondragon, A. Nefzaoui, & C. Saenz (Eds.), *Crop ecology, cultivation and uses of cactus pear* (pp. 105–113). Rome: FAO.

Pretti, L., Bazzu, G., Serra, P. A., & Nieddu, G. (2014). A novel method for the determination of ascorbic acid and antioxidant capacity in Opuntia ficus-indica using *in vivo* microdialysis. *Food Chemistry*, *147*, 131–137.

Rahimi, P., Abedimanesh, S., Mesbah-Namin, S. A., & Ostadrahimi, A. (2019). Betalains, the nature-inspired pigments, in health and diseases. *Critical Reviews in Food Science and Nutrition*, *59*, 2949–2978.

Ramadan, M. F., & Mörsel, J. T. (2003). Recovered lipids from prickly pear [*Opuntia ficus-indica (*L.) Mill] peel: A good source of polyunsaturated fatty acids, natural antioxidant vitamins and sterols. *Food Chemistry*, *83*(3), 447–456.

Ramadan, M.F., Moussa Ayoub T.E. & Rohn S. (eds.) (2021). *Opuntia spp.: Chemistry, bioactivity and industrial applications*. Springer Nature Switzerland AG. https://doi.org/10.1007/978-3-030-78,444-7.

Ramírez-Moreno, E., Cariño-Cortés, R., Cruz-Cansino, N. D. S., Delgado-Olivares, L., Ariza-Ortega, J. A., Montañez-Izquierdo, V. Y., ... Filardo-Kerstupp, T. (2017). Antioxidant and antimicrobial properties of cactus pear (*Opuntia*) seed oils. *Journal of Food Quality, 2017*, 3075907. https://doi.org/10.1155/2017/3075907

Ramírez-Moreno, E., Marques, C. D., Sánchez-Mata, M. C., & Goñi, I. (2011). *in vitro* calcium bioaccessibility in raw and cooked cladodes of prickly pear cactus (Opuntia ficus-indica L. Miller). *LWT-Food. Science and Technology, 44*(7), 1611–1615.

Rasoulpour, R., Afsharifar, A., & Izadpanah, K. (2018). Antiviral activity of prickly pear (Opuntia ficus-indica (L.) Miller) extract: Opuntin B, a second antiviral protein. *Crop Protection, 112*, 1–9.

Rasoulpour, R., Afsharifar, A., Izadpanah, K., & Aminlari, M. (2017). Purification and characterization of an antiviral protein from prickly pear (Opuntia ficus-indica (L.) Miller) cladode. *Crop Protection, 93*, 33–42.

Remes-Troche, J.M., Taboada-Liceaga, H., Gill, S., Amieva-Balmori, M., Rossi, M., Hernández-Ramírez, G., ... Whelan, K. (2021). Nopal fiber (Opuntia ficus-indica) improves symptoms in irritable bowel syndrome in the short term: A randomized controlled trial. *Neurogastroenterology and Motility*, 33: e13986. https://doi.org/10.1111/nmo.13986.

Ressaissi, A., Attia, N., Pacheco, R., Falé, P.L., & Serralheiro, M.L.M. (2020). Cholesterol transporter proteins in HepG2 cells can be modulated by phenolic compounds present in Opuntia ficus-indica aqueous solutions. *Journal of Functional Foods*, 64, 103,674. https://doi.org/10.1016/j.jff.2019.103674.

Reynoso-Camacho, R., & González de Mejía, E. (2008). Nopal (*Opuntia* spp.) and other traditional Mexican plants. In V. K. Pasupuleti & J. W. Anderson (Eds.), *Nutraceuticals, glycemic health and type 2 diabetes* (pp. 379–399). Ames, IA: Wiley-Blackwell.

Rocchetti, G., Pellizzoni, M., Montesano, D., & Lucini, L. (2018). Italian Opuntia ficus-indica cladodes as rich source of bioactive compounds with health-promoting properties. *Foods*, *7*(2), 24. https://doi.org/10.3390/foods7020024

Rodríguez-Rodríguez, C., Torres, N., Gutiérrez-Uribe, J. A., Noriega, L. G., Torre-Villalvazo, I., Leal-Díaz, A. M., ... Tovar, A. R. (2015). The effect of isorhamnetin glycosides extracted from Opuntia ficus-indica in a mouse model of diet induced obesity. *Food and Function*, *6*(3), 805–815.

Roome, T., Aziz, S., Razzak, A., Aslam, Z., Jamali, K. S., Sikandar, B., ... Shah, M. R. (2019). Opuntioside, opuntiol and its metallic nanoparticles attenuate adjuvant-induced arthritis: Novel suppressors of toll-like receptors-2 and-4. *Biomedicine and Pharmacotherapy*, 112, 108,624. https://doi.org/10.1016/j.biopha.2019.108624.

Rosales-Nieto, C. A., Rodríguez-Aguilar, M., Santiago-Hernandez, F., Cuevas-Reyes, V., Flores-Najera, M. J., Vázquez-García, J. M., ... Martin, G. B. (2021). Periconceptional nutrition with spineless cactus (Opuntia ficus-indica) improves metabolomic profiles and pregnancy outcomes in sheep. *Scientific Reports*, *11*(1), 1–14; https://doi.org/10.1038/s41598-021-86,653-w.

Sadowska-Bartosz, I. & Bartosz, G. (2021). Biological properties and applications of betalains. *Molecules*, 26, 2520; https://doi.org/10.3390/molecules26092520.

Sáenz, C. (2000). Processing technologies: An alternative for cactus pear (*Opuntia* spp.) fruits and cladodes. *Journal of Arid Environments*, *46*(3), 209–225.

Sáenz, C., Berger, H., Rodríguez-Félix, A., Galleti, L., Corrales-García, J., Sepúlveda, E., ... Rosell, C. (2013). *Agro-industrial utilization of cactus pear*. Rome: *Food and Agriculture Organization* (FAO). Available from http://www.fao.org/publications

Sáenz, C., Yoong, M., Figuerola, F., Chiffelle, I., & Estevez, A. M. (2012). Cactus pear cladodes powders as dietary fiber *source:* Purification and properties. *International Journal Food Science and Nutrition, 63*(3), 283–289.

Sáenz, C., Corrales, J., & Aquino, G. (2002). Nopalitos, mucilage, fiber, and cochineal. In P. S. Nobel (Ed.), *Cacti: Biology and uses* (pp. 211–234). Los Angeles, CA, USA: University of California Press.

Sakulnarmrat, K., & Konczak, I. (2012). Composition of native Australian herbs polyphenolic-rich fractions and *in vitro* inhibitory activities against key enzymes relevant to metabolic syndrome. *Food Chemistry, 134*, 1011–1019.

Sánchez-Tapia, M., Aguilar-López, M., Pérez-Cruz, C., Pichardo-Ontiveros, E., Wang, M., Donovan, S.M., ... Torres, M. (2017). Nopal (Opuntia ficus-indica) protects from metabolic endotoxemia by modifying gut microbiota in obese rats fed high fat/sucrose diet. *Scientific Reports* 7, 4,716. https://doi.org/10.1038/s41598-017-05096-4.

Santos-Zea, L., Gutiérrez-Uribe, J. A., & Serna-Saldivar, S. O. (2011). Comparative analyses of total phenols, antioxidant activity, and flavonol glycoside profile of cladode flours from different varieties of *Opuntia* spp. *Journal of Agricultural and Food Chemistry, 59*, 7054–7061.

Sciacca, F., Palumbo, M., Pagliaro, A., Di Stefano, V., Scandurra, S., Virzì, N., & Melilli, M. G. (2021). *Opuntia* cladodes as functional ingredient in durum wheat bread: Rheological, sensory, and chemical characterization. *CyTA-Journal of Food*, 19(1), 96–104. https://doi.org/10.1080/19476337.2020.1862918.

Serra, A. T., Poejo, J., Matias, A. A., Bronze, M. R., & Duarte, C. M. M. (2013). Evaluation of *Opuntia* spp. derived products as antiproliferative agents in human colon cancer cell line (HT29). *Food Research International*, 54(1), 892–901. https://doi.org/10.1016/j.foodres.2013.08.043.

Silva, M. A., Albuquerque, T. G., Pereira, P., Ramalho, R., Vicente, F., Oliveira, M. B. P., & Costa, H. S. (2021). Opuntia ficus-indica (L.) Mill.: A multi-benefit potential to be exploited. *Molecules*, 26(4), 951. https://doi.org/10.3390/molecules26040951.

Smeriglio, A., Bonasera, S., Germanò, M. P., D'Angelo, V., Barreca, D., Denaro, M., ... Trombetta, D. (2019). Opuntia ficus-indica (L.) Mill. fruit as source of betalains with antioxidant, cytoprotective, and anti-angiogenic properties. *Phytotherapy Research*, 33(5), 1526–1537. https://doi.org/10.1002/ptr.6345.

Standl, E., Khunti, K., Hansen, T.B., & Schnell, O. (2019). The global epidemics of diabetes in the twenty-first century: Current situation and perspectives. *European Journal of Preventive Cardiology*, 26(2S), 7–14. https://doi.org/10.1177/2047487319881021.

Stintzing, F.C., & Carle, R. (2005). Cactus stems (*Opuntia* spp.): A review on their chemistry, technology, and uses. *Molecular Nutrition and Food Research*, 49(2), 175–194. https://doi.org/10.1002/mnfr.200400071.

Stintzing, F.C., & Carle, R. (2007). Betalains–emerging prospects for food scientists. *Trends in Food Science and Technology*, 18(10), 514–525. https://doi.org/10.1016/j.tifs.2007.04.012.

Stintzing, F. C., Herbach, K. M., Mosshammer, M. R., Carle, R., Yi, W., Sellappan, S., ... Felker, P. (2005). Color, betalain pattern, and antioxidant properties of cactus pear (*Opuntia* spp.) clones. *Journal of Agricultural and Food Chemistry, 53*(2), 442–451.

Sutariya, B., & Saraf, M. (2017). Betanin, isolated from fruits of Opuntia elatior Mill. attenuates renal fibrosis in diabetic rats through regulating oxidative stress and TGF-β pathway. *Journal of Ethnopharmacology*, 198, 432–443. https://doi.org/10.1016/j.jep.2016.12.048.

Tesoriere, L., Butera, D., Pintaudi, A. M., Allegra, M., & Livrea, M. A. (2004). Supplementation with cactus pear (Opuntia ficus-indica) fruit decreases oxidative stress in healthy humans: A comparative study with vitamin C. *The American Journal of Clinical Nutrition, 80*, 391–395.

Tesoriere, L., Fazzari, M., Angileri, F., Gentile, C., & Livrea M. A. (2008). *in vitro* digestion of betalainic foods, stability and bioaccessibility of betaxanthins and betacyanins and antioxidative potential of food digesta. *Journal of Agricultural and Food Chemistry, 56*, 10487–10492. https://doi.org/10.1021/jf8017172.

Timpanaro, G., & Foti, V. T. (2014). The structural characteristics, economic performance and prospects for the Italian cactus pear industry. *Journal of the Professional Association for Cactus Development, 16*, 32–50.

Uebelhack, R., Busch, R., Alt, F., Beah, Z. M., & Chong, P. W. (2014). Effects of cactus fiber on the excretion of dietary fat in healthy subjects: A double blind, randomized, placebo-controlled, crossover clinical investigation. *Current Therapeutic Research*, 76, 39–44. http://dx.doi.org/https://doi.org/10.1016/j.curtheres.2014.02.001.

Yang, N., Zhao, M., Zhu, B., Yang, B., Chen, C., Cui, C., & Jiang, Y. (2008). Anti-diabetic effects of polysaccharides from Opuntia monacantha cladode in normal and streptozotocin-induced diabetic rats. *Innovative Food Science and Emerging Technologies*, 9(4), 570–574. https://doi.org/10.1016/j.ifset.2007.12.003.

Yoon, J. A., Hahm, S. W., Park, J. E., & Son, Y. S. (2009). Total polyphenol and flavonoid of fruit extract of Opuntia humifusa and its inhibitory effect on the growth of MCF-7 human breast cancer cells. *Journal of the Korean Society of Food Science and Nutrition, 38*(12), 1679–1684.

Zeghad, N., Ahmed, E., Belkhiri, A., Heyden, Y. V., & Demeyer, K. (2019). Antioxidant activity of Vitis vinifera, Punica granatum, Citrus aurantium and Opuntia ficus-indica fruits cultivated in Algeria. *Heliyon, 5*, e01575.

Zimmermann, H. (2017). Global invasion of cacti: Control, management and conflicts of interest. In P. Inglese, C. Mondragon, A. Nefzaoui, & C. Sáenz (Eds.), *Crop ecology, cultivation and uses of cactus pear* (pp. 172–185). Rome: FAO and ICARDA.

Zourgui, M. N., Hfaiedh, M., Brahmi, D., Affi, W., Gharsallah, N., Zourgui, L., & Amri, M. (2020). Phytochemical screening, antioxidant and antimicrobial activities of Opuntia streptacantha fruit skin. *Journal of Food Measurement and Characterization*, 14(5), 2721–2733; https://doi.org/10.1007/s11694-020-00518-w.

8

Molecular Mechanisms of Edible Macro- and Microalgae as Functional Foods or Sources of Nutraceuticals

Yvonne V. Yuan[1] *and B. Dave Oomah*[2]

[1] *School of Nutrition, Toronto Metropolitan University, Toronto, Ontario, Canada*
[2] *(Retired) Formerly with Summerland Research and Development Centre Agriculture and Agri-Food Canada Summerland, BC, Canada*

8.1 Introduction

Macro- and microalgae are perhaps the quintessential and original functional foods and sources of nutraceuticals, with their use by coastal peoples across the globe dating back to antiquity. For example, consumption of marine macroalgae was confirmed at the Monte Verde II archeological site in southern Chile dating approximately 14,000 years ago; a medicinal use for edible macroalgae is described in the Ebers papyrus of approximately 1,534 BCE Egypt, to treat or prevent breast cancer (Yuan, 2021); ethnobotanical studies indicate that the indigenous peoples of the Pacific coast of Canada (British Columbia) and the United States (Alaska) have a long history of macroalgae usage, particularly, *Porphyra abbottiae* (also known as *Pyropia abbottiae*) not only for nutritional and medical but also for trade purposes (Turner, 2003). The primacy of *P. abbottiae* and other *Porphyra* species to the indigenous cultures of the Pacific coast is manifested in the complex nomenclature of these red macroalgae among First Nations including Alutiiq and Dena'ina, Haida and Tlingit, Tshimshian, Wakashan, and Salish peoples (Turner, 2003). Microalgae such as *Spirulina platensis* (also known as *Arthrospira platensis*) was harvested by Aztecs from Lake Texcoco, in present day Mexico, as "Tecuitlatl" to consume with corn or tortillas; similarly, *S. platensis* is harvested from Lake Kossorom by the Kanembu women of Chad for consumption as "Dihé" (Yuan, 2021).

The importance of macro- and microalgae to coastal peoples across the globe is not surprising given that 71% of the Earth's surface is comprised by the oceans, albeit with less than 1% as freshwater. The marine macroalgal phyla or divisions of importance as traditional and functional foods, sources of nutraceuticals or ingredients, comprise the Ochrophyta, specifically the Phaeophyceae class (brown algae [B]); the Chlorophyta (i.e. the Chlorophyceae class (green algae [G]), and the Rhodophyta (i.e. the Rhodophyceae class (red algae [R]). Thus, marine macroalgae have a rich history of traditional food uses related to their protein content (particularly the Rhodophyta), long-chain ω-3

Molecular Mechanisms of Functional Food, First Edition. Edited by Rocio Campos-Vega and B. Dave Oomah.

polyunsaturated fatty acids (PUFAs; i.e. eicosapentaenoic acid [EPA] and docosahexaenoic acid [DHA]), minerals, vitamins, and dietary fibers. This is exemplified by the Phaeophyceae or kelps including "Arame" (*Eisenia bicyclis*), "Hijiki" (*Hizikia fusiformis*, also known as *Sargassum fusiforme*), "Kombu" (*Laminaria digitata*), "Wakame" in Japan, or "Sea Mustard" (*Undaria pinnatifida*) in Korea, "Sea Tangle", "Makonbu" or "Hai dai" (*Laminaria japonica* also known as *Saccharina japonica*) in Japan, Korea, or China, "Oarweed", "Tangleweed" or "Split laminaria" (*Laminaria setchelli*) in Japanese and Chinese soup stocks called Dashi, "Macro or Giant kelp" (*Macrocystis integrifolia*) or "Bull kelp" (*Nereocystis leutkeana*) in the Pacific Northwest (Yuan, Carrington, & Walsh, 2005; Yuan & Walsh, 2006); Chlorophyceae such as "Limu Pālahalaha" or "Limu papahapapa" (*Ulva fasciata*, also known as *Ulva lactuca*) in Hawai'i (also known as "Sea lettuces" comprised of the Ulva genus), "Latô and Arosep", "Latok", or "Umi-budō" meaning "Sea grapes" (*Caulerpa lentillifera*) in the Philippines, Malaysia, Okinawa, and Japan, respectively; Rhodophyceae such as "Nori" or "Kim/Gim" (*Pyropia tenera* also known as *Neopyropia tenera*) in Japan or Korea, "Zicai", "Susabi-nori or Amanori", "Kim" or "Open Sea Nori" (*Porphyra yezoensis*) in China, Japan, Korea, and Britain, respectively, "Ceylon moss" (*Gracilaria edulis*) in Southwest and Southeast Asia, "Sol" or "Dulse, Dillisk, Dilleasc, Creathnach" (*Palmaria palmata*) in Iceland, Scotland, Ireland, Norway, Atlantic Canada, or United States, "Jagalq'a or Dzagalggey", "Sgiw or Sgyuu," or "Leq'estén or Lhék'es or Lhak'us" (*P. abbottiae*) among Pacific Northwest First Nations including Dena'ina, Haida or Tshimshian, and Wakashan peoples (Tou, Oldaker, & Yuan, 2019; Turner, 2003; Yuan, 2021). Common among these coastal cultures is the incorporation of the various marine macroalgae throughout the diet whether sundried, fermented, compressed, roasted, or toasted in salads, soups, stews, sushi wraps, snacks, condiments, or seasonings, as well as in the form of aqueous extracts as hydrocolloids in gelled desserts (e.g. agar agar for almond jelly (Annin tofu) in China and agar jelly for Anmitsu in Japan; carrageenans for seaweed pie in Prince Edward Island, Canada), or beverages (e.g. carregeenans for an Irish moss drink in the Caribbean) (Rajauria & Yuan, 2021; Robledo and Freile Pelegrín, 1997; Tou et al., 2019; Yuan, 2008, 2007; Turner, 2003). Moreover, various marine macroalgae are wild-harvested or cultivated by the mariculture industry as sources of hydrocolloids (i.e. agar, alginates, carrageenans) as well as various nutritive and bioactive, and thereby, potentially nutraceutic, cosmeceutic, or pharmacognosic metabolites including the long-chain ω-3 PUFAs, carotenes and xanthophylls, tocols, mycosporine-like amino acids, scytonemins, fucoidans, polyphenols, phenolic acids, phlorotannins and lignans (Athukorala & Yuan, 2013; Athukorala, Trang, Kwok, & Yuan, 2016; Plaza, Cifuentes, & Ibáñez, 2008; Rajauria & Yuan, 2021; Tou et al., 2019; Yuan, 2008; 2007).

Various edible unicellular microalgae represent phyla or divisions including the Dinoflagellata (marine and freshwater dinoflagellates or plankton), Cryptophyta (freshwater microalgae, but also marine and brackish waters), Raphidophyta (marine and freshwater microalgae), and Cyanophyta (blue-green algae, e.g. Cyanophyceae) (Garcia, de Vicente, & Galán, 2017; Rajauria & Yuan, 2021; Tou et al., 2019). Microalgae not only played a role in the diets of ancient coastal cultures as identified above but also represent traditional foods in many others, including "Fah-tsai or Facai" (*Nostoc commune*; Cyanophyceae) in China, or "Tao" or "Water silk", "Pond silk" or "Mermaid's tree" (*Spirogyra varians*; Charophyta Division, Zygnematophyceae Class) in Thailand, and "Suizenji-nori"

(*Aphanotheca sacrum*; freshwater Cyanophyceae) in Japan (Garcia et al., 2017; Tou et al., 2019). Moreover, microalgae are central to macrobiotic diets and play a role as dietary supplements for nutritional (i.e. protein, long-chain ω-3PUFAs, carotenes and xanthophylls, carbohydrates) or functional food and nutraceutical benefits for today's consumers (Evans & Rabie, 2007; Garcia et al., 2017; Khan, Shin, & Kim, 2018; Lordan, Ross, & Stanton, 2011; Novel Foods Section, Food Directorate, Health Canada, 2018; Tou et al., 2019); examples include GREENS+™ which contains various blue-green microalgae, including *Spirulina sp.*, *Chlorella sp.*, *Dunaliella salina* as well as the macroalga Dulse, *P. palmata* (https://www.greensplus.com/superfood-powders), as well as numerous microalgal powders containing *Spirulina sp.* and *Chlorella sp.* such as Organika® Spirulina Powder (https://organika.com/products/organic-spirulina-powder), AlgaVia® Protein (https://www.canada.ca/en/health-canada/services/food-nutrition/genetically-modified-foods-other-novel-foods/approved-products/novel-food-information-whole-algal-protein.html) and SpiruZan™ (Evans & Rabie, 2007).

Eukaryotic or prokaryotic photosynthetic organisms, macro- and microalgae, respectively, are highly susceptible to oxidative stress due to growing environment variables including tidal zone habitats affecting water depths or exposure, nutrient availability, light irradiance and penetration, latitude, temperatures, and seasonality. Therefore, it is well recognized that the composition of wild-harvested algal specimens is highly variable, necessitating the provision of important details in research reports, including latitude and longitude of harvest locations, and time of year and date of harvest (Rajauria & Yuan, 2021). Ethnobotanical studies of First Nations peoples of the Pacific Northwest wild-harvesting *P. abbottiae* reveal specific months and stages of growth to obtain the "best-tasting" specimens, or even a "second growth" of more desirable algal fronds along with different preferences for "winter" or "spring" seaweed specimens (Turner, 2003). On the other hand, a robust global mariculture industry exists for the cultivation of many valorized macro- and microalgal species of interest. This technology enables sustainable cultivation of algal biomass with consistent morphology and composition by controlling seed stock, nutrients, light irradiance, temperature, agitation, and structures harvested (Athukorala et al., 2016; Evans & Rabie, 2007). An excellent example of a highly valorized cultivated macroalga is the Hana Tsunomata™ product of Acadian Seaplants (Nova Scotia, Canada) which is a cultivated Irish moss (*Chondrus crispus* [R]) sold to Japan; the product is particularly noted for its color palette of yellow, green, and fuschia pink algal fronds (https://www.acadianseaplants.com/hana-tsunomata). Moreover, cultivation of macroalgae is known to date back to seventeenth-century Japan wherein bamboo in shallow water was utilized as substrate for "Nori" (*P. tenera*) growth; more modern cultivation techniques use suspended ropes which are seeded with the filamentous growth phase of *P. tenera* (Turner, 2003). Cultivation of macroalgae can mitigate the exploitation and denudation of natural habitats and stocks, such as those with *C. crispus* beds in Prince Edward Island and Nova Scotia, Canada; in fact, the native *C. crispus* beds were ultimately overtaken by another macroalga, *Furcellaria lumbricalis* (R). Cultivation of microalgae is widely practiced not only for biomass production of macrobiotic products but also for the lucrative carotenoid and ω-3 PUFA nutraceutical markets (Evans & Rabie, 2007; Novel Foods Section, Food Directorate, Health Canada, 2018).

The past three decades have witnessed a steadily burgeoning body of knowledge in not only the scholarly literature but also the application of this basic and applied research to international patent applications pertaining to the nutritional, bioactive and thereby, functional food, nutraceutical, cosmeceutical, and pharmacognosic effects of macro- and microalgae and selected secondary metabolites as recently reviewed in Rajauria and Yuan (2021), Kim (2012, 2013), and Barrow and Shahidi (2008) among many other publications. This research has encompassed not only the potential bioactivities of comminuted macro- and microalgal biomasses, tissue extracts, and extractable constituents *in vitro*, but also *in vivo* animal model, case–control and epidemiological studies focused on reducing diet-related chronic disease risk factors linked to oxidative stress (Athukorala et al., 2016; Ceccoli, Costello, & Pillai, 2008; Choi et al., 2015; de la Coba, Aguilera, & López Figueroa, 2006; de la Coba 2019; Evans & Rabie, 2007; Grasa-López et al., 2016; Hitoe & Shimoda, 2017; Jo, Kim, & Cheong, 2019; Kim, Kim, Choi, & Lee, 2008; Mercurio et al., 2015; Chidambara Murthy et al., 2005; Nishihira, Nishimura, Sugawara, & Kudo, 2017; Orfanoudaki et al., 2020; Rastogi, Sonani, Madamwar, & Incharoensakdi, 2016; Shimabukuro et al., 2018; Skibola, Curry, VandeVoort, Conley, & Smith, 2005; Suh et al., 2017; Vega, Bonomi-Barufi, Gómez-Pinchetti, & Figueroa, 2020; Wang et al., 2016; Yang, Nam, Kong, & Kim, 2010; Yoshinaga, Maruya, Koikeda, & Nakano, 2018). While wet chemistry and *in vitro* cell culture studies are helpful in elucidating the molecular mechanisms of action of extracts and the constituent purified compounds, it is essential to determine *in vivo* bioaccessibility and bioavailability, metabolism, and target tissues of these molecules in understanding the potential efficacy to influence human health and potentially reduce diet-related chronic disease risks. An intriguing and potentially highly valorized application of algal extracts and secondary metabolites is the potential efficacy of UV-absorbing compounds from macro- and microalgae as potential nutraceuticals, antioxidants, and sunscreen molecules in cosmeceutical products (Aguilera, de la Coba, & López Figueroa, 2006; Bird, Fitzmaurice, Dunlap, Chalker, & Bandaranayake, 1987; Ceccoli et al., 2008; de la Coba et al., 2006, 2019; Schmid, Schurch, and Zulli, 2006; Vega et al., 2020). Many studies evaluated the efficacies of macro- and microalgae and constituents on diet-related chronic disease risk factors of continuing concern and cost to public health, including cardiovascular disease (CVD, e.g. atherosclerosis, hyperlipoproteinemia) (Choi et al., 2015; Grasa-López et al., 2016; Jo et al., 2019; Wang et al., 2016), liver cancer (Chidambara Murthy et al., 2005; Ranga Rao, Sarada, Shylaja, & Ravishankar, 2015), breast cancer (Skibola, 2004; Skibola et al., 2005; Yang et al., 2010), obesity (Hitoe & Shimoda, 2017), diabetes mellitus (Kim et al., 2008), and inflammation (Nishihara et al., 2017). Thus, this chapter reviews the evidence of the molecular mechanisms of action underlying the bioactivities of selected macro- and microalgal constituents in reducing diet-related chronic disease risk factors of concern to public health.

8.2 Cardioprotective Effects of Selected Macroalgae

The Global Burden of Disease Study (GBD) statistics for 2019 indicated that diet-related chronic disease risk factors accounted for approximately 34.66 million deaths for both females and males: elevated SBP (10.8 M; highest risk for females, second highest risk for males), dietary intake deficiencies or excesses (7.94 M; second highest risk for females, third highest risk for males), elevated blood glucose (6.5 M; third highest risk for females, fifth highest risk for males), elevated body mass index (BMI; 5.02 M; fifth highest risk for females, sixth highest risk for males), and elevated low-density lipoprotein (LDL) cholesterol (4.4 M; seventh highest risk for females and males) (Global Health Metrics, 2020). Moreover, the leading causes of "disability-adjusted life years" (DALYs) for males and females globally linked to diet-related chronic disease risk factors included (in descending order) ischemic heart disease (second leading cause), stroke (third leading cause), and diabetes (eighth leading cause). Thus, CVDs are clearly a leading cause of death across the globe (~17.9 M deaths in 2019), with 85% of these deaths attributable to heart attack (myocardial infarction) and stroke (WHO, 2020). CVD incidence was 1183.7 in "high-income" countries, compared to 823.2 in Southeast Asia, East Asia, and Oceania, 725.8 in South Asia, and 691.2 in Latin America and the Caribbean according to a trend analysis of CVD incidence per 100,000 using GBD data from 2017 with data analyzed according to Institute of Health Metrics and Evaluation (IHME) super regions (Amini, Zayeri, & Salehi, 2021). These regional epidemiological disease risk data suggest that other environmental variables may account for the highly variable habitual macro- and microalgal consumption patterns of these regions. For example, within a traditional Japanese diet, macroalgae may constitute between 10 and 25% of the diet, or approximately 3–13 g/day or as much as 3.5 kg per household (Turner, 2003; Yuan et al., 2005; Yuan & Walsh, 2006); similar dietary patterns can be expected in other East and Southeast Asian cultures given the variety of dietary algal species, preparation, and serving methods discussed above. On the other hand, macro- and microalgal consumption is generally very low in Western cultures, varying from 100% consumption by Japanese resident in Japan, compared to 12–20% by Hawai'ian residents of Japanese ancestry (Yuan et al., 2005; 2006); a notable exception are the First Nations peoples of the Pacific Northwest who consider edible seaweeds as an integral component of their "Traditional Ecological Knowledge" for nutrition, medicinal, and trade purposes, often harvesting as much 318 kg (wet wt) for family household and trade uses (Turner, 2003). Despite the emergence of seaweeds, also referred to as sea vegetables or superfoods, on North American restaurant menus and in specialty or health food grocery store shelves, consumption remains largely novel and episodic in nature (Rajauria & Yuan, 2021).

In epidemiological studies, regular seaweed consumption was inversely related to lower risk of hypertension and CVD. High seaweed consuming population (14.3 g fresh mass/adult/daily) like those adhering to the Japanese dietary guidelines had lower risk of death from all causes and CVD in a 15-year dietary patterns study of Japanese (79,594 men and women aged 45–75 years). *U. pinnatifida* powder (3.3 g/daily for 4 weeks) intake significantly reduced blood pressure and hypercholesterolemia in a dietary intervention trial of 36 elderly Japanese hypertensive outpatients. Blood pressure decreased similarly in 30 men and women diagnosed with metabolic syndrome consuming *U. pinnatifida* (6 g/daily for 1 month), and waist circumference decreased only in women compared to the placebo group (Shannon & Abu-Ghannam, 2019 and references therein).

The potential cardioprotective effects of dietary Phaeophyceae brown kelps such as *Ecklonia cava* was demonstrated in a randomized, double-blind, placebo-controlled 12-week clinical trial with 63 healthy female and male subjects, aged 19–80 years, with elevated total cholesterol (> 200 mg/dL or > 5.2 mmol/L, i.e. borderline high to high) or elevated LDL-cholesterol (> 110 mg/dL or > 2.8 mmol/L, i.e. near-optimal without heart disease or high with heart disease) (Choi et al., 2015). Subjects were administered 400 mg of *E. cava* ethanolic extract in tablet form (200 mg twice a day) or the placebo. The *E. cava* extract was analyzed by high pressure liquid chromatography (HPLC) to contain 8.2% of dieckol, a phlorotannin polymer of phloroglucinol (1,3,5-trihydroxybenzene; Figure 8.1). Participants were required to maintain their usual diet and physical activity and to refrain from consuming other functional foods or dietary supplements. After 12 weeks, LDL-cholesterol decreased significantly from 141.4 to 125.5 mg/dL in the *E. cava* treatment group compared to the placebo control, although there were no differences in plasma total cholesterol, high density lipoprotein (HDL)-cholesterol, triacylglycerol levels, or any anthropometric measures, including weight, BMI, or waist-to-hip ratio (Choi et al., 2015). Subjects tolerated the *E. cava* tablets well, suggesting that the ethanolic *E. cava* extract containing phlorotannins may help lower CVD risk in mildly hypercholesterolemic individuals (Choi et al., 2015).

Further evidence of the cardioprotective efficacies of polyphenol constituents of *E. cava* was reported in a randomized, double-blind, placebo-controlled 12-week clinical trial with 97 overweight female and male Korean adults (64 women and 33 men aged 19–55 years) and BMI between 24 and 29 kg/m^2 (Shin, Kim, Park, Lee, & Hwang, 2012). Subjects were assigned to one of three groups: low dose (LD) administered 72 mg of *E. cava* polyphenol

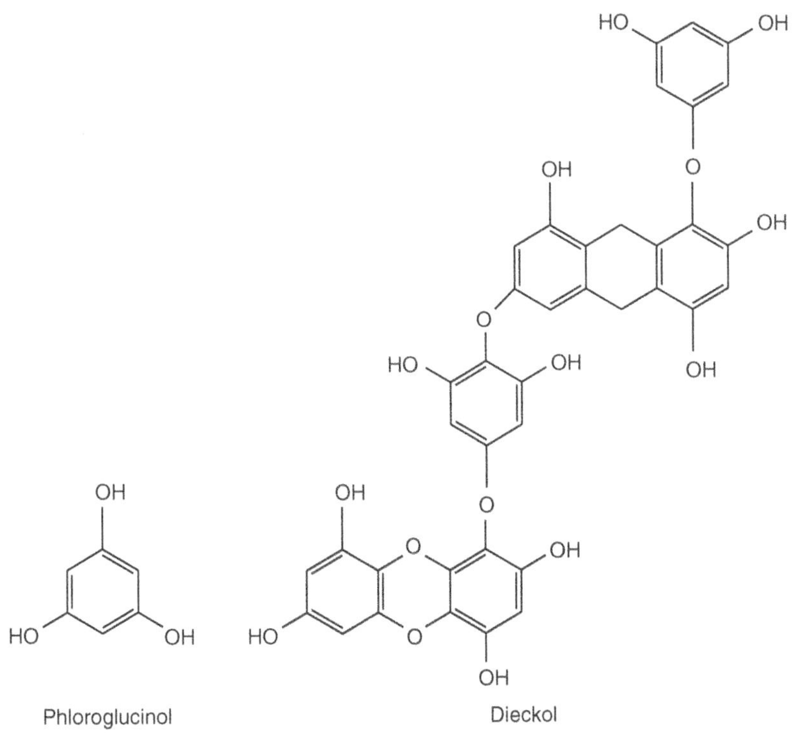

Figure 8.1 Phloroglucinol (1,3,5-trihydroxybenzene) and dieckol phlorotannin structures from brown kelp macroalgae.

extract (containing 98.5% phloroglucinol equivalents as dieckol, 8,8′-bieckol, 6,6′-bieckol, and phlorofurofucoeckol A; supplied by Livechem, Inc., Jeju, Korea) in a sweetened, canned beverage (246 mL); high dose (HD) administered 144 mg of *E. cava* polyphenol extract beverage or a placebo (PC) beverage. Subjects were asked to drink the test beverages between meals. Diets were not altered during the study, although food supplements or medications known to influence lipid or carbohydrate metabolism were to be avoided. As well, physical activity was not altered by subjects. The *E. cava* treatments (LD and HD) significantly reduced BMI, body fat ratio, waist circumference, and waist/hip ratio after the 12-week trial. In comparison to the PC group, only the HD subjects exhibited reduced body weight, BMI, body fat ratio, and waist circumference. Systolic blood pressure (SBP) and diastolic blood pressure (DBP) were not affected by LD treatment, but SBP was reduced for HD subjects after 12 weeks (Shin et al., 2012). Serum total cholesterol, LDL-cholesterol, total cholesterol/HDL-cholesterol, and atherogenic index (AI = [total-chol – HDL-chol]/HDL-chol) decreased in LD and HD groups compared to PC subjects, as well as compared to baseline levels. Serum HDL-cholesterol tended to increase in the LD group, and significantly so in the HD group after 12 weeks; HDL-cholesterol increased in HD compared to PC subjects. Serum triacylglycerols were unaffected by LD or HD treatments, while HD lowered serum glucose compared to baseline after 12 weeks but was not different between groups. Thus, the phlorotannin-rich *E. cava* extract may suppress fat deposition or upregulate its metabolism, via suppression of adipogenesis and lipogenesis (Shin et al., 2012). Alternatively, the phlorotannin-rich *E. cava* extract may also inhibit cholesterol absorption in reducing serum total and LDL-cholesterols in cardioprotection.

In other animal model studies of the cardioproctective effects of dietary Phaeophyceae brown kelps, researchers focused on other bioactive constituents, including alginates (linear polymers of 1,4-linked β-D-mannuronic acid and α-L-guluronic acids; Figure 8.2) as the major carbohydrate of kelps, such as *S. japonica* (Sea Tangle) (Jo et al., 2019), fucoidans (sulfated α-1,3 and 1,4 linked L-fucose (6-deoxy-L-galactose) polymers which frequently contain minor amounts of xylose, glucose, galactose, mannose, rhamnose as well as glucuronic acid, protein, phenolic compounds, phosphorus, and minerals such as Na^+, Mg^{2+},

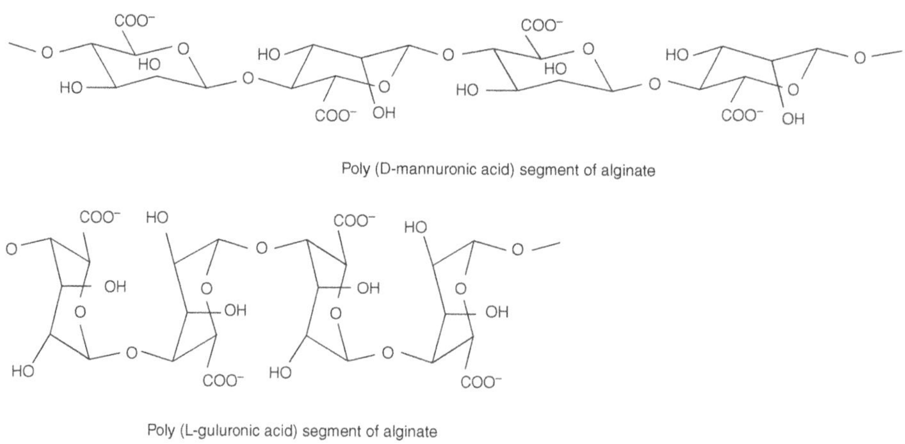

Figure 8.2 Generalized structures of alginate polymers from brown kelp macroalgae.

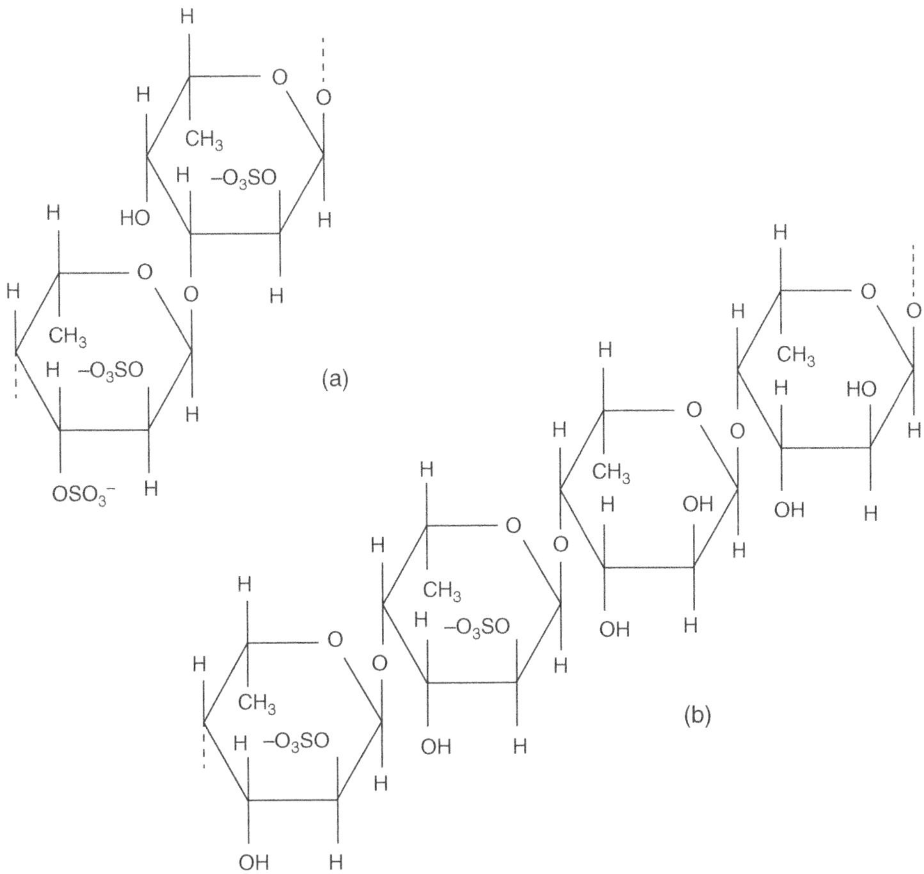

Figure 8.3 Structures of fucoidan and sulfated fucan from brown kelp macroalgae: (a), *Fucus vesiculosus*, disaccharide repeating unit [→ 4)-α-L-Fuc*p* -(2,3-OSO_3^-)-(1 → 3)-α-L-Fuc*p*-(2OSO_3^-)-(1→]; (b), *Arbacia lixula*, homogeneous, unbranched tetrasaccharide sulfated fucan repeating unit (Athukorala & Yuan, 2013).

Ca^{2+}, K^+, $Fe^{2+/3+}$, and Al^{3+}; Figure 8.3) of *S. japonica* (Wang et al., 2016) and fucoxanthin (oxygenated derivative of carotenes, a xanthophyll; Figure 8.4) of *U. pinnatifida* (Sea Mustard) (Grasa-López et al., 2016). Generally, fucoidans occur bound to alginates and are thus a convenient byproduct of alginate isolation and purification (Athukorala & Yuan, 2013). Jo et al. (2019) investigated 5-week-old male Sprague–Dawley rats fed on an AIN-93G basal diet containing 10% lard (BD), or a high-fat-high-cholesterol (HFHC; 20% lard with 1% cholesterol), or the HFHC diet supplemented with 1% or 2% powdered commercial Sea Tangle compressed and thinly sliced, designated "silk-shaped Sea Tangle (SST)" or HFHC diet supplemented with 1% or 2% powdered regular commercial Sea Tangle (CST) for 7 weeks. Both Sea Tangle specimens contained similar levels of water-soluble and alkali-soluble alginates 9.12–9.45% and 10.54–10.85%, respectively; however, the molecular weights of the component alginates were quite different between CST and SST at 12,500 and 6,400 Da, respectively. While rodent body wt gained was increased by the

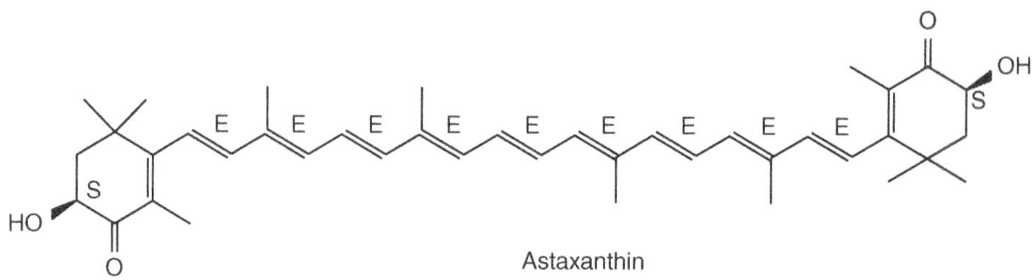

Astaxanthin

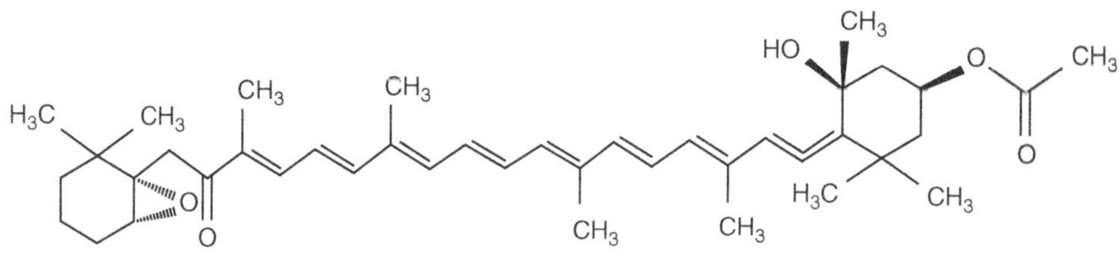

Fucoxanthin

Figure 8.4 Xanthophylls found in selected macro- and microalgae.

HFHC diet compared to the BD group, SST and CST supplementation mitigated that increase in a dose-dependent manner, resulting in significantly lower feed efficiency ratios in the SST and CST groups compared to the HFHC diet alone, which were ultimately similar to that of the BD-fed animals (Jo et al., 2019). Fat deposition (i.e. retroperitoneal, epididymal, and mesenteric) decreased dose-dependently in SST- and CST-fed animals, leading to significantly reduced total fat deposition in HFHC groups fed on 1 and 2% SST and CST (Jo et al., 2019). Serum triacylglycerols, LDL-cholesterol, total cholesterol, atherogenic index (AI = [total-chol – HDL-chol]/HDL-chol), and cardiac risk factor (CRF = total-chol/HDL-chol) values decreased dose-dependently with SST and CST supplementation of the HFHC diet, whereas HDL-cholesterol increased dose-dependently by the same treatments. SST supplementations were more effective than CST, particularly in lowering LDL-cholesterol. These researchers hypothesized that the serum lipid-lowering efficacy of SST supplementation may be associated with the lower molecular weight of the alginate content (Jo et al., 2019). On the other hand, the bioactivity of Sea Tangle fucoxanthin or fucoidan constituents as reported by other researchers may also be a factor in these results. Moreover, the additional processing of the SST supplement was intended to increase the solubility of constituents to efficiently improve blood lipid levels, antioxidant status, and lipid-regulating enzyme activities. Serum markers of endogenous antioxidant status, namely glutathione (GSH), GSH-reductase (GR), superoxide dismutase (SOD), and catalase (CAT) increased back to near-normal levels (similar to the BD group) in the SST- and CST-supplemented groups compared to the large reduction in the HFHC-fed group. Hepatic markers of endogenous antioxidant status such as GSH-peroxidase (GPx), GR, SOD, and CAT also increased back to near-normal levels (similar to the BD group) in the SST- and CST-supplemented groups compared to the large decreases in the HFHC-fed group. Thus, Jo et al. (2019) suggested that dietary Sea Tangle supplementation may effectively reduce oxidative stress with increased GSH and antioxidant enzyme levels in serum and liver tissues. Hepatic lipid-regulating enzyme activities, namely fatty acid synthase (FAS), 3-hydro-3-methylglutaryl CoA reductase (HMG-CoA reductase), and acyl-coenzyme A:cholesterol acyltransferase (ACAT) significantly decreased dose-dependently back to near-normal levels (similar to the BD group) in the SST- and CST-supplemented groups compared to the large increases in the HFHC-fed group, demonstrating a potent lipid-lowering effect of Sea Tangle mediated by suppression of lipogenic enzyme activities and hepatic HMG-CoA reductase and ACAT, key cholesterol-regulating enzymes.

Further evidence of the cardioprotective efficacies of Sea Tangle constituents such as the fucoidans was studied by Wang et al. (2016). Male LDL receptor ($LDLR^{-/-}$)-deficient rats, 6 weeks of age were fed on atherogenic diets (10% fat and 1.25% cholesterol) for 16 weeks with a daily gavage of 50 or 100 mg/kg fucoidan or solvent vehicle (0.5% carboxymethyl-cellulose sodium). The fucoidan was isolated from Sea Tangle cultured in Yantai, China. Serum triacylglycerols, total cholesterol, and LDL-cholesterol levels significantly decreased dose-dependently with fucoidan supplementation, while HDL-cholesterol levels substantially increased dose-dependently compared to the atherogenic diet-fed animals. Moreover, the fucoidan treatments significantly and dose-dependently reduced atherosclerotic plaque deposition in the aortic arch, descending thoracic aorta, and abdominal aorta of $LDLR^{-/-}$ rats compared to the atherogenic diet-fed animals (Wang et al., 2016). Immunohistochemistry of aortic plaque tissue revealed suppression of Mac-2-binding protein (a pro-inflammatory protein associated with atherosclerosis) and SM-22 smooth muscle protein expression in atherosclerotic plaque of fucoidan-treated animals as evidence of reduced macrophage infiltration and smooth muscle cell proliferation. Moreover, fucoidan treatments reduced superoxide reactive oxygen species (ROS) generation in aortic plaque as evidenced by reduced mRNA expression for ROS-generating enzymes and pro-inflammatory mediator expression. These results suggested that fucoidan was efficacious in reducing local aortic inflammation and reducing smooth muscle cell proliferation. Thus, fucoidan treatment exerted antioxidant activity *in vivo* by attenuating ROS production in aortic plaque. Further evidence of molecular mechanisms of action of fucoidan supplementation were elucidated with suppression of pro-inflammatory markers such as oxidized LDL receptor 1 (also known as lectin-type oxidized LDL receptor 1; LOX-1) in atherosclerotic plaque of $LDLR^{-/-}$ rats (Wang et al., 2016).

On the other hand, Grasa-López et al. (2016) studied another kelp commonly consumed in Korea and Japan, *U. pinnatifida* known as Wakame or Sea Mustard with a focus on the xanthophyll fucoxanthin in 7-week-old male Wistar rats fed on a standard rodent diet (CTL), or a high-fat diet (21% milk fat and 0.15% cholesterol; HF), or HF with a gavage of commercial ground Wakame (400 mg/kg body wt) in a H_2O:Tween 80 (100 : 1) suspension (UP group), or HF with a gavage of fucoxanthin (1 mg/kg body wt) in suspension (FU group) for 8 weeks. The commercial Wakame contained 0.98 mg of fucoxanthin/g dry wt by HPLC analysis. While the HF diet induced obesity as expected, UP-fed animals exhibited significantly decreased total body wt gained similar to the CTL group despite similar food consumption during the study. Interestingly, the total body wt gained by the FU-fed group was not different from the HF counterparts. Fat deposition, as epididymal, retroperitoneal, and mesenteric white adipose tissue depots significantly decreased by UP treatment compared to HF and FU groups. Total intra-abdominal fat masses for CTL, HF, UP, and FU groups were 5.10, 11.33, 7.11, and 9.35%, respectively. Serum profiles of animals revealed significant reductions in UP- and FU-fed groups versus HF for triacylglycerols, total cholesterol, LDL-/VLDL-cholesterol, glucose, insulin, and homeostatic model assessment (HOMA, a measure of β-cell function and insulin resistance), while HDL-cholesterol increased. Overall, the efficacy of UP in improving serum parameters was greater than that of FU treatment. UP and FU treatments improved blood glucose area under the curve (AUC) values back down to the level of the CTL group, compared to HF animals which exhibited insulin resistance (Grasa-López et al., 2016). The SBP and DBP of UP- and FU-fed animals were in the normal range of CTL and were both significantly lower than HF-fed animals. Serum levels of anti-inflammatory and pro-inflammatory markers such as adiponectin (plays a role in regulating glucose levels and fatty acid metabolism) were increased by UP and FU treatments similar to CTL levels which were significantly greater than for HF counterparts. Leptin (involved with food intake and energy homeostasis) and plasminogen activator inhibitor-1 (PA-1; elevated in abdominal obesity) levels were significantly reduced by UP and FU treatments compared to the elevated levels of HF counterparts. Adiponectin levels in retroperitoneal white adipose and hepatic tissues also increased in UP- and FU-fed animals compared to the low levels in HF animals, whereas UP and FU diets reduced leptin levels in retroperitoneal white adipose similar to CTL animals, compared to the large increase in HF animals. UP and FU supplementation of the HF diet of animals improved energy expenditure, fatty acid β-oxidation, and adipogenesis mediated by upregulation of peroxisome proliferator-activated receptor alpha (PPAR-α), peroxisome proliferator-activated receptor gamma (PPAR-γ), peroxisome proliferator-activated receptor gamma coactivator 1-α (PGC-1α), and thermogenin uncoupling protein 1 (UCP-1) in hepatic and retroperitoneal white adipose tissues, respectively. Taken together, these results indicate that dietary supplementation with Wakame and fucoxanthin are effective in reducing lipogenesis, CVD risk factors such as hypertension, elevated serum lipids, and glucose, insulin resistance, and inflammation (Grasa-López et al., 2016). The effect of the whole macroalga UP for many parameters was greater than for counterparts given isolated fucoxanthin, perhaps indicating the efficacy and/or synergy of other kelp constituents, including phlorotannins, fucoidans, and alginates. The anti-obesity properties of brown seaweed are attributed to the presence of fucoxanthin. For example, fucoxanthin (capsules 1 or 3 mg/daily for 4 weeks) significantly reduced BMI, visceral fat, basal metabolic rate, abdominal fat, and circumference (neck, arm, and thigh) in obese individuals ($n = 50$, aged 20–59 years, 26–30 kg/m^2 BMI). Resting energy expenditure increased with significant average weight loss (4.9 kg) in 115 obese, nondiabetic fatty liver (>11% liver fat) premenopausal women administered dried brown seaweed extract (300 mg containing 2.4 mg of fucoxanthin/daily for 16 weeks) (Hitoe and Shimoda 2017).

8.3 Antidiabetic Effects of Selected Macroalgae

Following up on the above discussion of diet-related chronic disease risks related to the leading CVD risk factors (i.e. ischemic heart disease and stroke), diabetes mellitus is notable as the eighth leading cause of "disability-adjusted-life-years" (DALYs) for males and females globally (Global Health Metrics, 2020). This is not surprising given the role of oxidative stress in types I and II diabetes mellitus. For example, in a randomized, placebo-controlled clinical trial with type 2 diabetes mellitus patients (9 men, 11 women; 40–70 years. old; BMI < 35 kg/m^2; fasting glucose >150 mg/dL (150–300 mg/dL)), treatment with 48 g/day equal parts *L. japonica* (Sea Tangle) and *U. pinnatifida* (Sea Mustard) kelps for 4 weeks significantly reduced fasting blood glucose (152 mg/dL), as well as postprandial blood glucose (203 mg/dL) levels (Kim et al., 2008). Kelp treatment lowered triacylglycerols and LDL-cholesterol levels and elevated HDL-cholesterol levels without changes in plasma total cholesterol. Kelp treatment of patients also reduced red blood cell lipid peroxidation products (TBARs) and greatly increased red blood cell antioxidant enzyme (CAT and GSH peroxidase) activities with no change in SOD activity (Kim et al., 2008).Thus, kelp-derived polysaccharides such as the laminarins (low-molecular-weight storage β-glucans, i.e. (1,3)-β-D-glucans comprised of (1,3)-β-D-glucopyranosyl residues with C6-O-branching in the main chain and some β-(1,6)-intrachain links), alginates, and fucans increased dietary fiber intake of the treatment group by 2.5-fold compared to the placebo group. Moreover, kelp supplementation positively benefited glycemic control in type 2 diabetes mellitus patients and was also efficacious in lowering plasma lipids and indices of oxidative stress.

A commercial phlorotannin extract InSea2® (https://insea2.com/insea2) produced from a hot water extraction of the kelps *Ascophyllum nodosum* and *Fucus vesiculosus* followed by filtration, ultrafiltration, and then spraydrying was evaluated for its efficacy *in vitro* and *in vivo* for digestive enzyme inhibition and slowing carbohydrate digestion and absorption, respectively (Roy et al., 2011). The kelp extract was treated to remove alginates and partially demineralized thereby containing a minimum of 10% polyphenols as chlorogenic acid equivalents with the remainder algal polysaccharides and minerals. The phlorotannin extract inhibited human salivary α-amylase and α-glucosidase dose-dependently with IC_{50} values of 2.8 μg/mL and 5.0 μg/mL, respectively, and near complete inhibition at sub-micromolar extract concentrations. Blood glucose elevation decreased (by 90%) after 30-min treatment with phlorotannin extract (7.5 mg/kg body wt), when male Wistar rats were gavaged with a 50:50 corn starch:safflower oil solution (2 mL/kg body wt) compared to the untreated placebo group. The blood glucose curves of phlorotannin-treated animals were similar to those given low glycemic index foods; for example, glucose absorption was prolonged for 6 hr compared to <2 hr with the placebo. Similarly, the peak plasma insulin levels decreased by 40% and the AUC by 22% in phlorotannin-treated animals compared to the placebo (Roy et al., 2011). The inhibitor constant (Ki) for α-amylase was 6.0×10^{-8} M indicating strong binding affinity of the phlorotannin extract to the enzyme. The binding affinity may involve multiple hydrogen bonds of phlorotannins with enzyme polypeptide carbonyl moieties. Roy et al. (2011) evaluating the strong hypoglycemic efficacy of the phlorotannin extract in this animal model hypothesized that algal polysaccharides as dietary fibers were only partly responsible for the effect, whereas phlorotannins were predominantly responsible for the hypoglycemic effect in modulating diabetes mellitus.

In a follow-up randomized crossover placebo-controlled clinical trial, healthy participants from the general population (12 women, 11 men, mean age 39.9 years; BMI 24.9 kg/m^2; not taking any medications, blood pressure in the normal range) participated in post-load plasma glucose and insulin concentrations testing with two 3-hr meal tolerance tests taken 1 week apart. Treatment with InSea2® phlorotannin extract (two 508-mg capsules containing 250 mg; innoVactiv Inc., Rimouski, Quebec, Canada) prior to a test meal of white bread (110 g, 50 g available carbohydrate) and water, reduced the insulin incremental

AUC by 12% compared to the placebo (Paradis, Couture, & Lamarche, 2011). The phlorotannin treatment also increased the Cederholm index of insulin sensitivity by 7.9%; this index indicates insulin sensitivity factoring in oral glucose tolerance and anthropometric variables: [50,000 + (fasting glucose—glucose at 2 hr) × 1.15 × 180 × 0.19 × body wt]/[120 × log (mean insulin) × mean glucose]. The single ingestion of 500 mg of phlorotannin extract had no significant effect on glucose response with no differences at any of the postprandial time points, or in the glucose incremental AUC, although there was a trend for reduced plasma glucose concentrations after the standard carbohydrate load with the phlorotannin extract treatment. The phlorotannin extract apparently altered the absorption rate of dietary carbohydrate within the gut resulting in less insulin for its subsequent processing within the plasma (Paradis et al., 2011). Thus, the InSea2® phlorotannin extract product with demonstrated *in vitro* α-amylase and α-glucosidase inhibitory activities (Roy et al., 2011) modulated insulin homeostasis in healthy adults after a carbohydrate-rich test meal (Paradis et al., 2011). The findings and conclusions of this post-load glucose challenge trial were limited by the lack of control of participants' diets prior to testing, as well as lack of data and control of participants' physical activity prior to the carbohydrate-load challenge tests.

Moreover, the phlorotannin fractions of kelps including *A. nodosum, Ecklonia bicyclis, E. cava, Ecklonia stolonifera, Ishige okamurae, Sargassum hystrix,* and *Sargassum polycystum* exhibited antidiabetic activities during *in vitro* and *in vivo* animal model studies through mechanisms such as inhibition of key enzymes (α-amylase, α-glucosidase, angiotensin-converting enzymes [ACEs], aldose reductase, dipeptidyl peptidase-4 [DPP-4], and protein tyrosine phosphatase 1B [PTP 1B] enzymes) and formation of advanced glycation end products (AGEs) as recently reviewed by Gunathilaka, Samarakoon, Ranasinghe, & Peiris (2020). Moreover, algal compounds modulate GLUT-4 and AMPK signaling pathway and trigger glucose tolerance. The antidiabetic activity of seaweed has been attributed to phlorotannins, fucoxanthins, polyphenolics, and polysaccharides which inhibit hepatic gluconeogenesis and reduce digestive enzyme activity (Pradhan et al., 2021).

Two compounds, fucoxanthin (fucosterol) and phlorotannins, account for the antidiabetic effects of brown algae primarily by inhibiting enzymes associated with reduction of blood glucose level and diabetic complications (Gunathilaka et al., 2020). For example, in human randomized controlled clinical trials, dieckol extract of brown algae (*Ecklonia cava,* 144 mg/day for 12 weeks) significantly reduced postprandial blood glucose level in 97 overweight adults. Furthermore, the same algae extract reduced oxidative stress-induced diabetic complications and exerted antihyperglycemic effects (reduction in total cholesterol, LDL-cholesterol, and their ratio and increase in HDL-cholesterol levels). Consumption of a commercial mixture (*InSea*^{2}) of the brown algea (*A. nodosum* and *F. vesiculosus*) inhibited α-amylase and α-glucosidase activity after 3 hr, although *F. vesiculosus* (500 and 2,000 mg) ingestion did not affect the postprandial blood glucose and insulin levels in healthy volunteers (Gunathilaka et al., 2020 and references therein). In type 2 diabetic patients (9 men and 11 women), supplementation of Laminaria (48 g/day, for 4 weeks) reduced fasting blood glucose and 2-hr postprandial glucose levels without affecting glycated hemoglobin level. Daily supplementation with *Undaria pinnatifida* and *Sacchariza polyschides* balanced blood glucose levels, reduced serum triglyceride levels, and increased high-density lipoprotein cholesterol in diabetic patients (Gunathilaka et al., 2020 and references therein). Consumption of seaweed extract (500 mg *InSea*[2]) before a carbohydrate meal (waffles and pure maple syrup) significantly enhanced postprandial cognitive performance in 33 women and 27 men (aged 18–65 years) with acute postprandial drowsiness (Shannon & Abu-Ghannam, 2019 and references therein). The effects were attributed to modulation of glucose metabolism by the high fiber and polyphenolic content (20% chlorogenic acid equivalent) of the seaweed extract.

The anti-obesity properties of brown seaweed are attributed to the presence of fucoxanthin. For example, fucoxanthin (capsules 1 or 3 mg/daily for 4 weeks) significantly reduced BMI, visceral fat, basal metabolic rate, abdominal fat, and circumference (neck, arm, and thigh) in obese individuals ($n = 50$, aged 20–59 years, 26–30 kg/m^2 BMI). Resting energy expenditure increased with significant average weight loss (4.9 kg) in 115 obese, nondiabetic fatty liver (>11% liver fat) premenopausal women administered dried brown seaweed extract (300 mg containing 2.4 mg of fucoxanthin/daily for 16 weeks) (Shannon & Abu-Ghannam, 2019 and references therein). *Himanthalia elongata* had the highest potential as ingredient to develop new functional foods for metabolic syndrome among seven commonly consumed seaweeds due to its high phenolic content (24 µmol GAE/g), antioxidant activity, angiotensin-converting enzyme-I inhibitory activity (IC_{50} = 65 µg/mL), and anti-inflammatory activity (Rico et al., 2018). Thus, these preclinical trials are evidence of the untapped potential of kelps and associated extracts in ameliorating diabetes mellitus risk reduction.

8.4 Anticarcinogenic Effects of Selected Macro- and Microalgae

GBD 2019 statistics indicate that among cancers, there were 2 M cases of breast cancer (99% of cases in females) with 701,000 deaths and 20.6 M DALYs globally; 2.2 M cases of colon and rectal cancers (42.1% in females and 57.6% in males) with 1.1 M deaths and 24.3 M DALYs globally (Global Health Metrics, 2020). Regionally, the 5-year prevalence of breast cancer per 100,000 in North America (622.87–640.31) and Western Europe (e.g. Germany, France, and United Kingdom; 663.43–707.38) is considerably greater than in East Asia (e.g. China and Japan; 197.04–507.88) (WHO, International Agency for Research on Cancer, 2020). The 5-year prevalence of colon and rectal cancers per 100,000 in North America (91.63–123.49 and 43.56–79.98, respectively) and Western Europe (131.08–49.34 and 75.84–80.16, respectively) are considerably greater than in China (51.75 and 45.21, respectively), but not Japan (228.74 and 130.41, respectively).

8.4.1 Breast Cancer Protective Effects

One of the first functional food or medicinal applications for edible macroalgae to treat or prevent breast cancer dates back to the Ebers papyrus (~1,534 BCE, Egypt). Indeed, average intake and frequency of Gim (*Porphyra* sp.) consumption was inversely associated with breast cancer risk in a case–control study with premenopausal and menopausal women (n = 362 for cases and control groups; 30–65 years old, 46 years mean age for both groups; 35% of women in each group were menopausal) in Korea (Yang et al., 2010). Moreover, these investigators found no such associations with Miyeok (*U. pinnatifida*) consumption and breast cancer disease risk. Food intake was estimated by a quantitative food frequency questionnaire with cases consuming a mean of 0.86 g of dry mass Gim/day, three times per week and 5.12 g wet wt Miyeok/day, 1.1 times per week; and the controls consuming a mean of 1.17 g dry mass Gim/day, 3.76 times per week and 5.45 g wet wt Miyeok/day, 1.15 times per week. Data analyses involved dividing the consumption amounts and frequencies into quintiles with Gim intake ranging from a median of 0.06 up to 2.00 g dry wt/day, with frequencies ranging from a median of 0.21–7 × per week; and Miyeok intake ranging from a median of 0.54–15.48 g wet wt/day, with frequencies ranging from a median of 0.21 × per month to 3.01 × per week. The inverse association of Gim intake with breast cancer risk was similar for both premenopausal and postmenopausal women (Yang et al., 2010). While these workers did not investigate potential molecular mechanisms of action of Gim constituents on breast cancer risk, other studies using animal models and cell culture assays may be useful in this regard.

Earlier animal studies evaluating the potential mammary cancer anticarcinogenic effects of marine red macroalgae demonstrated that diets containing 2% powdered *Porphyra tenera* inhibited the incidence of mammary tumors in rats treated with the carcinogen 7,12-dimethylbenz[a]anthracene (DMBA) (Yamamoto, Maruyama, & Moriguchi, 1987). Female Sprague–Dawley rats, aged 4 weeks, were fed a 2% *P. tenera* diet for 152 days, with animals dosed with DMBA (20 mg/kg body wt) on day 27. The control commercial diet fed to animals contained 4.4% fat and 3.5% fiber. The median time until the first detected palpable tumor was significantly delayed in *P. tenera*-fed rats (> 184 days) compared to control animals (126 days). Moreover, the number of *P. tenera*-fed rats with mammary tumors decreased (35 vs. 69% in the DMBA control group), while the tumor burden also decreased (9.3 vs. 13.6 g/rat in the DMBA control group). *P. tenera* diet supplementation can modulate DMBA toxicity and carcinogenicity potentially through immunopotentiation (Yamamoto et al., 1987).

On the other hand, the potential breast and mammary cancer anticarcinogenic effects of brown macroalgae have been relatively more intensely studied (Maruyama, Watanabe, & Yamamoto, 1991; Teas, Harbison, & Gelman, 1984; Teas, Vena, Cone, & Irhimeh, 2013; Yamamoto et al., 1987) in the wake of an initial hypothesis paper by Teas (1981). Teas hypothesized that dietary seaweeds may confer a protective effect based on the differences in breast cancer rates between Japanese and American women in concert with the clear difference in macroalgal consumption patterns between these two groups. The hypothesis paper proposed potential molecular mechanisms of action, including binding of bile acid sterols by macroalgal fiber components, modulation of intestinal flora involved with carcinogen metabolism, and stimulation of the immune system with respect to antioxidant properties. In a small, single-blinded placebo-controlled clinical trial with 15 healthy, postmenopausal American women (five controls, five who had been treated for estrogen receptor-positive (ER+) breast cancer, and five who had been treated for estrogen receptor-negative (ER-) breast cancer), studied over three consecutive arms of treatment: 1-month placebo; 1-month *U. pinnatifida* (5 g/day milled alga containing 8% fucoidan in capsules); and 1-month placebo washout, *U. pinnatifida* intake reduced urinary creatinine-standardized uPAR (human urokinase-type-plasminogen activator receptor) concentrations by 50%, which was reversed during the placebo washout (Teas et al., 2013). Generally, plasma uPAR concentrations increase after menopause and are associated with unfavorable breast cancer prognoses in patients. uPAR is an extracellular receptor involved in signaling pathways affecting the extracellular matrix, cell adhesion, inflammation, and tissue repair (Teas et al., 2013). Cancer progression is associated with increased shedding of uPAR from cell surfaces. The effects of *U. pinnatifida* supplementation were similar among control subjects as well as ER- and ER+ subjects. These workers also reported small increases in a serum protein, complement activation C3a des-arginine in all subjects during the *U. pinnatifida* arm of the trial; they discussed how this protein may indicate stimulation of an immune response, as well as reduction in an urinary progesterone-responsive protein as evidence of an influence on cell surface signaling receptors for steroid hormones (Teas et al., 2013). Thus, if *U. pinnatifida* supplementation is efficacious in decreasing uPAR levels, this may be one molecular mechanism underlying the carcinogenesis protective effects of dietary kelp fucoidan.

Diets containing 2% powdered brown kelps *Laminaria religiosa* or *L. japonica* significantly reduced the incidence of mammary tumors in female (4 weeks old) DMBA-treated Sprague–Dawley rats (Yamamoto et al., 1987). Rats were fed on the kelp-supplemented diets over 152 days, with DMBA (20 mg/kg body wt) administered on day 27. However, the median time until the first detected palpable tumor (> 184 days) was significantly delayed only in *L. religiosa*-fed rats compared to control animals (126 days), but not with *L. japonica*-fed animals. Also, the number of *L. religiosa* and *L. japonica*-fed rats with mammary tumors decreased at 35 and 50%, respectively, versus 69% in the DMBA control group, while the tumor burden also decreased (1.6 and 7.1 g/rat for *L. religiosa* and *L. japonica*-fed rats, respectively versus 13.6 g/rat in the DMBA control group). These investigators hypothesized that diet supplementation with kelps such as *L. religiosa* and *L. japonica* can modulate DMBA toxicity and carcinogenicity, potentially related to the ester sulfate content, that is, the fucoidans (Yamamoto et al., 1987). Fucoidans, as water-soluble sulfated polysaccharides, are absorbed by humans. One of the proposed mechanisms underlying the inhibitory effect of dietary kelp against chemically induced mammary carcinogenesis is the enhancement of antioxidant enzyme activity (i.e. GPx) and reduction in lipid peroxides in the livers of treated rats (Maruyama et al., 1991). These workers studied female Sprague–Dawley rats (aged 4 weeks) fed on a 2% *L. religiosa* diet and dosed with DMBA (20 mg/kg body wt) on day 27 of the start of the diet, with the study concluded on day 210 for a long-term study; another similarly designed short-term trial was concluded after 14 days of feeding. Hepatic lipoperoxides decreased in 2% *L. religiosa*-fed rats after 14 or 210 days compared to the DMBA control groups, whereas hepatic GPx decreased only in the 14-day study with the 2% *L. religiosa* group (Maruyama et al., 1991). The increased hepatic lipoperoxides of the DMBA control animals was ascribed to the reactivity of the carcinogen with PUFAs in the tissue membranes. Potential molecular mechanisms of the anticarcinogenic efficacy of *L. religiosa* diet supplementation included dietary fiber effects on DMBA absorption from the gastrointestinal tract and the antioxidant effect of carotenes and xanthophylls, namely β-carotene and fucoxanthin.

Further evidence of an anticarcinogenic effect of brown kelp constituents including fucoxanthin was reported by Teas et al. (1984) with 21-day-old female Sprague–Dawley rats fed a 5% *Laminaria angustata* diet with DMBA (5 mg) administered at 55 days of age and the feeding carried out over 26 weeks. Tumor appearance was significantly delayed in *L. angustata*-fed rats with a median time of 19 vs. 11 weeks in DMBA controls; rats fed on *L. angustata* exhibited fewer adenocarcinoma tumors per animal versus controls, and had fewer (63 vs. 76% in controls) mammary tumors (Teas et al., 1984). The dietary fiber component of *L. angustata*, specifically the cell wall polysaccharide fucoidan, may be responsible for the anticarcinogenic efficacy of the kelp-supplemented diet.

A pilot case study reported the effects of the brown kelp "Bladderwrack" (*F. vesiculosus*) intake (low dose of 700 mg or high dose of 1.4 g dry wt) on menstrual cycle length and hormonal status in three premenopausal women with abnormal menstrual cycle histories. The average menstrual cycle lengths significantly increased from a baseline of between 15 and 27 days to 25–32 days with 700 mg of supplementation and 31–36 days with 1.4 g of *F. vesiculosus* supplementation (Skibola, 2004). In addition, the average number of menstruating days decreased significantly in a dose-dependent manner from a baseline of between 6 and 9 days to 5 and 6 days with 700 mg of supplementation and 3–5 days with 1.4 g of *F. vesiculosus* supplementation. Moreover, plasma 17-β-estradiol levels significantly decreased dose-dependently, while progesterone levels increased: baseline 17-β-estradiol was 626 pg/mL compared with 164 and 92.5 pg/mL in the low and high dose *F. vesiculosus* groups, respectively; baseline progesterone was 0.58 ng/mL compared with 8.4 and 16.8 ng/mL in the low- and high-dose *F. vesiculosus* groups, respectively. Skibola (2004) concluded that dietary *F. vesiculosus* prolonged the menstrual cycle and exerted antiestrogenic effects in these premenopausal women. Thus, given the habitually longer menstrual cycle lengths of Japanese women (28 days) in combination with lower levels of circulating 17-β-estradiol, it is possible that an increased cycle length may lower the risk of estrogen-dependent diseases, including breast cancer. For example, shorter menstrual cycles and prolonged menstruation are associated with higher 17-β-estradiol and progesterone levels as well as breast cell proliferation rates; therefore, a reduction in menstrual cycles over a lifetime could decrease breast epithelial cell proliferation and thereby breast cancer risk (Skibola, 2004). In a follow-up study, Skibola et al. (2005) reported that when adult female Sprague–Dawley rats (200–250 g body wt) were fed on a basal control diet, low- or high-dose *F. vesiculosus* diets, 175 or 350 mg/kg/day, respectively, for 30 days, the estrous cycle was significantly affected in a dose-dependent manner, increasing from the control 4.3 days, to 5.4 and 5.9 days in the low- and high-dose groups, respectively. Dietary kelp supplementation increased the diestrus phase from 0.97 days for controls to 1.4 and 2.1 days in the low- and high-dose groups; the length of estrus, proestrus, and metestrus was not influenced by kelp supplementation. Similar to the case report above, rat serum 17-β-estradiol levels were decreased by the low-dose kelp supplementation from a baseline of 48.9 ng/L to 40.2 after 2 weeks feeding and to 36.7 ng/L after 4 weeks; progesterone levels did not differ with kelp supplementation (Skibola et al., 2005). When these workers exposed human-luteinized granulosa cells (hLGCs) to 25, 50, or 75 mmol/L *F. vesiculosus* aqueous ethanolic extract for 9 days, 17-β-estradiol in the cell culture medium dose-dependently decreased (23–35% reductions), while progresterone levels were unaffected. The antiestrogenic bioactivity of *F. vesiculosus* extracts was observed through competitive inhibition of hormone receptor binding assays, both ERα and ERβ:ERα receptor ligand binding decreased by 7, 21, and 52% and ERβ by 2, 18, and 58% with 0.5, 5, and 50 μmol/L kelp extract. Skibola et al. (2005) demonstrated antiestrogenic effects of dietary kelp and kelp extracts *in vitro* at realistic dose exposure levels, albeit these results were obtained with a brown kelp harvested in North America which is not habitually consumed by the Japanese. *F. vesiculosus* constituents which may be responsible may include the fucoidans as discussed above, or potentially fucoxanthin or polyphenolics such as the phlorotannins (Skibola et al., 2005).

Fucoxanthin induces more effective apoptosis than other carotenoids in several cancer cells (GOTO, HL-60, Caco-2, HepG-2, Neuro2a, DU145, PEL, PC-3, HeLa, H1299, HT-29, and DLD-1). It is quickly hydrolyzed to fucoxanthinol in the gastrointestinal tract by digestive enzymes (lipase and cholesterol esterase). Daily intake of cooked edible brown seaweed (*Undaria pinnafida* [Wakame]) (9.26 mmol fucoxanthin) for 1 week produced 0.8 pmol/mL of fucoxanthiol in human plasma (Miyashita & Hosokawa, 2015 and references therein). Fucoxanthinol has more potent antiproliferative effect on HL-60, MCF-7, and Caco-2 cancer cells than fucoxanthin. Fucoxanthin arrest cell cycle in the G0/G1 stage and induces apoptosis by altering gene and protein expressions in pathological conditions such as obesity and diabetes (of the MAPK [mitogen-activated protein kinase] and/or SAPK/JNK [stress-activated protein kinase/c-Jun N-terminal kinase] pathways).

8.4.2 Colon Cancer Protective Effects

In a case–control study, seaweed consumption was inversely correlated with colorectal cancer risk in Saitama Prefecture, Japan, with 79 colon and 102 rectal cancer cases and 653 controls, 40–69 years old, with females and males across all groups (Hoshiyama, Sekine, & Sasaba, 1993). The relative risk (RR) of colon cancer was significantly different between low seaweed ($\leq 4\times$ per week, RR = 1.0), intermediate (5–7$\times$ per week, RR = 0.5), and high consumption ($\geq 8\times$ per week, RR = 0.2). The RR of rectal cancer was also significantly different between low, intermediate, and high seaweed consumers with RR values of 1.0, 0.5, and 0.4, respectively. These workers discussed the potential anticarcinogenic effects of dietary fiber from the most commonly consumed dietary seaweeds in Japan, Wakame, and Nori instead of their antioxidant constituents such as ascorbic acid and β-carotene.

A previous animal study evaluated the potential anticarcinogenic effects of marine macroalgae (Yamamoto & Maruyama, 1985). The number of intestinal tumors decreased in 5-week-old male carcinogen-induced (1,2-dimethylhydrazine [DMH]; 20 mg/kg body wt/week for 12 weeks) Sprague–Dawley rats and fed on: diets containing 0.4% powdered or 2% *Eisenia bicyclis* (EB), *L. angustata* (LA) or *L. angustata* var. longissima (LL) for 12 weeks; or diets containing 2% *H. fusiformis* (HF), *L. japonica* var. ochotensis (LO), *Monostroma nitidum* (MN), or *P. tenera* (PT): 0.4% EB, LA, and LL diets (0.71, 0.57, and 0.29 tumors/rat, respectively) compared to the control group (1.33 tumors/rat); as well as in rats fed on 2.0% EB, LA, and LL diets (0.8, 0.6, and 0.5 tumors/rat, respectively), but not in the UP group (1.0 tumors/rat) versus the control group (1.5 tumors/rat); and in rats fed on 2% LO and PT (0.5 and 0.2 tumors/rat), but not in the HF (1.2 tumors/rat) or MN (1.6 tumors/rat) groups versus the control group (1.1 tumors/rat). The hot water extracts from the above macroalgae reduced the number of intestinal tumors in rats fed 0.5% hot water EB, LL, and LO extracts (0.6, 0.6, and 0.5 tumors/rat, respectively). Taken together, these results indicated that dietary fiber, sulfated ester, or brown macroalgal fucoidan content of diets may play a role in modulating intestinal carcinogenesis (Yamamoto & Maruyama, 1985).

Lee and Sung (2003) studied 4 week old male Sprague–Dawley rats (150–200 g body wt), dosed with the carcinogen azoxymethane (15 mg.kg/body wt/week at weeks 2 and 3) and fed a 7.5% fat basal diet (BD) or the BD supplemented with 15% Sea Tangle (*L. japonica*) for 11 weeks. The Sea Tangle diet was fed to one group of rats for 4 weeks prior to dosing with carcinogen to study effects on cancer initiation, another group was fed the Sea Tangle diet for 1 week following dosing with the carcinogen to study effects on cancer promotion, while the third group was fed Sea Tangle for the entire 11 weeks to study effects on initiation and promotion. Rats fed on Sea Tangle-supplemented diets exhibited significant reductions in the number of aberrant crypt foci (ACF)/colon, aberrant crypts(AC)/colon compared to the basal diet-fed group during the initiation phase (53 and 157 compared to 12 and 382, respectively). Sea Tangle-supplemented diets significantly reduced the number of ACF/colon and AC/colon compared to the basal diet-fed group during the promotion phase (54 and 156, respectively). Sea Tangle diets also significantly reduced the number of ACF/colon and AC/colon compared to the basal diet-fed group during initiation and promotion phases combined (49 and 142, respectively) (Lee & Sung, 2003). ACF are preneoplastic lesions that may be present in the colons of carcinogen-treated rats; and, when present in human colons, they are at high risk of developing into cancer. ACF in rat colons are known to proliferate and give rise to colon cancer. Moreover, crypt length (number of cells in each crypt), labeling index (number of labeled cells/crypt length), and proliferation zone (highest labeled cell/crypt length) all significantly decreased in Sea Tangle-fed rats during initiation, promotion, and initiation and promotion phases combined. This study revealed that Sea Tangle supplementation reduced the number of aberrant crypts and epithelial cell proliferation indicating inhibition of both initiation and promotion phases of colon carcinogenesis in this animal model. Sea Tangle may therefore exert anticarcinogenic bioactivities via direct binding of the carcinogen and its metabolites to macroalgal fibers; the fiber component may also have a direct fecal bulking effect to dilute the concentration of carcinogen *in situ*. Fucoidans and alginates may play a role in these anticarcinogenic effects (Lee & Sung, 2003).

Indeed, when Yoshinaga et al. (2018) studied the intestinal effects of a 2-week dietary supplementation with 4 g of commercial Wakame (*U. pinnatifida*) in 22 healthy 20–49 year old Japanese women, kelp supplementation increased total dietary fiber as expected, as well as defecation frequency from 4.6 to 5.6 × per week, defecation days from 4 to 5 days/week, and volume of stool from 9 to 13 pieces/week. Moreover, terminal restriction fragment length polymorphism analyses revealed that the fraction of bifidobacteria (*Bifidobacterium longum*) in feces increased with Wakame supplmentation and the relative abundance of Actinobacteria increased, while that of Bacteroides decreased in feces with Wakame supplementation (Yoshinaga et al., 2018). Thus, dietary kelp supplementation with Wakame exhibited bioactivity in laxation as discussed above, as well as being efficacious as a source of prebiotics, given that fucoidans can interact with the gut microbiome.

Several commercial algal (primarily from *Spirulina* and *Chlorella* sp) prebiotics products have been reported to be globally available for human health benefits upon consumption (Patel et al., 2021). Consumption (1 or 3 g/day for 4 weeks) of a proprietary wild-type strain of the green algae *Chlamydomonas reinhardii* (THN 6) improved gastrointestinal health of 51 adults (average age of 37–45 years) with self-reported gastrointestinal symptoms (Fields et al., 2020). Moreover, supplementation of C. *reinhardii* in diets of individuals with high-frequency gastrointestinal symptoms (e.g. irritable bowel syndrome) significantly reduced the occurrence of bowel discomfort, diarrhea, gas, and bloating, increased bowel movement regularity, and improved stool quality without changes in gut microbial composition. *Euglena gracilis* ingestion (2 g/day for 30 days) increased bowel movement frequency and stool volume stimulating *Faecalibacterium* (the most abundant butyrate-producing bacteria in the gastrointestinal tract) growth to improve digestive health, increase butyrate production, and promote defecation in 28 healthy adults (aged 40–60 years) (Nakashima et al., 2021). Moreover, increased abundance of butyrate-producing bacteria can improve insulin resistance and counteract type 2 diabetes. Fucoidan (*F. vesiculosus*/*Undaria pinnatifids* extract, 1 g/day) supplementation in male athletes ($n = 22$; average age of 25 years) for 7 days significantly improved mucosal health by increasing fecal lysozyme (gut markers of immunity and inflammation) indicating the potential to promote antimicrobial peptides secretion at the gut mucosa that contributes to mucosal immune health regulation (Cox, Cripps, Taylor, Fitton, & West, 2020).

Amaro, Barros, Guedes, Sousa-Pinto, and Malcata (2013) reviewed microalgal constituents of interest as anticarcinogens in the gastrointestinal tract. Microalgal compounds exhibiting anticarcinogenic bioactivities included PUFAs (i.e. DHA), carotenes (i.e. β-carotene), xanthophylls (i.e. the ketocarotenoid astaxanthin (3,3′-dihydroxy-β,β'-carotene-4,4′-dione (ASX), fucoxanthin), and sulfated polysaccharides (i.e. extracellular polysaccharide sheath (EPS) of cyanobacteria containing xylose, glucose, mannose, galactose, fructose, and fucose monomers) (Amaro et al., 2013; Rajauria & Yuan, 2021). van Beelen et al. (2009) reported reduced total ACF (44.3 and 48.2 vs. 92.8) in the colon of 4-week-old male Fischer 344 rats fed 5% corn oil with 20% menhaden fish oil [HFFO], 5% corn oil with 20% commercial [Martek life's DHA™] microalgal oil [HFMO]) compared to high-fat diets (25% corn oil [HFCO], for 8 weeks and dosed with azoxymethane (15 mg/kg body wt, at weeks 1 and 2). Moreover, ACF with a multiplicity of ≥4 crypts decreased as an indicator of tumor incidence in HFFO (4.8) and HFMO (5.2) groups versus HFCO controls (8.7). The microalgal oil was enriched with DHA (38.4% of fatty acids), while the menhaden oil contained smaller concentrations of EPA (9.17%) and DHA (13.9% of fatty acids). Lipid peroxidation in cecal contents and feces of rats was manifested by increased levels of thiobarbituric acid reactive substances (TBARS; produced only upon oxidation of PUFAs with ≥2 double bonds) in the cecum of HFFO (0.075 μmol/g) and HFMO (0.08 μmol/g) versus HFCO-fed controls (0.026 μmol/g) as well as in the feces: 0.191, 0.239, and 0.035 μmol/g feces, respectively. Thus, menhaden and microalgal oil supplementation of diets were equally effective in suppressing colon carcinogenesis of Fischer 344 rats (van Beelen et al., 2009). The anticarcinogenic effects of fish oil-supplemented diets include apoptosis (programmed type I cell death), induction in colonic crypts and oxidized metabolites of PUFAs that can induce chemopreventative phase II enzymes. A clear benefit of microalgal oil supplementation compared to fish oil is the absence of an undesirable fishy off-flavor.

8.4.3 Liver Cancer Protective Effects

ASX and ASX esters (ASXE) isolated from the freshwater Chlorophyta microalga *Haematococcus pluvialis* administered to Wistar rats (200–220 g body wt.) at 100 and 250 μg/kg body wt for 2 weeks followed by dosing with the liver carcinogen carbon tetrachloride

(2 g/kg body wt.), prevented biomarkers of CCl_4-induced liver damage (increased serum levels of glutamate pyruvate transaminase, glutamate oxaloacetate transaminase, and alkaline phosphatases) (Ranga Rao et al., 2015). Hepatic antioxidant enzymes (i.e. CAT, GPx, and SOD) and GSH which were reduced by 30–50% in CCl_4-treated controls were restored by ASX or even elevated by ASXE at the 250 μg/kg body wt treatment level. Thus, ASX and ASXE were hepatoprotective and exerted antioxidant protection against a known liver carcinogen (Ranga Rao et al., 2015). Chidambara Murthy et al. (2005) reported significantly increased endogenous hepatic antioxidant enzyme activities (CAT, GPx, and SOD) in a dose-dependent manner surpassing those of untreated controls in Wistar rats (180–220 g body wt) fed commercial diets supplemented with carotenoids isolated from the marine Chlorophyta microalga *Dunaliella salina* (125 or 250 μg/kg body wt. daily for 15 days) and dosed with carbon tetrachloride on day 14. *D. salina* contained 2.9% w/w carotenoids with β-carotene predominating at 86.5% with smaller amounts of α-carotene, lutein, and lycopene (Chidambara Murthy et al., 2005). Moreover, hepatic tissue TBARS increased by approximately 100% by the carcinogen, while *D. salina* carotenoid supplementation decreased TBARS levels below those of the untreated controls dose-dependently. β-carotene is well known as a singlet-oxygen quencher, indicating that the *D. salina* extract was an effective antioxidant scavenging free radicals and thereby restoring endogenous antioxidant enzyme levels inactivated by the free radicals generated by carbon tetrachloride-induced liver damage and carcinogenesis.

8.5 Antiviral Effects of Selected Macro- and Microalgae

Recent reviews (Alam et al., 2021; Ramos-Vega, Angulo, Bañuelos-Hernández, & Monreal-Escalante, 2021; Rosales-Mendoza et al., 2020) have focused on preclinical trials of microalgal-derived vaccines against human (e.g. Human immunodeficiency virus (HIV), Human papillomavirus (HPV), Influenza virus, Hepatitis B virus, Zaire Ebola virus, Zika virus, *Staphylococcus aureus*, and *Plasmodium falciparum*) and animal infectious diseases (Classical swine fever virus, Foot and mouth disease virus, White spot syndrome virus, infectious bursal disease virus, and Vibriosis). The main microalgae species used as production hosts and delivery vehicles of recombinant vaccines included *Chlamydomonas reinhardtii, D. salina*, *Schizochytrium* sp., *Thalassiosira pseudonana*, *Nannochloropsis* sp., and *Chlorella pyrenoidosa* (Ramos-Vega et al., 2021). After the selection of suitable bacterial, viral, or parasitic antigens, recombinant microalgal transformation can involve plasmids with constitutive or inducible promoters. Selection of the site of synthesis of the recombinant proteins depends on whether they will be purified, or if whole cell administration by oral route is possible, in which case the chloroplast or endoplasmic reticulum organelles are suitable. Several industrial processes have developed microalgae-derived edible vaccine; for example, *Arthrospira palatensis* phycocyanin that increase the specific IgA response to immunization, *H. pluvialis* astaxanthin (T and B lymphocytes increase in humans), and *Chlorella vulgaris* high-protein biomass (recombinant biopharmaceuticals) (Ramos-Vega et al., 2021 and references therein). Two algae species *C. reinhardtii* and *Schizochytrium* sp. have already been used to produce vaccines targeting HPV, H1N1 influenza virus, and Zika virus (Rosales-Mendoza et al., 2020).

Many algae bioactives such as agar, alginate, carrageenan, fucan, fucoidan, galactan, laminaran, and pheophorbide have high potential use against viral infectivity (Alam et al., 2021). The mechanism of action of these compounds involves inhibition of virus binding or internalization into host cells, blockage of virus adhesion and replication into host cells, inverse transcriptase inhibition in the RNA virus, inhibition of cell adhesion, reverse transcriptase blockage, cytopathic effect inhibition, partial blocking of the adhesion to endothelial cells, inhibition of adsorption, and invasion. Seaweed polysaccharides are able to inhibit virus attachment, reverse transcriptase, and protease activities.

Carrageenans (CGs), negatively charged polysaccharides isolated from red algae cell walls, have demonstrated efficient neutralization of many clinically relevant viruses, including HPV, HIV, herpes simplex virus (HSV), influenza virus, and dengue virus (DENV) (Lee, 2020). Moreover, CGs demonstrated anti-HIV efficacy in 11 clinical studies that recruited various cohorts (1,075 participants, predominantly women $n = 900$) based on HIV and sexual activity status. Carrageenans bind with viral glycoproteins blocking viral attachment and subsequent entry into host cell. Furthermore, CGs shields the positively charged sites of the HIV envelope glycoprotein (gp120) via electrostatic interactions, thereby interfering with the CD4-binding function of gp120 blocking HIV entry. CGs can also bind directly to the CD4 protein of T cells blocking/severing gp120 interaction, thereby preventing infection of CD4-positive T cells by HIV. GCs also block adsorption of respiratory viruses and demonstrate high-effective inhibitory activity against major clinically important respiratory infections including coronavirus. Another seaweed polysaccharide, dieckol (extracted from brown algae *Ecklonia cava*), is the most potent SARS-CoV3CLpro trans–/cis cleavage inhibitor in a dose-dependent and competitive manner with no toxicity (Alam et al., 2021).

Griffithsin (GRFT), a 121 amino acid lectin produced by the red macroalga *Griffithsia* sp., has broad spectrum antiviral activity with null toxicity and is considered a potential anti-SARS-CoV-1 and MERS-CoV agent (Rosales-Mendoza et al., 2020). GRFT antiviral activity is associated with homodimer complexes formation targeting the surface of pathogenic enveloped viruses such as the HIV, the severe acute respiratory syndrome (SARS-CoV-1), and the Middle East respiratory syndrome coronaviruses (MERS-CoV).The mechanism of action of GRFT involves impeding the interaction between gp120 and CD4 receptor-expressing cells thereby blocking viral fusion. Apparently, the GRFT high antiviral potency is derived from multivalent interaction of its three carbohydrate-binding domains targeting high-mannose-type oligosaccharides. GRFT exhibited anti-SARS-CoV-1 activity in vitro (Vero 76 cells) using four distinct strains. Furthermore, administration (oral, intravenous, and subcutaneous) of GRFT in Sprague–Dawley rats sustained therapeutic benefits up to 96 hr. Spirulina (*Arthrospira platensis*) extract (0.1 μg/mL) reduces LPS-activated macrophage and monocyte-induced TNF-α secretion (70 and 40%, respectively) that can attenuate hypercytokinemia, commonly known as cytokine storm mainly responsible for SARS-CoV-2 cases (Tzachor, Rozen, Khatib, Jensen, & Avni, 2021). This blue-green algae-based treatment is proposed (to serve as a specific TNF-α suppressor) as a novel anti-TNF-α therapy for viral cytokine storm and acute respiratory distress syndrome treatment. Consumption of *E. gracilis* fermentate (367 mg/day BetaVia™ Complete, a proprietary strain containing >50% 1,3 beta glucans and 20% protein) for 90 days reduced self-reported sick days, upper respiratory tract infections, and global severity of upper respiratory tract infection symptoms in 34 healthy, endurance-trained participants (Evans et al., 2019).

Preclinical studies in mice given a heat-stable vaccine orally protected infection by *S. aureus* (Dreesen, El Hamri, & Fussenegger, 2010). The vaccine, synthesized in the chloroplasts of the unicellular green alga *C. reinhardtii*, significantly reduced the pathogen load in the intestine of infected mice and conferred 80% protection against lethal doses of *S. aureus*. Induced mucosal immune response was evidenced by IgA quantified in fecal pellets collected each week during the vaccination program. Moreover, the IgA titers constantly increased during the entire vaccination program; similarly, the IgG titers of the systemic immune response increased throughout the vaccination trial (Dreesen et al., 2010).

In other studies, subcutaneous- and intramuscular-administered vaccines against HPV (Demurtas et al., 2013) and influenza in rodent animal models (Bayne et al., 2013), respectively, conferred immunity. In the latter study, induction of specific anti-E7 protein of the HPV type 16 IgGs and E7-specific T-cell proliferation were detected in C57BL/6 mice vaccinated with a *Chlamydomonas* extract and with affinity-purified protein. High levels of tumor protection were achieved in mice after challenge with a tumor cell line expressing the E7 protein (Demurtas et al., 2013). Bayne et al.(2013) expressed recombinant hemagglutinin proteins of the influenza virus in the microalga *Schizochytrium* sp. After a lethal challenge with homologous influenza virus, protective immunity was elicited with a single dose of 1.7, 5, or 15 μg recombinant hemagglutinin (rHA) with or without adjuvant at survival rates between 80 and 100%. Full protection (100%) was established at all dose levels with or without adjuvant when mice were given a second vaccination. Thus, these preclinical trials are paving the way to clinical trials, provided that researchers can optimize microalgal-derived recombinant protein production on a large scale (Ramos-Vega et al., 2021).

8.6 Current and Future Perspectives

From the above discussion, edible macroalgae of the Rhodophyceae, Chlorophyceae, and Phaeophyceae phyla as well as microalgae Cyanobacteria not only have a rich history of use as foods and medicines by coastal peoples across the globe but also that many of these macro- and microalgae may be efficacious in reducing the risk of diet-related chronic diseases of public health concern, including CVDs, and breast, colorectal, and liver cancers. Thus, macro- and microalgae and their bioactive constituents can a play a role as functional foods and sources of nutraceuticals or even pharmacognosics.

Fucoidan is probably the most investigated component of microalgae over the past three decades with more than 2,000 scientific papers (according to Pub Med) demonstrating its wide ranging physiological benefits to human health. Global research on fucoidan has focused primarily on immune support, gut and digestive health, antiviral, anticancer, anti-inflammation, and antiaging indications. This has resulted in increased fucoidan use in dietary supplements, functional foods and beverages, and skin care and dermatology products. Furthermore, high-quality organic fucoidan is generally easy to formulate, water-soluble, non-allergenic, and stable in various delivery systems.

Mozuku (*Nemacystus decipiens*) brown seaweeds fucoidan has excellent pharmacokinetics since it is well absorbed and detected in urine in 385 Okinawan subjects after a single oral intake (3 g) (Luthuli et al., 2019). Fucoidan supports immune function and restores beneficial microbial activity by improving the gut health of high-performance athletes.

Fucoidan supplementation (*Fucus vesicolosus/Undaria pinnatifida* extract 1 g/day for 7 days) increased fecal lysozyme (a marker of innate mucosal immunity) in professional team-sport athletes ($n = 22$, average age of 24.7 years) presumably by promoting antimicrobial peptides secretion (Cox et al., 2020).

Mekabu fecoidan (MF), a sulfated polysaccharide extracted from the sporophylls of *Undaria pinnatifida* supplementation (300 mg/day for 4 weeks) significantly increased antibody titers and natural killer cell activity against seasonal influenza vaccine in elderly (> 60 years) Japanese volunteers ($n = 70$) in a randomized, placebo-controlled, double-blind study (Negishi, Mori, Mori, & Yamori, 2013). The increased antibody production after vaccination can therefore prevent influenza epidemics in immunocompromised elderly. Furthermore, fucoidan is an effective adjuvant to increase vaccine efficacy (Kim & Joo, 2015). It upregulates expression of the activation markers major histocompatibility complex class II, cluster of differentiation (CD25 and CD69) in spleen cells and increases the viability and tumor necrosis factor-α (TNF-α) production of spleen cells. Fucoidan can potentially treat viral lung function and lung damage according to a recent review (Fitton, Park, Karpiniec, & Stringer, 2021). For example, some fucoidan fractions (primarily galactofucan) prevent binding of viral spike protein to the anticoagulant effects of heparin that directly inhibits SARS-CoV-2 in vitro, but not to ACE2.

Fucoidan extracts also reduce allergic responses after ingestion and topical application by suppressing IgE induction in human B cells, even after the onset of atopic dermatitis (Iwamoto et al., 2011). It has been proposed as an effective therapeutic agent for patients who already developed type I allergic diseases, since fucoidan increases galectin-9 mRNA expression that can remove IgE thereby suppressing antigen-induced mast cells degranulation (Mizuno, Sakaguchi, & Sakane, 2020). Furthermore, fucoidan extracts (particularly from *Undaria pinnatifida* and *F. vesiculosus*) have received regulatory approvals for food use and dietary supplements in several countries.

The seaweed cultivation market valued at USD 16.7 billion in 2020 is projected to reach USD 30.2 billion by 2025 (Mehra, 2021). Three product streams, essential red, essential green, and essential yellow, use *C. reinhardtii*, a single-celled freshwater green algae species for protein production according to Triton Algae Innovations (Watson, 2021). Ultra-violet light stimulates the green algae to turn red and produce heme by upregulating compounds found in the heme pathway known to provide meat-like properties. Essential red (spray-dried powder) contains 40–50% protein, minerals (calcium and potassium), and carotenoids (astaxanthin and β-carotene). Essential green (FDA GRAS) is intended for use in many food products (juices, snacks, smoothies, pasta, noodles, protein/energy bars or bites, breads, and crackers). *C. reinhardtii* has similar protein levels to chlorella and spirulina with digestible glycoprotein cell wall. Noblegen (Ontario, Canada) is commercializing products from *euglena gracilis* such as proteins (that behaves like animal proteins) and β-glucans (to replace methylcellulose and titanium dioxide). Genesea and Israeli start-up is producing protein isolates and other ingredients (particularly suitable for low pH products) from marine macroalgae for plant-based food manufacture and many food and beverage applications. The company is currently experimenting with common seaweed varieties already approved for human use (Southey, 2021).

Progress Biotech, the Dutch firm, is producing algal DHA for the global infant formula industry. Its algal DHA has neutral taste and color; EU approved this algal oil originating from *Schizochytrium limacinum* (strain WZU477) as novel food for food product use in 2014 and recently for infant formula. The *Schizochytrium sp.* has high DHA content (up to 49% of total fat content) and grows rapidly independent of weather conditions (Chu, 2021). The British feed producer Anpario has produced a free-flowing powder composed of algal oil targeted at breeding animals and producers supplying omega-3 enriched meat, milk, and eggs (Byrne, 2021). A specific (unnamed) microalgae species using waste sugarcane byproducts as energy produces high DHA content (37% of total fatty acids) by fermentation.

There remains a need for further mechanistic studies with macro- and microalgae of known composition with respect to bioactive constituents of interest as well as specific purified compounds to elucidate the molecular mechanisms underlying the effects observed in animal model studies or clinically. Moreover, because macro- and microalgae are photosynthetically active organisms and algal composition is known to vary seasonally, with harvest location and nutrient availability among other variables, there is potential for the cultivation and mariculture of species of interest to ensure consistency and the presence of bioactive molecules of interest. This approach may be instrumental in the valorization and widespread use of macro- and microalgae for their nutritive components and bioactives in reducing chronic disease risk factors of public health across the globe.

Acknowledgments

The work of the author was supported by funding from the Faculty of Community Services of Toronto Metropolitan University (TMU). The author also thanks TMU MHSc in Nutrition Communication program students Sarah Muncaster, Chloe Kavcic, Madison Brown, and Georgia Alton for assisting with literature searches for this book chapter.

References

Aguilera, J., de la Coba, F., & López Figueroa, F. (2006). Use of micosporine type amino acid (porphyra-334) as an antioxidant. Spanish Patent No. ES2301293B1.

Alam, M., Parra-Saldivar, R., Bilal, M., Afroze, C. A., Ahmed, M., Iqbal, H., & Xu, J. (2021). Algae-derived bioactive molecules for the potential treatment of SARS-CoV-2. *Molecules, 26*(8), 2134. https://doi.org/10.3390/molecules26082134

Amaro, H. M., Barros, R., Guedes, A. C., Sousa-Pinto, I., & Malcata, F. X. (2013). Microalgal compounds modulate carcinogenesis in the gastrointestinal tract. *Trends in Biotechnology, 31*, 92–98.

Amini, M., Zayeri, F., & Salehi, M. (2021). Trend analysis of cardiovascular disease mortality, incidence, and mortality-to-incidence ratio: results from global burden of disease study 2017. *BMC Public Health, 21*, 401. https://doi.org/10.1186/s12889-021-10429-0

Athukorala, Y., Trang, S., Kwok, C., & Yuan, Y. V. (2016). Antiproliferative and antioxidant activities and mycosporine-like amino acid profiles of wild-harvested and cultivated edible Canadian marine red macroalgae. *Molecules, 21*(1), 119. https://doi.org/10.3390/molecules21010119

Athukorala, Y., & Yuan, Y. V. (2013). Fucoidans from marine brown macroalgae: Isolation, identification and potential biological activities. Ch. 24. In S.-K. Kim (Ed.), *Marine nutraceuticals: Prospects and perspectives* (pp. 393–435). Boca Raton, FL: CRC Taylor & Francis Press Inc.

Barrow, C.&Shahidi, F. (2008). Marine nutraceuticals and functional foods. CRC Press, Taylor & Francis Group. Boca Raton, FL. 494 pgs.

Bayne, A. C. V., Boltz, D., Owen, C., Betz, Y., Maia, G., Azadi, P., . . . Lippmeier, J. C. (2013). Vaccination against influenza with recombinant hemagglutinin expressed by *Schizochytrium* sp. confers protective immunity. *PLoS One*. https://doi.org/10.1371/journal.pone.0061790

Bird, G., Fitzmaurice, N., Dunlap, W.C., Chalker, B.E. & Bandaranayake, W.M. (1987). Sunscreen compositions and compounds for use therein. U.S. Patent No. 5000946A.

Byrne, J. (2021). Anpario launches microalgae-derived, omega-3 fatty acid supplement. https://feednavigator.com/05-Oct-2021.

Ceccoli, J., Costello, B. & Pillai, S. (2008). Marine extracts and biofermentations for use in cosmetics. U.S. Patent No. 8,202,519 B2.

Chidambara Murthy, K. N., Vanitha, A., Rajesha, J., Mahadeva Swamy, M., Sowmya, P. R., & Ravishankar, G. A. (2005). In vivo antioxidant activity of carotenoids from Dunaliella salina – A green microalga. *Life Sciences*, *76*(12), 1381–1390. https://doi.org/10.1016/j.lfs.2004.10.015

Choi, E.-K., Park, S.-H., Ha, K.-C., Noh, S.-O., Jung, S.-J., Chae, H.-J., . . . Park, T.-S. (2015). Clinical trial of the hypolipidemic effects of a brown alga *Ecklonia cava* extract in patients with hypercholesterolemia. *International Journal of Pharmacology*, *11*, 798–805. https://doi.org/10.3923/ijp.2015.798.805

Chu, W. (2021). AAK inks deal to supply Progress Biotech's algae omega-3 oil for infant formula use. https://nutraingredients.com/05-Oct-2021.

Cox, A. J., Cripps, A. W., Taylor, P. A., Fitton, J. H., & West, N. P. (2020). Fucoidan supplementation restores fecal lysozyme concentrations in high-performance athletes: A pilot study. *Marine Drugs*, *18*(8), 412. http://dx.doi.org/10.3390/md18080412

de la Coba, F., Aguilera, J., Korbee, N., de Gálvez, M. V., Herrera-Ceballos, E., Álvarez-Gómez, F., & Figueroa, F. L. (2019). UVA and UVB photoprotective capabilities of topical formulations containing Mycosporine-like amino acids (MAAs) through different biological effective protection factors (BEPFs). *Marine Drugs*, *17*(1), 55. https://doi.org/10.3390/md17010055

de la Coba, F., Aguilera, J. & López Figueroa, F. (2006). Use of a mycosporin-type amino acid (shinorine) as an antioxidant. International Patent No. 2007/026038 A3.

Demurtas, O. C., Massa, S., Ferrante, P., Venuti, A., Franconi, R., & Giuliano, G. (2013). A Chlamydomonas-derived human papillomavirus 16 E7 vaccine induces specific tumor protection. *PLoS One*. https://doi.org/10.1371/journal.pone.0061473

Dreesen, I. A. J., El Hamri, G. C., & Fussenegger, M. (2010). Heat-stable oral alga-based vaccine protects mice from Staphylococcus aureus infection. *Journal of Biotechnology*. https://doi.org/10.1016/j.jbiotec.2009.12.006

Evans, D. & Rabie, M. (2007). Algal and algal extract dietary supplement composition. International Patent No. WO 2007/062274 A1.

Evans, M., Falcone, P. H., Crowley, D. C., Sulley, A. M., Campbell, M., Zakaria, N., . . . Herrlinger, K. A. (2019). Effect of a euglena gracilis fermentate on immune function in healthy, active adults: a randomized, double-blind, placebo-controlled trial. *Nutrients, 11*(12), 2926. http://dx.doi.org/10.3390/nu11122926

Fields, F. J., Lejzerowicz, F., Schroeder, D., Ngoi, S. M., Tran, M., McDonald, D., . . . Mayfield, S. (2020). Effects of the microalgae *Chlamydomonas* on gastrointestinal health. *Journal of Functional Foods, 65*, 103738. https://doi.org/10.1016/j.jff.2019.103738

Fitton, J. H., Park, A. Y., Karpiniec, S. S., & Stringer, D. N. (2021). Fucoidan and lung function: Value in viral infection. *Marine Drugs, 19*(1), 4. https://doi.org/10.3390/md19010004

Garcia, J. L., de Vicente, M., & Galán, B. (2017). Microalgae, old sustainable food and fashion nutraceuticals. *Microbial Biotechnology, 10*(5), 1017–1024.

Global Health Metrics. (2020). Global burden of 87 risk factors in 204 countries and territories, 1990–2019: A systematic analysis for the Global Burden of Disease Study 2019. *The Lancet, 396*(10258), P1123–P1249. https://doi.org/10.1016/S0140-6736(20)30752-2

Grasa-López, A., Miliar-García, Á., Quevedo-Corona, L., Paniagua-Castro, N., Escalona-Cardoso, G., Reyes-Maldonado, E., & Jaamillo-Flores, M.-E. (2016). *Undaria pinnatifida* and Fucoxanthin ameliorate lipogenesis and markers of both inflammation and cardiovascular dysfunction in an animal model of diet-Induced obesity. *Marine Drugs, 14*(8), 148. MDPI AG. Available fromhttp://dx.doi.org/10.3390/md14080148

Gunathilaka, T. L., Samarakoon, K., Ranasinghe, P., & Peiris, L. (2020). Antidiabetic potential of marine brown algae-a mini review. *Journal of Diabetes Research, 2020*, 1230218. https://doi.org/10.1155/2020/1230218

Hitoe, S., & Shimoda, H. (2017). Seaweed fucoxanthin supplementation improves obesity parameters in mildly obese Japanese subjects. *Functional Foods in Health and Disease, 7*(4), 246–262.

Hoshiyama, Y., Sekine, T., & Sasaba, T. (1993). A case–control study of colorectal cancer and its relation to diet, cigarettes, and alcohol consumption in Saitama Prefecture, Japan. *Tohuku Journal of Experimental Medicine, 171*, 153–165.

Iwamoto, K., Hiragun, T., Takahagi, S., Yanase, Y., Morioke, S., Mihara, S., . . . Hide, M. (2011). Fucoidan suppresses IgE production in peripheral blood mononuclear cells from patients with atopic dermatitis. *Archives of Dermatological Research, 303*(6), 425–431.

Jo, H. G., Kim, M. J., & Cheong, S. H. (2019). Sea tangle (*Saccharina japonica*), an edible brown seaweed, improves serum lipid profiles and antioxidant status in rats fed high-fat and high-cholesterol diets. *Journal of Applied Phycology, 31*(6), 3957–3967. https://doi.org/10.1007/s10811-019-01849-y

Khan, M. I., Shin, J. H., & Kim, J. D. (2018). The promising future of microalgae: current status, challenges, and optimization of a sustainable and renewable industry for biofuels, feed, and other products. *Microbial cell factories, 17*(1), 36. https://doi.org/10.1186/s12934-018-0879-x

Kim, M. S., Kim, J. Y., Choi, W. H., & Lee, S. S. (2008). Effects of seaweed supplementation on blood glucose concentration, lipid profile, and antioxidant enzyme activities in patients with type 2 diabetes mellitus. *Nutrition Research and Practice, 2*(2), 62–67.

Kim, S.-K. (2012). Marine cosmeceuticals: Trends and prospects. CRC Press, Taylor & Francis Group, Boca Raton, FL. 412 pgs.

Kim, S.-K. (2013). Marine nutraceuticals: Prospects and perspectives. CRC Press, Taylor & Francis Group, Boca Raton, FL. 444 pgs.

Kim, S. Y., & Joo, H. G. (2015). Evaluation of adjuvant effects of fucoidan for improving vaccine efficacy. *Journal of Veterinary Science, 16*(2), 145–150.

Lee, C. (2020). Carrageenans as broad-spectrum microbiocides: Current status and challenges. *Marine Drugs, 18*, 435. https://dx.doi.org/10.3390/md18090435

Lee, E.-J., & Sung, M.-K. (2003). Chemoprevention of azoxymethane induced rat colon carcinogenesis by seatangle, a fiber-rich seaweed. *Plant Foods for Human Nutrition, 58*, 1–8.

Lordan, S., Ross, R. P., & Stanton, C. (2011). Marine bioactives as functional food ingredients: potential to reduce the incidence of chronic diseases. *Marine Drugs, 9*, 1056–1100.

Luthuli, S.,Wu, S.,Cheng, Y.,Zheng, X.,Wu, M.,&Tong, H. (2019). Therapeutic effects of fucoidan: A review on recent studies. *Marine Drugs*,17(9):487; https://doi.org/10.3390/md17090487.

Maruyama, H., Watanabe, K., & Yamamoto, I. (1991). Effect of dietary kelp on lipid peroxidation and glutathione peroxidase activity in livers of rats given breast carcinogen DMBA. *Nutrition and Cancer, 15*, 221–228.

Mehra,A. (2021) Seaweed cultivation market worth $30.2 billion by 2025. https://www.marketsandmarkets.com/requestsampleNew.asp?id=152763701.

Mercurio, D. G., Wagemaker, T. A., Alves, V. M., Benevenuto, C. G., Gaspar, L. R., & Maia Campos, P. M. (2015). *In vivo* photoprotective effects of cosmetic formulations containing UV filters, vitamins, Ginkgo biloba and red algae extracts. *Journal of Photochemistry and Photobiology. B, Biology, 153*, 121–126. https://doi.org/10.1016/j.jphotobiol.2015.09.016

Miyashita, K., & Hosokawa, M. (2015). The beneficial health effects of fucoxanthin. In D. Bagchi, A. Swaroop, & M. Bagchi (Eds.), *Genomics, proteomics and metabolomics in nutraceuticals and functional foods* (pp. 122–134). Chichester, West Sussex, UK: John Wiley & Sons Ltd.

Mizuno, M., Sakaguchi, K., & Sakane, I. (2020). Oral administration of fucoidan can exert anti-allergic activity after allergen sensitization by enhancement of galectin-9 secretion in blood. *Biomolecules, 10*(2), 258. https://doi.org/10.3390/biom10020258

Nakashima, A., Sasaki, K., Sasaki, D., Yasuda, K., Suzuki, K., & Kondo, A. (2021). The alga Euglena gracilisstimulates *Faecalibacterium* in the gut and contributes to increased defecation. *Scientific Reports, 11*(1), 1–8. https://doi.org/10.1038/s41598-020-80306-0

Negishi, H., Mori, M., Mori, H., & Yamori, Y. (2013). Supplementation of elderly Japanese men and women with fucoidan from seaweed increases immune responses to seasonal influenza vaccination. *Journal of Nutrition, 43*(11), 1794–1798.

Nishihira, J., Nishimura, M., Sugawara, M., & Kudo, M. (2017). Double-blind, parallel group, placebo-controlled study of *Kjellmaniella crassifolia* Miyabe (Gagome) in human: The potential of Gagome to activate the immune system. *Functional Foods in Health and Disease, 7*(10), 758–772.

Novel Foods Section, Food Directorate, Health Canada. (2018). Novel food information – whole algal protein to be used as alternative protein source in unstandardized foods. Retrieved from https://www.canada.ca/en/health-canada/services/food-nutrition/genetically-modified-foods-other-novel-foods/approved-products/novel-food-information-whole-algal-protein.html.

Orfanoudaki, M., Hartmann, A., Alilou, M., Gelbrich, T., Planchenault, P., Derbré, S., . . . Ganzera, M. (2020). Absolute configuration of mycosporine-like amino acids, their wound healing properties and in vitro anti-aging effects. *Marine Drugs, 18*(1), 35. https://doi.org/10.3390/md18010035

Paradis, M.-E., Couture, P., & Lamarche, B. (2011). A randomized crossover placebo-controlled trial investigating the effect of brown seaweed (Ascophyllum nodosum and Fucus vesiculosus) on postchallenge plasma glucose and insulin levels in men and women. *Applied Physiology, Nutrition, and Metabolism*, *36*(6), 913–919. https://doi-org.ezproxy.lib.ryerson.ca/10.1139/h11-115

Patel, A. K., Singhania, R. R., Awasthi, M. K., Varjani, S., Bhatia, S. K., Tsai, M. L., . . . Dong, C. D. (2021). Emerging prospects of macro-and microalgae as prebiotic. *Microbial Cell Factories*, *20*(1), 1–16. https://doi.org/10.1186/s12934-021-01601-7

Plaza, M., Cifuentes, A., & Ibáñez, E. (2008). In the search of new functional food ingredients from algae. *Trends in Food Science and Technology*, *19*, 31–39.

Pradhan, B., Nayak, R., Patra, S., Jit, B. P., Ragusa, A., & Jena, M. (2021). Bioactive metabolites from marine algae as potent pharmacophores against oxidative stress-associated human diseases: a comprehensive review. *Molecules*, *26*(1), 37. https://dx.doi.org/10.3390/molecules26010037

Rajauria, G.&Yuan, Y.V. (2021). Recent advances in micro- and macroalgal processing: Food and health perspectives. John Wiley & Sons, Chichester, West Sussex, UK. 642 pgs.

Ramos-Vega, A., Angulo, C., Bañuelos-Hernández, B., & Monreal-Escalante, E. (2021). Microalgae-made vaccines against infectious diseases. *Algal Research*, *58*, 102408. https://doi.org/10.1016/j.alagal.2021.102408

Ranga Rao, A., Sarada, R., Shylaja, M. D., & Ravishankar, G. A. (2015). Evaluation of hepatoprotective and antioxidant activity of astaxanthin and astaxanthin esters from microalga-Haematococcus pluvialis. *Journal of Food Science Technology*, *52*, 6703–6710.

Rastogi, R. P., Sonani, R. R., Madamwar, D., & Incharoensakdi, A. (2016). Characterization and antioxidant functions of mycosporine-like amino acids in the cyanobacterium Nostoc sp. R76DM. *Algal Research*, *16*, 110–118.

Rico, D., Diana, A. B. M., Milton-Laskibar, I., Fernández-Quintela, A., Silván, J. M., Rai, D. K., . . . Martínez-Villaluenga, C. (2018). Characterization and *in vitro* evaluation of seaweed species as potential functional ingredients to ameliorate metabolic syndrome. *Journal of Functional Foods*, *46*, 185–194.

Robledo, D., & Freile Pelegrín, Y. (1997). Chemical and mineral composition of six potentially edible seaweed species of Yucatán. *Botanica Marina*, *40*, 301–306.

Rosales-Mendoza, S., García-Silva, I., González-Ortega, O., Sandoval-Vargas, J. M., Malla, A., & Vimolmangkang, S. (2020). The potential of algal biotechnology to produce antiviral compounds and biopharmaceuticals. *Molecules*, *25*(18), 4049. http://dx.doi.org/10.3390/molecules25184049

Roy, M.-C., Anguenot, R., Fillion, C., Beaulieu, M., Bérubé, J., & Richard, D. (2011). Effect of a commercially-available algal phlorotannins extract on digestive enzymes and carbohydrate absorption *in vivo*. *Food Research International*, ***44***(9), 3026–3029.

Schmid, D., Schurch, C., & Zulli, F. (2006). Mycosporine-like amino acids from red algae protect against premature skin-aging. *European Cosmetics*, *9*, 1–4.

Shannon, E., & Abu-Ghannam, N. (2019). Seaweeds as nutraceuticals for health and nutrition. *Phycologia*, *58*(5), 563–577.

Shimabukuro, J., Yogi, T., Nago, M., Teruya, T., Konishi, T., & Tako, M. (2018). Induction of apoptosis by fucoidan isolated from a traditional food, Saccharina longissima, in U937 human leukemia cells. *Functional Foods in Health and Disease*, *8*, 242–255.

Shin, H. C., Kim, S. H., Park, Y., Lee, B. H., & Hwang, H. J. (2012). Effects of 12-week oral supplementation of *Ecklonia cava* polyphenols on anthropometric and blood lipid parameters in overweight Korean individuals: A double-blind randomized clinical trial. *Phytotherapy Research: PTR*, *26*(3), 363–368. https://doi.org/10.1002/ptr.3559

Skibola, C. F. (2004). The effect of F. vesiculosus, an edible brown seaweed, upon menstrual cycle length and hormonal status in three pre-menopausal women: A case report. *BMC Complementary and Alternative Medicine*, *4*, 10. https://doi.org/10.1186/1472-6882-4-10

Skibola, C.F.,Curry, J.D.,VandeVoort, C.,Conley, A.&Smith, M.T. (2005). Brown kelp modulates endocrine hormones in female Sprague–Dawley rats and in human luteinized granulosa cells. *Nutrition and Cancer*, *135*, 296–300.

Southey, F. (2021). Making protein isolates from seaweed, not soy:"There is no reason to grow industrial ingredients on land". www.foodnavigator.com,19-Oct-2021.

Suh, S. S., Lee, S. G., Youn, U. J., Han, S. J., Kim, I. C., & Kim, S. (2017). Comprehensive expression profiling and functional network analysis of Porphyra-334, one Mycosporine-like amino acid (MAA), in human keratinocyte exposed with UV-radiation. *Marine Drugs*, *15*(7), 196. https://doi.org/10.3390/md15070196

Teas, J. (1981). The consumption of seaweed as a protective factor in the etiology of breast cancer. *Medical Hypotheses*, *7*, 601–613.

Teas, J., Harbison, M. L., & Gelman, R. S. (1984). Dietary seaweed (*Laminaria*) and mammary carcinogenesis in rats. *Cancer Reseach*, *44*, 2758–2761.

Teas, J., Vena, S., Cone, D. L., & Irhimeh, M. (2013). The consumption of seaweed as a protective factor in the etiology of breast cancer: Proof of principle. *Journal of Applied Phycology*, *25*, 771–779. https://doi.org/10.1007/s10811-012-9931-0

Tou, J., Oldaker, C., & Yuan, Y. V. (2019). Novel marine sources of nutraceuticals and functional foods: The marine crustacean Krill, macro- and microalgae. Ch. 119. In S.-K. Kim (Ed.), *Encyclopedia of marine biotechnology* (pp. 2651–2690). West Sussex, UK: John Wiley & Sons Ltd.

Turner, N. J. (2003). The ethnobotany of edible seaweed (*Porphyra abbottae* and related species; Rhodophyta: Bangiales) and its use by First Nations on the Pacific Coast of Canada. *Canadian Journal of Botany*, *81*(4), 283–293. https://doi.org/10.1139/b03-029

Tzachor, A., Rozen, O., Khatib, S., Jensen, S., & Avni, D. (2021). Photosynthetically controlled spirulina, but not solar spirulina, inhibits TNF-α secretion: potential implications for COVID-19-related cytokine storm therapy. *Marine Biotechnology*, *23*(1), 149–155.

van Beelen, V. A., Spenkelink, B., Mooibroek, H., Sijtsma, L., Bosch, D., Rietjens, I. M. C. M., & Alink, G. M. (2009). An n−3 PUFA-rich microalgal oil diet protects to a similar extent as a fish oil-rich diet against AOM-induced colonic aberrant crypt foci in F344 rats. *Food and Chemical Toxicology*, *47*, 316–320. https://doi.org/10.1016/j.fct.2008.11.014

Vega, J., Bonomi-Barufi, J., Gómez-Pinchetti, J. L., & Figueroa, F. L. (2020). Cyanobacteria and red macroalgae as potential sources of antioxidants and UV radiation-absorbing compounds for cosmeceutical applications. *Marine Drugs*, *18*(12), 659. https://doi.org/10.3390/md18120659

Wang, X., Pei, L., Liu, H., Qv, K., Xian, W., Liu, J., & Zhang, G. (2016). Fucoidan attenuates atherosclerosis in LDLR−/−mice through inhibition of inflammation and oxidative stress. *Marine Drugs*, *148*. https://doi:10.3390/md14080148

Watson, E. (2021). Tofurky teams up with Triton Algae Innovations for algae-fueled "meat innovations" set to launch in Q1, 2022; https://www.foodnavigator-usa.com; Article September 22, 2021.

World Health Organization (WHO), International Agency for Research on Cancer. (2020). https://gco.iarc.fr/today/home.

Yamamoto, I., & Maruyama, H. (1985). Effect of dietary seaweed preparations on 1,2-dimethylhydrazine-induced intestinal carcinogenesis in rats. *Cancer Letters, 26*, 241–251.

Yamamoto, I., Maruyama, H., & Moriguchi, M. (1987). The effect of dietary seaweeds on 7,12-dimethylbenz[a]anthracene-induced mammary tumorigenesis in rats. *Cancer Letters, 35*, 109–118.

Yang, Y. J., Nam, S.-J., Kong, G., & Kim, M. K. (2010). A case-control study on seaweed consumption and the risk of breast cancer. *British Journal of Nutrition, 103*, 1345–1353.

Yoshinaga, K., Maruya, R., Koikeda, T., & Nakano, T. (2018). Effectsof *Undaria pinnatifida* (wakame)on the human intestinal environment. *Functional Foods in Health and Disease, 8*(10), 477–493.

Yuan, Y. V. (2007). Antioxidants from edible seaweeds. Ch. 19. In F. Shahidi & C.-T. Ho (Eds.), *Antioxidant measurement and applications* (pp. 268–301). Washington, DC: ACS, Oxford University Press.

Yuan, Y. V. (2008). Marine algal constituents. Ch. 11. In C. Barrow & F. Shahidi (Eds.), *Marine nutraceuticals and functional foods* (pp. 259–296). Boca Raton, FL: CRC Taylor & Francis Press Inc.

Yuan, Y. V. (2021). Algae: A functional food with a rich history and future superfood. Ch. 1. In G. Rajauria & Y. V. Yuan (Eds.), *Recent advances in micro- and macroalgal processing: Food and health perspectives* (pp. 3–13). West Sussex, UK: John Wiley & Sons, Ltd.

Yuan, Y. V., Carrington, M. F., & Walsh, N. A. (2005). Extracts from dulse (P. palmata) are effective antioxidants and inhibitors of cell proliferation *in vitro*. *Food and Chemical Toxicology, 43*, 1073–1081.

Yuan, Y. V., & Walsh, N. A. (2006). Antioxidant and antiproliferative activities of extracts from a variety of edible seaweeds. *Food and Chemical Toxicology, 44*, 1144–1150.

9

Hempseed: A Functional Food Source

B. Dave Oomah

(Retired) Formerly with Summerland Research and Development Centre, Agriculture and Agri-Food Canada, Summerland, BC, Canada

9.1 Introduction

The evolutionary and domestication history of *Cannabis* is unclear because of its early origin (dating back to 6000 years). The *Cannabis* genus consists of three species, *C. sativa*, *C. indica*, and *Caecostenetroides ruderalis*. Although hemp is consistently classified as *C. sativa*, it is genetically related (high common alleles proportion) to *C. indica* type marijuana than to *C. sativa* strains from a genome perspective. Furthermore, the genetic differences between hemp and marijuana are not limited to genes involved in tetrahydrocannabinol [THC] production, since hemp typically possesses the B_D allele coding for cannabidiolic acid [CBDA] synthase (instead of the B_T allele coding for tetrahydrocannabinolic acid [THCA] synthase in marijuana) in cannabinoid biosynthesis (Sawler et al., 2015). Hemp cultivars are generally derived from a broad base with high level heterozygosity.

9.2 Hempseed Characteristics

Hempseeds (actually fruits, described botanically as "achenes") (Figure 9.1) are enclosed within the fruit wall (pericarp) representing the "hull" or "shell" (Citti, Pacchetti, Vandelli, Forni, & Cannazza, 2018; Small, 2017). The physical dimensions of the seed (Figure 9.2) depend on cultivar and moisture varying in length (3.3–5 mm), width (2.6–3.9 mm), and diameter (2.1–3.2 mm) (Oomah, Busson, Godfrey, & Drover, 2002; Sacilik, Öztürk, & Keskin, 2003; Taheri-Garavand, Nassiri, & Gharibzahedi, 2012). Seeds in monoecious varieties (for example, *cv. Futura 75*) are generally smaller than in dioecious ones (*cv. Finola*) grown in 2018 in the same environment (Valsavorie, Italian Alps; Pavlovic et al., 2019). The average weight (17–32 g) varied for 62 different accessions, comprising cultivars, seed bank accessions, breeding lines, and wild plants seed produced by open pollination in an experimental field in 1999 (Small & Marcus, 2000). The hull varied among these accessions

Molecular Mechanisms of Functional Food, First Edition. Edited by Rocio Campos-Vega and B. Dave Oomah.

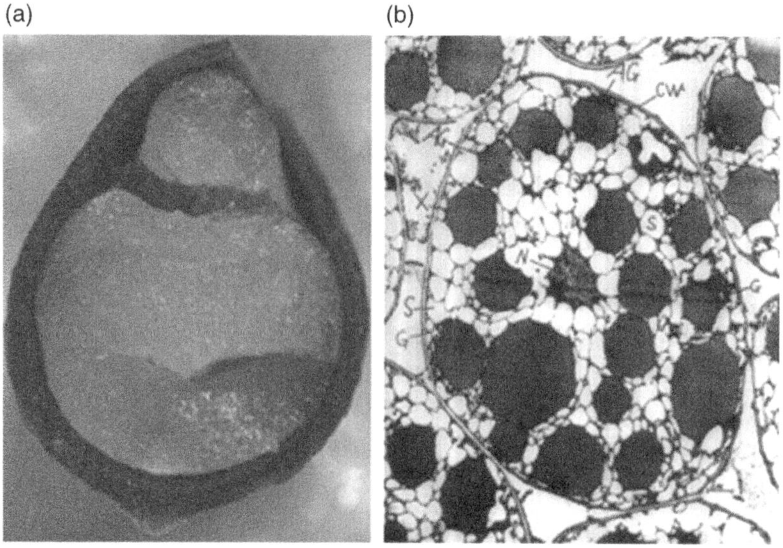

Figure 9.1 (a) Hempseed stained for lignin with phloroglucinol; hempseed hull contained red-stained lignin (17.6%; 16–19.5% range), cellulose (32%; 22–36.7% range), and xylan (5.7–17% based on xylose) as the predominant nondigestible dietary fiber [NDF] that influences seed protein digestibility Most of the seed is filled by an embryo rich in oils, proteins, and carbohydrates. *Source:* Schultz et al. (2020)/ELSEVIER. (b) Electron micrograph of a cross-section of hempseed fixed with potassium permanganate and post stained with lead citrate. Subcellular components consist of cell wall (CW); nucleus (N); spherosome (S); aleurone grains (AG); crystalloid (X); and globoid (G); the crystalloids occupy almost, all of the volume of the aleurone grains. The aleurone grains and crystalloids are 1–4 μ in diameter according to analytical ultracentrifuge pattern. Magnification: 12,000 X. *Source:* St. Angelo, Yatsu, and Altschul (1968)/ELSEVIER.

from~30 to 42% of the seed weight. A low proportion of hull (by weight) is desirable since the heart (cotyledons) attracts high value. Variation in heart % is generally low among hempseed cultivars (5.5% coefficient of variation for genotypic [G] or environmental [E] effects or their combination [G × E]). For example, hemp hearts averaged 60% (54–67% range) for 20 international representative industrial hemp varieties and breeding lines (including 10 of Australian origin). However, five hemp cultivars grown in Romania had higher heart proportion (59–70%) (Schultz et al., 2020 and references therein). Similar kernel yield (53–66%) was reported for eight hempseed cultivars native to China, although their thousand seed weight varied significantly (13–51 g) (Chen et al., 2010).

Hempseed consists of the primary macronutrients oil, protein, and carbohydrates (each representing approximately a quarter and often up to a third of the seed), together with minerals and other minor bioactive compounds. Seed composition (except ash, hemicellulose, and lignin) varied significantly ($P < 0.005$) among ten widely cultivated industrial Canadian hemp cultivars grown in 2012 in southern Quebec (45°25' N, 73°56'W) (Table 9.1); the cultivars included the high-yielding varieties (*CanMa*, *CFX-2*, *CFX-1*, *CRS-1*, and *Finola*) and dual-purpose (fiber/seed) varieties (*Alyssa*, *Anka*, *Delores*, *Jutta*, and *Yvonne*). Cultivar *Finola* bred primarily for seed production had the highest seed oil and crude protein contents (306 and 280 g/kg, respectively) (Vonapartis et al., 2015).

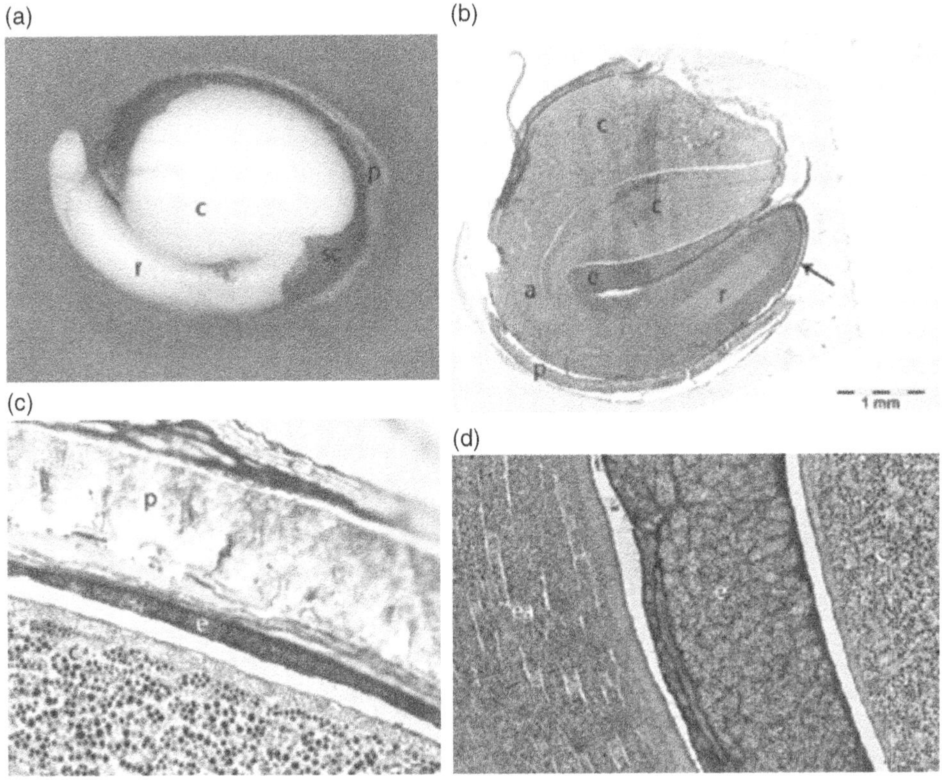

Figure 9.2 Stereomicroscope image (a) and semi-thin section stained with PAS reaction, (b) hemp seed revealing the embryo apex (a), cotyledons (c), endosperm (e) pericarp (p), root (r), and green seed coat (sc); (c, d) higher magnifications of hemp seed semi-thin section, a single layer endosperm envelops the embryo (C) and forms a multilayered structure in the portion between the axis/root of the embryo (ea) and cotyledons (C). *Source:* Reale and Cerri (2019)/The Society of Plant Reproductive Biologists.

Table 9.1 Hempseed composition (g/kg dry matter).

Composition	Mean	Range
Dry matter (%)	93.8	92.8–98.9
Protein (N × 6.25)	256	238–280
Oil	292	269–306
Saturated fatty acid (SFA %)	1.00	0.96–1.03
Unsaturated fatty acid (USFA %)	9.00	8.97–9.04
Ash	55	51–58
Neutral detergent fiber (NDF)	357	327–388
Acid detergent fiber (ADF)	278	259–298
Cellulose	165	141–181
Hemicellulose	80	63–94
Lignin	112	105–118

Source: Adapted from Vonapartis, Aubin, Seguin, Mustafa, and Charron (2015).

The geographic origin of hemp influences its seed composition even when grown under the same conditions. Thus, seed from 11 hemp cultivars and 9 accessions of different geographic origins varied (g/kg dry seed) in oil (285–360), protein (316–356), polyphenolic (5.9–10.6), and phytic acid (43–75) contents (Galasso et al., 2016). Protein content (average 337 g/kg dry weight [dw]) varied significantly ($P \leq 0.01$) among cultivars and accessions and correlated highly (r = 0.402 and 0.540; $P = 0.01$) with thousand seed weight and antioxidant activity, respectively. Oil content (average 319 g/kg) also varied significantly ($P = 0.01$) with significant differences for fatty acids, particularly linoleic acid [LA] (46–58% of total fatty acids) and alpha linolenic acid [ALA] (12–28%) among genotypes. ALA was positively correlated with oil content and inversely associated with LA (r = 0.386 and −0.749; $P = 0.01$, respectively) (Galasso et al., 2016). The average composition of hempseed cultivar *Fedora* grown in Italy was 92.7% moisture, 248, 245, 281, and 53 g/kg of protein, lipids, total carbohydrates (including cellulose, hemicellulose, lignin, soluble, and insoluble fibers), and ash, respectively. The seeds were collected during three consecutive years and refrigerated at −20 °C soon after harvest; seeds were mixed in equal proportions among the three harvests to average the composition over different harvests (Siano et al., 2019). The seed oil content (supercritical carbon dioxide fluid extraction [SCFE] 263–375 g/kg) was significantly ($P = 0.01$) influenced by genotype [G], climatic/environmental [E] conditions and their interaction [G × E] for 51 different hemp genotypes (25 varieties, 19 accessions, and 7 hemp samples from a Hungarian cross-breeding program), grown in Hohenthurm, Germany, for 2 years (2000 and 2001) (Kriese et al., 2004; Matthäus, Schumann, Brühl, & Kriese, 2006). However, variability was low (< 10% coefficient of variation [CV]) in seed protein (20–27%) and lipid (27–38%) contents among 20 international industrial hemp varieties and breeding lines (Schultz et al., 2020). The oil content ranged from 29.6 to 36.5% (average 32%) for 21 different hempseed samples (3 each from 7 provinces/sites) produced in north-western Turkey, predominant hempseed production region (Kiralan, Gül, & Kara, 2010).

Extraction methods can also affect oil content/yield; for example, enzyme (Protex 7 L, Viscozyme L, Kemzyme, Feedzyme, and Natuzyme)-assisted cold pressing produced higher ($P < 0.05$) oil content (28–33% vs 27%) than untreated hempseed without affecting protein, fiber, and ash contents (Latif & Anwar, 2009). Other methods such as microwave-assisted extraction [MAE] yield lower oil (34 vs 38%) with higher oxidation stability than Soxhlet extraction [SE] primarily due to higher tocopherol content (~12%) that enhanced antioxidant activity ([9%] or 72 vs 66% DPPH scavenging activity) (Rezvankhah, Emam-Djomeh, Safari, Askari, & Salami, 2019). Hempseed oil yield is dependent on the polarity of the extracting solvent; the higher the solvent polarity, the lower the oil yield, the highest yield obtained with the most apolar solvent (*n*-hexane) (Alonso-Esteban et al., 2020).

Hulls obtained after commercial hempseed processing contain twice the amount of neutral and acid detergent fibers (NDF and ADF), half the protein, a third of the oil and 80% of ash in the seed (House, Neufeld, & Leson, 2010). Therefore, shelling results in higher fat (1.5 times by weight) and protein content of dehulled hempseed. In another study, hull fractions (> 350 μm [33%] and > 250 μm [18.2%], total 51.2%) had fourfold crude fiber and half the oil content than the cotyledons (> 180 μm [39.4%] and < 180 μm [9.4% of fractions]; total 48.8%) generated by sieving hempseed meal after mechanical cold-pressing the

Table 9.2 Composition (g/kg dry matter) of hempseed and its derivatives.

	Seed				Meal	
	Whole	Dehulled	Hull	Cake	Regular	Defatted
Protein	245–256	320–359 (294)*	127–131	335	335	392
Lipid	337–355	325–467 (520)	103–115	140	106–111	10
Moisture	64–67	49 (54)	51–80	71	56–64	86
Carbohydrates	276–344	151 (57)	641	221	426–453	573
Ash	51–56	64 (49)	32–34	59	67–72	32–97
Total fibers	216–338	nr (27)	nr	144	426–492	454

Source: Adapted from Potin and Saurel (2020); Pihlanto, Nurmi, and Mäkinen (2020).
*Grijó, Pivaa, Osoriob, and Cardozo-Filho (2019).

seed (Pojić et al., 2014). The hull fractions were also abundant in Cannabisin B and N-*trans*-caffeoyltyramine (267–287 mg/kg), whereas catechin (312–744 mg/kg) and *p*-hydroxybenzoic acid (124–129 mg/kg) predominated in the cotyledon fractions. The cotyledons contained most of the antinutrients (trypsin inhibitors, phytic acid, glucosinolates, and condensed tannins) (Pojić et al., 2014).

The composition of hempseed and its derivatives (Table 9.2) indicates that protein and lipids (w/w, dry basis) are about 26 and 36%, respectively, in whole seeds, 35.7and 41.7% in hulled seeds (hemp hearts or cotyledons), and 13 and 11% in the hulls (Potin & Saurel, 2020). The seed hull is low in protein (8.8–16.3%, Mattila et al., 2018), but enriched in carbohydrates, mostly fibers (69% vs 32% vs 16% in whole seeds and hulled seeds, respectively). Hemp hearts, also known as dehulled seed, generally contain higher protein and digestible fiber than whole seed (Schultz et al., 2020). Dehulled hempseed typically contains 1½ times as much fat and protein in weight as the whole seeds. However, hemp hearts composition depends on the commercial dehulling process (Grijó et al., 2019). Furthermore, processing hempseed to meal increases the average crude protein (1.7 folds) and reduces the fat content from 30 to 10% (House et al., 2010). The kernel obtained using a laboratory huller varied significantly ($P < 0.05$) in oil (285–511 g/kg), protein (205–376 g/kg), ash (47–72 g/kg), and moisture (4.9–6.9%) contents for eight hempseed cultivars native to China (Chen et al., 2010).

The residual cake or meal obtained generally after cold-pressing hempseed is low in lipids (~11%) and rich in protein (30–50% dry matter) depending on hempseed variety, oil extraction method, and efficiency. Flour of hempseed (*cv. Fedora*) obtained after cold-pressing was enriched in carbohydrates, ash, and protein (9, 17 and 24%, respectively), but reduced (~38%) in lipids (Siano et al., 2019). The protein is further enriched (up to 44% increase) by delipidating the cake with organic solvent thereby adding value to this product for protein extraction. The minerals are concentrated in the dehulled seed as indicated by its ash content. Hemp hulls contain on average 32% (22–36.7%) crystalline cellulose, dietary fiber predominantly xylan (5.7–17.1%, based on xylose), and pectin; xylan produces prebiotic oligosaccharides; nondigestible food ingredients that promote bacteria growth in the colon and can improve health (Schultz et al., 2020).

9.3 Hempseed Oil and Fatty Acid Composition

The nutritional composition of hempseed is unique, particularly because of the health benefits of its fatty acid spectrum balanced in the essential omega-3 and omega-6 fatty acids. Furthermore, the omega-6 to omega-3 fatty acid ratio in hempseed oil is considered optimal for human health based on similar value found in Mediterranean and Japanese diets with historically low incidence of heart diseases (Russo & Reggiani, 2013). Moreover, a 4 : 1 ratio of omega-6/omega-3 has been associated with a 70% reduction in total mortality from cardiovascular disease in humans (Simopoulos, 2002). The most commonly reported range for individual fatty acids of hempseed oil (Table 9.3) indicates high variability (~2 folds) in monounsaturated and saturated fatty acids contents. Hempseed oil consists primarily of linoleic [LA] (46–62%), α-linolenic [ALA] (12–28%), oleic (8–17%), palmitic (4–9%), stearic (2–5%), and γ-linolenic [GLA] (0.5–4.5%) acids together with small amounts (≤ 2%) of other fatty acids.

Hempseed oil is among the few GLA-containing oils (black currant, borage, and evening primrose; 15–19, 23, and 3–15%, respectively (Ozen, Weiss, & Mauer, 2003). High GLA variation (64% coefficient of variation) among 20 international industrial hemp varieties and breeding lines indicates that the content of this fatty acid may be further improved by breeding (Schultz et al., 2020). GLA was also reported to be the most discriminating fatty acid (0.9–3.3% of total fatty acid; $P < 0.001$) among eight hempseed varieties (whole or hulled) from the Plant Variety Database of the European Commission (Alonso-Esteban et al., 2020). Stearidonic acid [SDA], one of the minor hempseed fatty acids, occurs only in a few seed oils such as black current and borage. The presence of both GLA and SDA in hempseed oil implies the occurrence of and their biosynthesis by the Δ6 desaturase enzyme (Guil-Guerrero, 2007). Both GLA and SDA present in hempseed oil are precursors for the rapid synthesis of longer chain fatty acids mediating many important biological processes. Some cultivars show interesting levels of GLA and SDA (>3% and ~ 1%, respectively); one cultivar (*Finola*) exhibited the highest GLA and SDA contents (4.5 and 1.5%, respectively) (Galasso et al., 2016).

Fatty acid content of hempseed oil depends on several factors. For example, the fatty acid composition varied widely among the largest number of investigated accessions (>500), including fiber-, drug-, and wild-type hempseed from plants around the world. Linoleic, linolenic, and oleic acids typically accounted for 55, 15, and 10% of the total hempseed oil fatty acid content, respectively. The minor fatty acids, GLA, varied significantly with the highest level (~6.8%) found in a variety from Ermaskovskaja, northern Russia. SDA was generally present up to 2.5% with only a few cultivars displaying less than 0.3% (Mölleken & Theimer, 1997a, 1997b). However, complete statistical analysis was not presented for these accessions, thereby limiting determination of variation due to genotype and/or geographic origin. Climate and latitude were previously considered responsible for *Cannabis indica* and *Cannabis sativa* cannabinoid profiles, although genetics is now known to govern these chemotypes. Similarly, fatty acid profiles are presumed to be under genetic control (Mölleken & Theimer, 1997a, 1997b). Genotypes differed significantly ($P = 0.01$) in linoleic acid (46–58%), α-linolenic acid (12–28.4%), GLA (0.5–4.5% of total fatty acid), and all other fatty acids (Galasso et al., 2016). GLA content varied

Table 9.3 Fatty acid contents of hempseed.

		Finola										
Fatty acids	**Range**	**A**	**B**	**C**	**D**	**E**	**F**	**G**	**H**	**I**	**J**	**K**
Palmitic acid	4.2–8.8	6.2–8.8	6–8.6	6.7–7	6.1–6.8	4.2–8.2	6	6.9	5	6.6	5.9	6.4
Palmitoleic acid	0.1–0.3	nd	0.12–0.24	nd	0.1–0.13	nd	0.14	nd	nd	nd	nd	nd
Stearic acid	1.95–4.6	2.3–3.3	2.3–4.6	2.1–2.8	2.3–2.7	1.95–2.6	2.3	2.1	2	2.7	2.7	3.2
Oleic acid	8.3–16.8	8.3–15.5	9.2–16.8	9.4–13	11.4–15.9	8.5–14.7	9.2	9.4	9	10.1	10.9	15.6
Linoleic acid	46.1–62	53.6–62	46.1–58.2	55.6–56.6	55.4–56.9	50.5–60.5	55	56.2	56	54.3	56.6	53.4
α-Linolenic acid	12.4–28.4	12.8–25.7	12.4–28.4	14.7–17.3	16.5–20.4	17.1–25.2	19.6	17.3	22	19.1	19.4	15.1
γ-Linolenic acid	0.5–4.5	0.5–4.1	0.5–4.5	2.6–4.5	0.6–1.1	0.5–2.4	4.5	4.5	4	3.6	2.0	2.7
Stearidonic acid	0.2–2	nd	0.2–1.5	nd	0.3–0.5	nd	1.5	nd	2	nd	nd	nd
Arachidic acid	0.5–2.5	0.6–0.9	0.5–1	0.6–0.8	0.6–0.8	0.5–0.7	1.0	0.8	nd	0.9	0.8	1.0
Gondoic acid	0.2–1.4	nd	0.2–0.4	nd	0.3–0.4	nd	0.3	nd	nd	nd	nd	nd
Eicosadienoic acid	0.04–1.4	0.1	nd	0.75–1.3	0.04–0.06	nd	nd	1.29	nd	nd	nd	nd
Behenic acid	0.15–0.42	nd	0.15–0.42	nd	0.21–0.25	nd	0.35	nd	nd	0.3	0.3	0.4
Erucic acid	0.12–0.18	nd	nd	1.31–2.55	0.12–0.18	nd	nd	1.62	nd	1.4	0.9	0.9
Linoleic/α-Linolenic	1.6–4.6	2.4–4.2	1.6–4.6	3.3–3.8	2.8–3.4	2.4–2.95	2.8	3.25	2.6	2.8	2.9	3.5
ΣSFA	6.7–13.2	9.8–13.2	8.9–14.7	9.4–10.6	9.4–10.6	6.7–11.6	9.6	9.8	7	10.5	9.7	11
ΣMUFA	8.5–17.1	9.5–17.1	9.5–17.4	10.7–15.6	12.1–16.4	8.5–14.7	9.7	11	9	10.1	10.9	15.6
ΣPUFA	67–91.9	67–91.9	71–80.7	73.6–79.6	73.2–78	76.3–82.8	85.3	79.2	82	77	78	71.2
PUFA/SFA	6.6–12	6.8–6.96	5.5–8	7.5–7.8	7.4–7.8	7.2–11.5	8.9	8.1	11.7	7.3	8	6.5
Oxidizability (Cox) Value	7.6–13.3	8.5–13	7.6–13.3	9.5–10.7	9.5–10.7	9.1–12.3	11	10.6	11.5	10.6	10.6	9.5
Atherogenic index (AI)	0.05–0.1	0.08	0.06–0.09	0.07–0.08	0.07	0.06–0.08	0.07	0.08	0.06	0.08	0.07	0.07
Thrombogenic index (TI)	0.08–0.2	0.10–0.12	0.10–0.13	0.11	0.09–0.1	0.08–0.1	0.1	0.1	0.07	0.1	0.09.	0.12

Source: Data for A adapted from Schultz et al. (2020); B and F from Galasso et al. (2016); C and G from Vonapartis et al. (2015); D and E from Kiralan et al. (2010) and Chen et al. (2010); I, H from Callaway (2004); J and K from Oomah et al. (2002).

significantly (0–41 g/kg oil; $P = 0.05$) by genotype and less by year of cultivation. Wider range of GLA (3.4–68 g/kg) has previously been reported depending on the field site (Matthäus et al., 2006 and references therein). High GLA content has been developed in the seed oil of Canadian hempseed cultivars (*Canda* > 3.5%, *Debby* > 5%, and *Joey* > 4%) (Salentijn, Zhang, Amaducci, Yang, & Trindade, 2015). SDA content varied among genotypes (4–16 g/kg) and was strongly correlated (r = 0.87) with GLA content indicating that seeds high in GLA also show high SDA because the same enzyme system is involved in their biosynthesis. Environmental conditions had minimal effect on linoleic (491–570 g/kg) and linolenic acids (169–249 g/kg) content with high correlation between the years (r = 0.80 and 0.86, respectively; $P = 0.01$). The total saturated (75–89 g/kg) and monounsaturated fatty acids [oleic acid] (75–155 g/kg) were genetically stable and independent of the year of cultivation (Matthäus et al., 2006).

The two essential fatty acids, LA and ALA, comprise up to 80% of total fatty acids of hempseed oil for 21 different hempseed samples produced in north-western Turkey (Kiralan et al., 2010). Supercritical [SFE] extracted oil from dehulled hempseeds of eight cultivars native to China ranged from 76 to 83% in PUFA, 56–63% LA, 13.5–23.4% ALA, and 0.5–2.4% GLA (Chen et al., 2010). The shell (hull) where phospholipids or glycolipids are important contains higher (38%) oleic acid, lower (34%) LA, AA, and gondoic acids than the seed, GLA is minimized (< 3%), and SDA is not detected (Mölleken & Theimer, 1997a). Moreover, the ratio of oleic acid (18 : 1) to stearic acid (18 : 0) in the shell was significantly higher (~4folds) than that in the seed (16.1vs 4.2) indicating the transcriptional upregulation of stearoyl-CoA desaturases 1, 2, 3, and 4 (*Scd1*, *−2*, *−3*, and *− 4*), genes that encode the enzymes catalyzing stearic acid desaturation.

Variation (coefficient of variation [CV]) in fatty acid of whole ground seed was determined for 20 international industrial varieties and breeding lines (Schultz et al., 2020). The essential fatty acids linoleic and α-linolenic acids (56 and 18% of total fatty acids) displayed low and high variability (3.2 and 16.8% CV), respectively. The other abundant fatty acids (>1%) in descending order were oleic (11%), palmitic (7%), and stearic (2.8%) acids with 14.7, 9.5, and 9.4% CV, respectively (Schultz et al., 2020). The fatty acid profiles differed significantly ($P < 0.001$) among eight hempseed varieties (whole or hulled) from the Plant Variety Database of the European Commission. However, the scatter (homoscedasticity) among whole (unhusked) hempseed varieties was high for oleic, stearidonic, and α-linolenic acids (0.52, 0.46, and 0.37, respectively) (Alonso-Esteban et al., 2020). Only extraction with *n*-hexane: acetone (1 : 1, v/v) showed no statistically significant differences ($P = 0.092$) among the fatty acid profiles of hempseed varieties. However, all extraction system (*n*-hexane, acetone, n-hexane: acetone (1 : 1, v/v), absolute ethanol, 96% ethanol, water) produced similar FA profiles for each hempseed variety. Only simultaneous oil saponification and extraction (MeOH/ KOH/ HCl to pH 1) reduced LA, ALA, GLA, and SDA contents, while increasing MUFA and SFA percentages instead of PUFA (Alonso-Esteban et al., 2020). Extraction techniques also produced variation in fatty acid content of hulled hempseed oil; LA (52–63.6%), ALA (18.6–24%), oleic acid (8.3–15.2%), SFA (8.9–11.3%), and LA/ALA ratio (2.25–3.42) (Mookerjee, 2020).

Genotypes differed significantly ($P = 0.01$) in linoleic acid (46–58%), α-linolenic acid (12–28.4%), and all other fatty acids. GLA content varied (0.5–4.5% of total fatty acid) among genotypes with *Finola* containing the highest GLA and SDA (1.5%) (Galasso et al., 2016). The overall average linoleic and α-linolenic acid contents were 544.7 and 213.6 g/kg, respectively, in the oil of 51 hempseed genotypes (Kriese et al., 2004). The fatty acid content (g/kg oil) for 51 hempseed genotypes was on average: myristic (0.3, range [R] 0–0.4), palmitic (58.4, R 53–65), linoleic (545, R 491–580), α-linolenic (214, R 169–249), γ-linolenic (20, R 6.7–40.8), and stearidonic (9.5, R 3.9–16). These values expressed in g/kg seed were myristic (0.1, range [R] 0–0.2), palmitic (18.8, R 15.7–22.6), linoleic (175.5, R 142–216), α-linolenic (68.7, R 50–87), γ-linolenic (6.4, R 2–13.4), and stearidonic (3.1, R 1.2–5.2). The fatty acid content depended more on environment than on genetics as indicated by the average broad-sense heritability of 0.23 (range 0 to 0.61 for arachidic and linoleic acids, respectively) (Kriese et al., 2004).

Fatty acid composition was cultivar and site dependent for three Italian hempseed cultivars (*Carmagnola, Futura, and Felina*) grown in two sites (*Treviglio* and *Cavriana*) in the *Lombardia* region in 3 years. Cultivars grown in Treviglio contained higher total unsaturated fatty acids than those grown in Cavriana (86–89 vs 75–82% based on HNMR). In addition, oleic, linoleic, and α-linolenic acids varied with the ranges 10.8–22.5, 31.9–63.1, and 11.3–20.9%, respectively (Lesma et al., 2014). *Finola* and *Futura 75* (monoecious and dioecious cultivar, respectively) grown in the same environment differed in the following fatty acid content: pentadecanoic acid (6.1 vs 0%), palmitic (0.1 vs 6.4%), oleic (9.4 vs 11.5%), γ-linolenic (4.2 vs1.86%), Cis-11-eicosenoic[C20 : 1] (9.3 vs 0.4%), linolenic (10 vs 18.4%), SFA (9.3 vs 10.3%; $P = 0.05$), MUFA (18.7 vs 12.1%: $P < 0.001$), PUFA (72 vs 77.6%; $P = 0.021$), omega-3 (10 vs 18.4%), omega-6 (62 vs 59%), and ω6/ω3 ratio (6.25 vs 3.21) (Pavlovic et al., 2019).

Hempseed oil composition depends on its origin and/or the extraction procedure. The simple screw press can extract 60–80% of hempseed (virgin) oil containing up to 7% γ-linolenic acid and 2.5% stearidonic acid (Matthäus & Brühl, 2008). Unrefined cold-pressed hempseed oil contained trace amounts of eicosadienoic and myristic acids and higher amounts of ALA and oleic acids than refined oils (20.3 vs 16.7 and 12 vs 9%, respectively) (Mikulcová, Kašpárková, Humpolíček, & Buňková, 2017). However, the fatty acid profile of refined hempseed oil (from Spain) (Montserrat-de la Paz, Marín-Aguilar, García-Giménez, & Fernández-Arche, 2014) was within the range for other hempseed oil (Table 9.3). Extraction techniques produced variation in fatty acid content of hulled hempseed oil; LA (52–63.6%), ALA (18.6–24%), oleic acid (8.3–15.2%), SFA (8.9–11.3%), and LA/ALA ratio (2.25–3.42). Moreover, the high UFA/SFA ratio (6.7–9.1) favors blood cholesterol reduction and cardiovascular disease prevention. Method of extraction techniques also affected β-carotene content (3.8–16.6 mg/ kg) (Mookerjee, 2020).

Seed fatty acid (except palmitic [$P < 0.05$], linoleic and erucic acids) content varied significantly ($P < 0.0001$) among 10 widely cultivated industrial Canadian hemp cultivars grown in 2012 in southern Quebec, Canada (Vonapartis et al., 2015). The average fatty acid content (%) was palmitic (6.8), stearic (2.5), oleic (11.8), linoleic (56.1), α-linolenic (16), γ-linolenic (3.2), arachidic (0.75), eicosadienoic (0.95), and erucic (2.1) acids; the total saturated and unsaturated fatty acids comprised 1and 9% of oil.

The most complete fatty acid profile consisted of 15 fatty acids identified in oil of hempseed (domestic origin) from seven north-western Turkish regions (sites) (Kiralan et al., 2010). The spectrum included the principal fatty' acids, palmitic (6.1–6.8% of total fatty acids), stearic (2.3–2.7%), linoleic (55.4–56.9%), α-linolenic (16.5–20.4%), oleic (11.4–15.9%) and minor (< 1%) ones; arachidic (0.6–0.76%), behenic (~ 0.2%), lignoceric (0.06–0.12%), myristic (~ 0.03%), γ-linolenic (0.6–1.1%), stearidonic (0.34–0.47%), gadoleic (0.32–0.39%), palmitoleic (~0.1%), eicosadienoic (~ 0.05%), and erucic (0.12–0.18%) acids. However, statistical analysis was not determined for differences among these hempseed oils (Kiralan et al., 2010).

Fatty acid composition varied widely even for the same hempseed cultivar "*Finola*" analyzed by several investigators (Callaway, 2004; Galasso et al., 2016; Vonapartis et al., 2015) (Table 9.3). *Finola* has one of the highest γ-linolenic and stearidonic acid contents (~ 4 and 2% of total fatty acids) with small differences in the principal fatty acids, palmitic (5–6.9%), stearic (2–2.3%), linoleic (55–56.2%), α-linolenic (17.3–22%), and oleic (9–9.4%) acids. Differences were higher in the minor fatty acids, presumably due to analytical methods since the earlier study (Callaway, 2004) identified only the major fatty acids.

Industrially cold-pressed commercial hempseed oils display similar fatty acid profile consisting of six major (palmitic, stearic, oleic, linoleic, α- and γ-linolenic acids) and three minor (≤ 1%) (arachidic, behenic, and erucic) fatty acids. Interestingly, these oils were devoid of palmitoleic, stearidonic, gondoic, and eicosadienoic acids, presumably lost during processing.

The most commonly reported ratio is within the1.6–4.6 range for the two essential fatty acids, linoleic and α-linolenic acids (Table 9.3). This low ratio of hempseed considered ideal for a healthy diet is presumably advantageous in reducing the risk/prevalence of many chronic diseases (Leyva, McCullough, & Pierce, 2011) and attributed to the health benefits of hempseed oil (Dunford, 2015). The WHO recommends a LA: ALA ratio of 5–10 : 1 in the diet, and a similar ratio (5–15 : 1) is required for industrially produced baby food by the European Society for Pediatric Gastroenterology and Nutrition (Elmadfa & Kornsteiner-Krenn, 2010 and references therein). The ratio of saturated (SFA): monounsaturated (MUFA): polyunsaturated (PUFA) is considered important to describe the health impact of edible oils with 1 : 1 : 1 ratio accepted as beneficial (Elmadfa & Kornsteiner-Krenn, 2010). This 1 : 1 SFA: MUFA ratio is maintained in hempseed and its oil; however, PUFA level (≥ 85%, 67–92% range) is considered the highest among vegetable oils (Pate & Ranali, 1999; Senila et al., 2020). Hempseed and its oil exhibit a favorable PUFA/SFA ratio (6.6–12) since both the British Department of Health and WHO/FAO recommend a ratio above 0.45 as guideline for a "balanced diet" (Lesma et al., 2014). The SFA: MUFA: PUFA ratio (8 : 10 : 82 to 11 : 14 : 75) of hempseed is unique among vegetable oils, although apparently hempseed and flaxseed oils display similar total saturated, monounsaturated, and polyunsaturated fatty acids and PUFAs/SFAs ratio (5.5 vs 5, 8.3 vs 6.3, 82.9 vs 84.9, and 15.1 vs 16.9, respectively) (Senila et al., 2020).

The oxidizability (Cox) value of the oils was calculated by the percentage of C18 fatty acids applying earlier proposed formula (Fatemi & Hammond, 1980):

$Cox = (1[C18{:}1\%] + 10.3\ [C18{:}2\%] + 21.6\ [C18{:}3\%])/100$. The high linoleic acid content primarily accounts for the Cox value (7.6–13.3) that was within the range of high linoleic safflower oils (Longoria-Sanchez et al., 2019). Moreover, the α-linolenic acid content further increases the oxidative stability rendering the oil unsuitable for frying (limited to 3% α-linolenic acid containing oil).

The atherogenic index (AI) and thrombogenic index (TI) were computed according to Ulbricht and Southgate (1991) as a ratio between some saturated and unsaturated fatty acids:

$$AI = \left(C12:0 + 4 \times C14:0 + C16:0 / \left[\Sigma MUFA + \Sigma(n-6) + \Sigma(n-3)\right]\right) 0.05 - 0.1$$

$$TI = \left(C14:0 + C16:0 + C18:0 / \left[0.5 \times \Sigma MUFA + 0.5 \times \Sigma(n-6) + 3 \times \Sigma(n-3) + \Sigma(n-3) / \Sigma(n-6)\right] 0.08 - 0.2\right.$$

The low atherogenic and thrombogenic indices (Table 9.3) may potentially favor reduction of serum cholesterol and atherosclerosis and heart disease prevention (Rudel, Kelly, Sawyer, Shah, & Wilso, 1998). The atherogenic index was similar to and the thrombogenic index lower than those reported for sunflower and olive oils (0.07 and 0.14 for atherogenic and 0.28 and 0.32 for thrombogenic index, respectively) (Ulbricht & Southgate, 1991).

9.3.1 Phytosterols

Phytosterols in addition to PUFAs are critically important for human health. Both PUFAs and phytosterols have been implicated in cardiovascular disease prevention due to their plasma cholesterol-lowering effects (Lee et al., 2011). Phytosterols benefit health by inhibiting cholesterol intestinal absorption, reducing blood low-density lipoprotein (LDL) and exerting anti-inflammatory and anticarcinogenic effects. In hyperlipidemic populations, phytosterol intake (1.5–2 g/day) reduced cholesterol (10–15%) within 3 weeks [EFSA, 2008]. In addition, β-sitosterol—the predominant plant phytosterol can inhibit various cancer (colon T-29, prostate LNCaP, and breast MDA-MB-23) cells (Islam et al., 2017 and references therein).

Hempseed (Canadian origin) contained campesterol, stigmasterol, β-sitosterol, and Δ^5-avenasterol (265, 51, 991, and 115 mg/kg seed, respectively) accounting for 18.6, 3.6, 69.6, and 8.1% of the total phytosterol, respectively. Similar proportion of β-sitosterol was found in chia seed, chickpea, and lentil (Islam et al., 2017). The campesterol, stigmasterol, β-sitosterol, and Δ^5-avenasterol contents (115, 25, 536, and 76 mg/kg, respectively) accounted for 15.3, 3.3, 71.3, and 10.1% of the total phytosterol, respectively, for hexane (Soxhlet) extracted oil from hempseed (cv. *Fedora*) grown in Italy and collected during three consecutive years. This phytosterol profile was consistently similar to those of cold-pressed oil and the residual cake flour oil obtained from *Fedora* hempseed (Siano et al., 2019). Matthäus and Brühl (2008) provided the most complete phytosterol profile of hempseed oil (Table 9.4) consisting primarily of β-sitosterol, campesterol, Δ^5-avenasterol, and stigmasterol (92% of total phytosterols), together with minor (< 2%) amounts of sitostanol, $\Delta^{5,24}$-stigmastadienol, and Δ^7-avenasterol and six other phytosterols each representing ≤1% of total phytosterols. The sitosterol and campesterol contents averaged 464 and 114 mg/kg (dry matter), respectively, for 13 industrial dual purpose hempseed accessions field grown in Northwestern Russia at the same location in 2017–2018 (Grigoryev, Illarinova, & Shelenga, 2020).

Refined hempseed oil contains 1.8–1.9% (~2%) unsaponifiable matter with sterols representing 15% of this fraction. β-Sitosterol and campesterol were predominant comprising 68 and 18% of the phytosterol fraction (Dunford, 2015; Montserrat-de la Paz et al., 2014).

Table 9.4 Phytosterol content (mg/kg) of hempseed oil, seed, and cake flour.

	A		B	C		
Phytosterols	**Average**	**Range**	**Refined HSO**	**Oil**	**Seed**	**Cake flour**
β-Sitosterol	3191	2704–4434	1905	530.4	536.1	528.4
Campesterol	709	257–1001	506	117.4	115.4	125.1
Δ^5-Avenasterol	336	209–572	143			
Stigmasterol	133	97–181	100	28.2	24.7	27.4
Sitostanol	76	39–127				
$\Delta^{5,24}$-Stigmastadienol	54	36–65	32			
Δ^7-Avenasterol	51	32–79	19.9	72.6	76.2	69.6
24-Methylene-cholesterol	40	25–81	7.1			
Clerosterol	36	23–74	58.4			
Cholesterol	34	22–74	0.9			
Δ^7-Stigmasterol	30	19–47	21.7			
Δ^7-Campesterol	26	12–39				
Campestanol	11	0–19				
Total phytosterols	4727	3922–6719	2794	748.6	752.4	750.5

Source: Data for A adapted from Matthäus and Brühl (2008); B from Montserrat-de la Paz et al. (2014); C from Siano et al. (2019).

The total sterols averaged 2.8 g/kg in the unsaponifiable fraction of virgin oil from 10 hempseed cultivars harvested in 2000. The hempseed oil contained 3.6–6.7 g phytosterols/kg oil consisting primarily of β-sitosterol, campesterol, and Δ^5-avenasterol (~70, 15, and 7% of total phytosterol, respectively). The oil contained limited amounts of Δ^7-phytosterols (~2% of total phytosterols) compared to other high linoleic acid oils (sunflower or safflower) (Matthäus & Brühl, 2008). In comparison to other common vegetable oils, the total phytosterols content in hempseed oil was similar to that found in canola oil (6.9 g/kg), but higher than in soybean (4.6 g/kg) and sunflower (4.1 g/kg) oils (Ghazani & Marangoni, 2013). Only four dietary phytosterols (β-sitosterol, campesterol, stigmasterol, and cycloartenol; 3160,706,114, and 169 mg/kg, respectively) were identified in commercial hempseed oil (Living Harvest) using APCI liquid chromatography–tandem mass spectrometry. Among 14 commercial vegetable oils, only sesame and canola oils have higher β-sitosterol and campesterol levels than hempseed oil (Mo, Dong, Hurst, & van Breemen, 2013).

9.3.2 Tocopherols

Tocopherol is an important minor bioactive compound of hempseed oil. Total tocopherols (average 880 mg/kg oil) varied significantly (600–1350 mg/kg oil; $P < 0.01$) among seed from 20 hemp cultivars of different geographical origins grown in pots under the same conditions for 1 year (Galasso et al., 2016). A lower range (413–1112 mg/kg oil or

Table 9.5 Tocopherol content (mg/kg) of hempseed oil.

Hempseed oil	α	β	γ	δ	Total	Vit E
HMSO[1]	32.3	8.1	734	28.7	803	106.0
HMSO[2]	25.6	6.0	598	39.7	669	85.8
HMSO[3]	7–30.5	0.6–2.5	126–294	5.1–24.9	139–352	
HMSO[4]	18.3–35.3		175–333	18.9–34.9	221–390	35.9–68.9
HMSO[5]	7–11	7–10	712–860	22–51	764–962	
Fasamo[6]	34	6	733	25	797	107.6
Hempola[6]	63	8	705	24	799	133.7
FOL[6]	52	10	791	16	868	131.3
FINOLA (cold-pressed)[7]	19.7	0.6	91.6	2.09	114	
SE[8]	7.9		793	31.9	832.6	87.5
MAE[8]	47.1		841	41.5	929.7	131.6
Control[9]	43		615	33.2	691.2	104.8
Enzyme[9]	35–53		654–725	24.1–35.8	724–788	101–126
Most Report	43.1	7.6	718.9	31.9	790.2	111.5
Range	7.9–63	6–10	175–841	16–42	221–930	36–134
Tocopherol (%)	3.8–7.9	0.8–1.2	88.2–92	1.8–5.9		

Source: Data for hempseed oil (HMSO) sourced from [1]Montserrat-de la Paz et al. (2014); [2]Uluate and Özdemir (2012); [3]Kriese et al. (2004); [4]Chen et al. (2010); [5]Blade, Ampong-Nyarko, and Przybylski (2006); [6]Oomah et al. (2002); [7]Smeriglio et al. (2016); [8]Rezvankhah et al. (2019); [9]Latif and Anwar (2009).

143–340 mg/kg seed) was reported for total tocopherol content of 51 hempseed genotypes grown in Hohenthurm, Germany, for 2 years; γ-tocopherol (362–967 mg/kg) accounted for 85% of the total tocopherols (Kriese et al., 2004; Matthäus et al., 2006) (Table 9.5). Oil (chloroform: methanol; 2:1, v/v extract) from four hempseed cultivars grown in Alberta, Canada, over 2 years had total tocopherols ranging from 764 to 964 mg/kg. The isomers α-, β-, γ-, and δ-tocopherol contents also varied (7–11, 7–10, 712–860, and 22–51 mg/kg, respectively) for these oils (Blade et al., 2006). The tocopherol isomers α: β: γ: δ ratio is 5:2:90:3 in hempseed oil; the α- and γ-isomers contributing primarily to vitamin E and antioxidant activities, respectively. However, the relatively high tocopherol content (800–1100 mg/kg; predominantly 85% γ-tocopherol) did not lead to high storage stability of virgin hempseed oil (Matthäus & Brühl, 2008).

The combination of high levels of γ- and α-tocopherols in hempseed oils may offer DNA damage protection, resulting in reduced cancer risk (Oomah et al., 2002 and references therein). Hempseed oil contains other minor tocopherol compounds, such as α-tocotrienols (15–31 mg/kg) present in oils from four hempseed cultivars grown in Alberta, Canada (Blade et al., 2006) and plastochromanol (0–15 mg/kg) detected in 51 hempseed genotypes grown in Hohenthurm, Germany (Matthäus et al., 2006). Hempseed oil had the highest γ- and δ-tocopherols (598 and 40 mg/kg oil) and oxidative stability (2-thiobarbituric acid-reactive substances [secondary oxidative products TBARS – 6.4 mmol malonaldehyde [ME] equivalent/g oil) compared to other unconventional (radish, terebinth, stinging nettle and laurel) oilseeds (Uluate & Özdemir, 2012).

Environmental factors (growing year) affect tocopherol levels in hempseed and its oil more than genotype (71 and 86 vs 24 and 10% of the total variance for γ- and δ-tocopherols, respectively) for seven industrial hempseed cultivars grown in Greece for three consecutive years (Irakli et al., 2019). The γ- and δ-tocopherols content varied significantly (46–113 and 6–13 mg/kg seed) among the cultivars during the 3 years (2016–2018). This contrast with a previous study indicating that total tocopherols (410–1110 mg/kg) and γ-tocopherols (360–970 mg/kg) contents of hempseed were stable and unaffected by year of cultivation (Matthäus et al., 2006). Tocopherol content in whole hempseed varied for total, α-, γ-, and δ-tocopherols (900, 3.5, 46–2949, and 6–1157 mg/kg seed, respectively). These values for the oil were 562–1350, 26–246. 6–8, 564–893, and 13–42 mg/kg oil for total, α-, β-, γ-, and δ-tocopherols, respectively (Farinon, Molinari, Constantini, & Merendino, 2020).

The total tocopherol can reach over 900 mg/kg of oil in hempseed depending on the extraction process. Microwave treatment of hempseed induces positive beneficial changes in oil quality such as elevated tocopherol concentration. For example, continuous microwave drying (950 W, 60 Hz, 6 min) of hempseed cultivar "*Fasamo*" increased (9.3 fold) β-tocopherol content of hexane extracted oil thereby increasing total tocopherol and vitamin E levels (8 and 19%, respectively) (Oomah et al., 2002).

Two- or threefold higher γ-tocopherol content can be recovered using alternative extraction process (supercritical fluid, ultrasound assisted [UAE], microwave assisted [MAE] extractions) than the traditional cold-pressing; MAE and UAE increased total tocopherol extraction (11 and 17%) (Aladić et al., 2015). Moreover, microwave-assisted extraction yielded significantly higher total tocopherol (930 vs 833 mg/kg), γ-tocopherol (841 vs 793 mg/kg), α-tocopherol (47 vs 7.9 mg/kg), and δ-tocopherol (41.5 vs 31.9 mg/kg) than Soxhlet extraction (Rezvankhah et al., 2019). Pressurized *n*-propane extraction yields higher (25–30%) total tocopherol of dehulled hempseed oil than conventional organic solvents (hexane and isopropanol, 3 : 2 v/v) (Grijó et al., 2019). Enzyme (Protex 7 L, Viscozyme L, Kemzyme, Feedzyme, and Natuzyme) hydrolysis (6 h, 40 °C) enhanced total tocopherol extraction (5–14%) with α-,γ-, and δ-tocopherols concentrations ranging from 35 to 53, 654–725, and 24.1–35.8 mg/kg, respectively (Latif & Anwar, 2009).

Commercial hempseed oils (*Fountain of Life* and *Hempola*) contained 62 and 95% higher α-tocopherol, respectively, thereby elevating their vitamin E content than those obtained from "*Fasamo*" seeds (Oomah et al., 2002). However, Italian Food grade oil cold-pressed from hempseed cultivar "*Finola*" had the lowest tocopherol content (Smeriglio et al., 2016). Vitamin E content of hempseed oil is similar to those of peanut, olive and soybean oils. A daily serving portion of hempseed (30 g) can contribute to satisfying up to 12 and 14% of the vitamin E average intake for men and women, respectively, based on EFSA recommendation (average daily intake of 11 and 13 mg/day of vitamin E as α-tocopherol for women and men, respectively) (Farinon et al., 2020).

9.3.3 Phospholipids

Phospholipids, by-products of oilseed processing are useful as emulsifiers, thickening agents, or antioxidants and are key molecules for nutritional properties in food and nutraceuticals applications. Several biological activities are attributed to phospholipids including anti-inflammatory, anti-atherosclerotic, anti-carcinogenic, and antioxidant properties, memory improvement and cardiovascular diseases risk reduction. Phospholipids in mechanically pressed hempseed cake consist mainly of lysophospholipid (phosphatidylinositol [PI-5], phosphatidylethanolamine [PE-5], phosphatidylcholine [PC-4]), and cardiolipin [CL-5] based on normal phase liquid chromatography mass spectrometry (Buré et al., 2016). The ALA-containing phospholipid species include phosphatidic acid [PA], phosphatidylglycerol [PG], phosphatidylinositol, phosphatidylethanolamine [PE], phosphatidylcholine [PC], and phosphatidylserine [PS]. Moreover, the phospholipid *n*−3 PUFA-rich ALA is in the *sn*-2 position of the glyceryl backbone—a valuable position for nutritional applications, except for PC 18:2–18:3 where both isomers were found in all cakes (Buré et al., 2016).

In a subsequent study, the total lipid content (120 g/kg dry matter) of the same mechanically pressed hempseed cake consisted of 4–6% of total phospholipids. Phosphatidylcholine and phosphatidylinositol, the predominant PL constitute ~45% and 21–25% of the total hempseed cake PLs, with a smaller proportion (9–12%) of phosphatidic acid (Cansell et al., 2017) (Table 9.6). Phosphatidylcholine is also the major constituent (45–51%) of the 51 phospholipids identified in methanol/chloroform extract (1:1, v/v) of hemp inflorescences along with ceramides (21%) and phosphoglycerols (10%) (Antonelli et al., 2020). The phospholipid distribution in hempseed cake compares favorably with those of sunflower crude seed oil (PC [30–52%], PE [17–26%], PI [16–30%], and PA [11–22%] (Carbone et al., 2009). Furthermore, the fatty acid composition of phospholipid differs considerably from those of hempseed cake oil. Thus, phospholipid of hempseed cake has higher (2.6x)

Table 9.6 Fatty acid composition (% weight) of total phospholipids and lipids and phospholipids distribution of hempseed cake.

Fatty acids	Phospholipids	Lipids	Phospholipids	Mol (%)	% Total
Palmitic	22.2	8.2	Phosphatidylcholine (PC)	44–45	51.8–54.2
Stearic	18.1	3.4	Phosphatidylinositol (PI)	21–25	25.3–29.4
Oleic	10.3	15.1	Phosphatidic acid (PA)	9–12	10.6–14.5
Vaccenic	1.5	0.9	Others	5–7	6–8.2
Linoleic	44.2	55.4			
α-Linolenic	7.9	13.1			
γ-Linolenic	nd	1.8			
Arachidic	0.7	0.9			
Eicosenoic	0.8	0.4			
Lignoceric	1.3	nd			
Others	2.9	0.8			
ΣSFA	33.6	12.9			
ΣMUFA	13.4	16.6			
ΣPUFA	53.0	70.6			
Σ*n*-6PUFA	45.1	57.5			
Σ*n*-3PUFA	7.9	13.1			

Source: Adapted from Cansell et al. (2017).

saturated fatty acids (primarily palmitic and oleic acids), lower PUFA (¾) and α-linolenic acid (40%) than those of the total lipids (Cansell et al., 2017).

Phospholipids are generally concentrated in hempseed hull (shell [pericarp + testa]) that also contains higher (38%) oleic acid and lower (34%) α-linolenic acid than cotyledon oil (Mölleken & Theimer, 1997a). The fatty acid composition of hempseed cake PL is comparable to that of rapeseed, particularly palmitic, linoleic, and α-linolenic acids (18–28%, 38–48%, and 7–9%, respectively) (Carbone et al., 2009). Hempseed (cv. *Fedora*) oil contains squalene (43.5 mg/kg) within the range (20–300 mg/kg) for most edible oils; flour and seed lipids contain 40 and 49 mg/kg, respectively. Squalene is known as a peroxyl radical scavenger that protects lipid peroxidation (Siano et al., 2019).

9.4 Proteins

In addition to high-quality lipids, hempseed contains protein whose nutritional quality is determined by its total protein content, amino acid profile, and digestibility. Protein is mostly located in hempseed endosperm often refereed to hempseed hearts or dehulled hempseed since the hull contains less (lower) protein, total amino acids, and fat (lipids) than the whole seed (14 vs 26, 8.7 vs 21.2, and 12.5 vs 35 g/100 g dw, respectively) (Mattila, Mäkinen, et al., 2018). Therefore, dehulling and defatting enrich protein content (28%) by carbohydrate reduction (38%) of the resulting meal that produces protein isolate with higher extraction yield (37%), protein recovery (19%), and crude protein content (91 vs 82%) compared to those extracted from whole seeds (Shen et al., 2020). The protein content of hempseed meal averaged 340 g/kg (dm) and differed significantly ($P < 0.01$) predominantly between years and among varieties. The meal was extracted from *n*-hexane defatted six hempseed varieties grown for two consecutive years (2011–2012) in two different locations in Lombardy (Italy) region (Russo & Reggiani, 2015). In the feed industry, the quality of hempseed meal protein is considerably superior to those of soybean or rapeseed protein because of its higher arginine and methionine contents and post-rumen availability (an excellent source of rumen undegraded protein (RUP) (Russo & Reggiani, 2015 and references therein).

Proteomic analysis identified 181 expressed proteins in hempseed (French cv. *Futura*), most involved in seed metabolic processes (41%), responses to stimulus (8%), and biological processes (7%). Biomolecular pathway maps attributed the protein data set, mainly ribosomal proteins (i.e. those involved in DNA transcription, translation, and replication and repair activities) to metabolic and genetic information processes (67 and 37%, respectively) (Aiello et al., 2016). However, protein recovery from hempseed press cake vary (34–51% yield) depending on extraction conditions (Potin & Saurel, 2020). The storage proteins in hempseed consist primarily of globulin (edestin) and albumin, although the actual amount varies; 65 and 33%, respectively (Maloma & Aluko, 2015), 67–75% 11S globulins and 25–37% 2S albumins (Aluko, 2017). The globulin and albumin were recovered from dialyzed and clarified supernatant obtained from protein extract (1 : 10 w/v 0.5 NaCl) of commercially defatted hempseed protein meal (~40% protein) (Maloma & Aluko, 2015). Higher globulin content (up to 82% of total hempseed protein) has also been reported in addition to albumin and a vicilin-like protein (7S – β-conglycinin) comprising 13% and up to 5% of total hempseed protein (Farinon et al., 2020 and references therein). Protein recovery was 37.8, 0.8 and 1.98% [47 : 1 : 2.5] relative to (dehulled acetone defatted hempseed) flour weight and 72.7, 1.3, and 3.9% [56 : 1 : 3] based on protein content of flour for 11S, 7S, and 2S, respectively. Moreover, 11S is significantly higher in crude protein content (97%) than 7S and 2S proteins (58 and 66%, respectively) (Ajibola, 2020).

Edestin is one of the earliest studied plant seed globulins obtained/prepared as a crystalline protein from hempseed (Docimo, Caruso, Ponzoni, Mattana, & Galasso, 2014; Ritthausen, 1881). It is obtained in high purity because of preparation ease and great stability and has been isolated from the aleurone grains and salt (sodium phosphate-sodium chloride, pH 7.8) soluble protein fraction of dehulled hempseeds. The total soluble protein fraction had sedimentation coefficients of 6S, 12S, and 18S according to ultracentifuge patterns (6S, 13S, and 20S for aleurone grains), whereas those from edestin crystalloids were 13S and 20S. Edestin (300,000 molecular weight) consisted of a hexamer (50,000 molecular weight subunit) and spherical particle (~80–85 Å diameter) (St. Angelo et al., 1968). Crystalline edestin (CAS Registry Number: 9007-57-2) is a face-centered cubic lattice ($ao = 114$ Å molecular surface configurations, 80 Å diameter, 1.35 density, and 300,000 molecular weight). The crystal structure of edestin (electron micrograph) is identical to that of Hubbard squash seed globulin (Hall, 1950).

Edestin yield was 49% from dehulled hempseed (143 mg/g hempseed) ground in liquid nitrogen and accounts for about 60–80% of the total protein content (Kim & Lee, 2011). This major 11S hempseed storage protein is a hexamer with a 48 and < 18.4 kDa protein components comparable to soy glycinin consisting of six identical subunits with the acidic and highly heterogeneous basic subunit (33 and 20 kDa, respectively) linked by one disulfide bond (Farinon et al., 2020; Tang, Ten, Wang, & Wang, 2006). It consists of two isoforms Ede1 and Ede2 containing significant amounts of essential amino acids, particularly sulfur amino acids and arginine. These divergent edestin types are encoded by seven cDNAs (*CsEde1A-D* [1536 bp] and *CsEde2A-C* [1476 bp]) isolated from *C. sativa* (cv. *Carmagnola*) (Docimo et al., 2014). Coding sequence similarity is high (~99%) among different forms of the same type, but lower (64%) between the different types. The edestin-type *CsEde1* is highly expressed than *CsEde2* (4.4x) in developing hempseed. Both edestin types are high in arginine (~12%), whereas edestin type 2 (*CsEde2*) is particularly rich in methionine (2.4 vs 0.8% for *CsEde1*). *CsEde1* and *CsEde2* have high coding sequence similarity (~75%) to 11S globulin precursor isoforms 1 (A & B) and 2 (A & B), respectively, and similar (52%) amino acid sequence (Docimo et al., 2014). The three edestin types (*CsEde1, CsEde2,* and *CsEde3*) (Figure 9.3) isolated from *C. sativa* (cv. *Futura*) are encoded by six genes; two genes each for the two isoforms code for types 1, 2, and 3 edestins (*CdEde1A* and *CdEde1B*; *CdEde2A* and *CdEde2B;* and *CdEde3A* and *CdEde3B*, respectively). The edestins coded by these 6 genes have a common N-terminal sequence of 23 amino acids containing 4 cysteine residues, three of which are located on the acid α subunit (and one on the basic β subunit) (Ponzoni, Brambilla, & Galasso, 2018; Potin & Saurel, 2020). Edestin 1, 2, and 3 consist of 511, 491, and 491 amino acids, respectively, according to the NCBI database.

Other hempseed storage proteins include the 7S globulin encoded by the single 7S-vicilin-like gene (*Cs7S* [1893 bp DNA fragment]) with a 48 kDa subunit partially associated through a disulfide bond with the edestin β basic subunit. The N-terminal sequence (21 AA) of the 7S polypeptide has 4-N-glycosylation sites and contains three cysteine residues

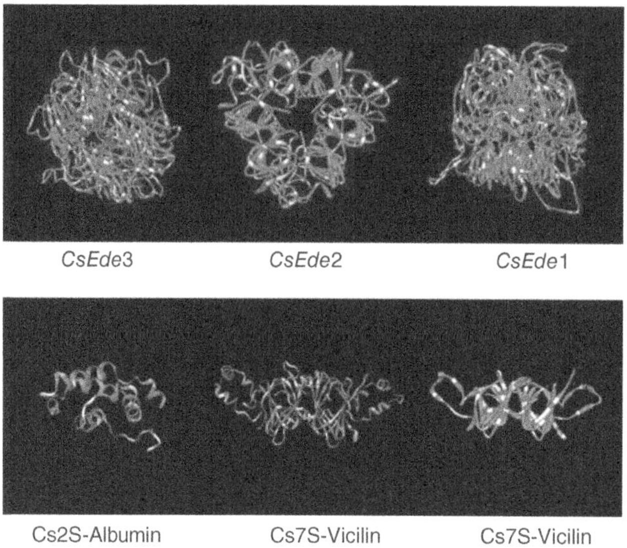

Figure 9.3 Hempseed proteins, edestin 1, 2, and 3 (*CsEde1, CsEde2,* and *CsEde3*), 7S-Vicilin-like proteins, and albumin (Cs2S). *Source:* Uniprot Swiss Model Repository.

(Ponzoni et al., 2018). The 7S-vicilin-like protein (*Cs7S* [A0A219D1T7_CANSA]) consists of 493 amino acids (55,890 Da) with sequence similarity to Korean pine (*Pinus korealensis*) vicilin crystal structure (4lej. 1. A) according to Swiss model (https://unitpro.org). A 2S albumin (10 kDa) has also been identified with two subunits of 7 and 3 kDa, respectively. Two genes (*Cs2S-1* and *Cs2S-2* [13738 bp]) encode for hempseed albumins consisting of seven polypeptides (6–35 kDa) with no interchain disulfide bond. Sulfur amino acids (18% of the total amino acids) abundant in the 2S albumin correspond to two cysteine and three methionine residues and six cysteine and five methionine residues in the small and large subunits, respectively (Ponzoni et al., 2018; Potin & Saurel, 2020). The 2S albumin (*Cs2S-1*[A0A219D1L6_CANSA]) protein consists of 142 amino acids (16,742 Da) and is considered a plant lipid transfer/seed storage protein/protease (trypsin-α-amylase) inhibitor according to Swiss model (https://unitpro.org). MS/MS protein (10% trichloroacetic acid [TCA] in cold acetone extract) analysis of two hempseed cultivars (monoecious cv. *Futura 75* and dioecious cv. *Finola*) revealed two edestin subunits (35 and 18 kDa) along with the heat shock protein (70 kDa), β-conglycinin (50 kDa), and vicilin (Pavlovic et al., 2019). In addition to storage proteins, six other proteins were identified: ATP synthase F1 subunit 1 (mitochondrion), cytochrome oxidase subunit 2 (mitochondrion), phenylalanine ammonia lyase, ribulose 1, 5-bisphosphate carboxylase/oxygenase large subunit (chloroplast), ribulose 1, 5-bisphosphate carboxylase/oxygenase large subunit partial (chloroplast), and Ycf1 (chloroplast) in cold-pressed hempseed cake (Kotecka-Majchrzak, Sumara, Fornal, & Montowska, 2021).

Albumin and globulin extracted from defatted (mechanical press solvent-free) hempseed protein meal differ in their structural and functional properties (Maloma & Aluko, 2015). The nonreducing sodium dodecyl sulfate-polyacrylamide gel electrophoresis (SDS-PAGE) profiles reveal seven (10–42 kDa) and three (6–35 kDa; [21, 34 and 45 kDa under reduced condition]) major polypeptides for albumin and globulin, respectively. Albumin has higher α-helix (5.9–7.6 vs 4.5–5.4%), lower β-strand (14.7–16.9 vs 17.8–18.8%), independent of pH (3, 7 or 9), and similar (~13%) β-turns compared to the secondary structure of globulin. This rigid conformational structure of the globulin modulates its functional properties (protein solubility and foaming capacity). Moreover, the globulin is considered nutritionally superior because of its higher content of hydrophobic, aromatic, and branched chain (12 vs 9%) and sulfur-containing amino acids relative to albumin (Maloma & Aluko, 2015).

9.4.1 Amino Acid Profile

The amino acid profile of hempseed and its derived products (Table 9.7) has recently been summarized (Potin & Saurel, 2020). Arginine and glutamic acids are most abundant in whole seeds (2.3–3.1% and 3.7–4.6% of whole seeds, respectively). Hempseed proteins generally have higher amount of sulfur-containing amino acids, but lower isoleucine, lysine, and phenylalanine contents than casein or soy proteins (Farinon et al., 2020). Hempseed hull has lower total amino acid content (8.7 vs 21.2 g/100 g dw) than the whole seed; its phenylalanine and tyrosine levels are lower than the recommended 6–9.3 g/16 g N (found in whole eggs) (Mattila, Mäkinen, et al., 2018). Moreover, commercial hempseed hulls displayed the lowest amount of total essential amino acids (26 vs 33 g/16 g N for the whole seed) among other plant protein sources (buckwheat, faba bean, flaxseed, lupin, rapeseed press cake, and quinoa) (Mattila, Mäkinen, et al., 2018). An earlier study (House et al., 2010) compared the protein content and quality and amino acid profile of hempseed and products derived from hempseed cultivars (one of four cultivars: *Finola*, *Crag*, and early maturing *USO-14* and *USO-31*) grown in Western Canada. Thirty samples of hemp seed products from two commercial hemp crushing plants included whole hemp seed (11), cold-pressed hempseed meal (10), shelled hemp seeds [hemp nuts] (6), and seed hulls (3). The scores of the limiting amino acid lysine (based on the FAO/WHO reference protein for children 2–5 years old) were 0.62, 0.61, 0.58, and 0.50 for hempseed, dehulled hempseed, hempseed meal, and hulls, respectively. Leucine and tryptophan are the second and third limiting amino acids depending on the product. However, amino acid scores for histidine, isoleucine, threonine, methionine + cysteine, phenylalanine + tyrosine, and valine were above 1 for hempseed products except hulls indicating these amino acids meet the requirement for dietary protein. Their methionine + cysteine scores were even higher than those of casein. Only histidine, phenylalanine + tyrosine and valine scores were above 1 for the hulls (House et al., 2010). The amino acid contents of hempseed hull obtained from hempseed grown in Starbuck (MB) in 2015 (Kim, Koo, Kim, & Nyachoti, 2018) were higher than reported previously and are included in the table.

The aromatic amino acids (AAA [tyrosine, phenylalanine, and tryptophan]) are associated with protein and other essential secondary metabolites synthesis. Some of these metabolites mediate nervous signals transmissions and quench reactive oxygen species in the brain (Parthasarathy et al., 2018). The phenylalanine + tyrosine score of hempseed products (Table 9.7) (>1) meets the dietary protein requirement, whereas that of tryptophan (<1) is below the intake requirement (House et al., 2010). Generally the hulls contain lower amounts of aromatic amino acids compared to the whole seed. For example, the hull

Table 9.7 Amino acid profiles of selected hempseed products.

	Amino acids (g/100 g of total amino acids)			
	Seed	**Hull**	**Dehulled**	**Cake**
Essential amino acids				
Histidine	2.7	2.6–3.7	2.9	3.0
Isoleucine	3.9	3.8–4.3	3.9	3.9
Leucine	6.9	7.3–7.7	6.6	6.9
Lysine	4.0	3.9–4.8	3.6	3.9
Methionine	2.5	2.0–2.4	2.2	2.4
Phenylalanine	4.7	5.0–5.7	4.4	4.7
Threonine	4.0	3.5–4.1	3.8	3.8
Tryptophan	0.97	0.71	1.1	1.1
Valine	5.2	5.3–6.3	5.1	5.1
Nonessential amino acids				
Alanine	4.7	4.5–4.7	4.6	4.4
Arginine	11.9	10.3–13.2	12.6	12.4
Aspartic acid	11.1	10.2–10.8	10.9	10.7
Cysteine	1.7	1.9	1.1	1.8
Glutamic acid	17.9	14.5–16.1	19.8	18.1
Glycine	4.8	4.9–5.1	4.7	4.8
Proline	4.3	5.0–6.5	4.5	4.5
Serine	5.3	4.9–5.1	5.2	5.4
Tyrosine	3.4	3.3–3.9	3.0	3.2
AAA	9.1	10.4–12.1	8.6	9.0
Antioxidant AA	17.9	20.0–21.4	17.3	18.1
Met + Cys	4.2	4.3–8.7	3.3	4.2
BCAA	16.0	12.1–18.3	15.6	15.9
Cytotoxic effect	19.8	21.8–28.7	19.4	20.0
EAA	34.9	28.9–37.7	33.6	34.8
Fisher ratio	1.76	1.0–1.8	1.84	1.77
Ile + Leu + Phe + Val	20.7	15.6–24	20.0	20.6
Arg + His + Glu	32.5	19.1–27.4	35.3	33.5
Lys/Arg ratio	0.34	0.2–0.4	0.29	0.31
PER1	2.26	1.3–2.5	2.11	2.25
PER2	2.31	1.5–2.6	2.21	2.33
PER3	2.01	0.95–2.1	2.07	2.22
SAT1	22.2	19.5–23.9	20.9	21.7
SAT2	7.4	5.3–8.0	6.8	7.0

Source: Adapted from Potin and Saurel (2020).

contains ~½, ~¼ and ~ 60% (as is basis) of phenylalanine, tryptophan, and tyrosine than those in the whole seeds (House et al., 2010). This difference was lower between the hulls and the whole seeds in phenylalanine and tyrosine levels (3.95 vs 3.41 g/16 g N and 3.04 vs 2.07 g/16 g N, respectively) (Mattila, Mäkinen, et al., 2018). The aromatic acid content (~8.4 g/100 g protein) of hempseed meal (Russo & Reggiani, 2015) and dehulled flour (Rodriguez-Martin et al., 2019) was similar to those of flaxseed meal and soy (Oomah, 2001; Oomah, 2020).

The antioxidant potency of most food derived peptides increases in the presence of some amino acids (histidine, tyrosine, methionine, lysine, tryptophan, and proline) (Girgih, Udenigwe, & Aluko, 2011). The sum of these amino acids (17–21 g/100 g total amino acids) was therefore considered to reflect the antioxidant amino acid (Antioxidant AA) of the hempseed products. Several crucial cellular mechanisms including redox balance and methylation depend on the sulfur amino acids (SAAs) cysteine and methionine concentrations (antioxidant level). However, SAAs restriction can benefit metabolic health since total cysteine and S-adenosylmethionine concentrations in human plasma have been associated with fat mass (Olsen et al., 2020 and references therein). A meal low in sulfur-amino acids (SAAs) methionine and cysteine and high in polyunsaturated fatty acids (PUFA) [Cys/Met_{low} + PUFA] reduces plasma concentrations of total, free, and reduced cysteine significantly in healthy individuals in a randomized postprandial pilot study (Olsen et al., 2020). The antioxidant level of hempseed meal and dehulled flour (4.1 and 3.4 g/100 g total amino acids, respectively) (Rodriguez-Martin et al., 2019; Russo & Reggiani, 2015) was higher than those of flax and soy proteins and commercial flaxseed meal (Oomah, 2020). The SAAs (Met+Cys) score for hempseed (whole, dehulled and meal) was higher (> 1.5) indicating that these products meet and exceed the dietary protein requirement for SAAs than that of casein (House et al., 2010). The SAAs level in whole hempseed was comparable to while those in hempseed hull (2.9 g/16 g N) was lower than those in whole egg protein (3.5–5.4 g/16 g N) (Mattila, Mäkinen, et al., 2018).

The branched-chain amino acids (BCAAs) [valine, leucine, and isoleucine] are major players in preventing future risks of type 2 diabetes (T2D) and other vascular or inflammatory stress-induced pathologies (Abolbaghaei, Oomah, Tavakoli, & Hosseinian, 2018). Elevated BCAAs cause insulin resistance and rise in plasma BCAAs, and their metabolites are independently associated with cardiovascular risk (Neinast, Murashige, & Arany, 2019). BCAA supplementation improves prognosis of liver cancer patient, whereas inhibiting its intake is beneficial in certain cancers (e.g. glioma) (Lieu, Nguyen, Rhyne, & Kim, 2020). BCAA is also positively associated with the risk of various diseases sharing etiological aspects with colorectal cancer (CRC). BCAA intake was inversely associated with CRC risk and may be related to a reduced risk of sigmoid colon cancer in a multicenter Italian case-control study including 1953 CRC cases (Rossi et al., 2020).

BCAAs constitute ~16% of amino acids in hempseed products with molar relative abundance of approximately 1.3 : 1.8 : 1[valine: leucine: isoleucine], except hempseed hull (12–18% of total amino acids) (Table 9.7). The average protein intake in males of 1.7 g/kg/day translates to approximately 88, 145, and 66 mg/kg/day of valine, leucine, and isoleucine (1.3 : 2.2 : 1 [Val:Leu:Ile]) (Neinast, Murashige, & Arany, 2019). The BCAA levels of dehulled flour and meal (16.6 and 16.3 g/100 g protein) (Rodriguez-Martin et al., 2019; Russo &

Reggiani, 2015) are comparable to those of commercial flaxseed meal and flaxseed and soy proteins (Oomah, 2001; Oomah, 2020). High proportion of aspartic acid (or asparagine), cysteine, histidine, and proline exert cytotoxic effects in in silico analyses (Avilés-Gaxiola, Gutiérrez-Grijalva, León-Felix, Angulo-Escalante, & Heredia, 2020). The sum of these amino acids accounts for almost 20% (⅕) of the total amino acids of hempseed products and even higher for hempseed hulls (Table 9.7).

Hemp proteins (from whole and hulled seeds, seed hull, and press cake) contain sufficient essential amino acids (over ⅓ [~36%] of the total amino acids), except the limiting amino acid lysine for adult nutrition (Potin & Saurel, 2020). The essential amino acid amount is comparable to common high-quality proteins (e.g. soy and casein) deemed sufficient for the FAO/WHO suggested requirements for 2–5 -year-old children (Wang & Xiong, 2019) and to flaxseed protein and commercial meal (Oomah, 2020). The low Fisher ratio (BCAA/AAA) of hempseed proteins (from whole and hulled seeds, seed hull, and press cake) is desirable traits for functional foods to meet specific physiological needs. These hempseed products displayed lower Fisher ratio (~1.8) than flaxseed protein, commercial flaxseed meal, and soy protein (Oomah, 2020). Elevated hypertriglyceridemia associated with increased plasma levels of isoleucine, leucine, phenylalanine, and valine predisposes to cardiovascular disease (Volk et al., 2020). The sum of these amino acids was considered as an index of hypertriglyceridemia (HyperTrig); ~20% of total amino acids for hempseed and its products (Table 9.7) comparable to those of rapeseed and soy protein isolates (16 and 20 g/100 g, respectively) (Volk et al., 2020). Hempseed and its products are excellent source of arginine, histidine, and glutamine, the three amino acids known to have strong effects on the immune functions in the body. These amino acids account for ~34% of the total amino acids, except for the hulls or 34 g/100 g protein for dehulled flour (Rodriguez-Martin et al., 2019) and comparable to those of flaxseed protein, commercial flaxseed meal, and soy protein (Oomah, 2020). The lysine/arginine ratio (≤ 0.3), a determinant of cholesterolemic and atherogenic effects of a protein, is low for hempseed proteins, suggesting that they are less lipidemic and atherogenic than soybean protein with a lysine/arginine ratio of 0.88 (Oomah, 2001 and references therein).

Protein efficiency ratio (PER) was calculated from the amino acid composition based on the equations proposed by Mir, Riar, and Singh (2019):

$$\text{PER1} = -0.684 + 0.456 \times \text{Leu} - 0.047 \times \text{Pro}$$

$$\text{PER2} = -0.468 + 0.454 \times \text{Leu} - 0.105 \times \text{Tyr}$$

$$\text{PER3} = -1.816 + 0.435 \times \text{Met} + 0.78 \times \text{Leu} + 0.211 \times \text{His} - 0.944 \times \text{Tyr}$$

The PER values were above two for hempseed products and considered typical of high-quality proteins (Mir et al., 2019), except for the hulls. Commercial hempseed hulls had the lowest PER2 value among plant protein sources (faba bean, lupin, rapeseed press cake, flaxseed, hempseed, buckwheat, and quinoa) (Mattila, Mäkinen, et al., 2018). The PER values of whole hempseed and hempseed hull were 1.8 and 1.4 (almost half of the standard casein 2.7) calculated based on leucine-tyrosine interaction (PER2). PER2 values were 1.5 and 1.8 calculated similarly for commercial hempseed hull and extruded hempseed hull, respectively (Kim et al., 2018). The net efficiency ratio is a surrogate measure of feed conversion efficiency in rats consuming hempseed compared to casein. This ratio consistent with PER value was higher for dehulled hempseed than whole hempseeds and hempseed meals (0.75 vs 0.65). Moreover, the higher net efficiency ratio/PER of dehulled hempseed increased its protein digestibility (11%) due to removal of the hull that depress digestibility (House et al., 2010). The digestible indispensable amino acid score [(DIAAS = 54) recommended by the Food and Agriculture Organization (FAO)] of hempseed protein was similar to those of faba bean and oat protein based on a single study on commercial hempseed cake (24.5% feed ingredient in pigs diet (Presto, Lyberg, & Lindberg, 2011). However, the pig ileal apparent digestibility of hempseed cake amino acids was higher than those of soybean meal products, and canola meal, although hempseed cake showed low tract digestibility of organic matter and energy (Herreman, Nommensen, Pennings, & Laus, 2020). The protein digestibility-corrected amino acid score (PDCAAS) for whole hempseeds, dehulled/shelled hempseeds, and hempseed meals (51, 61, and 46, respectively) is within the range of major pulses (lentils, pinto beans) and above cereal grains, particularly whole wheat (House et al., 2010). Hempseed proteins (whole seed and press cake) have high (~86%) *in vivo* digestibility that increases up to 95% similar to animal proteins on dehulling. The *in vitro* digestibility of hempseed meal proteins is similar to its *in vivo* values and that of isoelectric precipitated hemp protein isolate is even higher than that of soy protein isolate (88 vs 80%) (Potin & Saurel, 2020). Increased intake of leucine, isoleucine, lysine, threonine, and tyrosine (SAT1) elevates their plasma levels that are associated with increased satiety. Threonine and tyrosine (SAT2) particularly are presumed to account for this increased satiety (Volk et al., 2020). SAT1 and SAT2 account for approximately 20 and 7% of the total amino acids for the hempseed products, except the hulls.

The two divergent edestin types 1 and 2, isolated from *C. sativa* (cv *Carmagnola*) have similar amino acid sequence similarity (52%), are high in arginine (~12% of total amino acids), whereas type 2 is particularly rich in methionine (2.4 vs 0.8% for *CsEde1*) (Docimo et al., 2014). However, the methionine content of type 2 edestin is comparable to those of 11S, 7S, and 2S and the arginine level lower than those of globulin, albumin, and 11S (Table 9.8). Although albumins are sufficient in lysine, they are deficient in other essential amino acids (isoleucine, leucine, tryptophan, tyrosine, and phenylalanine).The 11S and 7S globulins have similar amino acid profiles, although the latter is limiting in sulfur amino acids (Potin & Saurel, 2020). Edestin (types 1 and 2) is rich in isoleucine, serine, and valine and low in cysteine, histidine, lysine, aspartic, and glutamic acids relative to other hempseed proteins (globulin, albumin, 11S, 7S, and 2S) (Table 9.8). These amino acid differences result in high BCAA, PER3 (type 2 only), hypertriglyceridemia index (HyperTrig), and low antioxidant potency (AAA), antioxidant level and immune function for edestin compared to the other hempseed proteins. Edestin type 2 exerts higher PER3 (~ 6×) [although PER1 and PER2 values were similar for both types] and antioxidant level almost double that of type 1 determined based on amino acid profile. Among all hempseed proteins, 2S displayed the most differences in amino acids profile resulting in the lowest amount of essential and branched chain amino acids, PER2, hypertriglyceridemia index HyperTrig, SAT1, lysine/arginine ratio, and cytotoxic effects. 2S also displayed the highest amount of aromatic amino acids, Fisher ratio, PER1, and SAT2 (Table 9.8).

2S protein is limiting in valine, isoleucine, leucine, and tryptophan; 7S protein is also limiting in tryptophan, but rich in lysine, whereas 11S is limiting in lysine. However, 11S, 7S, and 2S have high (88, 85, and 73%, respectively) *in vitro* protein digestibility (1.6 mg

Table 9.8 Amino acid profiles of hempseed proteins.

	Amino acids (g/100 g of total amino acids)						
	Edestin 1	**Edestin 2**	**Globulin**	**Albumin**	**11S**	**7S**	**2S**
Essential amino acids							
Histidine	2.0	2.4	3.9	3.9	2.9	3.1	3.2
Isoleucine	4.3	4.6	2.9	2.1	3.8	3.7	2.0
Leucine	6.3	6.6	5.6	4.2	6.6	6.3	4.0
Lysine	2.1	2.1	3.7	7.7	3.4	6.5	6.4
Methionine	0.8	2.4	4.1	1.8	2.5	2.5	2.2
Phenylalanine	3.5	4.3	3.3	1.4	4.5	3.7	1.4
Threonine	4.1	3.5	2.6	4.9	3.4	3.8	4.1
Tryptophan	1.0	0.9	0.3	0.2	1.2	0.7	0.2
Valine	7.4	6.4	3.4	3.0	4.7	4.9	3.1
Nonessential amino acids							
Alanine	5.9	6.9	2.8	4.1	5.2	5.5	5.9
Arginine	12.0	11.1	16.1	13.4	13.6	10.8	12.5
Aspartic acid	4.7	5.4	9.5	8.3	11.0	9.2	7.5
Cysteine	1.0	1.1	3.3	3.4	1.6	2.2	4.9
Glutamic acid	8.4	6.9	21.5	21.4	18.4	21.0	25.6
Glycine	7.0	7.1	4.1	8.7	4.1	4.2	5.8
Proline	3.8	4.3	3.9	4.0	3.7	3.9	3.9
Serine	8.2	6.9	5.7	5.4	5.6	5.0	5.0
Tyrosine	3.5	2.1	3.4	2.1	3.7	3.1	2.2
AAA	8.0	7.3	7.0	3.7	9.4	7.5	18.0
Antioxidant AA	13.1	14.1	19.3	19.7	17.5	19.8	19.5
Antioxidant level	1.8	3.4	7.4	5.2	4.0	4.8	7.1
BCAA	17.9	17.6	11.9	9.3	15.2	14.9	9.1
Cytotoxic effect	11.5	13.1	20.1	19.6	19.3	18.4	2.0
EAA	31.3	33.2	29.8	29.2	26.6	35.2	33.1
Fisher ratio	2.3	2.4	1.7	2.5	1.6	2.0	3.8
HyperTrig	21.4	22.0	15.2	10.7	19.7	18.6	0.9
Immune function	22.3	20.3	41.5	38.7	34.9	34.9	41.3
Lys/Arg ratio	0.2	0.2	0.2	0.6	0.3	0.6	0.5
PER1	2.0	2.1	1.7	1.0	2.2	2.0	2.4
PER2	2.1	2.3	1.7	1.2	2.1	2.1	1.0
PER3	0.5	2.9	2.0	1.1	1.5	1.9	1.1
SAT1	20.2	19.1	18.2	21.0	21.0	23.4	10.5
SAT2	7.6	5.6	6.0	7.0	7.1	6.9	18.7

Source: Adapted from Docimo et al. (2014) for edestin 1 and 2; Potin and Saurel (2020) for globulin and albumin; Ajibola (2020) for 11S, 7S, and 2S.

trypsin +3.1 mg chymotrypsin/mL, 37 °C, 10 min). The lower protein digestibility of 2S (albumin) compared to 11S and 7S proteins may be due to its relatively high cysteine content (3.7 vs 1.5 μmol total sulfhydryl/g). Furthermore, 2S has lower nutritional value and BCAAs compared to other hempseed storage proteins (Ajibola, 2020). All hempseed proteins have lower lysine/arginine ratio (0.2–0.6) than rapeseed or soy protein isolates (≥ 0.8) (Volk et al., 2020). The globulin can exert superior immune function comparable to 2S and cytotoxic effects similar to albumin and high antioxidant level (similar to 2S) than other hempseed proteins based on its amino acid profile (Table 9.8).

SDS-PAGE under nonreduced conditions reveals 6 subunits (163, 148, 114, 80, 19, and 15 kDa; 19, 7, 16, 19, 11, and 27% in relative abundance, respectively) for 2S; 4 subunits (155, 100, 49, and 15 kDa; 11, 11, 14, and 65%, respectively) for 7S and 5 subunits (163, 118, 83, 51, and 15 kDa; 6, 7, 16, 27, and 45%, respectively), for 11S. Under reduced conditions, the 2S consists of 5 subunits (126, 103, 80, 18, and 13 kDa; 4, 16, 28, 16, and 35% relative abundance, respectively); the 7S 5 subunits (157, 106, 85, 20, and 15 kDa; 8, 15, 24, 8, and 44%, respectively), and the 11S, 6 subunits (158, 103, 82, 64, 20, and 15 kDa; 16, 17, 7, 20, 8, and 31%, respectively) (Ajibola, 2020). Thus, the 11S, 7S, and 2S proteins have higher relative abundance of low (<100 kDa) than high (>100 kDa) molecular weight subunits under reduced SDS-PAGE (67–80 vs 20–33%). The prevalence of some amino acids also depends on the hydrolysing enzyme. For example, glutamic acid, glycine, phenylalanine, proline, tyrosine, and valine occur frequently in peptic and tryptic digest (37 °C, 16 h), whereas they are poorly expressed in the pancreatic hydrolysate of hempseed (cv. *Futura*) protein concentrate (HPC) obtained from defatted hempseed flour (Aiello, Lammi, Boschin, Zanoni, & Arnoldi, 2017).

9.4.2 Protein Hydrolysates

Protein hydrolysis generally depends on the protease, hempseed protein extract, and hydrolysis time. For example, acid-soluble protein isolate (AC-HPI) obtained from cold-pressed hempseed cake is significantly resistant to proteolysis than the alkali-soluble protein isolate (AL-HPI) (Teh, Bekhit, Carne, & Birch, 2016). Proteolytic concentrate (HT) among other proteases (AFP, Pro-G, actinidin, and zingibain) completely hydrolyzed the acidic and basic edestin subunits in AL-HPI in 1 and 2 hr, respectively, both subunits in AC-HPI were hydrolyzed in 1 hr. AL-HPI generated more antioxidant (ORAC) peptides (with highest ORAC 246 μM Trolox/g; 73 vs 69% DPPH scavenging and ACE-inhibitory activities 77 vs 65%) than AC-HPI after HT proteolysis. Moreover, the HT hydrolysates (2 hr) exhibited maximum ACE inhibitory activity (73–77%) with the lowest zeta potential (~ – 25 mv) corresponding to the highest proteolytic activity (Teh et al., 2016).

The extent of hempseed protein hydrolysis depends on the enzyme; enzymes with exo- and endo-proteinase activities (e.g. pepsin + pancreatin, flavourzyme) generate higher degree of hydrolysis (DH, 28%) than other enzymes. Hempseed protein isolate (obtained by alkali extraction/acid precipitation from hempseed protein meal [HPM, 37% protein, Hemp Oil Canada, St. Agathe, MB, Canada] was hydrolyzed with alcalase (50 °C, pH 8, 4 hr); papain (65 °C, pH 6, 4 hr); and PP (pepsin [37 °C, pH 2, 2 hr] followed by pancreatin [37 °C, pH 7.5, 4 hr]). Hydrolysis of hempseed protein isolate produces four peptides (300–9560 Da) and a minor polypeptide (~ 1 kDa) [alcalase and PP hydrolysates] indicating resistance to digestion (Malomo, Onuh, Girgih, & Aluko, 2015).

Hydrolyzed and unhydrolyzed (native) protein isolates had similar amino acid composition except for the low cysteine (0.12 vs 1.3%) and high proline content of the 2% pepsin hydrolysate relative to other samples. This 2% papain HPH exhibited the highest ACE-inhibitory activity (48%), whereas the other HPHs (except 2% papain) had high renin-inhibitory activities (>50%). However, combination of hydrolysates (1% alcalase +2 or 4% pepsin) provides effective systolic blood pressure lowering effect in spontaneously hypertensive rats (Malomo et al., 2015). Hydrolysate-simulating human gastrointestinal digestion has been prepared with (4% w/v, 37 °C) pepsin [2 hr] + pancreatin [4 hr] treatment of acetone decolorized hempseed protein isolate (HPI) (obtained by alkali solubilization/acid precipitation of defatted commercial hempseed meal) (Girgih, Udenigwe, & Aluko, 2011). The hempseed hydrolysate (HPH) was further separated by ultrafiltration membrane (UF – 1, 3, 5 and 10 kDa cutoff) into low (<1 and 1–3 kDa) and high (3–5 and 5–10 kDa) fractions (Table 9.9).

Amino acid content of HPH was similar to that of HPI, whereas those of low (<1 and 1–3 kDa) differed from the high (3–5 and 5–10 kDa) membrane fraction. Low fractions have relatively high leucine (7–10 vs 5–7%), phenylalanine (5–8 vs 3–4%), and valine (5.6 vs 4.6%) and low cysteine (0.3–0.7 vs 1.3–1.6%) levels. High fractions have low leucine and phenylalanine. The permeate (<1 kDa) has high content of hydrophobic (44 vs 33–38% of total amino acids) and aromatic amino acids (14 vs 7–10%) (Table 9.9). DPPH scavenging potency is higher in low than in high fractions (24 vs 19% for <1 and 5–10 kDa, respectively) presumably due to their higher hydrophobic amino acid contents (38–44 vs 33–36% of total amino acids) and very low in HPH (~4%) [presumably due to the low ratio of hydroxyl radical scavenging to nonscavenging peptides].

HPH exhibits high Fe^{2+} chelating capacity (72%) and protects linoleic acid peroxidation, and the ferric reducing antioxidant power (FRAP) is similar to those of chickpea protein hydrolysate (Girgih, Udenigwe, & Aluko, 2011). The HPH UF fractions exhibit low hydroxyl scavenging activities (16–24%) compared to glutathione (82%) that may affect their ability to modulate (attenuate) several ROS-induced disease conditions (aging, cancer, and diabetes). A subsequent study showed higher proportion of low (< 3 kDa) to high-molecular-weight (>3 kDa) hempseed hydrolysate fraction (74 vs 26%) suggesting potentially higher *in vivo* bioavailability due to their stronger resistance to gastrointestinal proteolysis (Girgih, Udenigwe, Li, Adebiyi, & Aluko, 2011). Moreover, HPH effectively inhibited ACE and renin (IC_{50} 0.67 and 0.81, respectively) and reduced SBP (−20 mmHg) significantly ($P < 0.05$) after only 2 hr of oral administration to SHRs and similar to the hypotensive effect of captopril after 4 hr. The low-molecular-weight fraction displayed lower antihypertensive effect than HPH (Girgih, Udenigwe, Li, Adebiyi, & Aluko, 2011).

In a later study, commercial defatted coarse hempseed protein meal (25% protein) was directly hydrolyzed by simulated human gastrointestinal protein digestion (pepsin [4% HPM, pH 2, 37 °C, 2 hr] + pancreatin [pH 7.5, 37 °C, 4 hr] sequential hydrolysis). This hempseed meal hydrolysate (HMH) substituted for casein (0.5 and 1% levels) significantly suppressed (> 2-folds) plasma ACE activity and renin attenuating thereby the normal increase

Table 9.9 Amino acid (%) profile of hempseed protein hydrolysates.

			Membrane fraction (kDa)						
	HPI	**HPH**	**<1**	**1–3**	**3–5**	**5–10**	**HMH**	**HPH20A**	**HPH60A+**
Essential amino acids									
Histidine	2.7	2.8	2.6	2.5	2.5	2.5	3.0	2.8	2.8
Isoleucine	3.7	3.8	4.2	4.2	4.0	3.9	4.0	3.6	3.9
Leucine	5.5	6.8	9.9	7.3	5.2	4.8	6.8	7.1	7.3
Lysine	3.0	3.0	3.2	2.9	2.6	2.5	3.9	3.1	3.2
Methionine	2.4	1.8	1.9	2.0	1.9	1.8	2.0	0.6	0.6
Phenylalanine	3.7	4.6	7.7	5.0	3.2	2.9	4.5	5.0	5.2
Threonine	3.5	3.7	3.6	3.8	4.0	4.0	3.7	3.9	4.0
Tryptophan	1.4	1.2	1.6	1.4	1.2	1.1	12.6	0.7	0.8
Valine	4.1	4.7	5.7	5.3	4.5	4.2	5.2	4.1	4.7
Non-essential amino acids									
Alanine	4.1	4.5	4.9	4.8	4.3	4.1	4.0	5.0	5.0
Arginine	13.9	14.1	13.9	13.0	13.6	13.3	2.1	13.3	13.8
Aspartic acid	11.8	11.4	9.5	11.7	12.8	12.7	10.8	12.4	12.1
Cysteine	1.5	1.3	0.3	0.7	1.3	1.6	1.2	1.8	1.7
Glutamic acid	22.4	20.1	15.2	19.3	22.7	22.9	18.1	18.8	18.6
Glycine	4.4	4.3	3.2	3.9	4.5	4.7	4.5	4.6	4.6
Proline	4.1	4.0	3.2	4.0	4.2	4.9	4.7	2.8	1.3
Serine	4.8	4.6	4.7	4.8	4.7	4.5	5.4	6.4	6.3
Tyrosine	3.1	3.5	4.8	3.5	3.1	3.6	3.6	4.0	4.1
AAA	8.1	9.3	14.0	10.0	7.4	7.6	20.7	9.7	10.1
Antioxidant AA	16.6	16.2	17.3	16.4	15.3	16.4	29.8	14.0	12.8
Antioxidant level	3.9	3.1	2.3	2.7	3.1	3.4	3.2	2.4	2.3
BCAA	13.3	15.3	19.7	16.7	13.6	13.0	16.0	14.7	15.9
Cytotoxic effect	20.1	19.5	15.6	18.9	20.8	21.7	19.7	19.8	17.8
EAA	29.9	32.3	40.3	34.3	28.8	27.7	45.6	30.8	32.6
Fisher ratio	1.6	1.6	1.4	1.7	1.8	1.7	0.8	1.5	1.6
HyperTrig	17.0	19.9	27.4	21.7	16.8	15.8	20.5	19.8	21.1
Immune function	39.0	36.9	31.7	34.7	38.8	38.7	23.2	35.0	35.2
Lys/Arg ratio	0.2	0.2	0.2	0.2	0.2	0.2	1.9	0.2	0.2
PER1	1.6	2.2	3.7	2.4	1.5	1.3	2.2	2.4	2.6
PER3	1.2	1.6	2.8	2.0	0.6	−0.2	1.6	0.8	0.9
SAT1	18.8	20.7	25.6	21.6	18.8	18.9	22.0	21.7	22.5
SAT2	6.6	7.1	8.4	7.3	7.1	7.6	7.3	7.9	8.1

in systolic blood pressure (SBP) when fed to spontaneously hypetensive rats (SHRs) [young rats for 8 weeks (preventive) or adult rats for 4 weeks (hypertension treatment) (Girgih, Alashi, He, Malomo, & Aluko, 2014). HMH inhibited ACE (70%) and renin (35%) *in vitro*; HMH (1%) significantly ($P<0.05$) reduced SBP (10%) in rats after 3rd week feeding, suppressing superoxide dismutase (SOD) and catalase expression together with lower total peroxide levels. Normotensive rats fed HMH and hempseed protein isolate [HPI] (1%) attenuated plasma renin levels significantly ($P<0.05$) compared to casein-only rat group. HMH peptides were either more active or had higher absorption characteristics than peptides from HPI within rats gastrointestinal tract. HMH diet but not HPI diet maintained the SHRs at normal blood pressure. SBP reductions were directly related to plasma ACE and renin levels suggesting the *in vivo* mechanism of action. SBP attenuation of hempseed protein hydrolysates was superior to those of soybean protein hydrolysate in SHRs. HMH had lower protein (55 vs 95%), higher fat (4 vs 0%), ash (30 vs 5%), and NaCl (22 vs 0.9%) compared to HPI and significantly higher total contents of hydrophobic and aromatic amino acids than HPI. HMH had higher tryptophan (12.6 vs 1.4%) and lower arginine (2.1 vs 13.9%) compared to HPI (Girgih et al., 2014). These investigators conclude that hempseed meal is a suitable raw material to develop (generate) peptide products due to the presence of desirable bioactive peptide components (high arginine residues and branched chain amino acids).

Hempseed protein hydrolysates have the highest renin inhibitory activities (IC_{50} values 0.08–0.81 mg/mL) among food-derived protein hydrolysates similar to the synthetic renin inhibitor Z-Arg-Arg-Pro-Phe-His-Sta-Ile-His-Lys-(Boc)-OMe. This superior renin inhibitory activity (0.054 mM) is due to the presence of the tripeptide trytophan-tyrosine-threonine [WYT] from hempseed protein (Pihlanto & Mäkinen, 2017 and references therein). Simulated gastrointestinal digest—pepsin (pH 2, 37 °C, 1 hr) followed by pancreatin (pH 7.5, 37 °C, 2 hr) of hempseed protein isolate (75% protein) reduced the distribution of high molecular weight (>10 kDa) and increased (18 vs 40% post digestion) those of low molecular (<1 kDa) peptides. Moreover, 85% of the peptides post digestion had molecular weight between <1 and < 10 kDa compared to 79% of the hempseed protein isolate >10 kDa (Lin, Pangloli, Meng, & Dia, 2020). Hydrolysis increased antioxidant activity (DPPH and ORAC), reduced colon cancer (HT-29 and HCT-116), cells proliferation (27 and 45%, respectively), significantly reduced LPS-induced NO production, and suppressed proinflammatory cytokines (IL-6 and TNF-α) levels of hempseed protein isolate indicating the anti-inflammatory potential and anticancer health benefits of the hydrolysate (Lin et al., 2020).

Hydrolysates have been produced from hempseed protein isolate (HPI) obtained by basic extraction (0.25% w/v Na_2SO_3, pH 10.5, 1 hr) isoelectric precipitation (pH 4.3) of defatted hempseed flour (Rodriguez-Martin et al., 2019). These hydrolysates were generated by fast alcalase (0.3 AU/g protein, pH 8, 50 °C, 20 min) treatment [HPH20A – 30% DH]) and its combination with flavourzyme (0.3 AU/g protein, pH 8, 50 °C, 1 hr + 60 leucine amino peptidase unit [LAPU]/g protein flavourzyme, pH 7, 15 min) [HPH 60A + 15AF -60% DH]. The amino acid profiles were similar for defatted hempseed flour, hempseed protein isolate, and their hydrolysates HPH20A and HPH60A + 15 AF, except for the lowest amount of essential amino acids (24 vs 34% of total amino acids) of the alcalase hydrolysate (Table 9.9). These hempseed products were rich in negatively charged amino acids (~12% aspartic acid + [and] asparagine; ~18% glutamine and glutamic acid) and arginine (~13%) similar to those reported for edestin. The hempseed protein isolate and their hydrolysates HPH20A and HPH60A + 15 AF had similar cytotoxic effects on BV-2 microglia cell viability. Hydrolysis increased DPPH scavenging activity (36 and 40% for HPH20A and HPH60A + 15 AF, respectively, compared to HPI. Both hydrolysates reduce proinflammatory mediators (TNF-α, IL-β, and IL-6) and increase anti-inflammatory mediator (cytokine IL-10) in LPS-stimulated BV-2 microglial cells, indicating that they can abate neuroinflammation (Rodriguez-Martin et al., 2019). The suppression of tumor necrosis factor-α (TNF-α) and upregulation of IL-10 also enhance insulin sensitivity in fat and muscle tissues to maintain normal glucose levels improving normal fat metabolic function in many metabolic diseases. The anti-inflammatory effects of these hydrolysates were further confirmed to regenerate M2 macrophage that attenuate inflammation in LPS-stimulated human primary monocytes (CD14 monocytes isolated from healthy nonsmokers male [25–35 years] blood donors) (Rodriguez-Martin et al., 2020). Both hydrolysates reduce proinflammatory mediators (TNF-α, IL-β, and IL-6), increase anti-inflammatory mediators (IL-10 and IL-4), upregulate anti-inflammatory gene expression (CD200R and MRC1), and suppress proinflammatory marker genes (CCR7 and iNOS) and the mRNA transcriptional activity of chemotaxis genes (CCR2 and CCL2) (Rodriguez-Martin et al., 2020).

Hempseed protein isolate (isoelectric precipitated from *n*-hexane defatted hempseed flour) was hydrolyzed (1 hr) with alcalase (pH 8.5, 55 °C), flavourzyme (pH 7, 55 °C), protamax (pH 7, 55 °C), pepsin (pH 2, 37 °C), or trypsin (pH 8, 37 °C). The~48 kDa polypeptide of the 7S-like globulin component in HPI degrades upon alcalase and protamax hydrolysis, but is relatively resistant to flavourzyme resulting in low flavourzyme – HPH solubility (28%). In spite of its low solubility, flavourzyme – HPH (metal-chelating peptide) exhibited the highest zinc-binding capacity (89%), thereby improving zinc bioavailability and concurrently expanding hempseed protein utilization. Pepsin hydrolysate of hempseed protein isolate contains low-molecular-weight peptides (<21.5 kDa); 83% of the peptides (<680 Da) consists of 5 or 6 amino acid residues. The 7S-like globulin (~48 kDa) is readily hydrolyzed by alcalase, papain, protomax, pepsin, and trypsin, but resistant to flavourzyme. Pepsin, papain, and trypsin are most effective in hydrolyzing hempseed protein isolate with pepsin generating (yielding) the most complex mixture of short peptides. Enzyme hydrolysis improves protein solubility with papain and pepsin hydrolysates exhibiting high solubility (≥ 90%) and flavourzyme hydrolysates low solubility improvement (28%). Pepsin is most efficient in producing Zn^{2+}-HPH complexes (Wang & Xiong, 2018). These zinc-binding hyrolysates/peptides can control red blood cell production claimed to be potentially beneficial in treating patient suffering from defective erythropoiesis associated diseases such as myelodysplastic syndromes (MDS), sickle cell anemia, and leukemia (Bresnick & Tanimura, 2019).

Among six commercial protease enzymes (flavourzyme, protomax, neutrase, papain, trypsin, and alcalase), only alcalase (0.8% w/w, pH 8.5, 65 °C, 3 hr) hydrolysate (HSP_{27}) of pretreated (90 °C, pH 8.5, 30 min) hempseed protein exhibited considerable α-glucosidase inhibitory activity (α-GIA) (58%) at 27% degree of hydrolysis (DH). The purified alcalase hydrolysate ($HSP_{27-75-3-9}$ with higher α-GIA than acarbose) (IC_{50} value 0.02 vs 0.09 mg/mL) had high EAA (50 vs 29 [$HSP_{27-75-3-9}$ vs HSP_{27}]), proline (19 vs 4), and methionine (5.3 vs

0.3) and BCAA (especially Leu, Met, Pro and Arg) (34 vs 15% of total amino acids for HSP) (Table 9.9). Two hypoglycaemic peptides (Leu-Arg [LR; 287 Da] and Pro-Leu-Met-Leu-Pro [PLMLP; 568 Da]) inhibit α-glucosidase that modulate glucose homeostasis with the hydrophobic amino acids (Pro and Leu) contributing to the α-glucosidase inhibitory activity (α-GLA). The biological activity of these peptides is maintained during simulated gastrointestinal tract (pepsin and pepsin + pancreatin) digestion (Ren et al., 2016).

Pepsin (pH 2, 37 °C, 16 hr) hydrolyzed (HP) hempseed (cv. *Futura*) protein obtained from Tris–HCl/0.5 M NaCl buffer (pH 8) extract of defatted hempseed flour inhibited *in vitro* dipeptidyl peptidase IV (DPPIV) activity almost twice (by 32 vs 17.5%) as effectively as trypsin hydrolysates (HT) (4 mg/mL in 1 mM HCl, 37 °C, 16 hr). Cellular (Caco-2 cell) model also demonstrated higher DPPIV inhibitory activity of HP relative to HT hydrolysates (by 22.5 vs 15.5%). The DPPIV inhibitory activities of the hydrolysates increased two- to fivefolds *in situ* (cellular) and *ex vivo* (human serum) 37% after RADA16 hydrogel encapsulation indicating their potential antidiabetic potential and therapeutic target for glycemic control promoting insulin secretion, reducing glucagon release and/or metabolic diseases (Lammi, Bollati, Gelain, Arnoldi, & Pugliese, 2019).

The peptic (16% yield, 20% DH) and tryptic (25% yield, 47% DH) hydrolysates of hempseed (cv. *Futura*) protein obtained from Tris–HCl/0.5 M NaCl buffer (pH 8) extract of defatted hempseed flour dose-dependently inhibited 3-hydroxy-3-methyl-glutaryl-coenzyme A reductase (HMGCoAR) activity, a key enzyme in cholesterol metabolism. These hydrolysates abundant in proline and valine (and hydrophobic amino acids) have similar amino acid composition according to hierarchical clustering analysis (HCA), but apparently differ in diverse structure reflecting their different (HMGCoAR) activity (Aiello et al., 2017). The pepsin hydrolysates (16 hr, 37 °C) of the same hempseed (French cv *Futura*) protein were investigated for their hypocholesterolemic properties using human hepatic HepG2 cell model system to elucidate their cholesterol-lowering mechanism/function relative to the low-density lipoprotein receptor (LDLR)-sterol-responsive element binding protein-2 (SREBP-2) pathway (Zanoni, Aiello, Arnoldi, & Lammi, 2017). The low-molecular-weight (<3000 Da [< 3 kDa]) pepsin hydrolysate increased SREBP-2 protein by upregulating LDLR protein levels and activated 5′-adenosine monophosphate-activated protein kinase (AMPK) pathway to modulate 3-hydroxy-3-methylglutaryl coenzyme A reductase (HMGCoAR) regulation similar to statins. Although SREBP-2 mainly regulates genes involved in cholesterol biosynthesis and homeostasis, it can also target/treat cancers with dysfunctional cholesterol metabolism. Thus, hempseed low-molecular-weight peptic hydrolysate alone like statin or in combination with statin can potentially be used to treat different cancers, including breast, prostate, lung, and hepatocellular cancer as demonstrated for statins (Xue et al., 2020). The peptic and tryptic hydrolysates also effectively inhibit 3-hydroxymethyl-coenzyme A reductase activity, a key cholesterol metabolism enzyme and the main target of statins (a competitive inhibitor of this enzyme) (Aiello et al., 2017).

Dipeptidyl peptidase IV (DPP-IV) inhibitory drugs (e.g. gliptins) have been developed to treat type 2 diabetes since DPP-IV inhibitors maintain insulinotropic activity during the postprandial phase. In addition to glycemic control administration of DPP-IV inhibitors is clinically effective in CVD patients due to their anti-atherogenic effects (Martínez-Sánchez, Gabaldón-Hernández, & Montoro-García, 2020). Hempseed was compared with other rich dietary source of tryptophan (pea, rice, and soy) to generate tryptophan-hydrolysates and peptides with DPP-IV inhibitory activity (Nogonierma & FitzGerald, 2015). Prehydrated (60 g/L, 50 °C, 1 hr) commercial hempseed protein isolate (47% w/w protein) was hydrolyzed with three proteo–/peptidolytic enzymes; (2%, 50 °C, 1, 2 or 4 hr) protamex, corolase L10 or promod 144MG. The isolate and hydrolysate (4 hr) were also hydrolyzed with pepsin (37 °C, pH 2, 90 min), followed by corolase PP (pH 7.5, 150 min) to simulate *in vitro* gastrointestinal digestion (SGID). All hempseed protein hydrolysates exhibited high DPP-IV inhibitory potency ($IC_{50} < 2$ mg protein equivalent/mL after 4 hr hydrolysis). The amino group content of the DPP-IV hydrolysate (after 4 hr) was ~15 mg/g N. Hempseed hydrolysates (after 4 hr and SGID) had high proportion (50–60% and 65% for H-SGID and ≥ 80% [4 hr hydrolysate obtained following SGID], respectively) of low-molecular-weight peptides (<1 kDa); SGID increased the proportion of short peptides in the hydrolysates. The IC_{50} value was 4.5 mg dw hydrolysate/mL for hydrolysate from simulated *in vitro* gastrointestinal digestion of hempseed protein isolate, whereas nonhydrolyzed protein isolates did not inhibit DPP-IV (Nogonierma & FitzGerald, 2015). Moreover, the protamex hydrolysate (2 hr) from hempseed, pea, rice, and soy protein isolates had low IC_{50} values (< 3, ~1, ~1.5, and ≤ 1.5 mg/mL, respectively) within the range (2–3 mg/mL) of protein hydrolysate enriched in DPP-IV-inhibiting peptides (Boots, 2013; Nogonierma & FitzGerald, 2015). An edestin 2 derived peptide; DIFNPRGG [Asp-Ile-Phe-Asn-Pro-Arg-Gly-Gly] also inhibited DPP-IV and angiotensin-converting enzyme I (ACE), in addition to exerting antioxidant and glucose uptake stimulating activities (Aiello et al., 2017).

9.4.3 Peptides

Specific proteins (edestin 1, 2, and 3, albumin, 7S vicilin-like protein, ATP synthase β subunit, prenyltransferase 5, and chloroplast RF21) were identified by UHPLC-Q-TOF-MS/MS after tryptic digest of acetone-extracted protein from cold-pressed hempseed oil. A total of 119 peptides (11 uniquely specific to hempseed [consisting of 9–27 amino acids] among 9 other investigated oils) were identified from these eight proteins (Table 9.10). The storage protein edestin generated most (89%) of these peptides; 51 peptides from edestin 3, 29 from edestin 2, and 26 from edestin 1. Six and four peptides were derived from albumin and the 7S vicilin-like protein, respectively, whereas only one peptide each was obtained from ATP synthase β subunit, prenyltransferase 5, and chloroplast RF21 (Kotecka-Majchrzak et al., 2020).

Peptides[1] indicate the presence of leucine and proline that promote synergy with non-peptide antioxidants and other antioxidant amino acids (alanine, cysteine, isoleucine, except for P1 [GTLDLVSPRL]) (Etemadian et al., 2021). Peptides[2] with leucine [L] in the terminal position exhibits antioxidant activity (Etemadian et al., 2021). Peptides[3] contain the memory improving amino acids glycine and proline [GP], threonine and proline [TP], threonine and lysine [TL], or lysine and histidine [LH] (Tanaka et al., 2020). Peptides[4] contain glutamine–glutamine that confers anti-inflammatory activity (Avilés-Gaxiola

Table 9.10 Unique peptides in tryptic digest of cold-pressed hempseed oil.

	PAN	Protein	Peptide	Length	HPB%
1	SNQ45160.1	Edestin 3	AMPEDVIANSYQISR[7]	15	47
2			FYIAGNPHEDFPQSR[7]	15	47
3			GFSVNLIQEAFNVDSETAR	19	42
3			GVLGTLFPGCAETFEEAQVSVGGGR[1,3]	25	40
5			NAMYAPHYNINAHSHYAIR[6]	20	40
6			TAVYGDQNECQLNR[3]	14	21
7			VECEGGMIESWNPNHEQFQCAGVALLR[1,5]	27	41
8	SNQ45158.1	Edestin 3	FYIAGNPHQEFPQSMMTQQGR[4,7]	22	27
9			LEACEPDHR[1,2]	9	33
10			QGQALTVPQNFAVVK	15	53
11	SNQ45153.2	7S Vicilin-like protein	EILSSQQEGPIVYIPDSR[3,4]	18	39
Peptides in tryptic digest of specific proteins					
P1	CDP79023.1	Edestin 1	GTLDLVSPLR[1,3,5]	10	50
P2			YTIQQNGLHLPSYTNTPQLVYIVK[1,4]	24	38
P3	CDP79028.1	Edestin 2	LQVVDDNGR[2]	9	33
P4	SNQ45160.1	Edestin 3	LTIQPNGLHLPSYTNGPQLIHVIR[1,2,3]	24	46
P5	SNQ45153.2	7S Vicilin-like protein	GPELAAAFGLSLER	14	57
P6			NNYGWSIALDEFSYSPLR[1,5,6,7]	18	39
P7			TLFLPQYLDSELTIFIR[1,3]	17	53

Source: Adapted from Kotecka-Majchrzak et al. (Kotecka-Majchrzak et al., 2021; Kotecka-Majchrzak, Sumara, Fornal, & Montowska, 2020); PAN-Protein Accession number; HPB% percent hydrophobicity.

et al., 2020). Peptides[5] have the hydrophobic hypoglycemic amino acids proline-leucine or leucine-arginine (Ren et al., 2016). Peptide[6] with asparagine [N] at the N-terminal contributes to antioxidant activity (Lu et al., 2010). Peptides[7] contain the isoleucine-alanine [IA] motif indicative of DPP-IV inhibitory activity (Nogonierma & FitzGerald, 2019).

Peptide 3 – the triglycine [GGG] forms part of "taste" peptide that can impart/enhance umami threshold in fermented soy sauce (Zhu, Sun-Waterhouse, Chen, Cui, & Wang, 2021); high glycine is also associated with elevated postprandial satiety (Camacho-Barcia et al., 2021). Peptide 5 – has the [AMYA] motif that forms part of the peptide sequence imparting acetylcholinesterase (AChE) – inhibitory activity according to Malomo and Aluko (2019). Peptides 8 and 10 may be considered glutamine-rich peptides with immune boosting properties and together in combination with leucine and phenylalanine can benefit glucose disposal in all body tissues attributed to their insulinotrophic effects (Steijns & Ward, 2004).The nonapeptides 9 and P5 contain glutamic acid, arginine, and cysteine (constituting 44 and 33% of the total amino acid residues) known as blood pressure reducing agent (Samsamikor, Mackay, Mollard, & Aluko, 2020). Proline and lysine in these peptides also increase their ACE inhibitory activity. Peptides P1, P2, and P3 were also identified in hempseed proteins (Aiello et al., 2017). Peptides contained between 6 and 43 amino acids; those with the highest frequency (89%) consisting of 6–20 amino acids or (74%) of 6–16 amino acids. The highest number of peptides contained ≤16 amino acids; 9, 10, 10, 12, and 8% consisting of 8, 9, 13, 15, and 16 amino acids, respectively.

Cold-pressed [<40 °C] hempseed cake generated 41 unique oilseed species-specific peptides from 22 proteins after trypsin hydrolysis (1 : 40, w/w [trypsin/protein ratio], 37 °C, overnight) followed by Sep-Pak C_{16} RP-HPLC purification) (Kotecka-Majchrzak et al., 2021). Thirteen of these peptides were derived from edestin 1, 12 from edestin 2, 12 from edestin 3, and 4 from 7S vicilin-like protein. The eleven identified proteins with their number of unique peptides were edestin 1 [53], edestin 2 [50], edestin 3 [46], 7S vicilin-like protein [23], ATP synthase F1 subunit 1 (mitochondrion) [17], albumin [10], ribulose 1, 5-bisphosphate carboxylase/oxygenase large subunit (chloroplast) [11], phenylalanine ammonia lyase [3], ribulose 1,5-bisphosphate carboxylase/oxygenase large subunit partial (chloroplast) [3], cytochrome oxidase subunit 2 (mitochondrion) [1], and Ycf1 (chloroplast) [3] (Kotecka-Majchrzak et al., 2021). These peptides contained between 8 and 30 amino acids consisting predominantly (81%) of 8–20 amino acids. One of the edestin 1 peptide GI**LGV**TFPGCPETFEESQR contained the LGV amino acid motive known to be a strong ACE inhibitor (Orio et al., 2017).

Peptides occurring with the highest frequency (21, 17, and 14%) contained 14, 10, and 20 amino acids, respectively. Thirty four (11 each from edestin 1, 2 and 3 and one from the 7S vicilin-like protein) of the 42 unique hempseed cake specific peptides were also identified in acetone extracted protein from hempseed oil (Kotecka-Majchrzak et al., 2020). Fifteen peptides (8 and 7 from edestin 1 and 2, with 10–27 and 9–15 amino acids, respectively) have also been reported in hydrolysed hempseed protein concentrate previously (Aiello et al., 2017); eight of these peptides (5 edestin 1 and 3 edestin 2) were later confirmed (Lammi et al., 2019). Hempseed protein is highly susceptible to proteolysis yielding the highest amount of peptide with high hydrophobicity with pancreatin (43%, 47.5% degree of hydrolysis [DH]), followed by trypsin (25%, 46.6% DH) and pepsin (16%, 19.7% DH) (Aiello et al., 2017; Farinon et al., 2020). Hempseed (cv. *Futura*) protein concentrate (HPC) was prepared from defatted hempseed flour using Tris–HCl/0.5 M NaCl buffer (pH 8) then hydrolyzed (37 °C, 16 hr), with pepsin (pH 2), trypsin (4 mg/mL in 1 mM HCl,) or pancreatin (4 mg/mL in H_2O) (Aiello et al., 2017).

Generally, the peptic hydrolysate is the richest source of identified peptides (Table 9.11) derived from numerous proteins, with a small prevalence of the two isoforms of edestin (6% each from edestins 1 and 2) as well as DNA-directed RNA polymerase subunit β (6%) and protein Ycf2 (6%). However, the percentage of edestin derived peptides increases greatly in the tryptic hydrolysate (40 and 24% from edestins 1and 2, respectively) (Aiello et al., 2017). Pepsin readily (1 min) hydrolyses edestin (acidic and basic subunits) from hempseed protein isolate releasing oligopeptides (< 10 kDa) that decreases in molecular weight with increasing incubation time. Trypsin digestion after peptic hydrolysis (37 °C, 2 hr each) further reduces molecular weight distribution of the oligopeptides.

Thirteen[1] of the 22 peptides (59%) are considered hypocholesterolemic hempseed peptides according to Zanoni et al. (2017). However, only 6 ([1a]) peptides are described as glucose uptake stimulating peptide. Fifty and 55% of the peptides ([2] and [3]) are considered

Table 9.11 ACE-inhibitor peptides.

	PAN	Protein	Sequence	HPB%	Length
1	H9A1V5	Acyl-activating enzyme 3 (peps)	ALSKNSMVKKFNLSSIKYIG[1,2,3,5,6,8]	40	20
2	H9A1V5	Acyl-activating enzyme 3 (peps)	NDVKKFIAGQVASFKRL[5]	47.1	17
3	H9A8L2	Acyl-activating enzyme 13 (peps)	SPIGGGPEQLVMFVVLKNGY[2,4,9]	50	20
4	H9A8L2	Acyl-activating enzyme 13 (peps)	GAVLNIAECCLLPTSYPRKDD[1a,2,8,9]	42.9	21
5	H9A8L3	Acyl-activating enzyme 14 (peps)	IPWTQLSPIRCAAESWAHM[1,2,7,8,9]	57.9	19
6	A6P6W0	Cannabidolic acid synthase-like 1	RIWGEKYFGKNFNRLVKVK[1a,2,3,8]	29.4	19
7	A6P6W0	Cannabidolic acid synthase-like 1	GEKYFGKNFNRLVKVKT[1a,2,3]	36.8	17
8	C6KI62	Chalcone synthase-like protein 1	PIGISDWNSLFWIVHP[3,9]	62.5	16
9	A0A0C5B2L0	Cytochrome c biogenesis protein CcsA	SSEKGMIATFCCITGLL	41.2	17
10	A0A0C5B2I8	Cytochrome f	ATGRIVCANCHLANKPVDIEVP[7,9]	50	22
11	A0A088MFF4	Delta 12 desaturase	PGRVLSLFVTLTLGWPLY[1a,2,4,6,8,9]	61.1	18
12	A0A090CXP8	Edestin 2	DIFNPRGG[3]	37.5	8
13	E5DKP2	MatR	YSIQKVFSAGRLVGGEKGPYSV[2,4,6,8]	36.4	22
14	A0A0C5APZ1	NAD(P)H-quinone	LPDTHGEAHYSTCMLLAGILLKMG[1a,2,7]	45.8	24
15	A0A0C5AUJ6	NADH-plastoquinone oxidoreductase subunit 5	GPTPISALIHAATM[1a, 4,6,9]	64.3	14
16	E5DL82	NADH-ubiquinone oxidoreductase chain 5	PRNSWISCNMRLNAITL[3]	47.1	17
17	E5DL82	NADH-ubiquinone oxidoreductase chain 5	LICILLFIGAVGKS[3,4]	64.3	14
18	A0A0C5ARZ4	Photosystem I P700 chlorophyll a apoprotein A1	SHLNWVCIFLGFHSFGLYI[1,2,3,8]	52.6	19
19	A0A0C5ARZ4	Photosystem I P700 chlorophyll a apoprotein A1	IPDKANLGFRFP[1,3,5,8,9]	58.3	12
20	A0A0C5ARQ8	RNA polymerase subunit beta	QIQFEGFCRF[1,3,8]	40	10
21	A0A0E3TIL1	THCA synthase	NPRENFLKCFSKHIPNNVA[1,2,3,8,9]	42.1	19
22	A7IZZ1	TPS1-(−)-limonene synthase chloroplastic	VRFEPQFSYFRI[1,3]	50	12

Source: Adapted from Aiello et al. (2017); PAN-Protein Accession number; HPB% percent hydrophobicity.

antioxidative and DPP-IV inhibitor peptides, respectively, whereas 5 and 3 peptides ([4] and [5]) are propyl endopeptidase and renin inhibitor, respectively. Four and three peptides ([6] and [7]) regulate the gastric (stomach) mucosal membrane activity and ubiquitin proteolysis activity (Aiello et al., 2017). Peptides ([8]) are DPP-IV inhibitors according to Lammi et al. (2019). 2P (two proline) peptides ([9]) are highly stable and can therefore maintain their bioactivities under various conditions (Graham et al., 2018).

The peptic and tryptic digestion/hydrolysis pattern is similar for hempseed protein isolate (HPI) obtained from commercial defatted hempseed flour/meal [prepared from dehulled, SCO_2 defatted hempseed], 11S and 7S proteins [extracted from HPI] (Wang, Tang, Yang, & Gao, 2008). The 22 ACE inhibitor peptides (Table 9.11) are multifunctional; 11 of those peptides (6, 7, 8, 12, 16, 17, 18, 19, 20, 21, and 22) also inhibit DPP-IV; 11 other peptides (3, 4, 6, 7, 8, 9, 10, 11, 13, 15, and17) stimulate glucose uptake; 5 inhibit propyl endopeptidase (3, 11, 13, 15, and 17), and 3 (peptides 1, 2, and 5) inhibit renin (Aiello et al., 2017; Lammi et al., 2019; Zanoni et al., 2017).

Peptide 2 (NDVKKFIAGQVASFKRL) with an N at the N-terminal contributes to antioxidant activity (Lu et al., 2010); the peptide also has leucine at the terminal position as well as two amino acids (alanine) indicating potential antioxidant activity (Etemadian et al., 2021). Peptide 3 has the satiating amino acids leucine, isoleucine, lysine, and tyrosine (LIKY) that reduce the feeling of hunger (Volk et al., 2020), and the triglycine [GGG] forms part of "taste" peptide that can impart/enhance umami threshold in fermented soy sauce (Zhu et al., 2021). Peptide 4 has the satiating amino acids leucine, threonine, and tyrosine [LTY] and therefore reduce hunger (Volk et al., 2020); it also contains cysteine, proline, aspartic acid, and histidine [CPDH] considered to be cytotoxic amino acids (Avilés-Gaxiola et al., 2020); and the memory improving amino acids tyrosine and proline [YP] (Tanaka et al., 2020). Peptides 6 and 18 contain the neuroantioxidant amino acids tyrosine, phenylalanine, and tryptophan [YFW] (Parthasarathy et al., 2018). Peptide 8 contains proline, aspartic acid, and histidine [PDH] considered cytotoxic amino acids (Avilés-Gaxiola et al., 2020). Peptides 9 and 16 have leucine in the terminal position and contain other antioxidant amino acids (alanine, cysteine, and isoleucine) (Etemadian et al., 2021). Peptide 11 (PGRVLSFVTLTTLGWPLY) has the hydrophobic amino acids Proline and Leucine that contribute to α-glucosidase (α-GLA) inhibitory activity (Ren et al., 2016); moreover, the two proline residues 2P enable the peptide to maintain its biological activity during simulated gastrointestinal tract (pepsin and pancreatin) digestion; the peptide also contains tyrosine, phenylalanine, and tryptophan [YFW] considered neuroantioxidant amino acids (Parthasarathy et al., 2018); leucine, threonine, and tyrosine [LTY] are satiating amino acids and therefore reduce hunger (Volk et al., 2020); it also has Leucine and Proline that promote synergy with nonpeptide antioxidants (Etemadian et al., 2021). Peptide 13 with valine at the C-terminal is an effective radical scavenger (Lu et al., 2010) and glycine-proline [GP] motif with memory improving potential (Tanaka et al., 2020). Peptides 14 (LPDTHGEAHYSTCMLLAGILLKMG) and 17 (LICILLFIGAVGKS) contain leucine and proline that promote synergy with nonpeptide antioxidants and other antioxidant amino acids (alanine, cysteine, and isoleucine) (Etemadian et al., 2021). Peptide 15 (GPTPISALIHAATM) contains the memory improving amino acids glycine and proline [GP] and threonine and proline [TP] (Tanaka et al., 2020); it may also exert antioxidant activity due to the presence of alanine and histidine with leucine and proline promoting synergy with nonpeptide antioxidants (Etemadian et al., 2021). Peptide 21 with asparagine [N] at the N-terminal contributes to antioxidant activity (Lu et al., 2010).

Peptic hydrolysate (pH 2, 37 °C, 2 hr) of alkali extracted acid precipitated hempseed protein isolate (HPI, 84% protein) obtained from commercial hempseed protein meal (HPM, 37% protein) was separated into 8 fractions (F1 —F8) by preparative reverse-phase-HPLC (RP-HPLC) (Malomo & Aluko, 2019). Only five fractions (F1, F3, F4, F7, and F8) inhibited acetylcholinesterase (AChE) with F7 (most hydrophobic) exerting the highest (98%) *in vitro Electrophorous electricus* (electric eel) *ee*AChE-inhibitory activity. Generally, peptide fractions normally exhibit superior inhibitory properties than unfractionated hydrolysates. F7 (<1200 Da) consisted predominantly of small peptides (3–8 amino acids) with one peptide containing 11 amino acids. AChE inhibitory peptides of hempseed derived from edestin (10, 4, and 5 peptides from edestin 1, 2, and 3, respectively) were within the 3–11 amino acid size range. However, almost half (47%) of the 19 AChE inhibitory peptides had 3 amino acids, whereas the frequency of the peptide length was 37% for 4 amino acids. The F7 peptides also exerted higher enzyme inhibitory potency of human *hm*AChE (kinetic constants [reduced V_{max}, CE, and K_i]) compared to unfractionated hydrolysate demonstrating its potential against neurodegenerative diseases (Malomo & Aluko, 2019). In animals, AChE modulates the major neurotransmitter acetylcholine (ACh) that degrades excessively at high AChE activity and can disrupt neurotransmission contributing to Alzheimer's disease (AD) pathogenesis.

Ninety peptides belonging to 33 hempseed proteins were identified in the peptic hydrolysate (37 °C, 16 hr) of hempseed protein isolate; the highest number (13) of peptides originated from edestin. These heterogeneous peptides consisted of 7–28 amino acids (860–3210 Da) with the most abundant (41%) containing predominantly basic residues and acid and neutral (32 and 27%, respectively) residues (Zanoni et al., 2017). These low-molecular-weight hempseed peptides (<3000 Da) increase SREBP2 protein by upregulating LDLR protein levels and activate AMPK pathway to modulate HMGCoAR regulation. Hempseed peptides exert statin-like activity by inhibiting HMGCoAR, the rate-limiting enzyme that leads to cholesterol biosynthesis and regulating transcription factors improving cellular LDL uptake (Bhullar & Wu, 2020). Pepsin hydrolyzed hempseed peptide mixtures dose-dependently inhibited (*in vitro*, direct enzyme binding) HMGCoAR activity resulting in significant endogenous cholesterol level reduction during hypercholesterolemia. The peptides also inactivated HMGCoAR activity in HepG2 hepatic cells via Ser-872 phosphorylation by phosphor-5′-adenosine monophosphate-activated protein kinase. SREBP2 is a key transcriptional factor in lipid metabolism controlling gene expression involved in cholesterol metabolism (Boachie, Yao, & Udenigwe, 2018). Another hempseed bioactive peptide resembled the casein glycomacropeptide-derived peptide (IPPKKNQDKTE) that reduces high glucose-induced insulin resistance and accompanying oxidative stress. It (the hempseed peptide) inhibited adipocyte lipid accumulation and increased glucose uptake in cells and mice (Bhullar & Wu, 2020).

9.4.4 Simulated Gastrointestinal Digestion

A simulated gastrointestinal digestion (GID) model was used with purified porcine intestinal brush border membrane (BBM) vesicles to demonstrate the molecular basis of digestibility for hempseed protein isolated from defatted hempseed (cv *Carmagnola*) meal.

This *in vitro* gastrointestinal digest (37 °C, with simulated salivary fluid (SSF [human salivary amylase], 2 min), simulated gastric fluid (SGF, pH 2.7, porcine pepsin, 2 hr), and simulated intestinal fluid (SIF [pancreatic lipase + trypsin + α-chymotrypsin + pancreatic α-amylase], 2 hr) released 41 peptides, 12 of which survived the action of BBM peptidases. Six proteins (edestin 1, 2, and 3, 7S vicilin-like protein, albumin, and ATP synthase subunit α) were identified by shotgun analysis of hempseed protein isolate (HPI). Thirty additional proteins were identified including metabolic enzymes, protease inhibitors, actin, and other proteins homolog to barley, maize, wheat, and other *Triticum* or cereal species. Peptides resistant to peptidase digestion contained acidic amino acid motives (LLR) corresponding to ACE-inhibitor sequence. DIFNPRGG (peptide 12 Table 9.11) was one of the 13 peptides identified after BBM gastrointestinal digest of HPI enabling it to be absorbed in the bloodstream (Mamone, Picariello, Ramondo, Nicolai, & Ferranti, 2019). *In vitro* gastrointestinal digested peptides of hempseed protein isolates were within the 6–32 amino acid size range. However, ⅔ (67%) of the 51 digested peptides had 6–14 amino acids with 10 amino acid containing peptides having the highest (22%) frequency. After BBM, *in vitro* gastrointestinal digested peptides of hempseed protein isolate were within the 8–28 amino acid size range (Mamone et al., 2019). However, almost ⅔ (62%) of the 13 digested peptides had 9–14 amino acids with 14 amino acid containing peptides having the highest (23%) frequency.

Eight (RP-HPLC) peptide fractions were obtained from simulated gastrointestinal tract (GIT) digest (4% w/v, protein basis 37 °C; pepsin, pH 2, 2 hr + pancreatin, pH 7.5, 4 hr) of hempseed protein isolate (Girgih, Udenigwe, & Aluko, 2013). Two of these fractions (F2 and F6) contained bioactive peptides with high superoxide, hydroxyl, ferric reducing antioxidant power (FRAP), and metal chelating antioxidant properties. Both fractions were rich in hydrophobic amino acids, predominantly isoleucine, leucine, and tyrosine in F2 and proline, cysteine, and methionine in F6 relative to other fractions (Girgih, Udenigwe, & Aluko, 2013). Hydrophobic amino acids act as antioxidants by increasing peptide solubility in nonpolar environments facilitating free radical interaction and terminating their reactivity. The presence of some amino acids (histidine, tyrosine, methionine, lysine, tryptophan, and proline) increases the antioxidant potency of most food derived peptides. Moreover, histidine, isoleucine, leucine, lysine, methionine, tyrosine, and tryptophan can contribute to the ferric reducing antioxidant power (FRAP) of protein hydrolysates (Girgih, Udenigwe, & Aluko, 2011). In an earlier study, peptides (<1, 1–3, 3–5, and 5–10 kDa) obtained by membrane ultrafiltration of the same simulated gastrointestinal tract hempseed protein digest exhibited significantly higher *in vitro* antioxidant properties (DPPH and hydroxyl radical scavenging activities and ferric reducing power) than the unfractionated hydrolysate (HPH). Moreover, HPH and peptide fractions >3 kDa exerted significantly higher metal chelation activity than glutathione (cellular physiological antioxidant peptide) (Girgih, Udenigwe, & Aluko, 2011). Peptides derived from the simulated gastrointestinal tract digest of hempseed protein isolate may account for the reduction in angiotensin-converting enzyme (ACE) activity and plasma renin resulting in reduced systolic blood pressure in spontaneous hypertensive rats (Girgih, Alashi, He, Malomo, & Aluko, 2014). Peptides with Proline-Serine-Leucine-Proline-Alanine (PSLPA) and Tryptophan-Valine-Tyrosine-Tyrosine (WVYY) motives exert the highest systemic blood pressure reduction. Peptides with the highest combined ACE and renin inhibition rates and highest systolic blood pressure reduction are Serine-Valine-Tyrosine-Threonine (SVYT) and Isoleucine-Proline-Alanine–Glycine-Valine (IPAGV). Hydrophobic amino acids play important roles in bioactive peptides; they can contribute to the radical-scavenging properties of the peptides and expected to favor oxidation inhibition (when the hydrophobic amino acids are distributed on both sides of the peptide chains) (Lu et al., 2010). Valine at the C-terminal can act as effective radical scavengers and Asparagine [N] at the N-terminal may also contribute to the peptide's antioxidant activity (Table 9.11). Dipeptides with Pro, such as Gly-Pro and Tyr-Pro [YP], can be transported intact across the blood–brain barrier (BBB) into the parenchyma of peptide-perfused mouse brain (Tanaka et al., 2020).

Simulated gastrointestinal (pepsin [2 hr] followed by pancreatin [4 hr]) hydrolyzed hempseed protein inhibited renin *in vitro* by binding to its active site (Aluko, 2019). The hydrolyzed hempseed protein (HPH) exhibited higher renin inhibitory activity resulting in significant SBP reduction than the ultrafiltered fractions (<1 and 1–3 kDa) (IC_{50} values 0.8 vs 2.5 and 1.9 mg/mL [0.05 vs 0.06 and 0.09 mM]). The hempseed peptides reduced renin production attenuating thereby SBP in spontaneously hypertensive rats (SHR). The tri-, quadra-, and penta-peptides (WYT, SVYT, and IPAGV) were the most active *in vitro* renin inhibitors (77, 86, and 74%, respectively) with uncompetitive inhibition pattern. However, IPAGV was most effective in reducing SBP in SHR than WYT; IPAGV and SVYT also produced longer lasting effect than the tripeptide WYT. Apparently, WYT was weakly bound to renin during *in vivo* interactions or highly susceptible to clearance by plasma peptidases. The tetra- and penta-peptides (SVYT and IPAGV) were resistant to plasma clearance thereby maintaining SBP-lowering effects (Aluko, 2019).

Two peptides (Asn-His-Ala-Val [NHAV] and His-Val-Arg-Glu-Thr-Ala-Leu-Val [HVRETALV]) exhibit effective antioxidant activity *in vitro* (FRAP, DPPH, super oxide, and hydroxyl radical scavenging activities) and on hydrogen peroxide-induced cell apoptosis in pheochromocytoma 12 (PC12 cells—classical *in vitro* neuronal model). Hence, these peptides are considered to exhibit neuroprotective effects. These peptides were purified from hydrolyzed (Alcalase, 50 : 1 w/w, pH 9.4, 50 °C, 1 hr) hempseed protein isolate (89% protein) obtained by base (alkali) extraction (pH 9, 40 °C, 70 min)/isoelectric precipitation (pH 5) from commercial defatted hempseed protein meal (Lu et al., 2010). Acid hydrolysis (6 M HCl, 110 °C, 6 hr) of hempseed protein isolate (HPI, 94% protein) obtained by alkali extraction/acid precipitation of defatted hempseed (cv. *Futura*, 35.8% protein) generated four fractions (PHPH1-PHPH4). ACE-inhibitory activity of HPH and PHPH3 (the most abundant fraction) was 45 and 85%, respectively. Peptides constituting the PHPH3 fraction were Glycine-Valine-Leucine-Tyrosine (GVLY), Leucine-Glycine-Valine (LGV), Arginine-Valine-Arginine (RVR), and Isoleucine-Glutamic acid-Glutamic acid (IEE), all from edestin sequence. The peptides GVLY, LGV, and RVR showed high ACE inhibition (>93%), although GVLY was the most active (IC_{50} 16 vs 145 and 526 μM for LGV and RVR, respectively). Both GVLY and LGV peptides were hydrophobic (1.78 and 0.94 Log P values). All active peptides contained a valine (hydrophobic branched chain amino acid) residue (Orio et al., 2017).

Edestin-derived bioactive peptides are produced by fermenting commercial hempseed protein extract with proprietary microbial (probiotic containing *Lactobacillus* and *Bifidobacterium* as well as kefir grains) proteases (Ruggiero & Pacini, 2020). This fermented hempseed protein extract (0.6 μg/well) significantly ($p < 0.05$) inhibits human colon HCT116 cell (bearing mutation of the KRAS gene) proliferation. This study demonstrates that edestin fermentation generates peptides exhibiting biological activity in human cell cultures. Edestin 1(A0A090DLH8- [YVLRGRARVQVVNHMGQKC]) has striking functional resemblance (similarity) to the human vitamin D-binding protein (Gc-protein; P02774), the precursor of vitamin D-binding protein-derived Macrophage Activating Factor (DBP-MAF). Apparently, fermentation increased acidity activating proteases resulting in the bioactive peptide VLANAFQ release from edestin identical to the DBP-MAF peptide TPTELAKLVNKRSE (Ruggiero & Pacini, 2020). Sourdough (*Saccharomyces cerevisiae* LBS) fermentation of commercial organic hempseed flour (31% protein, 49% fiber, 8% fat and 3% carbohydrates) also induces fast acidification with high acetic acid production in bread (Nissen, Bordoni, & Gianotti, 2020). Such fermentation can potentially release higher amount of these DBP-MAF bioactive peptide. This fermentation-derived peptide like DBP-MAF can potentially be considered as immunotherapy for prostate, metastatic breast, and colon cancer patients due to its inhibition of cancer cell proliferation, inhibition of tumor-induced angiogenesis, and tumoricidal macrophages activation (Pacini, Punzi, Morucci, Gulisano, & Ruggiero, 2012).

9.5 Carbohydrates

Carbohydrates are the least investigated hempseed macro components, although fiber and specialty carbohydrates are the 4th important nutraceutical ingredient type after probiotics, proteins, and amino acids according to global market analysis (Markets and Markets, 2021). Carbohydrates constitute 20–30% of hempseed with over twice that amount found in the hulls (697 vs 344 g/kg dry weight) and consist almost entirely of dietary fiber (98%). The rest is composed of sugars, fructose, glucose, and sucrose (5, 4 and 11 g/kg dry weight, respectively) (Mattila, Mäkinen, et al., 2018). Fiber content among hempseed cultivars ranges between 30.3 and 38.8% (Leonard, Zhang, Ying, & Fang, 2019). Components of dietary fiber, neutral detergent fiber (NDF), acid detergent fiber (ADF), and cellulose varied significantly ($P < 0.005$), whereas differences were insignificant for hemicellulose (63–94 g/kg) and lignin (105–118 g/kg) among ten widely cultivated industrial Canadian hemp cultivars (Table 9.1) (Vonapartis et al., 2015). Moreover, most of the NDF (cellulose, hemicellulose, and lignin) is present in the hempseed hull and influences seed protein digestibility.

9.5.1 Dietary Fiber

The ratio of water soluble to insoluble fiber is 20:80 in hempseed fiber mostly located in the hull (Callaway, 2004). Hempseed hull has twice the neutral detergent and acid detergent fibers (65 vs 32 and 50 vs 24% as is basis) than whole hempseed cultivars (one of four cultivars: *Finola*, *Crag*, and early maturing *USO-14* and *USO-31*) grown in Western Canada (House et al., 2010). The NDF and ADF contents of the seed meal derived from these hempseed cultivars are similar to those of the whole seed; the lowest contents occurred in dehulled seed (7.8 and 3.3% as is basis). Interestingly, the proportion of ADF/NDF is similar (~0.7) for the whole seed, seed meal, and hempseed hull and lower (0.4) for dehulled seed (House et al., 2010).

Table 9.12 Carbohydrate composition (g/kg dry matter) of hempseed products.

	Hempseed cake		
TC-Total carbohydrate	537.6[a]	430–667[b]	
SC-structural carbohydrate	464.1	392–562[b]	
NSC-nonstructural carb	73.5		
NDF	518.8	447–478[c]	
ADF	395.5	304–421	
ADL (acid detergent lignin)	63.1	117	
Starch	56.2	15–56	
Carb fractions (g/kg TC)			
CA-Sugars & Soluble fraction	31.2		
CB1-Starch & nonstarch polysaccharides	105.7		
CB2-Available cell wall fraction	581.9		
CC-Unavailable cell wall	281.2		
	Whole seed[d]	Seed meal[d]	HS[e]
Total dietary fiber	276–350	426	54
Soluble fiber	29–54	164	15.8
Nonsoluble fiber	222–309	262	32.6

Source: Adapted from: [a]Serrapica et al. (2019); [b]Mirpoor, Giosafatto, and Porta (2021); [c]Stastnik, Pavlata, and Mrkvicova (2020); [d]Pihlanto et al. (2020); [e]HS dehulled seed, Pastell, Putkonen, and Rita (2019).

Hempseed hulls contain on average 32% cellulose and 6–17% (w/w) xylan as the predominant noncellulosic polysaccharides (Schultz et al., 2020). Cellulose, lignin, and hemicellulose account for about 46, 31, and 23% of the total hempseed insoluble fiber, respectively (Vonapartis et al., 2015). Hemp hearts, also known as dehulled seed, generally contain lower dietary fiber (Table 9.12) than whole seed, but highly digestible (Schultz et al., 2020). Water-insoluble dietary fiber (IDF), water soluble polysaccharides (SDFP), and oligosaccharides (SDFS) represent 86, 12, and 2% of total dietary fiber of whole hempseed (350 g/kg fresh weight) and 60, 29, and 9% of total dietary fiber of dehulled (peeled) hempseed (54 g/kg fresh weight) (Pastell et al., 2019). Fat removal from hempseed results in hemp defatted flour containing 40.9% fiber (Rodriguez-Martin et al., 2019). However, fiber content (3.4%) is considerably low in high protein (86%) isolate obtained by isoelectric precipitation of defatted hemp meal (Das, 2015).

Hempseed cake is particularly rich in fiber and total carbohydrates relative to other oilseed cakes (soybean, rapeseed, cotton, groundnut, sesame, black cumin, and cardoon) (Mirpoor et al., 2021). The best and complete carbohydrate and dietary fiber composition has recently been reported for a South Italian commercial residual hempseed cake after oil extraction (screw press, 8 mm diameter, 22 rpm rotation speed, 70 °C from preheated seeds 50 °C) (Table 9.12) (Serrapica et al., 2019). The carbohydrate is fractionated and characterized according to the Cornell Net Carbohydrate and Protein System (CNCPS) into water-soluble carbohydrates ([CA] – mainly sugars and short oligosaccharides), medium degradable carbohydrates ([CB1] – starch and nonstarch polysaccharides), slow degradable carbohydrates ([CB2] – digestible cell walls), and undegradable and indigestible structural carbohydrates ([CC] – cell walls and lignin). The cell walls (available and unavailable [CB2 and CC]) were the most predominant carbohydrate fractions (582 and 281 g/kg total carbohydrates [TC], respectively) (Table 9.12). Hempseed cake has lower CA (31 vs 250 g/kg TC), and CC (281 vs 431 g/kg TC), but higher CB2 (582 vs 212 g/kg TC) than sunflower cake. Hempseed cake has similar TC (538 vs 595 g/kg dm), higher SC (464 vs 383 g/kg dm) and lower NSC (73.5 vs 212 g/kg dm) than sunflower cake (Serrapica et al., 2019).

Hempseed fiber provides 25 and 21 MJ/kg of gross energy from hempseed and hempseed cake, respectively. Furthermore, the fiber is microbially degraded in chicken caeca with increasing hempseed feed fiber content (10, 20, 30% hempseed cake); it releases short-chain fatty acids that may be absorbed and constitute approximately 8% of the total energy used in chicken (Kalmendal, 2008). Actually, hempseed cake or hempseed meal (up to 200 g/kg) does not affect performance parameters, feed consumption, feed conversion, egg quality, or mortality in chicken. Other studies report increases in tibia strength of broiler chicken fed hempseed up to 50 g/kg in diets for 35 days (Stastnik et al., 2020). However, rumen digestibility is lower for *in vitro* neutral detergent fiber of hempseed cake compared to sunflower cake (16 vs 35%) (Serrapica et al., 2019).

The fiber fraction of hempseed (~66% of the dietary fiber in the hempseed supplemented diet), although mostly insoluble, is apparently a good substrate for the gut microbiota. It considerably increases the cecal SCFA production that regulates energy metabolism and several metabolic functions in genetically obese male Zucker rats (Opyd, Jurgoński, Fotschki, & Juśkiewicz, 2020). Hempseed polysaccharide (42.1 kDa) consists primarily of xylose, arabinose, and galacturonic acid (35 : 25 : 22 M ratios) with smaller amounts (2–7 M ratio) of mannose, rhamnose, glucuronic acid, and galactose. The polysaccharide protects against cyclophosphamide-induced oxidative damage by enhancing antioxidant enzymes (superoxide dismutase [SOD], catalase [CAT], and glutathione peroxidase [GSH-Px]) through overexpression of their mRNA levels and reducing lipid peroxidation (malonaldehyde [MDA]) in mice intestine (Xue et al., 2020). The high arabinose content (60% molar ratio) of hempseed polysaccharide can positively benefit gut microbiota and attenuate metabolic disorders in overweight individuals by modulating colon fermentation and increasing short-chain fatty acid production in the gut (Kjølbæk et al., 2020).

Hempseed flour or protein (490 or 139 g/kg fiber) supplementation (10 or 20% starch replacement) increases fiber and dietary fiber (15–61 and 29–90 g/kg), primarily insoluble dietary fiber [63–354%] of starch-based gluten-free bread (Korus, Witczak, Ziobro, & Juszczak, 2017). However, during probiotic hempseed formulation, fiber is inversely associated with bacterial (*Bifidobacterium bifidum* & *Lactobacillus.plantarum*) growth and alcohol (cyclohexen-2-ol) production, but positively correlated with decanal and octanoic acid production (Nissen, Demircan, Taneyo-Saa, & Gianotti, 2019).

9.6 Phenolics

Total polyphenol content varied significantly (5.9–10.6 g/kg seed) among 20 hemp cultivars of different geographical origins grown in pots under the same conditions for 1 year (Galasso et al., 2016). Similar variation (46–80 mg gallic acid equivalent [GAE]/g of alcoholic extract from defatted seeds) occurred in total phenolic content among three Italian hempseed cultivars (*Carmagnola, Futura, and Felina*) and between two growing sites (*Treviglio* and *Cavriana*) in the *Lombardia* region in 3 years. Variation in flavonoids content (54–109 mg rutin equivalent [RE]/mg extract) was limited (Lesma et al., 2014). Seed of the popular Italian hemp cultivar "*Futura 75*" had 2.33 mg GAE/g dry weight [dw] of total polyphenols and 4.4 mg quercetin equivalent [QE]/g dw of total flavonoids exhibiting high antioxidant activity (134 μmol Trolox equivalent [TE]/g dw), antiradical activity (40% DPPH inhibition), *ex vivo* cellular antioxidant activity in red blood cells [CAA-RBC], and antihemolytic effect (82 CAA units and 52% hemolysis, respectively) (Frassinetti et al., 2018). Aqueous ethanol (80%, 1 : 10 w/v) extract of "*Futura 75*" exhibited selective antimicrobial activity against the pathogenic biofilm-producing enterobacteria and methicillin-resistant (MRSA) *Staphylococcus aureus* similar to vancomycin without affecting growth of the probiotic *Bifidobacteria* and *Lactobacillus* strains. The higher ratio of beneficial bacteria can promote a healthier gut microbiota by releasing bioactive compounds preventing pathogen adhesion to intestinal mucosa and competition with pathogenic strains for nutrients. This antimicrobial activity was presumably associated with the high polyphenol content of hempseeds, particularly caffeoyltyramine and cannabisin (Frassinetti, Gabriele, Moccia, Longo, & Di Gioia, 2020). The extract dose-dependently reduced *S. aureus* viability with a fivefold increase in dead cells after extract (1 mg/mL) exposure apparently due partly to membrane damage mechanisms and/or multiple antibacterial activity mechanisms (cell wall inhibition, DNA, lipid, and/or protein synthesis).

Total polyphenols were 767, 744, and 21 mg GAE/kg determined with Folin–Ciocalteu method and 1540, 1078, and 23.5 mg caffeic acid equivalent [CAE]/kg based on RP-HPLC semi-quantitation for seed, flour, and oil of hempseed cultivar "*Fedora*" (among the most popular EU hemp varieties), respectively (Siano et al., 2019). Aqueous (80%) methanol extracted polar compounds (922, 1709, and 192 mg GAE/kg total polyphenols) from these same seed, flour, and oil exhibited *in vitro* antioxidant activity (74, 67, and 22% DPPH inhibition, respectively). The typical flavonoid compounds, cannflavins A, B, and C were detected in the polar oil (aqueous 80% methanol) extract (Moccia et al., 2019). This extract also contained hydroxycinnamic acid derivatives and several phenylpropionamides (*p*-coumaroyltyramine, feruloyltyramine, and caffeoyloctopamine), cytotoxic (~90% cell death) to Caco-2, and HT-29 (human colorectal adenocarcinoma) cells (Moccia, Maria Russo, et al., 2019).

Using reversed phase HPLC, the dominant phenolic acids in hemp seed oil were *p*-hydroxybenzoic acid (60 μg/kg), followed by sinapic acid (30 μg/kg), vanillic acids (20 μg/kg), and *p*-coumaric acid (20 μg/kg) (Siger, Nogala-Kalucka, & Lampart-Szczapa, 2008). Polyphenols in hempseed oil (cv "*USO31*") investigated by UHPLC-HRMS and MS/MS (Faugno et al., 2019) showed the presence of dixyroxybenzoic acid, tyrosol hexoside, caffeoyl tartaric acid isomers, coumaric acid, dihydroxycoumarin hexoside [esculin], *N*-caffeoyltyramine, *N*-feruloyltyramine isomers, quercetin glycoside, dihydrokaemferol hexoside, kaempferol glycoside and isorhamnetin glycoside, quercetin derivatives (quercetin 3-O-[6′-O-deoxyhexosyl]-hexoside [rutin], quercetin hexoside, quercetin-3-O-hexuronide,quercetin-3-O-[6′-acetyl]- hexoside), flavonol kaempferol derivatives (kaempferol hexoside), and isorhamnetin-3-O-rutinoside. Absence of preseeding and low density (60 plants m^{-2}) ensured the highest phenols and polyphenols production. Seeds harvested in high-density fields with nonprefertilized soil and the use of mild temperature (40 °C) favored maximum flavonoids recovery. Several of these compounds are involved in plant defense and can be considered potential nutraceutical ingredients that can exert antifungal, antioxidant, anti-inflammatory, and anticancer properties. Caffeoyl tyramine constituted 35% of the phenolic component, whereas quercetin-*O*-hexuronide and kaempferol derivatives accounted for 51 and 20% of the flavonoid content (Faugno et al., 2019). Flavonoid content constituted only 2.8% of polyphenols (as total peak area) represented by seven flavonol glycosides (quercetin-7-*O*-acetyldeoxyhexose, quercetin-*O*-deoxyhexoside, quercetin pentoside, kaempferol *O*-deoxyhexoside 1, kaempferol *O*-deoxyhexoside 2, kaempferol pentoside 1, and kaempferol pentoside 2) in hempseed and its cold-pressed oil derived product (Faugno et al., 2019; Nigro et al., 2020). However, flavonoids accounted for almost all (98%) of the total phenolics of "*Finola*" oil (fresh weight) predominantly featuring flavanones, flavanols, and flavonols (Table 9.13). The oil contained 27.8 g quercetin equivalent of total flavonoid/kg fresh weight (Smeriglio et al., 2016).

Commercial hempseed hull has twice the total phenolic (1970 vs 960 mg GAE/kg dw; Folin-Ciocalteau method) and higher condensed tannin (procyanidins) contents (1440 vs 1050 mg/kg dw) than whole hempseed (Mattila et al., 2018). Proanthocyanidins in "*Finola*"

Table 9.13 Phenolic content (μg/kg fresh weight) of cold-pressed "*Finola*" oil.

Phenolic acids	μg/kg fw		μg/kg fw
Hydroxybenzoic acid		*Flavonols*	
Gallic acid	0.31	Quercetin-3-*O*-rutinoside	39.08
Protocatechuic acid	3.06	Quercetin-3-*O*-glucoside	8.4
4-Hydroxybenzoic acid	1.97	Kaempferol-3-*O*-glucoside	65.01
Vanillic acid	0.078	Kaempferol-3-*O*-rutinoside	34.3
Hydroxycinnamic acids		*Isoflavones*	
Chlorogenic acid	1.03	Daidzein	15.7
trans-*p*-Coumaric acid	2.78	Genistein	15.3
Flavonoids		*Flavones*	
Flavanols		Apigenin	12.34
Catechin	52.8		
Epicatechin	101.6	**%Total phenolics**	
		Flavanones	50.30
Flavanones		*Flavanols*	21.69
Eriodictyol-7-*O*-glucoside	5.3	*Flavonols*	20.62
Naringenin-7-*O*-glucoside	14.45	*Isoflavones*	4.36
Eriodictyol	9.29	*Flavones*	1.73
Naringenin	297.4	*Hydroxybenzoic acids*	0.76
Naringin	31.6	*Hydroxycinnamic acids*	0.54

Source: Adapted from Smeriglio et al. (2016).

hempseed cake (~2.5 g/kg) consisted predominantly of (epi) catechin polymers (Pap et al., 2020). Total phenolic contents (mg GAE/kg) of hempseed oils were 20–2670 (Izzo, Pacifico, et al., 2020), 440–1880 (Liang, Appukuttan Aachary, & Thiyam-Holländer, 2015), and "*Finola*" oil 2675 (fresh weight) (Smeriglio et al., 2016). Expressed in caffeic acid equivalents hempseed and "*Fedora*" oils contained ~24 mg CAE/kg total phenolics (Liang et al., 2015; Siano et al., 2019). Bleaching reduced the phenolic content (163 mg GAE/kg) of industrially cold-pressed hemp seed oil (Hemp Oil Canada Inc., Saint Agathe, Manitoba, Canada) up to one-third (Aachary, Liang, Hydamaka, Eskin, & Thiyam-Holländer, 2016). The total flavonoid content, detected as luteolin equivalent, was 195 mg/kg in cold-pressed hempseed oil from New Zealand (Teh & Birch, 2013).

The major phenolic compounds gallic acid, 3,4-dihydroxybenzoic acid, syringic acid, catechin, and 1,2-dihydrobenzene (Table 9.14) account for 85% of the total phenolic compounds in hempseed purchased from a local market in Turkey (Babiker et al., 2021). Moreover, the hydroxybenzoic acids represent 52.8% of the total phenolics. Roasting (160 °C) particularly for 14 min significantly increases (2.5 folds) the total phenolics (167 vs

Table 9.14 Phenolic content (mg/kg) of untreated and roasted hempseed.

		Roasted	
Phenolic compounds	**Untreated**	**14 min**	**7–21 min**
Hydroxybenzoic acids			
Gallic acid	49.8	220.4	37.6–220.4
3,4-Dihydroxybenzoic acid	25.5	36.2	31.5–72.3
Syringic acid	28.0	28.5	28.5–47.9
Hydroxycinnamic acids			
Caffeic acid	7.1	42.4	15.6–42.4
p-Coumaric acid	0.8	0.9	0.9–2.5
trans-Ferulic acid	4.5	3.8	3.8–9.9
trans-Cinnamic acid	0.9	1.8	0.7–3.1
Flavanoids			
Catechin	33.8	86.1	58.7–86.1
Quercetin	4.0	37.6	10.7–37.6
Isorhamnetin	3.2	12.1	4.2–12.1
Naringenin	1.9	4.2	0.9–4.2
Apigenin 7 glucoside	3.6	7.1	6.3–7.4
Rutin trihydrate	3.4	10.9	10.9–14.9
Others			
1,2-Dihydroxybenzene	28.6	66.2	54.9–74.5
Resveratrol	0.5	0.6	0.6–2.9

Source: Adapted from Babiker et al. (2021).

425 mg GAE/kg FW) and flavonoid (290 vs 750 mg CE/kg FW) contents of hempseeds. Roasting (14 min) increased gallic acid content over fourfolds, but reduced the major phenolic compounds (gallic acid, 3,4-dihydroxybenzoic acid, catechin, 1,2-dihydrobenzene, and syringic acids) to 78% of the total (compared to 85% in nonroasted hempseed).

However, the DPPH scavenging activity decreased (18 vs 11%) upon roasting hempseed for 14 min, but increased to 33% after extended roasting (21 min) [despite similar/comparable total phenolic and flavonoid contents at 7 min roasting]. This increase is likely due to phenolic and flavonoid release by thermal degradation of cellular structure (Babiker et al., 2021). Cold-pressed hempseed (Serbian variety *Helena*) meal has higher catechin, ferulic acid, and 3,4-dihydroxybenzoic acid (>10, 10, and 3-folds, respectively) in addition to protocatechuic and sinapic acids (28 and 22 mg/kg) (Pojić et al., 2014) compared to those reported by Babiker et al., 2021.

Production of the prenylflavonoids cannflavins A and B (CFA and CFB) is induced by sprouting particularly in the Italian saline-tolerant variety *Ermo* (7 and 5 μg/g of CFA and CFB, respectively), targeting their anti-inflammatory activity to microsomal prostaglandin E_2 synthase (mPGES-1) and 5-lipoxygenase (5-LO) (Werz et al., 2014). Cannflavins inhibit prostanoid formation complementing the anti-inflammatory activity of the ω-3 fatty acids. The anti-inflammatory potency of cannflavin is intermediate between that of aspirin and dexamethasone. However, CFA and CFB act as dual inhibitors of mPEG-1 and 5-LO, two crucial enzymes in the biosynthesis of the proinflammatory mediators PGE2 and leukotrienes [LTs], respectively. Moreover, dual mPEG-1 and 5-LO inhibition is considered a pharmacological strategy to intervene with inflammatory diseases, particularly when their intake is associated to a fatty matrix high in ω-3 fatty acids. CFA and CFB inhibit prostaglandin E2 (PGE2) production by cultured rheumatoid synovial cells exerting anti-inflammatory activity presumably due to cyclo-oxygenase and potent (IC_{50} 0.9 μM) concentration-dependent lipoxygenase inhibition. Cannflavin A, B, and C also exert anti-leishmanial activity (Pollastro, Minassi, & Fresu, 2018). A cannflavin B derivative, canflanone (FBL-03G), increased apoptosis in two pancreatic cell lines (Panc-02 and Ptf1/p48-Cre [KPC]). FBL-03G (4 μM) more effectively killed pancreatic cells than radiotherapy (4 Gy), delayed *in vivo* local and metastatic tumor progression in animal pancreatic cancer models (Wild-type C57Bl/6 mice) and increased their survival compared to control cohorts. This canflanone was granted orphan drug status for pancreatic cancer by the United States Food and Drug Administration in 2019 (Tomko, Whynot, Ellis, & Dupré, 2020).

9.6.1 Lignanamides

The polyphenol profile of hempseed is unique because of its prominent hydroxycinnamic acid amides and lignanamides. Phenylpropionamides are the most representative secondary metabolites responsible for the antioxidant and antiradical activity of hempseed polar extracts. Compounds of this class mainly include N-*trans*-caffeoyltyramine, N-*trans*-feruloyltyramine, and their various structural dimers called lignanamides. Lignanamides are naturally occurring secondary metabolites consisting of a lignan (phenylpropanoid [C8-C8′]) structure oxidatively coupled with an amide group and therefore considered a lignan subclass (often termed *lignan-amide*) (Leonard, Zhang, Ying, & Fang, 2020). Whole hempseed and hempseed hull contained 0.77 and 1.1–3.4 g/kg dry weight of total lignanamides, respectively, with higher dihydrobenzofuren-type (e.g. grossamide) than phenylpropylene-type (e.g. cannabisin F) lignanamide (Mattila, Pihlava, et al., 2018). The most abundant lignanamides in whole hempseed extract were cannabisin A, B, and M (74, 56 and 10 μg/mg, respectively) (Leonard et al., 2020 and references therein). SC-CO_2 extraction increased lignanamides (6 cannabisins, caffeoyltyramine, and 3 grossamides) content of hempseed (cv. *Finola*) cake (2.4 vs 1.9 g/kg dw), whereas total cannabisins content (1.4 g/kg) was unaffected (Pap et al., 2020).

Hempseed hull consisted primarily of cannabisin B and *N-trans*-caffeoyltyramine as the major phenolic compounds according to an earlier study (Sakakibara, Katsuhara, Ikeya, Hayashi, & Mitsuhashi, 1991). *N-trans*-caffeoyltyramine content of raw hempseed (Australian *cv. CRS1* grown in Tasmania) hull (128 μg caffeoyltyramine [CTE] equivalents/g sample) (Leonard, Zhang, Ying, Xiong, & Fang, 2021) (Table 9.15) is lower than previously reported values for hull fraction of hempseed meal or whole hempseed (287 mg *trans*-cinnamic acid equivalents/kg dry sample [152 mg *trans*-cinnamic acid equivalents/kg dry sample for commercial whole meal] and 25 mg/g, respectively) (Pojić et al., 2014; Zhou, Wang, Ji, Lou, & Fan, 2018). The total phenylpropionamide content of hempseed hull is higher than reported value for whole hempseed (400 μg CTE equivalents/g vs 234 μg/g [70% EtOH extract of petroleum ether defatted hempseed] for 14 phenylpropionamides). Extrusion increases the total phenylypropionamide content (25–78%) of hempseed hull consisting of 26 compounds including hydroxycinnamic acid amides and lignanamides (Leonard et al., 2021). Extrusion, particularly at low speed, significantly increases *in vitro* enzyme activities (α-glucosidase [41 vs 51–57%] and acetylcholinesterase [63 vs 67%] inhibition) of whole (extract) hempseed hull. Moreover, the free fraction of extruded hempseed hull has α-glucosidase inhibition comparable to acarbose indicating potential antidiabetic effect or activities.

Over 30 phenylpropionamides (including polyphenols, lignanamides, and caffeoyltyramine-like compounds) have been isolated from aqueous ethanol (95% EtOH) extract of *n*-hexane defatted hempseed. Most of these lignanamides inhibit the NF-κB signaling pathway thereby exerting good antineuroinflammatory activity.

These compounds can therefore be useful in preventing/attenuating neurodegenerative diseases and/or the development of Alzheimer's disease (pathological hallmark of neuroinflammation) (Zhou, Wang, Lou, & Fan, 2018). Lower aqueous ethanol (70% EtOH) extracted only fourteen phenylpropionamides (234 μg total phenylpropionamides [TPA]/mg extract) from petroleum ether defatted hempseed (Zhou, Wang, Ji, et al., 2018). The TPA included 12 lignanamides [N-*trans*-caffeoyloctopamine, N-*trans*-caffeoyltyramine, cannabisin (A, B, C, D, E, F, M, N), grossamide, and 3, 3′-demethyl-grossamide], and two phenylamides [N-*trans*-coumaroyltyramine and N-*trans*-feruloyltyramine]. Cannabisin A, B, and N-*trans*-caffeoyltyramine (~33, 22, and 11% of total TPA content, respectively) accounted for ~66% (⅔) of total TPA content (Table 9.15). The TPA extract alleviated LPS-induced hippocampal neuronal damage by reducing the primary inflammatory cytokines (IL-1β, IL-6, and TNF-α) levels thereby exerting neuroprotective effects (Zhou, Wang, Lou, & Fan, 2018). The extracts F03 (acetone, 50% EtOH) and F05 (75% EtOH, ethyl acetate fraction) obtained from *n*-hexane defatted hempseed showed comparatively low lignanamide contents (Table 9.15). Nevertheless, both F03 and F05 extract reduced inflammatory response (TNF-α and IL-6 cytokine expression) in activated primary human monocytes and suppressed the chemokine receptor CCR2 involved in inflammatory diseases of the

Table 9.15 Phenypropionamide content (mg/g) of raw and extruded hempseed hull, defatted hempseed, and extracts.

Compounds	Raw[1]	Extruded[1]	Defatted HS[2]	(%TPA)[2]	F03[3]	F05[3]
Hydroxycinnamic acid amides (HCA)						
N-trans-caffeoyltyramine	128.1	161–242	24.2–25.8	10.74	1.28	6.36
N-trans-caffeoyloctopamine	8.05	10.9–17.7	2.5–2.8	1.15	0.10	0.31
N-trans-feruloyltyramine	47.4	56–90	14.8–15.4	6.49	0.16	1.76
N-trans-coumaroyltyramine	11.9	15.1–19.9	9.96–10.99	4.51	0.04	0.45
Total HCA	195.5	244–370	52–55			
Lignanamides						
Cannabisin A	10.6	17.4–27.4	73–75	31.79	0.02	0.19
Cannabisin B	18.2	36–63	51–60	23.76	0.06	0.57
Cannabisin C	8.6	12.6–18.7	10.1–11.4	4.64		
Cannabisin D	21.8	18.3–27.1	9.9–12.2	4.63		
Cannabisin E	9.2	8.8–13	9.7–10.1	4.24		
cis-Cannabisin E	2.9	4.9–7.8				
Cannabisin F	10.3	17.6–28.6	7.2–7.7			
Cannabisin G	10.7	10.4–12.8				
Cannabisin M	48.2	51.4–55.7	10.0–10.6	4.42		
*Iso*Cannabisin N	1.16	1.8–2.6	1.0–1.1	0.45		
Cannabisin O	0.68	1.8–3.4				
Cannabisin Q	4.44	7.5–11.7				
3-Demethyl Cannabisin G	3.21	7.6–13.5				
Grossamide	44.2	45.9–49.4				
Grossamide K	8.3	7.2–8.7				
3′3-demethyl grossamide	1.05	1.0–2.2				
Rac-Sativamide A	1.24	1.0–1.5				
Total lignanamides	205	260–345	172–188		1.58	8.88
Total phenylpropanoids	400	503–715	223–243			

Values for raw and extruded hempseed hulls are expressed in μg caffeoyltyramine (CTE) equivalents/g sample; TPA – total phenylpropionamides; HS – hempseed; F03 and F05 – extracts; *Source:* adapted from [1]Leonard et al. (2021); [2]Zhou et al. (2018); [3]Martinez et al. (2020).

central nervous system (Martinez, Montserrat-de la Paz, De la Puerta, Garcia-Giménez, & Fernández-Arche, 2020). The extracts also exhibited strong scavenging activity; DPPH [248 and 309 μM TE/g extract] and ABTS [410 and 540 μM TE/g extract] for F03 and F05, respectively, due to their high phenolic and flavonoid contents 19 and 57 mg GAE/g and 5 and 18 mg QE/g extract (Martinez et al., 2020).

N-*trans*-caffeoyltyramine, N-*trans*-feruloyltyramine, cannabisin M, grossamide, and cannabisin D and B were the major phenylpronamides in raw hempseed hull accounting for 32, 12, 12, 11, 5.4, and 4.6% of the total phenylpropanoides content, respectively. The hydroxycinnamic acid amides accounted for approximately half of the total phenylpropanoids. Extrusion reduced the relative amount of cannabisin D, M and grossamide simultaneously increasing cannabisin B content (>three folds), particularly at high temperature and low speed (100 °C, 150 rpm) (Table 9.15). N-*trans*-caffeoyltyramine and cannabisin B were the most abundant phenolic compounds in the sieved (>250 μm) hempseed meal hull fractions (267–287 and 117–153 mg *trans*-cinnamic acid equivalents/kg dry sample, respectively (Pojić et al., 2014). These two compounds, *N-trans*-caffeoyltyramine and cannabisin B (66 and 142 mg extracted from 2 kg of hempseed hulls), exhibited high DPPH scavenging activity (IC_{50} of 9.4 and 11.2 μg/mL, respectively) and protected *in vitro* oxidation of human low-density lipoprotein (Chen et al., 2012). The total lignanamide (four containing cannabisin F and two grossamide compounds) content of hempseed was 77 mg/100 g dw (30 and 47 mg/100 g cannabisin F- and grossamide types), while that of the hulls was 111–336 mg/100 g dw (34–50 and 76–292 mg/100 g cannabisin F- and grossamide types) (Mattila, Pihlava, et al., 2018).

Cold-pressed (25–30 °C) hempseed (*cv. Fedora*) defatted with supercritical CO_2 (SFE), then extracted with SFE (100% CO_2, 15 g/min, 300 bar) and 20% ethanol as cosolvent contained 4 lignanamides (cannabisin A, B, and C representing 30% of the total peak area), 2 amides (*trans* and *cis* caffeoyltyramine accounting for 51% of the total peak area) (Michailidis, Angelis, Nikolaou, Mitakou, & Skaltsounis, 2021). The hempseed paste extracts exhibited collagenase inhibitory activity (>95% for SFE extracts vs 80% for UAE extracts at 600 μg/mL) presumably due to the presence of cannabisin A, B, and C and are potential candidates for nutritional and dermo-cosmetic products development.

Cannabisin A has stronger inhibitory effect on SARS-CoV-2 due to its higher binding affinity to SARS-CoV-2 Mpro (−12.76 kcal/mol) compared to darunavir (−12 kcal/mol) or ritonavir (−7.7 kcal/mol) (Figure 9.4). The logP value predicted for cannabisin A (5.2) is similar to the approved drug ritonavir (5.6) and higher than that of darunavir (2.2), thereby further supporting cannabisin A as a potential drug for SARS-CoV-2 (Ngo, Pham, Le, Pham, & Vu, 2020). Moreover, cannabisin A has larger pulling work W and Fmax (104 kcal/mol and 901 pN, respectively) values than the HIV-1 PR inhibitor darunavir (84 kcal/mol, 818 pN), the most reliable SARS-CoV-2 main protease (Mpro) inhibitor.

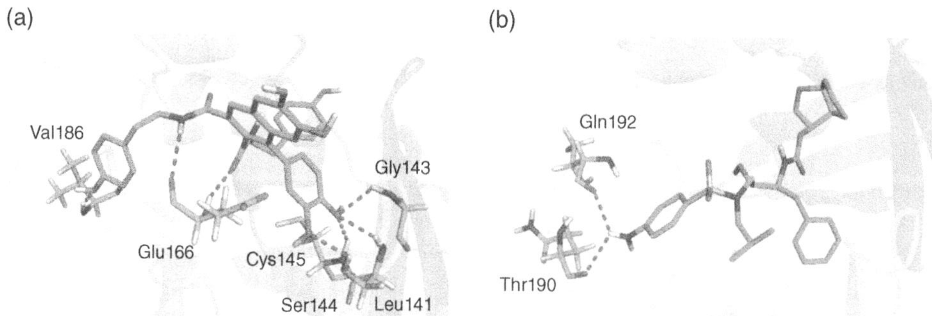

Figure 9.4 SARS-CoV-2 main protease complex conformation with (a) cannabisin A and (b) darunavir. *Source:* Ngo et al. (2020).

Lignanamides and their biosynthetic precursors, phenylamides, exert diverse biological activities including antifeedent, anti-inflammatory, and neuroprotective activities and insecticidal effects. Nigro et al. (2020) identified seven phenylamides (*N*-caffeoyloctopamine, *N*-*p*-coumaroyloctopamine, *N*-caffeoyltyramine, *N*-*p*-coumaroyltyramine, *N*-feruloyltyramine [2 isomers], and tri-*p*-coumaroylspermidine) and 25 lignanamides [*N*-caffeoyltyramine (*N*-caffeoyloctopamine dimers 1&2, *N*-feruloyltyramine dimer 1, *N*-caffeoyloctopamine dimer 2), *N*-caffeoyltyramine dimer hydroxyl derivative, grossamide K, demethylgrossamide, 3,3′-demethyl-heliotropamide, and 16 cannabisins (cannabisin A, B, B-isomer, C, C-isomer, D, D-isomer, E, E-isomer, F, H, H-isomers 1 &2, I, M and Q)] in ethanol extract of *n*-hexane defatted hempseed. Lignanamides-rich hempseed extract (LnHS) significantly inhibited U-87 malignant glioblastoma neuroepithelial cancer cells proliferation, reduced U-87 cell migration, inducing further DNA damage, and suppressed autophagic apoptosis. LnHS negatively and specifically regulates U-87 cell line survival and migration and may potentially cross the blood brain barrier thereby impacting brain health and cognition.

Lignanamides exert anti-inflammatory effects by reducing production, release or gene suppression of proinflammatory cytokines (tumor necrosis factor α [TNF-α], interlukin-6 [IL-6]) from immune cells. In a rat model, hempseed lignanamides (cannabisin-containing phenylpropanoid [TPA]) neutralized the lipopolysaccharide [LPS]-induced inflammatory effects by alleviating memory loss and reversing neuronal damage. Cannabisin B extracted from hempseed hull exhibited stronger DPPH scavenging ability than soybean isoflavones and flax secoisolarisiresinol diglucoside (SDG) (Chen et al., 2013). Hempseed lignanamides (cannabisins A, C, D, and M) are powerful antioxidants (>80% DPPH inhibition rate compared to quercetin), while others (grossamide, cannabisin D and G) display weak cytotoxic effect against human prostate cancer (LNCaP) cells (Leonard et al., 2020). Cannabisin-B exhibits strong antioxidative scavenging activity with anticancer efficacy inducing autophagic death in hepatoma cells by arresting the S phase cell cycle (Pollastro et al., 2018).

Three of the 11 lignanamides (*N-trans*-caffeoyltyramine, 3,3′-demethyl-heliotropamide, and 3,3′-demethylgrossamide) extracted from defatted (petroleum ether) hempseed inhibited acetylcholinesterase (83, 63, and 87%, respectively) *in vitro*, in addition to exhibiting high antioxidant (DPPH, ABTS, and ORAC [≥80%]) activities. The acetylcholinesterase inhibitory (AChE) activity of these lignanamides was associated with their similar favorable binding mode to galanthamine (a competitive pharmaceutical AChE inhibitor use to treat cognitive decline in mild to moderate Alzheimer's disease and other memory impairments) (Yan et al., 2015). Interestingly, *N-trans*-caffeoyltyramine, 3,3′-demethylgrossamide along with cannabisin B, exhibits the highest arginase inhibitory activity (>50% at 100 μM; IC_{50} values 21, 127 and 244 μM, respectively) among nine antioxidant lignanamides extracted from cold-pressed (20 °C) hempseed (*Fedora 17*) cake (Bourjot, Zedet, Demange, Pudlo, & Girard-Thernier, 2016). *N-trans*-caffeoyltyramine (because of its strong arginase inhibitory and antioxidant activities) can be a potential lead compound to develop new drugs to treat endothelial dysfunction associated with cardiovascular diseases (coronary artery disease, T2DM, and microvascular perfusion).

The two nor-lignanamides (Sativamide A and Sativamide B) derived from *N-trans*-caffeoyltyramine with an unprecedented skeleton have a unique benzo-angular triquinane ring system (Figure 9.5). Sativamide A and B mitigate tunicamycin (Tm)-induced

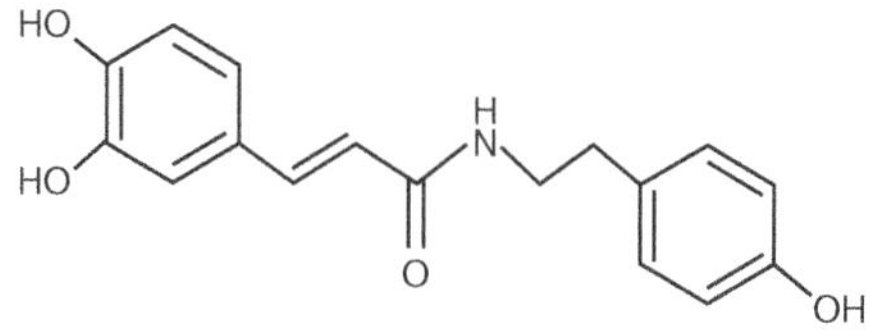

N-trans-caffeoyltyramine

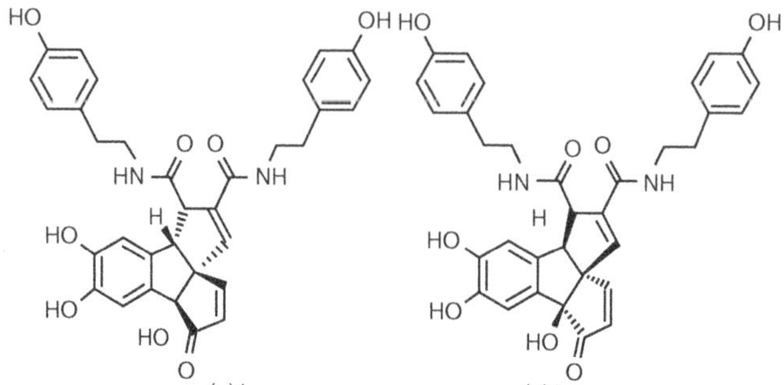

Figure 9.5 The two nor-lignanamides (Sativamide A and Sativamide B) derived from *N-trans*-caffeoyltyramine with an unprecedented skeleton have a unique benzo-angular triquinane ring system. *Source:* Zhu et al. (2018).

neuroblastoma PC-12 cell toxicity in a dose-dependent manner. These sativamides also dose-dependently suppress ER stress-mediated cytotoxicity in thapsigargin (Tg)-induced neuroblastoma PC-12 and SH-SY5Y cells. Furthermore, the cytoprotection of the sativamides (50 μM) is similar to that of salubrinal (25 μM), an inhibitor of endoplastic reticulum (ER) stress that plays an important role in neurodegenerative diseases (Zhu et al., 2018).

9.7 Minor Components

9.7.1 Minerals

Mineral contents in hempseed covered a wide range, varying >6 folds in Ca and K, 2–3 folds in Fe, Zn, Mg, and Cd, and 1.7x in Mn for five Romanian varieties (Three monoecious – *Denise*, *Diana* & *Zenit* and two dioecious – *Armanca* & *Silvana*) (Mihoc, Pop, Alexa, & Radulov, 2012). Heterogeneity was highest among mineral contents of 32 industrial hempseed cultivars registered in the European Union (Farinon et al., 2020). Potassium content of hempseed depends on accumulated soil Fe, whereas variety, fertilization, and their interaction affect Ca content. Both K and Ca compete with Mg soil availability and may therefore be responsible for its content in hempseed. *Zenit* had the highest K and Fe and the lowest Mn contents and average Cd content. However, agronomic practices (fertilization and distance between plants) have no or insignificant effects on Fe, Mn, Zn, and Cd contents of hempseed. Processing such as roasting (160 °C, 7, 14 or 21 min) reduces major mineral contents of hempseed except Ca that increases significantly to 1,009 mg/kg after

Table 9.16 Mineral contents (mg/kg) of hempseed and its products.

	Oil[1]	Flour[1]	Seed[1]	Seed[2]	Seed[3]	Seed[4]	Extruded[3]	Hulls[2]	Hearts[5]
Macro elements									
P				11,700	9,525		8,439–9,042	3,930	
K	20.7	5,064	2,517	9,210	7,558	2,520–28,210	7,541–7,724	5,290	3,479
Ca	53.9	1,907	944	1,270	823	900–9,550	925–1,009	1,660	298
Mg	199.1	2,311	2,682	4,960	4,687	2,370–6,940	4,335–4,667	2,210	1,975
Na	88.8	90.6	67.5			68–270			505
Micro elements									
Fe	1.7	152.5	98.0	43.8	101	40–2,400	102–124	30.6	99.7
Cu	1.5	11.9	5.0	18.9	0.88	8–20	0.89–0.99	18.0	9.6
Mn	0.8	94.7	44.4	105	93	40–150	96–98	121	
Zn	0.9	54.7	48.4	69.7	36.8	40–110	38.1–49.6	29.2	44.2
B	0.4	4.1	2.0		13.1		12.5–12.9		
Al	1.8	3.5	2.1						10.1
Mo	0.6	0.2	0.5						
Co	0.06	0.03	0.03						
Ni	0.03	0.85	0.79						
Pb	0.09	0.03	0.04						

Source: Data for [1]adapted from Siano et al. (2019); [2]Mattila, Pihlava, et al. (2018); [3]Babiker et al. (2021); [4]Farinon et al. (2020); [5]Senila et al. (2020).

14 min roasting; individual trace elements except boron increase with roasting time (Babiker et al., 2021) (Table 9.16).

Magnesium, potassium, and calcium are dominant minerals in hempseed (24.5% lipids), flour (13.6% lipids), and oil obtained from the cultivar *Fedora* (Siano et al., 2019). The flour, seed, and oil have the highest, moderate, and lowest contents of these macro elements, respectively (Table 9.16). The calcium and potassium contents in the flour are over twice that in the seed (1,907 vs 944 mg/kg and 5,064 vs 2,517 mg/kg, respectively). Similarly, Cu, Mn, and Ba contents in flour are ≥2 folds than in the seed.

Fe, Zn, Mn, and Cu are among the microelements in high concentration, whereas Co, Ni, and Pb are less abundant (<1%) in hempseed samples and below established safety threshold. The average Cu concentration in whole hempseed varies (7–13 mg/kg) within the values for seed and flour (5 and 12 mg/kg); Mo in the oil is higher than in seeds (0.57 vs 0.50 mg/kg) and higher than in flour (0.19 mg/kg). Fe concentration is higher (50%) in the flour than in the seed with low levels in the oil (1.7 mg/kg an order of magnitude lower than those reported earlier [Uluate & Özdemir, 2012]); similarly Zn content in the oil was lower than those reported earlier (0.9 vs 1.7 mg/kg). Moreover, Ca and Fe contents in cold-pressed hempseed oil from Turkey were > 6 folds that from Finland (Mattila, Pihlava, et al., 2018; Uluate & Özdemir, 2012). Some metals (Fe [11 mg/kg oil], Ni, Cu) accelerate autoxidation and significantly impact the shelf life of hempseed oil (Uluate &

Özdemir, 2012). This high Fe level of hempseed oil surpasses the maximum recommended levels (1.5 and 5 mg/kg) for refined and virgin oils, respectively. Moreover, hempseed oil is a good source of Zn due to its higher level than most commercial refined oils, peanut, sunflower, rice, soybean, and corn (Carbone et al., 2009). Hempseed hulls contain 30–65% less major elements (except Ca) and Zn than whole seeds; Cu and Mn are more evenly distributed in the whole seed. Hempseed and its hulls have higher and lower amounts of Mn (≥6×) and Cd (0.06×), respectively, compared to flaxseed (Mattila, Pihlava, et al., 2018).

Shelled hempseed compared to other oilseeds (shelled poppy, sunflower, sesame, and flaxseed) purchased from a Romanian supermarket contains the highest Mg content (1,975 vs 1,320–1,528 mg/kg), Fe (100 vs 35–53 mg/kg), K similar to flaxseed (~3,500 vs 1,823–2,833 mg/kg), Ca similar to sesame seed (298 and 257 mg/kg, respectively, but lower than other oilseeds [644–1,337 mg/kg]), Na similar to sunflower seed (~500 mg/kg, but lower than flaxseed [1,090 mg/kg]), similar Zn (44 vs 34–40 mg/kg), low Cu (10 vs 13–17 mg/kg), and high Al (10 vs 8–38 mg/kg) (Senila et al., 2020). Mineral (Ca, Mg, Fe, and Zn) content of cold-pressed hempseed oil is comparable to those of radish (*Raphanus sativus*) seed (Uluate & Özdemir, 2012). The reference daily intake (RDI) established by Food and Drug Administration (FDA) regulations is Ca (1,000 mg), Na (2,400 mg), Fe (15 mg), Cu (2 mg), K (3,500 mg), Mg (400 mg), Mn (2 mg), and Zn (15 mg). In commercial cold-pressed hempseed cake, Zn, Fe, and Ca sequester with phosphorous to form sparingly soluble inositol-1,2,3,4,5,6-hexakisphosphate ([$InsP_6$], 34 mg $InsP_6$/g press cake) complexes stored as phytate, often accounting >1% of dry weight. This phosphate (19.6 mg PO_4^{3-}/g press cake) can be easily released by phytase hydrolysis (*Escherichia coli* AppA phytase, 37 °C) in an ecofriendly process adding value to press cakes in a circular bioeconomy (Herrmann, Ruff, & Schwaneberg, 2020).

9.7.2 Phytic Acid

Phytic acid contents (43–75 g/kg; 58.4 g/kg average) varied significantly ($P < 0.01$) among hexane defatted hempseed from 20 hemp cultivars of different geographical origins grown under the same environment (Galasso et al., 2016). French hempseed cultivars have particularly high (≥7%) phytate content, higher than all Italian genotypes. However, variation in phytic acid of hexane defatted meal was significantly higher in year of production than hempseed genotypes (3 dioecius [*Carmagnola*, *CS* and *Fibranova*] and 3 monoecious [*Futura 75*, *Felina 32* and *Ferimon*]) (83 vs 57 – F values from ANOVA) (Russo & Reggiani, 2015). Moreover, the monoecious had significantly higher phytic acid content than dioecius genotypes (74–77 vs 62–65 g/kg). The phytic acid content of these 6 varieties was positively correlated ($r = 0.62–0.67$; $P < 0.01$) with other antinutritional factors (condensed tannins, trypsin inhibitors, and saponins) (Russo & Reggiani, 2015). Phytic acid content (15–31 g/kg; 26.7 g/kg average) varied significantly (16% CV) among 20 industrial hemp varieties and breeding lines currently representing international varieties including 10 of Australian origin (Schultz et al., 2020). Most of the antinutrients including phytic acid are present in the cotyledons (21 g/kg for >180 μM fraction) compared to the cold-pressed whole meal (23 g/kg) (Pojić et al., 2014) indicating that hempseed hulls have lower phytic acid content than the whole seed (21 vs 53 g/kg dw) (Mattila, Pihlava, et al., 2018).

Finola hempseed cake and SC-CO_2 defatted cake contain similar phytic acid (45 and 49 g/kg dry weight) mostly unaffected by enzyme (Flavorpro 750 MDP and Promod 439 L) hydrolysis (Pap et al., 2020).

9.7.3 Pigments

Pigments, primarily chlorophyll and carotenoids, impart light to dark green color of high-quality virgin (screw press) hempseed oil (Matthäus & Brühl, 2008). Cold-pressed hempseed oil from New Zealand has higher chlorophyll (75 vs 69 or 0.86 mg pheophytin *a*/kg) than flax or canola oils (Teh & Birch, 2013). Lower mean total (1.5 mg/kg) chlorophyll content (range 0.4–4.8 mg/kg) has been reported for thirteen commercial cold-pressed hempseed oils available in the Italian market (Izzo, Castaldo, et al., 2020). The optimal cold press condition (60 °C, 20 Hz frequency, 6 mm ID nozzle screw expeller) to obtain the highest hempseed oil recovery (23%) has total (99 mg/kg) chlorophyll content with 59 and 39 mg/kg of chlorophyll-*a* and -*b*, respectively. The residual oil in the press cake (10%) completely extractable by supercritical CO_2 has three times higher total chlorophyll content (229 mg/kg) with 194 and 35 mg/kg of chlorophyll-*a* and -*b*, respectively, compared to cold-pressed hempseed oil (Aladić et al., 2014). Canadian industrial cold-pressed hempseed oil contained 34–68 mg/kg total chlorophyll depending on extraction solvent (methanol or diethyl ether). Diethyl ether extracts the most chlorophyll (68, 49–53, and 15–19 mg/kg of total chlorophyll, chlorophyll-*a* and-*b*, respectively). Methanol extracts the least chlorophyll (34–43, 22–31, and 11 mg/kg of total chlorophyll, chlorophyll-*a* and-*b*, respectively) (Aachary et al., 2016).

Total chlorophyll and carotene contents of hempseed oil depend on the extraction method with cold pressed oil containing the highest amount (157 and 39 mg/kg), followed by Soxhlet [Soxterm, n-hexane, 180 °C, 175 min] (102 and 28 mg/kg) and SCO_2 (91 and 34 mg/kg) (Aladić et al., 2014). Moreover, total chlorophyll (10–179 mg/kg) and carotene (8–58 mg/kg) contents of SCO_2 extracted hempseed (*Fedora 17*) oil depend on processing (pressure, temperature, and time) conditions (300 and 400 bar pressure, 40 and 60 °C, constant 1.94 CO_2 kg/h, separator at 15 bar, 25 °C); the highest pigment concentration is obtained at high pressure (400 bar) and long extraction time (180 min) (Aladić et al., 2015). Canadian commercial cold-pressed hempseed oils have significantly higher β-carotene content (34 and 53 mg/kg oil) than those extracted from seed (*cv. Fasamo*, 20 mg/kg oil). However, microwave treatment (6 min) of hempseed increases pigments (chlorophyll and carotenoid) of the oil resulting in increased oxidation temperature indicating protective antioxidant effect. Moreover, continuous microwave drying (950 W, 60 Hz, 6 min) of hempseed cultivar "*Fasamo*" increased (9.3 fold) β-tocopherol content of hexane extracted oil thereby increasing total tocopherol and vitamin E levels (8 and 19%, respectively) (Oomah et al., 2002).

Hempseed oil obtained by extrusion is darker (17.4 vs 4.8, *b* color value) with higher chlorophyll *a*, *b*, and carotenoids (12 vs 3.8; 3.5 vs 0.5; and 1,475 vs 377 μg/mL, respectively) than cold-pressed oil (Moon et al., 2017). Notably, carotenoids, particularly β-carotene, synergistically improve vitamin E efficacy in attenuating cognitive impairment, consequently attenuating the pace of cognitive decline among middle-aged adults, delaying thereby the onset of AD and other forms of dementia in older age (Beydoun et al., 2020). Chlorophyll, a powerful prooxidant, interferes with oxidative stability and rancidity; it increases susceptibility to light-induced oxidation or photo-oxidation. Therefore, refining is necessary to reduce chlorophyll, metals and phospholipids improving hempseed oil quality. Chlorophyll is easily removed using a small bed of bleaching earth (1–5% or more appropriately ~2% by mass of extracted oil) to obtain a light yellow hempseed oil (Catchpole et al., 2009). Ultrasonic bleaching with industrial clay almost completely removes chlorophyll (~98% reduction) accompanied by low peroxide value and reduced phenolic content (118 vs 163 mg/kg GAE for control oil) of hempseed oil (Aachary et al., 2016).

9.7.4 Vitamins

Data on hempseed vitamins are limited. Hempseed has the most active antihemorrhagic factor (vitamin K) in White Leghorns chicks among other foods including cereals and seeds, vegetables, animal organs, fats and oils, and hen's egg (Dam, 1935). The fat-soluble vitamin occurs in the soluble nonsterol fraction of the unsaponifiable matter of hempseed. Whole hempseed (100 g) contains vitamin A (3,696 IU β-carotene), vitamin B-1 (0.9 mg thiamine), vitamin B-2 (1.1 mg riboflavin), vitamin B-3 (2.5 mg niacin), vitamin B-6 (0.3 mg pyridoxine), vitamin D (19 IU), and vitamin E (3 mg γ-tocopherol) (Garner-Wizard, 2013). *Finola* has 4 and 1 mg/kg of vitamins B1 and B2, respectively (Callaway, 2004). However, Johnson (2019) disclosed higher vitamin B-1and B-3 (13 and 92 mg/kg) and lower vitamin B-2 and B-6 (3 and 6 mg/kg) levels together with folate and vitamin A (1,100 and 10 μg/kg) in hempseed. Dietary intake of antioxidants such as β-carotene (as well as other carotenoids), vitamins A, C, and E reduce lipid peroxidation, ROS production, apoptosis, and protein and DNA oxidative damage (Beydoun et al., 2020). These antioxidants may also play a neuroprotective role.

9.8 Molecular Mechanism of Action

The health benefits of hempseed are often generally attributed to its oil and/or lipids constitution and fatty acid composition. The highly polyunsaturated (72–83%) hempseed oil, enriched in linoleic (*n*−6, 50–70%) and linolenic (*n*−3, 15–34%) fatty acids, favors serum cholesterol and atherosclerosis reduction and heart diseases prevention (Montserrat-de la Paz et al., 2014). Dietary linoleic acid intake was inversely associated with cardiovascular health disease (CHD) risk in a dose–response manner in prospective observational studies. Moreover, increasing polyunsaturated (PUFA) consumption instead of saturated fatty acids (SFAs) reduced CHD in a meta-analysis of intervention trials (Schwingshackl et al., 2018 and references therein). The *n*−6/*n*−3 (3 : 1) ratio of hempseed oil believed to be ideal for optimal health, particularly cardiovascular health has several medicinal and physiological effects, including cholesterol and high blood pressure reduction, glaucoma and cancer treatment, and anti-inflammatory benefits (Rezvankhah et al., 2019).

Virgin hempseed oil contains up to 7% γ-linolenic acid (GLA-18:3*n*−3) and 2.5% stearidonic acid (SDA-18:4*n*−3) (Matthäus & Brühl, 2008). GLA and SDA (respective biologic metabolites of LA and ALA) enable bypassing the first critical Δ-6 desaturase enzymatic step, thereby facilitating the conversion into the biologically active form of long-chain PUFAs. GLA exerts anti-inflammatory activity since it is rapidly converted to dihomo-γ-linolenic acid (DGLA; 20:3*n*-6) in the human body (Farinon et al., 2020). GLA benefits patients with rheumatoid arthritis and atopic dermatitis, whereas SDA can improve eicosanoids formation for people with deficient Δ-6-desaturase function (Matthäus & Brühl, 2008). GLA protects β-amyloid peptide fragment [$A\beta_{25-35}$]-induced neurotoxicity by reducing ROS and blocking caspase-3 activation, thereby attenuating apoptotic morphological alteration. GLA also inhibits proinflammatory cytokines (tumor necrosis factor-α [TNF-α] and prostaglandin E2 [PGE2]) production and blocks nuclear factor κB [NF-κB] subunit p65 activation by suppressing IκB-α degradation. The inhibitory effect of GLA is accompanied by reduced expression of mitogen-activated protein kinases (MAPKs) such as extracellular protein regulated protein kinase (ERK1/20) and *c*-Jun-N-terminal kinase (JNK) activity. Thus, GLA prevents Aβ25–35 damage via NF-κB signaling pathway (Youn, Lee, & Jun, 2018). SDA intake can also exert anti-inflammatory effects, effectively increase EPA composition of plasma and erythrocyte membrane and reduce serum insulin in humans (Gao, Fan, Tan, Zhang, & Wang, 2019). Eicosenoic acid (EA; 20:1, *cis*-11) or gondoic acid present in hempseed oil (< 0.3–1.5%) is detectable in comparable quantities in only a few oils (peanut, maize, sunflower, and soybean oils). EA is an intermediate in the human metabolic pathway from palmitic (16:0) to nervonic (24:1, cis-15) acid, the latter FA is important in the production of cerebrosides, characteristic components of nerve membranes and the "white substance" of the brain. Thus, this EA also contributes to hempseed oil beneficially important nutrient (Mölleken & Theimer, 1997a).

Commercially refined hempseed oil contains unsaponifiable matter (~1.9%) high in phytosterols (15% of unsaponifiable matter) consisting primarily of β-sitosterol (1,905 mg/kg oil) and campesterol (506 mg/kg oil). These phytosterols may be responsible for reducing/lowering low-density lipoprotein (LDL) levels in human blood, cholesterol levels, platelet aggregation, and endothelial damage. The major diterpene alcohol, phytol, present in hempseed oil (168 mg/kg oil) has been reported to exhibit anticancer and antioxidant properties (Dunford, 2015). Both PUFAs and phytosterols have been implicated in cardiovascular disease prevention due to their plasma cholesterol-lowering effects (Lee et al., 2011).

A recent interventional review investigated the increased intake of plant-based omega-3 fats for all-cause mortality, cardiovascular events, adiposity, and lipids including 162,796 adults at varying cardiovascular risk mainly from high-income countries participating in randomized controlled trials (12–18 months duration) (Abdelhamid et al., 2020). The review showed that increased ALA intake may slightly reduce the risk of cardiovascular events (5 RCTs, 19,327 participants, low certainty) and probably reduces the risk of arrhythmias (2 RCTs, 4,912 participants, moderate certainty). Increasing dietary ALA intake may be slightly protective of some heart and circulatory diseases (slightly reducing cardiovascular events and heart irregularities [moderate-or low-certainty evidence]) with minimal or no difference to all-cause cardiovascular or coronary deaths or coronary events (Abdelhamid et al., 2020).

9.8.1 Hempseed Oil and/or its Components

9.8.1.1 Cellular Effects

Hempseed, its derivative and/or constituents, has been the focus of some cellular studies (Table 9.17). In human rheumatoid arthritis fibroblast-like synovial (MH7A) cells, hempseed oil reduced cell survival rate, promoted apoptotic cell death time- and dose

Table 9.17 Effects of hempseed oil at the cellular level.

References	Benefits	Cells/organism
Jeong et al. (2014)	Hempseed oil (HO)-treated synovial cells accumulated lipid, produced ROS, and induced apoptosis	MH7A-human rheumatoid arthritis fibroblast-like synovial cells
Moccia, Maria Russo, et al. (2019)	Polar HO extract delayed cancer cell proliferation	Human colon adenocarcinoma (HT-29)
Moccia et al. (2019)	Polar HO extract reduced colorectal adenocarcinoma	Caco-2 and HT-29 (150 μg/mL, 24 hr, ~90% cell death)
Jin and Lee (2018)	Hexane hempseed extracts ameliorated lipogenesis in sebocytes and reduced inflammation	*Probionibacterium acnes*
Lee et al. (2011)	Hempseed meal intake protected neurodegeneration mediated by LA and ALA; inhibited cholesterol uptake	*Drosophila*

dependently, increased lipid accumulation, level of intracellular reactive oxygen species (ROS), and expression of major ER stress markers (glucose-regulated protein 78 and C/EBP homologous protein [CHOP]). CHOP functions as an antirheumatoid factor downstream of hempseed oil in MH7A cells (Jeong et al., 2014). Hempseed oil-treated synovial cells accumulated lipid, produced ROS, and induced apoptosis. This lipid peroxidation and ROS accumulation may potentially explain the hempseed oil-mediated cell death. Presumably, significant proportions of the accumulated lipids were converted by lipid peroxidation to ROS that mediated cell death through the involvement of the transition factor CHOP (Jeong et al., 2014).

A polar oil extract (POE) from cold-pressed hempseed (cv. *Codimono*) dose dependently induced significant (50%) delay in human colon adenocarcinoma (HT-29) cell growth, cytostatic autophagy, and cell cycle arrest (Moccia, Maria Russo, et al., 2019). In a follow-up study, polar oil (aqueous 80% methanol) extract (*cv. Fedora*) hempseed containing hydroxycinnamic acid derivatives and cannabinoids significantly reduced Caco-2 and HT-29 cells viability (Moccia, Siano, et al., 2019). This cytotoxic effect was associated with apoptotic cell death activation independent of antioxidant capacity of the oil extract. Caspase-3 activation mediated Caco-2 cells apoptosis, while caspase was not involved in HT-29 cells presumably suggesting POE specificity toward the malignant phenotype. Furthermore, POE from hemp seed and flour did not interfere with Caco-2 and HT-29 cell growth, although POE was rich in several compounds (N-*trans*-caffeoyltyramine, cannabisin A and B, *p*-coumaroyltyramine, feruloyltyramine, and caffeoyloctopamine) (Moccia, Siano, et al., 2019). It is therefore difficult to delineate the effects of hempseed oil from other POE components in these cell studies.

Hempseed hexane extracts (HSHE) exert antimicrobial, anti-inflammatory, antilipogenic, and collagen-promoting properties useful in treating acne vulgaris (Jin & Lee, 2018). HSHE reduced *Propionibacterium acnes*-induced iNOS, COX-2 (and the subsequent nitric oxide and prostaglandin [PGE_2] production), and inflammatory cytokines (IL-1β and IL-8) expressions. These anti-inflammatory effects result from regulating NF-κB (phosphorylated IKKα/β, IκBα, and NF-κB expressions) and MAPKs signaling, attenuating *p*-NF-κB translocation, inhibiting ERK and JNK phosphorylation and their downstream targets (c-Fos and c-Jun), and inhibiting activator protein 1(AP-1) promotor activity. HSHE-regulated inflammation and lipid biosynthesis (lipogenic signaling pathway by enhancing p-AMPKα expression and supressing p-mTOR, PPARγ, SREBP1, and FAS) by modulating AMPK and AKT/FoxO1 signaling and inhibiting 5-lipoxygenase and promoting *in vitro* collagen biosynthesis. HSHE anti-inflammatory activity is attributed to its major components, including linoleic and γ-linolenic acids. HSHE inhibited IKK, IκB, NF-κB, p38, JNK an ERK phosphorylation and regulating NF-κB and MAPK signal pathways in *P. acnes*-induced HaCaT cells (Jin & Lee, 2018).

The antioxidant activity of hempseed meal (HSM) is lost during digestion and absorption in the *Drosophila* model (Lee et al., 2011). HSM intake suppressed Aβ42 cytotoxicity (a model of *Drosophila* AD) due to LA and ALA protective neurodegenerative effects with no effects on (polyglutamate-induced) Parkinson disease (PD) and (induced JNK activated) Huntington disease (HD). LA and HSM inhibit cholesterol uptake with LA acting antagonistically with cholesterol during *Drosophila* development. LA and ALA mediate the neuroprotective effects of hempseed meal against amyloid since they exert cytotoxic effects on Aβ42. PUFAs (LA) suppress Niemann-Pick C1-like gene *NPC1L1*, a key intestinal cholesterol absorption protein (also critically important for cholesterol uptake that is effectively inhibited by β-sitosterol) (Lee et al., 2011).

9.8.1.2 Animal Studies

The effects of hempseed on platelet aggregation have been investigated in normal and hypercholesterolemic animals (Table 9.18), because high omega-6 fatty acids can produce eicosanoids and stimulates thromboxane A2 synthesis. Dietary hempseed (10% hempseed or 10% partially delipidated hempseed) induced limited beneficial effects (significant heart fibrillation reduction presumably due to elevated cardiac GLA level) after 8 weeks on cardiac function during ischemia/reperfusion challenge in male New Zealand white rabbits. The study did not support the use of dietary hempseed to protect the heart during ischemic insult, although it demonstrated the benefits of hempseed fatty acids (Prociuk et al., 2006). A similar hempseed treatment (10% hempseed) increased plasma PUFAs and GLA levels in male hypercholesterolemic New Zealand white rabbits after 8 weeks feeding. Hempseed supplementation to the diet normalized cholesterol-induced platelet aggregation, partly due to increased plasma GLA levels. Hempseed PUFAs can therefore potentially reduce cardiovascular disease by reducing platelet aggregation (Prociuk et al., 2008).

Hempseed (5 or 10% ground hempseed, w/w) supplemented in a regular rat chow diet for 12 weeks induced changes in plasma fatty acids (increased total plasma PUFAs significantly) of male Sprague–Dawley rats. ALA and LA levels increased significantly in hempseed-fed rats in a concentration-dependent manner. Hempseed supplementation significantly inhibited platelet aggregation and rate of aggregation due primarily to the specific fatty acid composition of hempseed and high plasma ALA. Hempseed may be beneficial in protecting against clot-induced strokes and myocardial infarctions (Richard et al., 2007). Platelet aggregation and rate of aggregation were significantly inhibited by

Table 9.18 Animal studies.

Reference	Treatment	Benefits
Prociuk et al. (2006)	10% hempseed or delipidated hempseed, 8 wk	Limited beneficial cardiac function in male rabbits
Prociuk et al. (2008)	10% hempseed supplementation, 8 weeks	Normalized platelet aggregation in male rabbits
Richard, Ganguly, Steigerwald, Al-Khalifa, and Pierce (2007)	5 and 10% hempseed, 12 weeks	Reduced platelet aggregation in rats
Richard et al. (2007)	5 and 10% hempseed, 12 weeks	Inhibited platelet and rate of aggregation in rats
Al-Khalifa et al. (2007)	5 and 10% hempseed supplementation, 12 wk	Conferred beneficial cardioprotection in rats
Kaushal, Dhadwal, and Kaur (2020)	10% hempseed supplementation, 1 mo	Antihypercholesterolemic effects via COX-2 in rats
Gavel et al. (2011)	10% hempseed supplementation, 8 wk	Atherogenesis and contractile function in hypercholesterolemic rabbits
Kaushal, Gupta, and Kulshreshtha (2020)	10% hempseed lipids, 2 mo	Hepato-protective effects; improved markers of fatty liver and modulated redox homeostasis in female rats
Hashemzaeh, Moghaddam, Karimi, Haghighi, and Hassanpour (2017)	Hempseed oil (8 mL/kg/d, 1 wk)	Modified lipid profile and atherogenic index positively in male rats
Pasqua et al. (2020)	Standard diet supplemented with 9.5% hempseed oil, 8 wk	Reduced plasma triglyceride and contracture/diastolic function in male Wistar rats
Opyd et al. (2020)	1% native or partially defatted hempseed, 8 wk in rats with metabolic disorders	Improved lipid metabolism and reduced triglyceride accumulation in hepatic lipid metabolism
Rezapour-Firouzi et al. (2018)	Hempseed/evening primrose oil (HSO/EPO) combination, 28 d	Reduced experimental encephalomyelitis (EAE) development in female C57BL/6 mice
Rezapour-Firouzi, Mohammadian, Sadeghzadeh, and Mazloomi (2020)	Evening primrose/hempseed oil (EPO/HSO) combination, 2 wk	Suppressed inflammation in chronic EAE-induced female C57BL/6 mice
Huang et al. (2021)	Cannabis seed oil (30 μL) by gavage, 8 wk in female $ApoE^{-}/^{-}$mice (C57/BL6) fed a high cholesterol diet	Reduced low-density lipoprotein cholesterol, serum necrosis factor α, and nitric oxide and exerted antiatherosclerotic effects

5 and 10% hempseed in male Sprague–Dawley rat diets after 12 weeks intervention (Richard et al., 2007). The fatty acid composition (most likely/presumably ALA) of hempseed was purportedly responsible for the beneficial platelet antiaggregation effects. Hempseed (20.6 and 24.4; 3.3 and 4.9; 0.3 and 0.6 mg/g LA, ALA, and GLA, respectively, for 5 and 10% hempseed in the diet) fed to male Sprague–Dawley rats with regular chow for 12 weeks significantly increased plasma ALA and GLA levels and only ALA levels in heart tissues (Al-Khalifa et al., 2007). Apparently, these omega-3 fatty acids are effectively absorbed and selectively deposited within the heart with hempseed addition to the diet. Hempseed-supplemented diets conferred beneficial cardioprotective effects (contractile performance; enhanced relaxation) in ischemia-repurfusion challenged hearts presumably induced by ALA. The hempseed dietary intervention did not affect eicosapentaenoic (EPA) and docosahexaenoic (DHA) acids, two physiologically important fatty acids (Al-Khalifa et al., 2007). These studies demonstrate the potential of dietary hempseed in preventing or treating cardiovascular diseases, like myocardial infarctions and strokes where thrombotic episodes are central feature of the pathology. Kaushal, Dhadwal, and Kaur (2020) compared the antioxidant capacity of protein (*n*-hexane defatted hempseed) and lipid fraction of hempseed (HS, 10%) supplemented to high-fat diets (HFD) induced hypercholesterolemic Wistar rats. HS scavenged free radical *in vitro*, ameliorated hypercholesterolemia (HC) [indicated by improved lipid profiles, aortic tissue and EEG patterns] and relieved COX-2 mediated inflammation. The anti-inflammatory effects of HS were attributed to PPARγ activation by COX-2-dependent prostaglandins (PGs) [PGDS and SeH] of the D- and J-series. The study demonstrates that the antihypercholesterolemic effects of HS are mediated through redox-sensitive modulation of inflammatory pathways. Dietary hempseed (10%, w/w) supplemented to the regular diet (125 g/daily for 8 weeks) was mildly beneficial against contractile dysfunction with atherosclerotic vessels in 0.5% cholesterol-fed male albino New Zealand white rabbits. However, hempseed did not protect against cholesterol-induced aortic plaque development (Gavel et al., 2011).

Hempseed lipids (10% + 90% pellet diet for 2 months) ameliorated markers of fatty liver (reduced hepatic fat accumulation, increased HDL, and changes in liver function markers), decreased levels of liver enzymes (serum glutamic oxaloacetic transaminase [SGOT], serum glutamic pyruvic transaminase [SGPT], and alkaline phosphatase [ALP]), modulated redox homeostasis (reversed HFD induced changes in inflammatory mediators COX-2, sEH, IL-4, and TNF-α expression and improved hPGDS expression), and increased antioxidant enzymes (catalase, GPx, GST) and reduced MDA levels in female Wistar rats (Kaushal, Gupta, & Kulshreshtha, 2020). Hempseed oil treatment (8 ml/kg/d for 1 week) reduced total cholesterol (TC), low-density lipoprotein cholesterol (LDL-C), very low-density lipoprotein cholesterol and triacylglycerols in male rats compared to normal control (distilled water 0.5 ml/kg), atherogenic index (AI) compared to carbon tetrachloride (CCl4, 0.5 ml/kg)-intoxicated (TCC) group. Hempseed oil can also reduce liver damage due to decreased plasma hepatic marker enzymes (ALT, AST, and ALP) and serum bilirubin levels (Hashemzaeh et al., 2017). Hempseed oil (9.5% w/w) supplemented to a standard diet for 8 weeks reduced plasma triglycerides and contracture/diastolic function (LVRDP) in male Wistar rats (Pasqua et al., 2020). Furthermore, hempseed oil lowered infarct size (area) without changes in oxidative stress, basal cardiac parameters, inflammatory process, and protein damage markers in I/R subjected hearts. However, phenolic-enriched hempseed oil (9.5 w/w % hempseed oil +0.5 w/w% soybean lecithin) with whey protein (13.3%) effectively mitigated cardiovascular risk by reducing and/or preventing cardiac infarction and cardiac dysfunction during ischemia/reperfusion (I/R) due to increased AMP-activated protein kinase [AMPK] activation (Pasqua et al., 2020). Dietary hempseed intake (1%, 8 weeks) improved some markers associated with lipid metabolism (blood plasma and liver), reducing triglyceride accumulation in hepatic lipid metabolism of Wistar rats with metabolic disorders induced by an unbalanced (high fat +cholesterol [HF]) diet (Opyd et al., 2020). Native hempseed consumption improved plasma cholesterol and increased hepatic PPARγ expression (adipogenesis stimulation and regulation of genes involved in cholesterol transport and metabolism). Native (1%) and partially defatted (presumably dehulled) hempseed supplementation stimulated cecal SCFA production with the defatted hempseed (15% dietary fiber) exerting a pronounced effect indicating it to be a good fermentation substrate for the gut microbiota. Hempseed supplementation also improved the antioxidant status of obese rats disturbed by HF dietary feeding (Opyd et al., 2020).

Hempseed/evening primrose oil combination (HSO/EPO [6–8%/8–12% GLA]) is used to treat inflammatory conditions since it can correct essential fatty acid (EFA) deficiency, repair the phospholipids of cell membrane lipids, and modulate immunological responses. HSO/EPO reduced experimental encephalomyelitis (EAE – an animal MS model) development, promoted remyelination, eliminated tissue damage, and downregulated interferon gamma (IFN-γ) in chronically EAE-induced female C57BL/6 mice (Rezapour-Firouzi et al., 2018). The anti-inflammatory effects of HSO/EPO correct essential fatty acid (EFA) deficiency, repair cell membrane phospholipids, and modulate immunological responses. Moreover, HSO/EPO treatment suppressed regulatory associated protein of TORC1 (*RAPTOR*) and IFN-γ gene expressions and regenerated the myelin sheath demonstrating potent downregulation of immune response for MS treatment (Rezapour-Firouzi et al., 2018). The same EAE model was used to evaluate the effects of evening primrose/hempseed oil (EPO/HSO) combination to improve the membrane fatty acids composition of spleen and blood cells and immunological factors relative to rapamycin (RAPA – inhibitor of mammalian target of rapamycin [mTOR] signaling in fatty acid metabolism and T cells activation) (Rezapour-Firouzi et al., 2020). EPO/HSO treatment significantly increased essential fatty acids content in cell membrane of the spleen and blood of chronically EAE-induced female C57BL/6 J mice. Moreover, overexpression of the interleukin (IL-4, IL-5, and IL-13) genes demonstrates repair of the cell membrane structure and inflammation suppression by EPO/HSO in EAE (Rezapour-Firouzi et al., 2020).

Cannabis seed oil administered by gavage for 8 weeks reduced serum triglycerides, low-density lipoprotein cholesterol, serum tumor necrosis factor α, and nitric oxide in female Apo $E^{-}/^{-}$ mice fed a high-cholesterol diet (Huang et al., 2021). Moreover, the oil treatment improved endothelial cell integrity and arterial lipid deposition and exerted antiatherosclerotic effects by modulating serum lipid profiles and inflammatory responses (reducing pro-inflammatory cytokines (C-reactive protein [CRP], TNF-α, and NO). The antiatherosclerosis effect of the seed oil is similar to that of seal oil. The seed oil similar to seal oil alleviated glucose-induced human microvascular endothelial (HMEC-1) cell damage without inhibiting oxidative stress (MDA-lipid peroxidation) in mice.

9.8.1.3 Human Studies

Human studies with hempseed oil can be segregated in two categories; one using hempseed oil and the other co-supplemented with evening primrose oil (Table 9.19). Cold-pressed hempseed oil (30 mL/day for 8 weeks with a 4-week washout period) was compared to cold-pressed extra virgin olive oil for their effects on plasma lipid profiles, transepidermal water loss, skin quality, and dermal medication usage in atopic dermatitis patients (1 male, 15 females; 38.1 ± 8.5 years old; 24.9 ± 4.5 kg/m^2 BMI) in a randomized single-blind controlled crossover study (Callaway et al., 2005). The hempseed oil intervention significantly modified plasma fatty acid profiles in all lipid fractions, particularly for linoleic acid, ALA, and GLA; LA, ALA, and GLA levels increased without affecting plasma arachidonic acid levels. Furthermore, clinical atopic symptoms (skin dryness and itchiness) decreased significantly ($P = 0.03$) after hempseed oil intervention. This efficacy of dietary hempseed oil intervention was apparently due to the exceptionally high PUFA (>80%) level of the oil, metabolically favorable n-6/n-3 fatty acid ratio (2 : 1), and the presence of GLA and SDA that enable bypassing the delta-6-desaturase rate-limiting enzymatic step. In short, hempseed oil stabilizes lipid homeostasis in atopic dermatitis. Gene expressions involved in lipid homeostasis and inflammation are modulated in an *in vitro* 3D model of atopic dermatitis. These gene expressions can be altered by stabilizing/improving cholesterol homeostatis through the upregulation of cholesterol efflux regulatory protein encoded by ATP-binding cassette transporter (ABCA1) gene in tissues impaired with barrier function due to inflammation (Park et al., 2021).

Table 9.19 Human studies.

References	Intervention	Effects
Callaway et al. (2005)	30 mL/d CPH oil, 8 wk intervention	Reduced clinical atopic symptoms effectively
Schwab et al. (2006)	30 mL/d hempseed oil, 4wk	Reduced CHD predictor; no hemostatic effect
Kaul et al. (2008)	2 g/d hempseed oil, 12 wk	No effects on platelet aggregation
Del Bo et al. (2019)	3 g/d hempseed oil, 8 wk	Increased PUFA intake; no effects on other lipid markers and LDL-C levels
Rezapour-Firouzi et al. (2013)	Hempseed oil cosupplemented with evening primrose oil (9 : 1, 18–21 g/d); 6 mo	Reduced inflammation by modulating cytokine production; increased IL-4 cytokine; increased general health
Rezapour-Firouzi et al. (2014)	Hempseed oil cosupplemented with evening primrose oil (9 : 1, 18–21 g/d); 6 mo	Reduced liver enzyme (ALT, AST, and GST) levels; Improved extended disability status score (EDSS)
Rezapour-Firouzi et al. (2015)	Hempseed oil cosupplemented with evening primrose oil (9 : 1, 18–21 g/d); 6 mo	Regulated surface-membrane enzymes (FADS2 & sPLA2)

Cold-pressed hempseed (cv. *Finola*) oil (2.5 : 1 $n-6/n-3$ ratio, 30 mL/day, 4 weeks) was compared to flaxseed oil (0.25 : 1 $n-6/n-3$ ratio) for their effects on the profile of serum lipids and fasting concentrations of serum total and lipoprotein lipids, plasma glucose and insulin, and hemostatic factors in healthy humans (8 M + 6 F, 45 years, 25 kg m^{-2} BMI) in a randomized, double-blind, crossover trial (Schwab et al., 2006). Hempseed oil supplementation for 4 weeks increased both LA and γ-linolenic acid in serum cholesterol esters and triglycerides and reduced total-to HDL cholesterol ratio, a coronary heart disease predictor with essentially no changes in LDL particles composition (Schwab et al., 2006). Hempseed oil (572, 186, and 26 mg LA, ALA, and GLA, respectively, in 1 g capsule) consumed twice daily did not affect plasma fatty acid concentration, lipid parameters (TC, HDL—C, LDL-C, and TG), LDL oxidation, inflammatory markers (CRP and TNFα), and collagen or thrombin-induced platelet aggregation in healthy volunteers (88, 54 women and 34 men; 35 years old; 24 kg m^{-2} BMI) after 3 months in a double-blind, placebo-controlled clinical study (Kaul et al., 2008). The nonsignificant effect of hempseed oil on plasma LA and ALA levels has also been similarly observed in rabbits presumably due to the low level administration of these fatty acids. Other PUFA-enriched oils, fish, and flaxseed oils used in the study showed effects similar to hempseed oil and placebo (sunflower oil) except for the plasma fatty acid profiles (Kaul et al., 2008). In an 8-week, pilot randomized parallel dietary intervention, hempseed oil (3 g/d providing 1.4 and 0.7 g/d of LA and ALA, respectively) reduced total saturated and monounsaturated fatty acid, increased total $n-3$ and n-6 PUFAs [LA, DGLA, and ARA], and the omega-3 index [total levels of EPA and DHA; an additional biomarker of CVD risk and a predictive morbidity and mortality CVD parameter] of red blood cells phospholipids in children and adolescents (23 M + 13 F, 6–15 years old) with primary hyperlipidemia without affecting other lipid markers and LDL-C levels (Del Bo et al., 2019).

The potential therapeutic benefits of hempseed oil cosupplemented with evening primrose oil (9 : 1, 18–21 g/d for 6 months) were investigated in relapsing–remitting type of multiple sclerosis (RRMS-a chronic inflammatory disorder) patients in a double-blind, randomized trial (N = 65, 23 M and 42 F, average age 34 years with 6.8 years disease duration) (Rezapour-Firouzi et al., 2013). The cosupplemented oil significantly increased the anti-inflammatory cytokine interleukin-4 (IL-4) concentration indicating reduced inflammation in young adults with MS. The dietary PUFA of the cosupplemented oils (particularly GLA and SDA levels) and their metabolites are presumed to affect/modulate the inflammatory functions and cytokines production. The HSO/EPO ratio (9 : 1) in the study provide a n-3/n-6 PUFAs ratio of 1 : 4 that can competitively inhibit dihomo-gamma linolenic acid (GLA) to arachidonic acid (AA) resulting in more anti-inflammatory prostaglandin E1 (PGE1). Furthermore, the GLA and SDA of HSO/EPO (2 : 1 ratio of n-6/n-3) enable to efficiently bypass the D6D (Delta-6-Desaturase; ALA conversion to SDA) enzymatic step (Rezapour-Firouzi et al., 2013).

In a follow-up study, the cosupplemented hempseed and evening primrose oils intervention significantly reduced liver enzyme alanine amino-transferase (ALT), aspartate-aminotransferase (AST), and gamma glutamyl transferase (GGT) levels and improved extended disability status score (EDSS) in the same patients (Rezapour-Firouzi et al., 2014). Moreover, these RRMS patients benefited from the immunomodulatory effects (for example, suppression of IFN-γ production) of the oil constituents. A later study showed that the improvement in clinical symptoms of RRMS patients was regulated by surface-membrane enzymes (FADS2 & sPLA2) (Rezapour-Firouzi et al., 2015). Secretory phospholipase-A2 (sPLA2) controls PUFA metabolism and apparently play a fundamental role in cell injury in the central nervous system (CNS), production of pro-inflammatory mediators, and MS pathogenesis. The sPLA2 hydrolyzes phospholipids to release arachidonic acid. Oil cosupplementation significantly increased blood cells PUFAs and reduced D6D and sPLA2 concentrations thereby reducing the mean EDSS. Furthermore, altering PUFAs rate suppresses sPLA2 (Rezapour-Firouzi et al., 2015).

9.8.2 Hempseed Protein and/or its Products

Hempseed protein and its hydrolysates and/or peptides have been less studied than hempseed oil. Recently, the effects of hempseed protein (20 or 40 g) consumption were evaluated on blood glucose (BG), insulin, and satiety of normoglycemic (<5.6 mmol/L) adults (14 M + 13 F; average age 22 years) compared to soybean protein in two acute randomized repeated-measures crossover experiments (Mollard et al., 2021). Hempseed protein ([Hemp Pro 70]), like soybean (HISOLATE 63), dose-dependently attenuates postprandial blood glucose and insulin concentrations, influences satiety and food intake compared to a carbohydrate control when consumed as part of a fruit smoothie. Hempseed protein is a viable alternative to soy protein for glucose regulation, although fat and fiber contents were different (Mollard et al., 2021). Simulated gastrointestinal (pepsin and pancreatin) hydrolyzed hempseed protein meal (HMH-64% protein, dry weight basis with high tryptophan 12.6% and low arginine 2.1%) was used as a source of antihypertensive peptides substituting casein (0.5 and 1%) fed to growing spontaneously hypertensive rats (SHRs) for 8 weeks or adults for 4 weeks (Table 9.20). HMH attenuated normal increases in systolic blood pressure (SBP) in growing (120 vs 158 mmHg for casein control) and adult rats (119 vs 150 mmHg for casein control). Moreover, HMH reduced plasma ACE activity and renin level indicating that HMH therapeutically prevents and treats hypertension in SHRs (Girgih, Alashi, He, Malomo, & Aluko, 2014). In a subsequent study, the antioxidant peptides in HMH (0.5 or 1% w/w, for 8 weeks) reduced lipid peroxidation rate and total peroxides, elevated plasma antioxidant enzyme (SOD and CAT) levels, and total antioxidant capacity in 20-week-old SHRs (Girgih, Alashi, He, Malomo, Raj, et al., 2014).

The antioxidant peptides attenuate plasma antioxidant stress markers (antioxidant enzymes [SOD and CAT] of the vascular system and lipid peroxides) in young (0.5 or 1% HMH w/w, for 8 weeks) and adult 20-week-old SHRs (4 weeks) spontaneously hypertensive rats. In normotensive WKY rats (1% HMH substituting casein for 4 weeks), reduced plasma peroxides indicating that under normal physiological conditions, HMH peptide consumption can reduce the risk of diseases associated with excessive radical compounds. Hempseed peptides upregulated antioxidant enzyme defense system, scavenged radical and metal ions, and inhibited lipid peroxidation (Girgih, Alashi, He, Malomo, Raj, et al., 2014). Single oral administration (200 mg/kg body weight [bw]) of alcalase (1%) hempseed hydrolysate reduced BP (−33 mg Hg) after 4 hr similar to the drug captopril (10 mg/kg bw). The pepsin (2 and 4%) hydrolysates produced the longest lasting BP-lowering effect (−23 mmHg) after 24 hr oral administration. However, the SBP-reducing effects were not directly related to *in vitro* ACE and renin inhibitions presumably

Table 9.20 Protein studies.

References	Hempseed protein products	Benefits
Girgih, Alashi, He, Malomo, and Aluko (2014)	Pepsin+pancreatin hydrolyzed hempseed protein meal (HMH) (0.5 and 1% HMH, 8 weeks) fed to SHRs	HMH reduces normal SBP increase; ACE and renin inhibitory activities
Girgih, Alashi, He, Malomo, Raj, et al. (2014)	Pepsin+pancreatin hydrolyzed hempseed protein meal (HMH) (0.5 and 1% HMH, 8 weeks) fed to SHRs	HMH peptides upregulates antioxidant enzymes and inhibits lipid peroxidation
Malomo et al. (2015)	Single oral dose (200 mg/kg bw) hempseed hydrolysate administered to SHRs	1% Alcalase hydrolysate reduces BP after 4 h; pepsin hydrolysates has long-lasting BP-lowering effect
Zanoni et al. (2017)	Pepsin hydrolyzed hempseed protein extract effects on human hepatic HepG2 cells	The hydrolysate exhibits statin-like hypocholesterolemic effects
Logarušić et al. (2019)	Neutrase hydrolyzed hempseed protein isolate effects on human keranocyte (HaCaT) cells	The hydrolysate promotes HaCaT cell growth and inhibits ROS formation
Rodriguez-Martin et al. (2019)	Alcalase and alcalase+flavourzyme hydrolyzed hempseed protein isolate effects on BV-2 microglial cells	The hydrolysate protects BV-2 cells from inflammation
Rodriguez-Martin et al. (2020)	Alcalase and alcalase+flavourzyme hydrolyzed hempseed protein isolate effects on primary human monocytes	The hydrolysate reduces proinflammatory and increases anti-inflammatory mediators
Lin et al. (2020)	Pepsin-pancreatin hydrolyzed hempseed protein isolate effects on colon cancer (HT-29 and HCT-116) and macrophage cells	The hydrolysate inhibits colon cancer cell proliferation, suppresses pro-inflammatory cytokines, and NO production
Wei et al. (2021)	Enzyme hydrolyzed hempseed peptide (<400 Da) effects on human liver cancer cells (Hep3B)	The hydrolysate increases apoptosis, reduces Hep3B viability, and proliferation

indicating reduced absorption rate and clearance after absorption from the gastrointestinal tract (Malomo, Onuh, Girgih & Aluko, 2015).

Pepsin hydrolyzed hempseed (cv. *Futura*) protein extract (0.25, 0.5, or 1 mg/mL containing 90 peptides) was investigated for its hypocholesterolemic properties using human hepatic HepG2 cells on the LDLR-SREBP2 pathway. The hypocholesterolemic effects were mediated by regulating the rate-controlling enzyme HMGCoAR of cellular cholesterol biosynthesis pathway (by activating the AMPK pathway), upregulating the LDLR protein levels through SREBP2 protein increase. The hydrolysate increases PCSK9 levels similar to the mechanism of action of statins coregulated by SREBP2 activation (Zanoni et al., 2017). Neutrase hydrolyzed hempseed protein isolate (1–10 mg/mL) stimulated normal human keranocyte (HaCaT) cell growth (53 and 62% at 1 and 2 mg/mL), but highly cytotoxic on HeLa cancer cells (at 5 and 10 mg/mL). Moreover, the hydrolyzed protein isolate (1 mg/mL) significantly inhibited ROS formation in HaCaT (25%) and H_2O_2 exposed HeLa (28%) cells and reduced ROS cells (16 and 18%, respectively) (Logarušić et al., 2019).

Alcalase and alcalase + flavourzyme hydrolyzed hempseed protein isolate (50 and 100 μg/mL) were evaluated for their neuroinflammatory activities on BV-2 microglial cells (Rodriguez-Martin et al., 2019). The hydrolysates (containing an oligopeptide 300–400 Da) suppressed mRNA transcription levels of inflammatory mediators (TNF-α, IL-1β, and IL-6) and upregulated anti-inflammatory cytokine IL-10 expression in LPS-stimulated BV-2 microglial cells. In a subsequent study, the anti-inflammatory effect of HPHs was evaluated on proinflammatory (TNF-α, IL-1β, and IL-6) and anti-inflammatory mediators (IL-10 and IL-4) mediators in LPS-stimulated primary human monocytes (Rodriguez-Martin et al., 2020). Hempseed protein hydrolysates (50 or 100 μg/mL) block the LPS-induced inflammatory state by reducing and increasing the transcriptional activity of proinflammatory and anti-inflammatory genes, respectively. Moreover, HPHs polarize primary human monocytes toward M2 (protective cytokines [IL-4 and IL-10]) poorly migratory anti-inflammatory macrophages important in tissue repair by suppressing C—C chemokine receptor 2 (CCR2) and C—C ligand 2 (CCL2) transcriptional activities (Rodriguez-Martin et al., 2020). Simulated gastrointestinal digest (1 hr pepsin followed by 2 hr pancreatin) of hempseed protein isolate significantly inhibited colon cancer (HT-29 and HCT-116) cells proliferation. The pepsin-pancreatin digest significantly reduced nitric oxide (NO) production and suppressed proinflammatory cytokines (IL-6 and TNF-α) levels in LPS-induced RAW 264.7 macrophage cells (Lin et al., 2020). Small molecular weight hempseed peptide (<400 Da) obtained by enzyme hydrolysis (protease and papain [3 and 0.4% w/v, protein basis] pH 5, 70 °C, 3 hr) increases apoptosis, reduces cell viability, proliferation, and cell migration in Hep3B human liver cancer cells without affecting normal liver cells. This anticancer effect due to increased cellular ROS levels is modulated by β-catenin via increased Akt and GSK-3β phosphorylation, upregulation of proapoptotic bad and cleaved caspase 3, and suppression of antiapoptotic Bcl-2 protein, subsequently suppressing β-catenin (Wei et al., 2021).

9.8.3 Hempseed Phytochemicals

Some hempseed components other than oil and protein have also been investigated for their health benefits (Table 9.21). Hempseed hull extract (60% ethanol) contained two antioxidant compounds *N-trans*-caffeoyltyramine and cannabisin B with high DPPH scavenging activity (IC_{50} of 9.4 and 11.2 μg/mL, respectively) and protected *in vitro* oxidation of human low-density lipoprotein (Chen et al., 2012). Cannabisin B purified from the same hempseed hull was investigated for anticancer activity in HepG2 human hepatoblastoma cells (Chen et al., 2013). It significantly inhibited HepG2 cells proliferation by inducing autophagic cell death. Furthermore, cannabisin B dose-dependently induced S phase cell arrest (presumably impacting DNA synthesis and/or a cyclin that regulates S phase) by inhibiting the AKT/mTOR (mammalian target of rapamycin) pathway. This considerable antiproliferating activity of cannabisin B can potentially be used as a promising chemopreventive agent against hepatoblastoma (liver) disease (Chen et al., 2013).

The potential effect of grossamide, a hempseed lignanamide and its underlying anti-neuroinflammatory mechanisms, was investigated against lipopolysaccharide (LPS)-induced inflammatory response in BV2 microglia cells (Luo et al., 2017). Grossamide isolated from defatted hempseed downregulated LPS-mediated pro-inflammatory mediators (IL-6 and TNF-α) thereby reducing their mRNA levels, protein production, and release

Table 9.21 Benefits of hempseed phytochemicals.

References	Hempseed phytochemicals	Mode of action
Chen et al. (2012)	Hempseed hull (60% EtOH extract)	High DPPH inhibition stronger than flax SDG
Chen et al. (2013)	Cannabisin B from hempseed hull	Induces autophagic cell death on HepG2 cells
Luo et al. (2017)	Hempseed grossamide	Inhibits IL-6 and TNF-α production in LPS-induced BV2 microglial cell models
Zhou, Wang, Ji, et al. (2018)	Phenylpropionamides from 70% EtOH hempseed extract	Reduces cytokine levels, alleviates hippocampal neuronal damage, and prevents nerve cell damage in mice
Wang, Luo, and Fan (2019)	Cannabisin F from hempseed	Upregulates SIRT1 expression and reduces ROS production in LPS-induced BV2 microglia cells
Wang et al. (2019)	Coumaroylaminobutanol glycopyranoside (GLC)	Upregulates AMPK expression and inhibits neuro-inflammation and oxidative stress through NF-κB and Nrf-2 signaling pathways
Wen et al. (2019)	Hempseed polysaccharides (HSP)	HSP activates Nrf2/Keap 1 pathway inducing antioxidant enzymes and phase II detoxifying enzymes in IPEC-1 cells to resist ROS

in BV2 micoglia cells. Grossamide is therefore a potential therapeutic candidate to inhibit inflammation in neurodegenerative diseases (Alzheimer's disease, Parkinson's disease, multiple sclerosis) (Luo et al., 2017). Hempseed extract (70% EtOH extract of petroleum ether defatted hempseed) containing fourteen phenylpropionamides (TPA −234 g/kg extract) alleviated LPS-induced hippocampal neuronal damage in male Kunming mice by reducing primary inflammatory cytokines (IL-1β, IL-6, and TNF-α) levels exerting thereby neuroprotective effects (Zhou, Wang, Ji, et al., 2018). The extract consisted primarily of cannabisin A, cannabisin B, and N-*trans*-caffeoyltyramine (~33% or⅓, 22 and 11% of total TPA content, respectively) accounting for ~66% (⅔) of total TPA content. TPA extract attenuated LPS-induced cognitive deficit similar to the nootropic memory enhancing drug piracetam. At low dose (1 g/kg) it prevented LPS-induced nerve cell damage in the CA3 hippocampal region demonstrating neuroprotective effects. NF-κB signaling pathway modulation has been proposed as the anti-neuroinflammatory mechanism based on grossamide, a representative phenylpropionamide compound (Zhou, Wang, Ji, et al., 2018). Thus, phenylpropionamide inhibits pro-inflammatory mediators and dose dependently reduces NF-κB subunit p65 phosphorylation and its translocation into the nucleus.

The antineuroinflammatory mechanisms of another hempseed lignanamide, cannabisin F was investigated against lipopolysaccharide (LPS)-induced inflammatory response and oxidative stress in BV2 microglia cells (Wang et al., 2019). Cannabisin F suppressed the production of LPS-induced pro-inflammatory mediators (IL-6 and TNF-α) and reduced their mRNA levels dose dependently in BV2 micoglia cells. Moreover, cannabisin F upregulated SIRT1 expression, suppressed NF-κB signaling pathway, reduced cellular ROS production, and promoted Nrf2 expression (Wang et al., 2019). Coumaroylaminobutanol glucopyranoside (GLC), yet another phenylpropionamide upregulated adenosine monophosphate-activated protein kinase (AMPK) expression, suppressed the nuclear factor-kappa B (NF-κB) signaling pathway (through IκBα and NF-κB p65 phosphorylation inhibition), and reduced proinflammatory cytokine levels in LPS-induced BV2 microglia (Wang et al., 2019). Moreover, GLC inhibits neuroinflammation through the TLR4/NF-κB and Nrf-2/HO-1 signaling pathways in concert with AMPK.

The antioxidant activity of hempseed polysaccharides (HSP) was evaluated on H_2O_2-stimulated intestinal porcine epithelial (IPEC-1) cells to determine its efficacy to prevent and treat oxidative damage, especially in the intestines (Wen et al., 2019). Moreover, HSP significantly enhanced cellular antioxidant defense (SOD, CAT and GSH-Px) protecting IPEC-1 cells from oxidative stress. HSP also upregulated intracellular transcription factor nuclear factor erythroid-2-related factor 2 (Nrf2) expressions and inhibited the level of Kelch-like ECH-associated protein 1 (Keap1) thereby activating Nrf2/Keap1 pathway.

9.8.4 Overall Mechanism of Action

The above studies indicate very complex mechanisms of action of hempseed and its individual components that benefit human health. However, some components (oil, protein, and lignanamides [represented by grossamide]) share a common mechanism – NF-κB signaling pathway (Figure 9.6) despite their different mode of action. Hempseed oil offers neuroprotection by inhibiting proinflammatory cytokines and cholesterol uptake, ameliorates hypercholesterolemia and fatty acid markers, exhibits immunomodulatory effects, and reduces cardiovascular disease risk. Hempseed protein, either as protein hydrolysate or peptide, modulates vascular tone by blood pressure-lowering effects, ACE and/or renin inhibition, nitric oxide-mediated vasodilation, increased antioxidant response, and anti-inflammatory effects. Other hempseed bioactives such as the lignanamides are known for their neuroprotection due to their anti-inflammatory effects. Apoptosis is another common mode of operation for hempseed oil, protein, and lignanamides. For example, hempseed oil promoted apoptotic cell death of rheumatoid arthritis fibroblast-like synovial (MH7A) cells (Jeong et al., 2014). The protein as protease-papain hydrolysate-induced apoptosis in human liver cancer cells by β-catenin ROS modulation (Wei et al., 2021). Lignanamides, on

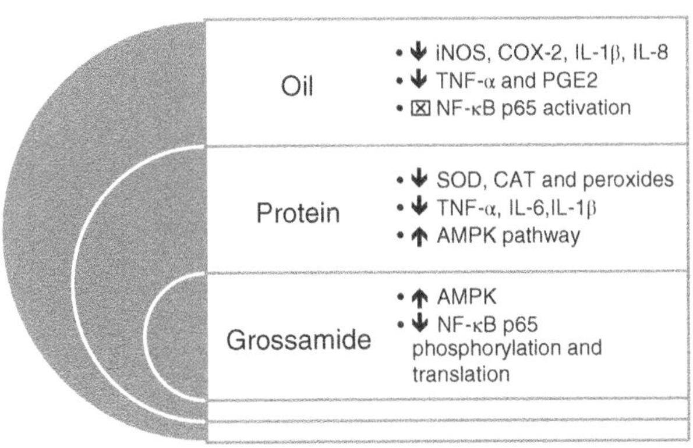

Figure 9.6 Common NF-κB signaling pathway for hempseed oil, protein, and lignanamides.

the other hand, promote HepG2 cells apoptosis by S phase cell arrest (Chen et al., 2013). The antioxidant defense system of hempseed polysaccharides activates the Nrf2/Keap1 pathway similar to Nrf2 upregulation by grossamide by increasing antioxidant enzymes (Wang et al., 2019; Wen et al., 2019).

The complex and multidirectional effects of hempseed intake are demonstrated by RNA sequencing of whole-blood transcriptome of ewes fed with a controlled diet (CTR) or 5% hempseed supplemented diet (HSG) (Iannaccone et al., 2019). HSG (5%) supplemented diet influenced the expression of 314 genes, 217 of which were upregulated. These overexpressed genes were highly associated with (1) [25 genes] – oxidative phosphorylation [genes related to NADH dehydrogenase, succinate-ubiquinone oxidoreductase (SDHA), and ATP-synthase complex], thermogenesis pathways [27 genes], (2) [11 genes] carbon metabolism; sugar (glycolysis/glucogenesis [11 genes]), galactose pathways, and (3) protein synthesis (ribosome and protein translation). HSG supplementation most affected pathways related to energy production and had no detrimental effect on either innate or adaptive immune system (Iannaccone et al., 2019). Figure 9.7 is a simple illustration of genes modulated by hempseed.

Recent studies suggest that several components in hempseed can suppress or alleviate severe acute respiratory syndrome coronavirus-2 (SARS-CoV-2). For example, Cannabisin A has stronger inhibitory effect on SARS-CoV-2 due to its higher binding affinity to SARS-CoV-2 Mpro compared to drugs such as darunavir or ritonavir (Ngo et al., 2020). Linoleic acid (the predominant fatty acid in hempseed) tightly binds to the structure of SARS-CoV-2

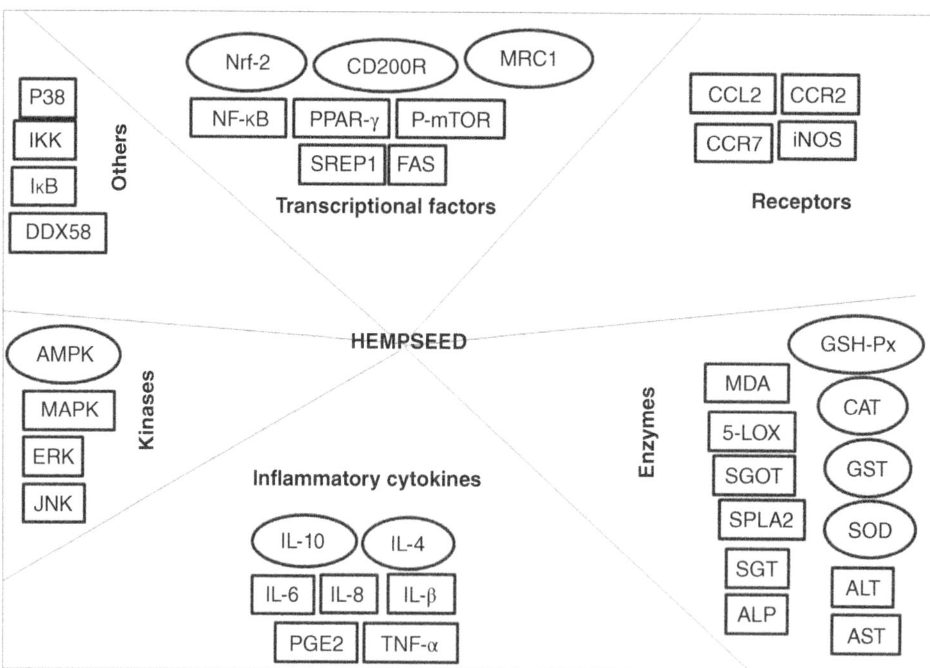

Figure 9.7 Molecular mechanisms of hempseed and its components. Upregulated molecular targets (oval). Down regulated molecular targets (rectangle).

spike (S) glycoprotein in three composite-binding pockets. The binding stabilizes a locked S conformation thereby reducing *in vitro* interaction with angiotensin-converting enzyme 2 (ACE2) (Toelzer et al., 2020). Moreover, SARS-CoV-2 replication is suppressed by LA synergy with the COVID-19 drug remdesivir indicating potential intervention strategies targeting LA binding by SARS-CoV-2. Other components present in hempseed, for example insoluble dietary fiber fermentation in the gut generates bacteria and other microorganisms suppresses the *DDX58* gene that play a central role in viral cell entry and the pro-inflammatory interferon-beta (IFN-β) that participates in the cytokine storm associated with severe SARS-CoV-2 cases (Hancocks, 2021). Another study (Das, 2021) demonstrates that polyunsaturated fatty acids, particularly linoleic and arachidonic acids (42–62 and 0.5–2.5% of total fatty acid, respectively, in hempseed) can inactivate SARS-CoV-2 by; (A) suppressing excess cytokine production (interleukin-6 [IL-6] and tumor necrosis factor-α [TNF-α] prevalent in COVID-19 cytokine storm; (B) regulating M1 and M2 macrophages generation required for timely inflammation and its resolution; and (C) mediating the actions of interferons needed for protection against viral infections (Das, 2021). Moreover, administration of linoleic acid and arachidonic acid and cofactors needed for optimum metabolism may benefit the prevention and management of COVID-19. Alpha-linoleic [ALA] and gamma-linoleic acids (12–28 and 0.5–4.5% of total fatty acid, respectively, in hempseed) along with other fatty acids are known to activate both gram-positive and gram-negetive bacteria, fungi, enveloped viruses including influenza, hepatitis B and C viruses (HBV and HCV). These fatty acids (LA, GLA, and AA) released by cytokine-activated macrophages exert antimicrobial action and can therefore prevent several respiratory infections (Das, 2021 and references therein). Moreover, the use of omega-3 fatty acid (12–28% of total fatty acid in hempseed) (supplements) was significantly associated with reduced risk (12% [95% CI {8–16%}]) of testing positive for SARS-CoV-2 in women (Louca et al., 2021). The protective effect of omega-3 fatty acid was even higher (16 and 21%) in the Swedish and US cohorts, respectively.

9.9 Current Trends and Perspectives

The licensed land to grow hemp is expected to increase to over 9 M acres by 2030 with hempseed grain leading the allocation increase (33 vs 66% for 2016 and 2030, respectively) according to Applied Food Sciences (Nutraingredients-usa.com, 2021). The industrial hemp market (including hempseed, hempseed oil, hemp fiber, and CBD hemp oil) is expected to grow to USD 26.6 billion by 2025 (34% CAGR 2019–2025), although the CBD hemp oil is projected to be the fastest growing segment during the forecast period (MarketsandMarkets, 2021). Nevertheless, two important products currently results from industrial hempseed processing, hempseed oil and protein. The protein is obtained from "hemp hearts", i.e. dehulled hempseed to improve sensory characteristics for favorable market adoption. In fact, V-70™ is claimed to be a clean CO_2 extracted 70% plant protein with near white color and neutral flavor according to Applied Food Sciences (Nutraingredients-usa.com, 2021). Thus, hempseed protein is expected to fill in the plant protein gap due to the increasing global annual demand for protein from plant source expected to increase from 140 to 155 megaton/year from 2020 to 2030 (Nutraingredients.

com, 2020). The plant-based protein market is projected to reach USD 14.5 billion by 2025 (7.1% CAGR) attributed mainly to technology innovation in plant-based protein extraction (MarketsandMarkets, 2020). The hempseed hull is a rich source of lignanamides that can be purified to develop products for neuroprotection. For example, Neoechulin A, an indole alkaloid has been isolated from 75% EtOH extract of petroleum ether defatted hempseed (Yan et al., 2016). It also inhibits β-amyloid protein (1–42)-induced microglial activation mainly attributed to NF-κB and p38 MAPK pathways inhibition in murine microglia BV2 cells (Sasaki-Hamaa et al., 2016). Neoechinulin A promoted silent information regulator two homolog 1(SIRT1), an important drug target for neurogenerative diseases, expression in HEK293 cell lines. SIRT1, a member of the sirutin family, regulates various cellular processes including energy metabolism, cell-cycle progression, and aging. SIRT1 activators or compounds promoting its expression can be candidates for aging and age-related diseases (Yan et al., 2016).

Hempseed protein is increasingly being used in many food products. For example, hemp protein [HMP] (50% protein, 10% fat, 5% carbohydrates); protein constituted 30% w/w of the final product (bar); bars without chocolate coating containing hempseed protein had high viscosity similar to those with whey protein concentrate and contained high amount of fiber, but low nutritional value compared to those made with other plant proteins (Maleki, Tomasevic, Djekic, & Solowiej, 2020). The V-70 clean hempseed CO_2 extracted 70% plant protein has applications in nutritional bars, protein beverage blends, bakery products, sports nutrition products, and dairy and meat analogs according to Applied Food Sciences. In fact, hempseed protein concentrate (71% protein, 15% carbohydrate, 7% fat, wet basis) can be used up to 60% (w/w) to replace soy protein isolate (91% protein, 1% carbohydrate, 3% fat, wet basis) in the production of high moisture meat analogues (Zahari et al., 2020). Hempseed protein concentrate absorbs less water and requires higher denaturation temperature compared to soy protein isolate and produces meat analog with comparable texture to soy protein isolate alone. However, few consumers are aware of hempseed as a dietary ALA source. In a study, 80% of participants (115 volunteers, 64 males and 51 females sampled 4 baked products [muffins and cookies] – two with and two without shelled hemp seed) liked or strongly liked muffins containing shelled hemp seed. Although 77% of the participants were aware of (has heard) omega-3 fatty acids, only 7% knew of hemp seed as a source (Grady et al., 2017).

Several components of hempseed have been proposed for novel product development. One such product Olimpiq StemxCell SL® contains hempseed extract as active component; it is specifically designed to treat diabetes mellitus by modulating blood sugar levels, enhancing regenerative capacity, and preventing diabetic complications (Doczi-Keresztesi et al., 2012). Edestin from hempseed has been proposed for the production of blood plasma substitute. Hempseeds is cold pressed, or CO_2 extracted, decolorized, filtered, extracted in salt water solution (20–60 °C), and filtered to obtain edestin (Horska, 2007). Hempseed protein fractions are proposed for use to form stable emulsions for lipid microencapsulation. The fraction (≥ 5,000 Da) is isolated from aqueous alkaline extraction of hempseed, isoelectric precipitation, and followed by separation. The isolated hempseed protein fractions form stable emulsions with lipids and form microencapsulated powders (Samaranayaka, Wanasundara, Ray, & Green, 2019). The benefits of hempseed fatty acids led to the development of a product comprising linoleic acid (LA) and α-linolenic acid

(ALA) in a LA/ALA ratio greater than 1:1 and less than 5:1 containing γ-linolenic acid (15–85%) to treat diseases including breast cancer, cardiovascular disease, and inflammation (Peskin, 2014). The lignanamides Sativamide A and B are proposed to treat neurodegenerative diseases including neuroblastoma, pheochromocytoma, Parkinson's disease, Alzheimer's disease, Huntington's disease, and prion diseases (Zhu et al., 2018). A consumable composition is proposed containing *N-trans*-caffeoyltyramine, *N-trans*-feruloyltyramine, or *p*-coumaroyltyramine to modulate metabolism addressing the pathogenesis of metabolic disorders such as fatty liver disease, nonalcoholic steatohepatitis, and type II diabetes mellitus (Heil, 2019). Moreover, the composition is formulated as a dietary supplement, food ingredient or additive, a medical food, nutraceutical or pharmaceutical composition. The composition is claimed to improve HNF4α activity, insulin-like growth factor, blood sugar, insulin, C peptide, triglyceride, free fatty acid and blood uric levels, glucose transporter expression, microalbumin, adinopectin, total serum cholesterol, high-density lipoprotein, and low-density lipoprotein levels. The composition improves metabolism, liver function, fasting plasma glucose and postprandial plasma glucose levels, glycosylated hemoglobin HbA1c, body weight, insulin sensitivity, or serum lipid profile (Heil, 2019).

Hemp is an ancient sustainable "green" crop that has an interrupted history due to the presence of phytocannabinoids. However, the hempseed revolution is advancing even though the CBD hemp oil is the fastest growing segment. Hempseed has not qualified for any health claims to date, although its high polyunsaturated fatty acid content fits in with the blood cholesterol-lowering effect. Moreover, the phytosterol content of hempseed further demonstrates its potential for blood cholesterol-lowering effect and its insoluble fiber can benefit digestive health. The fatty acid composition of hempseed oil, particularly α-linolenic acid along with γ-linolenic acid and stearidonic acid, can potentially improve cardiovascular health. Furthermore, the 3:1 n−6/n−3 ratio can benefit skin diseases and provide relief from eczema. The lignanamides from the underexploited hempseed hulls can potentially benefit cognitive health.

Novel hempseed products frequently appear in the news. For example, hempseed extract features in the new energy beverage along with ginseng, guarana, B vitamins, caffeine, sugar, and taurine launched recently in Germany. This energy drink represents innovation trends in hemp products (Arthur, 2021). Improved varieties of new oilseed crops including hemp are being developed for Eastern Canada. The project funded by Agriculture and Agri-Food Canada and implemented by Eastern Canada Oilseed Development Alliance (ECODA) also focuses on product innovation to boost market opportunities for oilseeds and value-added products (Demaree-Saddler, 2018). Meanwhile, the 2018 US Farm bill recognizes hemp as an agricultural commodity. Recently, the Association of American Feed Control Officials (AAFCO) received submission for hemp to become an approved feed ingredient. Montana bill "*HB 396*" already includes hempseed food ingredient as commercial feed, authorizes their use for pets, specialty pets, and horses, and approves hemp as an additive or defines ingredient in animal feed or medicated feed for livestock (feednavigator.com, 2021). Thailand is promoting hemp as an economic crop and the Ministry of Public Health has published (notification no. 425) specific requirements with imposed limits on THC and CBD for hempseed, its oil and protein products use (Koe, 2021). According to the Thai legal framework, hemp can be used as an ingredient in food products, including breakfast cereal, bakery products, beverages, snacks, butter, and food supplements. Moreover, hempseed protein can be used in supplements, protein powder, and general food products, whereas hempseed oil can be used in dairy fat spread, fat spreads, and salads. These developments and the physiological benefits of industrial hempseed and its components described in this chapter reassure us that many new value-added products will eventually emerge for use in the functional food, nutraceutical, and other industries.

References

Aachary, A. A., Liang, J., Hydamaka, A., Eskin, N. A. M., & Thiyam-Holländer, U. (2016). A new ultrasound-assisted bleaching technique for impacting chlorophyll content of cold-pressed hempseed oil. *LWT (Food Science and Technology Lebensmittel-Wissenschaft), 72*, 439–446.

Abdelhamid, A.S., Brown, T.J., Brainard, J.S., Biswas, P., Thorpe, G.C., Moore, H.J., Deane, K.H.O., Summerbell, C.D., Worthington, H.V., Song, F., and Hooper, L. (2020). Omega-3 fatty acids for the primary and secondary prevention of cardiovascular disease. *Cochrane Database of Systematic Reviews*, 3, https://doi.org//10.1002/14651858.CD003177.pub5.

Abolbaghaei, A., Oomah, B. D., Tavakoli, H., & Hosseinian, F. (2018). Influence of branched chain amino acids on insulin sensitivity and the mediator roles of short chain fatty acids and gut hormones: a review. *Journal of Food Bioactives, 2*, 1–15.

Aiello, G., Fasoli, E., Boschin, G., Lammi, C., Zanoni, C., Citterio, A., & Arnoldi, A. (2016). Proteomic characterization of hempseed (*C. sativa* L.). *Journal of Proteomics, 147*, 187–196.

Aiello, G., Lammi, C., Boschin, G., Zanoni, C., & Arnoldi, A. (2017). Exploration of potentially bioactive peptides generated from the enzymatic hydrolysis of hempseed proteins. *Journal of Agricultural and Food Chemistry, 65*, 10174–10184.

Ajibola, C.F. (2020). *Structural and functional properties of hemp seed storage proteins*. MSc thesis, University of Manitoba, Winnipeg, Manitoba, Canada, 113. http://hd http://l.handle.net/1993/34525.

Aladić, K., Jarni, K., Barbir, T., Vidović, S., Vladić, J., Bilić, M., & Jokić, S. (2015). Supercritical CO_2 extraction of hemp (*C. sativa*) seed oil. *Industrial Crops and Products, 76*, 472–478.

Aladić, K., Jokić, S., Moslavac, T., Tomas, S., Vidović, S., Vladić, J., & Šubarić, D. (2014). Cold pressing and supercritical CO_2 extraction of hemp (*C. sativa*) seed oil. *Chemical and Biochemical Engineering Quarterly, 28*(4), 481–490.

Al-Khalifa, A., Maddaford, T. G., Chahine, M. N., Austria, J. A., Edel, A. L., Richard, M. N., . . . Pierce, G. N. (2007). Effect of dietary hempseed intake on cardiac ischemia–reperfusion injury. *American Journal of Physiolology Regulatory. Integrative and Comparative Physiology, 292*, R1198–R1203.

Alonso-Esteban, J.I., González-Fernández, M.J., Fabrikov, D., Torija-Isasa, E., Sánchez-Mata, M.C. & Guil-Guerrero, J.L. (2020). Hemp (*C. sativa* L.) varieties: fatty acid profiles and upgrading of γ-linolenic acid-containing hemp seed oils. *European Journal of Lipid Science and Technology*, 122 (7) 1900445; https://doi.org/10.1002/ejlt.201900445.

Aluko, R. E. (2017). Hemp seed (*C. sativa* L.) proteins: composition, structure, enzymatic modification, and functional and bioactive properties. In S. R. Nadathur,

J. P. D. Wanasundara, & L. Scanlin (Eds.), *Sustainable Protein Sources* (pp. 121–132). London, UK: Academic Press.

Aluko, R.E. (2019). Food protein-derived renin-inhibitory peptides: *in vitro* and *in vivo* properties. *Journal of Food Biochemistry*, 43 (1) e12648; https://doi.org/10.1111/jfbc.12648.

Antonelli, M., Benedetti, B., Cannazza, G., Cerrato, A., Citti, C., Montone, C.M., Piovesana, S. & Laganá, A. (2020). New insights in hemp chemical composition: a comprehensive polar lipidome characterization by combining solid phase enrichment, high-resolution mass spectrometry, and cheminformatics. *Analytical and Bioanalytical Chemistry*, 412, 413–423; https://doi.org/10.1007/s00216-019-02247-6.

Arthur, R. (2021). PepsiCo launches Rockstar Energy + Hemp. Beveragedaily.com 08 Apr 2021.

Avilés-Gaxiola, S., Gutiérrez-Grijalva, E. P., León-Felix, J., Angulo-Escalante, M. A., & Heredia, J. B. (2020). Peptides in colon cancer: current state of knowledge. *Plant Foods for Human Nutrition, 75*, 467–476.

Babiker, E.E., Uslu, N., Juhaimi, F.A., Ahmed, I.A.M., Ghafoor, K., Özcan, M.M., and Almusallam, I.A. (2021). Effect of roasting on antioxidative properties, polyphenol profile and fatty acids composition of hemp (*C. sativa* L.) seeds. *LWT – Food Science and Technology*, 139, 110537; https://doi.org/10.1016.j.lwt.2020.110537.

Beydoun, M.A., Canas, J.A., Fanelli-Kuczmarski, M.T., Maldonado, A.I., Shaked, D., Kivimaki, M., Evans, M.K., and Zonderman, A.B. (2020). Association of antioxidant vitamins A, C, E and carotenoids with cognitive performance over time: A cohort study of middle-aged adults. *Nutrients*, 12, 3558; http://dx.doi.org/10.3390/nu12113558.

Bhullar, K. S., and Wu, J. (2020). Dietary peptides in aging: evidence and prospects. *Food Science and Human Wellness*, 9(1); https://doi.org/10.1016/j.fshw.2020.01.001.

Blade, S. F., Ampong-Nyarko, K., & Przybylski, R. (2006). Fatty acid and tocopherol profiles of industrial hemp cultivars grown in the high latitude prairie region of Canada. *Journal of Industrial Hemp, 10*(2), 33–43.

Boachie, R., Yao, S., & Udenigwe, C. C. (2018). Molecular mechanisms of cholesterol-lowering peptides derived food proteins. *Current Opinion in Food Science, 20*, 58–63.

Boots, J-W. P. (2013). Protein hydrolysate enriched in peptides inhibiting DPP-IV and their use. *US Patent* 2013/0096074 A1. Issued April 18, 2013.

Bourjot, M., Zedet, A., Demange, B., Pudlo, M., & Girard-Thernier, C. (2016). *In vitro* mammalian arginase inhibitory and antioxidant effects of amide derivatives isolated from the hempseed cakes (*C. sativa*). *Planta Medica International Open, 3*, e64–e67.

Bresnick, E.H., and Tanimura, N. (2019), Methods of controlling red blood cell production, *US Patent* US 2019/0270992 A1, Issued September 5, 2019.

Buré, C., Solgadi, A., Yen-Nicolaÿ, S., Bardeau, T., Libong, D., Abreu, S., . . . Cansell, M. (2016). Electrospray mass spectrometry as a tool to characterize phospholipid composition of plant cakes. *European Journal of Lipid Science and Technology, 118*, 1282–1292.

Callaway, J., Schwab, U., Harvima, I., Halonen, P., Mykkänen, O., Hyvönen, P., & Järvinen, T. (2005). Efficacy of dietary hempseed oil in patients with atopic dermatitis. *Journal of Dermatological Treatment, 16*, 87–94.

Callaway, J. C. (2004). Hempseed as a nutritional resource: an overview. *Euphytica, 140*, 65–72.

Camacho-Barcia, L., García-Gavilán, J., Papandreou, C., Hansen, T.T., Harrold, J.A., Finlayson, G., Blundell, J.E., Sjödin, A., Halford, J.C.G., and Bulló, M. (2021) Circulating metabolites associated with postprandial satiety in overweight/obese participants: The SATIN study. *Nutrients*, 13, 549; https://doi.org/10.3390/nu13020549.

Cansell, M., Bardeau, T., Morvan, E., Grélard, A., Buré, C., & Subra-Paternault, P. (2017). Phospholipid profiles of oleaginous pressed cakes using NMR and gas chromatography. *Journal of the American Oil Chemist's Society*, *94*(9), 1219–1223.

Carbone, V., Carelli, A.A., Carrin, M.E., Cozzolino, R., Crew, C., Aparicio, R., Ayadi, M., Baumler, E.R., and Penci, M.C. (2009). Assessment of chemical and physical–chemical properties of edible oils. In G.L. Russo (ed.) *Sixth Framework Programme Priority 5 Food Quality and Safety Priority*, Call 4-C, Specific Support Action, MAC-Oils Mapping and Comparing Oils, The Scientific Handbook, The European Communities, MAC-Oils Consortium. http://www.mac-oils.eu.

Catchpole, O. J., Tallon, S. J., Eltringham, W. E., Gray, J. B., Fenton, K. A., Vagi, E. M., . . . Zhu, J. R. Y. (2009). The extraction and fractionation of specialty lipids using near critical fluids. *Journal of Supercritical Fluids*, *47*, 591–597.

Chen, T., Hao, J., He, J., Zhang, J., Li, Y., Liu, R., & Li, L. (2013). Cannabisin B induces autophagic cell death by inhibiting the AKT/mTOR pathway and S phase cell cycle arrest in HepG2 cells. *Food Chemistry*, *138*(2–3), 1034–1041.

Chen, T., He, J., Zhang, J., Li, X., Zhang, H., Hao, J., & Li, L. (2012). The isolation and identification of two compounds with predominant radical scavenging activity in hempseed (seed of *C. sativa* L.). *Food Chemistry*, *134*, 1030–1037.

Chen, T., He, J., Zhang, J., Zhang, H., Qian, P., Hao, J., & Li, L. (2010). Analytical characterization of hempseed (seed of *C. sativa* L.) oil from eight regions in China. *Journal of Dietary Supplements*, *7*(2), 117–129.

Citti, C., Pacchetti, B., Vandelli, M. A., Forni, F., & Cannazza, G. (2018). Analysis of cannabinoids in commercial hemp seed oil and decarboxylation linetic studies of cannabidolic acid (CBDA). *Journal of Pharmaceutical and Biomedical Analysis*, *149*, 532–540.

Dam, H. (1935). The antihæmorrhagic vitamin of the chick: occurrence and chemical nature. *Nature*, *135*(3417), 652–653.

Das, R. (2015). Multienzyme modification of hemp protein for functional peptide synthesis. *Journal of Food Processing*, 738984; http://dx.doi.org/10.1155/2015/738984

Das, U. N. (2021). Arachidonic acid in (for) COVID-19. *Agro FOOD Industry Hi-Tech*, *32*(1), 62–65.

Del Bo, C., Deon, V., Abello, F., Massini, G., Porrini, M., Riso, P., & Guardamagna, O. (2019). Eight-week hempseed oil intervention improves the fatty acid composition of erythrocyte phospholipids and the omega-3 index, but does not affect the lipid profile in children and adolescents with primary hyperlipidemia. *Food Research International*, *119*, 469–476.

Demaree-Saddler, H. (2018).Canada invests in oilseeds. Retrieved April 9, 2021, from https://www.world-grain.com/authors/87-holly-demaree-saddler.

Docimo, T., Caruso, I., Ponzoni, E., Mattana, M., & Galasso, I. (2014). Molecular characterization of edestin gene family in *C. sativa* L. *Plant Physiology and Biochemistry*, *84*, 142–148.

Doczi-Keresztesi, Z., Jung, J., Kiss, I., Mezei, T., Szabo, L., & Ember, I. (2012). Retinal and renal vascular permeability changes caused by stem cell stimulation in alloxan-induced diabetic rats measured by extravasation of fluorescein. *In Vivo*, *26*, 427–436.

Dunford, N. T. (2015). Hemp and flaxseed oil: properties and applications for use in food. In G. Talbot (Ed.), *Specialty Oils and Fats in Food and Nutrition Properties, Processing and Application* (pp. 39–63). Amsterdam, Holland: Woodhead Publishing.

Elmadfa, I., & Kornsteiner-Krenn, M. (2010). Fat and nutrition. In D. Boskou & I. Elmadfa (Eds.), *Frying of Food: Oxidation, Nutrient and Non-Nutrient Antioxidants, Biologically Active Compounds, and High Temperatures* (second ed., pp. 1–22). Boca Raton, FL: CRC Press.

Etemadian, Y., Ghaemi, V., Shaviklo, A.R., Pourashouri, P., Mahoonak, A.R.S., and Rafipour, F. (2021). Development of animal/plant-base protein hydrolysate and its application in food, feed and nutraceutical industries: State of the art. *Journal of Cleaner Production*, 278, 123219; https://doi.org/10.1016/j.jclepro.2020.123219.

European Food Safety Authority (EFSA). (2008). Plant sterols and blood cholesterol—scientific substraction of a health claim related to plant sterols and lower/reduced blood cholesterol and reduced risk of (coronary) heart disease pursuant to Article 14 of Regulation (EC) No. 1924/2006, EFSA J 6(8): 781 –n/a.

Farinon, B., Molinari, R., Constantini, L., and Merendino, N. (2020). The seed of industrial hemp (*C. sativa* L.): Nutritional quality and potential functionality for human health and nutrition. *Nutrients*, 12, 1935; https://www.mdpi.com/2072-6643 (7 December1935).

Fatemi, S. H., & Hammond, E. G. (1980). Analysis of oleate, linoleate and linolenate hydroperoxides in oxidized ester mixtures. *Lipids*, *15*, 379–385.

Faugno, S., Piccolella, S., Sannino, M., Principio, L., Crescente, G., Baldi, G. M., . . . Pacifico, S. (2019). Can agronomic practices and cold-pressing extraction parameters affect phenols and polyphenols content in hempseed oils? *Industrial Crops and Products*, *130*, 511–519.

Feednavigator.com. (2021). Supporters hail Montana bill passage as milestone in development of US hemp grain industry. 16 April 2021; https://feednavigator.com.

Frassinetti, S., Gabriele, M., Moccia, E., Longo, V., and Di Gioia, D. (2020). Antimicrobial and antibiofilm activity of *C. sativa* L. seeds extract against *S. aureus* and growth effects on probiotic *Lactobacillus* spp. *LWT-Food Science and Technology*, 124, 109149; https://doi.org/10.1016/j.lwt.2020.109149.

Frassinetti, S., Moccia, E., Caltavuturo, L., Gabriela, M., Longo, V., Bellani, L., . . . Giorgetti, L. (2018). Nutraceutical potential of hemp (*C. sativa* L.) seeds and sprouts. *Food Chemistry*, *262*, 56–66.

Galasso, I., Russo, R., Mapelli, S., Ponzoni, E., Brambilla, I.M., Battelli, G., and Reggiani, R. (2016). Variability in seed traits in a collection of *C. sativa* L. genotypes. *Frontiers in Plant Science*, 7, 688; https://www.frontiersin.org/articles/10.3389/fpls.2016.00688/full.

Gao, Y.-X., Fan, C., Tan, X., Zhang, I., and Wang, J. (2019). Stearidonic acid promotes insulin secretion via stimulation of G protein-coupled receptor 40 in MIN6 pancreatic β-cells. *Journal of Functional Foods*, 60, 103450; https://doi.org/10.1016/j.jff.2019.103450.

Garner-Wizard, M. (2013). *Hempseed Food: The REAL Secret Ingredient for Health & Happiness*. Lulu.com.

Gavel, N., Edel, A., Bassett, C., Weber, A., Merchant, M., Rodriguez-Leyva, D., & Pierce, G. (2011). The effect of dietary hempseed on atherogenesis and contractile function in aortae from hypercholesterolemic rabbits. *Acta Physiologica Hungarica*, *98*(3), 273–283.

Ghazani, S. M., & Marangoni, A. G. (2013). Minor components in canola oil and effects of refining on these constituents: a review. *Journal of the American Oil Chemists' Society*, *90*(7), 923–932.

Girgih, A. T., Alashi, A., He, R., Malomo, S., & Aluko, R. E. (2014). Preventive and treatment effects of a hemp seed (*C. sativa* L.) meal protein hydrolysate against high blood pressure in spontaneously hypertensive rats. *European Journal of Nutrition*, *53*(5), 1237–1246.

Girgih, A. T., Alashi, A. M., He, R., Malomo, S. A., Raj, P., Netticadan, T., & Aluko, R. E. (2014). A novel hemp seed meal protein hydrolysate reduces oxidative stress factors in spontaneously hypertensive rats. *Nutrients*, *6*(12), 5652–5666.

Girgih, A. T., Udenigwe, C. C., & Aluko, R. E. (2011). *In vitro* antioxidant properties of hemp seed (*C. sativa* L.) protein hydrolysate fractions. *Journal of the American Oil Chemists Society*, *88*, 381–389.

Girgih, A. T., Udenigwe, C. C., & Aluko, R. E. (2013). Reverse-phase HPLC separation of hemp seed (*C. sativa* L.) protein hydrolysate produced peptide fractions with enhanced antioxidant capacity. *Plant Foods for Human Nutrition*, *68*, 39–46.

Girgih, A. T., Udenigwe, C. C., Li, H., Adebiyi, A. P., & Aluko, R. E. (2011). Kinetics of enzyme inhibition and antihypertensive effects of hemp seed (*C. sativa* L.) protein hydrolysates. *Journal of the American Oil Chemists Society*, *88*, 1767–1774.

Grady, A., Schmidt, M., Milar, P., Sadrija, A., Smith, K., & Erickson, D. (2017). Acceptability and knowledge of hemp seed as a dietary source of ALA. *Journal of the Academy of Nutrition and Dietetics*, *117*(9), A51.

Graham, B., Ward, A., Cottrell, C., Joyce, M.G., Kanekiyo, M., Pallesen, J. . . . Turner, H. (2018). Perfusion coronavirus spike proteins and their use. *World Patent WO 2018/081318 A1*, Issued May 3, 2018.

Grigoryev, S. V., Illarinova, K. V., & Shelenga, T. V. (2020). Hempseeds (*Cannabis* spp.) as a source of functional food ingredients, prebiotics and phytosterols. *Agricultural and Food Science*, *29*, 460–470.

Grijó, D. R., Pivaa, G. K., Osoriob, I. V., & Cardozo-Filho, L. (2019). Hemp (*C. sativa* L.) seed oil extraction with pressurized n-propane and supercritical carbon dioxide. *The Journal of Supercritical Fluids*, *143*, 268–274.

Guil-Guerrero, J. L. (2007). Stearidonic acid (18:4*n*−3): metabolism, nutritional importance, medical uses and natural sources. *European Journal of Lipid Science and Technology*, *109*(12), 1226–1236.

Hall, C. E. (1950). Electron microscopy of crystalline edestin. *Journal of Biological Chemistry*, *185*, 45–51.

Hancocks, N. (2021). High fiber diet could control COVID-19 inflammation, study suggests. *Nutraingredients.com*, April 1, 2021.

Hashemzaeh, M., Moghaddam, A., Karimi, I., Haghighi, Z. M. S., & Hassanpour, S. (2017). Semiprotective effects of hempseed oil on carbon tetrachloride-induced hepatotoxicity in male rats: an ultra-short toxicological intervention. *Journal of Chemical Health Risks.*, *7*(4), 319–330.

Heil, C.L. (2019). Composition for modulating metabolism. *World Patent WO 2019/140046 A1*, Issued July 18, 2019.

Herreman, L., Nommensen, P., Pennings, B., and Laus, M.C. (2020).Comprehensive overview of the quality of plant- and animal-sourced proteins based on the digestible indispensable amino acid score. *Food Science and Nutrition*, 1–13. https://doi.org/10.1002/fsn3.1809.

Herrmann, K. R., Ruff, A. J., & Schwaneberg, U. (2020). Phytase-based phosphorus recovery process for 20 distinct press cakes. *ACS Sustainable Chemistry and Engineering*, *8*(9), 3913–3921.

Horska, I. (2007). Agent for substitution of blood plasma and a method of its production. *U.S. Patent 2007/0196526 A1*, Issued August 23, 2007.

House, J. D., Neufeld, J., & Leson, G. (2010). Evaluating the quality of protein from hemp seed (*C. sativa* L.) products through the use of the protein digestibility-corrected amino acid score method. *Journal of Agricultural and Food Chemistry*, *58*, 11801–11807.

Huang, W., Zeng, Z., Lang, Y., Xiang, X., Qi, G., Lu, G., & Yang, X. (2021). Cannabis seed oil alleviates experimental atherosclerosis by ameliorating vascular inflammation in apolipoprotein-E-deficient mice. *Journal of Agricultural and Food Chemistry Article ASAP*; https://doi.org/10.1021/acs.jafc.0c07251.

Iannaccone, M., Lanni, A., Contali, F., Esposito, S., Martino, C., Bennato, F., De Angelis E., Grotta, L., Pomilio, F., Giansante, D. & Martino, G. (2019). Whole blood transcriptome analysis in ewes fed with hemp seed supplemented diet. *Scientific Reports*, 9 : 16192; https://doi.org/10.1038/s41598-0.19-52712-6.

Irakli, M., Tsaliki, E., Kalivas, A., Kleisiaris, F., Sarrou, E., and Cook, C.M. (2019). Effect of genotype and growing year on the nutritional, phytochemical, and antioxidant properties of industrial hemp (*C. sativa* L.) seeds. *Antioxidants*, 8 (10), 491; https://doi.org/10.3390/antiox8100491.

Islam, M. A., Jeong, B.-G., Jung, J., Shin, E.-C., Choi, S.-G., & Chun, J. (2017). Phytosterol determination and method validation for selected nuts and seeds. *Food Analytical Methods*, *10*, 3225–3234.

Izzo, L., Castaldo, L., Narváez, A., Graziani, G., Gaspari, A., Rodríguez-Carrasco, Y., and Ritieni, A. (2020). Analysis of phenolic compounds in commercial *C. sativa* L. inflorescences using UHPLC-Q-Orbitrap HRMS. *Molecules*, 25, 631; https://doi.org/10.3390/molecules25030631.

Izzo, L., Pacifico, S., Piccolella, S., Castaldo, L., Narvàez, A., Grosso, M. & Ritieni, A. (2020). Chemical analysis of minor bioactive components and cannabidolic acid in commercial hemp seed oil. *Molecules*, 25, 3710; http://dx.doi.org/10.3390/molecules25163710.

Jeong, M., Cho, J., Shin, J.-I., Jeon, Y.-J., Kim, J.-H., Lee, S.-J., . . . Lee, K. (2014). Hempseed oil induces reactive oxygen species-and C/EBP homologus protein-mediated apoptosis in MH7A human rheumatoid arthritis fibroblast-like synovial cells. *Journal of Ethnopharmacology*, *154*, 745–752.

Jin, S., and Lee, M-Y. (2018). The ameliorative effect of hemp seed hexane extracts on the *P. acnes*-induced inflammation and lipogenesis in sebocytes. *PLoS ONE*, 13(8):e0202933. https://doi.org/10.1371/journal.pone.0202933.

Johnson, M. (2019). Hempseed oil components. *WO 2019/23493 A1*, World Intellectual Property Organization, Issued December 5, 2019.

Kalmendal, R. (2008). Hemp seed cake fed to broilers. *MSc thesis*, Department of Animal Nutrition and Management, Swedish University of Agricultural Sciences (SLU), Uppsala, Sweden, 43.

Kaul, N., Kreml, R., Austria, J. A., Richard, M. N., Edel, A. L., Dibrov, E., . . . Pierce, G. N. (2008). A comparison of fish oil, flaxseed oil and hempseed oil supplementation on selected parameters of cardiovascular health in healthy volunteers. *Journal of the American College of Nutrition*, *27*(1), 51–58.

Kaushal, N., Dhadwal, S., & Kaur, P. (2020). Ameliorative effects of hempseed (*C. sativa*) against hypercholesterolemia associated cardiovascular changes. *Nutrition, Metabolism and Cardiovascular Diseases*, *30*(2), 330–338.

Kaushal N, Gupta M, and Kulshreshtha E. (2020). Hempseed (*C. sativa*) lipid fractions alleviate high-fat diet-induced fatty liver disease through regulation of inflammation and oxidative stress. *Heliyon* 6(7); https://doi.org/10.1016/j.heliyon.2020.e04422.

Kim, J.-J., & Lee, M.-Y. (2011). Isolation and characterization of edestin from Cheungsam hempseed. *Journal of Applied Biology and Chemistry*, *54*(2), 84–88.

Kim, J. W., Koo, B., Kim, I. H., & Nyachoti, C. M. (2018). Effects of extrusion and microbial phytase on the apparent an standardized total tract digestibility of phosphorus in hemp hulls fed to growing pigs. *Journal of Animal Science*, *96*, 1838–1845.

Kiralan, M., Gül, V., & Kara, Ş. M. (2010). Fatty acid composition of hempseed oils from different locations in Turkey. *Spanish Journal of Agricultural Research*, *8*(2), 385–390.

Kjølbæk, L., Benítez-Páez, A., Del Pulgar, E. M. G., Brahe, L. K., Liebisch, G., Matysik, S., . . . Sanz, Y. (2020). Arabinoxylan oligosaccharides and polyunsaturated fatty acid effects on gut microbiota and metabolic markers in overweight individuals with signs of metabolic syndrome: a randomized cross-over trial. *Clinical Nutrition*, *39*(1), 67–79.

Koe, T. (2021). Hemp and CBD in Thailand: Food and supplement companies rush into NPD as FDA approves draft regulation. https://www.nutraingredients-asia.com, March 29, 2021.

Korus, J., Witczak, M., Ziobro, R., & Juszczak, L. (2017). Hemp (*C. sativa* subsp. *sativa*) flour and protein preparation as natural nutrients and structure forming agents in starch based gluten-free bread. *LWT-Food Science and Technology*, *84*, 143–150.

Kotecka-Majchrzak, K., Sumara, A., Fornal, E. & Montowska, M. (2020). Identification of species-specific peptide markers in cold-pressed oils. *Scientific Reports*, **10**, 19971; https://doi.org/10.1038/s41598-020-76944-z.

Kotecka-Majchrzak, K., Sumara, A., Fornal, E., & Montowska, M. (2021). Proteomic analysis of oilseed cake: a comparative study of species-specific proteins and peptides extracted from ten seed species. *Journal of the Science of Food and Agriculture*, *101*, 297–306.

Kriese, U., Schumann, E., Weber, W. E., Beyer, M., Brühl, L., & Matthäus, B. (2004). Oil content, tocopherol composition and fatty acid patterns of the seeds of 51 *C. sativa* L. genotypes. *Euphytica*, *137*, 339–351.

Lammi, C., Bollati, C., Gelain, F., Arnoldi, A., and Pugliese, R. (2019). Enhancement of the stability and anti-DPPIV activity of hempseed hydrolysates through self-assembling peptide-based hydrogels. *Frontiers in Chemistry*, 6, 670; https://doi.org/10.3389/fchem.2018.00670.

Latif, S., & Anwar, F. (2009). Physicochemical studies of hemp (*C. sativa*) seed oil using enzyme-assisted cold-pressing. *European Journal of Lipid Science and Technology*, *111*, 1042–1048.

Lee, M. J., Park, S. H., Han, J. H., Hong, Y. K., Hwang, S., Lee, S., . . . Cho, K. S. (2011). The effects of hempseed meal intake and linoleic acid on *Drosophila* models of neurodegenerative diseases and hypercholesterolemia. *Molecules and Cells*, *31*, 337–342.

Leonard, W., Zhang, P., Ying, D., & Fang, Z. (2019). Hempseed in food industry: nutritional value, health benefits, and industrial applications. *Comprehensive Reviews in Food Science and Food Safety*, *19*(1), 282–308.

Leonard, W., Zhang, P., Ying, D., and Fang, Z. (2020) Lignanamides: sources, biosynthesis and potential health benefits—A minireview, *Critical Reviews in Food Science and Nutrition*; https://doi.org/10.1080/10408398.2020.1759025.

Leonard, W., Zhang, P., Ying, D., Xiong, Y., and Fang Z. (2021). Extrusion improves the phenolic profile and biological activities of hempseed (*C. sativa* L.) hull. *Food Chemistry*, 346 : 128606; https://doi.org/10.1016/j.foodchem.2020.128606.

Lesma, G., Consonni, R., Gambaro, V., Remuzzi, C., Roda, G., Silvani, A., . . . and Visconti, G.L. (2014). Cannabinoid-free *C. sativa* L. grown in the Po valley: evaluation of fatty acid profile, antioxidant capacity and metabolic content. *Natural Product Research*, *28*(21), 1801–1807.

Leyva, D. R., McCullough, R. S., & Pierce, G. N. (2011). Medicinal use of hempseeds (*C. sativa* L.): Effects on platelet aggregation. In V. R. Preedy, R. R. Watson, & V. B. Patel (Eds.), *Nuts and Seeds in Health and Disease Prevention* (pp. 637–646). San Diego, CA: Academic Press.

Liang, J., Appukuttan Aachary, A., & Thiyam-Holländer, U. (2015). Hemp seed oil: Minor components and oil quality. *Lipid Technology*, *27*(10), 231–233.

Lieu, E. L., Nguyen, T., Rhyne, S., & Kim, J. (2020). Amino acids in cancer. *Experimental and Molecular Medicine*, *52*, 15–30.

Lin, Y., Pangloli, P., Meng, X., and Dia, V.P. (2020). Effect of heating on the digestibility of isolated hempseed (*C. sativa* L.) protein and bioactivity of its pepsin-pancreatin digests. *Food Chemistry*, 314, 126198: https://doi.org/10.106/j.foodchem.2020.126198.

Logarušić, M., Slivac, I., Radošević, K., Bagović, M., Redovniković, I. R., & Srček, V. G. (2019). Hempseed protein hydrolysates' effects on the proliferation and induced oxidative stress in normal and cancer cell lines. *Molecular Biology Reports*, *46*, 6079–6085.

Longoria-Sanchez, A., Morales, M. V., Oomah, B. D., Medina-Godoy, S., Ochoa-Espinoza, X., Góngora-Gómez, A. M., & Espinosa-Alonso, L. G. (2019). Characteristics and antioxidant properties of cold pressed high oleic and linoleic oils from Mexican safflower varieties. *Emirates Journal of Food and Agriculture*, *31*(9), 679–687.

Louca P, Murray B, Klaser K, Mark S Graham, Mohsen Mazidi, Emily R Leeming, Ellen Thompson, Ruth Bowyer, David A Drew, Long H Nguyen, Jordi Merino, Maria Gomez, Olatz Mompeo, Ricardo Costeira, Carole H Sudre, Rachel Gibson, Claire J Steves, Jonathan Wolf, Paul W Franks, Sebastien Ourselin, Andrew T Chan, Sarah E Berry, Ana M Valdes, Philip C Calder, Tim D Spector, and Cristina Menni (2021). Modest effects of dietary supplements during the COVID-19 pandemic: insights from 445 850 users of the COVID-19 Symptom Study app. *BMJ Nutrition, Prevention and Health*; http://dx.doi.org/10.1136/bmjnph-2021-000250.

Lu, R. R., Qian, P., Sun, Z., Zhou, X. H., Chen, T. P., He, J. F., . . . Wu, J. (2010). Hempseed protein derived antioxidative peptides: purification, identification and protection from hydrogen peroxide-induced apoptosis in PC12 cells. *Food Chemistry*, *123*, 1210–1218.

Luo, Q., Yan, X., Bobrovskaya, L., Ji, M., Yuan, H., Lou, H., & Fan, P. (2017). Anti-neuroinflammatory effects of grossamide from hemp seed via suppression of TLR-4-mediated NF-κB signaling pathways in lipopolysaccharide-stimulated BV2 microglia cells. *Molecular and Cellular Biochemistry*, *428*(1–2), 129–137.

Maleki, J., Tomasevic, I., Djekic, I. & Solowiej, B.S. (2020). The effect of protein source on the physicochemical, nutritional properties and microstructure of high-protein bars intended for physically active people. *Foods*, 9, 1467; https://doi.org/10.3390/foods9101467.

Maloma, S. A., & Aluko, R. E. (2015). A comparative study of the structural and functional properties of isolated hemp seed (*C. sativa* L.) albumin and globulin fractions. *Food Hydrocolloids*, *43*, 743–752.

Malomo, S.A. & Aluko, R. E. (2019). Kinetics of acetylcholinesterase inhibition by hemp seed protein-derived peptides. *Journal of Food Biochemistry*, 43,7, e12867; https://doi.org/10.1111/jfbc.12897.

Malomo, S. A., Onuh, J. O., Girgih, A. T., & Aluko, R. E. (2015). Structural and antihypertensive properties of enzymatic hemp seed protein hydrolysates. *Nutrients*, *7*, 7616–7632.

Mamone, G., Picariello, G., Ramondo, A., Nicolai, M. A., & Ferranti, P. (2019). Production, digestibility and allergenicity of hemp (*C. sativa* L.) protein isolates. *Food Research International*, *115*, 562–571.

MarketsandMarkets. (2020). Plant-based market projected to reach $14.5 billion by 2025. www.marketsandmarkets.com.

MarketsandMarkets. (2021). Industrial hemp market projected to reach $26.6 billion by 2025. www.marketsandmarkets.com.

Martinez, J. R., Montserrat-de la Paz, S., De la Puerta, R., Garcia-Giménez, M. D., & Fernández-Arche, M. Á. (2020). Characterization of bioactive compounds in defatted hempseed (*C. sativa* L.) by UHPLC-HRMS/MS an anti-inflammatory activity in primary human monocytes. *Food and Function*, *11*, 4057–4066.

Martínez-Sánchez, S.M., Gabaldón-Hernández, J.A., and Montoro-García, S. (2020). Unraveling the molecular mechanisms associated with the role of food-derived bioactive peptides in promoting cardiovascular health. *Journal of Functional Foods*, 64, 103645; https://doi.org/10.1016/j.jff.2019.103645.

Matthäus, B., & Brühl, L. (2008). Virgin hemp seed oil: An interesting niche product. *European Journal of Lipid Science and Technology*, *110*(7), 655–661.

Matthäus, B., Schumann, E., Brühl, L., & Kriese, U. (2006). Hempseed oil—Influence of the genotype on the composition in a two-year study. *Journal of Industrial Hemp*, *10*(2), 45–65.

Mattila, P., Mäkinen, S., Eurola, M., Jalava, T., Pihlava, J.-M., Hellström, J., & Pihlanto, A. (2018). Nutritive value of commercial protein-rich plant products. *Plant Foods for Human Nutrition*, *73*, 108–115.

Mattila, P., Pihlava, J.-M., Hellström, J., Nurmi, M., Eurola, M., Mäkinen, S., . . . Pihlanto, A. (2018). Contents of phytochemicals and antinutritional factors in commercial protein-rich plant products. *Food Quality and Safety*, *2*, 213–219.

Michailidis, D., Angelis, A., Nikolaou, P.E., Mitakou, S., and Skaltsounis, A.L. (2021). Exploitation of *Vitis vinifera*, *Foeniculum vulgare*, *C. sativa* and *Punica granatum* by-product seeds as dermo-cosmetic agents. *Molecules*, 26, 731; https://doi.org/10.3390/molecules26030731.

Mihoc, M., Pop, G., Alexa, E., and Radulov, I. (2012). Nutritive quality of Romanian hemp varieties (*C. sativa* L.) with special focus on oil and metal contents of seeds. *Chemistry Central Journal*, 6, 122. http://journal.chemistrycentral.com/content/6/1/122.

Mikulcová, V., Kašpárková, V., Humpolíček, P., and Buňková. L. (2017). Formulation, characterization and properties of hemp seed oil and its emulsions. *Molecules*, 22, 700; https://doi.org/10.3390/molecules22050700.

Mir, N. A., Riar, C. S., & Singh, S. (2019). Effect of pH and holding time on the characteristics of protein isolates from *Chenopodium* seeds and study of their amino acid profile and scoring. *Food Chemistry*, *272*, 165–173.

Mirpoor, S. F., Giosafatto, C. V. L., & Porta, R. (2021). Biorefining of seed oil cakes as industrial co-streams for production of innovative bioplastics. A review. *Trends in Food Science and Technology*, *109*, 259–270.

Mo, S., Dong, L., Hurst, W. J., & van Breemen, R. B. (2013). Quantitative analysis of phytosterols in edible oils using APCI liquid chromatography–tandem mass spectrometry. *Lipids*, *48*(9), 949–956.

Moccia, S., Maria Russo, M., Spagnuolo, C., Tedesco, I., Picariello, G., Siano, F., . . . Russo, G. L. (2019). Activation of cytostatic autophagy by polar extract of hempseed oil (*C. sattiva* L.) in colorectal adenocarcinoma cell line. *Nutrition, Metabolism and Cardiovascular Diseases*, *29*, 871–885.

Moccia, S., Siano, F., Russo, G.L., Volpe, M.G., La Cara, F., Pacifico, S., Piccolella, S. and Picariello, G., (2019). Antiproliferative and antioxidant effect of polar hemp extracts (*C. sativa* L., Fedora cv.) in human colorectal cell lines. *International Journal of Food Sciences and Nutrition*, pp.1–14. https://doi.org/10.1080/09637486.2019.1666804.

Mollard, R. C., Johnston, A. J., Serrano Leon, A., Wang, H., Jones, P., & MacKay, D. S. (2021). Acute effects of hemp protein consumption on glycemic and satiety control: results of two randomized crossover trials. *Applied Physiology, Nutrition, and Metabolism*, (ja). https://doi-org.uml.idm.oclc.org/10.1139/apnm-2020-0907.

Mölleken, H., & Theimer, R. R. (1997a). Survey of minor fatty acids in *C. sativa* L. fruits of various origins. *Journal of the International Hemp Association*, *4*(1), 13–17.

Mölleken, H., & Theimer, R.R. (1997b) Evaluierung von *C. sativa* saatgutherkünften im Hinblick aufein verbesserte Ölqualität, pp. 485–499. In: *Bioresource Hemp 97*, Proceedings of the Symposium, Frankfurt, Germany. Nova-Institute, Köln, Germany.

Montserrat-de la Paz, S., Marín-Aguilar, F., García-Giménez, M. D., & Fernández-Arche, M. A. (2014). Hemp (*C. sativa* L.) seed oil: analytical and phytochemical characterization of the unsaponifiable fraction. *Journal of Agricultural and Food Chemistry*, *62*(5), 1105–1110.

Mookerjee, A. (2020). Investigation of different extraction methods for hemp seeds. *MSc thesis*, University of Saskatchewan, Saskatoon, Canada, 104.

Moon, Y.-H., Song, Y.-S., Kim, K.-S., Lee, J.-E., Yu, G.-D., Lee, Y.-H., . . . Cha, Y.-L. (2017). Comparison of color and major components of hempseed oils extracted with pressuring and extruded methods. *Journal of Oil and Applied Science*, *34*(3), 666–672.

Neinast, M., Murashige, D., & Arany, Z. (2019). Branched chain amino acids. *Annual Review of Physiology*, *81*, 139–164.

Ngo, S. T., Pham, N. Q. A., Le, L. T., Pham, D.-H., & Vu, V. V. (2020). Computational determination of potential inhibitors of SARS-CoV-2 main protease. *Journal of Chemical Information and Modeling*, *60*(12), 5771–5780.

Nigro, E., Crescente, G., Formato, M., Pecoraro, M.T., Mallardo, M., Piccolella, S., Daniele, A., and Pacifico, S. (2020). Hempseed lignanamides rich-fraction: chemical investigation and cytotoxicity toward U-87 glioblastoma cells. *Molecules*, 25, 1049; https://oi.org/10.3390/molecules25051049.

Nissen, L., Bordoni, A. & Gianotti, A. (2020) Shift of volatile organic compounds (VOCs) in gluten-free hemp-enriched sourdough bread: a metabolomic approach. *Nutrients*, 12 (4), 1050; https://doi,org/10.3390/nu12041050.

Nissen, L., Demircan, B., Taneyo-Saa, D.L., and Gianotti, A. (2019). Shift of aromatic profile in probiotic hemp drink formulations: a metabolomics approach. *Microorganisms*, 7(11), 509; http://dx.doi.org/10.3390/microorganisms7110509.

Nogonierma, A. B., & FitzGerald, R. J. (2015). Investigation of the potential of hemp, pea, rice and soy protein hydrolysates as a source of dipeptidyl peptidase (DPP-IV) inhibitory peptides. *Food Digestion: Research and Current Opinion*, *6*, 19–29.

Nogonierma, A.B., and FitzGerald, R.J. (2019). Features of dipeptidyl peptidase IV (DPP-IV) inhibitory peptides from dietary proteins. *Journal of Food Biochemistry*, 43 (1), e12451. First Published November 1, 2017; https://doi.org/10.1111/jfbc.12451.

Nutraingredients.com. (2020). AnPro yeast protein: finding a sustainable solution for the protein gap. Issue August 10, 2020; www.nutraingreients.com.

Nutraingredients.com. (2021). Anything but CBD – The real opportunity in hemp. Issue April 5, 2021; www.nutraingreients.com.

Olsen, T., Turner, C., Øvrebø, B., Bastani, N.E., Refsum, H. & Vinknes, K.J. (2020). Postprandial effects of a meal low in sulfur amino acids and high polyunsaturated fatty acids compared to a meal high in sulfur amino acids and saturated fatty acids on stearoyl-CoA-desaturase indices and plasma sulfur amino acids: a pilot study. *BMC Research Notes*, 13 : 379; https://doi.org/10.1186/s13104-020-05222-y.

Oomah, B. D. (2001). Flaxseed as a functional food source. *Journal of the Science of Food and Agriculture*, *81*, 889–894.

Oomah, B.D. (2020). Flaxseed By-products. Pages 267–289 *in* Campos-Vega, R., Oomah, B.D. and Vergara-Castañeda, H.A. (Eds.). *Food Wastes and By-products: Nutraceutical and Health Potential*. John Wiley & Sons Ltd. Chichester, UK, (ISBN: 9781119534105; DOI: 10.1002/9781119534167).

Oomah, B. D., Busson, M., Godfrey, D. V., & Drover, J. C. G. (2002). Characteristics of hemp (*C. sativa* L.) seed oil. *Food Chemistry*, *76*, 33–43.

Opyd, P. M., Jurgoński, A., Fotschki, B., & Juśkiewicz, J. (2020). Dietary hemp seeds more effectively attenuate disorders in genetically obese rats than their lipid fraction. *The Journal of Nutrition*, *150*(6), 1425–1433.

Orio, L. P., Boschin, G., Recca, T., Morelli, C. F., Ragona, L., Francescato, P., . . . Speranza, G. (2017). New ACE-inhibitory peptides from hempseed (*C. sativa* L.) proteins. *Journal of Agricultural and Food Chemistry*, *65*, 10482–10488.

Ozen, B. F., Weiss, I., & Mauer, L. J. (2003). Dietary supplement oil classification and detection of adulteration using Fourier Transform Infrared Spectroscopy. *Journal of Agricultural and Food Chemistry*, *51*, 5871–5876.

Pacini, S., Punzi, T., Morucci, G., Gulisano, M., & Ruggiero, M. (2012). Effects of vitamin D-binding protein-derived macrophage-activating factor on human breast cancer cells. *Anticancer Research*, *32*, 45–52.

Pap, N., Hamberg, L., Pihlava, J-M., Hellström, J., Mattila, P., Eurola, M. and Pihlanto, A. (2020). Impact of enzymatic hydrolysis on the nutrients, phytochemicals and sensory properties of oil hemp seed cake (*C. sativa* L. FINOLA variety). *Food Chemistry*, 320 : 126530: https://doi.org/10.1016/j.foodchem.2020.126530.

Park, A. Y., Bourtembourg, M., Chrétien, A., Hubaux, R., Lancelot, C., Salmon, M., & Fitton, J. H. (2021). Modulation of gene expression in a sterile atopic dermatitis model and inhibition of *S. aureus* adhesion by fucoidan. *Dermatopathology*, *8*, 69–83.

Parthasarathy, A., Cross, P.J., Dobson, R.C.J., Adams, L.E., Savka, M.A., and Hudson, A.O. (2018). The three-ring circus: metabolism of the three proteogenic aromatic amino acids and their role in the health of plants and animals. *Frontiers in Molecular Biosciences*, 5 : 29; https://doi.org/10.3389/fmolb.2018.00029.

Pasqua, T., Rocca, C., Lupi, F.R., Baldino, N., Amelio, D., Parisi, O.I., Granieri, M.C., De Bartolo, A., Lauria, A., Dattilo, M., Perrotta, I.D., Puoci, F., Cerra, M.C., Gabriele, D. & Angelone, T. (2020). Cardiac and metabolic impact of functional foods with antioxidant properties based on whey derived proteins enriched with hemp seed oil. *Antioxidants*, 9, 1066; http://dx.doi.org/10.3390/antiox9111066.

Pastell, H., Putkonen, T., & Rita, H. (2019). Dietary fiber in legumes, seeds, vegetables, fruits and mushrooms: Comparing traditional and semi-automated filtration techniques. *Journal of Food Composition and Analysis*, *75*, 1–7.

Pate, D. W., & Ranali, I. P. (1999). *Advances in Hemp Research* (pp. 243–255). Binghamton, New York: The Haworth Press.

Pavlovic. R., Panseri, S., Giupponi, L., Leoni, V., Citti, C., Cattaneo, C., Cavaletto, M., and Giorgi, A. (2019). Phytochemical and ecological analysis of two varieties of hemp (*C. sativa* L.) grown in a mountain environment of Italian Alps. *Frontiers in Plant Science*, 10 : 1265; https://doi.org/10.3389/fpls.2019.01265.

Peskin, B. (2014), Medicament for treatment of cancer, cardiovascular diseases and inflammation. *U.S. Patent 8,728,546 B1*, issued May 20, 2014.

Pihlanto, A., and Mäkinen, S. (2017). The function of renin and the role of food-derived peptides as direct renin inhibitors. In, A.N. Tolekova (ed.), *Renin-Angiotensin System—Past, Present and Future*, IntechOpen, London, UK, http://dx.doi.org/10.5772/intechopen.69513.

Pihlanto, A., Nurmi, M., & Mäkinen, S. (2020). Hempseed protein: processing and functional properties. In G. Crini & E. Lichtfouse (Eds.), *Sustainable Agriculture Reviews 42 Hemp Production and Applications* (pp. 223–237). Cham, Switzerland: Springer Nature.

Pojić, M., Mišan, A., Sakač, M., Hadnađev, D., Tamara Šarić, B., Milovanović, I., & Hadnađev, M. (2014). Characterization of byproducts originating from hemp oil processing. *Journal of Agricultural and Food Chemistry*, *62*(51), 12436–12442.

Pollastro, F., Minassi, A., & Fresu, L. G. (2018). Cannabis phenolics and their bioactivities. *Current Medicinal Chemistry*, *25*(10), 1160–1185.

Ponzoni, E., Brambilla, I. M., & Galasso, I. (2018). Genome-wide identification and organization and organization of seed storage protein genes of *C. sativa*. *Biologia Plantarum*, *62*(4), 693–702.

Potin, F., and Saurel, R. (2020). Hemp seed as a source of food proteins. In G. Crini & E. Lichtfouse (eds.), *Sustainable Agriculture Reviews 42*, Springer Nature Switzerland, pp. 265–2941: https://doi.org/10.1007/976-3-030-41384-2_9.

Presto, M. H., Lyberg, K., & Lindberg, J. E. (2011). Digestibility of amino acids in organically cultivated white-flowering faba bean and cake from cold-pressed rapeseed, linseed and hemp seed in growing pigs. *Archives of Animal Nutrition*, *65*(1), 21–33.

Prociuk, M. A., Edel, A. L., Gavel, N., Deniset, J., Ganguly, R., Austria, J. A., . . . Pierce, G. N. (2006). The effects of dietary hempseed on cardiac ischemia/reperfusion injury in hypercholesterolemic rabbits. *Experimental Clinical Cardiology*, *11*(3), 198–205.

Prociuk, M. A., Edel, A. L., Richard, M. N., Gavel, N. T., Ander, B. P., Dupasquier, C. M. C., & Pierce, G. N. (2008). Cholesterol-induce stimulation of platelet aggregation is prevented by a hempseed-enriched diet. *Canadian Journal of Physiology and Pharmacology*, *86*, 153–159.

Reale, L., & Cerri, M. (2019). Hemp (*C. sativa* L.) seeds: morpho-histological and histochemical characterization. *The International Journal of Plant Reproductive Biology*, *11*(1), 4–8.

Ren, Y., Liang, K., Jin, Y., Zhang, M., Chen, Y., Wu, H., & Lai, F. (2016). Identification and characterization of two novel α-glucosidase inhibitory oligopeptides from hemp (*C. sativa* L.) seed protein. *Journal of Functional Foods*, *26*, 439–450.

Rezapour-Firouzi, S., Arefhosseini, S. R., Ebrahimi-Mamaghani, M., Baradaran, B., Sadeghihokmabad, E., Torbati, M., . . . Zamani, F. (2014). Activity of liver enzymes in multiple sclerosis patients with hot-nature diet and co-supplemented hemp seed, evening primrose oils intervention. *Complementary Therapies in Medicine*, *22*(6), 986–993.

Rezapour-Firouzi, S., Arefhosseini, S. R., Mehdi, F., Mehrangiz, E.-M., Baradaran, B., Sadeghihokmabad, E., . . . Zamani, F. (2013). Immunomodulatory and therapeutic effects of hot-nature diet and co-supplemented hemp seed, evening prime rose oils intervention in multiple sclerosis patients. *Complementary Therapies in Medicine*, *21*, 473–480.

Rezapour-Firouzi, S., Arefhosseini, S. R., Mehrangiz, E.-M., Baradaran, B., Sadeghihokmabad, E., Mostafaei, S., . . . Chehreh, M. (2015). Alteration of delta-6-desaturase (FADS2), secretory phospholipase-A2 (sPLA2) enzymes by Hot-nature diet with co-supplemented hemp seed, evening prime rose oils intervention in multiple sclerosis patients. *Complementary Therapies in Medicine*, *23*, 652–657.

Rezapour-Firouzi, S., Mohammadian, M., Sadeghzadeh. M., and Mazloomi, E. (2020). Effects of co-administration of rapamycin and evening primrose/hemp seed oil supplement on immunologic factors and cell membrane fatty acids in experimental autoimmune encephalomyelitis. *Gene*, 759 : 144987; https://doi.org/10.1016/j.gene.2020.144987.

Rezapour-Firouzi, S., Shahabi, S., Mohammadzadeh, A., Tehrani, A. A., Kheradmand, F., & Mazloomi, E. (2018). The potential effects of hemp seed/evening primrose oils on the mammalian target of rapamycin complex 1 and interferon-gamma genes expression in experimental autoimmune encephalomyelitis. *Research in Pharmaceutical Sciences*, *13*(6), 523–532.

Rezvankhah, A., Emam-Djomeh, Z., Safari, M., Askari, G., & Salami, M. (2019). Microwave-assisted extraction of hempseed oil: studying and comparing of fatty acid composition, antioxidant activity, physicochemical and thermal properties with Soxhlet extraction. *Journal of Food Science and Technology*, *56*(9), 4198–4210.

Richard, M. N., Ganguly, R., Steigerwald, S. N., Al-Khalifa, A., & Pierce, G. N. (2007). Dietary hempseed reduces platelet aggregation. *Journal of Thrombosis and Hemostasis*, *5*, 424–425.

Ritthausen, K.H.(1881) www.encyclopedia.com.

Rodriguez-Martin, N. M., Montserrat-de la Paz, S., Toscano, R., Grao-Cruces, E., Villanueva, A., Pedroche, J., Millan, F., and Millan-Linares, M. C. (2020). Hemp (*C. sativa* L.) protein hydrolysates promote anti-inflammatory response in primary human monocytes. *Biomolecules*, *10*(5), 803; http://dx.doi.org/10.3390/biom10050803.

Rodriguez-Martin, N. M., Toscano, R., Villanueva, A., Pedroche, J., Millan, F. M.-d., la Paz, S., & Millan-Linares, M. C. (2019). Neuroprotective protein hydrolysates from hemp (*C. sativa* L.) seeds. *Food and Function*, *10*, 6732–6739.

Rossi, M., Mascaretti, F., Parpinel, M., Serraino, D., Crispo, A., Celentano, E., Glacosa, A., and La Vecchia, C. (2020). Dietary intake of branched-chain amino acids and colorectal cancer risk. *British Journal of Nutrition*, pp. 1–6; https://doi.org/10.1017/S0007114520003724.

Rudel, L. L., Kelly, K., Sawyer, J. K., Shah, R., & Wilso, M. D. (1998). Dietary monounsaturated fatty acids promote aorti atherosclerosis in LDL receptor-null ApoB100-overexpressing transgenic mice. *Arteriosclerosis, Thrombosis, and Vascular Biology*, *18*, 1818–1827.

Ruggiero, M., & Pacini, S. (2020). Fermentation of hemp seed proteins leads to formation of peptides that share sequence similarity with human vitamin D-binding protein. *Medical Research and Innovations*, *4*, 1–3. https://doi.org/10.15761/MRI.1000170

Russo, R., & Reggiani, R. (2013). Variability in antinutritional compounds in hempseed meal of Italian and French varieties. *Plant*, *1*(2), 25–29.

Russo, R., & Reggiani, R. (2015). Evaluation of protein concentration, amino acid profile and antinutritional compounds in hempseed meal from dioecious and monoecious varieties. *American.Journal of Plant Science*, *6*, 14–22.

Sacilik, K., Öztürk, R., & Keskin, R. (2003). Some physical properties of hemp seed. *Biosystems Engineering*, *86*(2), 191–198.

Sakakibara, I., Katsuhara, T., Ikeya, Y., Hayashi, K., & Mitsuhashi, H. (1991). Cannabisin A, an arylnaphthalene lignanamide from fruits of *C. sativa*. *Phytochemistry*, *30*, 3013–3016.

Salentijn, E. M. J., Zhang, Q., Amaducci, S., Yang, M., & Trindade, L. M. (2015). New developments in fiber hemp (*C. sativa* L.) breeding. *Industrial Crops and Products*, *68*, 32–41.

Samaranayaka, A.G.P., Wanasundara, U.N., Ray, M., & Green, R.C. (2019). Hemp protein and use for microencapsulation. *US Patent 2019/0335800 A1*, Issued November 7, 2019.

Samsamikor, M., Mackay, D., Mollard, R. C., & Aluko, R. E. (2020). A double-blind, randomized, crossover trial protocol of whole hemp seed protein and hemp seed protein hydrolysate consumption for hypertension. *Trials*, *21*, 1–13.

Sasaki-Hamaa, S., Hoshi, M., Niwa, Y., Ueda, Y., Kokaji, A., Kamisuki, S., . . . Oka, J.-I. (2016). Neoechinulin A induced memory improvements and antidepressant-like effects in mice. *Progress in Neuro-Psychopharmacology and Biological Psychiatry*, *71*, 155–161.

Sawler, J., Stout, J.M., Gardner, K.M., Hudson, D., Vidmar, J., Butler, L., Page, J.E. and Myles, S. (2015). The genetic structure of marijuana and hemp. *PLoS ONE*, 10(8): e0133292; https://doi.org/10.1371/journal.pone.0133292.

Schultz, C.J., Lim, W.L., Khor, S.F., Neumann, K.A., Schulz, J.M., Ansari, O., Skewes, M.A., and Burton, R.A. (2020). Consumer an health-related traits of seed from selected commercial and breeding lines of industrial hemp, *C. sativa* L. *Journal of Agriculture and Food Research*, 2, 100025; https://doi.org/10.1016/j.jafr.2020.100025.

Schwab, U. S., Callaway, J. C., Erkkilä, A. T., Gynther, J., Uusitupa, M. I. J., & Järvinen, T. (2006). Effects of hempseed and flaxseed oils on the profile of serum lipids, serum total and lipoprotein lipid concentrations and haemostatic factors. *European Journal of Nutrition*, *45*, 470–477.

Schwingshackl, L., Bogensberger, B., Benčič, A., Knüppel, S., Boeing, H., & Hoffmann, G. (2018). Effects of oils and solid fats on blood lipids: a systematic review and network meta-analysis. *Journal of. Lipid Research*, *59*, 1771–1782.

Senila, L., Neag, E., Cadar, O., Kovacs, M.H., Becze, A., and Senila, M. (2020). Chemical, nutritional and antioxidant characteristics of different food seeds. *Applied Sciences*, 10, 1569; http://dx.doi.org/10.3390/app10051589.

Serrapica, F., Masucci, F., Raffrenato, E., Sannino, M., Vastolo, A., Barone, C.M.A., and Di Francia, A. (2019). High fiber cakes from Mediterranean multipurpose oilseeds as protein sources for ruminants. *Animals*, *9*, 918. https://doi.org/10.3390/ani9110918.

Shen, P., Gao, Z., Xu, M., Ohm, J-B., Rao, J., and Chen, B. (2020). The impact of hempseed dehulling on chemical composition, structure properties and aromatic profile of hemp protein isolate. *Food Hydrocolloids*, 106; https://doi.org/10.1016/j.foodhyd.2020.105889.

Siano, F., Moccia, S., Picariello, G., Russo, G.L., Sorrentino, G., Di Stasio, M., La Cara, F., and Volpe, M.G. (2019).Comparative study of chemical, biochemical characteristic and ATR-FTIR analysis of seeds, oil and flour of the edible Fedora cultivar hemp (*C. sativa* L.). *Molecules*, 24 (1), 83; https://doi.org/10.3390/molecules24010083.

Siger, A., Nogala-Kalucka, M., & Lampart-Szczapa, E. (2008). The content and antioxidant activity of phenolic compounds in cold-pressed plant oils. *Journal of Food Lipids, 15*(2), 137–149.

Simopoulos, A. P. (2002). The importance of the ratio of omega-6/omega-3 essential fatty acids. *Biomedicine and Pharmacotherapy, 56*(8), 365–379.

Small, E. (2017). Classification of *C. sativa* L. in relation to agricultural, biotechnological, medical and recreational utilization. In S. Chandra, H. Lata, & M. A. ElSohly (Eds.), *C. sativa L. –Botany and Biotechnology* (pp. 1–61). Cham, Switzerland: Springer International Publishing AG.

Small. E., and Marcus, D. (2000) Hemp germplasm trials in Canada. Bioresource Hemp, *3rd International Symposium, Wolfsburg, Germany*. Online proceedings: http://www.hemphasis.com/files/publications/biorpap.htm.

Smeriglio, A., Galati, E. M., Monforte, M. T., Lanuzza, F., D'Angelo, V., & Circosta, C. (2016). Polyphenolic compounds and antioxidant activity of cold-pressed seed oil from Finola cultivar of *C. sativa* L. *Phytotherapy Research, 30*, 1298–1307.

St. Angelo, A. J., Yatsu, L. Y., & Altschul, A. A. (1968). Isolation of edestin from aleurone grains of *C. sativa*. *Archives of Biochemistry and Biophysics, 124*, 199–205.

Stastnik, O., Pavlata, L., and Mrkvicova, E. (2020). The milk thistle seed cakes and hempseed cakes are potential feed for poultry. *Animals*, 10, 1384; http://dx.doi.org/10.3390/ani10081384.

Steijns, J., & Ward, R. D. (2004). Glutamine rich peptides and their benefits for athletes and exercising people. *Agro Food Industry Hi Tech, 15*(6), 15–18.

Taheri-Garavand, A., Nassiri, A., & Gharibzahedi, S. M. T. (2012). Physical and mechanical properties of hemp seed. *International Agrophysics, 26*, 211–215.

Tanaka, M., Kiyohara, H., Yoshino, A., Takata, F., Dohgu, S., Katoka, Y., and Matsui, T. (2020). Brain-transportable soy dipeptide, Tyr-Pro, attenuates amyloid β peptide$_{25-35}$-inuced memory impairment in mice. *NPJ Science of Food*, 4, 7; https://doi.org/10.1038/s41538-020-0067-3.

Tang, C.-H., Ten, Z., Wang, X.-S., & Wang, X.-Q. (2006). Physicochemical and functional properties of hemp (*C. sativa* L.) protein isolate. *Journal of Agricultural and Food Chemistry, 54*, 8945–8950.

Teh, S.-S., Bekhit, A. E.-D., Carne, A., & Birch, J. (2016). Antioxidant and ACE-inhibitory activities of hemp (*C. sativa* L.) protein hydrolysates produced by the proteases AFP, HT, Pro-G, actidin and zingibain. *Food Chemistry, 203*, 199–206.

Teh, S. S., & Birch, J. (2013). Physicochemical and quality characteristics of cold-pressed hemp, flax and canola seed oils. *Journal of Food Composition and Analysis, 30*(1), 26–31.

Toelzer, C., Gupta, K., Yadav, S. K. N., Borucu, U., Davidson, A. D., Williamson, M. K., . . . Schaffitzel, C. (2020). Free fatty acid binding pocket in the locked structure of SARS-CoV-2 spike protein. *Science, 370*, 725–730.

Tomko, A.M., Whynot, E.G., Ellis, L.D. and Dupré, D.J. (2020). Anti-cancer potential of cannabinoids, terpenes, and flavonoids present in cannabis. *Cancers*, 12, 1985: http://dx.doi.org/10.3390/cancers12071985.

Ulbricht, T. L., & Southgate, D. A. (1991). Coronary heart disease: seven dietary factors. *Lancet*, *338*(8773), 985–992.

Uluate, S., & Özdemir, N. (2012). Antioxidant activities and oxidative stabilities of some unconventional oilseeds. *Journal of the American Oil Chemist's Society*, *89*, 551–559.

Volk, C., Brandsch, C., Schlegelmilch, U., Wensch-Dorendorf, M., Hirche, F., Simm, A., Gargum, O., Wiacek, C., Braun, P.G., Kopp, J.F., Schwerdtle, T., Treede, H., and Stangl, G.I. (2020). Postprandial metabolic response to rapeseed protein in healthy subjects. *Nutrients*, 12, 2270: https://doi.org/10.3390/nu12082270.

Vonapartis, E., Aubin, M.-P., Seguin, P., Mustafa, A. F., & Charron, J.-B. (2015). Seed composition of ten industrial hemp cultivars approved for production in Canada. *Journal of Food Composition*, *39*, 8–12.

Wang, Q., & Xiong, Y. L. (2018). Zinc-binding behavior of hemp protein hydrolysates: soluble versus insoluble zinc-peptide complexes. *Journal of Functional Foods*, *49*, 105–112.

Wang, Q., & Xiong, Y. L. (2019). Processing, nutrition, and functionality of hempseed protein: A review. *Comprehensive Reviews in Food Science and Food Safety*, *18*, 936–952.

Wang, S., Luo, Q., and Fan, P. (2019). Cannabisin F from hemp (*C. sativa*) seed suppresses lipopolysaccharide-induced inflammatory responses in BV2 microglia as SIRT1 modulator. *International Journal of Molecular Sciences*, *20*, 507. https://doi.org/10.3390/ijms20030507.

Wang, X.-S., Tang, C.-H., Yang, X.-Q., & Gao, W.-R. (2008). Characterization, amino acid composition and *in vitro* digestibility of hemp (*C. sativa* L.) proteins. *Food Chemistry*, *107*, 11–18.

Wei, L.-H., Dong, Y., Sun, Y.-F., Mei, X.-S., Ma, X.-S., Shi, J., Yang, Q.-L., Ji, Y.-R., Zhang, Z.-H., Sun, H.-N., Sun, X.-R., and Song, S. M. (2021). Anticancer property of hemp bioactive peptides in Hep3B liver cancer cells through Akt/GSK3β/β-catenin signaling pathway. *Food Science and Nutrition*, 00: 1–9; https://doi.org/10.1002/fsn3.1976.

Wen, Z. S., Xue, R., Du, M., Tang, Z., Xiang, X. W., Zheng, B., & Qu, Y. L. (2019). Hemp seed polysaccharides protect intestinal epithelial cells from hydrogen peroxide-induced oxidative stress. *International Journal of Biological Macromolecules*, *135*, 203–211.

Werz, O., Seegers, J., Schaible, A. M., Weinigel, C., Barz, D., Koeberle, A., . . . Appendino, G. (2014). Cannflavins from hemp sprouts, a novel cannabinoid-free hemp food product, target microsomal prostaglandin E_2 synthase-1 and 5-lipoxygenase. *PharmNutrition*, *2*, 53–60.

Xue, R., Du, M., Zhou, T., Ai, W., Zhang, Z., Xiang, X., Zhou, Y., and Wen, Z. (2020). Polysaccharides from hemp seed protect against cyclophosphamide-Induced Intestinal oxidative damage via Nrf2-Keap1 signaling pathway in mice. *Oxidative Medicine and Cellular Longevity*, https://doi.org/10.1155/2020/1813798.

Yan, X., Tang, J., dos Santos Passos, C., Nurisso, A., Simões-Pires, C. A., Ji, M., . . . Fan, P. (2015). Characterization of lignanamides from hemp (*C. sativa* L.) seed and their antioxidant and acetylcholinesterase inhibitory activities. *Journal of Agricultural and Food Chemistry*, *63*, 10611–10619.

Yan, X., Zhou, Y., Tang, J., Ji, M., Lou, H., & Fan, P. (2016). Diketopiperazine indole alkaloids from hemp seed. *Phytochemistry Letters*, *18*, 77–82.

Youn, K., Lee, S., & Jun, M. (2018). Gamma-linoleic acid ameliorates Aβ-induced neuroinflammation through NF-κB and MAPK signaling pathways. *Journal of Functional Foods*, *42*, 30–37.

Zahari, I., Ferawati, F., Helstad, A., Ahlström, C., Östbring, K., Rayner, M. & Purhagen, J.K. (2020). Development of high-moisture meat analogues with hemp and soy protein using extrusion cooking. *Foods 9*, 772; https://doi.org/10.3390/foods9060772.

Zanoni, C., Aiello, G., Arnoldi, A., & Lammi, C. (2017). Hempseed peptides exert hypocholesterolemic effects with a statin-like mechanism. *Journal of Agricultural and Food Chemistry*, *65*(40), 8829–8838.

Zhou, Y., Wang, S., Ji, J., Lou, H., & Fan, P. (2018). Hemp (*C. sativa* L.) seed phenylpropionamides composition and effects on memory dysfunction and biomarkers of neuroinflammation induced by lipopolysaccharide in mice. *ACS Omega*, *3*, 15988–15995.

Zhou, Y., Wang, S., Lou, H., & Fan, P. (2018). Chemical constituents of hemp (*C. sativa* L.) seed with potential anti-neuroinflammatory activity. *Phytochemistry Letters*, *23*, 57–61.

Zhu, G-Y., Jiang, Z-H., Yao, X-J., Liu, X., Fu, J., Bai, L-P., & Liu, L. (2018). Lignanamides and a method of treating neurodegenerative diseases by using the same. *US Patent 10,155,721 B2*, Issued December 18, 2018.

Zhu, X., Sun-Waterhouse, D., Chen, J., Cui, C., & Wang, W. (2021). Comparative study on the novel umami-active peptides of the whole soybeans and the defatted soybeans fermented soy sauce. *Journal of the Science of Food and Agriculture*, *101*(1), 158–166.

10

Phytochemicals and Functional Properties of Coffee: Molecular Mechanism of Action

Salvador A. Saldaña-Mendoza[1], Sandra Pacios-Michelena[1], Mónica L. Chávez-González[1], Leidy Johana Valencia-Hernández[1], A. Sócrates Palacios-Ponce[2], and Cristóbal N. Aguilar[1]

[1] *Food Research Department, School of Chemistry, Universidad Autónoma de Coahuila, Saltillo, Mexico*
[2] *ESPOL Polytechnic University, Escuela Superior Politécnica del Litoral, ESPOL, Facultad de Ingeniería en Mecánica y Ciencias de la Producción, Guayaquil, Ecuador*

10.1 Introduction

The philosopher Hippocrates, also known as the father of modern medicine, already recognized for 2,500 years the relationship that exists between the proper use of food, its therapeutic properties, and its health benefits, and this he affirmed through the phrase, "Let thy food be thy medicine and thy medicine be thy food" (Maurya, Chauhan, Yadav, Gangwar, & Maurya, 2021; Singh & Sinha, 2012). Since ancient times, old civilizations such as the Roman, Indian, Chinese, and Sumerian believed that the most effective way to avoid diseases was by preventing them mainly through a good diet (Maurya et al., 2021). Currently, the strong relationship between food and its physiological effect has increased consumers awareness of self-care through nutrition to prevent and treat diseases (Jiménez et al., 2015).

The term nutraceutical is the conjunction of the words "nutrition" and "pharmaceutical"; Stephen L DeFelice first introduced this term in 1989. This term is used to refer to any food or part of food that confers health benefits to those who consume it (Chiu, Venkatakrishnan, & Wang, 2020). Currently, nutraceuticals are considered as a food product or some of their secondary metabolites, mainly those phytochemicals that have bioactivities that can provide health benefits or that can be used to prevent diseases or act as adjuvants in the treatment of diseases (Lama et al., 2020). Nutraceuticals can be consumed directly by eating naturally occurring foods or through the consumption of capsules, tablets, powders, or liquids containing bioactive compounds. Nutraceuticals have been used in the management and prevention of various diseases such as diabetes, cardiovascular diseases, hypertension, neurodegenerative disorders (Ali, Ahmad, Juber Akhtar, & Khan, 2019; Ghaffari & Roshanravan, 2020), osteoporosis (Rajput, Wairkar, & Gaud, 2018), osteoarthritis (Wang, Leong, Cardoso, & Sun, 2018), arteriosclerosis (Moss, Williams, & Ramji, 2018),

Molecular Mechanisms of Functional Food, First Edition. Edited by Rocio Campos-Vega and B. Dave Oomah.

psoriasis (Raut & Wairkar, 2018), hypercholesterolemia (Catapano, Barrios, Cicero, & Pirro, 2019), among many others. Coffee is one of the most important food products in the world. It is considered the second most important product in commercial value (Fanali et al., 2020). Coffee is characterized by its unique flavor and thanks to its content rich in bioactive compounds such as methylxanthines and polyphenolic compounds with antioxidant activity such as caffeoylquinic acids, dicaffeoylquinic acids, feruloylquinic acids, and p-coumaroylquinic acids, among many others with nutraceutical properties, for which it is considered a functional food (Esquivel & Jiménez, 2012; Fanali et al., 2020)

10.2 Classification of Nutraceuticals

Nutraceuticals are classified in different ways, according to other authors. They can be classified based on the foods available in the market, on the food source from which they come, and their chemical nature (Table 10.1). Another interesting classification is based on origin, use, and method of preparation (Maurya et al., 2021) (Table 10.2).

Nutraceuticals perform different action mechanisms depending on the phytochemicals they contain, such as flavonoids, isoflavones, carotenes, fatty acids, and hydroxycinnamic acids, among others. The antioxidants are chemical compounds that, given their chemical structure, can stop free radicals. This capacity is used to treat and improve diseases caused by oxidative stress. There are many antioxidant compounds present in coffee, predominantly polyphenols such as tannins, flavonoids, chlorogenic acids, and procyanidins (Gemechu, 2020). The phenolic compounds have interesting biological properties such as being powerful antioxidants, excellent adjuvants to treat cancer, improve inflammatory processes, antihypertensive effects, among others. This group includes compounds such as catechin, epicatechin, gallic acid, ellagic acid, and ferulic acid (Samsonowicz, Regulska, Karpowicz, & Leśniewska, 2019). Alkaloids are interesting compounds with important activities; one of the most important alkaloids in coffee is trigonelline, to which antimicrobial, anticarcinogenic, and antihyperglycemic effects have been attributed (Montenegro et al., 2021). It possess neuroprotective and antidiabetic activities (de Abreu Pinheiro et al., 2021). Caffeine is another important alkaloid in coffee.

10.3 Regulation of Nutraceuticals

Nutraceuticals are governed under different regulations depending on each country since they have therapeutic properties without being considered drugs. For example, in the USA, as of 1994 food supplements were no longer considered additives due to the Dietary Supplement Law and the Health and Education Law. Therefore, food supplements require labeling for their commercialization indicating the safety of their consumption. In UK, supplements are considered food, exempt from the need for licenses and clinical trials for their commercialization (Rajput et al., 2018).

Codex Alimentarius, searching for safeguarding public health and guaranteeing fair commercial practices through the implementation of norms and recommendations, has established global standards for vitamins and minerals supplements based on purity and

Table 10.1 Nutraceuticals' classifications based on the foods available in the market, their source, and chemical nature.

Classification	Principals groups	References
Foods available in the market	• Traditional nutraceuticals promote nutritional benefits from the content of nutrients that they naturally possess. Those can be grouped based on 1) Their chemical constituents in a) nutrients b) herbal c) phytochemicals 2) Probiotic microorganisms and 3) Nutraceutical enzymes • Nontraditional: artificial foods prepared with the help of biotechnology 1) Fortified 2) Recombinant	y Postigo, Jacobo-Velázquez, Guajardo-Flores, Amezquita, and García-Cayuela (2021) Singh and Sinha (2012)
Food source	1) Dietary fiber 2) Probiotics 3) Prebiotics 4) Polyunsaturated fatty acids 5) Antioxidant vitamins 6) Polyphenols 7) Spices	Garima and Manoj (2016)
Chemical nature	1) Proteins 2) Carbohydrates 3) Lipids 4) Micronutrients	Jiménez et al. (2015).
Mechanism of action	Grouped based on the metabolic and physiological functions, classified as 1) Regulators of lipid metabolism 2) Antioxidants 3) Anti-inflammatory 4) Osteogenic 5) Immunoregulatory 6) Anticancer Grouped based on their origin, classified as 1) Animal origin 2) Plant 3) Mineral	Jiménez et al. (2015).

Table 10.2 Nutraceuticals classifications based on their origin, use, and method of preparation.

Classification	Basic Characteristics
Functional foods	Foods consumed regularly whose form has not been modified can be grouped into (a) processed and (b) fermented
Dietary supplements	The concentrated form of nutrients available in liquids, tablets, pills, chews, gels, powders, among other presentations for human consumption can be grouped into (a) natural, (b) semi-synthetic, and (c) synthetic
Medicinal foods	Designed to meet the nutritional needs of people with specific diseases or critical health conditions, are regulated by the food and drug administration and require medical supervision, contain all the nutrients of a balanced diet, and are free of contaminants
Dietary fibers	An edible and indigestible part of food that generally acts as prebiotics, these can be classified into a) soluble fiber and b) insoluble fiber
Farmaceuticals	The term is derived from "farm" and "pharmaceutical which come from genetically modified crops or animals to provide extra health benefits through their consumption
Probiotics and prebiotics	Probiotics and prebiotics refer to dietary supplements and drugs, as well as microorganisms added to foods for improving host health

Source: Based on Maurya et al. (2021).

bioavailability, establishing requirements for their packaging and labeling (Devla, Acharya, Acharya, & Kumar, 2011).

10.4 Current Landscape on Nutraceuticals

Functional foods and nutraceutical products help improve health, reduce health care costs, and support economic development in rural areas. Recently, the demand for functional foods has increased the number of projects financed by the European Commission (EC) and the United States. These studies have helped to understand the possible biologically active components in food, improving health and probably reducing disease risk while improving our general well-being (Daliri & Lee, 2015).

Country-specific regulations and justification for health claims are challenges that the functional food and nutraceutical market continue to face. Currently, the nutraceutical market is dominated by the United States and Europe by 37% and 33%, respectively, followed by Japan with 18% (y Postigo et al., 2021). The global nutraceutical market valued at $165.62 billion in 2014 (Ali et al., 2019) is projected to increase to USD 578.23 billion by 2025 (y Postigo et al., 2021).

10.5 Nutraceuticals and Health

Nutraceuticals have been of interest to many authors as an alternative to drugs for certain diseases, including allergies, cancer, diabetes, Parkinson's (Nasri, Baradaran, Shirzad, & Kopaei, 2014), and other important public health conditions such as Alzheimer's or cardiovascular diseases (CVDs) due to its high incidence and the lack of effective cures (Sadhukhan, Saha, Dutta, Mahalanobish, & Sil, 2018).

The growing trends in the use of nutraceuticals are due to their capacity to postpone or attenuate development of noncommunicable diseases of public health importance such as Alzheimer's (Nasri et al., 2014). Several publications attest to the positive effects of different nutraceutical plants, such as *Zizyphus jujube* and *Lavandula officinalis* (Rabiei, Rafieian-Kopaei, Heidarian, Saghaei, & Mokhtari, 2014; Rabiei, Rafieian-Kopaei, Mokhtari, & Shahrani, 2014). Different molecules with nutraceutical properties have been considered to eliminate this disease's detrimental effect in all its stages. These molecules have a protective action on neuronal cells against oxidative damage, the regulation of acetylcholinesterase (AChE) activity, and the attenuation of neurodegeneration (Sadhukhan et al., 2018). The flavonoids present in plants such as grapevine (*Vitis vinifera*), tea (*Camellia sinensis*), and cocoa (*Theobroma cacao*) modulate several neurological processes as shown by studies in which they interact with the neuronal-glial signaling pathways involved in neuronal survival. They modulate blood flow to the brain (Williams & Spencer, 2012), enzymes, and proteins involved in neuronal repair and synapse plasticity and inhibit neuropathological processes in brain regions often engaged in Alzheimer's pathogenesis (Wang et al., 2018).

Nutraceuticals in the form of vitamins, minerals, antioxidants, dietary fibers, and omega-3 polyunsaturated fatty acids, along with physical exercise, are recommended for the prevention and treatment of CVDs. Polyphenols alter cell metabolism and signaling, which is believed to reduce arterial disease (Nasri et al., 2014). Flavonoids block the cyclooxygenase enzyme that breaks down prostaglandins and prevents platelet aggregation. They also protect the vascular system that transports oxygen and nutrients to cells (Asgary, Keshvari, Sahebkar, Hashemi, & Rafieian-Kopaei, 2013). Phytosterols compete with dietary cholesterol by blocking absorption and at the same time facilitate its excretion from the body (Brufau, Canela, & Rafecas, 2008). Other compounds such as lycopene, β-carotenoids with antioxidant activity, saponins with antimutagenic effect, tannins, pectins, glucosinolates present in tomatoes, guava, pink grapefruit, watermelon, papaya, broccoli, soy, spinach, alfalfa, potato, citrus, and red fruits have been reported with antitumor and anti-inflammatory activities (Limer & Speirs, 2004; Stahl & Sies, 2005; Thomasset et al., 2007).

10.5.1 Coffee

Coffee is a plant of the genus *Coffea* belonging to the *Rubiaceae* family (Alves, Rodrigues, Antónia Nunes, Vinha, & Oliveira, 2017). It is native to Ethiopia, Africa, and is commonly cultivated in tropical and subtropical climates. It is one of the most important food products and the second most important commercial product after petroleum (Esquivel & Jiménez, 2012). The *Coffea arabica* mainly drives this crop's commercial importance, and *Coffea canephora* varieties are known as "arabica" and "robusta," respectively (Organización Internacional del Café, 2019).

World coffee production increased recently. Ten million tons of coffee was produced in 2019, led by Brazil (3.28 million tons) followed by Vietnam and Colombia (1.6 and 0.8 million tons, respectively) (Food and Agriculture Organization of the United Nations, 2021). Its polyphenols' content characterizes coffee, the contribution of antioxidants, and unique flavor (Saldaña-Mendoza et al., 2020; Zhang & Ku, 2019). The consumption of coffee to prevent some diseases has conferred the classification of "functional food."

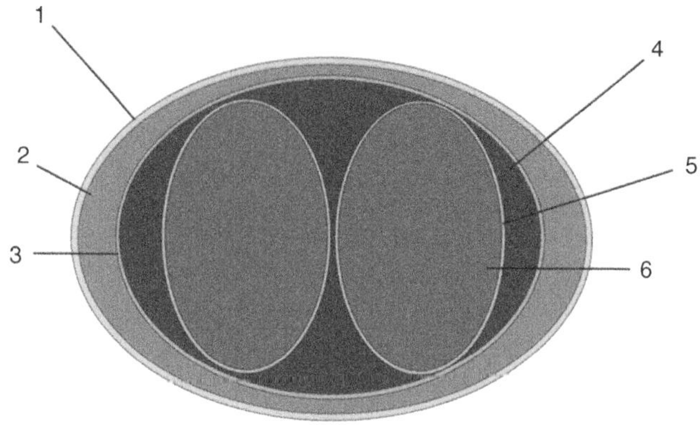

Figure 10.1 Structure of coffee cherry (1) outer skin (pericarp/exocarp), (2) pulp (mesocarp), (3) mucilage (endocarp), (4) parchment (endocarp), (5) silverskin, (6) bean (endosperm). *Source:* Based on Alves et al. (2017), Esquivel and Jiménez (2012), Vinícius de Melo Pereira et al. (2017).

These properties are derived from both green coffee beans composition and the changes they undergo during roasting (Esquivel & Jiménez, 2012).

10.5.2 The Coffee Bean

The coffee fruit (or cherry) is made up of five layers that cover the bean. The skin or pericarp is green when immature and turns red to purple at maturity. The pulp or mesocarp becomes a layer of viscous pectinaceous mucilage when ripe. Parchment is a thin yellow layer of polysaccharide, and the endosperm or silverskin covers the coffee bean (Alves et al., 2017; Esquivel & Jiménez, 2012; Vinícius de Melo Pereira, Soccol, Brar, Neto, & Soccol, 2017) (Figure 10.1).

10.6 Bioactive Compounds from Coffee

The effect of coffee on health and metabolic disorders has been studied and reported. Grosso et al. (2016) reported reduced risks of CVDs and mortality associated with cancer in nonsmoking patients. These effects emanate from coffee compounds with antitumor activity that acts on enzymes involved in tumor and metastasis genesis. Coffee consumption is associated with reduced risk of bladder, breast, oral and pharyngeal, colorectal, endometrial, esophageal, hepatocellular, leukemic, pancreatic, and prostate cancer (Yu, Bao, Zou, & Dong, 2011). Beneficial effects have been attributed to metabolic conditions (Shang, Li, & Jiang, 2016) such as type 2 diabetes, kidney stones and some liver conditions such as fibrous hepatitis and cirrhosis (Kennedy et al., 2016).

Espresso, boiled, and drip coffee extracts contribute to decreased platelet aggregation induced by ADP, collagen, arachidonic acid, and epinephrine. They also inhibit collagen-induced platelet aggregation as well as COX-1 and COX-2 cyclooxygenases more than chlorogenic acid or purified caffeine (Hutachok et al., 2020). Coffee consumption has been associated with reduced blood content of enzymes indicative of liver damage, such as aspartate aminotransferase (ASAT), γ-glutamyl transaminase (γ-GT), and alanine aminotransferase (ALAT). It also activates the transcription factor Nrf2 involved in the protection against oxidative and carcinogenic damage (Shaposhnikov et al., 2018). Consumption of coffee is related to reduction in cutaneous melanoma due to the suppressive effect of carcinogenesis promoted by 5-O-caffeoylquinic acid and caffeic acid that inhibit COX-2 expression. COX-2 plays a functional role in the development and expression of malignant melanoma. Moreover, nicotinaminic acid and niconinamide, generated during coffee roasting, have been associated with reduced carcinogenesis in skin induced by UVB radiation (Loftfield et al., 2015). The consumption of coffee also contributes to halitosis inhibition due to its antimicrobial activities, attributed to the content of compounds such as caffeine, α-dicarbonyl, glyoxal, methylglyoxal, and diacetyl (Alzoman, Alzahrani, Alwehaiby, Alanazi, & AlSarhan, 2021). Kuang et al. (2018) reported that coffee consumption reduces the levels of specific species of lysophosphatidylcholine, which have been implicated in arteriosclerosis and acute and chronic inflammation. Coffee consumption has also been associated with a lower probability of colorectal cancer. This effect is attributed to its antioxidant properties and the presence of melanoidins, which escape digestion and are metabolized by microbial species in the colon region, resulting in the release of antioxidants and associated low molecular weight substances (Castaldo et al., 2021). Moderate coffee consumption has beneficial effects against CVDs. Endothelial function is one of the predictors of CVDs, and oxidative stress is one of the most important risk factors for endothelial dysfunction due to decrease in NO bioavailability. Oxidative stress is generated by an imbalance in the production of NO and the overproduction of reactive oxygen species. Caffeine from coffee regulates NO metabolism and homeostasis through changes in the metabolism of asymmetric dimethyl-arginine (ADMA); the latter suppresses nitric oxide synthase which is metabolized and reduced by dimethyl arginine, dimethyl amino hydrolase 1 (DDAH1). Caffeine increases DDAH1 and decreases ADMA (Azad, Heshmati, Daneshzad, & Palmowski, 2021).

Kahweol and cafestol are other compounds of interest present in coffee. These compounds regulate angiogenesis, apoptosis, and inflammation processes. Coffee contains high antioxidants concentrations, widely described and shown to benefit health properties (Godos et al., 2014). Trigonelline has been reported with hypoglycemic and hypolipidemic activity. Another group of compound, melanoidins present in coffee exert antihypertensive and antiglycative activities (Rufián-Henares & Morales, 2007; Verzelloni, Tagliazucchi, Del Rio, Calani, & Conte, 2011).

The contents of soluble and bound hydroxycinnamates (HCAs) such as chlorogenic, caffeic, ferulic, and p-coumaric acids in coffee pulp (CP) in different stages of ripening have been analyzed and quantified. HCAs concentration peaks at the semi-ripe stage (7.4–25.5 mg/g CP, dw), suggesting that unripe or semi-ripe coffee cherries are potentially inexpensive sources of phenolic compounds (Rodríguez-Durán et al., 2014).

Methanolic extracts from roasted and spent coffee contain smaller quantities of phenolic compounds such as protocatechuic acid (4–13%) and gallic acid (4–24%) than other typical compounds chlorogenic acid (16–33% retained) and caffeine (23–36% retained) from spent coffee (Choi & Koh, 2017).

Coffee cherry pulp as by-product is another raw material with considerable amount of polyphenols such as chlorogenic acid, protocatechuic acid, gallic acid, and rutin

(4.9–9.2 mg gallic acid equivalents (GAE)/g DM); the pulp also contains caffeine in water extract (Heeger, Kosińska-Cagnazzo, Cantergiani, & Andlauer, 2017). Nonfermented coffee pulp extracts contain higher total hydroxycinnamic acids than fermented pulp (47.1 vs 30.9 g/kg) (Arellano-González, Ramírez-Coronel, Torres-Mancera, Pérez-Morales, & Saucedo-Castañeda, 2011)

Another bioactive compound of different subclasses had been reported due to extensive characterization and quantification of the minor bioactive compounds in extracts from coffee silverskin (CSS) containing alkaloids and polyphenols (phenolic acids, flavonoids, xanthones, and secoiridoids) using ultrasound-assisted extraction (UAE) with different solvents (methanol, ethanol, and water). At least 18 compounds were characterized including quinine, caffeic acid, ferulic acid, gallic acid, p-coumaric acid, syringic acid, trans-cinnamic acid, vanillic acid, hyperoside, kaempferol, quercetin dihydrate, quercitrin, rutin hydrate, and isogentisin (Nzekoue et al., 2020).

10.6.1 Molecular Mechanism of Action of Relevant Coffee Bioactive Compounds

10.6.1.1 Kahweol

Kahweol is a diterpene found in high amounts in *C. arabica* grains. Kahweol has antioxidant, anti-inflammatory, antiapoptotic, and antiproliferative properties (Lee, Choi, & Kang, 2021). It mediates apoptosis in lung adenocarcinoma cells (Kim, Hwang, & Jeong, 2009) and inhibits metastasis by modulating the expression of matrix metalloproteinases (MPPs) and growth factor endothelial vascular via the signal transducer and activator of protein transcription pathway (STAT3) (Kim et al., 2012). Kahweol significantly inhibits the expression of MMP-9 by inhibiting the signaling pathways Akt/p38 MAPK/JNK1/2 and (nuclear factor-kappa β) NF-κβ attenuating the metastasis of human fibrosarcoma cells (Choi et al., 2019). It also protects neurons against 6-OHDA-induced oxidative damage in SH-5Y5Y cells by upregulating heme oxygenase 1 through phosphatidylinositol 3-kinase (PI3K) and p38 mitogen-activated protein kinase (MAPK) pathways (Hwang & Jeong, 2008). It attenuates the oxidative hydrogen peroxide in induced damage in NIH3T3 cells (Lee & Jeong, 2007). Kahweol reduces the expression of connective tissue growth factor (CTGF) induced by angiotensin II (Ang II) in primary rat aortic smooth muscle cells and vascular smooth muscle cell (VSMC) line (A7r5) (Lee et al., 2021). This is achieved by inhibiting the expression of synthetic phenotypic markers of VSMC, which promote their migration to the intima zone of blood vessels stimulated by Ang II. Moreover, kahweol negatively regulates the expression of phosphorylated proteins extracellular signal regulated kinase (p-Erk), focal adhesion kinase (p-FAK), and Yes-associated protein (YEP) that increase in VSMC stimulated by Ang II, demonstrating that the use of kahweol can contribute to the development of treatments against arteriosclerosis.

10.6.1.2 Cafestol

Is a diterpene found in the grains of *C. arabica*. Like kahweol, it has anti-inflammatory, anticancer, proapoptotic properties and can increase glutathione transferase (GST) activity, modulate cellular detoxification mechanisms, as well as other classes of GST in mice and rats intestine and liver (Huber et al., 2002; Lima et al., 2017). In addition to GST enzymes induction, cafestol can elevate glutathione (GSH) levels by activating c-glutamylcysteine-synthetase (Huber, Scharf, et al., 2002). Cafestol-mediated induction of conjugating with antioxidant enzymes has also been observed in liver, intestinal, and kidney tissues (Huber et al., 2002; Huber, Scharf, et al., 2002). Lee and Jeong (2007) reported the protective effects of cafestol against oxidative stress induced by hydrogen peroxide and against DNA damage. However, its mechanism has not yet been elucidated. These effects can presumably explain some of the anticancer effects of coffee observed in humans and animal models. Cytotoxicity toward leukemic cells has also been observed, suggesting the need for preclinical studies regarding the anticancer properties of cafestol (Lima et al., 2017). Cafestol dose-dependently reduces cell viability and induces apoptosis in head and neck squamous carcinoma cell (HNSCC) (Kotowski et al., 2015). Cafestol also increases resistance to ABT-737 in Caki renal carcinoma cells through downregulation of the Mcl-1 protein and upregulation of Bim. ABT-737 markedly increases apoptosis in multiple myelomas, acute myeloid leukemia, chronic lymphatic leukemia, and lung cancer (Woo et al., 2014). However, resistance to ABT-737 in various cancer cells has been attributed to the high level of Mcl-1 expression. Cafestol may act as a potential agent to reduce cancer cell resistance to ABT-737 due to Mcl-1 suppression to increase ABT-737 sensitivity in multiple cancer cells (Woo et al., 2014).

10.6.1.3 Chlorogenic Acids

Chlorogenic acids (CGAs) are the most abundant polyphenols in coffee (Ludwig, Clifford, Lean, Ashihara, & Crozier, 2014). CGAs are a set of hydroxycinnamic esters together with quinic acid. These make up a family of esters similar to quinic acid and are grouped mainly in caffeic-quinic, di-caffeoylquinic, and feruloquinic acids, reaching up to 7% in dry weight in the arabica variety and 14% in canephora. Similar to other polyphenols, they provide beneficial health properties. They improve cognitive abilities (Mills, Oruna-Concha, Mottram, Gibson, & Spencer, 2013) in addition exerting antimicrobial and antiviral activities (Arauz, Ramos-Tovar, & Muriel, 2017). CGAs prevent diseases related to oxidative stress such as cancer, premature aging, strokes, Alzheimer's, Parkinson's, and other cardiovascular and neurodegenerative diseases (Gonçalves et al., 2017). The accumulation of β-amyloid peptides (Aβ), derived from β-amyloid precursor protein (APP), through sequential cleavages by β- and γ-secretase, is one of the most characteristic features of Alzheimer's disease. It is associated with inflammation, oxidative stress, and high expression of pro-apoptotic proteins resulting in brain cell dysfunction and death of the cerebral cortex (Fukutomi et al., 2021). The use of CGA's such as 5-caffeoylquinic acid has been reported to inhibit excessive AB production, besides upregulating the gene encoding low-density lipoprotein receptor-related protein 1 (LRP1), an Aβ efflux receptor. CGA's normalized the perivascular localization of aquaporin 4 facilitates Aβ clearance along the perivascular pathway. This suggests that 5-caffeoylquinic acid reduces Aβ deposition in the brain by modulating the Aβ clearance pathways and ameliorating cognitive decline and neuronal loss (Ishida et al., 2020). Furthermore, CGA's promote peroxisome proliferation-activated receptor (PPAR)-α expression (which is reduced in Alzheimer's disease). This activates the expression of α secretase, responsible for the nonamyloidogenic pathway of APP degradation and downregulates β-secretase (BACE-1). β-secretase is the primary enzyme responsible for Aβ peptide release in Alzheimer disease (Fukutomi et al., 2021). Therefore, CGAs may have beneficial effects on cognitive dysfunction in Alzheimer's disease. CGAs exhibit antioxidant effects by reducing the production of inflammatory mediators through mechanisms such as suppressing proinflammatory cytokine genes, inhibiting protein tyrosine phosphatase 1B, and modulating inflammatory nuclear factor kappa-light-chain-enhancer of activated B cells (NF-κB) activation via the redox-related tyrosine protein kinase/extracellular signal-regulated kinase and NF-kB– inducing kinase/IκB kinase pathways via oxidative stress reduction (Grosso, Godos, Galvano, & Giovannucci, 2017). Green coffee extracts have high CGAs content, from which their weight-reducing effect derives through AMP-activated protein kinase (AMPK) activation. The protein activation enhances GLUT4 translocation to the plasma membrane, increasing glucose transport to cells, resulting in peripheral glucose clearance. Furthermore, CGAs can inhibit glucose-6 phosphatase by limiting glucose production by glycogenesis or glycogenolysis (Gorji et al., 2019). For all these functions, CGAs are compounds of great interest in the food industry. However, they are unstable under high temperature and can undergo a transesterification reaction during food storage or processing, limiting their applications (Farah & Duarte, 2015). Besides, thermal processing of food products rich in chlorogenic acid can lead to acrylamide formation (Gonçalves et al., 2017).

10.6.1.4 Caffeine

Caffeine (1,3,7-trimethylxanthine) is an alkaloid that acts as a powerful stimulant of the nervous system. It is a common ingredient in pharmaceuticals, beverages, and foods (Li et al., 2020). It has been shown to benefit memory, learning, and delay cognitive decline (Ritchie et al., 2007). Its effects are mainly attributed to its antagonism of adenosine A2A receptors (A2AR), whose synergy with adenosine produces drowsiness (Haskell-Ramsay et al., 2018; Manalo & Medina, 2018). A2A receptor antagonists increase dopaminergic signaling by blocking the basal ganglia indirect pathway, which explains the symptomatic relief in Parkinson's disease (Jazayeri, Andrews, & Marshall, 2017). A2A receptor antagonists decrease the off-time and dyskinesias arising from prolonged L-DOPA use. Also, antagonism of presynaptic A1 receptors by caffeine favors the release of various other transmitters and negatively modulates A1-A2A receptor heterodimers internally (Borroto-Escuela & Fuxe, 2019). Caffeine changes dopaminergic systems by modulating adenosine receptors, attenuating certain neurological diseases by regulating dopaminergic pathways. The protective effect against dopaminergic neuronal loss by caffeine suggests that it may have therapeutic activity against neurodegenerative diseases such as Parkinson's (Alasmari, 2020). It also improves attention span and reaction time by releasing serotonin in the cerebral cortex and the interaction of methylxanthines with adenosine receptors. This mechanism inhibits the phosphodiesterase and adenyl cyclase enzymes by increasing adenosine monophosphate concentration, releasing catecholamines that induce positive inotropic and chronotropic, generating an alert state (Machado et al., 2020). Besides, caffeine can act as a protector against nitrite-induced methemoglobin (Das, Sen, & Sen, 2020). Caffeine also affects other biochemical pathways, for example: inhibiting PDEs (typically>100 μM), increasing cAMP; increasing the cytosolic calcium concentration (normally >500 μM), due to the blockade of Ca^{2+} reuptake by the sarcoplasmic reticulum; and activating the terianodine receptor to release Ca^{2+} stores (Jacobson, Gao, Matricon, Eddy, & Carlsson, 2020). Caffeine reduces histone deacetylase (HDAC) immunoreactivity in dopamine-depleted rats, which can contribute to its neuroprotection (Machado-Filho et al., 2014). Caffeine is a weak (mM) monoamine oxidase A (MAO-A) inhibitor and has served as a support for more potent MAO-A inhibitors (Petzer, Pienaar, & Petzer, 2013).

10.6.1.5 Trigonelline

Trigonelline belongs to the main components of coffee, where it is present in similar or smaller quantities than caffeine. Trigonelline is a derivative of vitamin B6 with a bitter taste. The content in green coffee beans is between 0.6 and 1%. When coffee is roasted at 230 °C, approximately 85% of the trigonelline decomposes into nicotinic acid, with few trigonelline molecules remaining in the roasted beans (Zhou & Zhou, 2012). Trigonelline demethylation during coffee roasting generates nicotinic acid, a water-soluble B vitamin also known as niacin. Therefore, coffee is an essential source of this vitamin in the diet.

Trigonelline has several pharmacological benefits, including hypoglycemic and hypolipidemic activities, sedative, antimigraine, antibacterial, antiviral, and antitumor effects that can improve memory retention and inhibit platelet aggregation (Zhou & Zhou, 2012). Trigonelline has two fundamental mechanisms to prevent cancer. The first is based on the prevention of cell invasion, an important means of controlling cancer cells. The other is to regulate the expression of the transcription factor Nrf2. The transcription factor Nrf2 is activated to protect invaded cells from damage (Hu, Wang, Zhang, & Qiu, 2019). Trigonelline inhibits Nrf2 expression in cancer cells (pancreatic carcinoma H6c7 and pancreatic duct cells) improving the effect of chemotherapy (Arlt et al., 2013). It interferes with the Nrf2 import mechanism but not with the export system. A study in pancreatic carcinoma cells (Panc1, Colo357, and MiaPaca2) showed that trigonelline efficiently decreases tert-Butylhydroquinone (tBHQ)-induced Nrf2 activity in all cell lines, thus reducing the accumulation of the protein Nrf2.

Along with Nrf2 inhibition, trigonelline blocks Nrf2-dependent proteasomal gene expression; these blocking effects were absent after Nrf2 treatment indicating that trigonelline inhibits Nrf2 activation (Arlt et al., 2013). Trigonelline has anti-invasive activity against cancer cells and is involved in the regeneration of dendrites and axons and enhances memory in animal models (Tohda, Kuboyama, & Komatsu, 2005). Nicotinic acid and trigonelline inhibit the invasion of AH109A cells (a rat hepatoma cell line) at low concentrations (2.5–40 μM). The invasive activity of AH109A cells is higher when the cells have been previously cultured in a ROS generating system. Nicotinic acid and trigonelline avoid the ROS-induced increase in cancer cells' invasiveness (Hirakawa, Okauchi, Miura, & Yagasaki, 2005). Trigonelline showed anticancer activity in P-388 leukemia cells in mice and also affects cervical and liver cancer (Zhou & Zhou, 2012). The analysis of antibacterial effect of trigonelline shows high activity against *Candida albicans* and *Candida parapsilosis*, using the microdilution method with 4 μg/ml concentrations (Özçelik, Kartal, & Orhan, 2011). It has antimicrobial activity against *Escherichia coli, Bacillus subtilis, Proteus mirabilis, Pseudomonas aeruginosa, Staphylococcus aureus, Acinetobacter baumannii, Enterococcus faecalis,* and *Klebsiella pneumoniae* (Zhou & Zhou, 2012). Besides, its antiviral activity against Herpes Simplex Type 1 has been observed with noncytotoxic concentrations of 1.6 μg/ml (Özçelik et al., 2011).

10.6.1.6 Sterols

Plant-based sterols, also called phytosterols, are common bioactive compounds in plants with physiological properties (Kamgang Nzekoue et al., 2020). Structurally, they have a cyclopentane-perhydro phenanthrene nucleus in their basic chemical structure with a hydroxyl group at carbon 3. They help reduce and control cholesterol levels, especially low-density lipoprotein (LDL-C), by inhibiting intestinal cholesterol absorption. The reduction of cholesterol is the therapeutic objective to prevent or treat cardiovascular disease (Meco López, Pascual Fuster, & Solà Alberich, 2016). Sterols possess antiteratogenic, anti-inflammatory, neuroprotector, neuromodulatory, anticancer, and hepatoprotective activities (Feng, Wang, Belwal, Li, & Luo, 2020; Randhir, Laird, Maker, Trengove, & Moheimani, 2020; Vivancos & Moreno, 2005). Therefore, food and pharmaceutical industries exploit them as bioactive ingredients in developing numerous products (Kamgang Nzekoue et al., 2020). The plant sterols entering the circulation are rapidly absorbed by different tissues and may affect various physiological processes (Plat et al., 2019). Sterols decrease cholesterol incorporation into mixed micelles, thereby lowering cholesterol bioavailability and absorption. However, other mechanisms have been proposed such as plant sterol-induced elevation in transintestinal cholesterol excretion (TICE) and stimulated efflux of cholesterol from the brush border membrane into the intestinal lumen (Nakano et al., 2016). During transport from the intestinal lumen into circulation, plant sterols can easily interfere with all types of immune cells present in the lymphoid follicles. Plant sterols are assimilated by immune cells in the circulation and incorporated into red blood cells and platelets. Plant sterols are also incorporated into other peripheral tissues, including the vascular wall, lungs, brain, and breast milk (Plat et al., 2019).

10.7 Extraction of Bioactive Compounds from Coffee

The technique or method to extract bioactive compounds from coffee depends on the raw material. Bioactive compounds can be obtained from roasted and green coffee beans or alternative sources of the coffee processing industry (Table 10.3) such as pulp, husk, silverskin, and used coffee (Acevedo et al., 2013; Murthy & Naidu, 2012).

These new technological extraction procedures have set a precedent for their feasibility in industrial applications compared to conventional processes. In the traditional method, hexane is used as a principal solvent through a simple soxhlet process over several hours; other solvents have also been used such as ethanol, methanol, and isopropanol (Ballesteros et al., 2017).

In the autohydrolysis process, water act as a principal solvent; however, water is used as the subcritical fluid acting as a nontoxic less corrosive solvent with modifiers such as nitrogen and carbon dioxide (Getachew & Chun, 2017; Mayanga-Torres et al., 2017)

Technology unlike previously described conventional processes are based on a combination between solvents and modifiers sometimes with an innovative technical approach for efficient, eco-friendly, and fast extraction process. In these processes common and essential factors are studied such as time, temperature, pressure, potency, field intensity, solvent/solid ratio, the particle size of the solid matrix, among others according to the technology to improve the yield and quality of the compounds (Al-Dhabi et al., 2017; Barbosa-Pereira

Table 10.3 Principal methods and techniques to extract bioactive compounds from coffee.

Method of extraction	Technology used	References
Traditional liquid–solid extraction	Organics solvents applied a simple or soxhlet process.	Acevedo et al. (2013); Araújo and Sandi (2007)
	Autohydrolysis process	Ballesteros, Ramirez, Orrego, Teixeira, and Mussatto (2017)
New technological procedures	Microwave-assisted extraction (MAE) Ultrasound-assisted extraction (UAE), Accelerated solvent extraction (ASE), Supercritical fluid extraction (SFE) Pressurized fluid extraction (PFE) Pulsed electric fields	Al-Dhabi, Ponmurugan, and Maran Jeganathan (2017) Belandria et al. (2016) Barbosa-Pereira, Guglielmetti, and Zeppa (2018)

Source: Based on Acevedo et al. (2013), Murthy and Naidu (2012).

et al., 2018; Belandria et al., 2016). Extraction process must evaluate such aspects as operative costs, reliability, eco- and climate-friendly, and attractiveness in terms of the products' quality attributes required for implementation at an industrial scale.

10.8 New Coffee-based Food Products, Nutraceuticals, and Supplements Commercialized, Patented, and in Progress

Coffee by-products have been explored as a source of antioxidants and phenolic compounds, due to their high content of flavonols, anthocyanins, chlorogenic acids, among others (Ramirez-Coronel et al., 2004). In 2014, coffee cascara was reported with a market price of US$13–17 per kg (Ciummo, 2014). Among the products available on the market are soluble powders, which can replace powdered drinks with high sugar content. Iriondo-DeHond et al. (2020) developed an instant beverage from powdered aqueous extract from coffee cascara with low sugar and fat content and high fiber and vitamin C content.

Another segment of this industry is the beverages made from dried coffee husks. Examples of such products are the nonalcoholic carbonated beverages "Caskai Sparkling Cascara Infusion," with tartaric acid and organic cane sugar as ingredients (Jelderks & Zimmermann, 2021). The Spanish company SUPRACAFÉ S.A. commercializes a coffee cherry hulls infusion called "Tabifruit," with combined flavors such as orange-cinnamon, and orange-passion fruit berry-flavored (Supracafé S.A., 2020). PectCof B.V. discloses a patent treating pulp residues to obtain bioactive compounds, mainly pectin with gelling, thickening, emulsifying, and valuable prebiotic properties for the food and pharmaceutical industries (Belalcazar Otalora, 2014; Pectcof B.V., 2021). This company has proposed obtaining fibers from coffee pulp residues; "Dutch Gum" being its main products. This fiber can be an alternative to the use of Arabic gum in soft drinks; however, it is currently presented as a work in progress, aimed at personal care, paint manufacturing, and dairy products (Pectcof B.V., 2021). Spent coffee beans have been the source for the development of high antioxidant fiber (Castillo et al., 2013).

Currently, coffee hulls are used as a food ingredient in gluten-free products such as bread and other foodstuffs to improve sensory and nutritional properties. The Coffee Cherry Company has invested in coffee pulp recycling processes as an example of coffee hulls in the manufacture of low-fat, gluten-free products rich in antioxidants, proteins, potassium, and dietary fiber (Belliveau, 2021). In Europe, the coffee husk is considered a novel food. Therefore, it has not been authorized by the European Food Safety Authority (EFSA) for use in food; however, in the USA, the coffee husk is widely used in the beverage industry. Currently, α-glucosidase activity of coffee hulls has been investigated as an appetite suppressant in obtaining dairy products with the addition of inulin (Iriondo-DeHond et al., 2020).

Green coffee is also considered a novel food due to its polysaccharides, proteins, amino acids, minerals, and polyphenolic compounds. This product is marketed unroasted for the preparation of infusions or to obtain extracts. The extracts have also been marketed as food supplements underlining their health benefits and have been used as beverages or chewing gum (Klingel et al., 2020; Macheiner, Schmidt, Schreiner, & Mayer, 2019; Marcason, 2013). There are products on the market containing chlorogenic acid, such as Svetol® and Green Coffee Extract, which have shown potential in regulating glucose, blood pressure, and body weight (Henry-Vitrac, Ibarra, Roller, Mérillon, & Vitrac, 2010; Svetol, 2021; Watanabe et al., 2006). The literature also reports a patented extract from coffee fruit, called NeuroFactor™, which allows for the upregulation of the memory neuroprotein BDNF (FutureCeuticals, 2021).

In West Sumatera, Indonesia, Yemen, and Etiopia, the leaves of coffee plants have been used to make hot beverages such as herbal teas with beneficial health effects. They have been called "kahwa daun," "giser," and "kuti." However, more research is needed to investigate their phytochemical aspects and bioactivity (Klingel et al., 2020; Novita, Kasim, Anggraini, & Putra, 2018). Coffee silverskin is another by-product under study; it has high protein and ash contents compared to whole wheat flour and wheat bran, as well as antioxidant activity related to dietary fiber content (Borrelli, Esposito, Napolitano, Ritieni, & Fogliano, 2004; Murthy & Naidu, 2012). Powdered coffee silverskin can be used as food preservative due to its antioxidant potential, in addition to its composition in chlorogenic acid and caffeine (del Castillo et al., 2013). Coffee silverskin is a promising low-cost functional ingredient (Bresciani, Calani, Bruni, Brighenti, & Del Rio, 2014) to develop caffeine-free nutraceuticals or food supplements, or decaffeinated products (Regazzoni et al., 2016). Thus, the market for products from coffee by-products has increased and is in the process of approval by the corresponding authorities.

10.9 Conclusion

Growing consumer awareness of the health benefits of food and its nutritional benefits for preventing potential illnesses and improving health is the current focus of the global functional foods and nutraceuticals market. Functional foods and nutraceuticals are helping improve health. Besides, the growing demand for functional foods also helps producers diversify their crops and promote research and innovation. Coffee is one of the most popular and consumed beverages globally, accounting for 75% of the regular soft drinks consumption. It is proven that coffee and its active compounds can provide beneficial effects against various neurological pathologies, cancer, psychoactive, metabolic disorders, and liver dysfunctions. These benefits have led to an increase in coffee consumption.

10.10 Acknowledgments

We thank the National Council of Science and Technology (CONACYT-Mexico) for financial support through the project: FOSEC_CB 2017–2018 A1-S-42515. Salvador A. Saldaña Mendoza thanks CONACYT-Mexico and the Autonomous University of Coahuila for the financial support and scholarship to develop his Ph.D. studies.

References

Acevedo, F., Rubilar, M., Scheuermann, E., Cancino, B., Uquiche, E., Garcés, M., . . . Shene, C. (2013). Spent coffee grounds as a renewable source of bioactive compounds. *Journal of Biobased Materials and Bioenergy*, *7*(3), 420–428. https://doi.org/10.1166/jbmb.2013.1369.

Al-Dhabi, N. A., Ponmurugan, K., & Maran Jeganathan, P. (2017). Development and validation of ultrasound-assisted solid–liquid extraction of phenolic compounds from waste spent coffee grounds. *Ultrasonics Sonochemistry*, *34*, 206–213. https://doi.org/10.1016/j.ultsonch.2016.05.005.

Alasmari, F. (2020). Caffeine induces neurobehavioral effects through modulating neurotransmitters. *Saudi Pharmaceutical Journal*, *28*(4), 445–451. https://doi.org/10.1016/j.jsps.2020.02.005.

Ali, A., Ahmad, U., Juber Akhtar, B., & Khan, M. M. (2019). Engineered nano scale formulation strategies to augment efficiency of nutraceuticals. *Journal of Functional Foods*, *62*(September), 103554. https://doi.org/10.1016/j.jff.2019.103554.

Alves, R. C., Rodrigues, F., Antónia Nunes, M., Vinha, A. F., & Oliveira, M. B. P. P. (2017). State of the art in coffee processing by-products. In C. M. Galanakis (Ed.), *Handbook of Coffee Processing By-Products* (pp. 1–26). Elsevier. https://doi.org/10.1016/B978-0-12-811290-8.00001-3

Alzoman, H., Alzahrani, A., Alwehaiby, K., Alanazi, W., & AlSarhan, M. (2021). Efficacy of arabic coffee and black tea in reducing halitosis: a randomized, double-blind, controlled, crossover clinical trial. *Healthcare*, *9*(3), 250. https://doi.org/10.3390/healthcare9030250.

Araújo, J. M. A., & Sandi, D. (2007). Extraction of coffee diterpenes and coffee oil using supercritical carbon dioxide. *Food Chemistry*, *101*(3), 1087–1094. https://doi.org/10.1016/j.foodchem.2006.03.008.

Arauz, J., Ramos-Tovar, E., & Muriel, P. (2017). Coffee and the liver. In Muriel, P. *Liver pathophysiology: Therapies and antioxidants*. Elsevier. https://doi.org/10.1016/B978-0-12-804,274-8.00048-5.

Arellano-González, M. A., Ramírez-Coronel, M. A., Torres-Mancera, M. T., Pérez-Morales, G. G., & Saucedo-Castañeda, G. (2011). Antioxidant activity of fermented and nonfermented coffee (*C. arabica*) pulp extracts. *Food Technology and Biotechnology*, *49*(3), 374–378.

Arlt, A., Sebens, S., Krebs, S., Geismann, C., Grossmann, M., Kruse, M.-L., . . . Schäfer, H. (2013). Inhibition of the Nrf2 transcription factor by the alkaloid trigonelline renders pancreatic cancer cells more susceptible to apoptosis through decreased proteasomal gene expression and proteasome activity. *Oncogene*, *32*(40), 4825–4835. https://doi.org/10.1038/onc.2012.493.

Asgary, S., Keshvari, M., Sahebkar, A., Hashemi, M., & Rafieian-Kopaei, M. (2013). Clinical investigation of the acute effects of pomegranate juice on blood pressure and endothelial function in hypertensive individuals. *ARYA Atherosclerosis*, *9*(6), 326–331. https://doi.org/10.22122/arya.v9i6.643

Azad, B. J., Heshmati, J., Daneshzad, E., & Palmowski, A. (2021). Effects of coffee consumption on arterial stiffness and endothelial function: a systematic review and meta-analysis of randomized clinical trials. *Critical Reviews in Food Science and Nutrition*, *61*(6), 1013–1026. https://doi.org/10.1080/10408398.2020.1750343.

Ballesteros, L. F., Ramirez, M. J., Orrego, C. E., Teixeira, J. A., & Mussatto, S. I. (2017). Optimization of autohydrolysis conditions to extract antioxidant phenolic compounds from spent coffee grounds. *Journal of Food Engineering*, *199*, 1–8. https://doi.org/10.1016/j.jfoodeng.2016.11.014.

Barbosa-Pereira, L., Guglielmetti, A., & Zeppa, G. (2018). Pulsed electric field assisted extraction of bioactive compounds from cocoa bean shell and coffee silverskin. *Food and Bioprocess Technology*, *11*(4), 818–835. https://doi.org/10.1007/s11947-017-2045-6.

Belalcazar Otalora, A. F. (2014). *Pectin extraction from coffee pulp*.

Belandria, V., Aparecida de Oliveira, P. M., Chartier, A., Rabi, J. A., de Oliveira, A. L., & Bostyn, S. (2016). Pressurized-fluid extraction of cafestol and kahweol diterpenes from green coffee. *Innovative Food Science and Emerging Technologies*, *37*, 145–152. https://doi.org/10.1016/j.ifset.2016.07.022.

Borrelli, R. C., Esposito, F., Napolitano, A., Ritieni, A., & Fogliano, V. (2004). Characterization of a new potential functional ingredient: coffee silverskin. *Journal of Agricultural and Food Chemistry*, *52*(5), 1338–1343. https://doi.org/10.1021/jf034974x.

Borroto-Escuela, D. O., & Fuxe, K. (2019). Adenosine heteroreceptor complexes in the basal ganglia are implicated in Parkinson's disease and its treatment. *Journal of Neural Transmission*, *126*(4), 455–471. https://doi.org/10.1007/s00702-019-01969-2.

Bresciani, L., Calani, L., Bruni, R., Brighenti, F., & Del Rio, D. (2014). Phenolic composition, caffeine content and antioxidant capacity of coffee silverskin. *Food Research International*, *61*, 196–201. https://doi.org/10.1016/j.foodres.2013.10.047.

Brufau, G., Canela, M. A., & Rafecas, M. (2008). Phytosterols: physiologic and metabolic aspects related to cholesterol-lowering properties. *Nutrition Research*, *28*(4), 217–225. https://doi.org/10.1016/j.nutres.2008.02.003.

Castaldo, L., Izzo, L., Narváez, A., Rodríguez-Carrasco, Y., Grosso, M., & Ritieni, A. (2021). Colon bioaccessibility under *in vitro* gastrointestinal digestion of different coffee brews chemically profiled through UHPLC-Q-orbitrap HRMS. *Foods*, *10*(1). https://doi.org/10.3390/foods10010179.

Castillo M, Ibáñez E, Amigo-Benavent M, Herrero M, Plaza M, Ullate M (2013) Application of products of coffee silverskin in anti-aging cosmetics and functional food. International patent application No. WO2014128320A1.

Catapano, A. L., Barrios, V., Cicero, A. F. G., & Pirro, M. (2019). Lifestyle interventions and nutraceuticals: Guideline-based approach to cardiovascular disease prevention. *Atherosclerosis: X*, *1*(February), 100003. https://doi.org/10.1016/j.athx.2019.100003

Chiu, H. F., Venkatakrishnan, K., & Wang, C. K. (2020). The role of nutraceuticals as a complementary therapy against various neurodegenerative diseases: A mini-review. *Journal of Traditional and Complementary Medicine*, *10*(5), 434–439. https://doi.org/10.1016/j.jtcme.2020.03.008.

Choi, B., & Koh, E. (2017). Spent coffee as a rich source of antioxidative compounds. *Food Science and Biotechnology*, *26*(4), 921–927. https://doi.org/10.1007/s10068-017-0144-9.

Choi, J. H., Hwang, Y. P., Jin, S. W., Lee, G. H., Kim, H. G., Han, E. H., . . . Jeong, H. G. (2019). Suppression of PMA-induced human fibrosarcoma HT-1080 invasion and metastasis by kahweol via inhibiting Akt/JNK1/2/p38 MAPK signal pathway and NF-κB dependent transcriptional activities. *Food and Chemical Toxicology*, *125*(August 2018), 1–9. https://doi.org/10.1016/j.fct.2018.12.038

Daliri, E. B.-M., & Lee, B. H. (2015). *Current Trends and Future Perspectives on Functional Foods and Nutraceuticals*. https://doi.org/10.1007/978-3-319-23,177-8_10.

Das, D., Sen, S., & Sen, K. (2020). Caffeine and catechin toward prevention of drug induced oxidation of hemoglobin: A spectroscopic study. *Spectrochimica Acta - Part A: Molecular and Biomolecular Spectroscopy*, *232*, 118,167. https://doi.org/10.1016/j.saa.2020.118167.

de Abreu Pinheiro, F., Ferreira Elias, L., de Jesus Filho, M., Uliana Modolo, M., de Gomes Rocha, J. C., Fumiere Lemos, M., . . . Soares Cardoso, W. (2021). Arabica and onilon coffee flowers: Bioactive compounds and antioxidant capacity under different processes. *Food Chemistry*, *336*, 127,701. https://doi.org/10.1016/j.foodchem.2020.127701.

Devla, M. N., Acharya, S. R., Acharya, N. S., & Kumar, V. (2011). Dietary supplements: A legal status in India and in foreign countries. *International Journal of Pharmacy and Pharmaceutical Sciences*, *3*(SUPPL. 3), 7–12.

Esquivel, P., & Jiménez, V. M. (2012). Functional properties of coffee and coffee by-products. *Food Research International*, *46*(2), 488–495. https://doi.org/10.1016/j.foodres.2011.05.028.

Fanali, C., Della Posta, S., Dugo, L., Gentili, A., Mondello, L., & De Gara, L. (2020). Choline-chloride and betaine-based deep eutectic solvents for green extraction of nutraceutical compounds from spent coffee ground. *Journal of Pharmaceutical and Biomedical Analysis*, *189*, 113,421. https://doi.org/10.1016/j.jpba.2020.113421.

Farah, A., & Duarte, G. (2015). Bioavailability and metabolism of chlorogenic acids from coffee. In Victor R. Preedy *Coffee in Health and Disease Prevention*. Elsevier. https://doi.org/10.1016/B978-0-12-409,517-5.00087-5.

Feng, S., Wang, L., Belwal, T., Li, L., & Luo, Z. (2020). Phytosterols extraction from hickory (Carya cathayensis Sarg.) husk with a green direct citric acid hydrolysis extraction method. *Food Chemistry*, *315*. https://doi.org/10.1016/j.foodchem.2020.126217.

Food and Agriculture Organization of the United Nations. (2021). FAOSTAT. Retrieved January 15, 2021, from http://www.fao.org/faostat/es/#data/QC/visualize.

Fukutomi, R., Ohishi, T., Koyama, Y., Pervin, M., Nakamura, Y., & Isemura, M. (2021). Beneficial effects of epigallocatechin-3-O-gallate, chlorogenic acid, resveratrol, and curcumin on neurodegenerative Diseases. *Molecules*, *26*(2), 415. https://doi.org/10.3390/molecules26020415.

Garima, V., & Manoj, K. M. (2016). A review on nutraceuticals:classification and its role in various diseases. *International Journal of Pharmacy and Therapeutics*, *7*(4), 152–160.

Gemechu, F. G. (2020). Embracing nutritional qualities, biological activities and technological properties of coffee byproducts in functional food formulation. *Trends in Food Science and Technology*, *104*(August), 235–261. https://doi.org/10.1016/j.tifs.2020.08.005.

Getachew, A. T., & Chun, B. S. (2017). Influence of pretreatment and modifiers on subcritical water liquefaction of spent coffee grounds: A green waste valorization approach. *Journal of Cleaner Production*, *142*, 3719–3727. https://doi.org/10.1016/j.jclepro.2016.10.096.

Ghaffari, S., & Roshanravan, N. (2020). The role of nutraceuticals in prevention and treatment of hypertension: An updated review of the literature. *Food Research International*, *128*(May 2019). https://doi.org/10.1016/j.foodres.2019.108749.

Godos, J., Pluchinotta, F. R., Marventano, S., Buscemi, S., Volti, G. L., Galvano, F., & Grosso, G. (2014). Coffee components and cardiovascular risk: Beneficial and detrimental effects. *International Journal of Food Sciences and Nutrition*, *65*(8), 925–936. https://doi.org/10.3109/09637486.2014.940287.

Gonçalves, B., Moeenfard, M., Rocha, F., Alves, A., Estevinho, B. N., & Santos, L. (2017). Microencapsulation of a natural antioxidant from coffee—chlorogenic acid (3-caffeoylquinic acid). *Food and Bioprocess Technology*, *10*(8), 1521–1530. https://doi.org/10.1007/s11947-017-1919-y.

Gorji, Z., Varkaneh, H. K., Talaei, S., Nazary-Vannani, A., Clark, C. C. T., Fatahi, S., . . . Zhang, Y. (2019). The effect of green-coffee extract supplementation on obesity: A systematic review and dose–response meta-analysis of randomized controlled trials. *Phytomedicine*, *63*(1). https://doi.org/10.1016/j.phymed.2019.153018.

Grosso, G., Godos, J., Galvano, F., & Giovannucci, E. L. (2017). Coffee, caffeine, and health outcomes: an umbrella review. *Annual Review of Nutrition*, *37*, 131–156. https://doi.org/10.1146/annurev-nutr-071816-064941.

Grosso, G., Micek, A., Godos, J., Sciacca, S., Pajak, A., Martínez-González, M. A., . . . Galvano, F. (2016). Coffee consumption and risk of all-cause, cardiovascular, and cancer mortality in smokers and non-smokers: a dose–response meta-analysis. *European Journal of Epidemiology*, *31*(12), 1191–1205. https://doi.org/10.1007/s10654-016-0202-2.

Haskell-Ramsay, C. F., Jackson, P. A., Forster, J. S., Dodd, F. L., Bowerbank, S. L., & Kennedy, D. O. (2018). The acute effects of caffeinated black coffee on cognition and mood in healthy young and older adults. *Nutrients*, *10*(10). https://doi.org/10.3390/nu10101386.

Heeger, A., Kosińska-Cagnazzo, A., Cantergiani, E., & Andlauer, W. (2017). Bioactives of coffee cherry pulp and its utilization for production of Cascara beverage. *Food Chemistry*, *221*, 969–975. https://doi.org/10.1016/j.foodchem.2016.11.067.

Henry-Vitrac, C., Ibarra, A., Roller, M., Mérillon, J. M., & Vitrac, X. (2010). Contribution of chlorogenic acids to the inhibition of human hepatic glucose-6-phosphatase activity *in vitro* by svetol, a standardized decaffeinated green coffee extract. *Journal of Agricultural and Food Chemistry*, *58*(7), 4141–4144. https://doi.org/10.1021/jf9044827.

Hirakawa, N., Okauchi, R., Miura, Y., & Yagasaki, K. (2005). Anti-invasive activity of niacin and trigonelline against cancer cells. *Bioscience, Biotechnology and Biochemistry*, *69*(3), 653–658. https://doi.org/10.1271/bbb.69.653.

Hu, G. L., Wang, X., Zhang, L., & Qiu, M. H. (2019). The sources and mechanisms of bioactive ingredients in coffee. *Food and Function, 10*(6), 3113–3126. https://doi.org/10.1039/C9FO00288J.

Huber, W. W., Prustomersky, S., Delbanco, E., Uhl, M., Scharf, G., Turesky, R. J., . . . Schulte-Hermann, R. (2002). Enhancement of the chemoprotective enzymes glucuronosyl transferase and glutathione transferase in specific organs of the rat by the coffee components kahweol and cafestol. *Archives of Toxicology, 76*(4), 209–217. https://doi.org/10.1007/s00204-002-0322-1.

Huber, W. W., Scharf, G., Rossmanith, W., Prustomersky, S., Grasl-Kraupp, B., Peter, B., . . . Schulte-Hermann, R. (2002). The coffee components kahweol and cafestol induce γ-glutamylcysteine synthetase, the rate limiting enzyme of chemoprotective glutathione synthesis, in several organs of the rat. *Archives of Toxicology, 75*(11), 685–694. https://doi.org/10.1007/s00204-001-0295-5.

Hutachok, N., Angkasith, P., Chumpun, C., Fucharoen, S., Mackie, I. J., Porter, J. B., & Srichairatanakool, S. (2020). Anti-platelet aggregation and anti-cyclooxygenase activities for a range of coffee extracts (*C. arabica*). *Molecules (Basel, Switzerland), 26*(1). https://doi.org/10.3390/molecules26010010.

Hwang, Y. P., & Jeong, H. G. (2008). The coffee diterpene kahweol induces heme oxygenase-1 via the PI3K and p38/Nrf2 pathway to protect human dopaminergic neurons from 6-hydroxydopamine-derived oxidative stress. *FEBS Letters, 582*(17), 2655–2662. https://doi.org/10.1016/j.febslet.2008.06.045.

Iriondo-DeHond, A., Elizondo, A. S., Iriondo-DeHond, M., Ríos, M. B., Mufari, R., Mendiola, J. A., . . . del Castillo, M. D. (2020). Assessment of healthy and harmful maillard reaction products in a novel coffee cascara beverage: melanoidins and acrylamide. *Foods, 9*(5), 620. https://doi.org/10.3390/foods9050620.

Iriondo-DeHond, M., Iriondo-DeHond, A., Herrera, T., Fernández-Fernández, A. M., Sorzano, C. O. S., Miguel, E., & del Castillo, M. D. (2020). Sensory acceptance, appetite control and gastrointestinal tolerance of yogurts containing coffee-cascara extract and inulin. *Nutrients, 12*(3), 627. https://doi.org/10.3390/nu12030627.

Ishida, K., Misawa, K., Nishimura, H., Hirata, T., Yamamoto, M., & Ota, N. (2020). 5-Caffeoylquinic acid ameliorates cognitive decline and reduces Aβ deposition by modulating Aβ clearance pathways in APP/PS2 Transgenic Mice. *Nutrients, 12*(2), 494. https://doi.org/10.3390/nu12020494.

Jacobson, K. A., Gao, Z., Matricon, P., Eddy, M. T., & Carlsson, J. (2020). Adenosine A 2A receptor antagonists: from caffeine to selective non-xanthines. *British Journal of Pharmacology*, (March), bph.15103. https://doi.org/10.1111/bph.15103.

Jazayeri, A., Andrews, S. P., & Marshall, F. H. (2017). Structurally enabled discovery of adenosine A 2A receptor antagonists. *Chemical Reviews, 117*(1), 21–37. https://doi.org/10.1021/acs.chemrev.6b00119.

Jiménez, S. R., Valle, J. S. L., Ocampo, A. U., Pérez, S. C., Hernández, N. P., & Cárdenas, J. S. M. (2015). Consumo de nutracéuticos, una alternativa en la prevención de las enfermedades crónicas no transmisibles. *Biosalud, 14*(2), 91–103.

Kamgang Nzekoue, F., Khamitova, G., Angeloni, S., Nácher Sempere, A., Tao, J., Maggi, F., . . . Caprioli, G. (2020). Spent coffee grounds: a potential commercial source of phytosterols. *Food Chemistry*, *325*(February), 126836. https://doi.org/10.1016/j.foodchem.2020.126836.

Kennedy, O. J., Roderick, P., Buchanan, R., Fallowfield, J. A., Hayes, P. C., & Parkes, J. (2016). Systematic review with meta-analysis: Coffee consumption and the risk of cirrhosis. *Alimentary Pharmacology and Therapeutics*, *43*(5), 562–574. https://doi.org/10.1111/apt.13523.

Kim, H. G., Hwang, Y. P., Han, E. H., Choi, J. H., Kwon, K. Il, Chung, Y. C., . . . Jeong, H. G. (2012). The coffee diterpene kahweol inhibits metastasis by modulating expressions of MMPs and VEGF via STAT3 inactivation. *Food Chemistry*, *133*(4), 1521–1529. https://doi.org/10.1016/j.foodchem.2012.02.043.

Kim, H. G., Hwang, Y. P., & Jeong, H. G. (2009). Kahweol blocks STAT3 phosphorylation and induces apoptosis in human lung adenocarcinoma A549 cells. *Toxicology Letters*, *187*(1), 28–34. https://doi.org/10.1016/j.toxlet.2009.01.022.

Klingel, T., Kremer, J. I., Gottstein, V., Rajcic de Rezende, T., Schwarz, S., & Lachenmeier, D. W. (2020). A review of coffee by-products including leaf, flower, cherry, husk, silver skin, and spent grounds as novel foods within the European Union. *Foods*, *9*(5), 665. https://doi.org/10.3390/foods9050665.

Kotowski, U., Heiduschka, G., Seemann, R., Eckl-Dorna, J., Schmid, R., Kranebitter, V., . . . Thurnher, D. (2015). Effect of the coffee ingredient cafestol on head and neck squamous cell carcinoma cell lines. *Strahlentherapie Und Onkologie*, *191*(6), 511–517. https://doi.org/10.1007/s00066-014-0807-x.

Kuang, A., Erlund, I., Herder, C., Westerhuis, J. A., Tuomilehto, J., & Cornelis, M. C. (2018). Lipidomic response to coffee consumption. *Nutrients*, *10*(12), 1–12. https://doi.org/10.3390/nu10121851.

Lama, A., Pirozzi, C., Avagliano, C., Annunziata, C., Mollica, M. P., Calignano, A., . . . Mattace Raso, G. (2020). Nutraceuticals: An integrative approach to starve Parkinson's disease. *Brain, Behavior, and Immunity—Health*, *2*(December 2019), 100037. https://doi.org/10.1016/j.bbih.2020.100037

Lee, J. H., Choi, S. T., & Kang, Y. J. (2021). Kahweol, a diterpenoid molecule, inhibits CTGF-dependent synthetic phenotype switching and migration in vascular smooth muscle cells. *Molecules*, *26*(3), 640. https://doi.org/10.3390/molecules26030640.

Lee, K. J., & Jeong, H. G. (2007). Protective effects of kahweol and cafestol against hydrogen peroxide-induced oxidative stress and DNA damage. *Toxicology Letters*, *173*(2), 80–87. https://doi.org/10.1016/j.toxlet.2007.06.008.

Li, S., Wen, J., He, B., Wang, J., Hu, X., & Liu, J. (2020). Occurrence of caffeine in the freshwater environment: Implications for ecopharmacovigilance. *Environmental Pollution*, *263*, 114,371. https://doi.org/10.1016/j.envpol.2020.114371.

Lima, C. S., Spindola, D. G., Bechara, A., Garcia, D. M., Palmeira-dos-Santos, C., Peixoto-da-Silva, J., . . . Bincoletto, C. (2017). Cafestol, a diterpene molecule found in coffee, induces leukemia cell death. *Biomedicine and Pharmacotherapy*, *92*, 1045–1054. https://doi.org/10.1016/j.biopha.2017.05.109.

Limer, J. L., & Speirs, V. (2004). Phyto-oestrogens and breast cancer chemoprevention. *Breast Cancer Research*, *6*(3), 119–127. https://doi.org/10.1186/bcr781.

Loftfield, E., Freedman, N. D., Graubard, B. I., Hollenbeck, A. R., Shebl, F. M., Mayne, S. T., & Sinha, R. (2015). Coffee drinking and cutaneous melanoma risk in the NIH-AARP diet and health study. *Journal of the National Cancer Institute, 107*(2), 1–9. https://doi.org/10.1093/jnci/dju421.

Ludwig, I. A., Clifford, M. N., Lean, M. E. J., Ashihara, H., & Crozier, A. (2014). Coffee: Biochemistry and potential impact on health. *Food and Function, 5*(8), 1695–1717. https://doi.org/10.1039/c4fo00042k.

Machado-Filho, J. A., Correia, A. O., Montenegro, A. B. A., Nobre, M. E. P., Cerqueira, G. S., Neves, K. R. T., . . . de Barros Viana, G. S. (2014). Caffeine neuroprotective effects on 6-OHDA-lesioned rats are mediated by several factors, including pro-inflammatory cytokines and histone deacetylase inhibitions. *Behavioral Brain Research, 264*, 116–125. https://doi.org/10.1016/j.bbr.2014.01.051.

Machado, S., Sá Filho, A. S., Campos, C., de Paula, C. C., Bernardes, F., Murillo-Rodriguez, E., . . . Lattari, E. (2020). Can caffeine intake combined with aerobic exercise lead to improvement in attentional and psychomotor performance in trained individuals? *IBRO Reports, 8*(January), 76–81. https://doi.org/10.1016/j.ibror.2020.01.002.

Macheiner, L., Schmidt, A., Schreiner, M., & Mayer, H. K. (2019). Green coffee infusion as a source of caffeine and chlorogenic acid. *Journal of Food Composition and Analysis, 84*, 103,307. https://doi.org/10.1016/j.jfca.2019.103307.

Manalo, R. V. M., & Medina, P. M. B. (2018). Caffeine protects dopaminergic neurons from dopamine-induced neurodegeneration via synergistic adenosine-dopamine D2-like receptor interactions in Transgenic *Caenorhabditis elegans. Frontiers in Neuroscience, 12*(MAR), 1–9. https://doi.org/10.3389/fnins.2018.00137.

Marcason, W. (2013). What is Green Coffee extract? *Journal of the Academy of Nutrition and Dietetics, 113*(2), 364. https://doi.org/10.1016/j.jand.2012.12.004.

Maurya, A. P., Chauhan, J., Yadav, D. K., Gangwar, R., & Maurya, V. K. (2021). Nutraceuticals and their impact on human health. In C. Egbuna, A. P. Mishra, & M. R. Goyal (Eds.), *Preparation of Phytopharmaceuticals for the Management of Disorders* (pp. 229–254). Elsevier. https://doi.org/10.1016/B978-0-12-820,284-5.00011-3

Mayanga-Torres, P. C., Lachos-Perez, D., Rezende, C. A., Prado, J. M., Ma, Z., Tompsett, G. T., . . . Forster-Carneiro, T. (2017). Valorization of coffee industry residues by subcritical water hydrolysis: Recovery of sugars and phenolic compounds. *Journal of Supercritical Fluids, 120*, 75–85. https://doi.org/10.1016/j.supflu.2016.10.015.

Meco López, J. F., Pascual Fuster, V., & Solà Alberich, R. (2016). La utilización de los esteroles vegetales en la práctica clínica: de la química a la clínica. *Clinica e Investigacion En Arteriosclerosis, 28*(6), 283–294. https://doi.org/10.1016/j.arteri.2016.04.001.

Mills, C. E., Oruna-Concha, M. J., Mottram, D. S., Gibson, G. R., & Spencer, J. P. E. (2013). The effect of processing on chlorogenic acid content of commercially available coffee. *Food Chemistry, 141*(4), 3335–3340. https://doi.org/10.1016/j.foodchem.2013.06.014.

Montenegro, J., dos Santos, L. S., de Souza, R. G. G., Lima, L. G. B., Mattos, D. S., Viana, B. P. P. B., . . . Teodoro, A. J. (2021). Bioactive compounds, antioxidant activity and antiproliferative effects in prostate cancer cells of green and roasted coffee extracts obtained by microwave-assisted extraction (MAE). *Food Research International, 140*(December 2020). https://doi.org/10.1016/j.foodres.2020.110014

Moss, J. W. E., Williams, J. O., & Ramji, D. P. (2018). Nutraceuticals as therapeutic agents for atherosclerosis. *Biochimica et Biophysica Acta—Molecular Basis of Disease, 1864*(5), 1562–1572. https://doi.org/10.1016/j.bbadis.2018.02.006.

Murthy, P. S., & Naidu, M. M. (2012). Recovery of phenolic antioxidants and functional compounds from coffee industry by-products. *Food and Bioprocess Technology*, *5*(3), 897–903. https://doi.org/10.1007/s11947-010-0363-z.

Nakano, T., Inoue, I., Takenaka, Y., Ono, H., Katayama, S., Awata, T., & Murakoshi, T. (2016). Ezetimibe promotes brush border membrane-to-lumen cholesterol efflux in the small intestine. *PLOS ONE*, *11*(3), e0152207. https://doi.org/10.1371/journal.pone.0152207.

Nasri, H., Baradaran, A., Shirzad, H., & Kopaei, M. R. (2014). New concepts in nutraceuticals as alternative for pharmaceuticals. *International Journal of Preventive Medicine*, *5*(12), 1487–1499.

Novita, R., Kasim, A., Anggraini, T., & Putra, D. P. (2018). Kahwa daun: traditional knowledge of a coffee leaf herbal tea from West Sumatera, Indonesia. *Journal of Ethnic Foods*, *5*(4), 286–291. https://doi.org/10.1016/j.jef.2018.11.005.

Nzekoue, F. K., Angeloni, S., Navarini, L., Angeloni, C., Freschi, M., Hrelia, S., . . . Caprioli, G. (2020). Coffee silverskin extracts: Quantification of 30 bioactive compounds by a new HPLC-MS/MS method and evaluation of their antioxidant and antibacterial activities. *Food Research International*, *133*(February). https://doi.org/10.1016/j.foodres.2020.109128.

Organización internacional del Café. (2019). *Informe del mercado de café enero 2019*.

Özçelik, B., Kartal, M., & Orhan, I. (2011). Cytotoxicity, antiviral and antimicrobial activities of alkaloids, flavonoids, and phenolic acids. *Pharmaceutical Biology*, *49*(4), 396–402. https://doi.org/10.3109/13880209.2010.519390.

Petzer, A., Pienaar, A., & Petzer, J. P. (2013). The interactions of caffeine with monoamine oxidase. *Life Sciences*, *93*(7), 283–287. https://doi.org/10.1016/j.lfs.2013.06.020.

Plat, J., Baumgartner, S., Vanmierlo, T., Lütjohann, D., Calkins, K. L., Burrin, D. G., . . . Mensink, R. P. (2019). Plant-based sterols and stanols in health and disease: "Consequences of human development in a plant-based environment?" *Progress in Lipid Research*, *74*(February), 87–102. https://doi.org/10.1016/j.plipres.2019.02.003.

Rabiei, Z., Rafieian-Kopaei, M., Heidarian, E., Saghaei, E., & Mokhtari, S. (2014). Effects of Zizyphus jujube extract on memory and learning impairment induced by bilateral electric lesions of the nucleus basalis of meynert in rat. *Neurochemical Research*, *39*(2), 353–360. https://doi.org/10.1007/s11064-013-1,232-8.

Rabiei, Z., Rafieian-Kopaei, M., Mokhtari, S., & Shahrani, M. (2014). Effect of dietary ethanolic extract of *L. officinalis* on serum lipids profile in rats. *Iranian Journal of Pharmaceutical Research*, *13*(4), 1295–1301. https://doi.org/10.22037/ijpr.2014.1586

Rajput, R., Wairkar, S., & Gaud, R. (2018). Nutraceuticals for better management of osteoporosis: An overview. *Journal of Functional Foods*, *47*(March), 480–490. https://doi.org/10.1016/j.jff.2018.06.013.

Ramirez-Coronel, M. A., Marnet, N., Kolli, V. S. K., Roussos, S., Guyot, S., & Augur, C. (2004). Characterization and estimation of proanthocyanidins and other phenolics in coffee pulp (*Coffea, arabica*) by thiolysis-Hhgh-oerformance liquid chromatography. *Journal of Agricultural and Food Chemistry*, *52*(5), 1344–1349. https://doi.org/10.1021/jf035208t.

Randhir, A., Laird, D. W., Maker, G., Trengove, R., & Moheimani, N. R. (2020). Microalgae: A potential sustainable commercial source of sterols. *Algal Research*, *46*(January), 101772. https://doi.org/10.1016/j.algal.2019.101772.

Raut, G., & Wairkar, S. (2018). Management of psoriasis with nutraceuticals: An update. *Complementary Therapies in Clinical Practice*, *31*, 25–30. https://doi.org/10.1016/j.ctcp.2018.01.007.

Regazzoni, L., Saligari, F., Marinello, C., Rossoni, G., Aldini, G., Carini, M., & Orioli, M. (2016). Coffee silver skin as a source of polyphenols: High resolution mass spectrometric profiling of components and antioxidant activity. *Journal of Functional Foods, 20*, 472–485. https://doi.org/10.1016/j.jff.2015.11.027.

Ritchie, K., Carrière, I., De Mendonça, A., Portet, F., Dartigues, J. F., Rouaud, O., . . . Ancelin, M. L. (2007). The neuroprotective effects of caffeine: A prospective population study (the Three City Study). *Neurology, 69*(6), 536–545. https://doi.org/10.1212/01.wnl.0000266670.35219.0c.

Rodríguez-Durán, L. V., Ramírez-Coronel, M. A., Aranda-Delgado, E., Nampoothiri, K. M., Favela-Torres, E., Aguilar, C. N., & Saucedo-Castañeda, G. (2014). Soluble and bound hydroxycinnamates in coffee pulp (*coffea arabica*) from seven cultivars at three ripening stages. *Journal of Agricultural and Food Chemistry, 62*(31), 7869–7876. https://doi.org/10.1021/jf5014956.

Rufián-Henares, J. A., & Morales, F. J. (2007). Effect of *in vitro* enzymatic digestion on antioxidant activity of coffee melanoidins and fractions. *Journal of Agricultural and Food Chemistry, 55*(24), 10016–10021. https://doi.org/10.1021/jf0718291.

Sadhukhan, P., Saha, S., Dutta, S., Mahalanobish, S., & Sil, P. C. (2018). Nutraceuticals: An emerging therapeutic approach against the pathogenesis of Alzheimer's disease. In *Pharmacological Research* (Vol. 129). https://doi.org/10.1016/j.phrs.2017.11.028.

Saldaña-Mendoza, S. A., Ascacio-Valdés, J. A., Palacios-Ponce, A. S., Contreras-Esquivel, J. C., Rodríguez-Herrera, R., Ruiz, H. A., . . . Aguilar, C. N. (2020). Use of wastes from the tea and coffee industries for the production of cellulases using fungi isolated from the Western Ghats of India. *Systems Microbiology and Biomanufacturing*. 0123456789). https://doi.org/10.1007/s43393-020-00001-z

Samsonowicz, M., Regulska, E., Karpowicz, D., & Leśniewska, B. (2019). Antioxidant properties of coffee substitutes rich in polyphenols and minerals. *Food Chemistry, 278*(May 2018), 101–109. https://doi.org/10.1016/j.foodchem.2018.11.057

Shang, F., Li, X., & Jiang, X. (2016). Coffee consumption and risk of the metabolic syndrome: A meta-analysis. *Diabetes and Metabolism, 42*(2), 80–87. https://doi.org/10.1016/j.diabet.2015.09.001.

Shaposhnikov, S., Hatzold, T., Yamani, N. El, Stavro, P. M., Lorenzo, Y., Dusinska, M., . . . Collins, A. (2018). Coffee and oxidative stress: a human intervention study. *European Journal of Nutrition, 57*(2), 533–544. https://doi.org/10.1007/s00394-016-1,336-4.

Singh, J., & Sinha, S. (2012). Classification, regulatory acts and applications of nutraceuticals for health. *International Journal of Pharmacy and Biological Sciences, 2*(1), 177–187.

Stahl, W., & Sies, H. (2005). Bioactivity and protective effects of natural carotenoids. *Biochimica et Biophysica Acta - Molecular Basis of Disease, 1740*(2), 101–107. https://doi.org/10.1016/j.bbadis.2004.12.006.

Thomasset, S. C., Berry, D. P., Garcea, G., Marczylo, T., Steward, W. P., & Gescher, A. J. (2007). Dietary polyphenolic phytochemicals - Promising cancer chemopreventive agents in humans? A review of their clinical properties. *International Journal of Cancer, 120*(3), 451–458. https://doi.org/10.1002/ijc.22419.

Tohda, C., Kuboyama, T., & Komatsu, K. (2005). Search for natural products related to regeneration of the neuronal network. *NeuroSignals, 14*(1–2), 34–45. https://doi.org/10.1159/000085384.

Verzelloni, E., Tagliazucchi, D., Del Rio, D., Calani, L., & Conte, A. (2011). Antiglycative and antioxidative properties of coffee fractions. *Food Chemistry, 124*(4), 1430–1435. https://doi.org/10.1016/j.foodchem.2010.07.103.

Vinícius de Melo Pereira, G., Soccol, V. T., Brar, S. K., Neto, E., & Soccol, C. R. (2017). Microbial ecology and starter culture technology in coffee processing. *Critical Reviews in Food Science and Nutrition, 57*(13), 2775–2788. https://doi.org/10.1080/10408398.2015.1067759.

Vivancos, M., & Moreno, J. J. (2005). β-Sitosterol modulates antioxidant enzyme response in RAW 264.7 macrophages. *Free Radical Biology and Medicine, 39*(1), 91–97. https://doi.org/10.1016/j.freeradbiomed.2005.02.025.

Wang, A., Leong, D. J., Cardoso, L., & Sun, H. B. (2018). Nutraceuticals and osteoarthritis pain. *Pharmacology and Therapeutics, 187*, 167–179. https://doi.org/10.1016/j.pharmthera.2018.02.015

Watanabe, T., Arai, Y., Mitsui, Y., Kusaura, T., Okawa, W., Kajihara, Y., & Saito, I. (2006). The blood pressure-lowering effect and safety of chlorogenic acid from green coffee bean extract in essential hypertension. *Clinical and Experimental Hypertension, 28*(5), 439–449. https://doi.org/10.1080/10641960600798655.

Williams, R. J., & Spencer, J. P. E. (2012). Flavonoids, cognition, and dementia: Actions, mechanisms, and potential therapeutic utility for Alzheimer disease. *Free Radical Biology and Medicine, 52*(1), 35–45. https://doi.org/10.1016/j.freeradbiomed.2011.09.010.

Woo, S. M., Min, K. J., Seo, B. R., Nam, J. O., Choi, K. S., Yoo, Y. H., & Kwon, T. K. (2014). Cafestol overcomes ABT-737 resistance in Mcl-1-overexpressed renal carcinoma Caki cells through downregulation of Mcl-1 expression and upregulation of Bim expression. *Cell Death and Disease, 5*(11). https://doi.org/10.1038/cddis.2014.472.

y Postigo, L. O. C., Jacobo-Velázquez, D. A., Guajardo-Flores, D., Amezquita, L. E. G., & García-Cayuela, T. (2021). Solid-state fermentation for enhancing the nutraceutical content of agrifood by-products: Recent advances and its industrial feasibility. *Food Bioscience, 100*, 926.

Yu, X., Bao, Z., Zou, J., & Dong, J. (2011). Coffee consumption and risk of cancers: A meta-analysis of cohort studies. *BMC Cancer, 11*(1), 96. https://doi.org/10.1186/1471-2,407-11-96.

Zhang, L., & Ku, K. M. (2019). Biomarkers-based classification between green teas and decaffeinated green teas using gas chromatography mass spectrometer coupled with in-tube extraction (ITEX). *Food Chemistry, 271*(June 2018), 450–456. https://doi.org/10.1016/j.foodchem.2018.07.137

Zhou, J. Y., & Zhou, S. W. (2012). Isorhynchophylline: A plant alkaloid with therapeutic potential for cardiovascular and central nervous system diseases. *Fitoterapia, 83*(4), 617–626. https://doi.org/10.1016/j.fitote.2012.02.010.

Web References

Belliveau D. (2021). The Coffee Cherry Co. Available from https://coffeecherryco.com (accessed April 11, 2021).

del Castillo, M.D., Ibañez, M.E., Amigo, M., Herrero, M., Plaza del Moral, M., & Ullate, M. (2013). Application of products of coffee silverskin in anti-aging cosmetics and functional food. WO 2013/004873 2013. Available from https://digital.csic.es/handle/10261/92745 (accessed April 8, 2021).

Ciummo, B., (2014). What is Cascara? *Fresh Cup Magazine*. Retrieved January 1, 2019, from https://www.freshcup.com/what-is-cascara.

Future Ceuticals. (2021). Available from https://www.futureceuticals.com (accessed April 11, 2021).

Jelderks J. & Zimmermann U. (2021). Caskai. Available from https://caskai.com (accessed April 11, 2021).

Pectcof B.V. (2021). Unlocking the Potential of Coffee Pulp. Available from https://pectcof.com (accessed April 11, 2021).

Supracafé, S.A. (2020). Tabi Fruit. Available from https://www.supracafe.com/conoces-la-infusion-de-cascara-decafe-supracafe-maximiza-su-sostenibilidad-con-tabifruit (accessed April 8, 2020).

Svetol®. (2021). The science behind Svetol®. Available from https://svetol.com (accessed April 11, 2021).

11

Coffee Proteins: Functional Food Ingredients with Molecular Effects for Sustainable Health

*Amaia Iriondo-DeHond[1], Maria del Pilar Salazar[2], Laura Guspí Domènech[1], Jose Edgar Zapata Montoya[2], and Maria Dolores del Castillo[1]**

[1] *Instituto de Investigación en Ciencias de la Alimentación (CIAL), CSIC-UAM, Calle Nicolás Cabrera 9, 28049 Madrid, Spain*
[2] *Universidad de Antioquia (UdeA), Medellín, Antioquia, Colombia*

11.1 Introduction

Proteins are macromolecules made up of one or more linear chains of amino acids organized in unique spatial structures for each type of protein. The human body uses proteins and amino acids obtained through the diet for the proper development of different vital functions (Wu et al., 2014). The nutritional quality of proteins depends on their amino acid profile (essential and nonessential amino acids (EAA)), in addition to other factors such as digestibility, bioaccessibility, or the presence of antinutritional compounds (Hoffer, 2016). Taking into account the amount of proteins, different nutrition, and health claims can be applied to food products. A nutrition claim is any declaration that affirms, suggests, or implies that a food has specific beneficial nutritional properties due to its energy intake, nutrients, or other substances (Regulation EC 1924, 2006). On the other hand, health claims are those that relate food or one of its components to health (The European Commission, 2012). Table 11.1 summarizes nutrition and health claims made on foods based on their protein content.

The recommended daily amount of protein is 0.80 g per kilogram of body weight for both men and women with modest levels of physical activity (The European Commission, 2020b). Considering the male average weight of 72 kg and the female average weight of 63 kg, 58 g, and 50 g of proteins should be consumed daily, respectively. Protein needs of the population can be met by sources of animal protein such as meat, fish, eggs, or dairy products; or vegetable protein sources such as cereals, grains, and seeds. In the last decade, the demand for food protein increased, especially animal-source protein. The consumption of animal-source protein is associated with increased greenhouse gas emissions and overutilization of water (Wu et al., 2014). Therefore, one of the main objectives of the current European Commission's Farm to Fork Strategy, and the initiatives under the new European Green Deal, is to increase the availability and source of alternative proteins such as plant,

Molecular Mechanisms of Functional Food, First Edition. Edited by Rocio Campos-Vega and B. Dave Oomah.

Table 11.1 Nutrition (Regulation (EC) No. 1924/2006) and health (Commission Regulation (EU) No. 432/2012) claims authorized by the European Commission related to proteins.

	Claim	Condition
Nutrition claims	Source of protein	A claim that a food is a source of protein, and any claim likely to have the same meaning for the consumer, may only be made where at least 12% of the energy value of the food is provided by protein.
	High protein	A claim that a food is high in protein, and any claim likely to have the same meaning for the consumer, may only be made where at least 20% of the energy value of the food is provided by protein.
Health claims	Protein contributes to a growth in muscle mass Protein contributes to the maintenance of muscle mass Protein contributes to the maintenance of normal bones	The claim may be used only for food which is at least a source of protein as referred to in the claim SOURCE OF PROTEIN as listed in the Annex to Regulation (EC) No. 1924/2006.

Source: Data from The European Commission (2012).

microbial, marine and insect-based proteins, and meat substitutes to search for new food sources to replace animal protein (The European Commission, 2020a).

Alternative proteins are estimated to account for 11% ($290bn) of the market by 2035, primarily due to plant-based proteins reaching price parity by 2023 with overall continuous growth in global protein market (worth trillions of dollars) (Watson, 2021). Moreover, plant-based protein consumption is expected to grow in three stages; 12, 16, and 7% CAGR (compound annual growth rate) for 2020–2025, 2025–2030, and 2030–2035, respectively (Watson, 2021). Alternatives are sought to meet the consumer's demands for a healthier diet, respectful for the environment and animal welfare, and to improve sustainability problems associated with meat production. These alternatives can be classified into two large groups:

1) Meat substitutes that claim to mimic texture, appearance, and organoleptic properties of meat products. This group includes foods formulated with plant sources such as soybeans, cereals, or legumes (Sha & Xiong, 2020); cultured meat produced *in vitro* using tissue-engineering technology (Li et al., 2020); or mycoproteins obtained by fungal fermentation (Hashempour-Baltork, Khosravi-Darani, Hosseini, Farshi, & Reihani, 2020).
2) Protein-rich foods that are not intended to imitate the texture of meat, such as vegetables, proteins extracted from algae and microalgae (Torres-Tiji, Fields, & Mayfield, 2020), proteins isolated from insects (Jantzen da Silva Lucas, Menegon de Oliveira, da Rocha, & Prentice, 2020), and the revalorization of proteins from by-products generated during agro-industrial processing (Gomes Almeida Sá, Moreno, & Carciofi, 2020).

Protein extraction from food enables production of concentrates and isolates with high protein content, which can be used in the preparation of food to improve its nutritional profile, or as a dietary supplement. Milk has been widely employed to obtain these type of ingredients (Meena, Singh, Panjagari, & Arora, 2017). However, it is still of animal-source origin and many consumers may have intolerances or allergies, or do not want to consume this animal source for ethical or ideological reasons (vegan population). Other natural protein sources commercially available are:

1) Daily diet vegetables, legumes, and grains. Concentrates and isolates of soy, rice, pea, and hemp proteins are commercially available for their use as food ingredient. Other plants such as potatoes, quinoa, corn, lentils, sesame, peanuts, chia seeds, and coffee beans can also be employed to obtain proteins.
2) Insects, algae, and microalgae. Although insects are an animal-source protein, they have attracted increasing attention as both human food and animal feed ingredient since they can be grown on low-value feeds and have a low carbon footprint (Hawkey, Lopez-Viso, Brameld, Parr, & Salter, 2021). The Farm to Fork Strategy also indicates that algae should become an important source of alternative protein for a sustainable food system and global food security (The European Commission, 2020a).
3) Food waste. Obtaining proteins from food waste minimizes waste disposal, maximizes the resources, and adds market value to different products. A potential sustainable source of proteins is oil extraction residues from seeds such as sunflower or pumpkin seeds (Gomes Almeida Sá et al., 2020).

Within the use of food waste as a source of proteins, coffee by-products may be a potential alternative to obtain protein enriched-extracts. Nowadays, there is a concentrated protein powder commercially available from green coffee beans called CoffeeProtein, and has excellent sensory profile, smooth texture, great capacity for mixing with different formulations, and is free of allergens such as gluten, lactose, egg, nuts, and soy (Applied Food Sciences, 2019). However, to the best of our knowledge, information is limited regarding the feasibility of coffee by-products as a sustainable source of proteins.

The coffee cherry is composed by an outer skin that covers the pulp (also known as "coffee cascara"") followed by a viscous and highly hydrated layer of mucilage, a thin yellowish endocarp, the parchment and, finally, the silverskin that covers each hemisphere of the green coffee bean (Esquivel & Jiménez, 2012). To obtain the green coffee bean, cascara, mucilage, and parchment are generated as by-products. Once the green coffee bean is obtained, it is roasted and the silverskin is released during the roasting process. Finally, spent coffee grounds (SCGs) are generated as a by-product when the coffee beverage is prepared at home or processed for soluble/instant coffee (Iriondo-DeHond, Iriondo-DeHond, & del Castillo, 2020). World consumption of coffee was projected at 166.3 million 60 kg bags for 2020/2021 (International Coffee Organization, 2021). The by-products generated during the whole process account for up to 90% of the dry weight of the cherry. This large volume of waste, which is a source of contamination and a risk for environmental sustainability, is composed by a massive number of nutrients and nonnutrients of great interest in nutrition and health such as proteins (de Melo Pereira et al., 2020; Iriondo-DeHond, Iriondo-DeHond, & del Castillo, 2020). The present chapter is an updated literature review of the potential of green coffee beans and coffee by-products as nutritious, safe, and sustainable sources of proteins for human consumption.

11.2 Coffee as an Alternative Protein Source

The green coffee bean and the by-products generated during the coffee cherry processing present different protein content (Figure 11.1a).

When the coffee cherry is depulped, 39 kg of cascara are produced from 100 kg of mature coffee cherries (Ferrão, 2009). Considering that the protein content of coffee cascara is around 9.55% (Iriondo-DeHond et al., 2020), this by-product can be a potential alternative source to animal-based proteins. On the other hand, the amount of proteins described in coffee parchment is lower than 1% (Iriondo-DeHond, Iriondo-DeHond, & del Castillo, 2020). Although the amount of coffee parchment produced is similar to that generated by coffee cascara (Ferrão, 2009), the application of this by-product as a protein source is not proposed because of its low protein content.

The green coffee bean has been recently proposed as an alternative plant-based protein source (Applied Food Sciences, 2019). Average protein content in *C. arabica* and *Coffea canephora* beans is approximately 11.5% dry basis (Clifford, 1985). Immature fruits at harvest lead to defective green beans that reduce the quality of the beverage (Mazzafera, 1999). Since the protein content of defective beans is similar to green beans from mature fruits (Mazzafera, 1999), instead of being discarded, these defective beans can be upcycled as a sustainable source of proteins.

Protein has been extracted from green coffee (*Coffea arabica*) beans (12.8% protein) by four different methods (CP-I-CP-IV); one of them simulating soy glycinin extraction (CP-I; 40.8% protein) and another without chlorogenic acid and fat removal (CP-III; 40.6% protein) (Rawel, Rohn, & Kroll, 2005). CP-III extraction resulted in reduced α-subunit content of the 11S proteins. Protein yield ranged between 7 and 45% with 40–50% protein content of the extracts, and protein solubility of CP-I-CP-IV was 45, 55, 50 and 98%, respectively, in

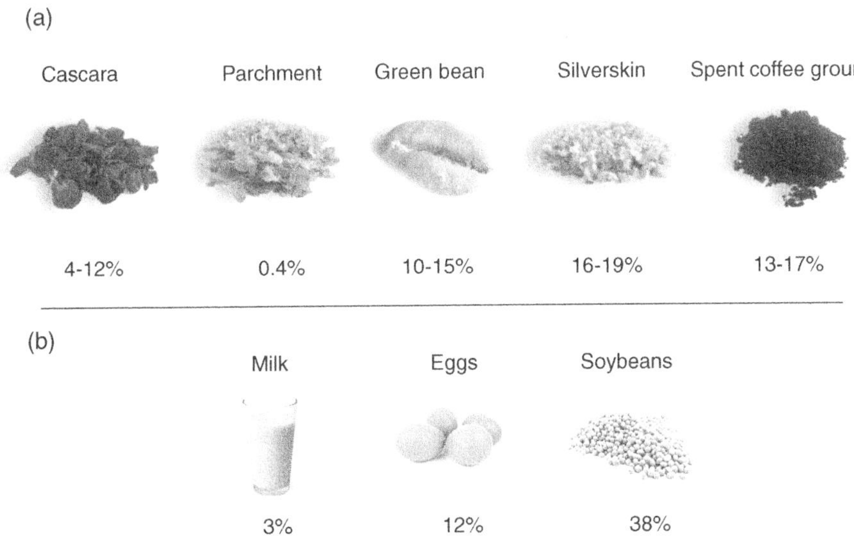

Figure 11.1 Protein content in green coffee beans and coffee by-products (a) and in conventional protein sources (b). Source of protein values of milk, egg, and soy: FoodData Central, USDA.

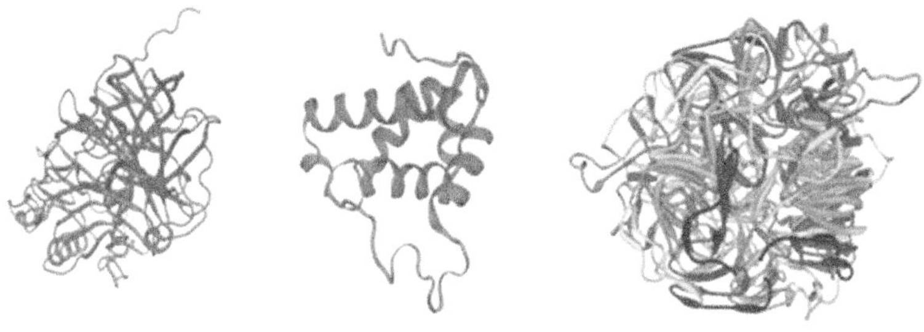

Figure 11.2 The 7S globulin-like (A0A6PWNJ1-1 [LOC113734078]) *Coffea arabica* protein consists of 440 amino acids (47,367 Da). The 2S seed storage protein (A0A6P6XN60 [LOC113743668]) *C. arabica* protein consists of 167amino acids (19,067 Da). The 11S storage protein (Q9ZNY2 [Q9ZNY2_COFAR]) *C. arabica* protein consists of 492 amino acids (55,119 Da) according to Swiss model (https://unitpro.org).

deionized water. In addition, all protein extracts exhibited antioxidant activity (5–35 mM Trolox/L).

Mature green coffee beans express maximum amount of the storage (11S globulin) proteins in the presence of other potential storage proteins (7S-like, sugar-dependent 1[SDP1 that encodes a patatin domain], SDP1-like, glutelin A2, A3, and patatin 6) (Cheng, Furtado, & Henry, 2018) (Figure 11.2).

The 11S endosperm protein of commercial coffee species (*Coffea Arabica* and *Coffea canephora*) accounts for approximately 45% of total grain protein, representing 5–7% of the coffee bean dry weight. The structure, typical of 11S plant (*Brassica napus, Vicia sativa, Oryza sativa,* and *Cucurbita maxima*) storage protein, consists of α (predominantly hydrophilic) and β (evenly hydrophilic and hydrophobic section) subunits (32 and 20 kDa, respectively) and 6 polypeptides (80, 52, 32, 22, 16, and 12 kDa) on 2DE SDS-PAGE (Rogers et al., 1999). The 11S coffee endosperm protein is encoded by the csp1gene or gene family expression.

Under nonreducing conditions, 11S green coffee storage protein displays two major protein bands (58 and 38 kDa) and four less abundant bands (170, 150, 16, and ≤ 14 kDa). The 58 kDa fraction reduces into high (acidic α-component, ~32–38 kDa) and low (basic β-isoform, ~ 20–22 kDa) molecular weight subunits (Abdelfatah Ali, 2013; Rawel et al., 2005). The surface-exposed α-chains of the 11S storage protein are the most reactive and susceptible to covalent reactions with phenolic compounds, whereas the β-chains located in the inner part of the 11S protein remain intact (Rawel et al., 2005). The 11S coffee storage protein (300–400 kDa) separates under nonreducing conditions (SS or urea) into 6 subunit (55–60 kDa) pairs connected by a disulfide link (Abdelfatah Ali, 2013). Circular dichroism (CD) spectroscopy revealed the predominance of the β-sheet (28–39%) conformation in the secondary structure, a molecular feature strongly influencing 11S globulin functional properties. The contents of α-helix, β-turn, and unordered secondary structure were 4.1–7.7, 18.6–20.3, and 36.2–48.4%, respectively, for *C. arabica* (Brazil and Guatemala) and *C. robusta* (Uganda and Indonesia) (Abdelfatah Ali, 2013). The low α-helix content indicates high digestibility of the green coffee 11S protein. Coffee 11S protein is acidic; 33–44% soluble at pH 2–3; and 56–79% at pH 4. The antioxidant capacity of 11S protein (~3 mM TE/g protein) is due to its strong covalent binding with polyphenols (hydroxycinnamic acid-caffeoylquinic acid [CQA]) since it decreases (~1.6 mM TE/g protein) when extracted in the presence of PVV. 11S coffee protein modification with CQA is responsible for its antioxidant capacity and physicochemical properties such as emulsion characteristics (emulsifying capacity and stability). CQA alters the structural (secondary structure α-helix, β-strand, β-turns or unordered) properties of 11S protein modifying its surface hydrophobicity that is responsible for physicochemical characteristics (thermal stability [DSC], emulsion characteristics) (Abdelfatah Ali, 2013). The coffee 11S proteins do not behave like soybean glycinin at low ionic strength (2 mM Tris, pH 8), but rather resembles other 11S proteins that do not dissociate under these conditions (Acuña et al., 1999).

Green coffee 11S seed storage protein, particularly the α-subunit, oxidatively fragments during aerobic incubation (37 °C, 0.02 M sodium borate, pH 8.5 containing trypsin, 48 hr) releasing free amino acids (Montavon, Duruz, Rumo, & Pratz, 2003). However, the 11S storage protein forms adducts with polyphenols prior to roasting and the protein subunits integrate into the melanoidins structure formed during roasting thereby participating in coffee aroma (Montavon, Mauron, & Duruz, 2003). Two-dimensional mapping of green beans revealed the expression of 16 and 5 (polypeptides) proteins in *C. canephora* and *C. arabica*, respectively. Moreover, another 8 proteins were upregulated and one suppressed in *C. canephora* compared to *C. arabica* (Gil-Agusti et al., 2005). Thirty proteins identified by reversed-phase-high performance liquid chromatography-electrospray ionization-tandem mass spectrometry RP-HPLC-ESI-MS/MS were differentially expressed; 3 peptides from proteins including glyceraldehyde-3-phosphate dehydrogenase (GAPD) [key enzyme in intermediary metabolism catalyzing the oxidation and subsequent phosphorylation of substrate aldehydes to acyl phosphate]; 4 peptides from phospholipid hydroperoxide glutathione peroxidase [protects cells and enzyme from oxidative damage]; two 11S-α chain peptides; two11S fragment peptides; one 11S-β chain peptide; and two 11S- α/β dimer peptides (Gil-Agusti et al., 2005).

Proteins, peptides, and amino acids participate in coffee aroma production due to their interaction with sugars in the Maillard reaction during roasting. Thirty peptides (4–10 kDa, pI range primarily within pH 4–6 and 7–9) have been identified in coffee beans accounting on average for 0.59 and 0.43% of dry mass (5.9 and 4.3 g/kg dwb) in *C. arabica* and *canephora*, respectively. *C. canephora* has fewer peptides in the 8–10 kDa than *C. arabica* (Mazzafera, Schimpl, & Kiyota, 2019). However, peptide content did not differ significantly between five *C. arabica* and four *C. robusta* types, although peptide composition differed significantly (Abdelfatah Ali, 2013). Peptide composition changed significantly, although peptide content remained unchanged during dry processing. Five to seven active protease enzymes (~⅓ of the total coffee bean protein) are present in green coffee and are stable in the intact beans for months because they are bound to arabinogalactans in the cell wall (Mazzafera et al., 2019).

Other proteins have also been reported in coffee beans. For example, a class III chitinase of *C. arabica* has been identified as the first known coffee allergen Cof a 1 (Manavski et al., 2012). The Cof a 1 cDNA encodes a 32 kDA (290 amino acids) *C. arabica* class III acidic endochitinase with high amino acid similarity to other class III plant chitinase (concavalin B and hevamine [major latex allergen]), a pathogenesis-related (PR) protein, presumably heat stable and therefore not destroyed by roasting green coffee beans and the digesting procedure. Another allergen in green coffee bean (18–40 kDa; ~12 mg/g roasted ground coffee) is probably a heterogeneous glycoprotein that can be precipitated at pH 4–4.5. An acid phosphatase and a trypsin inhibitor are also presumed to be part of the green coffee protein (Spiller, 1998).

By-products generated during the roasting and brewing process present the highest protein content among all the coffee by-products. Protein content in coffee silverskin ranges from 16.2 to 18.6% (Iriondo-DeHond, Iriondo-DeHond, & del Castillo, 2020). Although this by-product has a high protein content, only 20 kg of silverskin are produced from 1 ton of green coffee beans (Supracafé S.A., *pers. comm.*). In contrast, 650 kg of SCGs are generated from the processing of 1 ton of green coffee beans (Mussatto, Machado, Martins, & Teixeira, 2011). SCGs contain high amount of proteins, and the values reported by some authors were 6.7–9.9% (Lago, Antoniassi, & Freitas, 2001), 12.8–16.9% (Cruz et al., 2012), 13.6% (Campos-Vega, Loarca-Piña, Vergara-Castañeda, & Oomah, 2015), and 11.2% (Martinez-Saez et al., 2017). Other authors observed that the mean protein content is 13.6% after soluble coffee preparation (Mussatto, Ballesteros, Martins, & Teixeira, 2011; Silva, Nebra, Machado Silva, & Sanchez, 1998). SCGs protein content exceeds the amount present in coffee beans due to the concentration of nonextracted components in coffee residues during coffee preparation (Martinez-Saez et al., 2017).

Kjeldahl method has been normally used for the determination of protein content in coffee samples. Based on the interference of other nitrogen-containing compounds in this protein determination method, protein values of coffee samples present some limitations. Coffee beans and by-products contain nonprotein nitrogen compounds such as caffeine, trigonelline, niacin, purines, pyrimidines, and inorganic nitrogen (Delgado, Vignoli, Siikaaho, & Franco, 2008; Martinez-Saez et al., 2017). In fact, a study (Machado, Costa, Pimentel, Oliveira, and Alves, 2020) showed that one-quarter of the total nitrogen of coffee silverskin corresponded to the nonprotein fraction. Protein content also varies depending on the brewing and roasting processes, when the proteins can be fragmented, polymerized, and/or integrated into melanoidins (Bravo et al., 2012; Cruz et al., 2012).

In summary, taking into account the abundance and the protein content of each coffee by-product, besides green coffee beans, we propose the upcycling of cascara and SCGs as an alternative source of plant-based proteins. The amount of protein in these by-products is higher than protein levels in milk and similar to that found in eggs, which are conventional animal-based protein sources (Figure 11.1b). Coffee beans and by-products present lower protein values when compared to soybeans, which are one of the richest sources of protein, approximately 40% of dry weight (Singh, Meena, Kumar, Dubey, & Hassan, 2015). However, proteins from soy and coffee beans have similar structure. Most of the proteins in both products are seed storage proteins. In the green coffee bean, it is estimated that they can account for 40% of the total protein (López-Pedrouso, Bernal, Franco, & Zapata, 2014), 11S globulin being the principal component of these proteins (Acuña et al., 1999; Baú, Mazzafera, & Santoro, 2001; Mazzafera et al., 2019). Other proteins have also been identified in green coffee beans, such as 7S- and a 2S-type protein (Mazzafera et al., 2019), which are also present in soybeans (Preece, Hooshyar, & Zuidam, 2017; Singh et al., 2015). In the case of coffee by-products, information is limited on the molecular characterization of their proteins. Further research is needed to study the nature of proteins from coffee by-products, especially cascara, silverskin, and SCGs, in order to use them as an alternative source of animal-based proteins.

11.3 Protein Extraction from Coffee and Coffee By-products

Through different production processes, added value protein concentrates and isolates can be obtained for use in the food industry. The product resulting from this extraction can be classified according to its protein content into: (a) protein flours, which contain up to 65% protein; (b) protein concentrates, 65–90% protein; and (c) protein isolates, ≥ 90% protein. Protein extracts obtained from by-products of the food industry are proposed for their upcycling in the human diet to use as food ingredient or dietary supplement, or as techno-functional ingredients due to their emulsifying, gelling, foaming, and water absorption properties (Pojić, Mišan, & Tiwari, 2018).

Due to the similarity between soy and coffee proteins, processes commonly used for soy protein extraction can be applied on coffee samples. Large-scale processes for soybase preparation for protein extraction involve 4 steps: soaking the soybeans, grinding in cold water, filtering, and cooking at 93–100 °C for 30 minutes (Preece et al., 2017). Conventional soy protein extraction methods also involve alkali extraction using sodium hydroxide, lime, ammonia, or tri-basic sodium phosphate to bring the pH to slightly alkaline reaction for a better protein extraction (Berk, 1992). To date, the methodology described in the scientific literature for the extraction of coffee proteins from defective coffee beans and coffee by-products (silverskin and SCGs) is summarized in Table 11.2.

11.3.1 Protein Extraction from Defective Coffee Beans

Defective green coffee beans are discarded during the grading and selection of the coffee fruits to obtain a quality beverage. Proteins were extracted from defective green coffee beans using water extraction at different pH and precipitation by acidification or by adding

Table 11.2 Methods used to extract proteins from coffee by-products.

Sample	Extraction method	Recovered protein	Reference
Defective green coffee beans	Alkali extraction and isoelectric precipitation	55.9%	Castrillón-Monroy and Riaño-Luna (1999)
	Alkali extraction and HCl precipitation	25.2%	
Silverskin	Sequential extraction (alkaline—acid)	32.5%	Wen et al. (2020)
	UAE	14.0%	
	MAE	43.5%	
	Liquid:solid (ACN:water)	< 1%	Pérez-Míguez, Marina, and Castro-Puyana (2019)
	Subcritical water extraction	< 1%	
	HIFU	2.6%	
Spent coffee grounds (SCGs)	Tris buffer	0.3%	Valdés, Castro-Puyana, and Marina (2020)
	Urea buffer	2.9%	

UAE, ultrasound-assisted extraction; MAE, microwave-assisted extraction; HIFU, high intensity focused ultrasound.

salt (Castrillón-Monroy & Riaño-Luna, 1999). After studying different extraction conditions, the maximum protein was extracted at pH 9, 40 °C for 2 hr and extraction ratio of 1 : 8 (defective coffee beans:water). The extraction efficiency was 90.4% at lower volume (1 : 6.5 ratio), taking into account that the amount of protein in defective coffee beans was 34.95 mg per gram of coffee. Two methods were used for protein precipitation; maximum amount of precipitated protein was obtained by acidification (pH 3.4). On the other hand, isolates precipitated by HCl addition and with calcium chloride had 55.9 and 25.2% protein, respectively. The isolate with the highest protein content and the best properties was obtained by acid precipitation.

11.3.2 Extraction of Coffee Silverskin Proteins

Three different methods for protein extraction from coffee silverskin (14.6% protein) have recently been proposed (Wen, Zhang, et al., 2020): conventional extraction by pH variation and isoelectric precipitation, microwave-assisted extraction (MAE), and ultrasound-assisted extraction (UAE). The best yield using conventional extraction was obtained by sequential alkaline-acid extraction with 0.6 M NaOH +0.6 M HCl for 24 hours. In this case, increasing solvent concentration and time significantly increased extraction yield (Wen et al., 2020). The MAE (434.7 W, 20 min) with 0.6 M NaOH recovered significantly higher $(p < 0.05)$ percentage of protein compared to conventional extraction. These results indicate that microwaves allow a notable reduction in the extraction time due to high matrix rupture and components transfer. The UAE was carried out at a power of 38 W/cm^2 for 10 min and during the extraction by alkaline medium at 0.6 M. During the same period (10 min), UAE produced a threefold increase of protein extraction. In summary, both UAE and MAE are potentially rapid and effective tools for protein extraction from silverskin.

Pérez-Míguez et al. (2019) also extracted proteins from coffee silverskin to obtain protein hydrolysates. Three protocols were used to recover the proteins: (a) solid–liquid extraction using a mixture of water and an organic solvent as extractant, (b) subcritical water extraction, and (c) high-intensity focused ultrasound (HIFU) using different buffer solutions containing different additives as extracting solvents. However, protein recovery was low (Table 11.2).

Iriondo-DeHond et al. (2019) obtained a nutrient enriched coffee silverskin extract (CSE) by aqueous extraction easily applicable on an industrial scale. Although the main objective of this extraction was not to produce a protein-enriched extract, the protein content of CSE (16.7%) was obtained by stirring a mixture (50 g silverskin/L, 10 min, 100 °C), filtration and lyophilization.

11.3.3 Protein Extraction from SCGs

Two methods have been used to obtain protein isolates from SCGs using Tris or urea buffers (Valdés et al., 2020). Tris buffer and urea extractions yield <0.3 and 2.9 mg of protein/100 mg of SCGs, respectively. Proteins were obtained form SCGs derived from different brewing processes (espresso, French press, Italian, and paper filter filtered coffee). All the coffee brews employed to obtain the SCGs were treated with the same ratio of coffee and water, type of coffee, degree of roasting, and degree of grinding. For this reason, the only factors that differentiate the SCGs were the pressure applied, temperature, and time, which was different for each method of coffee preparation. As mentioned, the highest protein content was obtained in espresso coffee grounds, which were obtained by a higher pressure extraction but for a considerably shorter time. Therefore, proteins are retained in the SCGs and do not pass into the coffee brew.

11.4 Nutritional Value of Coffee Proteins

After carbohydrates and dietary fiber, protein is the most abundant macronutrient found in coffee by-products (Iriondo-DeHond, Iriondo-DeHond, & del Castillo, 2020). Depending on the origin, species, varietal, roasting, and brewing, a particular by-product could reach the nutritional claim of "source of proteins." As shown in Table 11.1, a food product can claim to be a "source of proteins" when at least 12% of the energy value of the food is provided by protein. This nutrition claim is applicable to cascara, green coffee beans, silverskin, and SCGs.

Protein quality is determined by its EAA composition, digestibility, and bioavailability of its amino acids. Table 11.3 shows the amino acid content of coffee by-products that can be a source of alternative plant-based proteins and also of the conventional protein sources (milk, egg, and soy). Of the total amino acids present in coffee cascara, 33% correspond to EAA. Major amino acids are aspartic acid, glycine, and glutamic acid. Tryptophan was not detected, but its presence could not be discarded since the acidic conditions necessary for the quantification provoke its hydrolysis (Iriondo-DeHond et al., 2020). Compared to other conventional protein sources, cascara presented lower values of total and EAA (Table 11.3).

Green coffee beans contain 44% EAA, values close to that reported for the conventional protein sources such as milk, egg, and soybeans. Glutamic and aspartic acids are the major amino acids of the coffee seed proteins (Table 11.3). In addition, amino acids such as leucine, alanine, lysine, and serine were also found in reasonable amounts (Shimizu & Mazzafera, 2000). The 11S protein is abundant in glycine, leucine, and glutamine and poor in sulfur-containing amino acids (cysteine and methionine) with 11 amino acids in high concentrations (5–11%) (Rogers et al., 1999). Free amino acid contents were 2.7–4.8 and 3.5–6 mg/kg (dry matter) for *Coffea arabica* and *C. robusta*, respectively (Abdelfatah Ali, 2013). However, *Coffea canephora* has higher free amino acids (0.15–2.5%) than *C. arabica* varieties (except for glutamate) (Mazzafera et al., 2019). Total free amino acid content differed between *C. arabica* and *C. robusta* (216–274 vs 159 mg/100 g of ground beans) due to differences in genotype, environment and postharvest treatment, maturation, and storage. Moreover, total free amino acids are higher in green than in roasted coffee (131 vs 86 mg/100 g) (Abdelfatah Ali, 2013). Interestingly, the amino acid content in green coffee beans changes not only during postharvest treatment but also during storage which can be attributed to the Maillard reaction (Arnold & Ludwig, 1996; Bagdonaite & Murkovic, 2018). Total amino acid content of green coffee beans was higher than that found for milk and similar to eggs.

Regarding the amino acid profile of silverskin, 46% of total amino acids correspond to EAA. Aspartic and glutamic acids followed by leucine, alanine, and glycine were the major amino acids found in this coffee by-product. Serine, asparagine, and arginine presented the highest values of free amino acids, Machado et al. (2020) observed that in coffee silverskin,

Table 11.3 Amino acid composition of coffee derivatives and conventional protein sources.

Amino acid (g/100 g)	Cascara[1]	Green bean[2]	Silverskin[3]	SCGs[4]	Milk[5]	Egg[6]	Soy[7]
Alanine	0.21	0.55	0.60	0.26	0.11	0.67	2.14
Arginine	0.05	0.42	0.50	0.01	0.13	0.79	2.94
Aspartic acid	0.28	1.22	1.02	0.57	0.29	1.27	3.76
Cysteine	0.02	0.33	0.04	0.02	0.04	0.39	0.77
Glutamic acid	0.21	2.28	0.92	0.46	0.80	1.63	6.80
Glycine	0.27	0.74	0.58	0.30	0.07	0.41	1.35
Histidine	0.05	0.32	0.21	0.04	0.1	0.28	0.95
Isoleucine	0.06	0.53	0.47	0.11	0.17	0.62	1.74
Leucine	0.12	1.01	0.76	0.28	0.34	1.05	3.06
Lysine	0.04	0.78	0.26	0.07	0.29	0.83	2.14
Methionine	0.03	0.17	0.08	0.03	0.09	0.49	0.49
Phenylalanine	0.10	0.66	0.53	0.13	0.16	0.66	2.15
Proline	0.16	0.76	0.52	0.18	0.33	0.56	2.13
Serine	0.19	0.68	0.54	0.06	0.20	0.92	2.07
Threonine	0.10	0.44	0.45	0.53	0.16	0.59	1.48
Tryptophan	N.A.	0.14	0.08	N.A.	0.04	0.17	0.45
Tyrosine	0.06	0.42	0.23	0.04	0.17	0.51	1.40
Valine	0.13	0.93	0.50	0.19	0.21	0.79	1.74
Total	2.08	12.38	8.29	3.28	3.70	12.54	37.55
Protein Quality Parameters (g/100 g amino acids)							
EAA	32.69	43.62	46.32	42.38	45.83	49.37	45.63
Fischer ratio	1.94	2.02	2.06	3.41	1.95	1.84	1.64
BCAA	14.90	19.95	20.87	17.68	19.46	19.48	17.41
AAA	7.69	9.85	10.13	5.18	10.00	10.61	10.65
PER1	1.59	2.75	3.20	2.95	3.09	2.90	2.76
PER2	1.85	2.88	3.40	3.28	3.22	2.88	2.84
PER3	1.10	2.49	3.67	4.35	2.64	3.01	2.12
SAT1	15.38	23.42	24.37	30.18	27.03	25.81	23.62
SAT2	4.81	4.68	6.39	16.16	5.41	6.02	5.14

AAA, aromatic amino acids; BCAA, branched-chain amino acids; EAA, essential amino acids; N.A., not analyzed; PER, protein efficiency ratio; SAT, satiety.
Source: Data from (1) Iriondo-DeHond et al. (2020), (2) Belitz, Grosch, and Schieberle (2009) and Iriondo-DeHond, Uranga, del Castillo, and Abalo (2020), (3) Machado et al. (2020) and Iriondo-DeHond et al. (2019), (4) Martinez-Saez et al. (2017), (5) FoodData Central, FDC ID: 746778, (6) FoodData Central, FDC ID: 748967, (7) FoodData Central, FDC ID: 1104766.

the major amino acids were aspartic and glutamic acids, and the BCAAs are found in the following amounts: leucine (7.6%), isoleucine (4.7%), and valine (5%) and also for proline (5.2%) and arginine (5%). They also found tryptophan, but in less quantity (0.8%). As for the green coffee beans, silverskin presented higher values of total amino acids compared to milk and similar proportion of EAA to milk, eggs, and soybeans (Table 11.3).

With regard to EAA, they comprise 42% of the total amino acids of SCGs (Martinez-Saez et al., 2017), values equivalent to those found in milk (Table 11.3). Aspartic acid, threonine, glutamic acid, glycine, and leucine presented the highest values among all the amino acids of SCGs (Campos-Vega et al., 2015; Martinez-Saez et al., 2017).

The branched chain amino acids (BCAA) valine, leucine, and isoleucine are important for the prevention and risk reduction of type 2 diabetes and other vascular or inflammatory stress-induced pathologies (Abolbaghaei, Oomah, Tavakoli, & Hosseinian, 2018). The green coffee bean, silverskin, and SCGs have values of BCAA similar to those of the conventional protein sources, ranging from 17 to 21 g/100 g of total amino acids (Table 11.3). On the other hand, the aromatic amino acids (AAA) tyrosine, phenylalanine, and tryptophan are associated with protein and other essential secondary metabolites synthesis involved in nervous signals transmissions and reducing reactive oxygen species in the brain (Parthasarathy et al., 2018). Cascara, the green bean, and SCGs have lower values of AAA compared to milk, egg, and soy. However, silverskin has AAA content comparable to milk (Table 11.3). Proteins with high BCAA, Fischer ratio (BCAA/AAA), and low content of AAA are searched for producing physiologically functional foods for specific needs, such as in patients with malnutrition associated with cancers, burns, trauma, and liver failure, and for nutritional support of children with chronic or acute diarrhea or milk protein allergies. SCGs have the highest Fischer ratio, even higher than milk, egg, and soy. The other coffee derivatives have similar Fischer ratios compared to the traditional protein sources (Table 11.3).

Protein efficiency ratio (PER) was calculated as another protein quality parameter, from the amino acid composition based on the following equations (Mir, Riar, & Singh, 2019):

$$PER1 = -0.684 + 0.456 \times Leu - 0.047 \times Pro$$
$$PER2 = -0.468 + 0.454 \times Leu - 0.105 \times Tyr$$
$$PER3 = -1.816 + 0.435 \times Met + 0.78 \times Leu + 0.211 \times His - 0.944 \times Tyr$$

PER values below 1.5 indicate low-quality proteins, and values above 2 are considered to be typical of high-quality proteins. Taking this information into account, as for the traditional protein sources (milk, egg, and soy), all coffee derivatives except cascara have PER values >2, showing the high protein quality of these products.

SAT1 and SAT2 parameters were calculated as satiety indicators. According to Volk et al. (2020), increased intake of leucine, isoleucine, lysine, threonine, and tryptophan (SAT1) elevates their plasma levels that are associated with increased satiety. Threonine and tryptophan (SAT2) might be the ones particularly responsible for this increased satiety (Volk et al., 2020). SAT1 and SAT2 values indicate that SCGs might have the most powelful hunger suppression effects after their intake. The green coffee bean and silverskin showed SAT1 and SAT2 values comparable to milk, egg, and soy (Table 11.3). Considering all the protein quality parameters calculated for coffee by-products, dietary supplements and/or food fortification based on these by-products may be an interesting approach to contribute to sustainable health of the global population.

11.5 Food Safety of Coffee Proteins

Some of the requirements that must be considered for approval of any novel food or ingredient according to regulation (EU) 2015/2283 are the absence of toxicity and food safety and low allergenic potential. Allergy caused by coffee protein isolates or concentrates has never been reported, and there is few available information on this matter. Studies performed by Applied Food Science, company that commercializes the CoffeeProtein green coffee concentrate, indicate that it is free of allergens such as gluten, lactose, eggs, nuts, and soy (Applied Food Sciences, 2019). Previous research by our group (Rios et al., 2020) also support the absence of gluten in cascara. In this study, the amount of gluten present in bread formulated with cascara extract and gluten-free flour was determined, and gluten levels were lower than 3 ppm. The limit established by the European Commission in Regulation (EC) 828/2014 for the labeling of products as "gluten-free" is 20 mg/kg (ppm). Therefore, the bread developed in this study was certified as safe for the population that has nutritional requirement for a gluten-free diet. To the best of our knowledge, the allergenicity of other coffee by-products has not been reported so far.

Considering the available data, the European Food Safety Authority (EFSA) has considered coffee cascara as a traditional food from a third country following Article 14 of Regulation (EU) 2015/2283 and does not raise any safety objections to placing this product on the market within the EU (European Food Safety Autority, 2021). However, coffee silverskin is considered a novel food and still needs to undergo EFSA's authorization procedure according to the Regulation (EU) 2015/2283 before it can be introduced into the European food market. Regarding SCGs, they have been recently considered not novel, and they can be used as food ingredients in the European market (European Parliament and of the Council, 2021).

11.6 Biological Properties of Coffee Proteins and Their Derivatives

The health claims ascribed to other food proteins (Table 11.1) may be applied to coffee proteins. Further investigation should be conducted to demonstrate this hypothesis. No previous studies on that field are available since the use of coffee as a source of proteins has been recently proposed (Applied Food Sciences, 2019). However, according to the increased requirements of new sources of food proteins, this is a line of research of great interest.

The EFSA Panel on Dietetic Products, Nutrition and Allergies (NDA) on the basis of the scientific data so far available, has concluded that there are insufficient evidences to establish a cause–effect relationship between consumption of food peptides and their biological effects *in vivo* (EFSA Panel on Dietetic Products, 2009). Consequently, EFSA has not authorized health claims for food peptides yet. More research should be performed in order to complete the scientific evidences on the health-promoting properties of peptides formed during the human digestion, hydrolysis with commercially available enzymatic preparations and chemicals. In order to achieve the goal for coffee proteins, an updated literature review on the bioactive properties of coffee peptides released by the activity of human digestive enzymes and other enzymes commonly used in food industry by *in silico* or *in vitro* approaches is provided in this section of the chapter (Table 11.4).

Table 11.4 Examples of peptides obtained by protease digestion from coffee or coffee by-products and their bioactivities.

Source	Assay	Digestion conditions	Peptides obtained with biological properties	Biological properties
Digestive enzymes approach				
Dehydrins from green coffee beans[1]	*In silico*	Pepsin	• YGGTTGT, YGGEGEQ, YGGTTGA, YGGEGVQ	• Opiod activity
11S globulin from green and roasted coffee beans and SCGs[2]	*In silico*	Pepsin, trypsin, and chymotrypsin	• GW, PL, GL, GF, VY, VF, VAF, VPK • AH, AL, VY, GL, TY • GL, AL	• DPP-IV and ACE inhibitor • Antioxidant • DPP-IV and ACE inhibitors and free radical scavengers
Protein extracts from silveskin[3]	*In vitro*	Pepsin and pancreatin	• TLNGV	• Antibacterial
Food industry enzymes approach				
Protein extracts from silveskin[3]	*In vitro*	Alcalase	• APGAGVY	• Antioxidant
Protein extracts from silveskin[3]	*In vitro*	Thermolysin	• LLNK • AVGVK, FASY, LLYQ, FDAVGVK, AFDAVGVK • LLSQ	• Antibacterial • ACE-inhibitor • Immunomodulator
Protein extracts from SCGs[4]	*In vitro*	Alcalase and thermolysin	• GGSMG, GVMDFQ, LDPIEF, GGGDL, YEAWL, NSLS, FPCDGP, GPVNIAY, GMAST, LGMAST DAHIPPG, DPHFGQPAVE, FGM, FLSRSD, IPNIH, LSDM, NLS, VDSFH, and NRGGY	• ACE-inhibitor and antioxidant

Letters indicate the corresponding amino acid. A = alanine, C = cysteine; H = histidine; I = isoleucine; M = methionine; S = serine; V = valine; G = glycine; L = leucine; P = proline; T = threonine; F = phenylalanine; R = arginine; Y = tyrosine; W = tryptophan; D = aspartic acid; N = asparagine; E = glutamic acid; Q = glutamine; K = lysine.
Source: Data from (1) Vinecky (2015), (2) Ribeiro et al. (2021), (3) Pérez-Míguez et al. (2019) and (4) Valdés et al. (2020).

Peptides obtained after hydrolysis with digestive enzymes may have health-promoting properties in the human body if they are bioavailable after digestion in physiological concentrations and remain intact until reaching the target organ (Segura-Camposa, Chel-Guerreroa, Betancur-Anconaa, & Hernandez-Escalante, 2011). The *in silico* study carried out by Vinecky (2015) indicated the possible release of peptides possessing opioid properties from dehydrin proteins from green coffee beans during the gastric digestion with pepsin (Table 11.4). In addition, according to the data generated by employing BIOPEP UWM database, peptides generated by digestion of the 11S globulin from green and roasted coffee beans and SCGs with digestive enzymes may inhibit dipeptidyl peptidase-IV (DPP-IV) and angiotensin I-converting enzyme (ACE) and possess antioxidant properties (Ribeiro, de Rocha, & Prudencio, 2021). Peptides from the protein extracts of silverskin samples submitted to different roasting process were obtained *in vitro* by enzymatic digestion employing pepsin and pancreatin (Pérez-Míguez et al., 2019). Of the 13 peptides obtained, only one (TLNGV) may be associated with potential antibacterial properties using the BIOPEP database. On the other hand, protein hydrolysates obtained using gastrointestinal digestive enzymes showed antioxidant properties and capacity to reduce micellar cholesterol solubility *in vitro* (Pérez-Míguez et al., 2019). Since these compounds can be generated *in vivo* after gastrointestinal digestion and possess antioxidant, hypocholesterolemic, and hypotensive properties, they can potentially help to reduce the risk of noncommunicable chronic diseases such as metabolic syndrome and cardiovascular diseases. These properties should be studied in clinical trials.

The enzymes commonly used in the food industry generate large amount of peptides from coffee by-products (Table 11.4). Pérez-Míguez et al. (2019) associated antioxidant and hypocholesterolemic properties to the peptidic fraction obtained by alcalase and thermolysin hydrolysis of coffee silverskin. Furthermore, the *in vitro* ACE-inhibitory effect has been evaluated in peptides obtained after the hydrolysis of SCG proteins with alcalase and thermolysin (Valdés et al., 2020). Results from this study showed that the hydrolysates possessed antioxidant properties and inhibited ACE, the latter being highest when SCGs samples were obtained from dark/medium roasted coffees, and after polyphenol removal. The studies by Pérez-Míguez et al. (2019) and Valdés et al. (2020) indicate that thermolysin activity allows to obtain peptides with better health-promoting properties from these by-products than alcalase. The antioxidant peptides obtained from coffee by-products using these enzymes can be used as food preservatives to avoid microbial growth and extend the shelf life of the food product. However, the study of the impact of adding these compounds in the right amount to achieve this effect must be considered in order to keep the sensory quality of the final food product.

Table 11.5 shows some biological parameters calculated from the amino acids present in coffee products. The antioxidant properties of most food derived peptides increase in the presence of some amino acids (histidine, tyrosine, methionine, lysine, tryptophan, and proline) (Girgih, Udenigwe, & Aluko, 2010). The sum of these amino acids reflects the antioxidant amino acids (antioxidant AA) of coffee derivatives (Table 11.5). The green bean has the highest values of antioxidant amino acids, in a similar amount to that of soy. On the other hand, redox balance and methylation depend on cysteine and methionine concentrations, which were also higher in the green coffee beans.

Table 11.5 Biological parameters of amino acids (g/100 g of total amino acids) from coffee derivatives and of conventional protein sources.

	Cascara	Green bean	Silverskin	SCGs	Milk	Egg	Soy
Antioxidant AA	16.35	20.92	16.65	10.98	27.57	22.49	20.13
Met + Cys	2.40	4.04	1.45	1.52	3.51	6.97	3.35
Ile + Leu + Phe + Val	19.71	23.51	26.30	33.84	23.78	24.15	21.35
Arg + His + Glu	14.90	24.39	19.66	15.55	27.84	21.38	28.46
Lys/Arg ratio	0.80	1.86	0.52	7.00	2.23	1.05	0.73

Increased plasma levels of isoleucine, leucine, valine, and phenylalanine were associated with hypertriglyceridemia. In this context, cascara showed the lowest values for the sum of these amino acids, lower than that for milk, egg, and soy (Table 11.5). Arginine, histidine, and glutamine are known to have strong effects on the immune functions in the human body, and the sum of these amino acids in the green bean is within the values of egg and milk. Last, the lysine to arginine ratio is used as an index to measure cholesterolomic and atherogenic effects of proteins, a lower ratio has lesser lipidemic and atherogenic effects (Mir et al., 2019). Silverskin showed the lowest Lys/Arg ratio, even lower than that for soy, followed by cascara, the green bean, and SCGs.

11.7 Conclusions

Coffee proteins are already commercially available (CoffeeProtein, Applied Food Sciences) for the formulation of foods for human consumption. CoffeeProtein consists of proteins (30%), dietary fiber, and other compounds (amino acids, antioxidants, vitamins, and minerals) with health-promoting properties. Similar products may be obtained from coffee by-products, which contain proteins in a range of concentration from 4 to 19%. Protein isolates and concentrates from both green coffee beans and coffee by-products (especially cascara, silverskin, and SCGs) can be obtained using extraction methods similar to those employed for soybean protein (water or alkali extraction) that are easily scalable at an industrial level. Green coffee beans and coffee by-products (cascara, silverskin, and SCGs) contain higher amounts of protein than milk (3%) and similar values when compared to eggs (12%). In addition, percentage of essential amino acids (32–46%) and other protein quality parameters are equivalent to that found for conventional protein sources such as milk, egg, and soybean. Therefore, coffee proteins can be used as a sustainable alternative to animal-based proteins. Peptides generated *in vitro* by enzymatic hydrolysis of isolated coffee proteins that have antioxidant properties can be used to better preserve the quality of foods and extend their shelf life. Coffee and its by-products are natural and sustainable sources of proteins in combination with other compounds of nutritional interest, such as dietary fiber. Novel ingredients obtained based on coffee proteins would be sustainable and hypoallergenic and suitable for all populations including those with special diets for health (celiac) or ideological (vegans) reasons.

Acknowledgments

The authors are grateful to the project "Nuevos conocimientos para la sostenibilidad del sector cafetero" funded by Consejo Superior de Investigaciones Científicas (CSIC) (201970E117) and "Novel coffee by-product beverages for an optimal health of the brain-gut axis (COFFEE4BGA)" funded by the Ministerio de Ciencia e Innovación (PID2019-111510RB-I00) that funded this work. The authors also acknowledge the financial support COLCIENCIAS, with its program of support to the national scientific community through the national doctoral program call no. 727 of 2015.

References

Abdelfatah Ali, M. K. (2013). *Interactions of food proteins with plant phenolics-modulation of structural, techno- and bio-functional properties of proteins*. Universität Potsdam Institut.

Abolbaghaei, A., Oomah, B. D., Tavakoli, H., & Hosseinian, F. (2018). Influence of branched chain amino acids on insulin sensitivity and the mediator roles of short chain fatty acids and gut hormones: a review. *Journal of Food Bioactives, 2*(August 2019). https://doi.org/10.31665/jfb.2018.2136

Acuña, R., Bassüner, R., Beilinson, V., Cortina, H., Cadena-Gómez, G., Montes, V., Nielsen, N. C., Baú, S. M. T., Mazzafera, P., & Santoro, L. G. (1999). Coffee seeds contain 11S storage proteins. *Physiologia Plantarum, 105*(1), 33–40. https://doi.org/10.1034/j.1399-3054.1999.105119.x.

Applied Food Sciences. (2019). *CoffeeProtein*. Organic Green Coffee Bean Protein. https://appliedfoods.com.

Arnold, U., & Ludwig, E. (1996). Analysis of free amino acids in green coffee beans. *Z Lebensm Unters Forsch, 203*(4), 379–384. https://doi.org/10.1007/BF01231078.

Bagdonaite, K., & Murkovic, M. (2018). Factors affecting the formation of acrylamide in coffee. *Czech Journal of Food Sciences, 22*(SI-Chem. Reactions in Foods V), S22–S24. https://doi.org/10.17221/10604-cjfs

Baú, S. M. T., Mazzafera, P., & Santoro, L. G. (2001). Seed storage proteins in coffee. *Revista Brasileira de Fisiologia Vegetal, 13*(1), 33–40. https://doi.org/10.1590/S0103-31312001000100004.

Belitz, H.-D., Grosch, W., & Schieberle, P. (2009). Coffee, tea, cocoa. In H.-D. Belitz, W. Grosch, & P. Schieberle (Eds.), *Food chemistry* (4th ed., pp. 938–951). Leipzig: Springer.

Berk, Z. (1992). Isolated soybean protein (ISP). In *Technology of Production of Edible Flours and Protein Products from Soybeans*. Food and Agriculture Organization of the United Nations. http://www.fao.org/3/t0532e/t0532e07.htm#:~:text=The basic principles of ISP,the solution%2C separated and dried.

Bravo, J., Juániz, I., Monente, C., Caemmerer, B., Kroh, L. W., De Peña, M. P., & Cid, C. (2012). Evaluation of spent coffee obtained from the most common coffeemakers as a source of hydrophilic bioactive compounds. *Journal of Agricultural and Food Chemistry, 60*(51), 12565–12573. https://doi.org/10.1021/jf3040594

Campos-Vega, R., Loarca-Piña, G., Vergara-Castañeda, H., & Oomah, B. D. (2015). Spent coffee grounds: A review on current research and future prospects. *Trends in Food Science and Technology, 45*(1), 24–36. https://doi.org/10.1016/j.tifs.2015.04.012.

Castrillón-Monroy, D., & Riaño-Luna, C. (1999). Extracción, preparación y almacenamiento de proteína obtenida de almendras de café defectuosas. *Cenicafé*, *50*(4), 299–312. https://www.cenicafe.org/es/publications/arc050%2804%29299-312.pdf

Cheng, B., Furtado, A., & Henry, R. J. (2018). The coffee bean transcriptome explains the accumulation of the major bean components through ripening. *Scientific Reports*, *8*(1), 1–11. https://doi.org/10.1038/s41598-018-29842-4.

Clifford, M. (1985). Coffee: Chemistry. In R. J. Clarke & R. Macrae (Eds.), *Elsevier applied science, London* (Vol. 4, Issue 3). Netherlands: Springer. https://doi.org/10.1007/978-94-009-4948-5

Cruz, R., Cardoso, M. M., Fernandes, L., Oliveira, M., Mendes, E., Baptista, P., Morais, S., & Casal, S. (2012). Espresso coffee residues: A valuable source of unextracted compounds. *Journal of Agricultural and Food Chemistry*, *60*(32), 7777–7784. https://doi.org/10.1021/jf3018854.

de Melo Pereira, G. V., de Carvalho Neto, D. P., Magalhães Júnior, A. I., do Prado, F. G., Pagnoncelli, M. G. B., Karp, S. G., & Soccol, C. R. (2020). Chemical composition and health properties of coffee and coffee by-products. In *Advances in food and nutrition research* (1st edition Vol. 91). Elsevier Inc. https://doi.org/10.1016/bs.afnr.2019.10.002.

Delgado, P. A., Vignoli, J. A., Siika-aho, M., & Franco, T. T. (2008). Sediments in coffee extracts: Composition and control by enzymatic hydrolysis. *Food Chemistry*, *110*(1), 168–176. https://doi.org/10.1016/j.foodchem.2008.01.029.

EFSA Panel on Dietetic Products, Nutrition and Allergies (NDA). (2009). Scientific Opinion on the substantiation of health claims related to isoleucine-proline-proline (IPP) and valine-proline-proline (VPP) and maintenance of normal blood pressure (ID 615, 661, 1831, 1832, 2891), and maintenance of the elastic properties of the arteries (ID 1832) pursuant to Article 13(1) of Regulation (EC) No 1924/2006. *EFSA Journal*, *7*(10), 1–18. https://doi.org/10.2903/j.efsa.2009.1259.

Esquivel, P., & Jiménez, V. M. (2012). Functional properties of coffee and coffee by-products. *Food Research International*, *46*(2), 488–495. https://doi.org/10.1016/j.foodres.2011.05.028.

European Food Safety Autority. (2021). Technical Report on the notification of cherry pulp from *C. arabica* L. and *C. canephora* Pierre ex A. Froehner as a traditional food from a third country following Article 14 of Regulation (EU) 2015/2283. *EFSA Supporting Publications*, *18*(5). https://doi.org/10.2903/sp.efsa.2021.EN-6657

European Parliament and of the Council. (2021). *Consultation request to determine the status of spent coffee grounds, defatted spent coffee grounds and defatted unused coffee grounds* (Vol. 4). European Parliament and of the Council.

Ferrão, J. E. M. (2009). *O Café, A Bebida Negra Dos Sonhos Claros*. Chaves Ferreira-Publicações S. A.

Gil-Agusti, M. T., Campostrini, N., Zolla, L., Ciambella, C., Invernizzi, C., & Righetti, P. G. (2005). Two-dimensional mapping as a tool for classification of green coffee bean species. *Proteomics*, *5*(3), 710–718. https://doi.org/10.1002/pmic.200401014.

Girgih, A. T., Udenigwe, C. C., & Aluko, R. E. (2010). In vitro antioxidant properties of Hemp Seed (Cannabis sativa L.) protein hydrolysate fractions. *Journal of the American Oil Chemists' Society*, *88*(3), 381–389. https://doi.org/10.1007/s11746-010-1686-7

Gomes Almeida Sá, A., Moreno, Y. M. F., & Carciofi, B. A. M. (2020). Plant proteins as high-quality nutritional source for human diet. *Trends in Food Science and Technology*, *97*(October 2019), 170–184. https://doi.org/10.1016/j.tifs.2020.01.011

Hashempour-Baltork, F., Khosravi-Darani, K., Hosseini, H., Farshi, P., & Reihani, S. F. S. (2020). Mycoproteins as safe meat substitutes. *Journal of Cleaner Production, 253*, 119958. https://doi.org/10.1016/j.jclepro.2020.119958.

Hawkey, K. J., Lopez-Viso, C., Brameld, J. M., Parr, T., & Salter, A. M. (2021). Insects: A potential source of protein and other nutrients for feed and food. *Annual Review of Animal Biosciences, 9*, 333–354. https://doi.org/10.1146/annurev-animal-021419-083930.

Hoffer, L. J. (2016). Human protein and amino acid requirements. *JPEN: Journal of Parenteral and Enteral Nutrition, 40*(4), 460–474. https://doi.org/10.1177/0148607115624084.

International Coffee Organization. (2021). *Coffee prices rallied in April 2021 amid an anticipated decline in production and signs of economic recovery from covid-19 pandemic* (Issue April). https://www.ico.org/Market-Report-20-21-e.asp.

Iriondo-DeHond, A., Elizondo, A. S., Iriondo-DeHond, M., Ríos, M. B., Mufari, R., Mendiola, J. A., Ibañez, E., & del Castillo, M. D. (2020). Assessment of healthy and harmful Maillard reaction products in a novel coffee cascara beverage: melanoidins and acrylamide. *Foods, 9*(5), 620. https://doi.org/10.3390/foods9050620.

Iriondo-DeHond, A., Iriondo-DeHond, M., & del Castillo, M. D. (2020). Applications of compounds from coffee processing by-products. *Biomolecules, 10*(9), 1219. https://doi.org/10.3390/biom10091219.

Iriondo-DeHond, A., Rios, M. B., Herrera, T., Rodriguez-Bertos, A., Nuñez, F., San Andres, M. I., . . . del Castillo, M. D. (2019). Coffee silverskin extract: nutritional value, safety and effect on key biological functions. *Nutrients, 11*(11), 2693. https://doi.org/10.3390/nu11112693

Iriondo-DeHond, A., Uranga, J. A., del Castillo, M. D., & Abalo, R. (2020). Effects of coffee and its components on the gastrointestinal tract and the brain–gut axis. *Nutrients, 13*(1), 88. https://doi.org/10.3390/nu13010088.

Jantzen da Silva Lucas, A., Menegon de Oliveira, L., da Rocha, M., & Prentice, C. (2020). Edible insects: An alternative of nutritional, functional and bioactive compounds. *Food Chemistry, 311*(July 2019), 126022. https://doi.org/10.1016/j.foodchem.2019.126022

Lago, R. C. A., Antoniassi, R., & Freitas, S. C. (2001). Centesimal composition and amino acids of raw, roasted and spent ground of soluble coffee. *II Simpósio de Pesquisa Dos Cafés Do Brasil Vitoria*, 1473–1477.

Li, X., Zhang, G., Zhao, X., Zhou, J., Du, G., & Chen, J. (2020). A conceptual air-lift reactor design for large scale animal cell cultivation in the context of *in vitro* meat production. *Chemical Engineering Science, 211*, 115269. https://doi.org/10.1016/j.ces.2019.115269.

López-Pedrouso, M., Bernal, J., Franco, D., & Zapata, C. (2014). Evaluating two-dimensional electrophoresis profiles of the protein phaseolin as markers of genetic differentiation and seed protein quality in common bean (*Phaseolus vulgaris* L.). *Journal of Agricultural and Food Chemistry, 62*(29), 7200–7208. https://doi.org/10.1021/jf502439u.

Machado, S., Costa, A. S. G., Pimentel B. F., Oliveira, M. B. P. P., & Alves, R. C. (2020). A study on the protein fraction of coffee silverskin: Protein/non-protein nitrogen and free and total amino acid profiles. *Food Chemistry, 326*(April), 126940. https://doi.org/10.1016/j.foodchem.2020.126940.

Manavski, N., Peters, U., Brettschneider, R., Oldenburg, M., Baur, X., & Bittner, C. (2012). Cof a 1: Identification, expression and immunoreactivity of the first coffee allergen. *International Archives of Allergy and Immunology, 159*(3), 235–242. https://doi.org/10.1159/000337461.

Martinez-Saez, N., Tamargo García, A., Domínguez Pérez, I., Rebollo-Hernanz, M., Mesías, M., Morales, F. J., Martín-Cabrejas, M. A., & del Castillo, M. D. (2017). Use of spent coffee grounds as food ingredient in bakery products. *Food Chemistry*, *216*, 114–122. https://doi.org/10.1016/j.foodchem.2016.07.173.

Mazzafera, P. (1999). Chemical composition of defective coffee beans. *Food Chemistry*. 64(4), 547–554. https://doi.org/10.1016/S0308-8146(98)00167-8.

Mazzafera, P., Schimpl, F., & Kiyota, E. (2019). Chapter 18. Proteins of coffee beans: recent advances. *Coffee*, *January*, 429–444. https://doi.org/10.1039/9781782622437-00429.

Meena, G. S., Singh, A. K., Panjagari, N. R., & Arora, S. (2017). Milk protein concentrates: opportunities and challenges. *Journal of Food Science and Technology*, *54*(10), 3010–3024. https://doi.org/10.1007/s13197-017-2796-0.

Mir, N. A., Riar, C. S., & Singh, S. (2019). Effect of pH and holding time on the characteristics of protein isolates from Chenopodium seeds and study of their amino acid profile and scoring. *Food Chemistry*, *272*(March 2018), 165–173. https://doi.org/10.1016/j.foodchem.2018.08.048

Montavon, P., Duruz, E., Rumo, G., & Pratz, G. (2003). Evolution of green coffee protein profiles with maturation and relationship to coffee cup quality. *Journal of Agricultural and Food Chemistry*, *51*(8), 2328–2334. https://doi.org/10.1021/jf020831j.

Montavon, P., Mauron, A. F., & Duruz, E. (2003). Changes in green coffee protein profiles during roasting. *Journal of Agricultural and Food Chemistry*, *51*(8), 2335–2343. https://doi.org/10.1021/jf020832b.

Mussatto, S. I., Ballesteros, L. F., Martins, S., & Teixeira, J. A. (2011). Extraction of antioxidant phenolic compounds from spent coffee grounds. *Separation and Purification Technology*, *83*(1), 173–179. https://doi.org/10.1016/j.seppur.2011.09.036.

Mussatto, S. I., Machado, E. M. S., Martins, S., & Teixeira, J. A. (2011). Production, composition, and application of coffee and its industrial residues. *Food and Bioprocess Technology*, *4*(5), 661–672. https://doi.org/10.1007/s11947-011-0565-z.

Parthasarathy, A., Cross, P. J., Dobson, R. C. J., Adams, L. E., Savka, M. A., & Hudson, A. O. (2018). A Three-Ring circus: Metabolism of the three proteogenic aromatic amino acids and their role in the health of plants and animals. *Frontiers in Molecular Biosciences*, *5*(APR), 1–30. https://doi.org/10.3389/fmolb.2018.00029.

Pérez-Míguez, R., Marina, M. L., & Castro-Puyana, M. (2019). High resolution liquid chromatography tandem mass spectrometry for the separation and identification of peptides in coffee silverskin protein hydrolysates. *Microchemical Journal*, *149*, 103951. https://doi.org/10.1016/j.microc.2019.05.051.

Pojić, M., Mišan, A., & Tiwari, B. (2018). Eco-innovative technologies for extraction of proteins for human consumption from renewable protein sources of plant origin. *Trends in Food Science and Technology*, *75*(March), 93–104. https://doi.org/10.1016/j.tifs.2018.03.010.

Preece, K. E., Hooshyar, N., & Zuidam, N. J. (2017). Whole soybean protein extraction processes: A review. *Innovative Food Science and Emerging Technologies*, *43*(October 2016), 163–172. https://doi.org/10.1016/j.ifset.2017.07.024

Rawel, H., Rohn, S., & Kroll, J. (2005). Characterization of 11S protein fractions and phenolic compounds from green coffee beans under special consideration of their interactions—A review. *Deutsche Lebensmittel-Rundschau*, *101*, 148–160. https://publishup.uni-potsdam.de/opus4-ubp/frontdoor/index/index/docId/13847

Regulation EC 1924. (2006). European Community (EC) No. 1924/2006 of the European Parliament and of the Council of 20 December 2006 on nutrition and health claims made on foods. *Official Journal of the European Union, L 404*, 9–25.

Ribeiro, E., de Rocha, T. S., & Prudencio, S. H. (2021). Potential of green and roasted coffee beans and spent coffee grounds to provide bioactive peptides. *Food Chemistry, 348*(November 2020), 129061. https://doi.org/10.1016/j.foodchem.2021.129061

Rios, M. B., Iriondo-deHond, A., Iriondo-deHond, M., Herrera, T., Velasco, D., Gomez-Alonso, S., Callejo, M. J., & del Castillo, M. D. (2020). Effect of coffee cascara dietary fiber on the physicochemical, nutritional and sensory properties of a gluten-free bread formulation. *Molecules, 25*, 1358. https://doi.org/10.3390/molecules25061358.

Rogers, W. J., Bézard, G., Deshayes, A., Meyer, I., Pétiard, V., & Marraccini, P. (1999). Biochemical and molecular characterization and expression of the 11S-type storage protein from *C. arabica* endosperm. *Plant Physiology and Biochemistry, 37*(4), 261–272. https://doi.org/10.1016/S0981-9428(99)80024-2.

Segura-Camposa, M., Chel-Guerreroa, L., Betancur-Anconaa, D., Hernandez-Escalante, V. M. (2011). Bioavailability of bioactive peptides. *Food Reviews International, 27*(3), 213–226. https://doi.org/10.1080/87559129.2011.563395.

Sha, L., & Xiong, Y. L. (2020). Plant protein-based alternatives of reconstructed meat: Science, technology, and challenges. *Trends in Food Science and Technology, 102*(February), 51–61. https://doi.org/10.1016/j.tifs.2020.05.022.

Shimizu, M. M., & Mazzafera, P. (2000). Compositional changes of proteins and amino acids in germinating coffee seeds. *Brazilian Archives of Biology and Technology, 43*(3), 259–265. https://doi.org/10.1590/S1516-89132000000300003.

Silva, M. A., Nebra, S. A., Machado Silva, M. J., & Sanchez, C. G. (1998). The use of biomass residues in the Brazilian soluble coffee industry. *Biomass and Bioenergy, 14*(5–6), 457–467. https://doi.org/10.1016/S0961-9534(97)10034-4.

Singh, A., Meena, M., Kumar, D., Dubey, A. K., & Hassan, M. I. (2015). Structural and functional analysis of various globulin proteins from soy seed. *Critical Reviews in Food Science and Nutrition, 55*(11), 1491–1502. https://doi.org/10.1080/10408398.2012.700340.

Spiller, M. A. (1998). The chemical components in coffee. In G. Spiller (Ed.), *Caffeine* (pp. 97–161). CRC Press/Taylor & Francis. https://doi.org/10.1201/9780429126789-2.

The European Commission. (2012). Commission Regulation (EU) No. 432/2012. *Official Journal of the European Union, 13*(1924). http://eur-lex.europa.eu/legal-content/EN/TXT/PDF/?uri=CELEX:32012R0432&from=EN.

The European Commission. (2020a). A Farm to Fork Strategy for a fair, healthy and environmentally-friendly food system. *Communication from the Commission to the European Parliament, the Council, the European Economic and Social Committee and the Committee of the Regions, 381*, 1–9.

The European Commission. (2020b). *Dietary Protein*. Health Promotion and Disease Prevention. https://ec.europa.eu/jrc/en/health-knowledge-gateway/promotion-prevention/nutrition/protein.

Torres-Tiji, Y., Fields, F. J., & Mayfield, S. P. (2020). Microalgae as a future food source. *Biotechnology Advances, 41*(February). https://doi.org/10.1016/j.biotechadv.2020.107536.

Valdés, A., Castro-Puyana, M., & Marina, M. L. (2020). Isolation of proteins from spent coffee grounds. Polyphenol removal and peptide identification in the protein hydrolysates by

RP-HPLC-ESI-Q-TOF. *Food Research International, 137*, 109368. https://doi.org/10.1016/j.foodres.2020.109368.

Vinecky, F. (2015). *Identificação e caracterização de peptídeos opioides presentes na fração proteica de grãos de C. arabica*. Universidad de Brasília.

Volk, C., Brandsch, C., Schlegelmilch, U., Wensch-Dorendorf, M., Hirche, F., Simm, A., Gargum, O., Wiacek, C., Braun, P. G., Kopp, J. F., Schwerdtle, T., Treede, H., & Stangl, G. I. (2020). Postprandial metabolic response to rapeseed protein in healthy subjects. *Nutrients, 12*(8), 1–22. https://doi.org/10.3390/nu12082270.

Watson, E. (2021). *"Alternative" proteins could account for 11% of global protein market by 2035, predicts report*. Food Navigator. https://www.foodnavigator-usa.com/Article/2021/03/24/Alternative-proteins-will-account-for-11-of-global-protein-market-by-2035-predicts-report.

Wen, L., Álvarez, C., Zhang, Z., Poojary, M. M., Lund, M. N., Sun, D. W., & Tiwari, B. K. (2020). Optimisation and characterization of protein extraction from coffee silverskin assisted by ultrasound or microwave techniques. *Biomass Conversion and Biorefinery*. https://doi.org/10.1007/s13399-020-00712-2.

Wen, L., Zhang, Z., Zhao, M., Senthamaraikannan, R., Padamati, R. B., Sun, D. W., & Tiwari, B. K. (2020). Green extraction of soluble dietary fiber from coffee silverskin: impact of ultrasound/microwave-assisted extraction. *International Journal of Food Science and Technology, 55*(5), 2242–2250. https://doi.org/10.1111/ijfs.14477.

Wu, G., Fanzo, J., Miller, D. D., Pingali, P., Post, M., Steiner, J. L., & Thalacker-Mercer, A. E. (2014). Production and supply of high-quality food protein for human consumption: Sustainability, challenges, and innovations. *Annals of the New York Academy of Sciences, 1321*(1), 1–19. https://doi.org/10.1111/nyas.12500.

12

Coffee, Ginger and Cinnamon: Molecular Mechanisms of Action on Human Health

Maria Bibiana Zapata Londoño[1], Diego Fernando Uribe Yunda[2], Johanny Aguillón Osma[3], and Maria Elena Maldonado-Celis[1]*

[1] *School of Nutrition and Dietetics, University of Antioquia, Medellín, Colombia*
[2] *Faculty of Exact and Applied Sciences, Instituto Tecnológico Metropolitano, Medellín, Colombia*
[3] *Faculty of Health Sciences, University of Quindío, Armenia, Colombia*

12.1 Introduction

Coffee, cinnamon, and ginger are rich in bioactive compounds which has led to renewed interest and increased experimental research to validate their health benefits. Therefore, this chapter presents information on the mechanisms of action involved in the prevention and/or control of chronic noncommunicable diseases associated with the antioxidant, antidiabetic, anti-inflammatory, cardioprotective, and hepatoprotective effects of coffee and these spices in order to understand their efficacy and safety as potential therapeutic agents. Coffee is one of the most popular and consumed beverages worldwide. The different forms of coffee preparation contain many phytochemicals whose content varies according to the preparation process, as well as the time of roasting. The bioactive compounds in coffee are responsible for the biological actions that will be elucidated in this chapter and whose mechanisms of action are associated with reduced risk of chronic noncommunicable diseases (Cheng, Sheen, Hu, & Hung, 2017; Guasch-Ferré, Merino, Sun, Fitó, & Salas-Salvadó, 2017).

Ginger (*Z. officinale*) is a perennial herbaceous plant native to South Asia; its rhizome is widely used as spice in food and beverages due to its characteristic pungent and spicy flavor by the gingerols and shogaols (Jafarzadeh et al., 2014; Srinivasan 2014). Its use in traditional eastern medicine dates back more than 2,500 years to treat diseases such as rheumatoid arthritis, digestion, and infections (Ali, Blunden, Tanira, & Nemmar, 2008; O'Hara, Kiefer, Farrell, & Kemper, 1998). More than 400 ginger compounds have been identified, the quantity and quality of which depend on the region of cultivation, the type and method of preparation (dry or fresh form) (Arablou & Aryaeian, 2018).

Cinnamon has been one of the most common spices obtained from the inner bark of the trees of the genus Cinnamomum, also from the leaf and the root. It is a tropical perennial plant of which about 250 species of this genus are known, the most economically

Molecular Mechanisms of Functional Food, First Edition. Edited by Rocio Campos-Vega and B. Dave Oomah.

important ones are *Cinnamomum verum* or *Cinnamomun zeylacum* (real Sri Lankan or Ceylon cinnamon), *Cinnamomum cassia* (Chinese cinnamon), *Cinnamomum burmannii* (Java or Indonesian cinnamon), and *Cinnamomum loureiroi* (Vietnamese or Saigon cinnamon) (Dorri, Hashemitabar, & Hosseinzadeh, 2018; Killday et al., 2011; Muhammad & Dewettinck, 2017; Vangalapati, Satya, Prakash, & Avanigadda, 2012). Preclinical and few human studies have demonstrated the beneficial health effects of cinnamon with anti-inflammatory and cardioprotective activities (Azimi, Ghiasvand, Feizi, Hariri, & Abbasi, 2014; Furlan et al., 2019; Qin, Dawson, Polansky, & Anderson, 2009). However, there are discrepancies among investigators due to the different forms of cinnamon used, dose, and duration of the interventions. Its pharmacological effects are attributed to the content of volatile oils such as cinnamaldehyde, (bark), eugenol (leaf) and camphor (root) (Gruenwald, Freder, & Armbruester, 2010), and procyanidins and catechin (bark) (Nonaka, Morimoto, & Nishioka, 1983).

Table 12.1 summarizes potential health benefits of the bioactive compounds present in coffee, ginger, and cinnamon to which their properties are attributed and discussed later.

12.2 Bioactive Compounds in Coffee, Ginger, and Cinnamon

Coffee is rich in many bioactive substances, and the concentration of caffeine, the most important component depends on genetic aspects and extrinsic factors (soil composition, climate, agricultural practices, storage conditions, roasting, and brewing times) (Farah & Donangelo, 2006; Gökcen & Sanlier, 2019; Górecki & Hallmann, 2020). Depending on the method of preparation and coffee type, the concentration of caffeine varies from 72 to 130 mg in 240 mL of brewed coffee and 58–76 mg in a single cup of espresso, and according to bean roasting this concentration changes from 152 to 232 mg/serving (Ludwig et al., 2014; McCusker, Goldberger, & Cone, 2003). Organic coffee contains significantly less caffeine (48 vs 58 mg/100 mL) than conventional coffee, longer roasting times significantly reducing caffeine level (73 vs 28 mg/100 mL) and during the first 3 min of coffee brewing a significantly higher concentration of caffeine was reported (Gökcen & Sanlier, 2019; Górecki & Hallmann, 2020). In addition, coffee has high content of phenolic acids such as the hydroxycinnamic acid family (caffeic, chlorogenic, *p*-coumaric, and ferulic acids); chlorogenic acid being the main phenolic acid followed by salicylic acid. Regular roasted ground coffee brews showed higher contents of chlorogenic acids (4,162 mg/100 g), chlorogenic acid lactones (779 mg/100 g), *p*-coumaric acid (2.4 mg/100 g), trigonelline (2,044 mg/100 g), nicotinic acid (100 mg/100 g), caffeine (4,565 mg/100 g), and theobromine (12.5 mg/100 g) than regular soluble coffee brews, even decaffeinated coffee brews presented comparable levels of these compounds. In addition, the boiling water pouring number and coffee ground size affect the content of chlorogenic acid (75–306 mg/serving) (Jeon et al., 2017; Ludwig et al., 2014), whereas the level of kahweol diterpene changes because of the characteristics of paper filter used (Yovana, Brígida, & Bragagnolo, 2018). Regarding polyphenol compounds, several flavonoids have been identified (epigallocatechin, quercetin-3-O-rutinoside [rutin], kaempferol-3-O-glucoside [astragalin], quercetin, quercetin-3-O-glucoside [isoquercetin], kaempferol). Their concentration is similar in conventional and organic coffee; however, the length of coffee

Table 12.1 Bioactive compounds and health benefits of coffee, ginger, and cinnamon.

Natural source	Bioactive compound	Effective concentration	Health Benefits	References
Coffee	Chlorogenic acid	0.1–100 μM	Induces cytoprotective genes expression	Boettler et al. (2011)
		1–10 mg/kg	Reduces cholesterol and fatty liver levels	Wan et al. (2013)
		15–60 mg/kg	Upregulates Nrf2 and nuclear cytoprotective genes expression mediated by Nrf2	Shi et al. (2018)
		30–90 mg/kg	Improves serum lipids and free fatty acid levels	Huang, Liang, Zhong, He, and Wang (2015)
		80 mg/kg	Reduces fasting plasma glucose, glycosylated hemoglobin levels, and the percentage of body fat	Jin et al. (2015)
		90 mg/kg	Improves fasting glucose, glucose tolerance, and fasting insulin concentrations	Peng, Qi, Zhong, Xu, and Zheng (2015)
		30–100 mg/kg	Reduces proinflammatory markers	Shan et al. (2009); Ma, Gao, and Liu (2015); Yuan et al. (2017)
		100 mg/kg	Decreases body weight and visceral fat accumulation; improves heart and liver structure and function	Bhandarkar, Brown, and Panchal (2019)
		0–2 mM	Immunomodulatory properties	Kang et al. (2012)
	Kahweol	0–5 μM	Immunomodulatory properties	Kim, Jung, and Jeong (2004)
		40 μM	Modulates liver inflammation	Seo et al. (2018)
		75 mg/kg	Modulates hepatic fibrosis	Seo et al. (2018)
	Cafestol	0–10 μM	Immunomodulatory properties	Kim et al. (2004)
		3–10 μM	Modulates endothelial dysfunction	Hao et al. (2018)
	Melanoidins	50–100 nM	Activates Nrf2	Sauer et al. (2013)

Ginger	6-Gingerol	10 μM	Alleviates the exaggerated vasoconstriction of diabetic aortae	Ghareib et al. (2015)
		50 μM	Prevents atherosclerosis, reduces cholesterol synthesis, and lipid accumulation	Kiyama (2020)
		100 μM	Induces relaxation in smooth muscle	Townsend et al. (2013); Townsend, Zhang, Xu, Wakita, and Emala (2014).
		0.05 or 0.2 mg/kg	Inhibits lipogenesis, upregulates fatty acid oxidation, reduces oxidative stress	Li et al. (2019)
	8-Gingerol	100 μM	Induces relaxation in smooth muscle	Townsend et al. (2013); Townsend et al. (2014).
	6-Shogaol	20 μM	Prevents oxidative stress	Chen et al. (2014)
		100 μM	Induces relaxation in smooth muscle	Townsend et al. (2013); Townsend et al. (2014).
Cinnamon	Cinnamaldehyde	10 M	Increases total glutathione	Wondrak et al. (2010)
		0.02%	Increases SOD activity and decreases intracellular ROS levels	Wang et al. (2020)
		20 mg/kg	Increases serum insulin, stimulates insulin release, and decreases fasting blood glucose and HbA1c	Anand, Murali, Tandon, Murthy, and Chandra (2010)
		20, 40 mg/kg	Improves cholesterol and lipid homeostasis	Huang, Yuan, Kim, Quan, and Chung (2011)
		40 mg/kg	Induces white adipose tissue browning during obesity	Zuo et al. (2017)
	Trans-cinnamaldehyde	30 M	Decreases intracellular ROS in lung fibroblasts	Choi (2020)
	Procyanidin C1	6.25, 12.5 and 25 μM	Enhances glucose entry into adipocytes	Sun et al. (2019)
	Methylhydroxychalcone polymer	0.1 mg/mL	Increases glycogen synthesis, glucose entry into adipocytes; promotes GS activity, and inhibits GSK-3 activities.	Jarvill-Taylor, Anderson, and Graves (2001)
	2-Methoxycinnamaldehyde	100 μg/kg	Protects against myocardial dysfunction	Hwa et al. (2012)
	Eugenol	10, 100 mg/kg	Reduces LDL and total cholesterol levels	Harb, Bustanji, Almasri, and Abdalla (2019)
	-(−)Epicatechin	10, 20, 40 mg/kg	Reduces total cholesterol, triglycerides and fat accumulation in the liver; decreases serum IL-2, IL-1β and IL-6 levels.	Cheng et al. (2017)

brewing changes their concentration. In short, brewing increases the levels of phenolic compounds and longer roasting time changes the ratio of glycosidic to aglycone forms (Górecki & Hallmann, 2020).

Over 400 different chemical compounds have been identified in ginger rhizome, including phenolic compounds, flavonoids, alkaloids, saponins, glycosides, steroids, terpenoids, and tannins (Dhanik, Arya, & Nand, 2017). The volatile oil confers the odor of ginger rhizome and consists of sesquiterpenes, such as zingiberene (36% w/w), curcumene (18%), and farnesene (10%) and monoterpenes, such as cineole (1.3%), linalool (1.3%), borneol (2.2%), geranial (citral ~1.4%), and neral (citral b, 0.8%) (Arablou & Aryaeian, 2018; Govindarajan, 1982a; Govindarajan, 1982b). The nonvolatile pungent components include major biologically active constituents such as gingerols, shogaols, paradols, and zingerone. Total flavonoid and phenolic contents were 3.66 ± 0.45 mg gallic acid/g and 10.22 ± 0.87 mg quercetin/g of dry rhizome weight (Mojani, Ghasemzadeh, Rahmat, Loh, & Ramasamy, 2014). The gingerols (23–25%) is a group of phenolic compounds that are the bioactive principle quantitatively more important, followed by shogaols (18–25%). Gingerols concentration is lower in dry compared to fresh ginger because of shogaols degradation rates which depend on pH changes and thermal effects with greatest stability at pH 4, whereas at 100 °C and pH 1, the reversible degradation is relatively rapid (Bhattarai, Tran, & Duke, 2001; Mojani et al., 2014; Prasad & Tyagi, 2015). After hydrogenation, shogaols can be transformed into paradols (Stoner, 2013). The gingerols are a group of chemical homologs differentiated by the length of their unbranched alkyl chains; the most abundant 6-gingerol is considered the principle responsible for the health benefits of ginger (Mojani et al., 2014).

The bioactive compounds in cinnamon are polyphenols, phenolic compounds, and essential oils. The main polyphenol compounds are catechin, (−)-epigallocatechin gallate, rutin, quercetin, kaempferol, isorhamentin, and A-and B-types procyanidins. The following phenolic compounds were identified protocatechuic, *p*-hydroxybenzoic, gallic, salicylic, syringic, vanillic, caffeic, chlorogenic, ferulic, *p*-coumaric, cinnamic, and caffeic acids, 3,4-dihydroxybenzaldehyde, *p*-hydroxybenzaldehyde, vanillin, and cinnamyl aldehydes (Anderson et al., 2004; Tanaka, Matsuo, Yamada, & Kouno, 2008). The main volatiles are the essential oils α-pinene, β-pinene, ρ-cymene, limonene, linalool, α-terpineol, benzaldehyde, cinnamaldehyde, eugenol, β-caryophyllene, cinnamyl acetate, α-humulene, acetyleugenol, benzyl benzoate, *trans*-cinnamaldehyde, caryophyllene oxide, l-borneol, l-bornyl acetate, E-nerolidol, camphene, α-phellandrene, α-terpinolene, and coumarin (Paranagama et al., 2010). The cinnamaldehyde, cinnamate, and cinnamic acid are the main components found in cinnamon and increase in concentration when cinnamon ages. The major compound in leaves is eugenol (70–95%); it gradually decreases with leaves development with a higher concentration of *trans*-cinnamaldehyde in leaves from 2-year-old branches, whereas bark is high in cinnamaldehyde content (65–80%) and antinutritional factors such as trypsin inhibitor (6–12 trypsin inhibitor unit/mg of protein), chymotrypsin inhibitor (2–5 chymotrypsin inhibitor unit/mg of protein), phytic acid (83–129 mg/100 g), and tannin content (0.65–2.18% of catechin equivalent) (Al-Numair, Ahmed, Ahmad, & Al-Assaf, 2007). In root bark camphor (60%) prevails, whereas the fruit is high in *trans* cinnamyl acetate (42–54%) concentration, the buds in terpene hydrocarbons (78%), and alpha-bergamotene (27%) (Vangalapati et al., 2012). The oil from cinnamon barks contains high levels of palmitic (31–33%), oleic (29–30%), linoleic (15–16%), and stearic acids (12–14%) (Al-Numair et al., 2007).

12.3 Molecular Mechanisms of Bioactive Compounds in Coffee, Ginger, and Cinnamon for Preventive Noncommunicable Chronic Diseases

12.3.1 Antioxidant

This section describes the molecular mechanisms involved in the antioxidant effects reported after use of these natural products in both *in vitro* and *in vivo* studies (Górecki & Hallmann, 2020; Harliansyah, Azian, Zurinah, & Yusof, 2005).

12.3.1.1 Coffee

Coffee constituents activate the Keap1-Nrf2 (Kelch-like ECH-associated protein 1 – nuclear factor erythroid 2-related factor 2) pathway (Figure 12.1), considered the master regulator of defense against oxidative stress by inducing cytoprotective genes expression (Baird & Yamamoto, 2020). The Nrf2 signaling pathway includes Cullin3 and Keap1 as negative regulators through complex formation with Nrf2. This pathway is activated when Keap1-thiol residues are modified by electrophiles or free radicals, which allow Nrf2 release, and its heterodimerization with Jun and small musculoaponeurotic fibrosarcoma (sMaf)

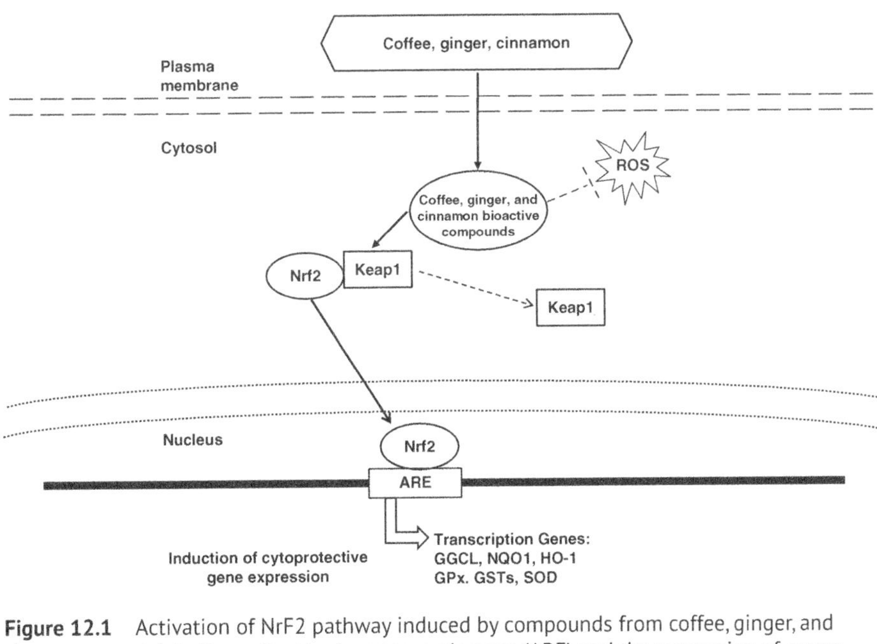

Figure 12.1 Activation of NrF2 pathway induced by compounds from coffee, ginger, and cinnamon modulate the antioxidant response element (ARE) and the expression of genes involved in endogenous antioxidant metabolism (GGCL, NQO1, HO-1, GPx, GSTs, and SOD). *Source:* Based on Li et al. (2019).

proteins of Nrf2 and its subsequent translocation to the nucleus, where it binds to antioxidant element response (ARE) or electrophile-response element (EpRE) at the promoters of Nrf2-induced genes. The Nrf2 has several conserved domains that play a functional role, among the most important are Neh-1 required for Nrf2 binding to DNA, Neh-2 to which Keap1 binds and Neh-3, Neh-4, and Neh-5 domains that can induce Nrf2 transactivation when oxidized (Da costa et al., 2019).

NrF2 is a basic leucine zipper transcription factor that regulates the response to oxidative damage by inducing expression of genes involved in preventing oxidative stress including γ-glutamyl cysteine synthetase (γ-GCS), heme oxygenase 1 (HO-1), glutathione reductase, glutathione peroxidase (GPx), thioredoxin-1, thioredoxin reductase, catalase (CAT), NAD(P)H:quinone oxidoreductase-1 (NQO1), superoxide dismutase (SOD), and peroxiredoxins (Da Costa et al., 2019; Wondrak et al., 2010).

In HT29 colon cancer cells, Nrf2 nuclear translocation after 3 hr chlorogenic acid ($\geq$ 10 nM) treatment significantly increases glutathione s-transferase A1 (GSTA1) (at $\geq$10 μM), gamma-glutamylcysteine ligase (GGCL) (at 100 μM), and HO-1 (at 0.1 μM) gene expression (Boettler et al., 2011). These results were confirmed on Sprague–Dawley rats that received 15–60 mg/kg of chlorogenic acid dissolved in water (once a day for 7 consecutive days), showed nuclear Nrf2 accumulation and significant upregulation of HO-1, GCL, and NQO1 gene expression, in a dose-dependent manner (Shi et al., 2018). Additionally, mixtures of cafestol and kahweol, two of the main diterpenes present in the lipid fraction of the coffee bean, stimulated Nrf2 and induced the expression of antioxidant proteins (Higgins, Cavin, Itoh, Yamamoto, & Hayes, 2008). Likewise, molecules produced during the roasting process such as N-methylpyridinium (NMP) from trigonelline and melanoidins from chlorogenic acid can also modulate the Keap1-Nrf2 pathway (Boettler et al., 2011). However, melanoidins (50–100 mM) and roasted coffee extracts (2–4 mg/ml), but not raw coffee extracts activated Nrf2 in macrophages, Caco-2 cells, and human gut tissue (Sauer et al., 2013).

12.3.1.2 Ginger

Cell models have been used to describe the mechanisms related to the antioxidant potential of ginger. For example, Ji et al. (2017) used human mesenchymal stem cells exposed to radiation and treated with a ginger extract at 0.1 mg/mL for 24 hours. This extract consisting mainly of gingerols and shogaols exhibited antioxidant protective effect by inducing Nrf2 translocation to the cell nucleus and thus activating cytoprotective genes expression encoding HO-1 and NQO-1 (Figure 12.1). Similar results were observed in human chondrocytes exposed to oxidative stress stimulated by interleukin 1β (IL-1β), where ginger extracts at 25 μg/mL stimulated the expression of several antioxidant enzymes and significantly reduced ROS production and lipid peroxidation (Mao et al., 2019). The bioactive components of ginger, particularly 6-shogaol, exhibit antioxidant activity by modulating the Nrf2 regulatory pathway (Mao et al., 2019). This signaling pathway controls the expression of genes such as HO-1, metallothionein 1 (MT1), aldo-keto reductase family 1 member B10 (AKR1B10), ferritin light chain (FTL), and γ-glutamyltransferase-like activity 4 (GGTLA4). Besides, 6-shogaol is also involved in regulating genes involved in glutathione biosynthesis, modulating the expression of catalytic and modifier subunits of the GCL enzyme. A more detailed analysis showed that 6-shogaol and its metabolites activate the Nrf2 pathway through alkylation of the cysteine residues of Keap1 protein (Chen et al., 2014).

12.3.1.3 Cinnamon

The antioxidant properties of cinnamon have been attributed to cinnamaldehyde and its analogues corresponding to *trans*-cinnamaldehyde, cinnamyl alcohol, cinnamic acid, sinapaldehyde, pinoresinol, and (−)-(2R, 3R)-5,7-dimethoxy-3′, 4′-methylenedioxy-flavan-3-ol (Godlewska-Żyłkiewicz et al., 2020; Li et al., 2019; Wang et al., 2020; Wondrak et al., 2010). *Trans*-Cinnamaldehyde and the mentioned phenolic compounds exert their antioxidant activity through the Nrf2 activation and consequently increase GCS, SOD, NQO-1, and HO-1 expression (Li et al., 2019) (Figure 12.1).

Treatment of human normal lung epithelial Beas-2B cell line with pinoresinol [PRO] (lignan) and (−)-(2R,3R)-5,7-dimethoxy-3′, 4′-methylenedioxy-flavan-3-ol [MFO] (flavonoid) for 18 hr induced Nrf2 activation and proteins encoded by its target genes (NQO1 and γ-GCS; Keap1) treated with PRO (25 M) and MFO (12.5 M), respectively, compared to untreated cells indicating that both phenols favor Nrf2 stabilization (Li et al., 2019). On the other hand, cinnamaldehyde and ethanol extract of *Cinnamomum cassia* bark (10 M) induced Nrf2 activation and increased HO-1, and γ-GCS protein levels in a primary culture of fetal colon cells after 4 hr treatment. Likewise, total glutathione increased after 24 hr treatment at 5 M and 10 M and genotoxicity-induced protection by hydrogen peroxide exposure in cells pretreated with the extract or with the pure compound at 2.5 M for 24 hr (Wondrak et al., 2010). Additionally, the effect of cinnamaldehyde on ROS, HO-1 and NQO-1 protein levels was abolished when Nrf2 gene expression was blocked suggesting that Nrf2 activation is essential for the antioxidant effect of cinnamaldehyde (Wang et al., 2020).

A structure–activity study revealed the possible mechanism by which cinnamaldehyde (aldehyde, unsaturated) present in cinnamon is able to activate the Nrf2 pathway. It acts as a Michael acceptor that enables spontaneous adducts formation with thiols, thereby modifying the Keap1 repressor cysteine residues and translocation of Nrf2 to the nucleus and induce target genes transcription (Wondrak et al., 2008). Similarly, *trans*-cinnamaldehyde can prevent oxidative stress through Nrf2 activation. Choi (2020) evaluated the role of the Nrf2 pathway in response to H_2O_2 (1 mM) induced oxidative damage of pre-treated (30 M *trans*-cinnamaldehyde, 1 hr) Chinese hamster lung fibroblasts for 24 hr. HO-1activity and phosphorylated Nrf2 levels increased and ROS decreased, but ROS increased when HO-1 was blocked using an inhibitor (ZnPP, 10 μM). Also, the expression of Keap1, the negative regulator of Nrf2, was lower after cinnamaldehyde treatment. These studies suggest that cinnamaldehyde and its analogs exert their antioxidant effect through the Nrf2 pathway, which positively regulates HO-1 enzyme expression.

12.3.2 Antidiabetes

Diabetes mellitus is characterized by the presence of chronic hyperglycemia, with alterations in the metabolism of macromolecules that result in limiting/obstructing insulin secretion or action (Lastra, Syed, Kurukulasuriya, Manrique, & Sowers, 2014). Insulin resistance and its secretion deficit should be considered in the etiology of T2DM (Lipscombe et al., 2016). Therefore, the antidiabetic effect of these plant products is reviewed accounting for the mechanisms of components that modulate these parameters.

12.3.2.1 Coffee

The molecular mechanisms involved in the antidiabetic potential of coffee is related to the modulation of glucose metabolizing-enzymes, increased glucose tolerance and insulin sensitivity, and reduced glucose storage (Figure 12.2). Hence, the regulation of key enzymes of glucose and lipid metabolism by trigonelline suppresses liver fatty acid synthase and increases glucokinase and carnitine palmitoyl activities in diabetic rats. This results in significant blood glucose, cholesterol, and triglycerides reduction and enhanced insulin sensitivity and antioxidant defense system in the pancreas of diabetic rats when treated with trigonelline (Zhou, Zhou, & Zeng, 2013). Likewise, chlorogenic acid (80 mg/kg for 12 weeks) modulates the activation of adenosine monophosphate (AMP) and activated protein kinase (AMPK) in the liver and skeletal muscle of diabetic mice (Jin et al., 2015). This regulates lipid and glucose metabolism by stimulating liver acetyl coenzyme A carboxylase and peroxisome proliferator-activated receptor alpha (PPAR-α), suppression and deactivation of glucose-6-phosphatase dehydrogenase, and the upregulation of the glucose transporter type 4 (GLUT4) in muscle cells. Thus, chlorogenic acid induces significant reduction of fasting plasma glucose, glycosylated hemoglobin, and the percentage of body fat in a diabetic mouse model (Jin et al., 2015). Similarly, chlorogenic acid (90 mg/kg/day) improved fasting glucose, glucose tolerance, and fasting serum insulin concentrations by upregulating sodium/glucose cotransporter 1 (SGLT1), glucose transporter type 2 (GLUT2), and

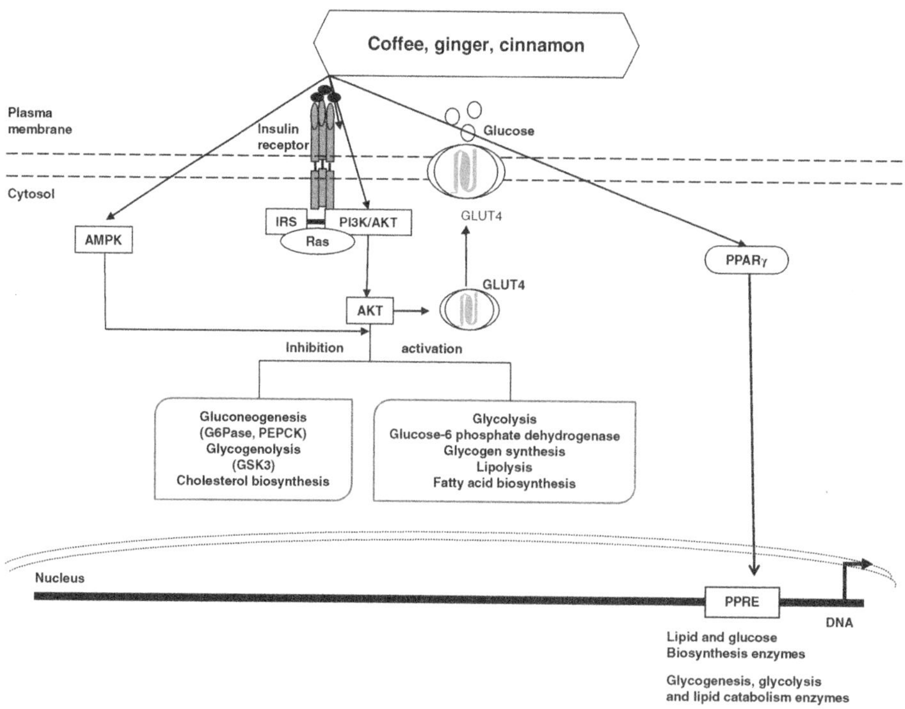

Figure 12.2 Antidiabetic mechanisms induced by phytochemicals of coffee, ginger, and cinnamon involve the insulin receptor, PI3K/AKT. and AMPK signaling pathways that reduce glucose level by inducing glucose uptake, glycolysis, glycogen synthesis, fatty acid biosynthesis, and by inhibiting carbohydrate and cholesterol biosynthesis.

proglucagon expression in the intestinal segments of a high-fat diet rat model (Peng et al., 2015). Finally, the residue obtained after coffee brewing, known as spent coffee grounds, exhibits antidiabetic potential by inhibiting α-glucosidase activity (Vázquez-Sánchez et al., 2018); likewise, the antioxidant and anti-inflammatory properties of spent coffee grounds are also important factors associated with its antidiabetic potential (Hernández-Arriaga, Oomah, & Campos-Vega, 2017; López-Barrera, Vázquez-Sánchez, Loarca-Piña, & Campos-Vega, 2016).

12.3.2.2 Ginger

The hypoglycemic effect of ginger is directly or indirectly related to the regulation of several signaling pathways related to insulin or insulin sensitivity (Figure 12.2). AMPK and phosphatidylinositol 3-kinase (PI3K)-protein kinase B (Akt) signaling pathway modulates insulin receptor expression, which significantly improves glucose uptake by activating the glucose transporter GLUT4 responsible for glucose entry into cells and circulating lipid levels reduction (Kord, Pourrajab, & Hekmatimoghaddam, 2020). Ginger also regulates the release/action of insulin (Mustafa, Eid, Jafri, El-Latif, & Ahmed, 2007), inhibits α-amylase and α-glucosidase synthesis (Wongsa, Chaiwarit, & Zamaludien, 2012), and enzymes related to gluconeogenesis and glycogenolysis (Jafarnejad et al., 2017).

12.3.2.3 Cinnamon

Cinnamon extract affects glycogen synthase, insulin receptor, GLUT4, glucose-6-phosphatase (G6Pase), glucose transporter 1 (GLUT1), PTP1B, pyruvate kinase (PK), phosphoenolpyruvate carboxy-kinase (PEPCK), and α-glucosidase *in vitro* (Cao, Polansky, & Anderson, 2007; Zhu et al., 2017).

The effect of cinnamon on GLUT4 and PTP1B was demonstrated in mouse adipocytes (3T3-L1) exposed to four polyphenol-rich fractions (1–10 μg/mL) and an aqueous extract (10 μg/mL) of cinnamon (*Cinnamomum burmannii*) (Cao et al., 2007). Insulin receptor levels increased after treatment with the fractions and GLUT4 levels increased after 3 hr treatment with one of the fractions and with the aqueous extract. The above indicates that cinnamon extracts and fractions favor glucose uptake by positively regulating the GLUT4 receptor and enabling greater insulin receptor phosphorylation by PTP1B inhibition and thus improving insulin sensitivity (Cao et al., 2007; Zhu et al., 2017).

The main cinnamon components exhibiting insulin-enhancing activity are pro-cyanidin type-A polymers present in several cinnamon species, corresponding to (+) catechin and (−) epicatechin (Anderson et al., 2004; Sun et al., 2019). Therefore, the effect of procyanidin C1 was evaluated on the regulation of insulin sensitivity of 3T3-L1 adipocytes. This polymer induced glucose entry into adipocytes at 6.25, 12.5, and 25 μM, and this effect was greater when combined with insulin through Akt phosphorylation and translocation of GLUT4 transporters (Sun et al., 2019).

Another cinnamon component, methylhydroxychalcone polymer is presumed to exert hypoglycemic action at 0.05 and 0.1 mg/mL. It induced insulin receptor autophosphorylation in 3T3-L1 adipocytes and elevated glucose entry, glucose incorporation, and glycogen synthesis in 3T3-L1 cells; this increase was enhanced when combined with insulin. Glycogen production decreased when cells were treated with a PI3K inhibitor, insulin, and polymer, indicating the role of this pathway in polymer-mediated effects. In addition, methylhydroxychalcone (0.1 mg/mL) stimulated GS activity and reduced the activity of GSK-3 enzyme, an inhibitor of GS phosphorylation (Jarvill-Taylor et al., 2001).

Anand et al. (2010) evaluated the mechanism of action of the antidiabetic effect of cinnamaldehyde in streptozotocin (STZ)-induced diabetic male Wistar rats and 7 days later received cinnamaldehyde at 20 mg/kg orally for 60 days. Serum insulin levels increased and fasting blood glucose and HbA1c (glycated hemoglobin) decreased in the treated group compared to the control group; it also stimulated insulin release in the isolated islets of the treated rats and extensively reduced glycogen in muscle and liver tissue compared to untreated mice. Cinnamaldehyde-induced GLUT4 translocation since GLUT4 levels decreased in the membrane of skeletal muscle tissue, the PK activity and mRNA increased in both liver and kidney from rats that received cinnamon compared to the control group, whereas the activity and expression of PEPCK enzyme decreased.

12.3.3 Cardiovascular Protection

Coffee, ginger, and cinnamon contain bioactive compounds that exert cardioprotective effects through antioxidant actions and blood lipid level-lowering effects. The mechanisms responsible remain somewhat unclear, although the available clinical and preclinical studies generally suggest a positive correlation between their intake and significant reduction in the cardiovascular risk factors (Wan et al., 2013).

12.3.3.1 Coffee

Molecular mechanism of chlorogenic acid (1–10 mg/kg) from coffee was demonstrated to play an important role in controlling cardiovascular disease (CVD) risk-related factors through cholesterol-lowering and fatty liver-attenuating properties by upregulating PPARα in hypercholesterolemic rats (Wan et al., 2013). Similarly, treatments with 30 to 90 mg/kg of 5-caffeoylquinic acid (5-CQA) modulate gene expression of proteins involved in hepatic lipid metabolism (LXR-α and PPARα) of Sprague–Dawley rats fed a high-fat diet, improving serum lipids and free fatty acids levels (Huang et al., 2015). Bhandarkar et al. (2019) evaluated 5-CQA potential to improve cardiovascular, liver, and metabolic responses in a rat model of metabolic syndrome; 5-CQA (100 mg/kg/day, 8 weeks) decreased body weight and visceral fat accumulation, and improved heart and liver structure and function.

The anti-inflammatory potential of coffee-derived compounds was evaluated using human umbilical vein endothelial cells subjected to cyclic strain and treated with 3 to 10 μM cafestol (Hao et al., 2018). Cafestol upregulated HO-1 and Sirtuin 1 (SIRT1) expression and, controlled ROS levels and MAPK phosphorylation, downregulating cyclic-strain-induced intracellular adhesion molecule 1 (ICAM-1), monocyte chemoattractant protein 1 (MCP-1), and IL-8 protein secretion (Hao et al., 2018).

12.3.3.2 Ginger

Ginger extracts inhibit hepatic fatty acids and thus novo triglyceride synthesis, reduce carbohydrate response element binding protein (ChREBP), lipogenic genes (acetyl-CoA carboxylase −1 [ACC], fatty acid synthase [FAS], stearoyl CoA desaturase −1 [SCD], and glucose 6-phosphatase [G6Pase]) and key enzymes in lipid metabolism (Lai et al., 2016).

Regarding cholesterol reduction, ginger increases bile production that favors cholesterol increase and phospholipids excretion (Arablou et al., 2014) by downregulating 3-hydroxy-3-methylglutaryl coenzyme-A reductase (HMGCR) expression (Wang et al., 2017). Ginger inhibits HMGCR action and activates LDL receptors (Srinivasan 2014), simultaneously reducing serum levels of PGE-2 and thromboxane B2 triggering an antithrombotic effect (Morvaridzadeh et al., 2020). Improvement in bile salt secretion and peroxisome proliferator-activated receptor delta factor (PPARδ) stimulation regulate peroxisome proliferator-activated receptor gamma factor (PPARγ) that contributes to glucose homeostasis, lipid metabolism regulation, and prevention of oxidative stress (De las Heras et al., 2017).

Ginger inhibits angiotensin converting enzyme (ACE) activity and prevents lipid peroxidation in the heart (Aludatt et al., 2016). Similarly, it lowers blood pressure by blocking calcium channels and prevents platelet aggregation (Marx et al., 2015). The bioactive compounds 6-gingerol, 8-gingerol, and 6-shogaol have the ability to lower blood pressure through various mechanisms, either independently or synergistically by blocking calcium channels (Cai et al., 2015; Townsend et al., 2013), inhibiting the renin-angiotensin system, particularly ACE activity (Bare, Mansur, Pili, & Helvina, 2020), relaxing activity on smooth muscle such as the aorta (Townsend et al., 2014), elevating nitric oxide production and fibrinolysis, synthetic thromboxane levels, and LDLs oxidation reduction (Ghareib et al., 2015). In addition, 6-gingerol affects key enzymes in lipid metabolism such as cholesteryl ester transfer protein (CEPT), FAS, ACC, HMGCR, lecithin cholesterol acyltransferase (LCAT), and lipoprotein lipase (Srinivasan 2014). It also prevents atherosclerosis by inducing autophagy via PI3K/Akt/mTOR and attenuates cholesterol synthesis and lipid accumulation via AMPK (Kiyama, 2020). On the other hand, 6-gingerol inhibits transforming growth factor beta (TGF-β) induced biglycan synthesis, suggesting atherosclerosis prevention (Micucci et al., 2020).

12.3.3.3 Cinnamon

Cinnamaldehyde, 2-methoxycinnamaldehyde, and epicatechin are the main cinnamon components that have been investigated for their cardioprotective effect and mechanism (Zuo et al., 2017; Hwa et al., 2012; Cheng et al., 2017).

Cinnamaldehyde (40 mg/kg) was administered for 8 weeks to induced (12 weeks high-fat diet) obese male C57BL/6J mice (Zuo et al., 2017). In these mice, cinnamaldehyde induced white adipose tissue (WAT) browning positively impacting obesity, and accordingly increased UCP1 expression, a marker protein of brown adipose tissue and PGC-1, a protein that promotes its expression. WAT transition to brown adipose tissue (BAT) was also promoted by cinnamaldehyde by inducing PPARα expression which is one of the main adipogenesis regulators and one of its coactivators corresponding to the protein PR-domain containing 16 (PRDM16). The antiadipogenic effects of cinnamaldehyde have been attributed to its ability to modulate PPAR-γ and the AMPK signaling pathway. This was demonstrated in high-fat diet induced obese mice receiving cinnamaldehyde (20 and 40 mg/kg for 1 month). Gene expression of PPAR-γ, C/EBPRα, SREBP1, and FAS decreased in epididymal fat tissue and phosphorylated AMPK levels increased compared to the control group. The antiobesity effect of cinnamaldehyde was purportedly due to the negative regulation that it exerts on PPAR-γ and C/EBPRα and to the AMPK induction pathway that participates in cholesterol and lipid homeostasis (Huang et al., 2011).

2-Methoxycinnamaldehyde isolated from *Cinnamomum cassia* was intravenously administered (100 µg/kg) to adult male rats; 10 min later ischemia was induced by blocking the coronary artery followed by induced reperfusion (Hwa et al., 2012). This compound protects against myocardial dysfunction by suppressing high mobility group box 1 (HMGB1) involved in inflammation and attenuating the levels of damage markers. Cinnamaldehyde (50 µM) also induced Nrf2 translocation and increased the levels of one of its target genes, HO-1, in endothelial cells that inhibits HMGB1 release by exerting both anti-inflammatory and antioxidant actions.

(−)-Epicatechin can also regulate blood lipid levels in hyperlipidemic rats by modulating the Insig-1-SREBP-cleavage-activating protein (SCAP) pathway, which regulates cellular lipid metabolism (Cheng et al., 2017). Epicatechin (10, 20 and 40 mg/kg) was administered for 12 weeks to male Sprague–Dawley with hyperlipidemia induced by a high-fat diet (Cheng et al., 2017). The levels of total cholesterol, triglycerides, and fat accumulation in the liver decreased in the group receiving this flavanol compared to the control group. The levels of SREBP-1c, Insig-1, and SCAP proteins in liver decreased in the eugenol-treated group compared to the control group; moreover, SIRT1, LXR-α expression decreased and FAS expression increased. LXR-α is important because it activates SREBP-1 transcription and can induce lipogenic transcription genes such as FAS and SIRT1 that inhibits SREBP-1c activity (Cheng et al., 2017).

12.3.4 Hepatoprotection

The liver is exposed to numerous compounds that can induce hepatocellular damage associated with oxidative stress, steatosis, cirrhosis, and jaundice that modify biochemical markers due to the components ability to modulate different signaling pathways that upregulate or downregulate gene expressions (Figure 12.3), depending on the cellular target (Ganesan, Jayachandran, & Xu, 2018).

12.3.4.1 Coffee

High coffee consumption is associated with reduced liver fibrosis markers such as serum levels of alpha-fetoprotein (AFP), aspartate aminotransferase/alanine aminotransferase ratio (AST/ALT), and the degree of liver steatosis. Likewise, the relative risks were 1.11 (95% CI: 0.76–1.61) for <1 cup per day, 0.70 (95% CI: 0.48–1.02) for 1–3 cups daily, and 0.47 (95% CI: 0.27–0.85) for ≥3 cups per day, suggesting a protective association between coffee consumption and the development of liver fibrosis (Freedman et al., 2009).

The hepatoprotective potential of coffee has been attributed mainly to caffeine, chlorogenic acid, and kahweol. Caffeine (0–10 nM) attenuates the progression of liver fibrosis in an *in vitro* model of hepatic stellate cells (HSC), by inhibiting the adhesion and activation of HSC through the regulation of procollagen type Ic and α-SMA expression. Likewise, in a mouse model of carbon tetrachloride-induced liver fibrosis, caffeine (37.5 mg/kg/day for 14 days) reduced ALT and AST, oxidative stress biomarkers, and CRP (Elmotasem, Farag, & Salama, 2018).

Chlorogenic acid treatment (60 mg/kg/day for 8 weeks) attenuates the degree of liver fibrosis by suppressing fibrogenic genes (α-SMA, type I collagen, type III collagen, and metalloproteinase inhibitor 1) and cytochrome P450 2E1 in rat models with carbon tetrachloride-induced liver fibrosis. Moreover, chlorogenic acid treatment increases GSH, SOD, and CAT levels and

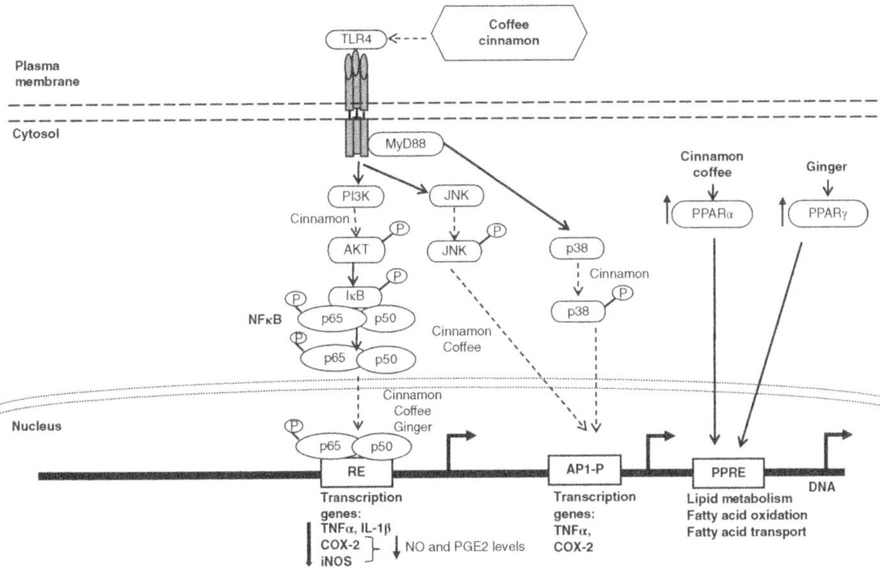

Figure 12.3 Cardiovascular protection, hepatoprotective, and immunomodulatory properties of bioactive compounds from coffee, ginger, and cinnamon modulate various cellular functions and signaling pathways such as PI3K/AKT, NF-kB, MAPKs (JNK, p38), and PPARs (α, γ). These signaling pathways modulate the transcription of genes involved in the inflammatory response such as the cytokines TNF-α and IL-1β, the pro-inflammatory enzymes (COX-2, iNOS), in lipid, fatty acid, and lipoprotein metabolism. *Source:* Based on Davari et al. (2020) and Kiyama (2020).

also stimulates nuclear Nrf2 translocation resulting in upregulating HO-1, NQO1, and GCLC genes (Shi et al., 2016). Likewise, chlorogenic acid counteracts acetaminophen-induced liver injury by downregulating cell-death related mechanism, maintenance of GSH and Trx (thioredoxin) homeostasis, and the modulation of signaling pathways, such as NF-κB, MAPK (mitogen-activated protein kinase), and TL3/4 (toll-like receptor) involved in the progression of hepatic alterations (Ji et al., 2013; Zheng, Sheng, Lu, & Ji, 2015) summarized in Figure 12.3.

Kahweol (40 μM) modulates NF-κB activation and proinflammatory cytokines expression in primary Kupffer cells and primary hepatocyte cultures exposed to lipopolysaccharide (LPS) (Seo et al., 2018). These results are supported by TGFβ and Jun N-terminal kinase (JNK) modulation (Figure 12.3) and reduced connective tissue growth factor and type I collagen in a thioacetamide (TAA)-induced hepatic fibrosis mouse model, treated with 75 mg/kg of kahweol for 8 weeks (Seo et al., 2018).

12.3.4.2 Ginger

Regarding the hepatoprotective effects, Lai et al. (2016) evaluated the protective effect and mechanism of action of ginger essential oil against nonalcoholic fatty liver disease (NAFLD), development in C57BL/6J mice administered high-fat diets supplemented with ginger essential oil (12.5 and 125 mg/kg) for 12 weeks. This treatment reduced glucose, insulin, free fatty acids, triglycerides, and total cholesterol levels. The molecular mechanism involved is suppression of different proteins of lipid metabolism such as sterol regulatory element-binding protein-1 (SERBP-1), ACC, FAS, HMGCR, and cytochrome p450-2E1

(CYP2E1). Furthermore, liver enzymes such as AST and ALT levels decreased by 44 and 86%, respectively. Histologically, consumption of this essential oil reduced the liver weight by 47%, indicating significant reduction in lipid accumulation. This study also showed significant increase in liver antioxidant capacity due to increased GSH, SOD, CAT, GPx, and glutathione reductase (GR) levels.

12.3.4.3 Cinnamon

Cinnamon can prevent NAFLD, alcoholic liver disease, steatosis, and protect against ischemia/reperfusion (Eidi, Mortazavi, Bazarga, & Zaringhalam, 2012; Golshahi, Araghi, Baghban, & Farzad-Mohajeri, 2019; Kanuri et al., 2009). Hepatoprotective benefit was observed for a commercial ethanol extract of cinnamon bark in steatosis (ethanol)-induced mice (Kanuri et al., 2009). Few studies have evaluated the mechanism by which cinnamon or its components exert their hepatoprotective effect, and until now the possible mechanism of 2-methoxycinnamaldehyde and (−)-epicatechin has been studied. Golshahi et al. (2019) administered 2-methoxycinnamaldehyde (1 mL/kg) in commercial powder and evaluated markers of cell death, liver function, and inflammation in 24 male rats with ischemia/reperfusion (IRI) induced liver. AST, ALT, malondialdehyde, and proinflammatory cytokines (tumor necrosis factor-alpha [TNF-α] and interleukin-6 [IL-6]) levels decreased compared to the group not receiving 2-methoxycinnamaldehyde. The histopathological analysis of rat liver revealed that unlike the control group, one of the markers of apoptosis, cleaved caspase 3 increased and NF-kB (targets genes involved in inflammation) levels decreased. This suggests that cinnamon protects against IRI induced damage by modulating cytokines and transcription factors involved in inflammation and apoptosis.

According to Cheng et al. (2017), (−)-epicatechin can explain the positive impact of cinnamon in animal models on hepatic steatosis. Hyperlipidemia-induced 72 male Sprague–Dawley rats were randomly assigned to six groups: the control group without hyperlipidemia, the group with hyperlipidemia that did not receive epicatechin, the other groups with hyperlipidemia received (10, 20 or 40 mg/kg) epicatechin for 12 weeks. The levels of proinflammatory cytokines: IL-2, IL-1β, and IL-6 in the serum and the expression of genes related to lipid metabolism including LXR-α and FAS were lower in the epicatechin groups than the control group; SIRT1 levels increased and hepatic fat accumulation decreased without significant morphological changes in liver cells in the epicatechin group compared to the control group.

12.3.5 Immunomodulation and Anti-Inflammation

This section focuses on different studies demonstrating the potential molecular signaling of anti-inflammatory effects induced with extracts or phytochemicals from coffee, ginger, and cinnamon that control several immune cells and involves inflammatory mediators such as IL-1α, IL-6, TNF-α, NO, and PGE-2 among others (Davari et al., 2020; Kiyama, 2020) summarized in Figure 12.3.

12.3.5.1 Coffee

Chlorogenic acid (60 mg/kg/day for 8 weeks) from coffee activated the TLR4/MyD88/NF-κB signaling pathway, by regulating NF-κB nuclear translocation and subsequently suppressing proinflammatory cytokines, in a rat model of carbon tetrachloride-induced liver fibrosis; this results in the downregulation of TLR4, myeloid differentiation factor 88 (MyD88), iNOS, and COX-2 expression (Shi et al., 2013). This confirms a previous *in vitro* model of LPS-induced inflammatory response using RAW 264.7 macrophage cells where chlorogenic acid (37.5 μg/mL) significantly decreased COX-2 expression, PGE2 release, and the activation of NF-κB and JNK/AP-1 pathways (Shan et al., 2009). Additionally, pretreatments with chlorogenic acid (30 mg/kg) suppressed serum TNF-α and IFN-γ production, the infiltration of hepatic macrophages, neutrophils and activated $CD4^+$ T lymphocytes, and TL4 signaling in Concanavalin A-induced hepatitis mouse models (Yuan et al., 2017). Likewise, intraperitoneal administration of chlorogenic acid (100 mg/kg) suppressed macrophage markers genes in fat tissue of high-fat diet obese mice, as well as proinflammatory molecules, such as MCP-1 and TNF-α (Ma et al., 2015); similarly, chlorogenic acid (1 mM administered orally for 15 days) improved colitis histologic scores and reduced macrophage inflammatory protein 2 and IL-1β after colitis induction with dextran sulfate sodium in C57BL/6 mice (Shan et al., 2009).

Other coffee constituents also exhibit immunomodulatory properties, including the modulation of NF-κB activation in LPS-induced macrophages by independent treatments with 0–2 mM caffeine, 0–5 μM kahweol, and 0–10 μM cafestol, suppressing the upregulation of iNOS, COX-2, and TNF-α expression induced by LPS (Kang et al., 2012; Kim et al., 2004). Additionally, Akt and extracellular signal-regulated kinase (ERK) phosphorylation inhibited by caffeine suggests that its anti-inflammatory properties are mediated by Akt-dependent NF-κB and ERK pathways (Kang et al., 2012).

12.3.5.2 Ginger

The constituents of ginger can regulate the most important step in the NF-κB signaling pathway through IKK activation, inhibition of IkBα degradation, and the nuclear transmission subunit p65 of NF-κB (Figure 12.3). Gingerol and 6-dihidroparadol inhibit iNOS and NO expressions in macrophages and downregulate the synthesis of proinflammatory markers such as TNFα, IL-6, IL-1β, and IL-12 (Kiyama, 2020). Another important component of ginger, paradol, is a potent inhibitor of COX-1 activity and exhibits significant antiplatelet effect compared to aspirin. This effect is attributed to the regulation of arachidonic acid metabolism at the COX-1 synthase/thromboxane system and to the inhibition of lipoxygenase activity (Ali et al., 2008; Srinivasan 2014; Tasneem, Liu, Li, Choudhary, & Wang, 2019).

A recent study demonstrated the safety and efficacy of a ginger (*Zingiber officinale*) rhizome (supercritical) extract, OLNP-06 against functional dyspepsia (FD) that considerably impacts quality of life, healthcare burden, and productivity loss (Panda, Nirvanashetty, Parachur, Krishnamoorthy, & Dey, 2020). OLNP-06 (standardized to 26% of total gingerols containing 6-, 8-, and 10-gingerol; 6-, 8-, and 10-shogoal) significantly reduced functional dyspeptic symptoms and improved all functional dyspepsia parameters in subjects suffering from FD as early as 14 days of supplementation compared to placebo in a randomized, double-blind, placebo-controlled, parallel-group, comparative clinical study ($n = 48$, 18–55 years old; 200 mg OLNP-06 twice daily for 28 days). Moreover, three major symptoms (postprandial fullness, upper abdominal bloating, and early satiation) improved significantly in the OLN-06 group compared to placebo (54 vs 13%) on day 28 and in total dyspeptic score (71–77 vs 28–38%) at the end of the study period (48 days).

12.3.5.3 Cinnamon

Very few studies have evaluated the effect of cinnamon supplementation on cytokines, transcription factors or other proteins involved in inflammation. Davari et al. (2020) randomly assigned 39 T2DM individuals to receive 3 g of cinnamon extract daily or a placebo for 8 weeks in a double-blind, randomized, controlled clinical trial. No significant changes occurred in levels of hs-PCR, TNF-α, NF-kB, IL-6, and the deacetylase (SIRT1) that regulates the half-life of several transcription factors, such as NF-kB.

The aqueous extract of cinnamon (*C. cassia*) bark (50 and 100 mg/mL) reduced IkBα phosphorylation, NF-kB inhibitor in LPS-stimulated macrophages, indicating that it does not enhance NF-kB activation through the induction of IkBα degradation. The evaluation of proteins downstream of the NF-kB pathway reduced JNK phosphorylation, p38, and ERK; it suggests that the anti-inflammatory effect of the extract is exerted through the regulation of the MAP kinase signaling pathway (Hong et al., 2012).

The immunomodulatory effect of the aqueous extract of *C. cassia* bark was evaluated in male BALB/c mice after 7 days treatment (Lee, Kim, Cho, Sohn, & Kang, 2011). The extract (0.2, 0.5, and 1 g/kg) reduced serum IFN-γ after T cells stimulation with anti-CD3 compared to the control. IFN-γ, IL-4, and IL-2 (mRNA and serum) levels decreased (except IL-2 increased in treated mice) in the splenocytes isolated from mice treated or not with the extract, and subsequently simulated with anti-CD3 mAb. The effect of the extract on proteins was evaluated since TCR/CD3 signaling leads to protein kinase C (PKC) activation which activates NF-kB and the pathway of the MAPK family. The extract inhibited the phosphorylation of JNK, p38, and ERK1/2 and reduced STAT4 phosphorylation, which could explain the suppression and IFN-γ levels. The results suggest that the extract can modulate the inflammatory response of T cells without affecting IL-2 production.

The immunomodulatory and anti-inflammatory effect of the aqueous extract of *C. cassia* bark was also studied in a model of experimental inflammatory bowel disease such as colitis, as well as in a macrophage cell line (Raw 264.7), antigen-presenting cells, and dendritic cells. Treatment with the extract (0.2 mg/mL) inhibited macrophage maturation, evidenced by reduced levels of activation markers corresponding to MHCII, and costimulatory molecules including B7.1, B7.2, and ICOS ligand. Likewise, the proinflammatory cytokines IL-1β and TNF-α and the COX-2 enzyme were suppressed compared to the control, with similar results in LPS-stimulated macrophages. This indicates that the anti-inflammatory effect of cinnamon extract is mediated by the inhibition of the activation and maturation of antigen-presenting cells (APC) (Kwon et al., 2011).

12.4 Conclusions

This chapter summarizes the antioxidant, antidiabetic, cardioprotective, hepatoprotective, anti-inflammatory, and immunomodulatory efficacy of coffee, ginger, and cinnamon. Coffee, ginger, and cinnamon can increase insulin sensitivity and regulate glucose metabolism; prevent cardiovascular disease by optimizing fatty acid, cholesterol, and lipoprotein levels, as well as blood pressure, which contributes to improve metabolic syndrome, prevent heart disease, and alleviate inflammatory processes associated with obesity, diabetes, arthritis, and inflammatory bowel diseases. The physiological effects and health benefits are consistent with the activation of molecular mechanisms triggered by the phytochemicals present in beverages, extracts, infusions, and/or essential oils prepared from roasted coffee beans, ginger rhizome, and cinnamon bark, which have been evidenced with *in vitro* and *in vivo* studies to validate what has been reported by traditional medicine and several clinical studies. This chapter, therefore, can serve as a starting point for further studies of the molecular mechanisms underlying the nutraceutical and functional potential of these natural products, as well as to validate their use alone or in combination with existing therapies.

References

Ali, B., Blunden, G., Tanira, M., & Nemmar, A. (2008). Some phytochemical, pharmacological and toxicological properties of ginger (*Z. officinale* Roscoe): a review of recent research. *Food Chemistry and Toxicology*, *46*(2), 409–420. https://doi.org/10.1016/j.fct.2007.09.085

Al-Numair, K. S., Ahmed, S., Ahmad, D., & Al-Assaf, A. H. (2007). Nutritive value, levels of polyphenols and anti-nutritional factors in Sri Lankan cinnamon (*C. zeylanicum*) and Chinese cinnamon (*C. cassia*). *Food Science Agricultural Research*, *154*, 5–21.

Aludatt, M. H., Rababah, T., Alhamad, M. N., Gammoh, S., Ereifej, K., Johargy, A., . . . Rawashdeh, M. (2016). Optimization of phenolic content, antioxidant, and inhibitory activities of α-glucosidase and angiotensin converting (ACE) enzymes from Z. officinale Z. *International Journal of Food Properties*, *19*(6), 1303–1316. https://doi.org/10.1080/10942912.2015.1063066

Anand, P., Murali, K. Y., Tandon, V., Murthy, P. S., & Chandra, R. (2010). Insulinotropic effect of cinnamaldehyde on transcriptional regulation of pyruvate kinase, phosphoenolpyruvate carboxykinase, and GLUT4 translocation in experimental diabetic rats. *Chemico-Biological Interactions*, *186*(1), 72–81. https://doi.org/10.1016/j.cbi.2010.03.044

Anderson, R. A., Broadhurst, C. L., Polansky, M. M., Schmidt, W. F., Khan, A., Flanagan, V. P., . . . Graves, D. J. (2004). Isolation and characterization of polyphenol type-A polymers from cinnamon with insulin-like biological activity. *Journal of Agricultural and Food Chemistry*, *52*(1), 65–70. https://doi.org/10.1021/jf034916b

Arablou, T., & Aryaeian, N. (2018). The effect of ginger (*Zingiber Officinale*) as an ancient medicinal plant on improving blood lipids. *Journal of Herbal Medicine*, *12*, 11–15. doi.org/10.1016/j.hermed.2017.09.005

Arablou, T., Aryaeian, N., Valizadeh, M., Sharifi, F., Hosseini, A., & Djalali, M. (2014). The effect of ginger consumption on glycemic status, lipid profile and some inflammatory markers in patients with type 2 diabetes mellitus. *International Journal of Food Sciences and Nutrition*, *65*(4), 515–520. https://doi.org/10.3109/09637486.2014.880671

Azimi, P., Ghiasvand, R., Feizi, A., Hariri, M., & Abbasi, B. (2014). Effects of cinnamon, cardamom, saffron, and ginger consumption on markers of glycemic control, lipid profile, oxidative stress, and inflammation in type 2 diabetes patients. *Review of Diabetic Studies: RDS*, *11*(3–4), 258–266. https://doi.org/10.1900/RDS.2014.11.258

Baird, L., & Yamamoto, M. (2020). The molecular mechanisms regulating the KEAP1-NRF2 pathway. *Molecular and Cellular Biology*, *40*(13), e00099–e00120. https://doi.org/10.1128/MCB.00099-20

Bare, Y., Mansur, S., Pili, A. P., & Helvina, M. (2020). Analysis of molecular interactions of 8-gingerol compounds in Ginger (*Z. officinale*) as ACE Inhibitor. *Bioeduscience*, *4*(2), 183–187. https://doi.org/10.22236/j.bes/424944

Bhandarkar, N. S., Brown, L., & Panchal, S. K. (2019). Chlorogenic acid attenuates high-carbohydrate, high-fat diet–induced cardiovascular, liver, and metabolic changes in rats. *Nutrition Research*, *62*, 78–88. https://doi.org/10.1016/j.nutres.2018.11.002

Bhattarai, S., Tran, V. H., & Duke, C. C. (2001). The stability of gingerol and shogaol in aqueous solutions. *Journal of Pharmaceutical Sciences*, *90*(10), 1658–1664. https://doi.org/10.1002/jps.1116

Boettler, U., Sommerfeld, K., Volz, N., Pahlke, G., Teller, N., Somoza, V., . . . Marko, D. (2011). Coffee constituents as modulators of Nrf2 nuclear translocation and ARE (EpRE)-dependent gene expression. *The Journal of Nutritional Biochemistry*, *22*(5), 426–440. https://doi.org/10.1016/j.jnutbio.2010.03.011

Cai, Z. X., Tang, X. D., Wang, F. Y., Duan, Z. J., Li, Y. C., Qiu, J. J., & Guo, H. S. (2015). Effect of gingerol on colonic motility via inhibition of calcium channel currents in rats. *World Journal of Gastroenterology*, *21*(48), 13466–13472. https://doi.org/10.3748/wjg.v21.i48.13466

Cao, H., Polansky, M. M., & Anderson, R. A. (2007). Cinnamon extract and polyphenols affect the expression of tristetraprolin, insulin receptor, and glucose transporter 4 in mouse 3T3-L1 adipocytes. *Archives of Biochemistry and Biophysics*, *459*(2), 214–222. https://doi.org/10.1016/j.abb.2006.12.034

Chen, H., Fu, J., Chen, H., Hu, Y., Soroka, D. N., Prigge, J. R., . . . Sang, S. (2014). Ginger compound [6]-shogaol and its cysteine-conjugated metabolite (m2) activate Nrf2 in colon epithelial cells *in vitro* and *in vivo*. *Chemical Research in Toxicology*, *27*(9), 1575–1585. https://doi.org/10.1021/tx500211x

Cheng, H., Xu, N., Zhao, W., Su, J., Liang, M., Xie, Z., . . . Li, Q. (2017). (−)-Epicatechin regulates blood lipids and attenuates hepatic steatosis in rats fed high-fat diet. *Molecular Nutrition and Food Research*, *61*(11). https://doi.org/10.1002/mnfr.201700303

Cheng, Y. C., Sheen, J. M., Hu, W. L., & Hung, Y. C. (2017). Polyphenols and oxidative stress in atherosclerosis-related ischemic heart disease and stroke. *Oxidative Medicine and Cellular Longevity*, *2017*, 8526438. https://doi.org/10.1155/2017/8526438

Choi, Y. H. (2020). Trans-cinnamaldehyde prevents oxidative stress-induced apoptosis in V79-4 Chinese hamster lung fibroblasts through the Nrf2-mediated HO-1 activation. *Biological and Pharmaceutical Bulletin*, *43*(11), 1707–1714. https://doi.org/10.1248/bpb.b20-00407

Da Costa, R. M., Rodrigues, D., Pereira, C. A., Silva, J. F., Alves, J. V., Lobato, N. S., & Tostes, R. C. (2019). Nrf2 as a potential mediator of cardiovascular risk in metabolic diseases. *Frontiers in Pharmacology*, *10*, 382. https://doi.org/10.3389/fphar.2019.00382

Davari, M., Hashemi, R., Mirmiran, P., Hedayati, M., Sahranavard, S., Bahreini, S., . . . Talaei, B. (2020). Effects of cinnamon supplementation on expression of systemic inflammation factors, NF-kB and Sirtuin-1 (SIRT1) in type 2 diabetes: a randomized, double blind, and controlled clinical trial. *Nutrition Journal*, *19*(1), 1. https://doi.org/10.1186/s12937-019-0518-3

De las Heras, N., Valero-Muñoz, M., Martín-Fernández, B., Ballesteros, S., López-Farré, A., Ruiz-Roso, B., & Lahera, V. (2017). Molecular factors involved in the hypolipidemic- and insulin-sensitizing effects of a ginger (*Z. officinale* Roscoe) extract in rats fed a high-fat diet.

Applied Physiology, Nutrition, and Metabolism, *42*(2), 209–215. https://doi.org/10.1139/apnm-2016-0374

Dhanik, J., Arya, N., & Nand, V. (2017). A Review on *Z. officinale*. *Journal of Pharmacognosy and Phytochemistry*, *6*(3), 174–184.

Dorri, M., Hashemitabar, S., & Hosseinzadeh, H. (2018). Cinnamon (*C. zeylanicum*) as an antidote or a protective agent against natural or chemical toxicities: a review. *Drug and Chemical Toxicology*, *41*(3), 338–351. https://doi.org/10.1080/01480545.2017.1417995

Eidi, A., Mortazavi, P., Bazargan, M., & Zaringhalam, J. (2012). Hepatoprotective activity of cinnamon ethanolic extract against CCI4-induced liver injury in rats. *EXCLI Journal*, *11*, 495–507.

Elmotasem, H., Farag, H. K., & Salama, A. A. (2018). *In vitro* and *In vivo* evaluation of an oral sustained release hepatoprotective caffeine loaded w/o Pickering emulsion formula–Containing wheat germ oil and stabilized by magnesium oxide nanoparticles. *International Journal of Pharmaceutics*, *547*(1–2), 83–96. https://doi.org/10.1016/j.ijpharm.2018.05.038

Farah, A., & Donangelo, C. M. (2006). Phenolic compounds in coffee. *Brazilian Journal of Plant Physiology*, *18*(1), 23–36. https://doi.org/10.1590/S1677-04202006000100003

Freedman, N. D., Everhart, J. E., Lindsay, K. L., Ghany, M. G., Curto, T. M., Shiffman, M. L., . . . HALT-C Trial Group. (2009). Coffee intake is associated with lower rates of liver disease progression in chronic hepatitis C. *Hepatology*, *50*(5), 1360–1369. https://doi.org/10.1002/hep.23162

Furlan, C. P. B., Valle, S. C., Marostica Junior, M. R., Ostman, E., Bjorck, I., & Tovar, J. (2019). Effect of bilberries, lingonberries and cinnamon on cardiometabolic risk-associated markers following a hypercaloric-hyperlipidic breakfast. *Journal of Functional Food*, *60*, 103443. https://doi.org/10.1016/j.jff.2019.103443

Ganesan, K., Jayachandran, M., & Xu, B. (2018). A critical review on hepatoprotective effects of bioactive food components. *Critical Reviews in Food Science Nutrition*, *58*(7), 1165–1229. https://doi.org/10.1080/10408398.2016.1244154

Ghareib, S. A., El-Bassossy, H. M., Elberry, A. A., Azhar, A., Watson, M. L., & Banjar, Z. M. (2015). 6-Gingerol alleviates exaggerated vasoconstriction in diabetic rat aorta through direct vasodilation and nitric oxide generation. *Drug Design, Development and Therapy*, *9*, 6019–6026. https://doi.org/10.2147/DDDT.S94346

Godlewska-Żyłkiewicz, B., Świsłocka, R., Kalinowska, M., Golonko, A., Świderski, G., Arciszewska, Ż., . . . Lewandowski, W. (2020). Biologically active compounds of plants: structure-related antioxidant, microbiological and cytotoxic activity of selected carboxylic acids. *Materials (Basel)*, *13*(19), 4454. https://doi.org/10.3390/ma13194454

Gökcen, B. B., & Sanlier, N. (2019). Coffee consumption and disease correlations. *Critical Reviews in Food Science and Nutrition*, *59*(2), 336–348. https://doi.org/10.1080/10408398.2017.1369391

Golshahi, H., Araghi, A., Baghban, F., & Farzad-Mohajeri, S. (2019). Protective effects of 2-methoxycinnamaldehyde an active ingredient of *C. cassia* on warm hepatic ischemia reperfusion injury in rat model. *Iranian Journal of Basic Medical Sciences*, *22*(12), 1400–1407. https://doi.org/10.22038/IJBMS.2019.13987

Górecki, M., & Hallmann, E. (2020). The antioxidant content of coffee and its *in vitro* activity as an effect of its production method and roasting and brewing time. *Antioxidants*, *10*(4), 308. https://doi.org/10.3390/antiox9040308

Govindarajan, V. S. (1982a). Ginger – chemistry, technology, and quality evaluation: part 1. *Critical Reviews in Food Science and Nutrition, 17*(1), 1–96. https://doi.org/10.1080/10408398209527347

Govindarajan, V. S. (1982b). Ginger – chemistry, technology, and quality evaluation: part 2. *Critical Reviews in Food Science and Nutrition, 17*(3), 189–258. https://doi.org/10.1080/10408398209527348

Gruenwald, J., Freder, J., & Armbruester, N. (2010). Cinnamon and health. *Critical Reviews in Food Science and Nutrition, 50*(9), 822–834. https://doi.org/10.1080/10408390902773052

Guasch-Ferré, M., Merino, J., Sun, Q., Fitó, M., & Salas-Salvadó, J. (2017). Dietary polyphenols, Mediterranean Diet, prediabetes, and Type 2 Diabetes: A narrative review of the evidence. *Oxidative Medicine and Cellular Longevity, 2017*, 6723931. https://doi.org/10.1155/2017/6723931

Hao, W. R., Sung, L. C., Chen, C. C., Chen, P. Y., Cheng, T. H., Chao, H. H., . . . Chen, J. J. (2018). Cafestol inhibits cyclic-strain-induced interleukin-8, intercellular adhesion molecule-1, and monocyte chemoattractant protein-1 production in vascular endothelial cells. *Oxidative Medicine and Cellular Longevity, 2018*, 7861518. https://doi.org/10.1155/2018/7861518

Harb, A. A., Bustanji, Y. K., Almasri, I. M., & Abdalla, S. S. (2019). Eugenol reduces LDL cholesterol and hepatic steatosis in hypercholesterolemic rats by modulating TRPV1 receptor. *Scientific Reports, 9*(1), 14003. https://doi.org/10.1038/s41598-019-50352-4

Harliansyah, A., Azian, M., Zurinah, W., & Yusof, Y. (2005). Effects of Z. officinale on superoxide dismutase glutathione peroxidase, catalase and glutathione content in HEPG2 cell line. *Malaysian Journal of Biochemistry and Molecular Biology, 11*(1), 36–41.

Hernández-Arriaga, A. M., Oomah, B. D., & Campos-Vega, R. (2017). Microbiota source impact *in vitro* metabolite colonic production and anti-proliferative effect of spent coffee grounds on human colon cancer cells (HT-29). *Food Research International, 97*, 191–198. https://doi.org/10.1016/j.foodres.2017.04.004

Higgins, L. G., Cavin, C., Itoh, K., Yamamoto, M., & Hayes, J. D. (2008). Induction of cancer chemopreventive enzymes by coffee is mediated by transcription factor Nrf2. Evidence that the coffee-specific diterpenes cafestol and kahweol confer protection against acrolein. *Toxicology and Applied Pharmacology, 226*(3), 328–337. https://doi.org/10.1016/j.taap.2007.09.018

Hong, J. W., Yang, G. E., Kim, Y. B., Eom, S. H., Lew, J. H., & Kang, H. (2012). Anti-inflammatory activity of cinnamon water extracts *in vivo* and *in vitro* LPS-induced models. *BMC Complementary and Alternative Medicine, 12*, 237. https://doi.org/10.1186/1472-6882-12-237

Huang, B., Yuan, H. D., Kim, D. Y., Quan, H. Y., & Chung, S. H. (2011). Cinnamaldehyde prevents adipocyte differentiation and adipogenesis via regulation of peroxisome proliferator-activated receptor-γ (PPARγ) and AMP-activated protein kinase (AMPK) pathways. *Journal of Agricultural and Food Chemistry, 59*(8), 3666–3673. https://doi.org/10.1021/jf104814t

Huang, K., Liang, X. C., Zhong, Y. L., He, W. Y., & Wang, Z. (2015). 5-Caffeoylquinic acid decreases diet-induced obesity in rats by modulating PPARα and LXRα transcription. *Journal of the Science of Food and Agriculture, 95*(9), 1903–1910. https://doi.org/10.1002/jsfa.6896

Hwa, J. S., Jin, Y. C., Leeb, Y. S., Kob, Y. S., Kim, Y. M., Shif, L. Y., . . . Chang, K. C. (2012). 2-Methoxycinnamaldehyde from *C. cassia* reduces rat myocardial ischemia and reperfusion injury *in vivo* due to HO-1 induction. *Journal of Ethnopharmacology, 139*(2012), 605–615. https://doi.org/10.1016/j.jep.2011.12.001

Jafarnejad, S., Keshavarz, S. A., Mahbubi, S., Sxaremi, S., Arab, A., Abbasi, S., & Djafarian, K. (2017). Effect of ginger (*Z. officinale*) on blood glucose and lipid concentrations in diabetic and hyperlipidemic subjects: A meta-analysis of randomized controlled trials. *Journal of Functional Foods, 29*, 127–134. https://doi.org/10.1016/j.jff.2016.12.006

Jafarzadeh, A., Mohammadi-Kordkhayli, M., Ahangar-Parvin, R., Azizi, V., Khoramdel-Azad, H., Shamsizadeh, A., . . . Khaksari, M. (2014). Ginger extracts influence the expression of IL-27 and IL-33 in the central nervous system in experimental autoimmune encephalomyelitis and ameliorates the clinical symptoms of disease. *Journal of Neuroimmunology, 276*(1–2), 80–88. https://doi.org/10.1016/j.jneuroim.2014.08.614

Jarvill-Taylor, K. J., Anderson, R. A., & Graves, D. J. (2001). A hydroxychalcone derived from cinnamon functions as a mimetic for insulin in 3T3-L1 adipocytes. *Journal of the American College of Nutrition, 20*(4), 327–336. https://doi.org/10.1080/07315724.2001.10719053

Jeon, J. S., Kim, H. T., Jeong, I. H., Hong, S. R., Oh, M. S., Park, K. H., . . . El-Aty, A. M. (2017). Determination of chlorogenic acids and caffeine in homemade brewed coffee prepared under various conditions. *Journal of Chromatography B Analytical Technologies in the Biomedical and Life Sciences, 1064*, 115–123. https://doi.org/10.1016/j.jchromb.2017.08.041

Ji, K., Fang, L., Zhao, H., Li, Q., Shi, Y., Xu, C., . . . Liu, Q. (2017). Ginger oleoresin alleviated γ-ray irradiation-induced reactive oxygen species via the Nrf2 protective response in human mesenchymal stem cells. *Oxidative Medicine and Cellular Longevity, 2017*. https://doi.org/10.1155/2017/1480294

Ji, L., Jiang, P., Lu, B., Sheng, Y., Wang, X., & Wang, Z. (2013). Chlorogenic acid, a dietary polyphenol, protects acetaminophen-induced liver injury and its mechanism. *The Journal of Nutritional Biochemistry, 24*(11), 1911–1919. https://doi.org/10.1016/j.jnutbio.2013.05.007

Jin, S., Chang, C., Zhang, L., Liu, Y., Huang, X., & Chen, Z. (2015). Chlorogenic acid improves late diabetes through adiponectin receptor signaling pathways in db/db mice. *PLOS One, 10*(4), e0120842. https://doi.org/10.1371/journal.pone.0120842

Kang, C. H., Jayasooriya, R. G. P. T., Dilshara, M. G., Choi, Y. H., Jeong, Y. K., Kim, N. D., & Kim, G. Y. (2012). Caffeine suppresses lipopolysaccharide-stimulated BV2 microglial cells by suppressing Akt-mediated NF-κB activation and ERK phosphorylation. *Food and Chemical Toxicology, 50*(12), 4270–4276. https://doi.org/10.1016/j.fct.2012.08.041

Kanuri, G., Weber, S., Volynets, V., Spruss, A., Bischoff, S. C., & Bergheim, I. (2009). Cinnamon extract protects against acute alcohol-induced liver steatosis in mice. *The Journal of Nutrition, 139*(3), 482–487. https://doi.org/10.3945/jn.108.100495

Killday, K. B., Davey, M. H., Glinski, J. A., Duan, P., Veluri, R., Proni, G., . . . Tempesta, M. S. (2011). Bioactive A-type proanthocyanidins from *C. cassia*. *Journal of Natural Products, 74*(9), 1833–1841. https://doi.org/10.1021/np1007944

Kim, J. Y., Jung, K. S., & Jeong, H. G. (2004). Suppressive effects of the kahweol and cafestol on cyclooxygenase-2 expression in macrophages. *FEBS Letters, 569*(1–3), 321–326. https://doi.org/10.1016/j.febslet.2004.05.070

Kiyama, R. (2020). Nutritional implications of ginger: chemistry, biological activities and signaling pathways. *Journal of Nutritional Biochemistry*, *86*, 108486. https://doi.org/10.1016/j.jnutbio.2020.108486

Kord, M. T., Pourrajab, F., & Hekmatimoghaddam, S. (2020). Ginger extract increases GLUT-4 expression preferentially through AMPK than PI3K signalling pathways in C2C12 muscle cells. *Diabetes, Metabolic Syndrome and Obesity*, *13*, 3231–3238. https://doi.org/10.2147/DMSO.S260224

Kwon, H. K., Hwang, J. S., Lee, C. G., So, J. S., Sahoo, A., Im, C. R., . . . Im, S. H. (2011). Cinnamon extract suppresses experimental colitis through modulation of antigen-presenting cells. *World Journal of Gastroenterology*, *17*(8), 976–986. https://doi.org/10.3748/wjg.v17.i8.976

Lai, Y. S., Lee, W. C., Lin, Y. E., Ho, C. T., Lu, K. H., Lin, S. H., . . . Sheen, L. Y. (2016). Ginger essential oil ameliorates hepatic injury and lipid accumulation in high fat diet-induced nonalcoholic fatty liver disease. *Journal of Agricultural and Food Chemistry*, *64*(10), 2062–2071. https://doi.org/10.1021/acs.jafc.5b06159

Lastra, G., Syed, S., Kurukulasuriya, L. R., Manrique, C., & Sowers, J. R. (2014). Type 2 diabetes mellitus and hypertension: An update. *Endocrinology and Metabolism Clinics of North America*, *43*(1), 103–122. https://doi.org/10.1016/j.ecl.2013.09.005

Lee, B. J., Kim, Y. J., Cho, D. H., Sohn, N. W., & Kang, H. (2011). Immunomodulatory effect of water extract of cinnamon on anti-CD3-induced cytokine responses and p38, JNK, ERK1/2, and STAT4 activation. *Immunopharmacology and Immunotoxicology*, *33*(4), 714–722. https://doi.org/10.3109/08923973.2011.564185

Li, A. L., Li, G. H., Li, Y. R., Wu, X. Y., Ren, D. M., Lou, H. X., . . . Shen, T. (2019). Lignan and flavonoid support the prevention of cinnamon against oxidative stress related diseases. *Phytomedicine*, *53*, 143–153. https://doi.org/10.1016/j.phymed.2018.09.022

Li, J., Wang, S., Yao, L., Ma, P., Chen, Z., Han, T. L., . . . Wang, J. (2019). 6-gingerol ameliorates age-related hepatic steatosis: Association with regulating lipogenesis, fatty acid oxidation, oxidative stress and mitochondrial dysfunction. *Toxicology and Applied Pharmacology*, *362*, 125–135. https://doi.org/10.1016/j.taap.2018.11.001

Lipscombe, L., Butalia, S., Dasgupta, K., Eurich, D., Goldenberg, R., Khan, N., . . . Bscphm, S. S. (2016). Pharmacologic management of type 2 diabetes: 2016 interim update. *Canadian Journal of Diabetes*, *40*(3), 193–195. https://doi.org/10.1016/j.jcjd.2016.02.006

López-Barrera, D. M., Vázquez-Sánchez, K., Loarca-Piña, M. G. F., & Campos-Vega, R. (2016). Spent coffee grounds, an innovative source of colonic fermentable compounds, inhibit inflammatory mediators *in vitro*. *Food Chemistry*, *212*, 282–290. https://doi.org/10.1016/j.foodchem.2016.05.175

Ludwig, I. A., Mena, P., Calani, L., Cid, C., Del Rio, D., Lean, M. E. J., & Crozier, A. (2014). Variations in caffeine and chlorogenic acid contents of coffees: What are we drinking? *Food Functional*, *5*(8), 1718–1726. https://doi.org/10.1007/s11032-017-0636-6

Ma, Y., Gao, M., & Liu, D. (2015). Chlorogenic acid improves high fat diet-induced hepatic steatosis and insulin resistance in mice. *Pharmaceutical Research*, *32*(4), 1200–1209. https://doi.org/10.1007/s11095-014-1526-9

Mao, Q. Q., Xu, X. Y., Cao, S. Y., Gan, R. Y., Corke, H., Beta, T., & Li, H. B. (2019). Bioactive compounds and bioactivities of ginger (*Z. officinale* roscoe). *Foods*, *8*(6), 1–21. https://doi.org/10.3390/foods8060185

Marx, W., McKavanagh, D., McCarthy, A. L., Bird, R., Ried, K., Chan, A., & Isenring, L. (2015). The effect of ginger (Z. officinale) on platelet aggregation: A systematic literature review. *PLOS One, 10*(10), 1–13. https://doi.org/10.1371/journal.pone.0141119

McCusker, R. R., Goldberger, B. A., & Cone, E. J. (2003). Caffeine content of specialty coffees. *Journal of Analytical Toxicology, 27*(7), 520–522. https://doi.org/10.1093/jat/27.7.520

Micucci, M., Bolchi, C., Budriesi, R., Cevenini, M., Maroni, L., Capozza, S., . . . Angeletti, A. (2020). Antihypertensive phytocomplexes of proven efficacy and well-established use: Mode of action and individual characterization of the active constituents. *Phytochemistry, 170*, 112222. https://doi.org/10.1016/j.phytochem.2019.112222

Mojani, M. S., Ghasemzadeh, A., Rahmat, A., Loh, S. P., & Ramasamy, R. (2014). Assessment of bioactive compounds, nutritional composition and antioxidant activity of Malaysian young ginger (*Z. officinale* Roscoe). *International Food Research Journal, 21*(5), 1931–1935.

Morvaridzadeh, M., Fazelian, S., Agah, S., Khazdouz, M., Rahimlou, M., Fahimeh, A., . . . Heshmati, J. (2020). Effect of ginger (Zingiber officinale) on inflammatory markers: A systematic review and meta-analysis of randomized controlled trials. *Cytokine, 135*, 155224. https://doi.org/10.1016/j.cyto.2020.155224

Muhammad, D. R. A., & Dewettinck, K. (2017). Cinnamon and its derivatives as potential ingredient in functional food—A review. *International Journal of Food Properties, 20*(Suppl. 2), 2237–2263. https://doi.org/10.1080/10942912.2017.1369102

Mustafa, S. S. S., Eid, N. I., Jafri, S. A., El-Latif, H. A. A., & Ahmed, H. M. S. (2007). Insulinotropic effect of aqueous ginger extract and aqueous garlic extract on the isolated perfused pancreas of streptozotocin induced diabetic rats. *Pakistan Journal of Zoology, 39*(5), 279–284.

Nonaka, G. I., Morimoto, S., & Nishioka, I. (1983). Tannins and related compounds. Part 13. Isolation and structures of trimeric, tetrameric, and pentameric proanthocyanidins from cinnamon. *Journal of Chemistry Society, Perkin Transactions, 1*, 2139–2145. https://doi.org/10.1039/P19830002139

O'Hara, M., Kiefer, D., Farrell, K., & Kemper, K. (1998). A review of 12 commonly used medicinal herbs. *Archives of Family Medicine, 7*(6), 523–536. https://doi.org/10.1001/archfami.7.6.523

Panda, S. K., Nirvanashetty, S., Parachur, V. A., Krishnamoorthy, C., & Dey, S. (2020). A randomized, double-blind, placebo controlled, parallel-group, comparative clinical study to evaluate the efficacy and safety of OLNP-06 versus placebo in subjects with functional dyspepsia. *Journal of Dietary Supplements*. https://doi.org/10.1080/19390211.2020.1856996

Paranagama, P. A., Wimalasena, S., Jayatilake, G. S., Jayawardena, A. L., Senayake, U. M., & Mubarak, A. M. (2010). A comparison of essential oil constituents of bark, leaf, root and fruit of cinnamon (C. zeylanicum Blum) grown in Sri Lanka. *Journal of the National Science Foundation of SriLanka, 29*(3–4), 147–153. https://doi.org/10.4038/jnsfsr.v29i3-4.2613

Peng, B. J., Qi, Z. H. U., Zhong, Y. L., Xu, S. H., & Zheng, W. A. N. G. (2015). Chlorogenic acid maintains glucose homeostasis through modulating the expression of SGLT-1, GLUT-2, and PLG in different intestinal segments of Sprague–Dawley rats fed a high-fat diet. *Biomedical and Environmental Sciences, 28*(12), 894–903. https://doi.org/10.3967/bes2015.123

Prasad, S., & Tyagi, A. K. (2015). Ginger and its constituents: Role in prevention and treatment of gastrointestinal cancer. *Gastroenterology Research and Practice, 2015*, 11. https://doi.org/10.1155/2015/142979

Qin, B., Dawson, H., Polansky, M. M., & Anderson, R. A. (2009). Cinnamon extract attenuates TNF alpha- induced intestinal lipoprotein ApoB48 overproduction by regulating inflammatory, insulin, and lipoprotein pathways in enterocytes. *Hormone and Metabolic Research*, *41*(7), 516–522. https://doi.org/10.1055/s-0029-1202813

Sauer, T., Raithel, M., Kressel, J., Münch, G., & Pischetsrieder, M. (2013). Activation of the transcription factor Nrf2 in macrophages, Caco-2 cells and intact human gut tissue by Maillard reaction products and coffee. *Amino Acids*, *44*(6), 1427–1439. https://doi.org/10.1007/s00726-012-1222-1

Seo, H. Y., Kim, M. K., Lee, S. H., Hwang, J. S., Park, K. G., & Jang, B. K. (2018). Kahweol ameliorates the Liver Inflammation through the Inhibition of NF-κB and STAT3 activation in primary kupffer cells and primary hepatocytes. *Nutrients*, *10*(7), 863. https://doi.org/10.3390/nu10070863

Shan, J., Fu, J., Zhao, Z., Kong, X., Huang, H., Luo, L., & Yin, Z. (2009). Chlorogenic acid inhibits lipopolysaccharide-induced cyclooxygenase-2 expression in RAW264.7 cells through suppressing NF-κB and JNK/AP-1 activation. *International Immunopharmacology*, *9*(9), 1042–1048. https://doi.org/10.1016/j.intimp.2009.04.011

Shi, A., Shi, H., Wang, Y., Liu, X., Cheng, Y., Li, H., . . . Dong, L. (2018). Activation of Nrf2 pathway and inhibition of NLRP3 inflammasome activation contribute to the protective effect of chlorogenic acid on acute liver injury. *International Immunopharmacology*, *54*, 125–130. https://doi.org/10.1016/j.intimp.2017.11.007

Shi, H., Dong, L., Jiang, J., Zhao, J., Zhao, G., Dang, X., . . . Jia, M. (2013). Chlorogenic acid reduces liver inflammation and fibrosis through inhibition of toll-like receptor 4 signaling pathway. *Toxicology*, *303*, 107–114. https://doi.org/10.1016/j.tox.2012.10.025

Shi, H., Shi, A., Dong, L., Lu, X., Wang, Y., Zhao, J., . . . Guo, X. (2016). Chlorogenic acid protects against liver fibrosis in vivo and in vitro through inhibition of oxidative stress. *Clinical Nutrition*, *35*(6), 1366–1373. https://doi.org/10.1016/j.clnu.2016.03.002

Srinivasan, K. (2014). Antioxidant potential of spices and their active constituents. *Critical Reviews in Food Science and Nutrition*, *54*(3), 352–372. https://doi.org/10.1080/10408398.2011.585525

Stoner, G. D. (2013). Ginger: Is it ready for prime time? *Cancer Prevention Research*, *6*(4), 257–262. https://doi.org/10.1158/1940-6207.CAPR-13-0055

Sun, P., Li, K., Wang, T., Ji, J., Wang, Y., Chen, K. X., . . . Wang, H. Y. (2019). Procyanidin C1, a component of cinnamon extracts, is a potential insulin sensitizer that targets adipocytes. *Journal of Agricultural and Food Chemistry*, *67*(32), 8839–8846. https://doi.org/10.1021/acs.jafc.9b02932

Tanaka, T., Matsuo, Y., Yamada, Y., & Kouno, I. (2008). Structure of polymeric polyphenols of cinnamon bark deduced from condensation products of cinnamaldehyde with catechin and procyanidins. *Journal of Agricultural and Food Chemistry*, *56*(14), 5864–5870. https://doi.org/10.1021/jf800921r

Tasneem, S., Liu, B., Li, B., Choudhary, M. I., & Wang, W. (2019). Molecular pharmacology of inflammation: Medicinal plants as anti-inflammatory agents. *Pharmacological Research*, *139*, 126–140. https://doi.org/10.1016/j.phrs.2018.11.001

Townsend, E. A., Siviski, M. E., Zhang, Y., Xu, C., Hoonjan, B., & Emala, C. W. (2013). Effects of ginger and its constituents on airway smooth muscle relaxation and calcium regulation.

American Journal of Respiratory Cell and Molecular Biology, *48*(2), 157–163. https://doi.org/10.1165/rcmb.2012-0231OC

Townsend, E. A., Zhang, Y., Xu, C., Wakita, R., & Emala, C. W. (2014). Active components of ginger potentiate β-agonist-induced relaxation of airway smooth muscle by modulating cytoskeletal regulatory proteins. *American Journal of Respiratory Cell and Molecular Biology*, *50*(1), 115–124. https://doi.org/10.1165/rcmb.2013-0133OC

Vangalapati, M., Satya, N. S., Prakash, D. S., & Avanigadda, S. (2012). A review on pharmacological activities and clinical effects of cinnamon species. *Research Journal of Pharmaceutical, Biological and Chemical Sciences*, *3*(1), 653–663.

Vázquez-Sánchez, K., Martinez-Saez, N., Rebollo-Hernanz, M., Del Castillo, M. D., Gaytán-Martínez, M., & Campos-Vega, R. (2018). *In vitro* health promoting properties of antioxidant dietary fiber extracted from spent coffee (*Coffee arabica* L.) grounds. *Food Chemistry*, *261*, 253–259. https://doi.org/10.1016/j.foodchem.2018.04.064

Wan, C. W., Wong, C. N. Y., Pin, W. K., Wong, M. H. Y., Kwok, C. Y., Chan, R. Y. K., . . . Chan, S. W. (2013). Chlorogenic acid exhibits cholesterol lowering and fatty liver attenuating properties by up-regulating the gene expression of PPAR-α in hypercholesterolemic rats induced with a high-cholesterol diet. *Phytotherapy Research*, *27*(4), 545–551. https://doi.org/10.1002/ptr.4751

Wang, P., Yang, Y., Wang, D., Yang, Q., Wan, J., Liu, S., . . . Yang, Y. (2020). Cinnamaldehyde ameliorates vascular dysfunction in diabetic mice by activating Nrf2. *American Journal of Hypertension*, *33*(7), 610–619. https://doi.org/10.1093/ajh/hpaa024

Wang, Y., Yu, H., Zhang, X., Feng, Q., Guo, X., Li, S., . . . Ma, Y. (2017). Evaluation of daily ginger consumption for the prevention of chronic diseases in adults: A cross-sectional study. *Nutrition*, *36*, 79–84. https://doi.org/10.1016/j.nut.2016.05.009

Wondrak, G. T., Cabello, C. M., Villeneuve, N. F., Zhang, S., Ley, S., Li, Y., . . . Zhang, D. D. (2008). Cinnamoyl-based Nrf2-activators targeting human skin cell photo-oxidative stress. *Free Radical Biology and Medicine*, *45*(4), 385–395. https://doi.org/10.1016/j.freeradbiomed.2008.04.023

Wondrak, G. T., Villeneuve, N. F., Lamore, S. D., Bause, A. S., Jiang, T., & Zhang, D. D. (2010). The cinnamon-derived dietary factor cinnamic aldehyde activates the Nrf2-dependent antioxidant response in human epithelial colon cells. *Molecules*, *15*(5), 3338–3355. https://doi.org/10.3390/molecules15053338

Wongsa, P., Chaiwarit, J., & Zamaludien, A. (2012). *In vitro* screening of phenolic compounds, potential inhibition against α-amylase and α-glucosidase of culinary herbs in Thailand. *Food Chemistry*, *131*(3), 964–971. https://doi.org/10.1016/j.foodchem.2011.09.088

Yovana, M., Brígida, M., & Bragagnolo, N. (2018). Physical characteristics of the paper filter and low cafestol content filter coffee brews. *Food Research International*, *108*, 280–285. https://doi.org/10.1016/j.foodres.2018.03.041

Yuan, Y., Gong, X., Zhang, L., Jiang, R., Yang, J., Wang, B., & Wan, J. (2017). Chlorogenic acid ameliorated concanavalin A-induced hepatitis by suppression of Toll-like receptor 4 signaling in mice. *International Immunopharmacology*, *44*, 97–104. https://doi.org/10.1016/j.intimp.2017.01.017

Zheng, Z., Sheng, Y., Lu, B., & Ji, L. (2015). The therapeutic detoxification of chlorogenic acid against acetaminophen-induced liver injury by ameliorating hepatic inflammation. *Chemico-Biological Interactions*, *238*, 93–101. https://doi.org/10.1016/j.cbi.2015.05.023

Zhou, J., Zhou, S., & Zeng, S. (2013). Experimental diabetes treated with trigonelline: effect on β cell and pancreatic oxidative parameters. *Fundamental and Clinical Pharmacology*, *27*(3), 279–287. https://doi.org/10.1111/j.1472-8206.2011.01022.x

Zhu, R., Liu, H., Liu, C., Wang, L., Ma, R., Chen, B., . . . Gao, S. (2017). Cinnamaldehyde in diabetes: A review of pharmacology, pharmacokinetics and safety. *Pharmacology Research*, *122*, 78–89. https://doi.org/10.1016/j.phrs.2017.05.019

Zuo, J., Zhao, D., Yu, N., Fang, X., Mu, Q., Ma, Y., . . . Gao, S. (2017). Cinnamaldehyde ameliorates diet-induced obesity in mice by inducing browning of white adipose tissue. *Cellular Physiology and Biochemistry*, *42*, 1514–1525. https://doi.org/10.1159/000479268

13

Chontaduro (*Bactris gasipaes* Kunth): A Sustainable Source of Healthy Bioactive Compounds

Ivan Luzardo-Ocampo[1,2], *Mario E. Rodríguez-García*[3], *and Olga Lucia Torres-Vargas*[4]

[1] *Institute of Neurobiology, Universidad Nacional Autónoma de México (UNAM), Queretaro, Mexico*
[2] *Research and Graduate Program in Food Science, Universidad Autonoma de Queretaro, Queretaro, Mexico*
[3] *Departamento de Nanotecnología, Centro de Física Aplicada y Tecnología Avanzada, Universidad Nacional Autónoma de México (UNAM)-Campus Juriquilla, Queretaro, Mexico*
[4] *Research Group on Agroindustrial Sciences, Interdisciplinary Science Institute, Universidad del Quindio, Armenia, Colombia*

13.1 Introduction

Palm products comprise a varied amount of agricultural products, mainly located in the Amazonia region. About 150 native palm species have been reported in the region due to the high diversity of environmental conditions, englobed by the Amazonian river and the Andes. Still, the list of species increases if the coastal regions of Central and South America are also considered. The diversity of these plants can be associated with their economic and cultural importance for the people involved with their growth, harvest, and use. Palm fruits are one of the primary food sources for Indigenous people of the Amazonian region, and they are sold in specialized markets due to their potential sensorial and health benefits (Smith, 2015b). Chontaduro (*B. gasipaes* Kunth) or peach palm occupies a high position within palm fruits due to its chemical composition and consumption-derived health benefits. However, the lack of extensive studies evaluating this fruit urges the need for providing more information and stimulates intensive research focus on this product. Hence, this chapter reviews the potential beneficial effects of chontaduro consumption targeting its bioactive compounds and their assessment in disease models.

13.2 Agronomical Description and Cultural Aspects

Chontaduro (*B. gasipaes* Kunth) are known by several names depending on the growth or production region. For instance, it is called peach palm in english; chima or chonta (Bolivia), pupunha (Brazil), chontaduro or pijiwao (Colombia); chonta dura or chunda (Ecuador); parépou (French Guiana); pijuayo or pifuayo (Peru); pijiguaio or pifuayo (Venezuela); and pejibaye (Costa Rica, Honduras, and Nicaragua) (Smith, 2015a). To date,

Molecular Mechanisms of Functional Food, First Edition. Edited by Rocio Campos-Vega and B. Dave Oomah.

the genetic difference has been confirmed between the cultivated chontaduro (*B. gasipaes* Kunth, var. gasipaes) and a wild not extensively used form (*B. gasipaes* Kunth var. chichagi) (Graefe, Dufour, van Zonneveld, Rodriguez, & Gonzalez, 2013). Although no recent statistical information can be found about its production levels, the Food and Agriculture Organization (FAO) categorizes all of these products within "palm products," indicating that Indonesia, Malaysia, Thailand, Brazil, and Colombia are the primary producers. Nonetheless, "palm products" include palm oil and palm kernel, which are unrelated to chontaduro fruit (FAO, 2018; Franco et al., 2019).

As an agricultural product, chontaduro is a drupe with occasional striations and mixed peel color (greenish-yellow to yellow, orange, dark red, and reddish-green, depending on the cultivar), whereas its mesocarp can range from creamy to white (Figure 13.1a). The fruit exhibits a mammiform, rounder, angular, or truncated tip (Figure 13.1b), whereas the endocarp contains a seed, usually located at the center of the fruit. The whole fruit varies considerably in shape (ovoid, elliptic, round, or cuneiform) and displays a wide range of dimensions (length: 2–7 cm, width: 2–8 cm) and weight (4–186 g) (Graefe et al., 2013). Chontaduro-producing palms reach up to 20 m in height and produce branches attaching bunches of fruits. Palms usually remain productive for several years after which they are abandoned by producers, but trees continue being productive producing fruits periodically, contributing to sustainable and long-lasting production (Smith, 2015a). Moreover, peach

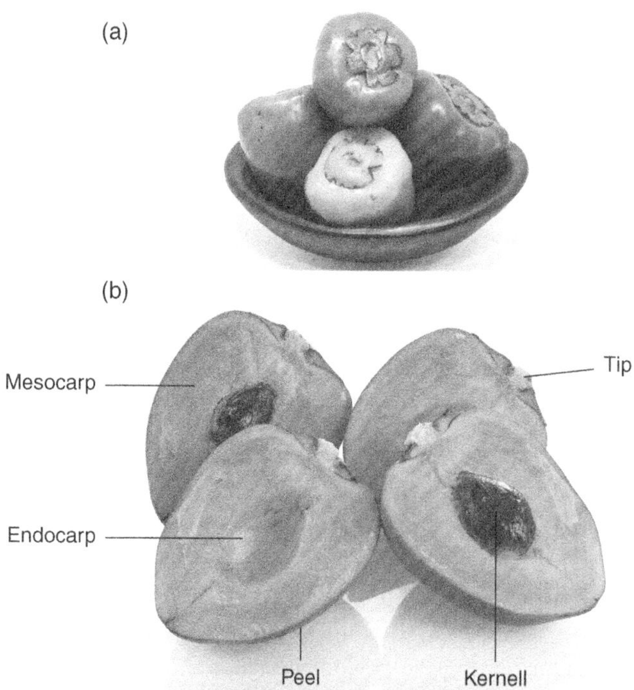

Figure 13.1 (a) Chontaduro (*Bactris gasipaes* Kunth) fruits color diversity. *Source:* anamejia18/Adobe Stock; (b) Main fruit parts. *Source:* anamejia18/Adobe Stock.

palms act as an important carbon sink in land-use systems, and the crops are less affected by environmental factors than other Amazonian crops (Graefe et al., 2013).

The fruit requires boiling prior to consumption primarily in many Latin-American cities along the continent, particularly in San José (Costa Rica), Cali (Colombia), and Manaus and Belém (Brazil). As popular dips, salt, lemon, mayonnaise, sweet ingredients (e.g. condensed milk), or a sauce made from a tomato substitute (*Solanum sessiliflorum*) are used, depending on the country (Smith, 2015a). The considerable carbohydrates and lipids composition of the fruit make it a very energetic food product used in school lunches in Northern Brazil, or it has been regionally incorporated as breakfast, afternoon snack, or sophisticated culinary presentation (Bezerra & da Silva, 2016). Since ancient times, parts of this plant (roots and leaves) have been used by Amerindian populations to combat hepatitis, malaria, parasites, and pro-inflammatory conditions such as gastritis and ear pain. However, these properties have not yet been evaluated through proper *in vitro* or *in vivo* assays (Peixoto Araujo et al., 2021).

13.2.1 Proximal Composition

Similar to other fruits, varieties differ in proximal components. A recent study (Torres-Vargas, Luzardo-Ocampo, Hernandez-Becerra, & Rodríguez-García, 2021) highlighted the proximal differences between unripe and ripe chontaduro fruits from Colombia, indicating that ripe fruits contain more protein, less lipids and minerals, and higher total carbohydrate contents than unripe fruits. Particularly for the mineral composition, unripe fruit flours were the richest in potassium. Still, ripe flours have higher amounts of sulfur and calcium (up to 5 mg/100 g), iron (~7 mg/100 g), and strontium (~5.5 mg/100 g). Table 13.1 depicts the proximal composition of ripe and unripe chontaduro. Most chontaduro components are highly digestible, accounting for an average of 51–95, 91, 87, and 96% of protein, dry matter, energy, and starch digestibility, respectively (Graefe et al., 2013).

Chontaduro fruits contain relatively low protein, but their mesocarp contains all essential amino acids (EAAs) (Smith, 2015a). Torres-Vargas et al. (2021) reported a higher (12.64–23.87%) protein content in the ripe fruits and attributed this difference to the potential biological protein synthesis during the fruit's development. The most important amino acids in chontaduro are lysine, methionine, cysteine, threonine, and tryptophan (16–49, 8–13, 19, 27–39, and 4.5–7 g/kg of chontaduro, respectively) (Graefe et al., 2013). The highest protein digestibility corrected amino acid scores (PDCAAS) in Chontaduro fruit are from methionine-cysteine (117%), valine (103%), threonine (99%), and isoleucine (90%) (Leterme et al., 2005). Table 13.2 shows the amino acids profile of yellow and red chontaduro from Colombia, together with the calculated values for aromatic amino acids (AAAs), antioxidant amino acids, branched-chain amino acids (BCAAs), Fisher ratio, and the adapted equations for the protein efficiency ratio (PER) (Mir, Riar, & Singh, 2019).

The amino acids profile of chontaduro fruit is considered of low quality from a nutritional viewpoint, although the PDCAAS values are better than cereals such as maize or wheat and comparable to other tropical crop seeds (37–88%) (Leterme et al., 2005). Values obtained for methionine + cysteine and isoleucine + leucine + phenylalanine + valine were higher than 1, indicating that at least the minimum dietary protein requirements are fulfilled by chontaduro protein since sulfur-containing aminoacids are covered (Tome, 2012).

Table 13.1 Proximal composition (%) of whole chontaduro (*B. gasipaes* Kunth) fruit at two ripeness stages (unripe/ripe) and heating procedure (uncooked and cooked).

Ripeness stage	Heating procedure	Protein	Lipids	Ash	Moisture	Starch	Dietary Fiber	Soluble dietary fiber	Insoluble dietary fiber	Total carbohydrates
Unripe	Uncooked	6.76–6.96	25.21–34.62	1.37–2.41	58.47–59.39	63.50–64.73	4.53–4.65	–	–	45.04–47.29
	Cooked	9.07	36.04	2.62	–	–	5.41	–	–	43.69–45.48
Ripe	Uncooked	6.13–8.20	22.99–23.63	2.59–2.63	62.41–63.11	68.89–70.61	4.88–4.97	–	–	55.82–56.17
	Cooked	10.53–10.62	28.86–29.66	2.60–2.76	–	–	10.28–10.34	0.8–1.6	4.3–7.9	47.63–48.43

Source: Adapted from Espinosa-Pardo, Martinez, and Martinez-Correa (2014), Torres-Vargas et al. (2021), and Yuyama et al. (2003).

Table 13.2 Amino acid profile (g/100 g protein) of yellow and red chontaduro.

	Yellow	Red	PDCAAS (%)
Essential amino acids (EAAs)			
Histidine	1.9	1.8	–
Isoleucine	2.5	2.8	90
Leucine	5.4	5.5	78
Lysine	5.0	4.7	79
Methionine	1.2	1.3	–
Phenylalanine	2.7	2.7	–
Threonine	3.9	3.8	99
Tryptophan	7.0	7.0	60
Valine	3.7	3.9	103
Nonessential amino acids			
Alanine	6.9	5.9	–
Arginine	5.8	5.1	–
Aspartic acid	9.0	7.7	–
Cysteine	2.0	1.7	–
Glutamic acid	8.3	7.9	–
Glycine	4.6	4.5	–
Proline	3.6	3.6	–
Serine	5.8	5.5	–
Tyrosine	1.9	2.2	–
AAA	5.30	5.60	–
Antioxidant AA	14.30	14.30	–
Met + Cys	3.20	3.00	–
BCAA	11.60	12.20	–
Cytotoxic effect	16.50	14.80	–
EAA	28.20	28.70	–
Fischer ratio	2.19	2.18	–
Ile + Leu + Phe + Val	14.30	14.90	–
Arg + His + Glu	16.00	14.80	–
Lys/Arg ratio	0.86	0.92	–
PER1	1.61	1.65	–
PER2	1.78	1.80	–
PER3	1.53	1.34	–
SAT1	18.70	19.00	–
SAT2	5.80	6.00	–

Source: Adapted from Leterme et al. (2005).
AAA: aromatic amino acids; **BCAA:** branched chain amino acids; **PDCAAS:** protein digestibility-corrected amino acids score; **PER:** protein efficiency ratio.

BCAA consumption has been linked to reduced overweight risk, while some circulant plasmatic amino acids such as BCAA and AAAs have been associated with glycemia and insulin resistance (Ottosson et al., 2016). Values obtained from chontaduro fruit showed BCAA (15.48–17.11% of total amino acids) and AAA (7.08–7.85% of total amino acids) profile meeting the minimum protein requirements, while BCAA profile could potentially indicate the ability of chontaduro to protect from obesity and overweight risk, but detailed *in vitro* and *in vivo* tests should confirm this hypothesis. Most of these amino acids account for antioxidant amino acids, where values were the same for both types of chontaduro.

Overall, chontaduro has low Fischer ratio (BCAA/AAA), suggesting the need for BCAA supplementation, which is highly desirable since BCAA are taken by the muscle to inhibit AAA transport toward the blood–brain barrier, providing the peripheral metabolism for other beneficial purposes involved in hepatic metabolism (Udenigwe & Aluko, 2010). Calculations from PER showed low-quality proteins ($PER < 1.0$), a confirmed trend *in vivo* (2.89 and 2.30 for 200 and 400 g/kg body weight (BW), respectively).

The lipid-rich fraction of the fruit contains higher levels of unsaturated fatty acids than common cereal crops such as corn, sorghum, or soybean, contributing to the reputed high-energy source of this fruit and elevated contents of vitamin A (Smith, 2015a). The main components of the fruit's oil are omega-3 (linolenic acid), omega 6 (linoleic acid), and omega 9 (oleic acid) fatty acids, but lauric and myristic acids are the most common saturated acids (Graefe et al., 2013). The total saturated fatty acids (SFAs) ranged between 29.6 and 46.3% of total fatty acids, except for one report ($\sum SA = 85.5\%$ total fatty acid) for peach palm from French Guiana due to high lauric, myristic, and palmitic acids (60.6, 18.9, and 6% of total fatty acid, respectively) (Graefe et al., 2013). Mechanical expeller press (25–30 °C, 60 rpm, 8 mm extruder diameter) oil has higher saturated/unsaturated (17/83%) than solvent (petroleum benzene) extraction (13.4/86.6%) (Cardona Jaramillo, Carrillo Bautista, Alvarez Solano, Achenie, & González Barrios, 2019). Torres-Vargas et al. (2021) informed the presence of some saturated acids in the fruit's oil from Colombian chontaduro, such as stearic, palmitic, and oleic acids, based on their infrared spectra peaks from their main functional groups (Table 13.3). Several sterols have also been reported in the oil, particularly sitosterol, that can be used for cosmetic and emollients purposes (Bezerra & da Silva, 2016).

Overall, chontaduro has low polyunsaturated fatty acids (PUFAs) profile compared to the monounsaturated fatty acids (MUFAs) content, exhibiting values higher than other Amazonian fruits such as buriti (34.20%) and brazil nuts (*Bertholletia excelsa* HBK) (38.78%); the latter considered an energy-dense nut with a proportional SFAs/MUFAs/PUFAs ratio (Dufoo-Hurtado et al., 2021). Due to its fatty acids profile, chontaduro has been ranked as an intermediate source of ω-6 fatty acids represented by linoleic and linolenic acids, as their contents account for 5–30% of the total fatty acids according to the proposed ranking for Amazonian fruits (dos Santos, Goncalves et al., 2020). The atherogenic index (AI) and thrombogenic index (TI) are inversely related to the amount of anti-atherogenic fatty acids in certain oil, indicating a potential effect of the food source to cause platelet aggregation. These indexes were calculated for the fatty acids profile of chontaduro using the suggested equations of Ulbricht and Southgate (1991).

The calculated values from Table 13.3 are close to AI reported for chontaduro (1.10), while calculated TI was higher (2.04) (dos Santos, Soares et al., 2020) and suggested the ability of chontaduro to prevent cardiovascular diseases further. Regarding oxidizability

Table 13.3 Fatty acid composition of chontaduro mesocarp (%).

	Content (%)					
Fatty acids	Range	*Bactris gasipaes* Kunth	*B. gasipaes chichagui*	*B. gasipaes gasipaes* (green)	*B. gasipaes gasipaes* (red)	*B. gasipaes gasipaes* (yellow)
Saturated fatty acids (SFAs)						
Lauric (12:0)	0.01–60.6	60.6	0.01–0.02	0.02	0.02–0.03	0.01
Myristic (14:0)	0.1–18.9	18.9	0.10–0.14	0.11–0.12	0.15	0.14
Palmitic (16:0)	6.0–44.8	6.0–44.8	34.8–34.9	34.9–35.3	34.0–38.6	39.9–40.8
Stearic (18:0)	0.4–3.5	0.4–3.5	1.1–1.5	0.9–1.0	1.5–1.6	1.2–1.4
Arachidic (20:0)	2.5	2.5	–	–	–	–
Monounsaturated fatty acids (MUFAs)						
Palmitoleic (16:1)	3.5–11.5	3.5–10.5	7.9–9.3	10.8–11.5	8.3–8.4	9.5–10.3
Oleic (18:1)	12.9–60.8	12.9–60.8	48.0–51.9	45.1–46.3	45.8–46.0	36.2–38.0
Polyunsaturated fatty acids (PUFAs)						
Linoleic (18:2)	1.3–21.1	1.4–21.1	2.4–4.9	5.3–5.4	4.0–8.0	8.6–8.8
Linolenic (18:3)	0.0–5.5	0.0–5.5	0.2–1.1	0.8–0.9	0.2–0.9	1.5–1.6
ΣSFAs	29.6–46.3*	29.6–46.3	36.0–36.6	35.9–36.4	35.5–40.4	41.4–42.3
ΣMUFAs	16.4–71.3	16.4–71.3	55.9–61.2	55.9–57.8	54.1–54.4	45.7–48.3
ΣPUFAs	1.4–26.6	1.4–26.6	2.6–6	6.1–6.3	4.2–8.9	10.1–10.4
PUFA/SFA	0.02–0.2	0.02–0.2	0.07–0.2	0.17	0.1–0.2	0.25
Oxidizability (Cox value)	6.4–67.2	6.4–67.2	36.0–36.5	35.9–36.4	35.5–40.4	41.1–42.3
Atherogenic index (AI)	0.5–3.7	1.8–3.7	0.5–0.6	0.6	0.5–0.6	0.7
Thrombogenic index (TI)	0.7–0.9	0.7–0.9	0.9	0.9	0.8–0.9	0.8

Source: Adapted from Graefe et al. (2013) and Restrepo, Estupiñán, and Colmenares (2016).

(calculated oxidizability or COX value), the presence of chemical groups such as alcohols, esters, ethers, and carboxylic acids confirms a high proportion of unsaturated fatty acids and the resistance to oxidative processes (dos Santos, Goncalves et al., 2020).

Wild Ecuadorian *B. gasipaes* (hexane extracted) seed oil (11.5%, w/w yield) had high amount of SFAs (75.7%), particularly lauric, myristic, and palmitic acids (33.3, 27.8, and 9.6%, respectively) (Radice et al., 2014). Beta-sitosterol, squalene, and triterpene cycloartenol were the main components (51, 17, and 9%, respectively, of the unsaponifiable matter [3%]). The abundant β-sitosterol and squalene provide *B. gasipaes* oil wide range of physiological functions such as anticancer and anticholestrolemia and an attractive resource for functional foods, supplements, and even pharmaceuticals. Squalene is recognized for skin benefits (hydration and emollient) and as an ingredient in topically applied lipid emulsions and nanostructured lipid carriers (Radice et al., 2014 and references therein).

Regarding the total carbohydrates, the fruit is rich in dietary fiber and starch, the main feature characterizing the fruit, but most of these components depend highly on the ripeness state. Although there is little information about the impact of tropical fruits on glycemix index (GI), Jiménez, Gómez, Pérez, and Blanco-Metzler (2012) reported its estimation from chontaduro cooked fruits and baked chips using 12 healthy male and female volunteers (18–51 years old), where the fruit and chips showed a similar glucose peak, but a longer decrease ($p > 0.05$) compared to white bread control (21 ± 10 mg/dL vs. 18 ± 12 mg/dL). The consideration of the total area under the curve (AUC) indicated that the cooked fruits displayed the lowest GI (35 ± 6), whereas the baked chips were considered moderate GI (60 ± 7). Potential gelatinization processes on starch when excess water is used for cooking chontaduro could induce further resistant starch formation and soluble fiber release, resulting in lower GI (Garsetti et al., 2005). Moreover, the fat content from chontaduro can delay gastric emptying and interacts with amylose and amylopectin, contributing to V-type-resistant starch formation, contributing to GI reduction.

A functional analysis of the chontaduro starch based on the ripeness state and cooking procedure (uncooked and cooked flours) displayed the uncooked flours as those with the highest potential to be used in viscous food formula. In contrast, the cooked flours showed the lowest apparent viscosity (Torres-Vargas et al., 2021). The authors suggested the feasibility of enriching viscosity-required food lines (e.g. infant products, since starches are easy to swallow). Chontaduro mesocarp exhibits a low glycemic index (35 mg/dL) compared to white bread, making this fruit proper for type II diabetes and coronary conditions (Graefe et al., 2013). Peach palm fruit starch can be used in products where slow and smooth retrogradation is desired such as breads, soups, chowder, and porridges without emulsifier or fat use (Felisberto et al., 2020).

13.3 Bioactive Compounds

Chontaduro is one of the most balanced fruit due to its antioxidant bioactive compounds (mainly carotenoids and polyphenols), low sodium content, high dietary fiber, complete amino acid profile, and unsaturated fatty acids (de Souza Mesquita et al., 2019; Torres-Vargas et al., 2021). The main reported bioactive compounds from *B. gasipaes* are shown in Table 13.3.

13.3.1 Dietary Fiber

Targeted as one of the main bioactive compounds, dietary fiber from chontaduro fruit is composed of insoluble fiber, highly esterified homogalacturonan (70% esterification), linear methyl, and minor contents of xylogalacturonan and rhamnogalacturonan (Avila-Sosa et al., 2019) that can be used as fermentable carbohydrate substrates by gut microbiota. Several monomeric fractions have been identified from dietary fiber in chontaduro by-products (sheaths and stems flours) such as fucose, arabinose, galactose, glucose, mannose, xylose, and uronic acids, most of them proposed as inexpensive sources of fiber components of food products (Bolanho, Danesi, & Beléia, 2015). Beyond their antioxidant potential (Vieira et al., 2021), the biological potential of chontaduro fiber might reside in the nondigestible fraction as gut microbiota can use it as a fermentable source to produce beneficial compounds such as short-chain fatty acids (SCFAs) (butyrate, ethyl, and propionate) (Bezerra & da Silva, 2016). The extraction of pectic polysaccharides from ripe chontaduro fruits yielded several dietary fiber components such as rhamnose, arabinose, xylose, galactose, and uronic acid with fermentative potential (Cantu-Jungles, Cipriani, Iacomini, Hamaker, & Cordeiro, 2017)

13.3.2 Phenolic Compounds

Phenolic compounds have not been the main focus of research regarding bioactive compounds from chontaduro. Anthocyanins (1 mg/100 g) and yellow flavonoids (17 mg/100 g) were the main identified phenolics in the pulp (lightly orange color) after a methanolic extraction, column chromatography purification, and liquid chromatography coupled to mass spectrometry analysis (LC–MS) (M. dos Santos, Mamede, Rufino, de Brito, & Alves, 2015). The same authors quantified 30 mg of gallic acid equivalents (GAE)/100 g fruits as total phenolic compounds. Caffeic and chlorogenic acids are absent in *B. gasipaes* fruits (Meinhart et al., 2019).

13.3.3 Carotenoids

The oily nature of chontaduro contributes to the fruit's richness in carotenoids, bioactive compounds with an impressive potential to be used in food formulations since all of them are Generally Recognized as Safe (GRAS) by most regulatory agencies (Luzardo-Ocampo, Ramírez-Jiménez, Yañez, Mojica, & Luna-Vital, 2021). As carotenoids usually exhibit low bioaccessibility and bioavailability due to their lipophilic nature, chontaduro can serve as a food matrix to potentially increase these digestive parameters.

Many carotenoids have been reported in chontaduro (Table 13.4), with average levels ranging between 0.7 and 13.9 mg/100 g (110–202 mg of β-carotene equivalents/g dry matter for Costa Rican varieties of red/orange chontaduroripe cooked fruits) (Rojas-Garbanzo et al., 2016), composed predominantly by β-carotene, two (Z)-isomers of γ-carotene, and one isomer of (Z)-lycopene. Most carotenes are provitamin A carotenoids with diverse retinol activity equivalents (RAEs) (37–609 μg/100 g). Chontaduro consumption might comply with public health programs covering vitamin A deficiencies in developing countries since its carotene contents (89–94% of provitamin A active β-, γ-, and δ-carotenes)

Table 13.4 Bioactive compounds of chontaduro (*B. gasipaes* Kunth) fruit pulp.

Bioactive compound	Amount		References
Dietary fiber components (%)	**Pulp**	**Peel**	
Rhamnose	3.9	–	Cantu-Jungles et al. (2017)
Arabinose	4.1–14.5	–	
Galactose	3.6	–	
Xylose	2.6–6.5	–	
Uronic acid	71.5	–	
Phenolic compounds (mg/100 g)	**Pulp**	**Peel**	
Total anthocyanins	1	–	dos Santos et al. (2015)
Yellow flavonoids	17	–	
Total phenolic compounds	54–104	–	Rojas-Garbanzo, Pérez, Vaillant, and Pineda-Castro (2016)
Carotenoids (mg/100 g)	**Pulp**	**Peel**	
β-carotene	0.40–5.40	7.30 ± 0.40	de Rosso and Mercadante (2007); Hempel et al. (2014); Noronha Matos, Praia Lima, Pereira Barbosa, Zerlotti Mercadante, and Campos Chisté (2019)
Trans-β-carotene	55.51	–	
(Z)-γ-carotene	0.10–3.90	1.80 ± 0.60	
Trans-γ-carotene	35.43	–	
Cis-γ-carotene	28.35	–	
Cis-δ-carotene	5.22	–	
E-δ-carotene	–	1.74 ± 0.04	
Trans-δ-carotene	45.77	–	
E-γ-carotene	–	4.10 ± 0.70	
(Z)-lycopene	0.04–0.83	–	
Trans-α-cryptoxanthin	0.12	–	
Total	3.18–3.24	15.00 ± 1.70	Matos et al. (2019)

are similar to carrot juice and two times higher than mango (Jatunov, Quesada, Díaz, & Murillo, 2010). Most carotenoids from chontaduro are found in the peel (91.34%, 33.69 vs. 3.24 mg/100 g), although this proportion depends on the cultivar and variety. The total carotenoids contents in chontaduro pulp are higher than other Amazonian fruits recognized for their carotenoids, such as bacaba (*Oenocarpus bacaba*), inajá (*Attalea maripa*), bacuri (*Platonia insignis*), murici (*Byrsonima crassifolia*), and pequia (*Caryocar villosum*) (+78.39, 87.65, 96.91, and 62.96%, respectively). However, total carotenoids in chontaduro are lower than two important carotenoids-rich fruits such as pajurá (*Couepia bracteosa*) (13 mg/100 g) and buriti (*Mauritia vinifera*) (14.2 mg/100 g) (Noronha Matos et al., 2019).

13.4 *In vitro* and *in vivo* Studies Featuring the Biological Potential of Chontaduro

Most *in vitro* and *in vivo* studies assessing the impact of chontaduro on disease models or selective mechanisms are usually focused on a targeted bioactive compound. Hence, these chemical biological studies (Table 13.5) have been divided by their main studied bioactive compound or group of components.

13.4.1 Dietary Fiber

Only few studies have evaluated dietary fiber from chontaduro or its by-products. Vieira et al. (2021) evaluating the antioxidant of xylooligosaccharides isolated from the chontaduro inner sheath and peel waste reported 14.82–44.14 mmol Trolox equivalents/mg xylo-oligosaccharides (ORAC) and 2.15–3.88 mg Trolox equivalents/mL as IC_{50} values for DPPH inhibition. The authors indicated that most of this antioxidant capacity depends on the depolymerization method from the original raw materials, solubility, the nature and molecular weight of the monosaccharide constituents, and their glycosidic bonds.

The evaluation of the chemical composition and *in vivo* fecal digestibility of solid-state fermented chontaduro fruit in pigs showed that crude fiber was the least digestible component (75–77%) compared to dry matter (90–92%), crude protein (92–93%), and the nitrogen-free extract (91–93%) (Caicedo et al., 2019). The *in vitro* fecal fermentation of extracted and dialyzed polysaccharides from ripe chontaduro using a human fecal inoculum showed the ability of these polysaccharides to induce significant pH reductions and increased SCFAs levels (acetate, propionate, and butyrate) and total branched-chain fatty acids (BCFAs) (Cantu-Jungles et al., 2017). Surprisingly, propionate and butyrate levels were higher indicating higher bifidogenic effect than commercial fructooligosaccharides subjected to the same procedure.

13.4.2 Carotenoids

Carotenoids from chontaduro have been one of its main research topics due to the oily nature of this fruit. Supercritical CO_2 extracts 20 mg/g carotenoids from this fruit with antioxidant capacities up to ~25% (based on the β-carotene bleaching procedure), was similar to quercetin during the same assay (26%) (Espinosa-Pardo et al., 2014). This value was lower than the reported β-carotene bleaching from an earlier report of the pulp (33.7%) (Contreras-Calderón et al., 2011), but it was placed among some of the highest values compared to other exotic fruits from Colombia. The β-carotene composition of the fruit may be associated with solid antioxidant properties linked to the elimination of reactive oxygen species (ROS) such as singlet oxygen ($^1O_2$), superoxide anion radical ($\cdot O_2^-$), hydroxy radical ($\cdot OH$), peroxy radical ($RCOO\cdot$), and nitric oxide (NO). Moreover, carotenoids can stimulate enzymatic and nonenzymatic mechanisms to avoid mitochondrial proapoptotic processes (Kawata, Murakami, Suzuki, & Fujisawa, 2018). Some of these biological properties from *B. gasipaes* carotenoids might explain a protective effect against colon cancer as suggested by epidemiologic surveys conducted in Ecuador, but *in vitro* and *in vivo* studies

Table 13.5 Biological properties of chontaduro (*B. gasipaes* Kunth) bioactive compounds.

Targeted bioactive compounds/ components	Treatment	Characteristics of the study	Main outcomes	Reference
***In vitro* studies**				
Dietary fiber	*Bactris gasipaes* wastes (inner sheath and peel)	Xylooligosaccharides from dietary fiber were extracted using an alkaline solution (2 M NaOH, 5 hr, 25 °C), neutralized (pH 6), and precipitated (cold ethanol, 14–16 hr). Resulting solutions were centrifuged (2,544 g, 15 min, 4 °C), washed (ethanol/acetone), and evaporated.	Extracted xylans (xylose, arabinose, acetyl-xylose, xylobiose, xylotetraose, and xylopentaose) showed antioxidant activity values of 14.82 mmol TE/mg XOS (ORAC method) and DPPH IC_{50}:2.15–3.88 mg/mL.	Vieira et al. (2021)
	Ripe chontaduro fruits	Fruits were peeled, and the pulp was manually extracted, freeze-dried, and milled (265 g). The dried pulp powder was defatted with chloroform-methanol (1 : 1), and polysaccharides were extracted with water (100 °C, 2 hr). The extracts were centrifuged (3,860 g, 20 min, 25 °C), and polysaccharides were precipitated with ethanol, centrifuged (3,860 g, 20 min, 4 °C), and freeze-dried. Starch was removed with α-amylase and pullulanase, followed by dialysis. Fiber-rich polysaccharides were fermented *in vitro* using a batch fecal fermentation procedure. Human fecal samples were placed in an anaerobic chamber, homogenized with buffer, strained, and inoculated with the extracted carbohydrates. Fermentation was conducted for 4, 8, 12, and 24 hr.	Polysaccharides extracted from *B. gasipaes* increased gas amounts (9.6–12.4 mL) and lowered pH values (7.4–6.9) after 24-hr fermentation. SCFAs (μmol/50 mg carbohydrates) such as acetate (34.7–51.4), propionate (6.6–10.3), and butyrate (4.1–6.6) also increased during this period.	Cantu-Jungles et al. (2017)

Phenolic compounds	Ripe *B. gasipaes* fruits	Ripe chontaduro fruits were lyophilized and stored at −70 °C. Subsequently, 5 g of the powder was mixed with 20 mL methanol:water (50 : 50 v/v), rested at 25 °C for 1 hr, and centrifuged (15,000 rpm, 15 min), and supernatants were collected. Then, 20 mL of water:acetone (70 : 30 v/v) were added for 1 hr at 25 °C and centrifuged. Both extracts were combined, 50 mL of water were added, and analyses were carried out.	Antioxidant capacity: 92% β-carotene/linoleic acid bleaching method and ORAC: 94 ± 1 g/g μM Trolox/g.	dos Santos et al. (2015)
	B. gasipaes fruits from Costa Rica (Tucurrique and Pérez-Zeledón varieties)	Batches of fruits were selected in color (red/orange), disinfected (100 ppm NaClO, 15 min), homogenized, and frozen (−20 °C)	Total carotenoids content: 110–202 μg equivalents of β-carotene equilvalents/g dry weight Total phenolic compounds: 54–104 mg GAE/g dry sample. ORAC antioxidant capacity: 32–50 μmol TE/g dry weight	Rojas-Garbanzo et al. (2016)
Carotenoids	*B. gasipaes* carotenoids extracted by Supercritical CO_2	Total bioactive compounds were extracted by supercritical CO_2 using several conditions (40, 50, and 60 °C; 100, 200, and 300 bar).	Extracted carotenoids (up to 20 mg/g extract) exhibited β-carotene inhibitions up to 25%	Espinosa-Pardo et al. (2014)
	B. gasipaes fruits rich in carotenoids oil in a mayonnaise emulsion	Fruits were ground, homogenized, lyophilized, and stored (−40 °C). Carotenoids were recovered in sunflower oil after ultrasound-assisted extraction (1 : 6 solid:liquid ratio, 5 min, 400 W, 40 kHz, 27 °C) and vacuum filtered (−25 mmHg, 27 °C). The oil (38.70%) was incorporated to a O/W mayonnaise emulsion, together with water (55.27%), lemon juice (3.96%), salt (1.1%), xanthan gum (0.83%), and natural garlic (0.14%)	Mayonnaise increased the bioaccessibility of carotenoids compared to lyophilized fruit (11-folds, $p < 0.05$). *trans*-Lycopenes, *trans*-y-lycopenes, and *trans*-β-carotenes were the most bioaccessible carotenoids	de Souza Mesquita, Neves, Pisani, and de Rosso (2020)

(*Continued*)

Table 13.5 (Continued)

Targeted bioactive compounds/ components	Treatment	Characteristics of the study	Main outcomes	Reference
	B. gasipaes carotenoid extract	*B. gasipaes* fruits were lyophilized (−40 °C, 48 hr) and extracted using a conventional ultrasound-assisted acetone and 1-butyl-3-methylimidazodium tetrafluoroborate ionic liquid extraction. Both extracts were digested *in vitro*.	Ion-liquid extracted carotenoids were more bioaccessible at the end of the digestion procedure (33.25% vs. 26.84%), particularly all-*trans*- lycopene (37.22%), *cis*-δ-carotene (3) (37.21%), all-*trans*-α-carotene (34.80%), and all-*trans*-β-carotene (34.13%). From these carotenoids, all-*trans*-β-carotene and all-*trans*-lycopene showed the highest Caco-2 uptakes (up to 100 ng carotenoids/ng cell protein)	de Souza Mesquita et al. (2021)
Whole fruit	*B. gasipaes* pulp	Fresh fruit was cleaned with tap water and separated into peel, pulp, and seeds. The pulp was homogenized for 8–10 min. Then, extracts were prepared by homogenizing fruit with an acidic methanol–water (50 : 50) solution, vortexed (1 min), centrifuged (2,879 *g*, 15 min), and the recovered supernatants were mixed with acetone:water (70 : 30) and diluted with water. Extracts were stored at −20 °C for 24 hr until analysis.	ABTS: 14.1 μmol TE/g FW; FRAP: 3.98 μmol TE/g FW; TPC: 65.7 mg GAE/100 g FW	Contreras-Calderón, Calderón-Jaimes, Guerra-Hernández, and García-Villanova (2011)
	Ripe *B. gasipaes* aqueous and ethanolic extracts	Ripe chontaduro mesocarp was dried (40 °C, 24 hr), powdered, and mixed (20 mg/mL) with saline solution (0.85%) for 48 hr. For the ethanolic extracts, the dried powder was mixed with ethanol (50 mL) for 48 hr, the solvent was evaporated, and the resulting extract was resuspended in DMSO with saline solution (0.85%) (1 : 9 v/v proportion). The extracts were then diluted up to 20 mg/mL. Spermatozoons from healthy male sperm samples were challenged against the extracts for up to 120 min.	Treatment (2 hr) of both *B. gasipaes* extracts induced cell agglutination significantly ($p < 0.05$) reducing motility, compared to the initial state without extracts. The authors suggested the potential use of *B. gasipaes* extracts as contraceptives.	Angulo-Paternina, Ospina-Medina, Arango-Valencia, and Cardona-Maya (2017)

***In vivo* studies**				
Dietary fiber	Solid-state fermented ripe chontaduro fruit	Ripe chontaduro fruits were washed, the seeds were removed, and the pulp was ground, sieved (1 cm particle size), and mixed with molasses, wheat powder, mineral salt, calcium carbonate, and plain yogurt. The resulting blend was fermented up to 15 days at 25 °C, and administered to male pigs (32 ± 2 kg)	Crude fiber fecal digestibility was the lowest (75.06–77.04%), followed by dry matter (90.23–91.65%), the free-nitrogen extract (91.24–93.30%), and crude protein (92.24–93.23%). The overall nutrients' digestibility was not affected by the chontaduro inclusion (up to 40%) in the pigs' diet.	Caicedo et al. (2019)
Carotenes	Cooked ripe whole fruit	Fruits (230–240 g) were administered to two healthy nonsmoking adults (male/female, 26 and 30 years, BMI < 25). Fruits were cooked by boiling (1.5 kg in 5 L water containing 20 g salt, 30 min). Fruits were provided with mayonnaise (50 g, 78% fat)	After 6 h, γ-carotene (Z)-isomers (20.1–81.4 nmol/L plasma), β-carotene (24.3–98.7 nmol/L plasma), and total lycopene levels ((Z)-lycopene isomers and E-lycopene) (6.9–47.0 nmol/plasma) were detected. RAE depended on the tested variety: orange varieties (378–583 μg/100 FW) were higher than yellow varieties (42–88 μg/100 FW)	Hempel et al. (2014)
Whole fruit	Red chontaduro	Pellets made from dry red chontaduro mixed with casein and ground commercial rat food (LABINA®) were administered to four groups of female Wistar adult rats: control lactating rats (n : 10), control post-lactating rats (n : 8), red chontaduro-lactating rats (n : 7), and red chontaduro-post-lactating rats (n : 8) during 30 days.	Red chontaduro consumption significantly ($p < 0.01$) increased BW gain (9.53 ± 1.30%) and total cholesterol (36.38 ± 10.76%) but decreased HDL-cholesterol (71.57 ± 8.97%) in lactating rat groups. For post-lactating rats, only HDL-cholesterol increased ($p < 0.01$) by chontaduro administration (10.77 ± 11.43%).	Carvalho et al. (2013)
	17 varieties of chontaduro	Chontaduro diets (200–800 g/kg BW) were administered to 10 Sprague–Dawley rats (150 ± 12 g) for 28 days	Starch was the most digestible component of the diets (91.9–97.7%), followed by the dry matter (80.9–92.7%), and nitrogen (apparent: 73.6–79.9%). Rats' weight increased up to ~175 g.	Leterme et al. (2005)

ABTS: 2,2′-azino-bis(3-ethylbenzothiazoline-6-sulfonic acid); **BMI:** body mass index; **BW:** body weight; **DPPH:** 2,2-diphenyl-1-picrylhydrazyl; **FRAP:** ferric ion-reducing antioxidant power; **FW:** fresh weight; **GAE:** gallic acid equivalents; **HDL:** high-density lipoprotein; **IC_{50}:** half inhibitory concentration; **ORAC:** oxygen radical absorbance capacity; **RAE:** retinol activity equivalents; **SCFAs:** short-chain fatty acids; **TE:** Trolox equivalents; **TPC:** total phenolic compounds; **XOS:** xylooligosaccharides.

are required to support this claim (Neira-Mosquera, Pérez-Rodríguez, Sánchez-Llaguno, & Moreno-Rojas, 2013).

An *in vivo* study reported the first proof of the carotenoids' potential bioavailability after human consumption of the fruit (Hempel et al., 2014). Whole fruit (230–240 g of cooked orange chontaduro with 50 g of mayonnaise) was administered to two healthy nonsmoking adults (26 and 30 years; body mass index (BMI) <25 kg/m^2), and blood samples were collected. The triglyceride-rich lipoprotein blood fraction was dried, mixed with methyl terbutylic ether (MTBE) and 0.1 methanol, and quantified by high-performance liquid chromatography coupled to photodiode array detection and atmospheric pressure chemical ionization-mass spectrometry (HPLC-PDA-APCI-MS). No β-carotene was detected at 0 hr in plasma after consumption, but values of 20.1–81.4 nmol/L plasma were identified at 6 hr. γ-carotene concentrations also increased after 6 hr consumption. The biological value of these carotenes, measured in RAEs, ranged from 42 to 378 μg/100 g fresh weight (FW).

Two studies recently evaluated the bioaccessibility of chontaduro carotenoids depending on the extraction method. An emulsion with mayonnaise from a carotenoids-rich oil showed increased bioaccessibility of *trans*-β-carotenes, *trans*-lycopenes, *trans*-γ-carotenes, and total carotenoids from the emulsion compared to the lyophilized fruit (up to 11-folds, values ranging from 30 to 60% of micellarization efficiency) (de Souza Mesquita et al., 2020). The authors also reported the bioaccessibility of carotenoids extracted by two different methods (conventional ultrasound-assisted acetone extraction and ionic liquid-mediated extraction) (de Souza Mesquita et al., 2021). After conducting an *in vitro* digestion, the intestinal phase was simulated with polarized Caco-2 cells, and the ion liquid extracted carotenoids were more bioaccessible and exhibited higher Caco-2 uptakes.

Carotenoids were extracted from *B. gasipaes* feedstocks mediated by ethanolic solution (50%, w/w) of an imidazolium-based ionic liquid (IL). The IL-extracted carotenoids supplementation (gavaged with 1 mg/kg bw) improved serum parameters (lowered serum glucose, alanine transaminase [ALT], and aspartate transaminase [AST] and increased high-density lipoprotein [HDL]), hepatic antioxidant response (increased glutathione s-transferase [GST]), and metabolic conditions (increased adiponectin, inhibited body mass gain) of adult male Wistar rats, even when fed with high-fat diet. The extract also decreased pro-inflammatory cytokines (IL1β and IL-6) and increased the anti-inflammatory cytokine IL-10 levels that act against liver damage mediated by oxidative stress (de Souza Mesquita et al., 2021).

13.4.3 Whole Fruit Components

One of the oldest investigation reports *in vivo* evaluation of 17 whole chontaduro fruit samples [yellow (2), yellow-orange, orange-green (2), red-orange, red (2), orange, orange-green (3), yellow-green (4), red-green] from several origins (rain forest and the Colombian Andes region) administered to 10 Sprague–Dawley rats (150 ± 12 g) for 28 days (Leterme et al., 2005). Peach palm (200, 400, 600, and 800 g/kg BW) was supplemented with casein + methionine, maize starch, oil, sucrose, minerals and vitamins, and ground rice hulls. Starch was the most digestible component of the diets, followed by dry matter and nitrogen. Compared to the control diet, the rats showed the highest feed intake and growth, particularly with the 800 g/kg BW.

The energetic composition of chontaduro can be helpful during pregnancy, aiming to provide nutrients and energy to both the lactating mother and children. Carvalho et al. (2013) provided evidence about the anti-obesogenic effects of red chontaduro consumption in lactating and post-lactating Wistar rats. The rats (female adult rats) were randomly assigned to four groups: control lactating rats, control post-lactating rats, chontaduro-lactating rats, and chontaduro-post-lactating rats. Unfortunately, the authors did not report the amount of chontaduro administered to rats (30-day experiment), but the fruit was mixed with the rats' basal diet, pelletized, and dried (60 °C). For the lactating groups, there were no differences ($p < 0.01$) for the food consumption between red chontaduro- and control-lactating rats, but red chontaduro consumption reduced the total BW gain, total cholesterol, and increased HDL-cholesterol ($p < 0.01$). Chontaduro significantly increased HDL-cholesterol in the post-lactating rats. The increase in plasma lipids and triglycerides in both post-lactating groups was associated with changes in the adipose tissue metabolism during lactation. According to the authors, the obtained HDL-cholesterol value at the end of the study (2.59 mmol/L) was within the optimal level recommended by the American Heart Association.

The allegedly aphrodisiac properties of chontaduro have stimulated research aiming to provide scientific evidence to this claim. An *in vitro* assay evaluating the impact of a chontaduro extract (aqueous and ethanolic extracts from dried mesocarp) on spermatozoon motility from male sperm samples showed an agglutination effect on these cells (20 mg dried chontaduro/mL extract) without significant kinematic effect (Angulo-Paternina et al., 2017). Hence, the authors proposed the use of chontaduro extract for contraceptives purposes.

13.5 Conclusions and Final Remarks

Several constituents and/or products of *B. gasipaes* have been proposed for use in the food industry. For example, the fruit peels flour containing high dietary fiber (48%) and protein (9%) exhibit good emulsion activity and stability (≥50%) and can therefore be potentially used as natural additive (dietary fiber or emulsifier source) for the food industry (Martínez-Girón, Osorio, & Ordoñez-Santos, 2021). Cookies produced from whole *B. gasipaes* fruit (pulp + peel) and pulp flour had low moisture (5–6%), high lipid content and total carbohydrate (~26 and 60%, respectively) with high total energy value (~502 kcal/100 g). Furthermore, the cookies exhibited good sensory acceptance (>70%) and high purchase intention (85%) demonstrating the use of the fruit peel (a by-product) in developing new food products (Ribeiro, Monteiro, do Carmo, da Silva Pena, & Chisté, 2021). Snacks have also been developed with peach palm that had synergistic effect in combination with tara gum and carboxymethyl cellulose on the sensory characteristics highlighting natural peach flavor and improving crunchiness. The snack has high purchase intention (85%) and potential as functional food due to its composition (9 g fat, 140 g dietary fiber, and 15.5 mg carotenoids/100 g dry basis) and antioxidant capacity (4,700 μmol TE/100 g dry weight) (López-Calvo, Pérez, Ivankovich Guillén, Calderón Villaplana, & Pineda Castro, 2015).

Chontaduro is a sustainable source of bioactive compounds since crops can be exploited for several years, continuously providing edible fruits with promising biological potential. This potential is even higher in those communities largely linked to the product's harvesting, distribution, and consumption and can be some of the main drivers to increase research focused on this fruit. As chontaduro contains all EAAs, bioavailable carotenoids, and dietary fiber components associated with high antioxidant and nutritional effects, the fruit can potentially be used as a nutrition-rich ingredient to formulate common food products. Surprisingly, despite its reported biological properties, only a few *in vitro* and *in vivo* studies have assayed the impact of fruit consumption. Hence, more research to valorize this exotic fruit is needed to significantly impact communities with nutritional deficiencies, particularly in underdeveloped countries.

References

Angulo-Paternina, R., Ospina-Medina, L., Arango-Valencia, V., & Cardona-Maya, W. D. (2017). *In vitro* effect of three aphrodisiac plants on human sperm motility. *Revista Cubana de Plantas Medicinales*, *22*(4), 1-1.

Avila-Sosa, R., Montero-Rodríguez, A. F., Aguilar-Alonso, P., Vera-López, O., Lazcano-Hernández, M., Morales-Medina, J. C., & Navarro-Cruz, A. R. (2019). Antioxidant properties of Amazonian fruits: A mini review of *in vivo* and *in vitro* studies. *Oxidative Medicine and Cellular Longevity*, *2019*, 1–11. https://doi.org/10.1155/2019/8204129.

Bezerra, C. V., & da Silva, L. H. M. (2016). Pupunha (*B. gasipaes*): General and consumption aspects. In K. Kristbergsson & J. Oliveira (Eds.), *Traditional foods. Integrating food science and engineering knowledge into the food chain* (First, pp. 399–405). SpringerUS. https://doi.org/10.1007/978-1-4899-7648-2_33.

Bolanho, B. C., Danesi, E. D. G., & Beléia, A. D. P. (2015). Carbohydrate composition of peach palm (*B. gasipaes* Kunth) by-products flours. *Carbohydrate Polymers*, *124*, 196–200. https://doi.org/10.1016/j.carbpol.2015.02.021.

Caicedo, W., Ferreira, F. N., Viáfara, D., Guaman, A., Socola, C., & Moyano, J. C. (2019). Chemical composition and fecal digestibility in pigs of fermented Chontaduro (*B. gasipaes* Kunth) fruit. *Livestock Research for Rural Development*, *31*(9), 1–10.

Cantu-Jungles, T. M., Cipriani, T. R., Iacomini, M., Hamaker, B. R., & Cordeiro, L. M. C. (2017). A pectic polysaccharide from peach palm fruits (*B. gasipaes*) and its fermentation profile by the human gut microbiota *in vitro*. *Bioactive Carbohydrates and Dietary Fiber*, *9*, 1–6. https://doi.org/10.1016/j.bcdf.2016.11.005.

Cardona Jaramillo, J.E.C., Carrillo Bautista, M.P., Alvarez Solano, O.A., Achenie, L.E.K. & González Barrios, A.F. (2019). Impact of the mode of extraction on the lipidomic profile of oils obtained from selected Amazonian fruits. *Biomolecules*, *9*, 329; https://doi.org/10.3390/biom9080329.

Carvalho, R. P., Lemos, J. R. G., de Aquino Sales, R. S., Martins, M. G., Nascimento, C. H., Bayona, M., . . . Monteiro, J. B. (2013). The consumption of red pupunha (*B. gasipaes* Kunth) increases HDL cholesterol and reduces weight gain of lactating and post-lactating Wistar rats. *The Journal of Aging Research and Clinical Practice*, *2*(3), 257–260. http://www.ncbi.nlm.nih.gov/pubmed/25580386

Contreras-Calderón, J., Calderón-Jaimes, L., Guerra-Hernández, E.,& García-Villanova, B. (2011). Antioxidant capacity, phenolic content and vitamin C in pulp, peel and seed from 24 exotic fruits from Colombia. *Food Research International, 44*(7), 2047–2053. https://doi.org/10.1016/j.foodres.2010.11.003.

de Rosso, V. V., & Mercadante, A. Z. (2007). Identification and quantification of carotenoids, by HPLC-PDA-MS/MS, from Amazonian Fruits. *Journal of Agricultural and Food Chemistry, 55*(13), 5062–5072. https://doi.org/10.1021/jf0705421.

de Souza Mesquita, L. M., Murador, D. C., Neves, B. V., Braga, A. R. C., Pisani, L. P., & de Rosso, V. V. (2021). Bioaccessibility and cellular uptake of carotenoids extracted from *bactris gasipaes* fruit: differences between conventional and Ionic liquid-mediated extraction. *Molecules, 26*(13), 3989. https://doi.org/10.3390/molecules26133989.

de Souza Mesquita, L. M., Neves, B. V., Pisani, L. P., & de Rosso, V. V. (2020). Mayonnaise as a model food for improving the bioaccessibility of carotenoids from *B. gasipaes* fruits. *LWT, 122*, 109022. https://doi.org/10.1016/j.lwt.2020.109022.

de Souza Mesquita, L. M., Ventura, S. P. M., Braga, A. R. C., Pisani, L. P., Dias, A. C. R. V., & de Rosso, V. V. (2019). Ionic liquid-high performance extractive approach to recover carotenoids from *B. gasipaes* fruits. *Green Chemistry, 21*(9), 2380–2391. https://doi.org/10.1039/C8GC03283A.

dos Santos, M., Mamede, R., Rufino, M., de Brito, E., & Alves, R. (2015). Amazonian native palm fruits as sources of antioxidant bioactive compounds. *Antioxidants, 4*(3), 591–602. https://doi.org/10.3390/antiox4030591.

dos Santos, O. V., Gonçalves, B. S., da Macêdo, C. S., da Conceição, L. R. V., Costa, C. E. F., Monteiro Júnior, O. V., de Souza, A. L. G., & da Lannes, S. C. S. (2020). Evaluation of quality parameters and chromatographic, spectroscopic, and thermogravimetric profile of Patauá oil (Oenocarpus bataua). *Food Science and Technology, 40*(suppl 1), 76–82. https://doi.org/10.1590/fst.01619.

dos Santos, O. V., Soares, S. D., Dias, P. C. S., de Duarte, S. P. A., dos Santos, M. P. L., & das do Nascimento, F. C. A. (2020). Chromatographic profile and bioactive compounds found in the composition of pupunha oil (*B. gasipaes* Kunth): implications for human health. *Revista de Nutrição, 33*. https://doi.org/10.1590/1678-9805202033e190146.

Dufoo-Hurtado, E., Luzardo-Ocampo, I., Ceballos-Duque, S. M., Oomah, B. D., Maldonado-Celis, M. E., & Campos-Vega, R. (2021). Nuts by-products: the Latin American contribution. In R. Bhat (Ed.), *Valorization of agri-food wastes and by-products* (First (pp. 289–315). Elsevier. https://doi.org/10.1016/B978-0-12-824044-1.00025-8

Espinosa-Pardo, F. A., Martinez, J., & Martinez-Correa, H. A. (2014). Extraction of bioactive compounds from peach palm pulp (*B. gasipaes*) using supercritical CO_2. *The Journal of Supercritical Fluids, 93*, 2–6. https://doi.org/10.1016/j.supflu.2014.05.010.

FAO. (2018). *Worldwide palm kernels producers*. FAOSTAT. http://www.fao.org/faostat/es/#data/QCL.

Felisberto, M.H.F., Costa, M.S., Boas, F.V., Leivas, C.L., Franco, C.M.L., de Souza, S.M., Clerici, M.T.P.S. & Cordeiro, L.M.C. (2020). Characterization and technological properties of peach palm (*B. gasipaes* var. gasipaes) fruit starch. *Food Research International*, 136 : 109569; https://doi.org/10.1016/j.foodres.2020.109569.

Franco, T. S., Potulski, D. C., Viana, L. C., Forville, E., de Andrade, A. S., & de Muniz, G. I. B. (2019). Nanocellulose obtained from residues of peach palm extraction (*B. gasipaes*). *Carbohydrate Polymers, 218*, 8–19. https://doi.org/10.1016/j.carbpol.2019.04.035.

Garsetti, M., Vinoy, S., Lang, V., Holt, S., Loyer, S., & Brand-Miller, J. C. (2005). The glycemic and insulinemic index of plain sweet biscuits: relationships to *in vitro* starch digestibility. *Journal of the American College of Nutrition*, *24*(6), 441–447. https://doi.org/10.1080/07315724.2005.10719489.

Graefe, S., Dufour, D., van Zonneveld, M., Rodriguez, F., & Gonzalez, A. (2013). Peach palm (*B. gasipaes*) in tropical Latin America: Implications for biodiversity conservation, natural resource management and human nutrition. *Biodiversity and Conservation*, *22*(2), 269–300. https://doi.org/10.1007/s10531-012-0402-3.

Hempel, J., Amrehn, E., Quesada, S., Esquivel, P., Jiménez, V. M., Heller, A., . . . Schweiggert, R. M. (2014). Lipid-dissolved γ-carotene, β-carotene, and lycopene in globular chromoplasts of peach palm (*B. gasipaes* Kunth) fruits. *Planta*, *240*(5), 1037–1050. https://doi.org/10.1007/s00425-014-2121-3

Jatunov, S., Quesada, S., Díaz, C., & Murillo, E. (2010). Carotenoid composition and antioxidant activity of the raw and boiled fruit mesocarp of six varieties of *B. gasipaes*. *Archivos Latinoamericanos de Nutricion*, *60*(1), 99–104. http://www.ncbi.nlm.nih.gov/pubmed/21090177

Jiménez, G., Gómez, G., Pérez, A. M., & Blanco-Metzler, A. (2012). Estimation of glycaemix index of peach palm (*B. gasipaes*) cooked fruits and chips, and pitahaya (Hylocereus spp.) pulp. *Archivos Latinoamericanos de Nutrición*, *62*(3), 242–249.

Kawata, A., Murakami, Y., Suzuki, S., & Fujisawa, S. (2018). Anti-inflammatory activity of β-Carotene, lycopene and tri-n-butylborane, a scavenger of reactive oxygen species. *In vivo*, *32*(2). https://doi.org/10.21873/invivo.11232

Leterme, P., García, M.-F., Londoño, A.-M., Rojas, M.-G., Buldgen, A., & Souffrant, W.-B. (2005). Chemical composition and nutritive value of peach palm (*B. gasipaes* Kunth) in rats. *Journal of the Science of Food and Agriculture*, *85*(9), 1505–1512. https://doi.org/10.1002/jsfa.2146.

López-Calvo, R., Pérez, A. M., Ivankovich Guillén, C., Calderón Villaplana, S., & Pineda Castro, M. L. (2015). Evaluación de la aceptación por consumidores de un bocadillo de pejibaye (*B. gasipaes*) y estudio de su potencial como alimento funcional. *Archivos Latinoamericanos de Nutrición*, *65*(1), 51–58.

Luzardo-Ocampo, I., Ramírez-Jiménez, A. K., Yañez, J., Mojica, L., & Luna-Vital, D. A. (2021). Technological applications of natural colorants in food systems: A review. *Foods*, *10*(3), 634. https://doi.org/10.3390/foods10030634.

Martínez-Girón, J., Osorio, C. & Ordoñez-Santos, L.E. (2021). Effect of temperature and particle size on physicochemical and techno-functional properties of peach palm peel flour (*B. gasipaes*, red and yellow ecotypes). *Food Science and Technology International*, July 2; https://doi.org/10.1177/10820132211025133.

Matos, K. A. N., Lima, D. P., Barbosa, A. P. P., Mercadante, A. Z., & Chisté, R. C. (2019). Peels of tucumã (Astrocaryum vulgare) and peach palm (*B. gasipaes*) are by-products classified as very high carotenoid sources. *Food Chemistry*, *272*, 216–221. https://doi.org/10.1016/j.foodchem.2018.08.053.

Meinhart, A. D., Damin, F. M., Caldeirão, L., de Jesus Filho, M., da Silva, L. C., da Silva Constant, L., Filho, J. T., Wagner, R., & Godoy, H. T. (2019). Chlorogenic and caffeic acids in 64 fruits consumed in Brazil. *Food Chemistry*, *286*, 51–63. https://doi.org/10.1016/j.foodchem.2019.02.004.

Mir, N. A., Riar, C. S., & Singh, S. (2019). Effect of pH and holding time on the characteristics of protein isolates from Chenopodium seeds and study of their amino acid profile and scoring. *Food Chemistry, 272*, 165–173. https://doi.org/10.1016/j.foodchem.2018.08.048.

Neira-Mosquera, J. A., Pérez-Rodríguez, F., Sánchez-Llaguno, S., & Moreno-Rojas, R. (2013). Study o the mortality in Ecuador related to dietary factors. *Nutrición Hospitalaria, 28*(5), 1732–1740. https://doi.org/10.3305/nh.2013.28.5.6755.

Ottosson, F., Ericson, U., Almgren, P., Nilsson, J., Magnusson, M., Fernandez, C., & Melander, O. (2016). Postprandial levels of branch chained and aromatic amino acids associate with fasting glycaemia. *Journal of Amino Acids, 2016*, 1–9. https://doi.org/10.1155/2016/8576730.

Peixoto Araujo, N. M., Arruda, H. S., Marques, D. R. P., de Oliveira, W. Q., Pereira, G. A., & Pastore, G. M. (2021). Functional and nutritional properties of selected Amazon fruits: A review. *Food Research International, 147*, 110520. https://doi.org/10.1016/j.foodres.2021.110520.

Radice, M., Viafara, D., Neill, D., Asanza, M., Sacchetti, G., Guerrini, A., & Maietti, S. (2014). Chemical characterization and antioxidant activity of Amazonian (Ecuador) *Caryodendron orinocense* Karst. and *B. gasipaes* Kunth seed oils. *Journal of Oleo Science, 63*(12), 1243–1250.

Restrepo, J., Estupiñán, J. A., & Colmenares, A. J. (2016). Estudio comparativo de las fracciones lipídicas de B. gasipaes Kunth (chontaduro) obtenidas por extracción soxhlet y por extracción con CO2 supercrítico. *Revista Colombiana de Química, 45*(1), 5. https://doi.org/10.15446/rev.colomb.quim.v45n1.57199

Ribeiro, G. S., Monteiro, M. K. C., do Carmo, J. R., da Silva Pena, R., & Chisté, R. C. (2021). Peach palm flour: production, hygroscopic behavior and application in cookies. *Heliyon, 7*(5); https://doi.org/10.1016/j.heliyon.2021.e07062.

Rojas-Garbanzo, C., Pérez, A. M., Vaillant, F., & Pineda-Castro, M. L. (2016). Physicochemical and antioxidant composition of fresh peach palm (*B. gasipaes* Kunth) fruits in Costa Rica. *Brazilian Journal of Food Technology, 19*. https://doi.org/10.1590/1981-6723.9715.

Smith, Nigel. (2015a). *B. gasipaes*. In Nigel Smith (Ed.), *Palms and people in the amazon* (First, pp. 177–194). Springer International Publishing. https://doi.org/10.1007/978-3-319-05509-1.

Smith, Nigel. (2015b). Palms and cultural landscapes. In N. Smith (Ed.), *Palms and people in the amazon* (First, pp. 1–8). Springer International Publishing. https://doi.org/10.1007/978-3-319-05509-1.

Tome, D. (2012). Criteria and markers for protein quality assessment—A review. *British Journal of Nutrition, 108*(S2), S222–S229. https://doi.org/10.1017/S0007114512002565.

Torres-Vargas, O. L., Luzardo-Ocampo, I., Hernandez-Becerra, E., & Rodríguez-García, M. E. (2021). Physicochemical characterization of unripe and ripe chontaduro (*B. gasipaes* Kunth) fruit flours and starches. *Starch - Stärke, 73*(7–8), 2000242. https://doi.org/10.1002/star.202000242.

Udenigwe, C. C., & Aluko, R. E. (2010). Antioxidant and angiotensin converting enzyme-inhibitory properties of a flaxseed protein-derived high fischer ratio peptide mixture. *Journal of Agricultural and Food Chemistry, 58*(8), 4762–4768. https://doi.org/10.1021/jf100149w.

Ulbricht, T. L. V., & Southgate, D. A. T. (1991). Coronary heart disease: seven dietary factors. *The Lancet, 338*(8773), 985–992. https://doi.org/10.1016/0140-6736(91)91846-M.

Vieira, T. F., Corrêa, R. C. G., de Fatima Peralta Muniz Moreira, R., Peralta, R. A., de Lima, E. A., Helm, C. V., Garcia, J. A. A., Bracht, A., & Peralta, R. M. (2021). Valorization of peach palm (*B. gasipaes* Kunth) waste: production of antioxidant xylooligosaccharides. *Waste and Biomass Valorization, In press.* https://doi.org/10.1007/s12649-021-01457-3.

Yuyama, L. K. O., Aguiar, J. P. L., Yuyama, K., Clement, C. R., Macedo, S. H. M., Fávaro, D. I. T., Afonso, C., Vasconcellos, M. B. A., Pimentel, S. A., Badolato, E. S. G., & Vannucchi, H. (2003). Chemical composition of the fruit mesocarp of three peach palm (*B. gasipaes*) populations grown in Central Amazonia, Brazil. *International Journal of Food Sciences and Nutrition, 54*(1), 49–56. https://doi.org/10.1080/0963748031000061994.

14

Saffron: A Functional Food with Potential Molecular Effects

Yousef Rasmi[1,2], Elihud Salazar[3], Ena Gupta[4], Behrokh Daei-Hasani[5], and Martín Calderón-Juárez[6]

[1] Department of Biochemistry, School of Medicine, Urmia University of Medical Sciences, Urmia, Iran
[2] Cellular and Molecular Research Center, Urmia University of Medical Sciences, Urmia, Iran
[3] Centro Universitario de la Costa, University of Guadalajara, Puerto Vallarta, Mexico
[4] Department of Home Science, University of Allahabad, Allahabad, India
[5] Department of Biology, Payame Noor University, Tehran, Iran
[6] Plan de Estudios Combinados en Medicina, National Autonomous University of Mexico, Faculty of Medicine, Mexico City, Mexico

14.1 Introduction

Crocus sativus L. (saffron) is a perennial herb belonging to the Iridaceae family usually called "red gold," "autumn crocus," or "saffron crocus." This stemless herb is cultivated all around the world especially in Irano-Turanian, East Asia (Kashmir), and Western Mediterranean (Spain) regions. The term "saffron" originated from the French word Safran, which means "yellow" in Arabic (Zargari, 1990). According to some experts, saffron was first recognized in the seventh century BCE, under Assyrian botanical reference compiled by Ashurbanipal, whereas use of saffron and its cultivation by humans spans over 3,500 years that extends across civilizations, cultures, and continents for the treatment of around 90 illnesses (Deo, 2003). Saffron is more acclimatized to mountain valleys, hill sides, and plateaus extending in altitudes from 600 to 1700 m and growth are also observed at approximately 2000 m altitude above sea level (Delgado, Aramburu, & Alonso, 2006).

A total of 300 t/yr saffron is produced annually around the world 80% of which contributed by Iran (Fernández, 2004). Saffron is considered as world's most precious and expensive spice derived from the three dried stigmas (25–30 mm long) of the *C. sativus* (saffron crocus).

The plant is not propagated by seed; hence, underground parts (bulbs or corms) are used to produce new plant. The three yellow stamens are also present in the flower, which is generally not collected as they do not contain active compounds. From each bulb one to seven light purple colored flowers are produced, with reddish-colored thread-like stigma of the flower that is treasured equally as natural colorant and as a spice. Around 70,000 flowers are used to yield about 200,000 dried stigmas producing 500 g of pure saffron. In the American market, it costs around $30 per ounce as only once a year saffron blooms and

Molecular Mechanisms of Functional Food, First Edition. Edited by Rocio Campos-Vega and B. Dave Oomah.

Table 14.1 The plant profile (systematic classification) of *C. Sativus*.

Kingdom	Plantae
Subkingdom	Tracheobionta, vascular plant
Superdivision	Spermatophyta, seed plants
Division	Spermatophyte
Clade	Angiosperms
Clade	Monocots
Order	Asparagales
Family	Iridaceae
Genus	*Crocus*
Species	*C. sativus*
Botanical name	*Crocus sativus* L.

should be collected in that short period (Schmidt, Betti, & Hensel, 2007; Wani, Hamza, & Mohiddin, 2011). The systematic classification of the plant is listed below (Table 14.1).

Saffron (stigmas of *C. sativus* L.) is full of significant nutrients and minerals, which is necessary for maintaining and regulating diverse metabolic processes in the body. The predominant natural bioactive constituents present in saffron are safranal, crocin, and picrocrocin (Abdullaev, 2002). About 28 volatile and aroma-yielding chemical compounds are present in saffron along with nonvolatile phytochemicals such as α- and β-carotenes, carotenoids, lycopene, zeaxanthin, and conjugated xanthocarotenoid. The major active compounds responsible for saffron aroma are dihydrooxophorone and ketoisophorone, whereas its characteristic color is due to crocetinglycosyl esters (Dhar, Sharma, Bhat, Chrungoo, & Kaul, 2017). Significant proportions of these compounds are present in saffron stigmas. The other important nutrients responsible for antioxidant property of saffron include proteins, thiamine, riboflavin, manganese, potassium, sodium, iron, zinc, and copper (Nehvi, Salwee, & Arshid-Gowhar, 2011). In Mesopotamia, many ancient cultures used saffron primarily as an herb or condiment, in traditional medicine its tinctures and extracts have been used as an expectorant, aphrodisiac, sedative, analgesic, diaphoretic, stomachic, carminative, antispasmodic, emenagogue, and eupeptic agent (Ghaffari & Roshanravan, 2019).

Bioactive compounds present in saffron can provide potential preventive and therapeutic applications. Its molecular and cellular effects are primarily attributed to its pharmacological effects. Research suggests that saffron and its bioactive constituents exert antioxidant (Samarghandian, Farkhondeh, & Zeinali, 2020), anti-inflammatory (Zeinali, Zirak, Rezaee, Karimi, & Hosseinzadeh, 2019), anticarcinogenic, antigenotoxic, antimutagenic, antitumoural and cytotoxic activities (Abdullaev et al., 2003; Tavakol-Afshari, Boskabady, & Salari, 2020), antihypertensive (Nasiri, Sameni, Vakili, Jarrahi, & Khorasani, 2015), antiatherogenic (Hemmati et al., 2015), antihyperlipidemic (Lahmass et al., 2017), antidiabetic (Milajerdi et al., 2018), anticonvulsion (Hosseinzadeh & Khosravan, 2002), neuroprotective (Bian, Zhao, & Lee, 2020), and antidepression (Orio, Alen, Ballesta, Martin, & Gomez de Heras, 2020) properties. Saffron and its chemical components are recommended for use either as an additive in food products or in the free form. The potential beneficial effect of saffron depends on its bioavailability and amount consumed, though amount of saffron consumed differs significantly between various types of culinary diet and different regions. This chapter reviews the recent systematic literature to bridge the gap among researchers by summarizing the chemistry of bioactive components, absorption, bioavailability, metabolism, degradation, and therapeutic applications of saffron.

14.2 Botanical Review

C. sativus (Saffron) is a perennial herbaceous plant that emerges from the ground and grows up to 40 cm in length from its bulbs (Figure 14.1). The bulbs are covered with many concentric spathes as its structure is massive varying in forms and sizes. The propagation of saffron is through bulbs, also known as corms, and new bulbs are formed by each corm that helps in plant multiplication. Thin and fibrous roots are at the bottom of the mother bulb, whereas contractile roots are formed at the base of the lateral bulbs. There is no seed propagation in case of saffron plant and new plants are produced from underground parts like bulbs or corms (Khalesi & Behboodi, 2003). Each bud has about 5 to 11 dark green narrow leaves ranging in the middle from 1.5 and 2.5 mm and 20 to 60 cm in length. The flowers of crocus are hermaphrodite, and the pollination is through butterflies and bees. At the beginning of autumn, the flowers start to bloom which are terminal, solitary, and purplish in color consisting of six petals (three internal and three external), three stamens, and a style ending with a single thread-like stigma comprising three filaments of deep

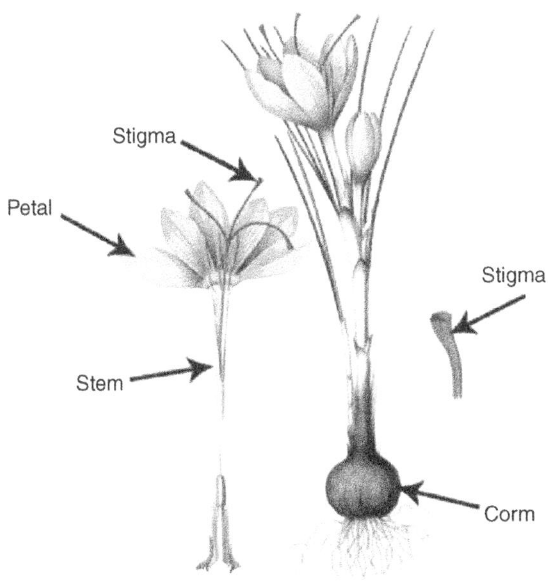

Figure 14.1 Different parts of *Crocus sativus* L. (Saffron) Plant (Shahi, Assadpour, & Jafari, 2016).

red-orange color of 2.5 to 3 cm long (Molina, García-Luis, Valero, Navarro, & Guardiola, 2003). Only once in a year saffron blooms mainly during 3–4 weeks in October–November and has to be collected within that short period as they are open for only few days. Germination occurs at one to 6 months at 18 °C and around 3 years are required for plants to flower from seeds (Nadkarni, 1996). Totally, 120,000 crocus flowers are harvested to produce 1 kg of dry saffron that is dried till they lose four-fifths of their weight.

14.3 Chemistry of Saffron

C. sativus is a remarkable herb of medicinal value because of its natural yellow color, flavor, and strong odor. More than 150 components are present in the saffron stigmas (Samarghandian & Borji, 2014). The other nutritional elements that offer health benefit include carbohydrates, proteins, fats, fiber, flavonoids, anthocyanins, vitamins (thiamine and riboflavin), and minerals. The classification of chemical compounds present in saffron is based on the presence of volatile compounds (safranal, terpene, terpene esters, and terpene alcohol) and nonvolatile compounds (carotenoids, alpha- and beta- carotenes, crocin, picrocrocin, α-crocin, zeaxanthin, and lycopene). De-glycosylation of picrocrocin produces the carboxaldehyde volatile compound safranal the main bioactive component of *C. sativus* (Himeno & Sano, 1987).

Research studies revealed that distinct flavor and aroma in saffron are primarily due to chemical compounds safranal, while picrocrocin is responsible for its bitter taste. Over 65% of the total aroma components are represented by safranal that is present in the world's best quality saffron, "Azafrán de La Mancha" (Carmona, Zalacain, Salinas, & Alonso, 2007). In the *Crocus* genus, the safranal precursor is picrocrocin (more stable substance than the crocetin esters). Dehydration and hydrolysis of picrocrocin also generate safranal, during storage, dehydration, and manipulation of saffron (José Bagur et al., 2017).

The yellow golden hue arises from naturally degraded carotenoid compounds such as crocin and crocetin. Crocin derived from the word "crocos" meaning saffron in German was first isolated in 1818 as the main carotenoid (Valle Garcia-Rodriguez, Serrano-Diaz, Lopez-Corcoles, Carmona, & Alonso, 2014). Its molecular formula and structure were established by Karrer and Salomon (1928), while its glycosidic nature was reported later (Decker, 1914), and its systematic name is 8,8-diapo-8,8-carotenoic acid. The dicarboxylic carotenoid (crocetin) derives the name crocin after glycosylation, as its chemical structure forms the central core of crocin. It is a natural apocarotenoid dicarboxylic acid with systematic name 8,8′-diapocarotene dioic acid. Crocins at high concentrations in the aqueous extract pool together and aggregate (Naess et al., 2006). Many crocins have been identified through high-performance liquid chromatography-UV–visible photodiode-array detection-mass spectrometry (Tarantilis, Tsoupras, & Polissiou, 1995). The odor and color active compounds in saffron emanate from bio-oxidative cleavage of zeaxanthin (Bolhassani, Khavari, & Bathaie, 2014). Kaempferol 3-*O*-alpha-(2,3-di-*O*-beta-D-glucopyranosyl) rhamnopyranoside was identified as a novel flavonol glycoside (Norbak & Kondo, 1999). Worldwide ISO 3632 standards regulate saffron quality by standardizing the saffron classification. The chemical identification and main chemical identifiers for the active components of the genus *Crocus* are presented in Table 14.2.

Table 14.2 The chemical identity of active components in the *Crocus* genus.

Chemical structure	IUPAC name	Chemical formula	Molecular weight	C.A.S. number
Crocin 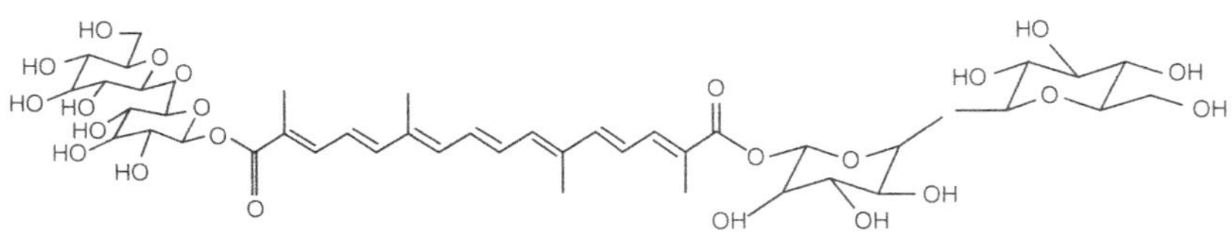	bis[(2*S*,3*R*,4*S*,5*S*,6*R*)-3,4,5-trihydroxy-6-[[(2*R*,3*R*,4*S*,5*S*,6*R*)-3,4,5-trihydroxy-6-(hydroxymethyl)oxan-2-yl]oxymethyl] oxan-2-yl] (2*E*,4*E*,6*E*,8*E*,10*E*,12*E*,14*E*)-2,6,11,15-tetramethylhexadeca-2,4,6,8,10,12,14-heptaenedioate	$C_{44}H_{64}O_{24}$	977 g/mol	42,553–65-1
Crocetin 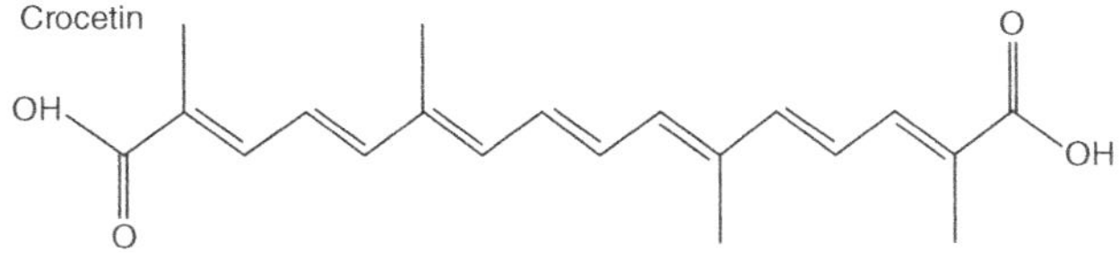	(2*E*,4*E*,6*E*,8*E*,10*E*,12*E*,14*E*)-2,6,11,15-tetramethylhexadeca-2,4,6,8,10,12,14-heptaenedioic acid	$C_{20}H_{24}O_4$	328.408 g/mol	27,876–94-4
Picrocrocin 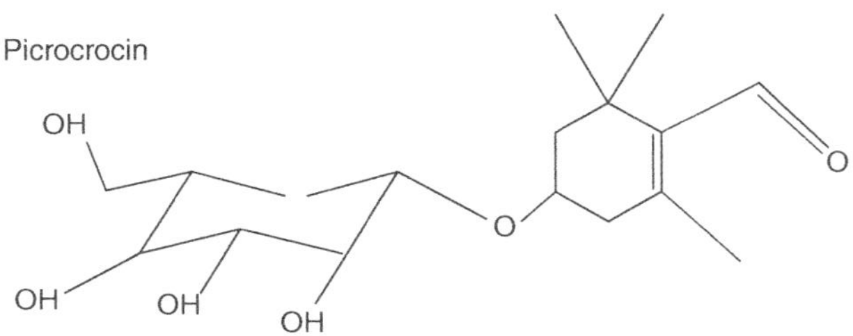	2,6,6-trimethyl-4-{[3,4,5-trihydroxy-6-(hydroxymethyl)oxan-2-yl]oxy} cyclohex-1-ene-1-carbaldehyde	$C_{16}H_{26}O_7$	330.37 g/mol	138–55-6
Safranal 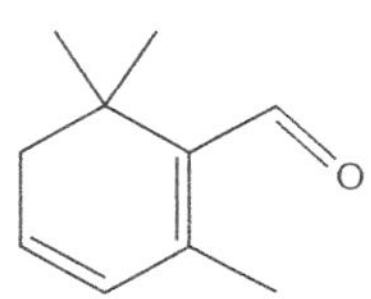	2,6,6-trimethylcyclohexa-1,3-diene-1-carbaldehyde	$C_{10}H_{14}O$	150.22 g/mol	116–26-7

14.4 Bioavailability of Saffron

The rate and extent of an active drug compound absorbed and commonly administered orally can be available to its targeted tissue (Chow, 2014). Several compounds can be found in saffron, such as carbohydrates, proteins, amino acids, minerals, mucilage, gums, and vitamins (riboflavin and thiamine) (Shahi et al., 2016). However, the biological active principal components of the saffron stigma are crocin, crocetin, picrocrocin, and safranal (Moratalla-López, Bagur, Lorenzo, Salinas, & Alonso, 2019); these compounds are responsible for the smell, taste, color, and pharmacological properties of saffron. Saffron (5 g) can provoke severe adverse reactions and 12–20 g/d is potentially fatal (Chryssanthi, Lamari, Georgakopoulos, & Cordopatis, 2011). Little is known about the bioavailability of saffron components. The absorption of saffron's metabolites are based on experimental data in animal models, most commonly rodents. The most relevant information about this topic is presented in this section.

Crocin is a carotenoid, different from other compounds of its class, because of its solubility in water due to its glycosidic esters. In spite of the continuous efforts to understand the absorption of crocins, the exact mechanism of absorption and metabolism remains unknown. Once ingested, crocins remain in the gastrointestinal tract and are excreted with feces (Xi, Qian, Du, & Fu, 2007), it is likely that crocins are hydrolyzed to crocetin in the intestinal lumen by enzymes located in the enterocytes epithelium such as β-glucosidases (Németh et al., 2003) and to a lesser extent by bacteria in the intestinal tract.

Rats were divided in two groups (Shakya, Nepal, Kang, & Jeong, 2020), one treated with a mix of oral antibiotics; cefadroxil (a cephalosporin that inhibits the enzymatic reactions necessary to produce bacterial cell wall, specifically β-lactamases in Gram-negative bacteria) (Tanrisever & Santella, 1986), erythromycin (a macrolide antibiotic with bacteriostatic effect against Gram-positive, Gram-negative bacteria by inhibiting the peptide production through the blockade of 23S ribosomal RNA in the 50s subunit in the ribosomes) (Farzam, Nessel, & Quick, 2020), and oxytetracycline hydrochloride (a broad-spectrum antibiotic that inhibits protein synthesis in bacteria by binding to the 30S ribosomal unit) (Pickens & Tang, 2010) and a control group. The bacterial enzymatic activity decreased in antibiotic administered rats and thus the intestinal microbiota ability to convert crocin to crocetin is expected to be compromised. The study showed sevenfold reduction in cumulative concentration of crocetin in plasma of rats treated with antibiotics, implying that intestinal microbiota plays a major role in crocetin pharmacokinetics.

One of the most important investigations (Chryssanthi et al., 2011) has contributed to the determination of crocetin absorption after oral consumption. In that study, a saffron infusion (200 mg in 150 mL of hot [80 °C] water rested for 5 min) was administered to 4 healthy volunteers, and blood samples were obtained before, 2, and 24 hours after infusion. Plasma concentration of *trans*-crocetin and *cis*-crocetin was measured by solid phase extraction (SPE) and high-performance liquid chromatography (HPLC), a reliable detection method for quantifying drugs and metabolites in biological fluids, including blood samples (Matuszewski, Constanzer, & Chavez-Eng, 2003). Interestingly, these experiments showed a steep increase (1.27–3.67 μM) in total crocetin during the first 2 hr in all participants, and even a non-negligible plasma concentration (0.10–0.24 μM) after 24 hours. Most of the total crocetin concentration was attributed to *trans*-crocetin, which is consistent with the previous concept that *trans*-crocetin is the main absorbed crocetin isomer (Moratalla-López et al., 2019). Moreover, *cis*-crocetin as a percentage of detectable crocetin doubled (24.5–49.4%) in the first 2 hr, and surprisingly, the *cis*-crocetin isomer was undetected in plasma samples after 24 hours ingestion (Chryssanthi et al., 2011).

Other major study that addresses the problem of crocetin absorption in humans was conducted by Umigai et al. (2011). In that study, plasma crocetin concentration was measured in 10 healthy individuals with a male–female proportion of 1:1by high-performance liquid chromatography, using solid-phase extraction technique (SPE-HPLC). The crocetin concentration was quantified before and after a single administration of 7.5, 15, and 22.5 mg of crocetin. Maximum plasma concentration was achieved after approximately 4–4.8 hours of ingestion. The mean maximum concentration increased from 100.9 ng to 279.7 ng when 7.5 mg and 22.5 mg of crocetin were ingested, respectively, and total clearance of crocetin from plasma after oral administration was between 12.8 (l/h) and 13.8 (l/h) without statistical difference among oral doses.

Once crocetin reaches the bloodstream, it is transported to the liver via the enterohepatic system and metabolized to crocetin-monoglucuronide and -diglucuronide (both glucuronide conjugates) (Asai, Nakano, Takahashi, & Nagao, 2005). Crocetin permeates the blood brain barrier, implying this compound has potential effects on the central nervous system (Lautenschläger et al., 2015). Abiotic *in vitro* digestion of 28 mg of saffron extract in three stages (oral, gastric, and duodenal digestion) showed that each component undergoes several modifications in its bioavailability (Almodóvar, Briskey, Rao, Prodanov, & Inarejos-García, 2020). For instance, bioaccessibility of crocin isomers is greatly reduced after the gastric phase of *in vitro* digestion, particularly *trans*-crocin-4. However, after all the *in vitro* digestion steps, crocetin concentration was not reduced.

After crocin, picrocrocin is the most abundant ionapocarotenoid, however, little is known about its absorption mechanism and biological effects. Experiments have shown that the intestinal epithelium directly absorbs picrocrocin *in vitro* is unlikely, because this compound needs to be further metabolized probably by either enterocytes epithelium or bacteria in the intestine lumen to be absorbed (Kyriakoudi, O'Callaghan, Galvin, Tsimidou, & O'Brien, 2015). Safranal has several effects on the central nervous system, sexual behavior, respiratory, metabolic, and antihypertensive activities, although little is known about its oral bioavailability (Rezaee & Hosseinzadeh, 2013).

14.5 Bioactivity of *C. sativus* L. Compounds

Saffron is a source of natural carotenoids and pigments with a specific chemical structure of the conjugated double bond system and is responsible for the processes of color, reaction, and energy transfer (Britton, 1995). Saffron can be considered a functional food or nutraceutical because it is rich in antioxidant compounds. Several scientific studies highlight the biomedical and medicinal properties of saffron or some of its metabolites in the prevention and modulation of various chronic diseases (Bolhassani et al., 2014)

During the last decade, several studies have described the bioactivity of saffron compounds. However, bioactive metabolites are also present in floral bioresidues (Hosseini, Razavi, & Hosseinzadeh, 2018) and *C. sativus* L. whole flowers. The effect of bioactive substances is not related to the dose consumed but to their bioavailability. Similarly, bioavailable doses can lead to different effects (Holst & Williamson, 2008).

14.5.1 Bioactivity of the Three Main Saffron Compounds

The three main ingredients of saffron include crocetin/crocins, picrocrocin, and safranal with potential therapeutic value and also present in the stigma (Alavizadeh & Hosseinzadeh, 2014). Other compounds confer its nutritional and health beneficial properties such as other carotenoids, carbohydrates, fiber, proteins, fats, vitamins (riboflavin and thiamine), minerals, as well as phenolic compounds such as anthocyanins and kaempferol glycosides, in addition to many other nourishing factors (Alavizadeh & Hosseinzadeh, 2014; Serrano-Díaz, Sánchez, Martínez-Tomé, Winterhalter, & Alonso, 2013). Saffron's main carotenoids are glycosidic esters responsible for their unique solubility in water. Their absorptions and metabolisms still remain unresolved (Chryssanthi et al., 2011). Transcrocetin is the only active metabolite that, if used with pure crocetin or saffron extract, crosses the blood brain barrier (BBB) and reaches the central nervous system (Yoshino, Yoshida, Umigai, Kubo, & Lee, 2011).

Safranal induces diverse bioactivities and are less known than crocins (Kyriakoudi et al., 2015). It counteracts oxidative stress in nerve cells by eliminating free radicals (Kanakis, Tarantilis, Tajmir-Riahi, & Polissiou, 2007). Safranal is a particular antidepressant agent (Hosseinzadeh & Sadeghnia, 2007) known to have anticonvulsant effects compared to crocins (Hosseinzadeh & Sadeghnia, 2007; Hosseinzadeh & Talebzadeh, 2005). Picrocrocin showed antiproliferative activity in human cancer cell lines (Kyriakoudi et al., 2015).

Saffron is linked through the anti-inflammatory (Hosseinzadeh & Younesi, 2002), anticonvulsants (Hosseinzadeh & Talebzadeh, 2005), satiating (Gout, Bourges, & Paineau-Dubreuil, 2010), antihistamines, anti-asthmatic, genoprotective, antitussive, protection of gastric mucosa, and sexual dysfunction activities (Licón, Carmona, Llorens, Berruga, & Alonso, 2010). Neuroprotective activity of saffron has been shown in an experimental model of cerebral ischemia (Hosseinzadeh & Sadeghnia, 2007; Zheng, Liu, Wang, & Xu, 2007). Saffron also has a neuroprotective effect in age-related macular degeneration. Thus, in a clinical trial, oral saffron supplementation (20 mg/d over 3 months) improved retinal flicker sensitivity and protected against retinal damage in 25 patients with age-related macular degeneration (Falsini et al., 2010; Piccardi et al., 2012).

It is useful for facilitating and controlling menstruation and relieves back pain associated with menstruation (Agha-Hosseini et al., 2008; Ulbricht et al., 2011). Moreover, crocetin administered to experimental animals after a hemorrhagic shock significantly increased survival and reduced apoptosis (Yang et al., 2011). Saffron can have antiamyloidogenic activity that has a positive effect on cognitive function. Amyloid β-peptide fibrillogenesis or acetylcholinesterase activity inhibitory effects are effective in Alzheimer's disease (Akhondzadeh et al., 2010; Alavizadeh & Hosseinzadeh, 2014). Crocetin can help prevent Parkinson's disease (Ahmad et al., 2005). Saffron and its metabolites are effective in various models of mental disorders, including depression and anxiety compared to fluoxetine or imipramine (Akhondzadeh et al., 2005; Akhondzadeh, Fallah-Pour, Afkham, Jamshidi, & Khalighi-Cigaroudi, 2004). In fact a meta-analysis confirmed saffron's beneficial effects on depression in adults with major depressive disorders (Hausenblas, Saha, Dubyak, & Anton, 2013). Saffron supplementation (150 mg of pure saffron from powdered dry saffron leaves [19.7 mg crocin and 0.25 mg safranal/g saffron]) combined with resistance training for 6 weeks improved concentration of blood markers implicated in depression in untrained young (average age 24 years old) men in a randomized, double-blind, and parallel study (Moghadam et al., 2021). Supplementation increased markers of antidepressive effects (dopamine), the endocannabinoid system markers [2-arachidonoglycerol (2-AG), and anandamide (AEA)] without changes in tryptophan serum concentrations.

Some studies claim potential weight loss upon saffron consumption (Gout et al., 2010), via preventing fat digestion from the diet by inhibiting pancreatic lipase, inhibiting inflammatory cytokines and differentiating fat cells, increasing satiety by decreased food intake, or increased neurotransmitter or hormonal function for feeling full, and increase in glucose and lipid metabolism.

Antitumor effects of saffron metabolites were observed in rats and cancer cells. Several studies have shown that saffron and/or its metabolites function as an apoptotic activator in cancer cells, but does not affect normal cells (Escribano, Alonso, Coca-Prados, & Fernández, 1996; Zahra, Reyhane, Hamidreza, & Majid, 2013). Recent *in vitro* and *in vivo* studies show that saffron components, especially crocins and crocetin, may have anticancer effects on breast, lung, and pancreatic cancer cells (Samarghandian & Borji, 2014). The mechanism of action is unknown to date, but its activity may be related to the antioxidant function of crocins, which can affect cell growth regulation and modulate gene expression and immune responses (Samarghandian, Tavakkol, & Davoodi, 2011). Crocetin is responsible for inhibiting nucleic acid synthesis, improving the activity of endogenous antioxidant systems, inducing apoptosis, and disrupting the signaling pathways of tumor growth factors (William, Gregory, Amitabha, Shrikant, & Animesh, 2012). Picrocrocin reduces the proliferation of adenocarcinoma and liver cancer cells in humans (Kyriakoudi et al., 2015).

Effect of liver protection *in vivo* studies has shown that crocin reduces caspases activation, essential mediators of apoptosis, and reduces the ratio between proapoptotic protein bax and antiapoptotic Bcl-2 (bax/Bcl-2), which decrease liver toxicity (Lari et al., 2015). Over the past decade, numerous studies (Table 14.3) have been conducted on the biomedical and medicinal properties of saffron and its bioactive molecules. The possible effects of saffron and its bioactive compounds depend on the amount of their consumption and bioavailability.

14.5.2 Bioactivity of Phenolic Compounds of *C. sativus* L. Flower

In addition to the bioactivity of saffron's primary ingredients mentioned above, the main by-product of saffron production, the tepals of *C. sativus* L. flower, contains alkaloids, minerals, anthocyanins, kaempferol, and their glycosides. This combination provides antioxidant features and potential remedial applications of saffron (Hosseini et al., 2018; Termentzi & Kokkalou, 2008). High phenolic contents are found in floral bioresidues, whole flowers, and tepals (Goupy, Vian, Chemat, & Caris-Veyrat, 2013; Serrano-Díaz, Sánchez, Martínez-Tomé, Winterhalter, & Alonso, 2014). Five kaempferol derivatives are known in this spice (Carmona, Sanchez et al., 2007). The highest amount of phenolic components is mainly in tepals (Serrano-Díaz et al., 2012). In traditional medicine, the tepals of *C. sativus* L. flower were used as an antispasmodic, anxiolytic, antitumor, and antidepressant (Hosseini et al., 2018). Several studies attribute the antioxidant activity of tepals to their

Table 14.3 Summary of the main bioactivity of saffron and its bioactive molecules.

Bioactive Compound	Bioactivity (Reference)	Model	Dose
trans-crocetin	Cross the blood–brain barrier and reach the central nervous system (Linardaki, Orkoula, Kokkosis, Lamari, & Margarity, 2013; Yoshino et al., 2011)	Rats	Oral administration (100 mg/kg)
Crocetin	Neuroprotection (Ahmad et al., 2005)	Hemi-Parkinson rats	Peripheral administration (25, 50, and 75 μg/kg body weight)
	Improved post-shock survival and reduced apoptosis (Yang et al., 2011)	Rats	Bolus injection (2 mg/kg body weight)
	Cardioprotective effects (after myocardial ischemia reperfusion injury) (Wang, Sun, Liu, & Fang, 2014)	Adult male Wistar rats	Intragastric administration (50 mg/kg/day)
Crocins	Hepatoprotective effects (Lari et al., 2015)	Rats	Intraperitoneally (25 mg/ kg body weight/ day for 4 weeks)
Safranal	Antidepressant (Hosseinzadeh & Sadeghnia, 2007)	Rats	Peripheral administration (15.5 mg/kg body weight)
	Anticonvulsant (Hosseinzadeh & Talebzadeh, 2005)	Mice	Injected (0.15 and 0.35 mg/kg)
Picrocrocin	Antitumor effects (Hosseinzadeh & Sadeghnia, 2007)	Human colon adenocarcinoma (Caco-2-cell model)	8–24 μM
Saffron extracts	Satiating (Gout et al., 2010)	Human (randomized, double-blind, placebo-controlled, parallel group)	Oral administration (capsule: 176.5 mg extract/day for 8 weeks)
	Reduce cognitive deterioration (Alzheimer's disease) (Farokhnia et al., 2014)	Patients (randomized double-blind parallel group)	Oral administration (capsule: 30 mg/d for 12 months)
Saffron	Premenstrual syndrome (Agha-Hosseini et al., 2008)	Women (double-blind, randomized and placebo-controlled trial)	Oral administration (capsule: 30 mg/d for 6 months)
	Neuroprotection (macular degeneration) (Bisti, Maccarone, & Falsini, 2014)	Albino rats with light-induced photoreceptors degenerations	Oral administration (1 mg/kg/day for 6 weeks)
	Improve the symptoms of children with deficit hyperactivity disorder (Baziar et al., 2019)	Children	Oral administration (20–30 mg/ day for 6 weeks)
	Effective treatment in depression and anxiety (Ghajar et al., 2017)	Patients (double-blind-controlled clinical trial)	Oral administration (30 mg/day per 6 weeks)

Source: Adapted from Moratalla-López et al. (2019).

phenolic content (Serrano-Díaz et al., 2012; Termentzi & Kokkalou, 2008). The stamens and perianth have significant antifungal, cytotoxic, and antioxidant activities (Zheng, Li, Ma, & Han, 2011). Kaempferol, kaempferide, and isorhamnetin have been detected in the pollen, which are methylated glycosides with antityrosinase activity (Li & Wu, 2002).

Tepal extract of *C. sativus* L. flower exerts hepatoprotective effect because it improves acute liver damage, returns blood parameters and liver histopathology to almost normal (Omidi, Riahinia, Montazer Torbati, & Behdani, 2014), and improves renal function (Arash Omidi & Totrabi, 2016) (Table 14.4). Another study on the efficacy of tepals extracts on the parameters of blood, immune system, and spleen histology showed increased antibody response without changes in spleen histology and hematological parameters (Babaei et al., 2014). Therefore, tepal extracts can be proposed for biomedical use. In addition, since tepal extract can play a role in the aging process, it has been used as a substance in luxury cosmetics to develop new skin treatments (De Mesa, Muñoz-Mingarro, & Pons, 2018).

Table 14.4 Bioactivities of phenolic compounds of saffron and floral bioresidues of *C. sativus* L. flower.

Bioactive Compound	Bioactivity (Reference)	Model	Dose
Tepal (ethanol; 80%) extracts of *C.s.* flower	Hepatoprotective effects (Omidi et al., 2014)	Rats	Administration by oral gavage (20 mg/kg body weight for 6 days)
	Ameliorative effects on kidney failures (Omidi & Totrabi, 2016)	Rats	Intraperitoneal injection (40 mg/kg body weight for 7/13 days)
	Increase antibody response (Babaei, Arshami, Haghparast, & Mesgaran, 2014)	Rats	Intraperitoneal injection (75 mg/kg body weight for 14 days)
Kaempferol aglycone	Antitumor effects (Riahi-Chebbi et al., 2019)	Colon cancer cells	75 μM
	Anti-inflammatory effects (Bian et al., 2019)	Cellular (rat intestinal microvascular endothelial cells [RIMVECs]) model of intestinal inflammation in rats	12.5, 25 and 50 μM
	Anti-inflammatory effects on *Helicobacter pylori*-induced inflammation (Yeon et al., 2019)	Gastric adenocarcinoma cell	6.25, 12.5, and 25 μM
	Antidepressant effects (Gao, Wang, Peng, & Deng, 2019)	Chronic social defeat stress mouse model	Intraperitoneal injection (20 mg/kg body weight)
	Wound healing effects (Özay et al., 2019)	Incisional and excisional wound models on diabetic and nondiabetic rats	Topically applied (1% weight/weight for 14 days)

(*Continued*)

Table 14.4 (Continued)

Bioactive Compound	Bioactivity (Reference)	Model	Dose
Kaempferol 3-*O*-*β*-sophoroside	Anti-inflammatory effects (Kim, Ku, & Bae, 2012)	Human endothelial cells	>0.05 μM
	Analgesic effects (Palanichamy & Nagarajan, 1990)	Mice	Intraperitoneal injection (50 mg/kg body weight)
Quercetin	Chemopreventive effects. Inhibit cell growth and invasion/migration of the cells (Chen et al., 2013)	Cultured oral squamous carcinoma cells	2 mg/mL
	Chemopreventive effects (Sharmila et al., 2014)	Male Sprague Dawley rats	Oral administration (200 mg/kg body weigh/thrice a week for 16 weeks)
Delphinidin 3-*O*-*β*-glucoside	Prevent tumor progress by inhibiting angiogenesis and cell migration (Viegas et al., 2019)	Breast cancer cells	200 μM

Source: Adapted from Moratalla-López et al. (2019).

Antioxidant, anti-inflammatory, sunscreen, and anticancer activities have been attributed to kaempferol (Calderon-Montano, Burgos-Moron, Perez-Guerrero, & Lopez-Lazaro, 2011; Colombo et al., 2019; Mohebbati, Khazdair, & Hedayati, 2017). Inhibition of inflammatory mediator expression has been demonstrated in rat intestinal inflammatory cell model (Bian et al., 2019). In addition, an anti-inflammatory effect was observed in *Helicobacter pylori*-induced inflammation (Gao et al., 2019). Kaempferol may inhibit endotoxin-induced inflammatory responses in acute lung injury in mice as well as cellular models (Qian et al., 2019). Further, the wound healing improvement in diabetic and nondiabetic rats (Özay et al., 2019) and antidepressant effect (Gao et al., 2019) are also attributed to kaempferol. The kaempferol 3-*O*-β-sophoroside, one of the main components in floral bioresidues, has proven anti-inflammatory activity in human endothelial cells, which can be useful in the treatment of inflammatory vascular diseases (Kim et al., 2012). Analgesic action has also been attributed to that in tests with mice (Palanichamy & Nagarajan, 1990).

In addition to kaempferol, quercetin 3-*O*-β-sophoroside is another flavonol present in tepals of *C. sativus* L. (Serrano-Díaz et al., 2014). Quercetin is a classic candidate to design anticancer drugs, primarily because it has the opposite effect on other signaling networks to prevent cancer progression (Priyadarsini et al., 2010). Therefore, this flavonol can be a potential bioactive compound for disease prevention and treatment (Bischoff, 2008). Delphinidin glucosides, along with other anthocyanins, can inhibit tumor progression because of their ability to inhibit blood vessel formation (Viegas et al., 2019). The main bioactivities of saffron phenolic compounds and floral bioresidues from *C. sativus* L. flowers mentioned above are shown in Table 14.4.

Consequently, investigation of the bioactivity has recently been evaluated for saffron's main components and phenolic compounds in saffron; flower tepal extracts, all its flavones, especially kaempferol and quercetin, as well as anthocyanins for their potential antioxidant activities and medicinal characteristics. However, more research is needed to understand the health effects of these phenolic compounds in floral bioresidues, whole *C. sativus* L. flower, and saffron.

14.6 Metabolism of Saffron

Cytochrome P450 is a very important enzymatic system involved in the metabolism of several drugs and compounds, including xenobiotics (McDonnell & Dang, 2013). Xenobiotics are a set of chemicals foreign to the body and have sundry physiological effects (Pollak, 1998). Cytochrome P450 is located in the hepatocytes, it is therefore no surprise that the liver plays a major role in the metabolism and detoxification of drugs, by two-phase xenobiotics biotransformation. Briefly, phase I reactions transform xenobiotics into hydrophilic solutes mainly using Cytochrome P450 pathways and phase II conjugates these metabolites to glucuronate, glutathione, and sulfate (Kalra & Tuma, 2018).

As mentioned earlier, saffron compounds are mainly metabolized in the liver after ingestion. Saffron metabolites significantly reduced CYP1A2 (13.7%) in thirty-four healthy volunteers after ingesting an infusion of *C. sativus* L. for 6 days (Begas et al., 2019). This study proposes that these metabolites do not interfere in other drugs metabolisms, but rather have some pharmacological interaction with other compounds metabolized by CYP1A2, particularly in males.

However, the metabolites derived from saffron greatly depends on environmental factors and thus saffron harvested in different parts of the world is expected to display significant variability. Cossignani et al. (2014) studied the metabolites in saffron from central Italy and indicated that in addition to environmental factors drying process also affect saffron composition. It is likely that depending on the origin of this spice, its metabolism would be modified in some way.

Albumin, the most abundant protein in plasma, transports several substances such as bilirubin, calcium, fatty acids, and some drugs (from anticoagulants and anesthetics) (Moman & Varacallo, 2019). The albumin transport functions are also extended to saffron metabolites. For instance, crocetin, one of the most studied apocarotenoids and steadfast bioavailable, is known to also bind to plasma albumin (Miller, Willett, Moss, Miller, & Belinka, 1982). The experimental data supported by this study further shows that crocetin might even bind to the same sites as fatty acids.

There are still many gaps remaining in saffron metabolism, particularly at the cellular level. Kyriakoudi et al. (2015) investigated cellular transport of picrocrocin, a relatively abundant compound in saffron. Picrocrocin is the most bioavailable substance among all apocarotenoids, although its *in vitro* cellular transport is similar to other apocarotenoids.

14.7 Analytical Methods for Saffron

In international trade agreements, the quality of saffron is determined in accordance with ISO 3632 : 2011 standard (ISO, 2003), which divides saffron into three categories in terms of physical and chemical properties such as color, aroma, and taste. Among them, color is the most important quality (Carmona, Zalacain, & Alonso, 2006) compared to aroma and taste.

A group of carotenoids derived from crocetin are responsible for saffron's color. Glucose, gentiobiose, and neapolitanose are sugars that esterify the crocetin end, providing crocetin esters, mainly as *trans* geometric isomers, and minor amounts of *cis* geometric isomers (Carmona, Zalacain, & Alonso, 2006). Crocetin esters content is 16%–28% of saffron (García-Rodríguez, Moratalla-López, López-Córcoles, & Alonso, 2021). Crocins are another way to find these carotenoids. The kinetics of these carotenoids have been investigated in saffron aqueous extracts during heat treatment in the dark (Sánchez, Carmona, Ordoudi, Tsimidou, & Alonso, 2008).

Picrocrocin is thought to have the greatest effect on the bitter taste of saffron (Carmona, Zalacain, & Alonso, 2006; Chrysanthou, Pouliou, Kyriakoudi, & Tsimidou, 2016; Maggi, Carmona, Sánchez, & Alonso, 2010). This compound is found only in the genus Crocus, of which the only edible species is *C. sativus* L. Thus, picrocrocin is a molecular marker of this spice that cannot mimic the taste of other spices, and it may play a role in identifying real saffron (Sánchez, Maggi, Carmona, & Alonso, 2011). The picrocrocin content is 7–16% in the dry stigma of *C. sativus* L. (Priscila del Campo et al., 2010). During heat treatment, picrocrocin's stability in aqueous extracts of saffron was measured and shown to be more stable than crocetin esters (Sánchez, Carmona, Jarén-Galán, Mosquera, & Alonso, 2011).

Safranal, a picrocrocin aglycone, is the main constituent of volatile saffron compounds and responsible for its aroma (Carmona, Zalacain, & Alonso, 2006; Carmona, Zalacain et al., 2007; Condurso, Cincotta, Tripodi, & Verzera, 2017; Maggi et al., 2009). Safranal accounts for ~65% of the total aroma, expressed as a percentage of the total volatile content, determined by gas chromatography (GC-SPME) (Moratalla-López et al., 2019; Urbani, Blasi, Chiesi, Maurizi, & Cossignani, 2015). The concentration of safranal is typically 0.1–0.6% of saffron (García-Rodríguez et al., 2017).

Nowadays, the quality of saffron is evaluated according to the ISO 3632 : 2011 standard (ISO, 2003), in which color, taste, and aroma are determined by UV–vis spectrophotometry. The color intensity (440 nm) is directly related to the ability of transmitting color to foodstuffs to which saffron is added (Alonso, Salinas, & Saez, 1998), depending on the content of crocetin ester in the dried *C. sativus* L stigma. Other wavelengths 257 nm and 330 nm evaluate picrocrocin and safranal, respectively. However, other substances, such as *trans*-crocetin esters and *cis*-crocetin esters, also adsorb at 250 and 330 nm (García-Rodríguez et al., 2021). Therefore, more accurate methods for measuring the crocetin esters, picrocrocin, and safranal have been developed and are constantly being updated.

The quality of saffron should be determined in such ways as to provide detailed amount of crocetin esters, picrocrocin, safranal, and other compounds of interest. Today, various analytical techniques are available to identify and quantify the key components of saffron (Table 14.5). Also, these methods have generally been validated using the standard metabolite of interest, so that compounds can be measured very accurately (Magnusson, 2014).

Table 14.5 Analytical techniques for saffron analysis.

Analytical technique (Reference)	Indicative data or analyte	Information
Spectrophotometry UV–vis (ISO, 2003)	Coloring strength, E1%1 cm257 nm and E1%1 cm330 nm	ISO 3632:2011
HPLC–DAD (Vahedi et al., 2018; Valle Garcia-Rodriguez et al., 2014; Valle García-Rodríguez et al., 2014)	Crocins, picrocrocin, and safranal	Saffron quality
LC/DAD/MS/MS (Carmona, Sanchez et al., 2007)	Crocins, picrocrocin, and flavonoids	Identification of saffron metabolites
UHPLC–MS/MS (Rocchi et al., 2018)	Crocins	Differentiation of process for obtaining saffron
NMR (Yilmaz, Nyberg, Mølgaard, Asili, & Jaroszewski, 2010).	Saffron compounds	Metabolic fingerprinting
DHS–GC–MS (Maggi et al., 2009)	Safranal and other volatile compounds	Aroma quality
e-Nose (Carmona, Martinez et al., 2006)	Volatiles of saffron as a whole	Determination of geographical origin
PTR–TOFMS (Nenadis, Heenan, Tsimidou, & Van Ruth, 2016)	Volatile compounds	Quality of aroma
Raman spectroscopy (Anastasaki, Kanakis, Pappas, Maggi, Zalacain, et al., 2010)	Sum of crocins and coloring strength	Saffron quality
NIR spectroscopy (Zalacain et al., 2005)	Saffron quality control	Saffron quality/ geographic origin
Derivatization–HPLC–DAD (Del Campo et al., 2009)	Free amino acids and ammonium	Determination of geographic origin
MIR spectroscopy (Anastasaki, Kanakis, Pappas, Maggi, Del Campo, et al., 2010)	FT-IR spectra saffron filaments	Determination of geographic origin
Tristimulus colorimetry (Alonso, Sanchez-Fernandez, Sáez, Zalacain, & Salinas, 2003)	Color	Saffron quality

Source: Adapted from Moratalla-López et al. (2019).

Various analytical methods such as gas chromatography/mass spectrometry (GC–MS) and olfactometry measurements (Culleré, San-Juan, & Cacho, 2011), GC–MS (Aliakbarzadeh, Sereshti, & Parastar, 2016; Maggi et al., 2009), HPLC–DAD (Kanakis, Daferera, Tarantilis, & Polissiou, 2004; Kuhn & Winterstein, 1934), HPLC–MS/MS (Rubert, Lacina, Zachariasova, & Hajslova, 2016), or ultrasound-assisted extraction by UV–vis (Maggi et al., 2011), are used to determine the volatile saffron fraction. To evaluate the crocetin esters and picrocrocin, liquid chromatography equipment with custom diode detector (HPLC-DAD) can be used in saffron aqueous extracts (Vahedi et al., 2018; Valle García-Rodríguez et al., 2014). HPLC-DAD is recommended to identify and measure safranal (Valle García-Rodríguez et al., 2014).

Therefore, the fingerprint of the saffron metabolite can be used to distinguish the original saffron from other saffrons, also, nuclear magnetic resonance (NMR) is used today with particular emphasis on geographical origin characteristics, aging determination, and fraud detection (Consonni & Cagliani, 2010; Consonni, Ordoudi, Cagliani, Tsiangali, & Tsimidou, 2016). In saffron, the NMR-based metabolomics approach was first adopted to obtain fingerprints from the Iranian saffron metabolites (Yilmaz et al., 2010).

According to analytical method of *C. sativus* L., in cold-pressed *C. sativus* L. juices, floral bioresidues, kaempferol derivatives, and anthocyanins were measured by liquid chromatographed with photodiode array detection (LC-DAD) using high-resolution mass spectrometry (LC-ESI- (HR) MS^n) (Tuberoso, Rosa, Montoro, Fenu, & Pizza, 2016). Goupy et al. (2013) identified flavonols and anthocyanins in tepals by ultraperformance liquid chromatography combined with diode array detection (UPLC-DAD) and ion trap mass spectrometry by electrospray ionization (ESI-MS^n). In addition, the HPLC-DAD method is described for the analysis of floral bioresidues of *C. sativus* L. flower (Serrano-Díaz et al., 2014). Kaempferols were extracted from *C. sativus* L. tepals by flash column chromatography and identified using thin-layer chromatography (TLC), infrared (IR), HPLC-DAD, and NMR (Zeka et al., 2015). The flavonoid fraction of saffron was evaluated by liquid chromatography with photodiode array detection and mass spectrometer (LC-DAD-MS/MS) (Carmona, Sanchez et al., 2007).

Hence, several analytical techniques are described to evaluate all bioactive compounds of saffron, residues of flowers, and whole *C. sativus* L flowers. Saffron quality, ISO 3632 : 2011, is the most appropriate standard that needs to be revised, with other more accurate analytical methods. Proven methods can be used to ensure the optimal and accurate evaluation of the compounds of interest. The growing interest in bioactive molecules of floral bioresidues and *C. sativus* L. whole flowers has become the focus of new research and new analytical methods.

14.8 Antioxidant Properties of Saffron

Saffron is known for its natural potent antioxidant and anti-inflammatory properties. Carotenoids, flavonoids, and anthocyanins have been proposed as the major components of saffron extract responsible for its antioxidant and anti- inflammatory activities. The antioxidant activities of methanol and water–methanol (50 : 50 v/v) extracts of *crocus sativus*' stigma were superior than those of carrots and tomatoes. Numerous studies claimed that among carotenoids constituents of saffron, crocin, and crocetin showed high radical scavenging activities followed by safranal. High radical scavenging activity of these constituents is presumably due to their ability to donate single hydrogen atom to the free radicals.

In vitro experiments and animal studies have shown that saffron extracts (especially crocetin and crocin) decrease plasma malondialdehyde (MDA) level as an index of lipid peroxidation, which is induced by reactive oxygen species (ROS) (Ghaffari & Roshanravan, 2019).

However, the synergistic effect of all the bioactive constituents produces significant antioxidant activity in saffron. These compounds can protect DNA and tRNA from harmful chemical reaction through the formation of ligand–polynucleotide complexes (Christodoulou, Kadoglou, Kostomitsopoulos, & Valsami, 2015). Saffron was touted by researchers for its antitumor benefits and cancer prevention properties, due in part to its high carotenoids content

(Abdullaev, 2002). According to many studies, strong evidence indicates that saffron and its components possess anticarcinogenic and antitumour effects *in vitro* and *in vivo*.

However, research is still ongoing to address the particular mechanism of action of each saffron component against malignancies. The antitumor activity is mainly attributed to inhibition of DNA and RNA synthesis, inhibition or suppression of cancer cell proliferation, apoptosis, inhibition of metastasis and angiogenesis, and changes in oncogenes' expression pattern or tumor-suppressive genes (Lambrianidou, Koutsougianni, Papapostolou, & Dimas, 2020).

Saffron extract, either applied topically or administered per os or intravenously, is effective against skin neoplasia (retardation or even inhibition). The anticarcinogenic properties of saffron are attributed to the presence of crocin, crocetin, picrocrocin, and safranal at concentrations ranging from 2 μM to 3 mM (Christodoulou et al., 2015). The anticancer effect of crocin on three human colorectal cancer cells (HCT-116, SW-480, and HT-29) has been recently reported. Crocin reduced the rate of cell proliferation, but HCT-116 showed higher sensitivity to crocin than the other two cell lines (Lambrianidou et al., 2020).

Hepatoprotective effects have been described in *in vivo* studies, whereby crocetin esters attenuated the activation of caspases, essential mediators of apoptosis or cell death, and the ratio between the proapoptotic bax and antiapoptotic BCL-2 (bax/Bcl-2) proteins, which reduces hepatotoxicity (Khorasany & Hosseinzadeh, 2016).

Despite the promising results that have emerged, there is a lack of human clinical trials.

Diffusion Pharmaceuticals LLC (Charlottesville, VA, USA) conducted a clinical trial with *trans*-sodium crocetinate (TSC) as adjuvant in radiotherapy and chemotherapy treatments for resistant solid tumors such as glioblastoma, in which hypoxia is the result of its rapid growth. The results strongly suggest that adding TSC during radiation therapy is beneficial for the treatment of newly diagnosed glioblastoma (Bagur et al., 2020).

Another clinical study demonstrated the anticancer effect of saffron in combination with chemotherapy in cancer patients suffering from liver metastasis. Thirteen participants had primary cancer, including esophagus, stomach, colon, ovarian, and breast cancers, and consumed capsules containing 50 mg of dried saffron stigma. From the total patients included in the study, six patients quit, in saffron-treated group, two patients showed partial and complete response (50%), whereas in placebo group, no response was observed (Lambrianidou et al., 2020). Saffron with its antioxidant properties has the potential to act as a preventive agent of gastric mucosa damage through enriching glutathione levels and reducing lipid peroxidation. It can inhibit ulcers that are induced by stress and histamines (Ashktorab et al., 2019).

Saffron's compounds are reported to be a safe and effective treatment to reduce the toxic side effects of some conventional chemotherapeutic drugs such as cisplatin, tamoxifen, and doxorubicin. Several studies have highlighted that the combined treatment of saffron extracts with chemotherapeutic drugs had synergistic effects, enhancing the outcome of the applied treatment (Lambrianidou et al., 2020).

In a recent study, the effect of a saffron infusion intake was assessed on the lipid profile of multiple sclerosis (MS) patients. A total of 35 MS patients took a mid-afternoon infusion for 96 days; 16 of them completed the treatment. Total blood cholesterol levels significantly decreased from 199.5 to 179.5 mg/dL, and triglyceride levels decreased from 124 to 101 mg/dL (Bagur et al., 2020).

14.9 Health Benefits of Saffron

Many studies have been published recently highlighting the biomedical and pharmacological properties of saffron or its metabolites; these are attributed mainly to the carotenoids: crocetin and crocins. Apart from these, there are other macro and micronutrients, carotenoids, and phenolic compounds such as anthocyanins, kaempferol glycosides, that confer saffron various nutritional and beneficial properties for health. These function on many body systems, such as the gastrointestinal, cardiovascular, endocrine, nervous, and immune systems.

Safranal and crocin isomers, as well as crocetin, are related to anxiolytic, neuroprotective, antidepressant, and anti-Alzheimer properties, which have been proven in several clinical trials (Orio et al., 2020). Recent systematic reviews and meta-analyses conclude that saffron is significantly more effective than placebo in reducing the severity of depression and exhibits similar or equal efficacy to antidepressants. Some well-designed clinical trials have indicated the efficacy of saffron in mild and moderate depression, more effective or at least equivalent to therapeutic doses of imipramine and fluoxetine in depression using Hamilton Depression Rating Scale. In a 4 week, double-blind, prospective, placebo-controlled clinical trial, the addition of crocin tablets (30 mg/day, 15 mg BID) amplified the effects of SSRIs in treatment of patients with mild to moderate depression. Also, due to the absence of substantial side effects, crocin was a particularly effective therapeutic adjuvant (Talaei, Moghadam, Tabassi, & Mohajeri, 2015). Postpartum depression is a subtype of depression. In a randomized, double-blind placebo-controlled trial on 60 women suffering from postpartum depression, 96% of patients in the saffron group were in remission compared to 43% in placebo group (Tabeshpour et al., 2017).

There is evidence that saffron dosing can reduce anxiety. A 6-week study involving 66 patients suffering from major depression accompanied with anxiety compared the anxiolytic effects of saffron (30 mg/day) and citalopram (40 mg/day). Obvious improvement of anxiety symptoms was observed. Two clinical trials were designed to investigate the anxiolytic effect of Affron® (a novel saffron extract). In the first study, Affron® (28 mg/day for 4 weeks) notably improved anxiety-like symptoms in healthy adults (Kell et al., 2017). In the second trial, Affron® (14 mg, b.i.d.) administered for 8 weeks improved anxiety and depressive symptoms in youths (aged 12–16 years) with mild-to-moderate anxiety or depression symptoms (Lopresti, Drummond, Inarejos-García, & Prodanov, 2018).

The sleep-enhancing effects of Affron® a standardized saffron extract derived from *C. sativus* stigma were demonstrated in a recent 3-arm, parallel-group, double-blind-controlled trial including 120 adults (n = 40 each for placebo [12 M + 28F], 14 [13 M + 27F] or 28 [10 M + 30F] mg Affron® groups, average age 50–55 years old) with unsatisfactory sleep (Lopresti, Smith, & Drummond, 2021). Supplementation of Affron® (containing over 3.5% lepticrosalides bioactive saffron compound) 1 hr before bed for 28 days improved sleep quality, mood after awakening, reduced insomnia (24 vs 6% rates compared to placebo) and increased evening salivary melatonin levels independent of the 14 or 28 mg daily dose. The study indicates that saffron sleep enhancing effects may be due to melatonin that regulates many biological functions including sleep, circadian rhythms, immunity, and reproduction and exerts antioxidant and anti-inflammatory effects. Saffron can influence melatonin by gut microbiota increased tryptophan bioavailability which is then converted to serotonin consequently producing melatonin and/or modulating inflammatory cytokines (interferon-γ and interlukin-1β) and cytochrome P450 enzyme activity according to Lopresti et al. (2021).

Recently, the neuroprotective role of dietary saffron has generated great interest. Extracts of *C. sativus* and its components can exert neuroprotection in neurodegenerative diseases, such as Alzheimer's and Parkinson's diseases, because they can interact with the glutamatergic, cholinergic, and dopaminergic systems. Oral intake of saffron improved cognitive function and functional status (community and home life and personal care duties) for subjects with neurodegenerative conditions such as Alzheimer's disease and amnesic mild cognitive impairment, which is suspected of being an early stage in neurodegenerative progression to dementia (Singletary, 2020).

Other mind–body conditions including premenstrual syndrome (PMS) may also respond to saffron therapy. In a trial 30 mg/day of saffron administered over the course of two menstrual cycle's significantly improved PMS discomfort in a group of women aged 20 to 45 years (Agha-Hosseini et al., 2008).

The efficacy of Satiereal® (a novel extract of saffron stigma) was evaluated in 60 mildly overweight women in a randomized, placebo-controlled, double-blind study as a supplement to reduce body weight over an 8-week period. The secondary variable, the snacking frequency, was assessed by daily self-recording of episodes by the subjects in a nutrition diary. Subjects consumed 176.5 mg extract per day ($n = 31$) or a matching placebo ($n = 29$). Results indicate that Satiereal consumption reduces snacking and creates a satiating effect that can contribute to body weight loss. The combination of an adequate diet with the supplement might help subjects engage in a weight loss program (Gout et al., 2010).

Saffron supplementation can potentially affect ocular neurodegenerative diseases, such as glaucoma, age-related macular degeneration, diabetic retinopathy, and retinitis pigmentosa. Saffron may hold the key to preventing the loss of sight (Heitmar, Brown, & Kyrou, 2019). In a trial by researchers at the University of Sydney, the participants mean BCVA (best corrected visual acuity) improved 0.69 letters ($p = 0.001$) and mean-pooled mfERG (multifocal electroretinogram) latency reduced 0.17 ms ($p = 0.04$) after taking a saffron dietary supplement tablet for 3 months (Broadhead et al., 2019). Some studies using saffron in glaucoma patients focus on analyzing its possible ocular hypotensive effect and have obtained contradictory results (Heitmar et al., 2019). The only existing clinical study in patients with primary open angle glaucoma (POAG) reported that oral saffron supplementation (30 mg daily for 1 month) can significantly reduce the intra ocular pressure (IOP) after 3 weeks. Compared to placebo, the addition of daily oral saffron supplementation for 1 month resulted in significant IOP reduction after the third week, but this effect reverted after a 4-week wash-out period (Bonyadi, Yazdani, & Saadat, 2014).

The gastroprotective effect of crocin in ethanol-induced gastric injury was demonstrated by increased gastric juice levels after crocin pretreatment. The mucosal levels of glutathione, malondialdehyde, and superoxide dismutase activity were restored with crocin treatment. The study showed that crocin protects rat gastric mucosa against ethanol-induced injury by displaying anti-inflammatory, anti-oxidative, anti-apoptotic, and mucin-secretagogue mechanisms, probably mediated by increased PGE2 release (El-Maraghy, Rizk, & Shahin, 2015). With its antioxidant properties, saffron has the potential to act as a preventive agent of gastric mucosa damage by increasing glutathione levels and reducing lipid peroxidation. Saffron can inhibit ulcers that are induced by stress and histamines (Inoue et al., 2005).

14.10 Conclusion and Future Perspectives

Saffron is one of the most important culinary plants and has various uses as a preservative, colorant, food ingredient, herbal remedies, and traditional medicine. It is taken from the dry stigmas of the flowering *C. sativus* L. plant. Nowadays, together with the bioactive compounds in the stigma, the floral bioresidues of *C. sativus* L. are an interesting source of phenolic compounds that make this flower a powerful source of antioxidant compounds. Bioactive phenolic components of floral bioresidues of *C. sativus* L. have been less studied than saffron. Flavonols, chiefly kaempferol, quercetin, and their glycosides, are noted for their antioxidant activity, which shows beneficial medicinal properties. High amounts of kaempferols can be extracted from *C. sativus* L. tepals, which are powerful anti-inflammatory agents; however, more research is needed on its health effects and bioavailability.

The antioxidant and anti-inflammatory properties of some spices have long been known to protect against chronic diseases. Today, the tendency to use special diets that contain spices or herbs to help the treatment of various diseases has increased. The potential health benefits of saffron depend on the amount consumed and its bioavailability. Studies on the bioavailability of spices are essential to determine their pharmacological and toxicological properties. There are studies on the pharmacokinetics of orally used crocins, but none determine the pharmacokinetics of saffron as a spice. In most reports studying the bioactivity of saffron, there is no description of the prototype, so the content of the main metabolites is unknown.

Also, there are many reports that do not obviously explain the methods of preparing different forms of saffron. This is a very important issue because it has the greatest effect on the concentration and absorption of metabolites. The most important of these is the ISO 3632:2011 standard, but many studies show that this standard need to be revised, especially for safranal analysis, and other more accurate analysis techniques such as HPLC-DAD or NMR. Meanwhile, there is currently an important trend in determining bioactive compounds of the floral bioresidues and whole flower of *C. sativus* L. In this case, the analytical methods used are appropriate and the content of these bioactive molecules can be evaluated. With this growing interest, new techniques are being developed. In addition, while the stability of these bioactive compounds has been investigated, other storage conditions of interest in the industry can also be studied (Bagur et al., 2020; Moratalla-López et al., 2019).

In brief, a wide range of flavonoids, including anthocyanins, kaempferol, quercetin, and their glycosides, as well as large amounts of crocetin, safranal, picrocrocin, and crocin as active ingredients in *C. sativus* L. are considered as dietary supplements, beverages, and functional foods; and in pharmaceutical industry, cosmetic materials, and as *C. sativus* L. based ingredients in herbal remedies.

References

Abdullaev, F. I., Riveron-Negrete, L., Caballero-Ortega, H., Hernández, J. M., Perez-Lopez, I., Pereda-Miranda, R., & Espinosa-Aguirre, J. J. (2003). Use of *in vitro* assays to assess the potential antigenotoxic and cytotoxic effects of saffron (*C. sativus* L.). *Toxicology in vitro*, *17*(5–6), 731–736.

Abdullaev, F. I. (2002). Cancer chemopreventive and tumoricidal properties of saffron (*C. sativus* L.). *Experimental Biology and Medicine, 227*(1), 20–25.

Agha-Hosseini, M., Kashani, L., Aleyaseen, A., Ghoreishi, A., Rahmanpour, H., Zarrinara, A. R., & Akhondzadeh, S. (2008). *C. sativus* L. (saffron) in the treatment of premenstrual syndrome: a double-blind, randomized and placebo-controlled trial. *BJOG: An International Journal of Obstetrics & Gynecology, 115*(4), 515–519. https://doi.org/10.1111/j.1471-0528.2007.01652.x

Ahmad, A. S., Ansari, M. A., Ahmad, M., Saleem, S., Yousuf, S., Hoda, M. N., & Islam, F. (2005). Neuroprotection by crocetin in a hemi-parkinsonian rat model. *Pharmacology Biochemistry and Behavior, 81*(4), 805–813.

Akhondzadeh, S., Fallah-Pour, H., Afkham, K., Jamshidi, A.-H., & Khalighi-Cigaroudi, F. (2004). Comparison of *C. sativus* L. and imipramine in the treatment of mild to moderate depression: A pilot double-blind randomized trial [ISRCTN45683816]. *BMC Complementary and Alternative Medicine, 4*(1), 12. https://doi.org/10.1186/1472-6882-4-12

Akhondzadeh, S., Sabet, M. S., Harirchian, M. H., Togha, M., Cheraghmakani, H., Razeghi, S., . . . Moradi, A. (2010). Saffron in the treatment of patients with mild to moderate Alzheimer's disease: a 16-week, randomized and placebo-controlled trial. *Journal of Clinical Pharmacy and Therapeutics, 35*(5), 581–588. https://doi.org/10.1111/j.1365-2710.2009.01133.x

Akhondzadeh, S., Tahmacebi-Pour, N., Noorbala, A-A., Amini, H., Fallah-Pour, H., Jamshidi, A-H, & Khani, M. (2005). *C. sativus* L. in the treatment of mild to moderate depression: a double-blind, randomized and placebo-controlled trial. *Phytotherapy Research, 19*(2), 148–151. doi: https://doi.org/10.1002/ptr.1647.

Alavizadeh, S. H., & Hosseinzadeh, H. (2014). Bioactivity assessment and toxicity of crocin: A comprehensive review. *Food and Chemical Toxicology, 64*, 65–80. https://doi.org/10.1016/j.fct.2013.11.016

Aliakbarzadeh, G., Sereshti, H., & Parastar, H. (2016). Pattern recognition analysis of chromatographic fingerprints of *C. sativus* L. secondary metabolites toward source identification and quality control. *Analytical and Bioanalytical Chemistry, 408*(12), 3295–3307. https://doi.org/10.1007/s00216-016-9400-8

Almodóvar, P., Briskey, D., Rao, A., Prodanov, M., & Inarejos-García, A. M. (2020). Bioaccessibility and pharmacokinetics of a commercial saffron (*C. sativus* L.) extract. *Evidence-Based Complementary and Alternative Medicine*, 1575730. https://doi.org/10.1155/2020/1575730

Alonso, G. L., Salinas, M. R., & Saez, J. R. (1998). Crocin as coloring in the food industry. *Recent Research and Development in Agricultural and Food Chemistry, 2*, 141–153.

Alonso, G. L., Sanchez-Fernandez, M. A., Sáez, J. R., Zalacain, A., & Salinas, M. R. (2003). Evaluation of the color of Spanish saffron using tristimulus colorimetry. *Italian Journal of Food Science, 15*(2), 249–258.

Anastasaki, E., Kanakis, C., Pappas, C., Maggi, L., Del Campo, C. P., Carmona, M., . . . Polissiou, M. G. (2010). Differentiation of saffron from four countries by mid-infrared spectroscopy and multivariate analysis. *European Food Research and Technology, 230*(4), 571–577.

Anastasaki, E. G., Kanakis, C., Pappas, C., Maggi, L., Zalacain, A., Carmona, M., . . . Polissiou, M. G. (2010). Quantification of crocetin esters in saffron (*C. sativus* L.) using Raman spectroscopy and chemometrics. *Journal of Agricultural and Food Chemistry, 58*(10), 6011–6017.

Asai, A., Nakano, T., Takahashi, M., & Nagao, A. (2005). Orally administered crocetin and crocins are absorbed into blood plasma as crocetin and its glucuronide conjugates in mice. *Journal of Agricultural and Food Chemistry*, *53*, 7302–7306. https://doi.org/10.1021/jf0509355

Ashktorab, H., Soleimani, A., Singh, G., Amr, A., Tabtabaei, S., Latella, G., . . . Brim, H. (2019). Saffron: The golden spice with therapeutic properties on digestive diseases. *Nutrients*, *11*(5). https://doi.org/10.3390/nu11050943

Babaei, A., Arshami, J., Haghparast, A., & Mesgaran, M. D. (2014). Effects of saffron (*C. sativus*) petal ethanolic extract on hematology, antibody response, and spleen histology in rats. *Avicenna Journal of. Phytomedicine*, *4*(2), 103–109.

Bagur, M. J., Alonso, S., Gonzalo, L., Jiménez-Monreal, A. M., Serrano-Heras, G., Martínez-Tomé, M., & Alonso, G. L. (2020). Effect of daily intake of a saffron infusion on blood cholesterol levels. *Applied Sciences*, *10*(14), 4763. https://doi.org/10.3390/app10144763

Baziar, S., Aqamolaei, A., Khadem, E., Mortazavi, S. H., Naderi, S., Sahebolzamani, E., . . . Shahmirzadi, M. (2019). *C. sativus* L. versus methylphenidate in treatment of children with attention-deficit/hyperactivity disorder: A randomized, double-blind pilot study. *Journal of Child and Adolescent Psychopharmacology*, *29*(3), 205–212.

Begas, E., Bounitsi, M., Kilindris, T., Kouvaras, E., Makaritsis, K., Kouretas, D., & Asprodini, E. K. (2019). Effects of short-term saffron (*C. sativus* L.) intake on the *in vivo* activities of xenobiotic metabolizing enzymes in healthy volunteers. *Food and Chemical Toxicology*, *130*, 32–43. https://doi.org/10.1016/j.fct.2019.05.013

Bian, Y., Liu, P., Zhong, J., Hu, Y., Fan, Y., Zhuang, S., & Liu, Z. (2019). Kaempferol inhibits multiple pathways involved in the secretion of inflammatory mediators from LPS-induced rat intestinal microvascular endothelial cells. *Molecular Medicine Reports*, *19*(3), 1958–1964. https://doi.org/10.3892/mmr.2018.9777

Bian, Y., Zhao, C., & Lee, S. M.-Y. (2020). Neuroprotective potency of saffron against neuropsychiatric diseases, neurodegenerative diseases, and other brain disorders: from bench to bedside. *Frontiers in Pharmacology*, *11*, 1602. https://doi.org/10.3389/fphar.2020.579052

Bischoff, S. C. (2008). Quercetin: potentials in the prevention and therapy of disease. *Current Opinion in Clinical Nutrition and Metabolic Care*, *11*(6), 733–740.

Bisti, S., Maccarone, R., & Falsini, B. (2014). Saffron and retina: Neuroprotection and pharmacokinetics. *Visual Neuroscience*, *31*(4–5), 355–361.

Bolhassani, A., Khavari, A., & Bathaie, S. Z. (2014). Saffron and natural carotenoids: Biochemical activities and anti-tumor effects. *Biochemistry Biophyisics Acta*, *1845*(1), 20–30. https://doi.org/10.1016/j.bbcan.2013.11.001

Bonyadi, M. H. J., Yazdani, S., & Saadat, S. (2014). The ocular hypotensive effect of saffron extract in primary open angle glaucoma: a pilot study. *BMC Complementary and Alternative Medicine*, *14*(1), 399. https://doi.org/10.1186/1472-6882-14-399

Britton, G. (1995). Structure and properties of carotenoids in relation to function. *The FASEB Journal*, *9*(15), 1551–1558. https://doi.org/10.1096/fasebj.9.15.8529834

Broadhead, G. K., Grigg, J. R., McCluskey, P., Hong, T., Schlub, T. E., & Chang, A. A. (2019). Saffron therapy for the treatment of mild/moderate age-related macular degeneration: a randomized clinical trial. *Graefes Arch Clinical and Experimental Ophthalmology*, *257*(1), 31–40. https://doi.org/10.1007/s00417-018-4163-x

Calderon-Montano, J. M., Burgos-Moron, E., Perez-Guerrero, C., & Lopez-Lazaro, M. (2011). A review on the dietary flavonoid kaempferol. *Mini-Reviews in Medicinal Chemistry*, *11*(4), 298–344. https://doi.org/10.2174/138955711795305335

Carmona, M., Martínez, J., Zalacain, A., Rodriguez-Mendez, M. L., de Saja, J. A., & Alonso, G. L. (2006). Analysis of saffron volatile fraction by TD–GC–MS and e-nose. *European Food Research and Technology*, *223*(1), 96–101.

Carmona, M., Sánchez, A. M., Ferreres, F., Zalacain, A., Tomás-Barberán, F., & Alonso, G. L. (2007). Identification of the flavonoid fraction in saffron spice by LC/DAD/MS/MS: Comparative study of samples from different geographical origins. *Food Chemistry*, *100*(2), 445–450.

Carmona, M., Zalacain, A., & Alonso, G.L. (2006). The chemical composition of saffron: color, taste and aroma. *Bomarzo SL*.

Carmona, M., Zalacain, A., Salinas, M. R., & Alonso, G. L. (2007). A new approach to saffron aroma. *Critical Reviews in Food Science and Nutrition*, *47*(2), 145–159. https://doi.org/10.1080/10408390600626511

Carmona, M., Zalacain, A., Sánchez, A. M., Novella, J. L., & Alonso, G. L. (2006). Crocetin esters, picrocrocin and its related compounds present in *C. sativus* stigmas and *Gardenia jasminoides* fruits. Tentative identification of seven new compounds by LC-ESI-MS. *Journal of Agricultural and Food Chemistry*, *54*(3), 973–979. https://doi.org/10.1021/jf052297w

Chen, S.-F., Nien, S., Wu, C.-H., Liu, C.-L., Chang, Y.-C., & Lin, Y.-S. (2013). Reappraisal of the anticancer efficacy of quercetin in oral cancer cells. *Journal of the Chinese Medical Association*, *76*(3), 146–152.

Chow, S.-C. (2014). Bioavailability and bioequivalence in drug development. *Wiley Interdisciplinary Reviews: Computational Statistics*, *6*, 304–312. https://doi.org/10.1002/wics.1310

Christodoulou, E., Kadoglou, N. P., Kostomitsopoulos, N., & Valsami, G. (2015). Saffron: a natural product with potential pharmaceutical applications. *Journal of Pharmacy and Pharmacology*, *67*(12), 1634–1649. https://doi.org/10.1111/jphp.12456

Chrysanthou, A., Pouliou, E., Kyriakoudi, A., & Tsimidou, M. Z. (2016). Sensory threshold studies of picrocrocin, the major bitter compound of saffron. *Journal of Food Science*, *81*(1), S189–S198. https://doi.org/10.1111/1750-3841.13152

Chryssanthi, D. G., Lamari, F. N., Georgakopoulos, C. D., & Cordopatis, P. (2011). A new validated SPE-HPLC method for monitoring crocetin in human plasma--application after saffron tea consumption. *Journal of Pharmaceutical and Biomedical Analysis*, *55*(3), 563–568. https://doi.org/10.1016/j.jpba.2011.02.018

Colombo, M., de Lima Melchiades, G., Michels, L. R., Figueiró, F., Bassani, V. L., Teixeira, H. F., & Koester, L. S. (2019). Solid dispersion of kaempferol: formulation development, characterization, and oral bioavailability assessment. *AAPS PharmSciTech*, *20*(3), 106. https://doi.org/10.1208/s12249-019-1318-y

Condurso, C., Cincotta, F., Tripodi, G., & Verzera, A. (2017). Bioactive volatiles in Sicilian (South Italy) saffron: safranal and its related compounds. *Journal of Essential Oil Research*, *29*(3), 221–227. https://doi.org/10.1080/10412905.2016.1244115

Consonni, R., & Cagliani, L. R. (2010). Chapter 4—Nuclear magnetic resonance and chemometrics to assess geographical origin and quality of traditional food products. In S. L. Taylor (Ed.), *Advances in food and nutrition research* (Vol. 59, pp. 87–165). Academic Press.

Consonni, R., Ordoudi, S. A., Cagliani, L. R., Tsiangali, M., & Tsimidou, M. Z. (2016). On the traceability of commercial saffron samples using 1H-NMR and FT-IR metabolomics. *Molecules, 21*(3), 286. https://doi.org/10.3390/molecules21030286

Cossignani, L., Urbani, E., Simonetti, M. S., Maurizi, A., Chiesi, C., & Blasi, F. (2014). Characterisation of secondary metabolites in saffron from central Italy (Cascia, Umbria). *Food Chemistry, 143*, 446–451. https://doi.org/10.1016/j.foodchem.2013.08.020

Culleré, L., San-Juan, F., & Cacho, J. (2011). Characterization of aroma active compounds of Spanish saffron by gas chromatography–olfactometry: Quantitative evaluation of the most relevant aromatic compounds. *Food Chemistry, 127*(4), 1866–1871. https://doi.org/10.1016/j.foodchem.2011.02.015

De Mesa, N. A., Muñoz-Mingarro, D., & Pons, E. B. (2018). Biowaste saffron extracts as active ingredients of cosmetic products antioxidants. *Spain patent ES2646415B1*.

Decker, F. (1914). Beiträge zur Kenntnis des Crocetins. *Archiv der Pharmazie, 252*(1–2), 139–160. https://doi.org/10.1002/ardp.19142520113

del Campo, C. P., Carmona, M., Maggi, L., Kanakis, C. D., Anastasaki, E. G., Tarantilis, P. A., . . . Alonso, G. L. (2010). Picrocrocin content and quality categories in different (345) worldwide samples of saffron (*C. sativus* L.). *Journal of Agricultural and Food Chemistry, 58*(2), 1305–1312. https://doi.org/10.1021/jf903336t

Del Campo, C. P., Garde-Cerdán, T., Sánchez, A. M., Maggi, L., Carmona, M., & Alonso, G. L. (2009). Determination of free amino acids and ammonium ion in saffron (C. sativus L.) from different geographical origins. *Food Chemistry, 114*(4), 1542–1548.

Delgado, M. C., Aramburu, A. Z., & Alonso, G. L. (2006). *The chemical composition of saffron: Color, taste and aroma*. Bomarzo.

Deo, B. (2003). *Growing Saffron–The world's most expensive spice*. Crop & Food Research (New Zealand Institute for Crop & Food Research). no. 20

Dhar, M. K., Sharma, M., Bhat, A., Chrungoo, N. K., & Kaul, S. (2017). Functional genomics of apocarotenoids in saffron: insights from chemistry, molecular biology and therapeutic applications. *Briefings in Functional Genomics, 16*(6), 336–347.

El-Maraghy, S. A., Rizk, S. M., & Shahin, N. N. (2015). Gastroprotective effect of crocin in ethanol-induced gastric injury in rats. *Chemico-Biological Interaction, 229*, 26–35. https://doi.org/10.1016/j.cbi.2015.01.015

Escribano, J., Alonso, G. L., Coca-Prados, M., & Fernández, J.-A. (1996). Crocin, safranal and picrocrocin from saffron (*C. sativus* L.) inhibit the growth of human cancer cells *in vitro*. *Cancer Letters, 100*(1), 23–30. https://doi.org/10.1016/0304-3835(95)04067-6

Falsini, B., Piccardi, M., Minnella, A., Savastano, C., Capoluongo, E., Fadda, A., . . . Bisti, S. (2010). Influence of saffron supplementation on retinal flicker sensitivity in early age-related macular degeneration. *Investigative Ophthalmology and Visual Science, 51*(12), 6118–6124. https://doi.org/10.1167/iovs.09-4995

Farokhnia, M., Shafiee Sabet, M., Iranpour, N., Gougol, A., Yekehtaz, H., Alimardani, R., . . . Akhondzadeh, S. (2014). Comparing the efficacy and safety of *C. sativus* L. with memantine in patients with moderate to severe Alzheimer's disease: a double-blind randomized clinical trial. *Human Psychopharmacology: Clinical and Experimental, 29*(4), 351–359.

Farzam, K., Nessel, T. A., & Quick, J. (2020). *Erythromycin. StatPearls Publishing*.

Fernández, J.-A. (2004). Biology, biotechnology and biomedicine of saffron. *Recent Research and Developments in Plant Science., 2*, 127–159.

Gao, W., Wang, W., Peng, Y., & Deng, Z. (2019). Antidepressive effects of kaempferol mediated by reduction of oxidative stress, proinflammatory cytokines and up-regulation of AKT/β-catenin cascade. *Metabolic Brain Disease*, *34*(2), 485–494.

García-Rodríguez, M. V., López-Córcoles, H., Alonso, G. L., Pappas, C. S., Polissiou, M. G., & Tarantilis, P. A. (2017). Comparative evaluation of an ISO 3632 method and an HPLC-DAD method for safranal quantity determination in saffron. *Food Chemistry*, *221*, 838–843. https://doi.org/10.1016/j.foodchem.2016.11.089

García-Rodríguez, M. V., Moratalla-López, N., López-Córcoles, H., & Alonso, G. L. (2021). Saffron quality obtained under different forcing conditions, considering various vegetative stages of corms. *Scientia Horticulturae*, *277*, 109811. https://doi.org/10.1016/j.scienta.2020.109811

Ghaffari, S., & Roshanravan, N. (2019). Saffron; An updated review on biological properties with special focus on cardiovascular effects. *Biomed Pharmacother*, *109*, 21–27. https://doi.org/10.1016/j.biopha.2018.10.031

Ghajar, A., Neishabouri, S. M., Velayati, N., Jahangard, L., Matinnia, N., Haghighi, M., . . . Meysamie, A. (2017). *C. sativus* L. versus citalopram in the treatment of major depressive disorder with anxious distress: a double-blind, controlled clinical trial. *Pharmacopsychiatry*, *50*(04), 152–160.

Goupy, P., Vian, M. A., Chemat, F., & Caris-Veyrat, C. (2013). Identification and quantification of flavonols, anthocyanins and lutein diesters in tepals of *C. sativus* by ultra performance liquid chromatography coupled to diode array and ion trap mass spectrometry detections. *Industrial Crops and Products*, *44*, 496–510. https://doi.org/10.1016/j.indcrop.2012.10.004

Gout, B., Bourges, C., & Paineau-Dubreuil, S. (2010). Satiereal, a *C. sativus* L extract, reduces snacking and increases satiety in a randomized placebo-controlled study of mildly overweight, healthy women. *Nutrion Research*, *30*(5), 305–313. https://doi.org/10.1016/j.nutres.2010.04.008

Hausenblas, H. A., Saha, D., Dubyak, P. J., & Anton, S. D. (2013). Saffron (*C. sativus* L.) and major depressive disorder: a meta-analysis of randomized clinical trials. *Journal of Integrative Medicine*, *11*(6), 377–383.

Heitmar, R., Brown, J., & Kyrou, I. (2019). Saffron (*C. sativus* L.) in ocular diseases: A narrative review of the existing evidence from clinical studies. *Nutrients*, *11*(3). https://doi.org/10.3390/nu11030649

Hemmati, M., Zohoori, E., Mehrpour, O., Karamian, M., Asghari, S., Zarban, A., & Nasouti, R. (2015). Anti-atherogenic potential of jujube, saffron and barberry: anti-diabetic and antioxidant actions. *EXCLI Journal*, *14*, 908–915.

Himeno, H., & Sano, K. (1987). Synthesis of crocin, picrocrocin and safranal by saffron stigma-like structures proliferated *in vitro*. *Agricultural and Biological Chemistry*, *51*(9), 2395–2400. https://doi.org/10.1080/00021369.1987.10868396

Holst, B., & Williamson, G. (2008). Nutrients and phytochemicals: From bioavailability to bioefficacy beyond antioxidants. *Current Opinion in Biotechnology*, *19*(2), 73–82. https://doi.org/10.1016/j.copbio.2008.03.003

Hosseini, A., Razavi, B. M., & Hosseinzadeh, H. (2018). Saffron (*C. sativus*) petal as a new pharmacological target: a review. *Iranian Journal of Basic Medical Sciences*, *21*(11), 1091–1099. https://doi.org/10.22038/IJBMS.2018.31243.7529

Hosseinzadeh, H., & Khosravan, V. (2002). Anticonvulsant effects of aqueous and ethanolic extracts of *C. sativus* L stigmas in mice. *Archives of Iranian Medicine*, *5*(1), 44–47.

Hosseinzadeh, H., & Sadeghnia, H. R. (2007). Protective effect of safranal on pentylenetetrazol-induced seizures in the rat: involvement of GABAergic and opioids systems. *Phytomedicine, 14*(4), 256–262. https://doi.org/10.1016/j.phymed.2006.03.007

Hosseinzadeh, H., & Talebzadeh, F. (2005). Anticonvulsant evaluation of safranal and crocin from *C. sativus* in mice. *Fitoterapia, 76*(7–8), 722–724. https://doi.org/10.1016/j.fitote.2005.07.008

Hosseinzadeh, H., & Younesi, H. M. (2002). Antinociceptive and anti-inflammatory effects of *C. sativus* L. stigma and petal extracts in mice. *BMC Pharmacology, 2*(1), 7. https://doi.org/10.1186/1471-2210-2-7

Inoue, E., Shimizu, Y., Shoji, M., Tsuchida, H., Sano, Y., & Ito, C. (2005). Pharmacological properties of N-095, a drug containing red ginseng, polygala root, saffron, antelope horn and aloe wood. *The American Journal of Chinese Medicine, 33*(1), 49–60.

ISO. (2003). *Saffron (C. sativus L) – Part 1: Specification and Part 2: Test methods.* Switzerland: International Organization for Standardization Geneva.

José Bagur, M., Alonso Salinas, G. L., Jiménez-Monreal, A. M., Chaouqi, S., Llorens, S., Martínez-Tomé, M., & Alonso, G. L. (2017). Saffron: An old medicinal plant and a potential novel functional food. *Molecules, 23*(1). https://doi.org/10.3390/molecules23010030

Kalra, A., & Tuma, F. (2018). *Physiology*. Liver: *StatPearls Publishing*.

Kanakis, C. D., Daferera, D. J., Tarantilis, P. A., & Polissiou, M. G. (2004). Qualitative determination of volatile compounds and quantitative evaluation of safranal and 4-hydroxy-2,6,6-trimethyl-1-cyclohexene-1-carboxaldehyde (HTCC) in Greek saffron. *Journal of Agricultural and Food Chemistry, 52*(14), 4515–4521. https://doi.org/10.1021/jf049808j

Kanakis, C. D., Tarantilis, P. A., Tajmir-Riahi, H. A., & Polissiou, M. G. (2007). Crocetin, dimethylcrocetin, and safranal bind human serum albumin: stability and antioxidative properties. *Journal of Agricultural and Food Chemistry, 55*(3), 970–977. https://doi.org/10.1021/jf062638l

Karrer, P., & Salomon, H. (1928). Zur Kenntnis der Safranfarbstoffe III. *Helvetica Chimica Acta, 11*(1), 711–713. https://doi.org/10.1002/hlca.19280110181

Kell, G., Rao, A., Beccaria, G., Clayton, P., Inarejos-García, A. M., & Prodanov, M. (2017). Affron® a novel saffron extract (*C. sativus* L.) improves mood in healthy adults over 4 weeks in a double-blind, parallel, randomized, placebo-controlled clinical trial. *Complementary Therapies in Medicine, 33*, 58–64.

Khalesi, M, & Behboodi, B. (2003). Development and Contraction of Contractile Roots in *C. sativus*. In *International Symposium on Saffron Biology and Biotechnology* 650, 55–58.

Khorasany, A. R., & Hosseinzadeh, H. (2016). Therapeutic effects of saffron (*C. sativus* L.) in digestive disorders: a review. *Iranian Journal of Basic Medical Sciences, 19*(5), 455–469.

Kim, T. H., Ku, S.-K., & Bae, J.-S. (2012). Inhibitory effects of kaempferol-3-O-sophoroside on HMGB1-mediated proinflammatory responses. *Food and Chemical Toxicology, 50*(3–4), 1118–1123.

Kuhn, R., & Winterstein, A. (1934). Über die Konstitution des Pikro-crocins und seine Beziehung zu den Carotin-Farbstoffen des Safrans. *Berichte der deutschen chemischen Gesellschaft (A and B Series), 67*(2), 344–357. https://doi.org/10.1002/cber.19340670239

Kyriakoudi, A., O'Callaghan, Y. C., Galvin, K., Tsimidou, M. Z., & O'Brien, N. M. (2015). Cellular transport and bioactivity of a major saffron apocarotenoid, picrocrocin (4-(β-d-glucopyranosyloxy)-2,6,6-trimethyl-1-cyclohexene-1-carboxaldehyde). *Journal of Agricultural and Food Chemistry, 63*(39), 8662–8668. https://doi.org/10.1021/acs.jafc.5b03363

Lahmass, I., Ouahhoud, S., Sabouni, A., Elyoubi, M., Benabbas, R., Elmoussaoui, R., . . . Saalaoui, E. (2017). Antihyperlipidemic effect of crude extract of saffron (*C. sativus*) stigma in healthy male rats. *Journal of Medical and Allied Sciences, 7*(1), 20.

Lambrianidou, A., Koutsougianni, F., Papapostolou, I., & Dimas, K. (2020). Recent advances on the anticancer properties of saffron (*C. sativus* L.) and its major constituents. *Molecules, 26*(1). https://doi.org/10.3390/molecules26010086

Lari, P., Abnous, K., Imenshahidi, M., Rashedinia, M., Razavi, M., & Hosseinzadeh, H. (2015). Evaluation of diazinon-induced hepatotoxicity and protective effects of crocin. *Toxicolology and Industrial Health, 31*(4), 367–376. https://doi.org/10.1177/0748233713475519

Lautenschläger, M., Sendker, J., Hüwel, S., Galla, H. J., Brandt, S., Düfer, M., . . . Hensel, A. (2015). Intestinal formation of trans-crocetin from saffron extract (C. sativus L.) and *in vitro* permeation through intestinal and blood brain barrier. *Phytomedicine, 22*, 36–44. https://doi.org/10.1016/j.phymed.2014.10.009

Li, C.-Y., & Wu, T.-S. (2002). Constituents of the pollen of *C. sativus* L. and their tyrosinase inhibitory activity. *Chemical and Pharmaceutical Bulletin, 50*(10), 1305–1309. https://doi.org/10.1248/cpb.50.1305

Licón, C., Carmona, M., Llorens, S., Berruga, M. I., & Alonso, G. L. (2010). Potential healthy effects of saffron spice (*C. sativus L. stigmas*) consumption. *Functional Plant Science and Biotechnology, 4*, 64–73.

Linardaki, Z. I., Orkoula, M. G., Kokkosis, A. G., Lamari, F. N., & Margarity, M. (2013). Investigation of the neuroprotective action of saffron (C. sativus L.) in aluminum-exposed adult mice through behavioral and neurobiochemical assessment. *Food and Chemical Toxicology, 52*, 163–170.

Lopresti, A. L., Drummond, P. D., Inarejos-García, A. M., & Prodanov, M. (2018). Affron®, a standardized extract from saffron (*C. sativus* L.) for the treatment of youth anxiety and depressive symptoms: a randomized, double-blind, placebo-controlled study. *Journal of Affective Disorders, 232*, 349–357.

Lopresti, A. L., Smith, S. J., & Drummond, P. D. (2021). An investigation into an evening intake of a saffron extract (affron®) on sleep quality, cortisol, and melatonin concentrations in adults with poor sleep: a randomized, double-blind, placebo-controlled, multi-dose study. *Sleep Medicine, 86*, 7–18.

Maggi, L., Carmona, M., del Campo, C. P., Kanakis, C. D., Anastasaki, E., Tarantilis, P. A., . . . Alonso, G. L. (2009). Worldwide market screening of saffron volatile composition. *Journal of the Science of Food and Agriculture, 89*(11), 1950–1954. https://doi.org/10.1002/jsfa.3679

Maggi, L., Carmona, M., Sánchez, A. M., & Alonso, G. L. (2010). Saffron flavor: Compounds involved, biogenesis and human perception. *Funct Plant. Science and Technology, 4*(Special Issue), 45–55.

Maggi, L., Sánchez, A. M., Carmona, M., Kanakis, C. D., Anastasaki, E., Tarantilis, P. A., . . . Alonso, G. L. (2011). Rapid determination of safranal in the quality control of saffron spice (C. sativus L.). *Food Chemistry, 127*(1), 369–373. https://doi.org/10.1016/j.foodchem.2011.01.028

Magnusson, B. (2014). *The fitness for purpose of analytical methods: a laboratory guide to method validation and related topics* (2nd edition 2014): Eurachem; 2014, 57 p. Available from: http://um.kb.se/resolve?urn = urn:nbn:se:ri:diva-731.

Matuszewski, B. K., Constanzer, M. L., & Chavez-Eng, C. M. (2003). Strategies for the assessment of matrix effect in quantitative bioanalytical methods based on HPLC-MS/MS. *Analytical Chemistry*, *75*(13), 3019–3030. https://doi.org/10.1021/ac020361s

McDonnell, A. M., & Dang, C. H. (2013). Basic review of the cytochrome P450 system. *Journal of the Advanced Practitioner in Oncology*, *4*(4), 263–268. https://doi.org/10.6004/jadpro.2013.4.4.7

Milajerdi, A., Jazayeri, S., Hashemzadeh, N., Shirzadi, E., Derakhshan, Z., Djazayeri, A., & Akhondzadeh, S. (2018). The effect of saffron (C. sativus L.) hydroalcoholic extract on metabolic control in type 2 diabetes mellitus: A triple-blinded randomized clinical trial. *Journal of Research in Medical Sciences: the Official Journal of Isfahan University of Medical Sciences*, *23*, 16. https://doi.org/10.4102/jrms.JRMS_286_17

Miller, T. L., Willett, S. L., Moss, M. E., Miller, J., & Belinka, B. A. (1982). Binding of crocetin to plasma albumin. *Journal of Pharmaceutical Sciences*, *71*, 173–177. https://doi.org/10.1002/jps.2600710209

Moghadam, B. H., Bagheri, R., Roozbeh, B., Ashtary-Larky, D., Gaeini, A. A., Dutheil, F., & Wong, A. (2021). Impact of saffron (*Crocus Sativus* Linn) supplementation and resistance training on markers implicated in depression and happiness levels in untrained young males. *Physiology and Behavior*, *233*, 113352. https://doi.org/10.1016/j.physbeh.2021.113352

Mohebbati, R., Khazdair, M. R., & Hedayati, M. (2017). Neuroprotective effects of medicinal plants and their constituents neuroprotective effects of medicinal plants and their constituents on different induced neurotoxicity methods: A review. *A Journal of Reports in Pharmaceutical Sciences*, *6*(1), 34–50.

Molina, R.V., García-Luis, A., Valero, M., Navarro, Y., & Guardiola, J.L. (2003). *Extending the harvest period of saffron*. In *1 International Symposium on Saffron Biology and Biotechnology* 650, 219–225.

Moman, R. N., & Varacallo, M. (2019). *Physiology*. Albumin: *StatPearls Publishing*.

Moratalla-López, N., Bagur, M. J., Lorenzo, C., Salinas, M., & Alonso, G. L. (2019). Bioactivity and bioavailability of the major metabolites of C. sativus L. flower. *Molecules*, *24*(15). https://doi.org/10.3390/molecules24152827

Nadkarni, K. M. (1996). *[Indian materia medica]; Dr. KM Nadkarni's Indian materia medica: with Ayurvedic, Unani-Tibbi, Siddha, allopathic, homeopathic, naturopathic and home remedies, appendices and indexes. 1* (Vol. 1). Popular Prakashan.

Naess, S. N., Elgsaeter, A., Foss, B. J., Li, B., Sliwka, H.-R., Partali, V., . . . Naqvi, K. R. (2006). Hydrophilic carotenoids: surface properties and aggregation of crocin as a biosurfactant. *Helvetica Chimica Acta*, *89*(1), 45–53. https://doi.org/10.1002/hlca.200690011

Nasiri, Z., Sameni, H. R., Vakili, A., Jarrahi, M., & Khorasani, M. Z. (2015). Dietary saffron reduced the blood pressure and prevented remodeling of the aorta in L-NAME-induced hypertensive rats. *Iranian Journal of Basic Medical Sciences*, *18*(11), 1143–1146.

Nehvi, F. A., Salwee, Y. A., & Arshid-Gowhar, A. (2011). Chemical composition, utility and processing of saffron under pack house concept [training manual]. *Advances in Saffron Biology, Production and Quality Improvement" for Afghan Delegation*.

Németh, K., Plumb, G. W., Berrin, J. G., Juge, N., Jacob, R., Naim, H. Y., . . . Kroon, P. A. (2003). Deglycosylation by small intestinal epithelial cell β-glucosidases is a critical step in the absorption and metabolism of dietary flavonoid glycosides in humans. *European Journal of Nutrition*, *42*, 29–42. https://doi.org/10.1007/s00394-003-0397-3

Nenadis, N., Heenan, S., Tsimidou, M. Z., & Van Ruth, S. (2016). Applicability of PTR-MS in the quality control of saffron. *Food Chemistry, 196*, 961–967.

Norbak, R., & Kondo, T. (1999). Flavonol glycosides from flowers of Crocus speciosus and C. antalyensis. *Phytochemistry, 51*(8), 1113–1119. https://doi.org/10.1016/S0031-9422(99)00109-0

Omidi, A., Riahinia, N., Montazer Torbati, M. B., & Behdani, M. A. (2014). Hepatoprotective effect of *C. sativus* (saffron) petals extract against acetaminophen toxicity in male Wistar rats. *Avicenna Journal of Phytomedicine, 4*(5), 330–336.

Omidi, A., & Totrabi, Z. (2016). The protective role of saffron petal extracts on gentamicininduced nephrotoxicity in rats. *Veterinary Science Development, 6*(6045), 9–12.

Orio, L., Alen, F., Ballesta, A., Martin, R., & Gomez de Heras, R. (2020). Antianhedonic and antidepressant effects of Affron®, a standardized saffron (*C. sativus* L.) extract. *Molecules, 25*(14), 3207. https://doi.org/10.3390/molecules25143207

Özay, Y., Güzel, S., Yumrutaş, Ö., Pehlivanoğlu, B., Erdoğdu, İ. H., Yildirim, Z., . . . Darcan, S. (2019). Wound healing effect of kaempferol in diabetic and nondiabetic rats. *Journal of Surgical Research, 233*, 284–296.

Palanichamy, S., & Nagarajan, S. (1990). Analgesic activity of *Cassia alata* leaf extract and kaempferol 3-O-sophoroside. *Journal of Ethnopharmacology, 29*(1), 73–78. https://doi.org/10.1016/0378-8741(90)90099-F

Piccardi, M., Marangoni, D., Minnella, A. M., Savastano, M. C., Valentini, P., Ambrosio, L., . . . Falsini, B. (2012). A Longitudinal follow-up study of saffron supplementation in early age-related macular degeneration: sustained benefits to central retinal function. *Evidence-Based Complementary and Alternative Medicine, 429124*. https://doi.org/10.1155/2012/429124

Pickens, L. B., & Tang, Y. (2010). Oxytetracycline biosynthesis. *American Society for Biochemistry and Molecular Biology Inc, 285*, P27509–P27515.

Pollak, J. K. (1998). A short review of the problems posed by xenobiotics in chemical mixtures and the role of mixed function oxidases. *International Journal of Environmental Health Research, 8*, 157–163. https://doi.org/10.1080/09603129873598

Priyadarsini, R. V., Murugan, R. S., Maitreyi, S., Ramalingam, K., Karunagaran, D., & Nagini, S. (2010). The flavonoid quercetin induces cell cycle arrest and mitochondria-mediated apoptosis in human cervical cancer (HeLa) cells through p53 induction and NF-κB inhibition. *European Journal of Pharmacology, 649*(1–3), 84–91.

Qian, J., Chen, X., Chen, X., Sun, C., Jiang, Y., Qian, Y., . . . Zheng, C. (2019). Kaempferol reduces K63-linked polyubiquitination to inhibit nuclear factor-κB and inflammatory responses in acute lung injury in mice. *Toxicology Letters, 306*, 53–60. https://doi.org/10.1016/j.toxlet.2019.02.005

Rezaee, R., & Hosseinzadeh, H. (2013). Safranal: From an aromatic natural product to a rewarding pharmacological agent. *Mashhad University of Medical Sciences, 16*, 12–26.

Riahi-Chebbi, I., Souid, S., Othman, H., Haoues, M., Karoui, H., Morel, A., . . . Essafi-Benkhadir, K. (2019). The phenolic compound Kaempferol overcomes 5-fluorouracil resistance in human resistant LS174 colon cancer cells. *Scientific Reports, 9*(1), 1–20.

Rocchi, R., Mascini, M., Sergi, M., Compagnone, D., Mastrocola, D., & Pittia, P. (2018). Crocins pattern in saffron detected by UHPLC–MS/MS as marker of quality, process and traceability. *Food Chemistry, 264*, 241–249.

Rubert, J., Lacina, O., Zachariasova, M., & Hajslova, J. (2016). Saffron authentication based on liquid chromatography high resolution tandem mass spectrometry and multivariate data analysis. *Food Chemistry, 204*, 201–209. https://doi.org/10.1016/j.foodchem.2016.01.003

Samarghandian, S., & Borji, A. (2014). Anticarcinogenic effect of saffron (*C. sativus* L.) and its ingredients. *Pharmacognosy Research, 6*(2), 99–107. https://doi.org/10.4103/0974-8490.128963

Samarghandian, S., Farkhondeh, T., & Zeinali, T. (2020). *C. sativus L.(saffron) extract antioxidant potential and use in aging* (pp. 193–200). Elsevier.

Samarghandian, S., Tavakkol, A. J., & Davoodi, S. (2011). Suppression of pulmonary tumor promotion and induction of apoptosis by *C. sativus* L. extraction. *Applied Biochemistry and Biotechnology, 164*(2), 238–247. https://doi.org/10.1007/s12010-010-9130-x

Sánchez, A. M., Carmona, M., Jarén-Galán, M., Mosquera, M. I., & Alonso, G. L. (2011). Picrocrocin kinetics in aqueous saffron spice extracts (*C. sativus* L.) upon thermal treatment. *Journal of Agricultural and Food Chemistry, 59*(1), 249–255. https://doi.org/10.1021/jf102828v

Sánchez, A. M., Carmona, M., Ordoudi, S. A., Tsimidou, M. Z., & Alonso, G. L. (2008). Kinetics of individual crocetin ester degradation in aqueous extracts of saffron (*C. sativus* L.) upon thermal treatment in the dark. *Journal of Agricultural and Food Chemistry, 56*(5), 1627–1637. https://doi.org/10.1021/jf0730993

Sánchez, A. M., Maggi, L., Carmona, M., & Alonso, G. L. (2011). *Authentication of saffron spice (C. sativus L.) Progress in Authentication of Food and Wine* (Vol. 1081, pp. 309–331). American Chemical Society.

Schmidt, M., Betti, G., & Hensel, A. (2007). Saffron in phytotherapy: pharmacology and clinical uses. *Wiener Medizinische Wochenschrift, 157*(13), 315–319.

Serrano-Díaz, J., Sánchez, A. M., Maggi, L., Martínez-Tomé, M., García-Diz, L., Murcia, M. A., & Alonso, G. L. (2012). Increasing the applications of *C. sativus* flowers as natural antioxidants. *Journal of Food Science, 77*(11), C1162–C1168. https://doi.org/10.1111/j.1750-3841.2012.02926.x

Serrano-Díaz, J., Sánchez, A. M., Martínez-Tomé, M., Winterhalter, P., & Alonso, G. L. (2013). A contribution to nutritional studies on *C. sativus* flowers and their value as food. *Journal of Food Composition and Analysis, 31*(1), 101–108. https://doi.org/10.1016/j.jfca.2013.03.009

Serrano-Díaz, J., Sánchez, A. M., Martínez-Tomé, M., Winterhalter, P., & Alonso, G. L. (2014). Flavonoid determination in the quality control of floral bioresidues from C. sativus L. *Journal of Agricultural and Food Chemistry, 62*(14), 3125–3133. https://doi.org/10.1021/jf4057023

Shahi, T., Assadpour, E., & Jafari, S. M. (2016). Main chemical compounds and pharmacological activities of stigmas and tepals of "red gold"; saffron. *Trends in Food Science and Technology, 58*, Elsevier Ltd., 69–78.

Shakya, R., Nepal, M. R., Kang, M. J., & Jeong, T. C. (2020). Effects of intestinal microbiota on pharmacokinetics of crocin and crocetin in male sprague–dawley rats. *Metabolites, 10*, 1–10. https://doi.org/10.3390/metabo10110424

Sharmila, G., Bhat, F. A., Arunkumar, R., Elumalai, P., Singh, P. R., Senthilkumar, K., & Arunakaran, J. (2014). Chemopreventive effect of quercetin, a natural dietary flavonoid on prostate cancer in *in vivo* model. *Clinical Nutrition, 33*(4), 718–726.

Singletary, K. (2020). Saffron: potential health benefits. *Nutrition Today, 55*(6), 294–303. https://doi.org/10.1097/nt.0000000000000449

Tabeshpour, J., Sobhani, F., Sadjadi, S. A., Hosseinzadeh, H., Mohajeri, S. A., Rajabi, O., . . . Eslami, S. (2017). A double-blind, randomized, placebo-controlled trial of saffron stigma (*C. sativus* L.) in mothers suffering from mild-to-moderate postpartum depression. *Phytomedicine*, *36*, 145–152.

Talaei, A., Moghadam, M. H., Tabassi, S. A. S., & Mohajeri, S. A. (2015). Crocin, the main active saffron constituent, as an adjunctive treatment in major depressive disorder: a randomized, double-blind, placebo-controlled, pilot clinical trial. *Journal of Affected Disorders*, *174*, 51–56. https://doi.org/10.1016/j.jad.2014.11.035

Tanrisever, B., & Santella, P. J. (1986). Cefadroxil: A review of its antibacterial, pharmacokinetic and therapeutic properties in comparison with Cephalexin and Cephradine. *Drugs*, *32*, 1–16. https://doi.org/10.2165/00003495-198600323-00003

Tarantilis, P. A., Tsoupras, G., & Polissiou, M. (1995). Determination of saffron (*C. sativus* L.) components in crude plant extract using high-performance liquid chromatography-UV–visible photodiode-array detection-mass spectrometry. *Journal of Chromatography A*, *699*(1–2), 107–118. https://doi.org/10.1016/0021-9673(95)00044-N

Tavakol-Afshari, J., Boskabady, M.-H., & Salari, R. (2020). *Anticancer properties of saffron* (pp. 485–492). Elsevier.

Termentzi, A., & Kokkalou, E. (2008). LC-DAD-MS (ESI+) Analysis and antioxidant capacity of *C. sativus* petal extracts. *Planta Medica*, *74*(05), 573–581.

Tuberoso, C. I. G., Rosa, A., Montoro, P., Fenu, M. A., & Pizza, C. (2016). Antioxidant activity, cytotoxic activity and metabolic profiling of juices obtained from saffron (C. sativus L.) floral by-products. *Food Chemistry*, *199*, 18–27. https://doi.org/10.1016/j.foodchem.2015.11.115

Ulbricht, C., Conquer, J., Costa, D., Hollands, W., Iannuzzi, C., Isaac, R., . . . Varghese, M. (2011). An evidence-based systematic review of saffron (*C. sativus*) by the Natural Standard Research Collaboration. *Journal of Dietary Supplements*, *8*(1), 58–114. https://doi.org/10.3109/19390211.2011.547666

Umigai, N., Murakami, K., Ulit, M. V., Antonio, L. S., Shirotori, M., Morikawa, H., & Nakano, T. (2011). The pharmacokinetic profile of crocetin in healthy adult human volunteers after a single oral administration. *Phytomedicine*, *18*, 575–578. https://doi.org/10.1016/j.phymed.2010.10.019

Urbani, E., Blasi, F., Chiesi, C., Maurizi, A., & Cossignani, L. (2015). Characterization of volatile fraction of saffron from Central Italy (Cascia, Umbria). *International Journal of Food Properties*, *18*(10), 2223–2230. https://doi.org/10.1080/10942912.2014.968787

Vahedi, M., Kabiri, M., Salami, S. A., Rezadoost, H., Mirzaie, M., & Kanani, M. R. (2018). Quantitative HPLC-based metabolomics of some Iranian saffron (*C. sativus* L.) accessions. *Industrial Crops and Products*, *118*, 26–29. https://doi.org/10.1016/j.indcrop.2018.03.024

Valle Garcia-Rodriguez, M, Serrano-Diaz, J., Tarantilis, P.A., Lopez-Corcoles, H., Carmona, M., & Alonso, G.L. (2014). Determination of saffron quality by high-performance liquid chromatography. *Journal of Agricultural and Food Chemistry*, *62*(32), 8068–8074.

Viegas, O., Faria, M. A., Sousa, J. B., Vojtek, M., Gonçalves-Monteiro, S., Suliburska, J., . . . Ferreira, I. M. P. L. V. O. (2019). Delphinidin-3-O-glucoside inhibits angiogenesis via VEGFR2 downregulation and migration through actin disruption. *Journal of Functional Foods*, *54*, 393–402. https://doi.org/10.1016/j.jff.2019.01.039

Wang, Y., Sun, J., Liu, C., & Fang, C. (2014). Protective effects of crocetin pretreatment on myocardial injury in an ischemia/reperfusion rat model. *European Journal of Pharmacology*, *741*, 290–296.

Wani, B. A., Hamza, A. K. R., & Mohiddin, F. A. (2011). Saffron: A repository of medicinal properties. *Journal of Medicinal Plants Research*, *5*(11), 2131–2135.

William, G. G., Gregory, R., Amitabha, R., Shrikant, A., & Animesh, D. (2012). Crocetin: an agent derived from saffron for prevention and therapy for cancer. *Current Pharmaceutical Biotechnology*, *13*(1), 173–179. http://dx.doi.org/10.2174/138920112798868566

Xi, L., Qian, Z., Du, P., & Fu, J. (2007). Pharmacokinetic properties of crocin (crocetin digentiobiose ester) following oral administration in rats. *Phytomedicine*, *14*, 633–636. https://doi.org/10.1016/j.phymed.2006.11.028

Yang, R., Vernon, K., Thomas, A., Morrison, D., Qureshi, N., & Van Way, I. I. I.,. C. W. (2011). Crocetin reduces activation of hepatic apoptotic pathways and improves survival in experimental hemorrhagic shock. *Journal of Parenteral and Enteral Nutrition*, *35*(1), 107–113.

Yeon, M. J., Lee, M. H., Kim, D. H., Yang, J. Y., Woo, H. J., Kwon, H. J., . . . Kim, J.-B. (2019). Anti-inflammatory effects of kaempferol on *Helicobacter pylori*-induced inflammation. *Bioscience, Biotechnology, and Biochemistry*, *83*(1), 166–173.

Yilmaz, A., Nyberg, N. T., Mølgaard, P., Asili, J., & Jaroszewski, J. W. (2010). 1H NMR metabolic fingerprinting of saffron extracts. *Metabolomics*, *6*(4), 511–517. https://doi.org/10.1007/s11306-010-0221-z

Yoshino, F., Yoshida, A., Umigai, N., Kubo, K., & Lee, M. C. (2011). Crocetin reduces the oxidative stress induced reactive oxygen species in the stroke-prone spontaneously hypertensive rats (SHRSPs) brain. *Journal of Clinical Biochemistry and Nutrition*, *49*(3), 182–187. https://doi.org/10.3164/jcbn.11-01

Zahra, B.S., Reyhane, H., Hamidreza, M., & Majid, S. (2013). Anticancer effects of crocetin in both human adenocarcinoma gastric cancer cells and rat model of gastric cancer. *Biochemistry and Cell Biology*, *91*(6), 397–403. doi: https://doi.org/10.1139/bcb-2013-0014%M 24219281.

Zalacain, A., Ordoudi, S. A., Díaz-Plaza, E. M., Carmona, M., Blázquez, I., Tsimidou, M. Z., & Alonso, G. L. (2005). Near-infrared spectroscopy in saffron quality control: determination of chemical composition and geographical origin. *Journal of Agricultural and Food Chemistry*, *53*(24), 9337–9341.

Zargari, A. (1990). *Herbal medicines* (pp. 83). Tehran: *Publication of Tehran University*.

Zeinali, M., Zirak, M. R., Rezaee, S. A., Karimi, G., & Hosseinzadeh, H. (2019). Immunoregulatory and anti-inflammatory properties of C. sativus (Saffron) and its main active constituents: A review. *Iranian Journal of Basic Medical Sciences*, *22*(4), 334–344.

Zeka, K., Ruparelia, K. C., Continenza, M. A., Stagos, D., Vegliò, F., & Arroo, R. R. J. (2015). Petals of *C. sativus* L. as a potential source of the antioxidants crocin and kaempferol. *Fitoterapia*, *107*, 128–134. https://doi.org/10.1016/j.fitote.2015.05.014

Zheng, C.-J., Li, L., Ma, W.-H., Han, T., & Qin, L.-P. (2011). Chemical constituents and bioactivities of the liposoluble fraction from different medicinal parts of *C. sativus*. *Pharmaceutical Biology*, *49*(7), 756–763. https://doi.org/10.3109/13880209.2010.547206

Zheng, Y.-Q., Liu, J.-X., Wang, J.-N., & Xu, L. (2007). Effects of crocin on reperfusion-induced oxidative/nitrative injury to cerebral microvessels after global cerebral ischemia. *Brain Research*, *1138*, 86–94. https://doi.org/10.1016/j.brainres.2006.12.064

15

Cocoa Shell: Source of Novel Bioactive Ingredients for the Prevention of Cardiometabolic Diseases

Miguel Rebollo-Hernanz[1,2], Silvia Cañas[1,2], Cheyenne Braojos[1,2], Paz Cano-Muñoz[1,2], and Maria A. Martín-Cabrejas[1,2]

[1] *Department of Agricultural Chemistry and Food Science, Universidad Autónoma de Madrid, Madrid, Spain*
[2] *Institute of Food Science Research, CIAL (UAM-CSIC), Madrid, Spain*

15.1 Introduction

Cocoa (*Theobroma cacao* L.) is the only commercially cultivated among the twenty-two species of the *Theobroma* genus (Kim, Lee, & Lee, 2011). Cocoa processing produces a significant amount of waste. In fact, just 10% of cocoa fruit's total weight is used for commercialization, while the remaining 90% is discarded as waste or by-products (Bhat & Sukumaran, 2012). Recently, cocoa by-products' revalorization has gained increased attention owing to the lack of natural resources and severe environmental issues associated with cocoa processing wastes disposal. The research proposes using these by-products in applications beyond traditional uses (fertilizer or feedstuff), such as food or biotechnological applications (Jahurul et al., 2013; Yusof, Khanahmadi, Amid, & Mahmod, 2016). The biorefinery of cocoa processing residues into valuable products is gaining interest, and as a result, developed countries are preparing strategic policies to evolve toward a circular bioeconomy (Kowalska, Czajkowska, Cichowska, & Lenart, 2017; Ravindran & Jaiswal, 2016). Hence, cocoa by-products valorization strategies and new applications are appearing in different fields.

Furthermore, a plethora of studies reports multiple health benefits from cancer prevention to metabolic syndrome regulation. Cocoa health benefits are associated with specific groups of phytochemicals, namely alkaloids (theobromine and caffeine) and phenolic compounds based on *in vivo* and epidemiological studies (Katz, Doughty, & Ali, 2011). Lesser attention has been given to the dietary fiber fraction of cocoa, which is present in their by-products but involves just a marginal fraction of the cocoa products. Notwithstanding this knowledge gap, the effects of consuming cocoa by-products may be similar to that of cocoa beans due to their comparable phytochemical composition (Rebollo-Hernanz, Zhang, Aguilera, Martín-Cabrejas, & de Mejia, 2019b). Then, research should focus on further investigating the chemical composition of cocoa by-products, their biological activity,

Molecular Mechanisms of Functional Food, First Edition. Edited by Rocio Campos-Vega and B. Dave Oomah.

and potential toxicity, and the mechanism underlying the observed effects. Hence, this chapter aims to review the literature and the main findings around the cocoa shell's health benefits and the molecular mechanisms of its bioactive compounds.

15.2 Cocoa Shell: From Field to Functional Foods Industry

15.2.1 Cocoa Processing and Cocoa Shell Production

The cocoa beans are processed to obtain cocoa products such as liquid chocolate, cocoa powder, and cocoa butter (Beg, Ahmad, Jan, & Bashir, 2017). The processing of the cocoa beans is carried out in two stages. The first stage involves processing in the field: harvesting, seeds and pulp separation from the pod, fermentation, and cocoa beans drying, storage, and export (Figure 15.1a). The second stage involves cleaning, roasting, grinding, and further steps related to the manufacture of chocolate and derivatives (Predan, Lazăr, & Lungu, 2019). Then, the cocoa beans are processed to obtain liquid chocolate, cocoa powder, and cocoa butter (Beg et al., 2017). Cocoa by-products are generated during cocoa processing (Figure 15.1).

Cocoa pod husk is the first by-product obtained from cocoa processing when the cocoa beans are removed from the pod. The cocoa shell (**Figure 15.1**b–d) is removed from the cocoa bean during roasting, obtaining the second by-product. The cocoa shell is separated along with the germ, and the cotyledon fragments, known as nibs, free of the shell are used in chocolate manufacture (Beckett, 2008). Hence, only two by-products are obtained during cocoa processing: the cocoa pod husk and the cocoa shell (Vásquez et al., 2019; Wollgast & Anklam, 2000).

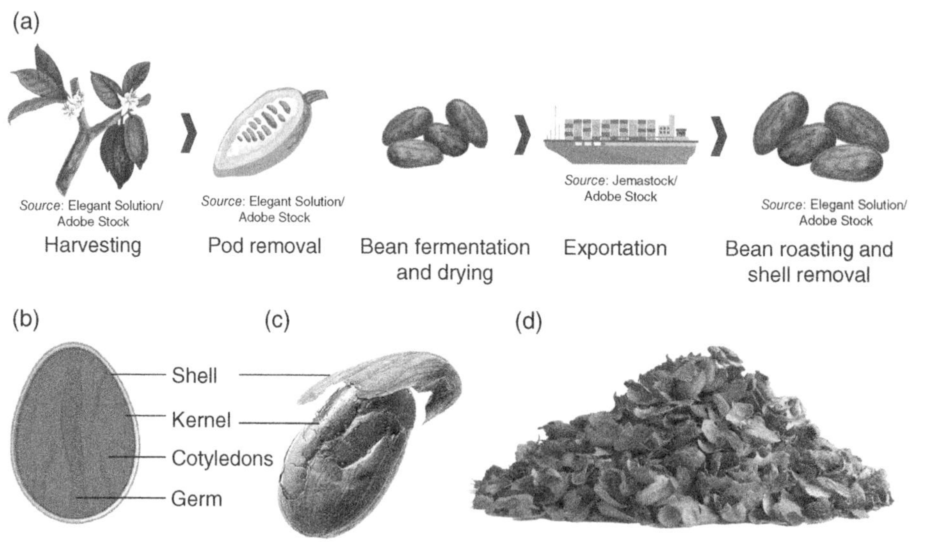

Figure 15.1 Cocoa processing stages (a) from harvesting to roasting, including the two steps where the cocoa pod husk and the cocoa shell are separated; the cocoa bean's anatomical structure (**b, c**), including (from the outside to the inside) shell, kernel, and cotyledons (composing the nibs), and germ; the appearance of the cocoa shell once separated from the cotyledons (d).

The generation of the cocoa shell is very significant in cocoa production, representing 12–20% of the cocoa bean (Okiyama, Navarro, & Rodrigues, 2017). Considering the world cocoa production in 2019/2020 (4.2 million tons), large amounts of the cocoa shell that can be revalorized are discarded every year: based on the weight percentage of the cocoa shell and the mentioned cocoa production data, 700 thousand tons of cocoa shells are produced worldwide. Hence, producing one kg of chocolate would generate almost 100 g of the cocoa shell. Therefore, the high cocoa shell generation warrants further evaluation of its feasibility as a potential food ingredient or nutraceutical. This by-product has attracted the attention of researchers mainly due to its nutritional properties and, to its economic recovery.

15.2.2 Nutritional Characterization of the Cocoa Shell

The cocoa shell is composed mainly of dietary fiber (40–64%), protein (12–33%), fat (5–19%), moisture (~5–10%), and minerals (~7–11%) (Arlorio, Coisson, Restani, & Martelli, 2001; Lecumberri, Mateos et al., 2007; Martín-Cabrejas, Valiente, Esteban, Mollá, & Waldron, 1994; Martínez et al., 2012) (Table 15.1). The dietary fiber fraction of the cell wall contains pectic polysaccharides (~45%), comprising a heterogeneous mixture of rhamnogalacturonans with variable branching degrees, hemicelluloses (~20%), including xyloglucans, galactoglucomannans, and glucuronoarabinoxylans, cellulose (~35%), and lignin (Redgwell et al., 2003). Cocoa shell also contains other minor compounds such as theobromine (1.2%), caffeine (0.2%), and polyphenols (0.2–5.8%) (Jokić et al., 2018; Lecumberri, Mateos et al., 2007). Protocatechuic and vanillic acids and mono- and polymeric procyanidins are the main phenolic compounds, among other minor identified compounds (de la Cádiz-Gurrea et al., 2020).

The composition of the cocoa shell varies considerably and is influenced by factors such as the cocoa's origin and processing. Regardless of its variability, the distinct composition of the cocoa shell, rich in methylxanthines, phenolic compounds, and dietary fiber, makes it a nutraceutical product of interest to many researchers. The cocoa shell can have multiple applications in the food industry due to its versatility as a food ingredient since it is a source of dietary fiber and can be used to produce extracts rich in methylxanthines and phenolic compounds. The cocoa shell has been recognized as a valuable by-product with many potential applications from this comprehensive nutritional and phytochemical perspective. Nonetheless, particular attention must be made to the cocoa shell's food safety aspect to transform it into a food ingredient.

15.2.3 Food Safety Aspects of the Cocoa Shell

Cocoa shell possesses excellent potential as a food ingredient since it contains many bioactive compounds, but before incorporation into food products, undesirable components that may be present in the cocoa shell must be eliminated, including microorganisms, mycotoxins, heavy metals, and polycyclic aromatic hydrocarbons (Barišić, Jozinović et al., 2020).

Microorganisms frequently contaminate the cocoa shell since birds and insects responsible for transmitting *Escherichia coli* and *Salmonella* contact the cocoa beans during the drying and storage stages (do Nascimento et al., 2010). Fungal contamination may arise during different cocoa bean processing stages. Most fungi, including *Aspergillus* species, *Eurotium*, and *Absidia* species, may be present on the cocoa shell. Fungal contamination

Table 15.1 Nutritional and chemical composition of the cocoa shell.

Compound	Concentration range	References
Energy (kcal/100 g)	121.8	Adamafio (2013)
Moisture (%)	4.3–10.1	Agus, Mohamad, and Hussain (2018); Arlorio et al. (2001); Barbosa-Pereira, Rojo-Poveda, Ferrocino, Giordano, and Zeppa (2019); Bonvehí and Coll (1999); Bonvehí and Jordà (1998); Grillo et al. (2019); Knapp and Coward (1935); Martín-Cabrejas et al. (1994); Martínez et al. (2012); Mazzutti, Rodrigues, Mezzomo, Venturi, and Ferreira (2018); Okiyama et al. (2018); Okiyama, Soares, Toda, Oliveira, and Rodrigues (2019); Adamafio (2013); Agus et al. (2018); Arlorio et al. (2001); Bonvehí and Benería (1998); Bonvehí and Coll (1999); Bonvehí and Jordà (1998); Grillo et al. (2019); Lecumberri et al. (2007); Martín-Cabrejas et al. (1994); Martínez et al. (2012); Okiyama et al. (2018, 2019); Rojo-Poveda, Mateus-Reguengo, and Stévigny (2019); Vojvodić, Komes, Vovk, Belščak-Cvitanović, and Bušić (2016)
Ash (%)	6.3–12.0	
Proteins (%)	11.6–21.6	
Fats (%)	1.8–18.5	
Carbohydrates (%)	7.9–25.8	Agus et al. (2018); Martínez et al. (2012); Rojo-Poveda et al. (2019)
Starch (%)	0.9–1.4	Bonvehí and Benería (1998); Bonvehí and Coll (1999)
Soluble sugars (%)	0.4	Lecumberri, Mateos et al. (2007)
Total sugars (%)	1.3–2.2	Bonvehí and Coll (1999)
Dietary fiber (%)		
Total dietary fiber	39.9–63.6	Arlorio et al. (2001); Bonvehí and Benería (1998); Lecumberri, Mateos et al. (2007); Martín-Cabrejas et al. (1994); Martínez et al. (2012); Okiyama et al. (2019); Redgwell et al. (2003); Rojo-Poveda et al. (2019)
Soluble dietary fiber	7.0–18.6	
Insoluble dietary fiber	24.2–51.9	
Methylxanthines		
Theobromine (%)	0.5–4.8	Arlorio et al. (2001); Barbosa-Pereira, Guglielmetti, and Zeppa (2018); Bonvehí and Jordà (1998); Jokić, Gagić, Knez, Šubarić, and Škerget (2018); Okiyama et al. (2018)
Caffeine (%)	0.1–0.4	Barbosa-Pereira et al. (2018); Bonvehí and Jordà (1998); Okiyama et al. (2018)
Polyphenols (mg/g)	1.9–58.0	Arlorio et al. (2001); Barbosa-Pereira et al. (2018); Hernández-Hernández, Viera-Alcaide, Morales-Sillero, Fernández-Bolaños, and Rodríguez-Gutiérrez (2018); Lecumberri, Mateos et al. (2007); Mazzutti et al. (2018)

Source: Based on Arlorio et al. (2001); Lecumberri, Goya et al. (2007); Martín-Cabrejas et al. (1994); Martínez et al. (2012).

alters the sensory properties of cocoa beans, and in addition can form mycotoxins (Okiyama et al., 2017). Ochratoxin A (OTA) is the commonest mycotoxin found in cocoa shells. OTA can appear at all stages of cocoa processing, but it is highest during drying and storage (Copetti, Pereira, Iamanaka, Pitt, & Taniwaki, 2010). Aflatoxins B1, B2, G1, and G2 are also present in cocoa by-products, primarily in the cocoa shell (Copetti et al., 2012). Numerous studies have reported the importance of removing contaminants such as OTA from cocoa shells by different methods (Amézqueta, González-Peñas, Lizarraga, Murillo-Arbizu, & López De Cerain, 2008; Arlorio et al., 2001; Barišić, Jozinović et al., 2020). OTA concentration in this ingredient generally ranges between 0.1 and 2.0 μg/kg, which is, however, considered acceptable by the European Commission for cocoa beans (less than 2 μg/kg) (Copetti, Iamanaka, Nester, Efraim, & Taniwaki, 2013; Rojo-Poveda, Barbosa-Pereira, Zeppa et al., 2020). Since the cocoa shell is a roasted by-product, some undesirable components such as acrylamide and 5-hydroxymethylfurfural may be present, albeit in nontoxic concentrations (Barišić, Flanjak et al., 2020). The drying and roasting stages can even produce polycyclic aromatic hydrocarbons. During the roasting stage, these compounds can migrate from the cocoa shell to the cocoa beans. Although some methods attempt to remove polycyclic aromatic hydrocarbons from cocoa shells, more studies are needed to eliminate these compounds successfully (Barišić, Jozinović et al., 2020).

Fertilizers, pesticides, or insecticides used during cocoa production and the contact with metal devices during cocoa processing may affect the content of heavy metals in cocoa, especially in cocoa shells, due to its high absorption capacity (Barišić, Jozinović et al., 2020). In order to use the cocoa shells as food ingredients, it is necessary to remove the heavy metals, so the cocoa shells are subjected to an alkalinization process with subsequent washing (Chronopoulos, Zuurbier, Brandstetter, & Jung, 2011).

Numerous studies have shown that cocoa shell has an enormous potential to be used as part of the animal diet. However, the high concentration of theobromine in the cocoa shell may produce adverse effects (Adamafio, 2013; Panak Balentić et al., 2018). The inclusion of the cocoa shell in the diet of laying hens is not recommended unless it is boiled since it can decrease egg production (Emiola, Ojebiyi, & Akande, 2011; Olubamiwa, Ikyo, Adebowale, Omojola, & Hamzat, 2006). Nonetheless, the European Food Safety Authority (EFSA) has limited cocoa shell's use in animal feed due to their theobromine content. The limit imposed by EFSA is 300 mg/kg of theobromine in animal feed and 700 mg/kg in feed for adult cattle (Alexander et al., 2008). For this reason, several methods have been proposed to remove theobromine from cocoa shells, including boiling the material (Olubamiwa et al., 2006) or the fermentation of cocoa shells by some fungi to remove theobromine (Adeyemo, Ajayi, Longe, & Olubamiwa, 2015; Aromolaran & Ogunsakin, 2019). Recently, bio-dethrobromianation has been achieved with a filamentous fungus *Talaromyces verruculosus* KY697103 (TvTD) treatment (Oduro-Mensah et al., 2020). This bio-detheobrominated cocoa pod husk (1.5 mg theobromine/g husks) at 30% substitution of isonitrogenous and isocaloric diets did not impact negatively on growth performance, physiology, and reproduction of male Sprague–Dawley rats. Therefore, the detheobrominated cocoa pod husk can safely replace up to 30% maize in animal feed formulations (Oduro-Mensah et al., 2020). The cocoa shell has recently been validated as a novel antioxidant and dietary fiber-rich food ingredient, free of contaminants, and exerting no adverse effects (Rebollo-Hernanz et al., 2020).

15.2.4 Cocoa Shell as a Functional Food Ingredient

Conventionally, the cocoa shell has been used to prepare infusions (Otero, 2018) or just employed as mulch, but due to its high content of bioactive compounds, it has gained interest as a new food-grade ingredient with health-promoting properties (Oddoye, Agyente-Badu, & Gyedu-Akoto, 2013). The cocoa shell exhibits high antioxidant capacity due to its high content in phenolic compounds (Azizah, Nik Ruslawati, & Swee Tee, 1999; Martínez et al., 2012). Cocoa shell's phytochemical composition and antioxidant capacity vary according to their geographical origin. Genotype, maturity, climate, type, and quality of the soil or stress are additional factors that can influence the content of antioxidant compounds in the cocoa shell (Bruna, Eichholz, Rohn, Kroh, & Huyskens-Keil, 2009). During cocoa bean processing, and specifically during the fermentation stage, significant changes occur in the concentration of phenolic compounds and antioxidant capacity (Mazor Jolić, Radojčic Redovnikovic, Marković, Ivanec Šipušić, & Delonga, 2011) since some phenolic compounds present in the cotyledon migrate toward the shell, which enrichers the by-product in polyphenols after fermentation (Hernández-Hernández et al., 2018; Kim & Keeney, 1984).

Previous reports also have investigated other green extraction processes: supercritical CO_2 (Mazzutti et al., 2018), pressurized liquid extraction (ethanolic) (Mazzutti et al., 2018), subcritical water extraction (Jokić et al., 2018), microwave-assisted extractions (aqueous and deep eutectic solvents) (Mellinas, Jiménez, & Garrigós, 2020; Pavlović, Jokić, Jakovljević, Blažić, & Molnar, 2020). The concentration of obtained phytochemicals varied among studies because of the technique used per se and the intrinsic differences among cocoa shell samples. Nevertheless, high theobromine, caffeine, and flavanols were extracted independent of the study. One of the first investigations showed that methanol was the preferred solvent for extracting antioxidant compounds from cocoa shell as a source of phenolic-rich ingredients (Azizah et al., 1999). Similarly, optimized CO_2 extracts exhibited increased antioxidant activity when redissolved in methanol and contained epicatechin and *p*-hydroxybenzoic acid as principal phenolic compounds (Arlorio et al., 2005). Another study in contrast showed better phenolic compounds extraction capabilities with acetone than other solvents such as ethanol, methanol, or water (Nsor-Atindana, Zhong, Mothibe, Bangoura, & Lagnika, 2012). Recent investigations have concluded that a combination of methanol and acetone appears to be the most effective extraction solvent (Hernández-Hernández et al., 2019).

Innovative and more sustainable alternatives are being implemented in food processing, cosmetic, and pharmaceutical industries. Following green chemistry and green engineering principles (Anastas & Eghbali, 2010), the term "green extraction" was defined, and six principles for green extraction were established (Chemat, Vian, & Cravotto, 2012). Consequently, green extractions are based on the design of extraction processes that reduce energy consumption, allow the use of alternative solvents, and ensure a safe and high-quality extract/product (Chemat et al., 2012). Based on these principles, green extractions from the cocoa shell following a biorefinery approach (**Figure 15.2**) involve obtaining the targeted compounds from the food matrix, valorizing the potentially generated co-products, preferably using water, reducing toxic compounds and enhancing safety to high added-value extracts. Recently, an optimized process has been described to obtain aqueous extracts rich in phenolic compounds from cocoa shell (Rebollo-Hernanz et al., 2021c). According to the study, the optimal conditions to achieve a higher extraction of phenolic compounds were 100 °C, 90 min, 0% citric acid, and 0.02 g/mL solid-to-solvent ratio. The main compounds identified

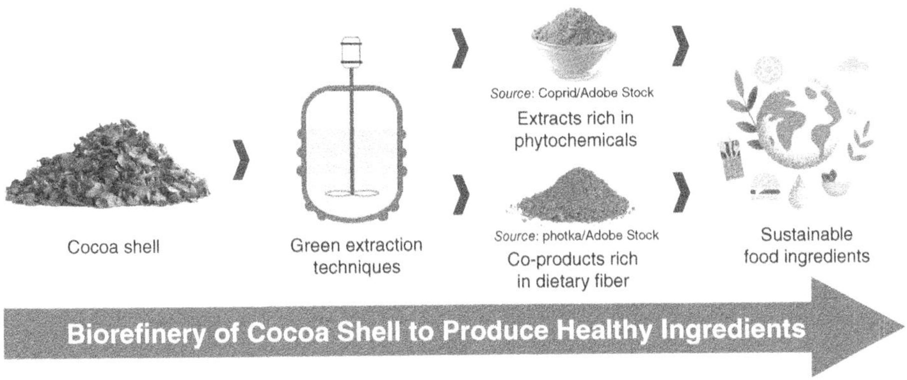

Figure 15.2 Biorefinery approach in the revalorization of the cocoa shell through green extraction techniques.

in this study were protocatechuic acid, procyanidin B2, (−)-epicatechin, and (+)-catechin, associated with the higher antioxidant capacity of the extracts.

The high amount of dietary fiber in the cocoa shell, and the insoluble/soluble dietary fiber ratio, together with the bioactive compounds present in the cocoa shell (free and bound to the dietary fiber), guides the sustainable use of this by-product both as an extract or as a fiber-rich ingredient (flour or residue from the phytochemicals green extraction). The use of the cocoa shell as a food ingredient has been widely studied in recent years (Panak Balentić et al., 2018). Recent studies have developed extruded corn snacks with the cocoa shell to improve their nutritional properties. The snacks acquired a pleasant brown color for the consumer because it resembles chocolate products with reduced resistant starch and increased total phenolic contents (Jozinović et al., 2019). Research has also shown that replacing wheat flour in breadmaking by up to 6% with cocoa shell soluble dietary fiber results in sensory-acceptable products (Collar, Rosell, Muguerza, & Moulay, 2009). The food industry has also paid attention to the use of the cocoa shell as a food ingredient. In particular, Kraft Foods have patented a process that allows the extraction of phenolic compounds and theobromine from the cocoa shell (Bradbury & Kopp, 2006). In addition, incorporation of the cocoa shell increases viscosity and accentuates the chocolate flavor in food products containing cocoa (Chronopoulos et al., 2011). The inclusion of the cocoa shell in biscuits can prevent the degradation of bioactive compounds naturally present in the food matrix through *in vitro* gastrointestinal digestion (Rojo-Poveda, Barbosa-Pereira, El Khattabi et al., 2020). Nevertheless, other studies claim that some bioactive compounds naturally present in the cocoa shell can degrade during production and storage due to their labile nature and low stability. Techniques such as microencapsulation can provide stability to phenolic compounds. Spray-drying a cocoa shell extract with maltodextrins generates microspheres capable of protecting phenolic compounds during storage (90 days) and cooking (in a biscuit model) (Papillo et al., 2019). The incorporation of phenolic compounds extracted from cocoa shells improves oils stability. The polyphenols incorporated in frying oils can act as antioxidants, delaying the generation of free fatty acids, keeping a low peroxide index, and increasing their antioxidant capacity without affecting the oil color (Manzano et al., 2017). Recently, the protective effect of cocoa shell-made ice creams against intestinal damage has been demonstrated (Rossin et al., 2021).

15.3 Molecular Mechanisms of Bioactive Compounds from Cocoa Shell in Preventing Cardiometabolic Diseases

Noncommunicable diseases have been framed as a global emergency. These chronic diseases comprise various conditions; this chapter focuses on those considered cardiometabolic, tightly associated with nutrition: obesity, diabetes, and cardiovascular disease (Miranda et al., 2019). Globally, 1.9 billion adults are overweight and 609 million adults obese, representing approximately 39% of the world's population (Chooi, Ding, & Magkos, 2019). Linked to obesity, type II diabetes is prevalent in 9% of the world's population, around 422 million people (Matsuzawa, 2014). High blood pressure and high cholesterol affect 59 and 39 million people, respectively (Mozaffarian et al., 2016). Therefore, cardiovascular disease has become the leading cause of death in western countries (Mozaffarian et al., 2016).

Despite numerous efforts to develop cocoa shell as a food ingredient (nutritional and phytochemical characterization, ingredients extraction/purification, and product development), there is still a lack of solid research validating the health effects of these ingredients to prevent chronic diseases, especially cardiometabolic diseases. Building a strong and reliable science foundation for functional food claims requires proving the bio-efficacy of functional food components *in vivo*, not just *in vitro* experiments (FAO, 2007). *In vitro* and animal models cannot determine cause-and-effect in humans, but they provide evidence for functionality associated with the active substance (Granato et al., 2020). The validation of by-products as food ingredients needs a thorough evaluation to confirm their functionality *in vitro* and using cell culture models, animals, and humans (preclinical and clinical trials) (Granato, Nunes, & Barba, 2017). Therefore, a molecular nutrition approach is needed to assess the effect of the cocoa shell's bioactive compounds on whole-body physiology and health status at the molecular and cellular levels. The precise identification of the molecular pathways underlying human metabolism is essential for promoting health preventing disease (Norheim et al., 2012). This section will cover the molecular effects of the main three groups of bioactive compounds from the cocoa shell (methylxanthines, phenolic compounds, and dietary fiber) compiled to date.

15.3.1 Molecular Mechanisms of Methylxanthines from Cocoa Shell

Methylxanthines are purine alkaloids commonly present in the diet. The number and positions of methyl substitutes in the xanthine ring determine the differences between the most-wellknown methylxanthines, the trimethylxanthine caffeine (1,3,7-trimethylxanthine), and the dimethylxanthines theobromine (3,7-dimethylxanthine) and theophylline (1,3-dimethylxanthine) (Martínez-López et al., 2014). Methylxanthines are one of the main bioactive compounds in cocoa shell (Franco, Oñatibia-Astibia, & Martínez-Pinilla, 2013; Panak Balentić et al., 2018). During cocoa bean fermentation, methylxanthines migrate from the bean to the cocoa shell. Three methylxanthines are primarily found in the cocoa shell: theobromine, caffeine, and theophylline (Beckett, 2009). Literature reports that methylxanthines show different beneficial effects in animal and human models, mainly antidiabetic, anti-obesity, and antihypertensive effects, as well as anti-inflammatory, anticancer, and neuroprotective functions (Jang et al., 2015; Jang, Kang, Mukherjee, & Yun, 2018; Mitani et al., 2017; Mitchell et al., 2011; Papadimitriou et al., 2015; Sugimoto et al., 2014). This section focuses on the effects of methylxanthines from the cocoa shell and their underlying mechanisms of action to prevent cardiometabolic diseases.

15.3.1.1 Methylxanthines and Obesity

Caffeine and theobromine have also been shown to suppress adipogenesis and stimulate thermogenesis and lipolysis, resulting in control of body weight gain (Acheson et al., 2004; Carrageta et al., 2018; Jang et al., 2015; Jang et al., 2018; Mitani et al., 2017). The main pharmacological effects of methylxanthines include adenosine receptor antagonism and phosphodiesterase inhibition, mainly due to their structural similarity to adenosine (Martínez-Pinilla, Oñatibia-Astibia, & Franco, 2015). This structural similarity triggers methylxanthines nonspecific binding to the adenosine receptor that stimulates lipolysis in adipocytes. These effects are known through the 1-methyl group of methylxanthines, such as caffeine or theophylline, but not theobromine. However, theobromine suppresses adipogenesis by degrading C/EBPβ via the adenosine A1 receptor (Green & Stanberry, 1977; Mitani et al., 2017).

On the other hand, methylxanthines present agonism with β-adrenergic receptors stimulating thermogenesis and consequently increasing the lipolytic cascade activity in adipocytes through two different mechanisms of action: by increasing cAMP-dependent protein kinase (PKA)-dependent cAMP levels and through nonselective inhibition of phosphodiesterases, enzymes responsible for cAMP hydrolysis. In this regard, evidence suggests that theobromine inhibits the cAMP-phosphodiesterase complex in different cell lines (Acheson et al., 2004; Jang et al., 2018; Rasouli & Zahraie, 2006; Sugimoto et al., 2014; Wu & Rajagopalan, 2016).

Theobromine inhibits differentiation and adipogenesis in 3T3-L1 adipocyte cell lines via AMPK activation and suppression in ERK and JNK signaling pathways. Furthermore, theobromine plays an essential role in browning in 3T3-L1 adipocytes mainly through β-adrenergic signaling and AMPK activation, as well as in the regulation of thermogenesis related genes, increasing their expression in both brown and white adipocytes (Jang et al., 2015; Jang et al., 2018). In turn, caffeine has been associated with reduced fatty acid synthesis and FASN expression in 3T3-L1 adipocytes in rats and higher lipid mobilization (Rebollo-Hernanz et al., 2019b; Zapata, Rebollo-Hernanz, Novakofski, Nakamura, & Gonzalez de Mejia, 2020).

On the other hand, caffeine is known to increase lipolysis through stimulation of the sympathetic nervous system, leading to increased release of noradrenaline and consequent activation of β-adrenergic receptors (Acheson et al., 2004). Moreover, the activation of the sympathetic nervous system suppresses hunger, enhances satiety, and stimulates energy expenditure, partly due to increased fat oxidation (Astrup, Toubro, Cannon, Hein, & Madsen, 1991). Human studies have shown that caffeine consumption in young trained and untrained individuals after an overnight fast increases resting metabolic rate, with increased plasma free fatty acid levels and increased lipolysis (LeBlanc, Jobin, Cote, Samson, & Labrie, 1985; Schubert et al., 2014). Other studies have shown that caffeine consumption favors increased energy expenditure and fat loss and also associated with reducing energy intake (Dulloo, Geissler, Collins, & Miller, 1989). On the other hand, caffeine consumption improved weight loss and prevented weight regain (Davoodi et al., 2014).

Recently, the microbiota-mediated antiobesity effects have emerged as another method to curb obesity, and the role of methylxanthines has also been tested. Theobromine modifies the intestinal microbiota in animal models, alleviating weight loss, although its interaction with other compounds present in the cocoa shell, such as fiber or polyphenols, attenuates or potentiates its effects (Martín-Peláez et al., 2017).

15.3.1.2 Methylxanthines and Diabetes

Methylxanthines also exert antidiabetic functions (Cherniack, Buslach, & Lee, 2018). Studies in animal models suggest that the consumption of coffee and caffeine increases glucose tolerance and insulin sensitivity (Matsuda et al., 2011; Morakinyo, Adekunbi, Dada, & Adegoke, 2013; Yamauchi et al., 2010).

In the case of caffeine consumption in humans, several studies indicate that regular caffeine consumption apparently improves insulin sensitivity and glucose tolerance and reduces blood glucose levels (Agardh et al., 2004; Ärnlöv, Vessby, & Risérus, 2004; Loopstra-Masters, Liese, Haffner, Wagenknecht, & Hanley, 2011). However, there is controversy regarding acute coffee consumption, although combined caffeine and carbohydrate consumption may apparently increase blood glucose levels for extended periods and decrease hypoglycemia episodes (Olateju, Begley, Green, & Kerr, 2015; Richardson, Thomas, Ryder, & Kerr, 2005; Watson, Jenkins, Hamilton, Lunt, & Kerr, 2000). On the other hand, few studies investigate long-term caffeine/coffee consumption, although we found meta-analyses where their consumption is associated with reduced risk (up to 23%) of diabetes (Yarmolinsky et al., 2015).

The antidiabetic role of theobromine has been proven in cell lines and animal models through sirtuin-1 activation and reduced extracellular matrix accumulation in diabetic kidney, which can have therapeutic potential for diabetic nephropathy (Papadimitriou et al., 2015). Recent studies in animal models also highlight the critical role of theobromine in the cocoa shell in glucose metabolism, mainly in the secretion of hormones related to appetite and insulin (Camps-Bossacoma et al., 2019). Cocoa shell theobromine has been associated with enhanced glucose tolerance in insulin-resistant hepatocytes by improving glucose and energy metabolism (Rebollo-Hernanz, Aguilera, Martin-Cabrejas, & Gonzalez de Mejia, 2021b).

Regarding human epidemiological studies of cocoa and theobromine, chocolate consumption has been associated with lower insulin resistance, although the effect seems to be mainly due to flavonoids (Alkerwi, Sauvageot, Crichton, Elias, & Stranges, 2016). Nevertheless, studies indicate that theobromine increases postprandial glucose and insulin, suggesting insulin sensitivity reduction presumably due to cAMP-phosphodiesterase inhibition (Smolders, Mensink, Boekschoten, de Ridder, & Plat, 2018).

15.3.1.3 Methylxanthines and Cardiovascular Disease

There is some controversy regarding caffeine consumption and protection against cardiovascular disease (Turnbull, Rodricks, Mariano, & Chowdhury, 2017). Moderate caffeine consumption reduces and even appears to protect against cardiovascular disease, although there are no observational studies with pure caffeine exposures, so harmful or protective effects due to other components cannot be ruled out (Ding et al., 2015; Paganini-Hill, 2011; Turnbull et al., 2017). Furthermore, moderate caffeine consumption is not associated with increased risk of arrhythmias, heart failure, changes in blood pressure, heart rate, cardiac output, or heart rate variability in healthy humans. However, studies are inconsistent regarding its role and the risk of ischemic heart disease, electrocardiogram parameters, or myocardial infarction (Turnbull et al., 2017). The population with hypertension or at risk of hypertension may be more sensitive to some of its effects (Turnbull et al., 2017).

On the other hand, clinical trials in humans suggest a cause–effect relationship between cocoa consumption and the correct maintenance of blood flow related to cardiovascular health (Arranz et al., 2013; Hooper et al., 2012; Kwok et al., 2015). The effects of dark chocolate consumption on cardioprotection are also highlighted, as this chocolate is rich in theobromine, so it is reasonable to deduce that theobromine can provide cardioprotective effect (Hooper et al., 2012; Lessens & Rakel, 2011; Tokede, Gaziano, & Djoussé, 2011). Theobromine modifies plasma lipid profile in animal models, reducing LDL and triglyceride levels and increasing HDL levels. However, in healthy humans, theobromine increases HDL, although it reduces all LDL types (Eteng & Ettarh, 2000; Jacobs et al., 2017). In addition, clinical trials have shown that theobromine consumption increases fasting and postprandial arterial diameter and decreases the augmentation index in obese subjects, although it does not improve endothelial function (Smolders et al., 2018). These studies suggest that theobromine can be, along with others, one of the compounds that contribute to the cardioprotective function of cocoa.

15.3.2 Molecular Mechanisms of Phenolic Compounds from Cocoa Shell

Phenolic compounds, along with fiber, are the most intriguing and researched components in the cocoa shell. They are the primary phytochemicals responsible for the health benefits attributed to this cocoa by-product. Flavonoids are a subclass of phenolic compounds that include flavanols, the most abundant phenolics in cocoa and cocoa by-products (Wollgast & Anklam, 2000). The cocoa shell's phenolic compounds are bioaccessible and susceptible to be absorbed after *in vitro* digestion and can prevent oxidative stress in the gut (Cañas et al., 2021; Rojo-Poveda, Barbosa-Pereira, El Khattabi et al., 2020). Similarly, recent studies have demonstrated that the anti-inflammatory properties of the phytochemicals from the cocoa shell protect intestinal cells from inflammation, preventing intestinal damage (Rossin et al., 2021). Several studies have also demonstrated antiproliferative and anticarcinogenic effects in different cell lines (Zainal, Abdah, Taufiq-Yap, Roslida, & Rosmin, 2014;. Lee et al., 2004; Ki, Hwang, Nam, Kyoung, & Hyong, 2005).

Notwithstanding the multiple studies evaluating the extraction of phenolic compounds and their *in vitro* antioxidant properties, few investigations unravel the biological activity of these compounds in cell and animal models or human clinical studies. Nonetheless, owing to the similar composition to cocoa and chocolate, we can estimate those bioactivities and the implied molecular mechanisms. The following paragraphs explore the molecular mechanisms and effects of the phenolic compounds from the cocoa shell on the prevention of obesity, diabetes, and cardiovascular diseases.

15.3.2.1 Phenolic Compounds and Obesity

Both interventional and observational studies have linked cocoa to antiobesity effect (Greenberg & Buijsse, 2013). Cocoa regulates lipid metabolism by suppressing the expression of proteins involved in lipogenesis and fatty acid synthesis while increasing lipolysis-related proteins (Ali, Ismail, Esa, & Pei, 2015; Ali, Ismail, & Kersten, 2014). Cocoa reduces obesity-related inflammation and insulin resistance primarily by downregulating proinflammatory genes (Gu, Yu, & Lambert, 2014). Most of these effects are associated with the phenolic fraction of cocoa (Halib, Ismail, Mohd Yusof, Osakabe, & Daud, 2020).

The administration of a cocoa procyanidin-rich extract to high-fat-fed C57BL/6J mice stimulated AMPK-α and UCP-1 expression in the adipose tissue, resulting in reduced obesity (Tomaru et al., 2007). In addition, cocoa-phenolic-rich powder consumption lowered plasma IL-6, MCP-1, and adiponectin levels. These alterations were associated with reduced obesity-derived inflammation and amelioration of fatty liver disease in high-fat-fed C57BL/6J mice (Gu et al., 2014). Likewise, cocoa phenolic extracts have been linked to decreased adipogenesis and increased β-oxidation and energy expenditure in the adipose tissue of high-fat-diet induced obese Wistar rats. These effects were associated with increased expression of PPARγ, PPARα, PGC1-α, SIRT1, and UCP-1 (Rabadan-Chávez, Quevedo-Corona, Garcia, Reyes-Maldonado, & Jaramillo-Flores, 2016). In cell culture, cocoa phenolic compounds also alleviated the high glucose-driven altered lipid values in liver cells. Reduction in fatty acid synthesis (SREBF1 and FASN downregulation) and increase in its oxidation (PPAR-α upregulation) were responsible for the lipid-lowering impact. AMPK, AKT, and PKCζ were also involved in these effects (Cordero-Herrera, Martin, Fernandez-Millan et al., 2015).

The limited studies carried out with the cocoa shell instead of cocoa per se indicate that cocoa shell phenolic compounds can reduce adipogenesis *in vitro* by reducing de novo lipogenesis (decreasing FASN activity) and stimulating lipolysis (Rebollo-Hernanz, Zhang, Aguilera, Martín-Cabrejas, & de Mejia, 2019a; Rebollo-Hernanz et al., 2021b). Additionally, cocoa shell phenolics serve as adipose tissue browning elicitors, stimulating mitochondrial biogenesis and mitochondrial activity via upregulation of PGC1-α and UCP-1 proteins (Rebollo-Hernanz et al., 2019a). Cocoa shell phenolics also modulate the inflammatory microenvironment produced in the adipose tissue in obesity conditions. These phytochemicals arrest the paracrine interplay macrophages-adipocytes by reducing TNF-α, MCP-1, IL-6 secretion, iNOS, and COX-2 expression and increasing adiponectin production (Rebollo-Hernanz et al., 2019a). In the liver, cocoa shell phenolic compounds reduce the fatty acid synthesis and stimulate β-oxidation by triggering CPT-1 activity. In this regard, cocoa shell's phenolics have shown potential in modulating hepatocytes metabolism via insulin/AKT/ERK pathways and stimulating mitochondrial function (Rebollo-Hernanz, Aguilera, Martin-Cabrejas, & Gonzalez de Mejia, 2021a).

15.3.2.2 Phenolic Compounds and Diabetes

Epidemiological studies have demonstrated that long-term cocoa and chocolate intake can reduce the risk of diabetes (Greenberg, 2015; Matsumoto, Petrone, Sesso, Gaziano, & Djoussé, 2015). Clinical intervention trials have shown promising effects of cocoa, including improving insulin sensitivity in medicated diabetes patients (Martin, Goya, & Ramos, 2016). Experimental animal models validated these results, suggesting that cocoa and its primary phenolic compounds behave as antidiabetic agents by protecting pancreatic beta cells and improving insulin sensitivity in tissues, including the liver, adipose tissue, and skeletal muscle (Ramos, Martín, & Goya, 2017). Regarding the underlying molecular mechanism of action, multiple *in vitro* experiments have demonstrated that cocoa flavanols can regulate insulin secretion in beta-pancreatic cells (Fernández-Millán et al., 2014) and target insulin-sensitive tissues due to their insulin-mimicking behavior or by regulating main proteins in the insulin signaling pathway (Greenberg & Buijsse, 2013).

Cocoa phenolics reduced hyperglycemia, enhanced insulin sensitivity, and increased β-cell mass and function in diabetic Zucker fatty rats (Fernández-Millán et al., 2015). Correspondingly, cocoa phenolic compounds improved insulin resistance and reverted hepatic oxidative stress (Cordero-Herrera, Martin, Escriva et al., 2015; Cordero-Herrera, Martín, Goya, & Ramos, 2015). Administering procyanidin-rich cocoa extract to high-fat C57BL/6J mice increases both muscle and adipose tissues AMPK-α signaling and GLUT4 translocation, leading to reduced hyperglycemia and insulin resistance (Tomaru et al., 2007). Comparably, administration of phenolic-rich cocoa powder increased the HOMA-IR in obese C57BL/6J mice given a high-fat diet (Gu et al., 2014).

Cocoa polyphenols and their microbial-derived metabolites 3,4-dihydroxyphenylacetic acid and 3-hydroxyphenylpropionic acid enhance glucose-stimulated insulin secretion (Fernández-Millán et al., 2014; Martín, Ramos, Cordero-Herrero, Bravo, & Goya, 2013). These effects are presumably mediated by increases in mitochondrial respiration (upregulation of mitochondrial complex III, IV, and V components and increases in cellular ATP production) (Rowley et al., 2017). Cocoa phenolic compounds can also enhance the insulin receptor cascade and activate the PI3K/AKT pathway in HepG2 cells (Cordero-Herrera, Martín, Bravo, Goya, & Ramos, 2013; Cordero-Herrera, Martín, Goya, & Ramos, 2014). Similarly, cocoa phenolics also modulated hepatic gluconeogenic and PEPCK expression through AKT and AMPK and increased GLUT-2 levels, therefore modulating glucose uptake (Cordero-Herrera et al., 2013).

The antidiabetic effects of cocoa shell phenolic components per se or phenolic-rich extract from the by-product have received little attention. However, effects are similar to those observed in cocoa extracts. Cocoa shell phenolic compounds can prevent adipocyte insulin resistance by stimulating GLUT4 translocation and glucose uptake via activation of insulin/AKT/PI3K signaling pathways (Rebollo-Hernanz et al., 2019a). In HepG2 cells, phenolic compounds from the cocoa shell reduced gluconeogenesis by inhibiting PEPCK and inducing glucose metabolism via glucokinase activation (Rebollo-Hernanz et al., 2021b). Functional beverages based on cocoa shell phenolic extracts have revealed α-glucosidase-inhibiting properties, reducing intestinal glucose absorption (Rojo-Poveda et al., 2019).

15.3.2.3 Phenolic Compounds and Cardiovascular Disease

Cocoa flavanols positively affect cardiovascular disease biomarkers such as LDL oxidation, the lipid profile, nitric oxide, endothelial dysfunction, and blood pressure (Khan et al., 2014; Lin et al., 2016). These effects are mediated by vasodilatory, anti-inflammatory, antithrombotic, and antiatherogenic processes (Arranz et al., 2013).

Cocoa procyanidins stimulate endothelial relaxation and activate endothelial NO synthase, the enzyme necessary for NO production (Karim, McCormick, & Tissa Kappagoda, 2000).

The European Food Safety Authority (EFSA) has determined a cause-and-effect link between cocoa flavanol consumption and the preservation of normal endothelium-dependent vasodilation (EFSA, 2014). Likewise, cocoa flavanols may reduce blood pressure by acting as ACE inhibitors (Actis-Goretta, Ottaviani, Keen, & Fraga, 2003).Two clinical trials demonstrate the cardioprotective effects of cocoa flavonols. Sustained (2× daily for one month) of cocoa extract containing 130 mg (−epicatechin) and 560 mg procyanidins increased flow-mediated vasodilation and concentration of epicatechin metabolites in the circulatory system. The treatment attenuated pulse wave velocity and blood pressure and lowered total cholesterol in healthy participants in a randomized double-blind parallel clinical trial (Heiss & Kelm, 2016). The therapeutic effects of high-dose cocoa flavonols (enriched in procyanidin dimers (192 mg) and trimers to decamers (649 mg)) were evaluated in patients (70 years average age; 80% male; 30 kg/m^2 BMI) with chronic heart failure in a single-center randomized double-blind placebo controlled cross-over trial (De Palma et al., 2016). Cocoa flavonols together with guideline-directed medical therapy improved cardiac function indicated by reduction in N-terminal pro-B-type natriuretic peptide (NT-ProBNP) in chronic heart failure, altered blood pressure, and reduced flow-mediated dilation.

Cocoa's antioxidant properties can even aid in the prevention of thrombosis and atherosclerosis. Cocoa shell phenolic compounds can decrease LDL oxidation, avoiding atherosclerosis progression (Vinson et al., 2006). Cocoa shell phenolics also influence inflammatory mediators such as adhesion molecules, cytokines, chemokines, growth factors, and enzymes (metalloproteases, cyclooxygenases, and lipoxygenases) (Goya et al., 2016). The cocoa shell can exert immunomodulatory effects by reducing ROS production and cytokines secretion (Mao, Van de Water, Keen, Schmitz, & Gershwin, 2002; Rebollo-Hernanz et al., 2019a), which are related to the progression of atherosclerosis and have been identified as potential predictors of cardiovascular disease risk. Hence, the EFSA has claimed that cocoa flavanols can protect lipids from oxidative damage while protecting cardiovascular health (EFSA, 2010).

Multiple animal studies feeding rats with high-fat-high-cholesterol diets have concluded that cocoa shell's phenolic compounds can attenuate lipid peroxidation in the serum and liver (Lecumberri, Goya et al., 2007; Ramos et al., 2008). Similarly, *in vitro* studies have demonstrated reduced absorption of fat, cholesterol, and bile salts during *in vitro* digestion associated with soluble dietary fiber and the phenolic content of the cocoa shell. Phenolic compounds can inhibit pancreatic lipase, destabilize cholesterol micelles, and bind bile salts, potentially reducing serum lipid concentration (Braojos et al., 2020; Rebollo-Hernanz et al., 2020).

15.3.3 Molecular Mechanisms of Dietary Fiber from Cocoa Shell

The dietary fiber content of cocoa shells has been extensively studied (Arlorio et al., 2001; Bonvehí & Benería, 1998; Lecumberri, Mateos et al., 2007; Martín-Cabrejas et al., 1994; Martínez et al., 2012; Okiyama et al., 2019; Redgwell et al., 2003; Rojo-Poveda et al., 2019). Cocoa shell fiber can reduce cardiovascular risks by lowering cholesterol and triacylglycerol levels and reduce diabetic symptoms by slowing glucose absorption (Lecumberri, Goya et al., 2007; Nsor-Atindana, Zhong, & Mothibe, 2012). Cocoa shells contain significant amounts of phenolic compounds that are adsorbed or bound to the dietary fiber enabling it to act as antioxidants and help reduce oxidative stress and inflammation in the gut (Rossin et al., 2019). This section summarizes the main effects and mechanisms of the dietary fiber from the cocoa shell in preventing cardiometabolic diseases.

15.3.3.1 Dietary Fiber acts as a Prebiotic Modulating Gut Microbiota

Studies in animals fed cocoa shell have revealed the potential prebiotic effect of its fiber due to changes in the microbiota community with an increase in *Lactobacillus* and *Bifidobacterium* genera and the short-chain fatty acids (SCFAs) profile (Massot-Cladera et al., 2015). The main SCFAs are acetic, propionic, and butyric acid, whose considerable beneficial effects on health range from immune system improvement to obesity, cancer, and others (Gibson et al., 2017; Wenzel, Gates, Ranger, & Klegeris, 2020). However, human studies are needed to confirm its prebiotic potential. Furthermore, human studies have shown an increase in lactobacilli and bifidobacterial colonies and clostridia reduction after cocoa shell's administration (Tzounis et al., 2011).

On the other hand, preliminary studies in pigs demonstrate the cocoa shell's ability to modulate gut microbiota, increasing the *Bacteroides-Prevotella* group population (Magistrelli et al., 2016). These studies reveal that the consumption of the cocoa shell (7.5%) does not modify the *Bifidobacterium* content, although it can reduce the *Firmicutes/Bacteroidetes* ratio (Magistrelli et al., 2016). However, more investigations are needed to evaluate the possible prebiotic effect of dietary fiber from the cocoa shell since that can add value to this by-product.

15.3.3.2 Dietary Fiber and Obesity

Studies highlight the effect of cocoa consumption in diminishing the risk of obesity by modulating the expression of genes related to lipid metabolism (Matsui et al., 2005). The effects of dietary fiber on fat absorption presumably account for the lower risk of obesity. Thus, cocoa shells' high dietary fiber content may decrease the risk of developing obesity (Lecumberri, Mateos et al., 2007). Supplementation with cocoa shell soluble fiber provides beneficial effects, decreasing food intake in high-fat diet fed rats due to the satiety potential of the soluble fiber (Ramos et al., 2008). Weight gain and food consumption in Zucker obese rats fed the cocoa shell soluble fiber fraction were lower than those fed the control diet (Sánchez et al., 2010).

The mechanism of action of dietary fiber on satiety and decreased energy intake is apparently due to gastric distension caused by increased secretion of saliva and gastric juices from increased chewing (Howarth, Saltzman, & Roberts, 2001). Recently, dietary fiber consumption has been linked to decreased appetite-stimulating hormones such as ghrelin and an increase in anorexigenic hormones such as glucagon-like peptides 1 and 2, cholecystokinin, and peptide YY (Baldassano et al., 2018; Reimer et al., 2012). However, comparative studies conducted with cocoa and cocoa fiber demonstrated an increase in ghrelin concentration after cocoa consumption (although not an increase in food intake) and significant decrease in leptin content after cocoa fiber consumption by rats (Massot-Cladera et al., 2017). Likewise, studies have shown a positive correlation between *Bifidobacterium* and *Lactobacillus* and leptin levels and an increase in butyrate and propionate leading to reduced ghrelin levels (Queipo-Ortuño et al., 2013). Therefore, butyrate and propionate can protect against obesity by preventing food intake due to stimulatory effects on anorexigenic gut hormones (Lin et al., 2012). This finding opens a research door on the potential effects that fermentation of cocoa shells dietary fiber can have on modifying the microbiota community, production of short-chain fatty acids, and consequently, on appetite-regulating hormones to reduce the risk of obesity.

15.3.3.3 Dietary Fiber and Diabetes

The principal mechanism of action of the cocoa shell's dietary fiber diabetes prevention potential is based on the difficulty of dietary fiber in diffusing nutrients from the lumen to the gut mucosa. Consequently, it requires longer small intestine and reduced glucose absorption (Benítez et al., 2021; Jenkins, Taylor, & Wolever, 1982). This lower glucose absorption decreases the response of certain glucose-related hormones such as insulin and the gastric inhibitory polypeptide. This effect appears to be greater after consuming viscous fibers (Jenkins & Jenkins, 1985; Kay, Grobin, & Track, 1981). Cocoa shell exhibits hypoglycemic properties in *in vitro* studies (Rebollo-Hernanz et al., 2020) and dietary fiber-rich fractions/ingredients (Lecumberri, Mateos et al., 2007; Nsor-Atindana, Zhong, & Mothibe, 2012).

Furthermore, when taken as part of a diet, the cocoa shell dietary fiber fractions (primarily the soluble fraction) can inhibit α-amylase (an enzyme involved in glucose hydrolysis), absorb the produced glucose, and retard its diffusion through the intestinal wall contributing to antidiabetic effects (Nsor-Atindana, Zhong, & Mothibe, 2012). The intake of soluble dietary fiber from the cocoa shell lowered plasma glucose and insulin levels in Zuker fatty rats, resulting in reduced insulin resistance (Sánchez et al., 2010). Moreover, studies on diabetic rats with cocoa extract (rich in soluble dietary fiber and phenolics) also outlined a hypoglycemic effect. These results can be due to the dietary fiber from this by-product and the antioxidant compounds' action (Ruzaidi, Amin, Nawalyah, Hamid, & Faizul, 2005).

The cocoa shell dietary fiber fermentation in the gut can control insulin and glucagon secretion by increasing glucagon-like peptide-1 (GLP-1) levels (Mandaliya & Seshadri, 2019; Pino, Mujica, & Arredondo, 2021). Hence, understanding cocoa shell's dietary fiber and its metabolization by the microbiota is critical to explain its antidiabetic properties and mechanisms of action.

15.3.3.4 Dietary Fiber and Cardiovascular Disease

The cocoa shell's dietary fiber contributes to cardiovascular health as it passes through the digestive system. Cocoa shells have been linked to improved cardiovascular health by reducing atherogenic fat absorption (Rojo-Poveda, Barbosa-Pereira, Zeppa, & Stévigny, 2020). *In vitro* studies have demonstrated the effects on the absorption of fat, cholesterol, and bile salts. Fat absorption reduction is associated with insoluble dietary fiber, whereas soluble fiber is related to cholesterol and bile salts-binding properties (Nsor-Atindana, Zhong, & Mothibe, 2012). Moreover, the cocoa shell dietary fiber can reduce pancreatic lipase activity by binding either lipids or the enzyme or destabilizing the micelles where pancreatic lipase hydrolyzes triglycerides before intestinal absorption (Braojos et al., 2020; Rebollo-Hernanz et al., 2020).

In vivo experiments have demonstrated improvements in the lipid profile of hypercholesterolemic rats that consumed cocoa shell's dietary fiber due to reduced total cholesterol,

LDL, triglycerides, and the atherogenic index. In contrast, HDL and free fatty acids remained unchanged (Lecumberri, Goya et al., 2007). The administration of cocoa shell flour also increased fecal fat excretion of rats fed a cholesterol-rich diet. When the rats were fed the soluble dietary fiber from the cocoa shell, the serum lipid profile improved due to increase in HDL values in conjunction with decrease in LDL, triglycerides levels, and lipid peroxidation (Ramos et al., 2008). The cocoa shell also improved lipid markers (LDL, HDL) in diabetic rats (Ruzaidi et al., 2005). Zucker fatty rats fed the cocoa shell (5%) diet gained less weight and consumed less food than those on the regular diet. This group also had lower total cholesterol, triglyceride, and blood pressure levels than the control group (Sánchez et al., 2010).

The primary mechanism of action of dietary fiber in lowering cholesterol is due to the action of fiber on bile acids. Fiber modifies the balance of bile acids by binding them; consequently, dietary fiber and bile salts are excreted together in feces. On the other hand, the adsorption of bile salts to dietary fiber results in the inability of cholesterol to solubilize in bile salts and the excretion of cholesterol and other lipids (Eastwood & Boyd, 1967; Leveille & Sauberlich, 1966; Story, 1985; Vahouny, Tombes, Cassidy, Kritchevsky, & Gallo, 1980). Later, since the synthesis of the primary bile acids requires cholesterol, the liver synthesizes more bile acids from cholesterol due to the imbalance in bile salts, thus decreasing cholesterol in the body (Björkhem, 1985). In addition, colonic bacteria convert fermentable fibers into SCFAs. Propionic acid can be absorbed and inhibits the liver's HMG-CoA reductase, which is responsible for limiting cholesterol production (Chen, Ma, Liang, Peng, & Zuo, 2011). Finally, dietary fiber intake can lower arterial blood pressure by reducing glucose absorption and decreasing insulin secretion (Aleixandre & Miguel, 2016).

15.4 Future Perspectives on the Preventive Effects of Cocoa Shell on Cardiometabolic Diseases

Research has demonstrated the potential of cocoa and the cocoa shell on the prevention of cardiometabolic diseases. Nonetheless, understanding the molecular mechanisms of action that govern each of the functions triggered by cocoa shell's bioactive compounds will require well-designed experiments and analysis. To date, the current investigations just focus on the chemical composition of the cocoa shell. Recent chemometric analysis has been able to classify the cocoa shell according to composition and origin (Barbosa-Pereira, Belviso, Ferrocino, Rojo-Poveda, & Zeppa, 2021; Rojo-Poveda, Zeppa, Ferrocino, Stévigny, & Barbosa-Pereira, 2021). Likewise, minor active compounds should also be investigated, as in the case of vitamin D. Cocoa shell may be a bioavailable source of vitamin D (Kon & Henry, 1935), which is synthetized during sun drying from the ergosterol derived from the yeast that ferments the cocoa bean (Knapp & Coward, 1935). Recent studies indicate that vitamin D can reduce cognitive decline in humans and therefore cocoa shell may be helpful in this regard (Du Preez et al., 2021). Moreover, biogenic amines are also found in the cocoa shell after fermentation (Delgado-Ospina et al., 2020) and can exhibit beneficial or detrimental effects to health, which should be still investigated (Gomez-Gomez, Borges, Minatel, Luvizon, & Lima, 2018). The health-promoting effects of the cocoa shell methylxanthines and phenolic compounds are affected by the amount ingested and their bioavailability (absorption, distribution, metabolism, and elimination) and the chemical structure and the biological activities of the metabolites produced (Williamson & Manach, 2005). Microbiota will play a critical role in phenolic compounds and dietary fiber functions (Aravind, Wichienchot, Tsao, Ramakrishnan, & Chakkaravarthi, 2021; Corrêa, Rogero, Hassimotto, & Lajolo, 2019; Makki, Deehan, Walter, & Bäckhed, 2018; Ozdal et al., 2016). The food matrix can also influence cocoa methylxanthines and phenolic compounds absorption. Other dietary components may influence enzymes and transporters involved in phytochemical absorption and metabolism. For example, polysaccharides of cacao pod husk exert various physiological functions and are referred as biological response modifiers. Pectin extracted from cacao pod husk containing 61% carbohydrates, 10.6% protein, 21% ash, and 36.8 mg gallic acid equivalents/g of total polyphenols was supplemented to diets (0, 1, 3, or 6 g/kg) calculated every week within 56 days of feeding white cultured shrimp (*Litopenaeus vannamei*) (Kuo, Chang, & Cheng, 2022). Pectin supplementation (3 g/kg) promoted shrimps growth (increased weight and length gain rates, and feed efficiency after 56 days. Moreover, the survival rate also increased for stress challenged shrimp (with *Vibrio alginolyticus* pathogen or hypothermal stress [14 °C]). Pectin (3 g/kg) supplemented diet regulated neuroendocrine and enhanced humoral and cellular immune responses by significantly increasing total and differential hemocyte counts, phenololoxidase and phagocytic activities, respiratory bursts, and clearance efficiency to *V. alginolyticus*, attenuating stress-induced dopamine and norepinephrine secretions, and promoting aerobic carbohydrates metabolism providing energy to enhance antihypothermal stress responses and disease resistance in shrimp (Kuo et al., 2022).

Furthermore, direct interactions between phytochemicals and specific dietary components may influence absorption. Generally, cocoa shell's dietary fiber interacts with phenolic compounds modifying their properties. Accordingly, the absorption of these phytochemicals may be affected by indirect effects of food on different gut physiology variables (pH, intestinal fermentation, bile salts excretion, transit duration, among others) (Domínguez-Avila et al., 2017; Quirós-Sauceda et al., 2014). Therefore, future research should focus on investigating the effects of gastrointestinal digestion on cocoa shell biological activity and then exploring the impact of microbiota. Then, cell and animal models could be used to validate the bioactivity and safety of the cocoa shell-based ingredients and uncover their molecular mechanisms of action. Multiomic experiments will help to comprehensively associate the targeted metabolites derived from the cocoa shell with the biological effects related to the prevention of cardiometabolic diseases. Finally, double-blind, cross-over, placebo-control clinical human studies will provide the final results needed to claim the beneficial effects of the cocoa shell and derived ingredients.

15.5 Concluding Remarks

Evidence has demonstrated the nutraceutical potential of the cocoa shell to obtain novel bioactive ingredients for preventing cardiometabolic diseases. Research on cocoa or isolated pure bioactive compounds has proven the efficacy of the components present in the cocoa shell and enabled investigation of the associated molecular mechanisms. This information provides a basis for understanding the cocoa shell's bioactivity and mechanisms of action as a functional ingredient. More *in vitro*, *in vivo*, and human studies are required to comprehend these ingredient's biological activity and molecular mechanisms in depth. Methylxanthines, phenolic compounds, and dietary fiber may act synergistically in the organism to reduce the risk of obesity, diabetes, and cardiovascular disease. There is still much to learn about the metabolism of the cocoa shell bioactive compound in the gastrointestinal tract, their metabolism, and their biological activity in different organs. Furthermore, clinical investigations on the effectiveness of the cocoa shell bioactive compounds and their metabolites will be required to validate this cocoa by-product as a novel food.

References

Acheson, K. J., Gremaud, G., Meirim, I., Montigon, F., Krebs, Y., Fay, L. B., . . . Tappy, L. (2004). Metabolic effects of caffeine in humans: Lipid oxidation or futile cycling? *American Journal of Clinical Nutrition*, *79*(1), 40–46. https://doi.org/10.1093/ajcn/79.1.40

Actis-Goretta, L., Ottaviani, J. I., Keen, C. L., & Fraga, C. G. (2003). Inhibition of angiotensin converting enzyme (ACE) activity by flavan-3-ols and procyanidins. *FEBS Letters*, *555*(3), 597–600. https://doi.org/10.1016/S0014-5793(03)01355-3

Adamafio, N. A. (2013). Theobromine toxicity and remediation of cocoa by-products: An overview. *Journal of Biological Sciences*, *13*(7), 570–576. https://doi.org/10.3923/jbs.2013.570.576

Adeyemo, G., Ajayi, A., Longe, O., & Olubamiwa, O. (2015). Gut morphology and釘internl organs of broiler birds fed graded levels of bio-detheobrominized cocoa bean shell (CBS) based diets. *American Journal of Experimental Agriculture*, *5*(2), 172–177. https://doi.org/10.9734/ajea/2015/10582

Agardh, E. E., Carlsson, S., Ahlbom, A., Efendic, S., Grill, V., Hammar, N., . . . Östenson, C. G. (2004). Coffee consumption, type 2 diabetes and impaired glucose tolerance in Swedish men and women. *Journal of Internal Medicine*, *255*(6), 645–652. https://doi.org/10.1111/j.1365-2796.2004.01331.x

Agus, B. A. P., Mohamad, N. N., & Hussain, N. (2018). Composition of unfermented, unroasted, roasted cocoa beans and cocoa shells from Peninsular Malaysia. *Journal of Food Measurement and Characterization*, *12*(4), 2581–2589. https://doi.org/10.1007/s11694-018-9875-4

Aleixandre, A., & Miguel, M. (2016). Dietary fiber and blood pressure control. *Food and Function*, *7*(4), 1864–1871. https://doi.org/10.1039/c5fo00950b

Alexander, J., Benford, D., Cockburn, A., Cravedi, J.-P., Dogliotti, E., Domenico, A. Di, Luisa Férnandez-Cruz, M., Fürst, P., Fink-Gremmels, J., Corrado, L., Galli, P., Grandjean, J., Gzyl, G., Heinemeyer, N., Johansson, A., Mutti, J., Schlatter, R., Van Leeuwen, C. Van, & Peteghem, P. V. (2008). Theobromine as undesirable substances in animal feed - Scientific opinion of the Panel on Contaminants in the Food Chain. *EFSA Journal*, *6*(9), 725. https://doi.org/10.2903/j.efsa.2008.725.

Ali, F., Ismail, A., Esa, N. M., & Pei, C. P. (2015). Transcriptomics expression analysis to unveil the molecular mechanisms underlying the cocoa polyphenol treatment in diet-induced obesity rats. *Genomics*, *105*(1), 23–30. https://doi.org/10.1016/j.ygeno.2014.11.002

Ali, F., Ismail, A., & Kersten, S. (2014). Molecular mechanisms underlying the potential antiobesity-related diseases effect of cocoa polyphenols. *Molecular Nutrition and Food Research*, *58*(1), 33–48. https://doi.org/10.1002/mnfr.201300277

Alkerwi, A., Sauvageot, N., Crichton, G. E., Elias, M. F., & Stranges, S. (2016). Daily chocolate consumption is inversely associated with insulin resistance and liver enzymes in the Observation of Cardiovascular Risk Factors in Luxembourg study. *British Journal of Nutrition*, *115*(9), 1661–1668. https://doi.org/10.1017/S0007114516000702

Amézqueta, S., González-Peñas, E., Lizarraga, T., Murillo-Arbizu, M., & López De Cerain, A. (2008). A simple chemical method reduces ochratoxin A in contaminated cocoa shells. *Journal of Food Protection*, *71*(7), 1422–1426. https://doi.org/10.4315/0362-028X-71.7.1422

Anastas, P., & Eghbali, N. (2010). Green Chemistry: Principles and Practice. *Chemical Society Reviews*, *39*(1), 301–312. https://doi.org/10.1039/b918763b

Aravind, S. M., Wichienchot, S., Tsao, R., Ramakrishnan, S., & Chakkaravarthi, S. (2021). Role of dietary polyphenols on gut microbiota, their metabolites and health benefits. *Food Research International*, *142*, 110189. https://doi.org/10.1016/J.FOODRES.2021.110189

Arlorio, M., Coisson, J. D., Restani, P., & Martelli, A. (2001). Characterization of pectins and some secondary compounds from *T. cacao* hulls. *Journal of Food Science*, *66*(5), 653–656. https://doi.org/10.1111/j.1365-2621.2001.tb04616.x

Arlorio, M., Coïsson, J. D., Travaglia, F., Varsaldi, F., Miglio, G., Lombardi, G., & Martelli, A. (2005). Antioxidant and biological activity of phenolic pigments from *T. cacao* hulls extracted with supercritical CO_2. *Food Research International*, *38*(8–9), 1009–1014. https://doi.org/10.1016/J.FOODRES.2005.03.012

Ärnlöv, J., Vessby, B., & Risérus, U. (2004). Coffee consumption and insulin sensitivity. *Journal of the American Medical Association*, *291*(10), 1199–1201. https://doi.org/10.1001/jama.291.10.1199-b

Aromolaran, O., & Ogunsakin, F. M. (2019). Degradation of Theobromine in Cocoa (*T. cacao*) by-products by Fermentation with Aspergillus niger. *South Asian Journal of Research in Microbiology*, 1–6. https://doi.org/10.9734/sajrm/2018/v2i329265

Arranz, S., Valderas-Martinez, P., Chiva-Blanch, G., Casas, R., Urpi-Sarda, M., Lamuela-Raventos, R. M., & Estruch, R. (2013). Cardioprotective effects of cocoa: Clinical evidence from randomized clinical intervention trials in humans. *Molecular Nutrition and Food Research*, *57*(6), 936–947. https://doi.org/10.1002/mnfr.201200595

Astrup, A., Toubro, S., Cannon, S., Hein, P., & Madsen, J. (1991). Thermogenic synergism between ephedrine and caffeine in healthy volunteers: A double-blind, placebo-controlled study. *Metabolism*, *40*(3), 323–329. https://doi.org/10.1016/0026-0495(91)90117-F

Azizah, A. H., Nik Ruslawati, N. M., & Swee Tee, T. (1999). Extraction and characterization of antioxidant from cocoa by-products. *Food Chemistry*, *64*(2), 199–202. https://doi.org/10.1016/S0308-8146(98)00121-6

Baldassano, S., Accardi, G., Aiello, A., Buscemi, S., Di Miceli, G., Galimberti, D., . . . Vasto, S. (2018). Fibers as functional foods and the effects on gut hormones: The example of β-glucans in a single arm pilot study. *Journal of Functional Foods*, *47*, 264–269. https://doi.org/10.1016/j.jff.2018.05.059

Barbosa-Pereira, L., Belviso, S., Ferrocino, I., Rojo-Poveda, O., & Zeppa, G. (2021). Characterization and classification of cocoa bean shells from different regions of venezuela using hplc-pda-ms/ms and spectrophotometric techniques coupled to chemometric analysis. *Foods, 10*(8), 1791. https://doi.org/10.3390/foods10081791

Barbosa-Pereira, L., Guglielmetti, A., & Zeppa, G. (2018). Pulsed electric field assisted extraction of bioactive compounds from cocoa bean shell and coffee silverskin. *Food and Bioprocess Technology, 11*(4), 818–835. https://doi.org/10.1007/s11947-017-2045-6

Barbosa-Pereira, L., Rojo-Poveda, O., Ferrocino, I., Giordano, M., & Zeppa, G. (2019). Analytical dataset on volatile compounds of cocoa bean shells from different cultivars and geographical origins. *Data in Brief, 25*. https://doi.org/10.1016/j.dib.2019.104268

Barišić, V., Flanjak, I., Tot, A., Budeč, M., Benšić, M., Jozinović, A., . . . Ačkar, Đ. (2020). 5-Hydroxymethylfurfural and acrylamide content of cocoa shell treated with high voltage electrical discharge. *Food Control, 110*, 107043. https://doi.org/10.1016/j.foodcont.2019.107043

Barišić, V., Jozinović, A., Flanjak, I., Šubarić, D., Babić, J., Milicević, B., . . . Ackar, D. (2020). Difficulties with use of cocoa bean shell in food production and high voltage electrical discharge as a possible solution. *Sustainability (Switzerland), 12*(10), 3981. https://doi.org/10.3390/SU12103981

Beckett, S. T. (2009). Industrial Chocolate. In S. T. Beckett (Ed.), *Industrial Chocolate Manufacture and and Use*. Chichester, West Sussex, UK: John Wiley & Sons Ltd.

Beckett, S. T. (2008). The Science of Chocolate. In *The Science of Chocolate* (2nd ed., pp. 240). Cambridge, UK: The Royal Society of Chemistry. https://doi.org/10.1039/9781847558053

Beg, M. S., Ahmad, S., Jan, K., & Bashir, K. (2017). Status, supply chain and processing of cocoa—A review. *Trends in Food Science and Technology, 66*, 108–116. https://doi.org/10.1016/j.tifs.2017.06.007

Benítez, V., Rebollo-Hernanz, M., Aguilera, Y., Bejerano, S., Cañas, S., & Martín-Cabrejas, M. A. (2021). Extruded coffee parchment shows enhanced antioxidant, hypoglycaemic, and hypolipidemic properties by releasing phenolic compounds from the fiber matrix. *Food and Function, 12*(3), 1097. https://doi.org/10.1039/d0fo02295k

Bhat, S. G., & Sukumaran, R. K. (2012). Enzyme technologies for bioconversion of food processing by-products. In *Valorization of Food Processing By-Products* (pp. 233–266). https://doi.org/10.1201/b12816

Björkhem, I. (1985). Mechanism of bile acid biosynthesis in mammalian liver. *New Comprehensive Biochemistry, 12*(C), 231–278. https://doi.org/10.1016/S0167-7306(08)60685-7

Bonvehí, J. S., & Benería, M. A. (1998). Composition of dietary fiber in cocoa husk. *European Food Research and Technology, 207*(2), 105–109. https://doi.org/10.1007/s002170050303

Bonvehí, J. S., & Coll, F. V. (1999). Protein quality assessment in cocoa husk. *Food Research International, 32*(3), 201–208. https://doi.org/10.1016/S0963-9969(99)00088-5

Bonvehí, J. S., & Jordà, R. E. (1998). Constituents of cocoa husks. *Zeitschrift Fur Naturforschung - Section C Journal of Biosciences, 53*(9–10), 785–792. https://doi.org/10.1515/znc-1998-9-1002

Bradbury, A., & Kopp, G. (2006). Polyphenol-enriched composition from cocoa shell extraction. In *EP Patent 1 728 434 A1*. https://patents.google.com/patent/US20060269633/en.

Braojos, C., Benitez, V., Rebollo-Hernanz, M., Cañas, S., Aguilera, Y., Arribas, S. M., & Martin-Cabrejas, M. A. (2020). Evaluation of the hypolipidemic properties of cocoa shell after simulated digestion using *in vitro* techniques and a cell culture model of non-alcoholic fatty liver disease. *Proceedings, 70*(1), 58. https://doi.org/10.3390/foods_2020-07669

Bruna, C., Eichholz, I., Rohn, S., Kroh, L. W., & Huyskens-Keil, S. (2009). Bioactive compounds in cocoa hulls (*T. cacao* L.) from different origins. *Journal of Applied Botany and Food Quality*, *83*, 9–13.

Camps-Bossacoma, M., Garcia-Aloy, M., Saldana-Ruiz, S., Cambras, T., González-Domínguez, R., Franch, À., . . . Castell, M. (2019). Role of theobromine in cocoa's metabolic properties in healthy rats. *Journal of Agricultural and Food Chemistry*, *67*(13), 3605–3614. https://doi.org/10.1021/acs.jafc.8b07248

Cañas, S., Rebollo-Hernanz, M., Aguilera, Y., Braojos, C., Benitez, V., Arribas, S., & Martin-Cabrejas, M. (2021). Bioaccessibility of phenolic compounds from cocoa shell subjected to *in vitro* digestion and its antioxidant activity in intestinal and hepatic cells. *Proceedings Paper. Medical Sciences Forum*. https://doi.org/10.3390/cahd2020-08612

Carrageta, D. F., Dias, T. R., Alves, M. G., Oliveira, P. F., Monteiro, M. P., & Silva, B. M. (2018). Anti-obesity potential of natural methylxanthines. *Journal of Functional Foods*, *43*, 84–94. https://doi.org/10.1016/J.JFF.2018.02.001

Chemat, F., Vian, M. A., & Cravotto, G. (2012). Green extraction of natural products: Concept and principles. *International Journal of Molecular Sciences*, *13*(7), 8615–8627. https://doi.org/10.3390/ijms13078615

Chen, Z. Y., Ma, K. Y., Liang, Y., Peng, C., & Zuo, Y. (2011). Role and classification of cholesterol-lowering functional foods. *Journal of Functional Foods*, *3*(2), 61–69. https://doi.org/10.1016/J.JFF.2011.02.003

Cherniack, E. P., Buslach, N., & Lee, H. F. (2018). The potential effects of caffeinated beverages on insulin sensitivity. *Journal of the American College of Nutrition*, *37*(2), 161–167. https://doi.org/10.1080/07315724.2017.1372822

Chooi, Y. C., Ding, C., & Magkos, F. (2019). The epidemiology of obesity. *Metabolism: Clinical and Experimental*, *92*, 6–10. https://doi.org/10.1016/j.metabol.2018.09.005

Chronopoulos, D., Zuurbier, R., Brandstetter, B., & Jung, C. (2011). Food comprising alkalized cocoa shells and method therefor. *Google Patents*, *1*(19), 14. https://patents.google.com/patent/US20110151098A1/en

Collar, C., Rosell, C. M., Muguerza, B., & Moulay, L. (2009). Breadmaking performance and keeping behavior of cocoa-soluble fiber-enriched wheat breads. *Food Science and Technology International*, *15*(1), 79–87. https://doi.org/10.1177/1082013208102643

Copetti, M. V., Iamanaka, B. T., Nester, M. A., Efraim, P., & Taniwaki, M. H. (2013). Occurrence of ochratoxin A in cocoa by-products and determination of its reduction during chocolate manufacture. *Food Chemistry*, *136*(1), 100–104. https://doi.org/10.1016/j.foodchem.2012.07.093

Copetti, M. V., Iamanaka, B. T., Pereira, J. L., Lemes, D. P., Nakano, F., & Taniwaki, M. H. (2012). Determination of aflatoxins in by-products of industrial processing of cocoa beans. *Food Additives and Contaminants - Part A Chemistry, Analysis, Control, Exposure and Risk Assessment*, *29*(6), 972–978. https://doi.org/10.1080/19440049.2012.660657

Copetti, M. V., Pereira, J. L., Iamanaka, B. T., Pitt, J. I., & Taniwaki, M. H. (2010). Ochratoxigenic fungi and ochratoxin A in cocoa during farm processing. *International Journal of Food Microbiology*, *143*(1–2), 67–70. https://doi.org/10.1016/j.ijfoodmicro.2010.07.031

Cordero-Herrera, I., Martín, M. A., Bravo, L., Goya, L., & Ramos, S. (2013). Cocoa flavonoids improve insulin signaling and modulate glucose production via AKT and AMPK in HepG2 cells. *Molecular Nutrition and Food Research*, *57*(6), 974–985. https://doi.org/10.1002/MNFR.201200500

Cordero-Herrera, I., Martín, M. Á., Escrivá, F., Álvarez, C., Goya, L., & Ramos, S. (2015). Cocoa-rich diet ameliorates hepatic insulin resistance by modulating insulin signaling and glucose homeostasis in Zucker diabetic fatty rats. *The Journal of Nutritional Biochemistry*, *26*(7), 704–712. https://doi.org/10.1016/J.JNUTBIO.2015.01.009

Cordero-Herrera, I., Martín, M. Á., Fernández-Millán, E., Álvarez, C., Goya, L., & Ramos, S. (2015). Cocoa and cocoa flavanol epicatechin improve hepatic lipid metabolism in *in vivo* and *in vitro* models. Role of PKCζ. *Journal of Functional Foods*, *17*, 761–773. https://doi.org/10.1016/J.JFF.2015.06.033

Cordero-Herrera, I., Martín, M. Á., Goya, L., & Ramos, S. (2014). Cocoa flavonoids attenuate high glucose-induced insulin signaling blockade and modulate glucose uptake and production in human HepG2 cells. *Food and Chemical Toxicology*, *64*, 10–19.

Cordero-Herrera, I., Martín, M. Á., Goya, L., & Ramos, S. (2015). Cocoa intake ameliorates hepatic oxidative stress in young Zucker diabetic fatty rats. *Food Research International*, *69*, 194–201. https://doi.org/10.1016/J.FOODRES.2014.12.039

Corrêa, T. A. F., Rogero, M. M., Hassimotto, N. M. A., & Lajolo, F. M. (2019). The two-way polyphenols-microbiota interactions and their effects on obesity and related metabolic diseases. *Frontiers in Nutrition*, *0*, 188. https://doi.org/10.3389/FNUT.2019.00188

Davoodi, S. H., Hajimiresmaiel, S. J., Ajami, M., Mohseni-Bandpei, A., Ayatollahi, S. A., Dowlatshahi, K., . . . Pazoki-Toroudi, H. (2014). Caffeine treatment prevented from weight regain after calorie shifting diet induced weight loss. *Iranian Journal of Pharmaceutical Research*, *13*(2), 707–718. https://doi.org/10.22037/ijpr.2014.1516

de la Cádiz-Gurrea, M. L., Fernández-Ochoa, Á., Leyva-Jiménez, F. J., Guerrero-Muñoz, N., del Villegas-Aguilar, M. C., Pimentel-Moral, S., . . . Segura-Carretero, A. (2020). LC–MS and spectrophotometric approaches for evaluation of bioactive compounds from Peru cocoa by-products for commercial applications. *Molecules*, *25*(14), 3177. https://doi.org/10.3390/molecules25143177

De Palma, R., Sotto, I., Wood, E. G., Khan, N. Q., Butler, J., Johnston, A., . . . Corder, R. (2016). Cocoa flavanols reduce N-terminal pro-B-type natriuretic peptide in patients with chronic heart failure. *ESC Heart Failure*, *3*, 97–106.

Delgado-Ospina, J., Di Mattia, C. D., Paparella, A., Mastrocola, D., Martuscelli, M., & Chaves-Lopez, C. (2020). Effect of fermentation, drying and roasting on biogenic amines and other biocompounds in Colombian criollo cocoa beans and shells. *Foods*, *9*(4). https://doi.org/10.3390/foods9040520

Ding, M., Satija, A., Bhupathiraju, S. N., Hu, Y., Sun, Q., Han, J., . . . Hu, F. B. (2015). Association of coffee consumption with total and cause-specific mortality in 3 large prospective cohorts. *Circulation*, *132*(24), 2305–2315. https://doi.org/10.1161/CIRCULATIONAHA.115.017341

do Nascimento, M. D. S., da Silva, N., da Silva, I. F., da Silva de, J. C., Marques, É. R., & Barbosa Santos, A. R. (2010). Enteropathogens in cocoa pre-processing. *Food Control*, *21*(4), 408–411. https://doi.org/10.1016/j.foodcont.2009.06.015

Domínguez-Avila, J. A., Wall-Medrano, A., Velderrain-Rodríguez, G. R., Chen, C. Y. O., Salazar-López, N. J., Robles-Sánchez, M., & González-Aguilar, G. A. (2017). Gastrointestinal interactions, absorption, splanchnic metabolism and pharmacokinetics of orally ingested phenolic compounds. *Food and Function*, *8*(1), 15–38. https://doi.org/10.1039/c6fo01475e

Du Preez, A., Lefèvre-Arbogast, S., Houghton, V., Lucia, C. de, Low, D. Y., Helmer, C., Féart, C., Delcourt, C., Proust-Lima, C., Pallàs, M., Ruigrok, S. R., Altendorfer, B., González-Domínguez, R., Sánchez-Pla, A., Urpi-Sardà, M., Andres-Lacueva, C., Aigner, L., Lucassen, P. J., Korosi, A., . . . Thuret, S. (2021). The serum metabolome mediates the concert of diet, exercise, and neurogenesis, determining the risk for cognitive decline and dementia. *Alzheimer's and Dementia*. https://doi.org/10.1002/ALZ.12428.

Dulloo, A. G., Geissler, C. A., Collins, A., & Miller, D. S. (1989). Normal caffeine consumption: Influence on thermogenesis and daily energy expenditure in lean and postobese human volunteers. *American Journal of Clinical Nutrition*, *49*(1), 44–50. https://doi.org/10.1093/ajcn/49.1.44

Eastwood, M. A., & Boyd, G. S. (1967). The distribution of bile salts along the small intestine of rats. *Biochimica et Biophysica Acta (BBA)/Lipids and Lipid Metabolism*, *137*(2), 393–396. https://doi.org/10.1016/0005-2760(67)90116-6

EFSA. (2010). Scientific opinion on the substantiation of health claims related to cocoa flavanols and protection of lipids from oxidative damage (ID 652, 1372, 1506, 3143), and maintenance of normal blood pressure (ID 1507) pursuant to Article 13(1) of Regulation (EC). *EFSA Journal*, *8*(10). https://doi.org/10.2903/j.efsa.2010.1792

EFSA. (2014). Scientific Opinion on the modification of the authorisation of a health claim related to cocoa flavanols and maintenance of normal endothelium-dependent vasodilation pursuant to Article 13(5) of Regulation (EC) No. 1924/2006 following a request in accordan. *EFSA Journal*, *12*(5). https://doi.org/10.2903/j.efsa.2014.3654

Emiola, I. A., Ojebiyi, O. O., & Akande, T. O. (2011). Performance and organ weights of laying hens fed diets containing graded levels of sun-dried cocoa bean shell (CBS). *International Journal of Poultry Science*, *10*, 986–989. https://doi.org/10.3923/ijps.2011.987.990

Eteng, M. U., & Ettarh, R. R. (2000). Comparative effects of theobromine and cocoa extract on lipid profile in rats. *Nutrition Research*, *20*(10), 1513–1517. https://doi.org/10.1016/s0271-5317(00)80032-9

FAO. (2007). Report on Functional Foods. *Food Quality and Standards Service*. https://scholar.google.com/scholar_lookup?title=Report on functional foods&publication_year=2007&author=Food and Agriculture Organization (FAO).

Fernández-Millán, E., Cordero-Herrera, I., Ramos, S., Escrivá, F., Alvarez, C., Goya, L., & Martín, M. A. (2015). Cocoa-rich diet attenuates beta cell mass loss and function in young Zucker diabetic fatty rats by preventing oxidative stress and beta cell apoptosis. *Molecular Nutrition and Food Research*, *59*(4), 820–824. https://doi.org/10.1002/MNFR.201400746

Fernández-Millán, E., Ramos, S., Alvarez, C., Bravo, L., Goya, L., & Martín, M. Á. (2014). Microbial phenolic metabolites improve glucose-stimulated insulin secretion and protect pancreatic beta cells against tert-butyl hydroperoxide-induced toxicity via ERKs and PKC pathways. *Food and Chemical Toxicology*, *66*, 245–253.

Franco, R., Oñatibia-Astibia, A., & Martínez-Pinilla, E. (2013). Health benefits of methylxanthines in cacao and chocolate. *Nutrients*, *5*(10), 4159–4173. https://doi.org/10.3390/nu5104159

Gibson, G. R., Hutkins, R., Sanders, M. E., Prescott, S. L., Reimer, R. A., Salminen, S. J., . . . Reid, G. (2017). Expert consensus document: The International Scientific Association for Probiotics and Prebiotics (ISAPP) consensus statement on the definition and scope of prebiotics. *Nature Reviews Gastroenterology and Hepatology*, *14*(8), 491–502. https://doi.org/10.1038/nrgastro.2017.75

Gomez-Gomez, H. A., Borges, C. V., Minatel, I. O., Luvizon, A. C., & Lima, G. P. P. (2018). Health benefits of dietary phenolic compounds and biogenic amines. In *Bioactive Molecules in Food* (pp. 1–25). Cham: Springer. https://doi.org/10.1007/978-3-319-54528-8_27-2

Goya, L., Martín, M. Á., Sarriá, B., Ramos, S., Mateos, R., & Bravo, L. (2016). Effect of cocoa and its flavonoids on biomarkers of inflammation: studies of cell culture, animals and humans. *Nutrients, 8*(4), 212. https://doi.org/10.3390/nu8040212

Granato, D., Barba, F. J., Bursać Kovačević, D., Lorenzo, J. M., Cruz, A. G., & Putnik, P. (2020). Functional foods: product development, technological trends, efficacy testing, and safety. *Annual Review of Food Science and Technology, 11*(1), 93–118. https://doi.org/10.1146/annurev-food-032519-051708

Granato, D., Nunes, D. S., & Barba, F. J. (2017). An integrated strategy between food chemistry, biology, nutrition, pharmacology, and statistics in the development of functional foods: A proposal. In *Trends in Food Science and Technology* (Vol. 62, pp. 13–22). Elsevier Ltd https://doi.org/10.1016/j.tifs.2016.12.010.

Green, R. D., & Stanberry, L. R. (1977). Elevation of cyclic amp in C-1300 murine neuroblastoma by adenosine and related compounds and the antagonism of this response by methylxanthines. *Biochemical Pharmacology, 26*(1), 37–43. https://doi.org/10.1016/0006-2952(77)90127-7

Greenberg, J. A. (2015). Chocolate intake and diabetes risk. *Clinical Nutrition, 34*(1), 129–133. https://doi.org/10.1016/j.clnu.2014.02.005

Greenberg, J. A., & Buijsse, B. (2013). Habitual chocolate consumption may increase body weight in a dose–response manner. *PLoS ONE, 8*(8), 597–609. https://doi.org/10.1371/journal.pone.0070271

Grillo, G., Boffa, L., Binello, A., Mantegna, S., Cravotto, G., Chemat, F., . . . Telysheva, G. (2019). Cocoa bean shell waste valorisation; extraction from lab to pilot-scale cavitational reactors. *Food Research International, 115*, 200–208. https://doi.org/10.1016/j.foodres.2018.08.057

Gu, Y., Yu, S., & Lambert, J. D. (2014). Dietary cocoa ameliorates obesity-related inflammation in high fat-fed mice. *European Journal of Nutrition, 53*(1), 149–158. https://doi.org/10.1007/s00394-013-0510-1

Halib, H., Ismail, A., Mohd Yusof, B. N., Osakabe, N., & Daud, Z. A. M. (2020). Effects of cocoa polyphenols and dark chocolate on obese adults: A scoping review. *Nutrients, 12*(12), 1–19. https://doi.org/10.3390/nu12123695

Heiss C, Kelm M (2016) Investigation on the effects of cocoa procyanidins on vascular function in healthy individuals. https://clinicaltrials.gov/ct2/show/NCT02728466.

Hernández-Hernández, C., Morales-Sillero, A., Fernández-Bolaños, J., Bermúdez-Oria, A., Morales, A. A., & Rodríguez-Gutiérrez, G. (2019). Cocoa bean husk: industrial source of antioxidant phenolic extract. *Journal of the Science of Food and Agriculture, 99*(1), 325–333. https://doi.org/10.1002/jsfa.9191

Hernández-Hernández, C., Viera-Alcaide, I., Morales-Sillero, A. M., Fernández-Bolaños, J., & Rodríguez-Gutiérrez, G. (2018). Bioactive compounds in Mexican genotypes of cocoa cotyledon and husk. *Food Chemistry, 240*, 831–839. https://doi.org/10.1016/J.FOODCHEM.2017.08.018

Hooper, L., Kay, C., Abdelhamid, A., Kroon, P. A., Cohn, J. S., Rimm, E. B., & Cassidy, A. (2012). Effects of chocolate, cocoa, and flavan-3-ols on cardiovascular health: a systematic review and meta-analysis of randomized trials. *The American Journal of Clinical Nutrition, 95*(3), 740–751. https://doi.org/10.3945/ajcn.111.023457

Howarth, N. C., Saltzman, E., & Roberts, S. B. (2001). Dietary fiber and weight regulation. *Nutrition Reviews, 59*(5), 129–139.

Jacobs, D. M., Smolders, L., Lin, Y., de Roo, N., Trautwein, E. A., van Duynhoven, J., . . . Mihaleva, V. V. (2017). Effect of theobromine consumption on serum lipoprotein profiles in apparently healthy humans with low HDL-cholesterol concentrations. *Frontiers in Molecular Biosciences, 4*(59), 1–11. https://doi.org/10.3389/fmolb.2017.00059

Jahurul, M. H. A., Zaidul, I. S. M., Norulaini, N. A. N., Sahena, F., Jinap, S., Azmir, J., . . . Omar, A. K. M. (2013). Cocoa butter fats and possibilities of substitution in food products concerning cocoa varieties, alternative sources, extraction methods, composition, and characteristics. *Journal of Food Engineering, 117*(4), 467–476. https://doi.org/10.1016/J.JFOODENG.2012.09.024

Jang, M. H., Kang, N. H., Mukherjee, S., & Yun, J. W. (2018). Theobromine, a methylxanthine in cocoa bean, stimulates thermogenesis by inducing white fat browning and activating brown adipocytes. *Biotechnology and Bioprocess Engineering, 23*(6), 617–626. https://doi.org/10.1007/s12257-018-0434-y

Jang, Y. J., Koo, H. J., Sohn, E. H., Kang, S. C., Rhee, D. K., & Pyo, S. (2015). Theobromine inhibits differentiation of 3T3-L1 cells during the early stage of adipogenesis via AMPK and MAPK signaling pathways. *Food and Function, 6*(7), 2365–2374. https://doi.org/10.1039/c5fo00397k

Jenkins, D. J. A., Taylor, R. H., & Wolever, T. M. S. (1982). The diabetic diet, dietary carbohydrate and differences in digestibility. *Diabetologia, 23*(6), 477–484.

Jenkins, D. J. A., & Jenkins, A. L. (1985). Dietary fiber and the glycemic response. *Proceedings of the Society for Experimental Biology and Medicine, 180*(3), 422–431. https://doi.org/10.3181/00379727-180-42199

Jokić, S., Gagić, T., Knez, Ž., Šubarić, D., & Škerget, M. (2018). Separation of active compounds from food by-product (cocoa shell) using subcritical water extraction. *Molecules, 23*(6), 1408. https://doi.org/10.3390/molecules23061408

Jozinović, A., Panak Balentić, J., Ačkar, Đ., Babić, J., Pajin, B., Miličević, B., . . . Šubarić, D. (2019). Cocoa husk application in the enrichment of extruded snack products. *Journal of Food Processing and Preservation, 43*(2). https://doi.org/10.1111/jfpp.13866

Karim, M., McCormick, K., & Tissa Kappagoda, C. (2000). Effects of cocoa extracts on endothelium-dependent relaxation. *Journal of Nutrition, 130*(8 SUPPL). https://doi.org/10.1093/jn/130.8.2105s

Katz, D. L., Doughty, K., & Ali, A. (2011). Cocoa and chocolate in human health and disease. *Antioxidants and Redox Signaling, 15*(10), 2779–2811. https://doi.org/10.1089/ars.2010.3697

Kay, R. M., Grobin, W., & Track, N. S. (1981). Diets rich in natural fiber improve carbohydrate tolerance in maturity-onset, non-insulin dependent diabetics. *Diabetologia, 20*(1), 18–21. https://doi.org/10.1007/BF00253811

Khan, N., Khymenets, O., Urpí-Sardà, M., Tulipani, S., Garcia-Aloy, M., Monagas, M., . . . Andres-Lacueva, C. (2014). Cocoa polyphenols and inflammatory markers of cardiovascular disease. *Nutrients, 6*(2), 844–880. https://doi.org/10.3390/nu6020844

Ki, W. L., Hwang, E. S., Nam, J. K., Kyoung, H. K., & Hyong, J. L. (2005). Extraction and chromatographic separation of anticarcinogenic fractions from cacao bean husk. *BioFactors, 23*(3), 141–150. https://doi.org/10.1002/biof.5520230303

Kim, H., & Keeney, P. G. (1984). (−)-Epicatechin content in fermented and unfermented cocoa beans. *Journal of Food Science, 49*(4), 1090–1092. https://doi.org/10.1111/j.1365-2621.1984.tb10400.x

Kim, J., Lee, K. W., & Lee, H. J. (2011). Cocoa (*T. cacao*) Seeds and Phytochemicals in Human Health. In *Nuts and Seeds in Health and Disease Prevention* (pp. 351–360). Elsevier Inc. https://doi.org/10.1016/B978-0-12-375688-6.10042-8

Knapp, A. W., & Coward, K. H. (1935). The vitamin D activity of cacao shell. *Biochemical Journal, 29*(12), 2728–2735. https://doi.org/10.1042/bj0292728

Kon, S. K., & Henry, K. M. (1935). The effect of feeding cacao shell to cows on the vitamin D content of butter (milk). *Biochemical Journal, 29*(9), 2051–2056. https://doi.org/10.1042/bj0292051

Kowalska, H., Czajkowska, K., Cichowska, J., & Lenart, A. (2017). What's new in biopotential of fruit and vegetable by-products applied in the food processing industry. *Trends in Food Science and Technology, 67*, 150–159. https://doi.org/10.1016/j.tifs.2017.06.016

Kuo, H. W., Chang, C. C., & Cheng, W. (2022). Pectin from dry cacao pod husk mediates growth performance, immune resistance responses and carbohydrate metabolism of *Litopenaeus vannamei* through dietary administration. *Aquaculture, 548*, 737613. https://doi.org/10.1016/j.aquaculture.2021.737613

Kwok, C. S., Boekholdt, S. M., Lentjes, M. A. H., Loke, Y. K., Luben, R. N., Yeong, J. K., . . . Khaw, K. T. (2015). Habitual chocolate consumption and risk of cardiovascular disease among healthy men and women. *Heart, 101*(16), 1279–1287. https://doi.org/10.1136/heartjnl-2014-307050

LeBlanc, J., Jobin, M., Cote, J., Samson, P., & Labrie, A. (1985). Enhanced metabolic response to caffeine in exercise-trained human subjects. *Journal of Applied Physiology, 59*(3), 832–837. https://doi.org/10.1152/jappl.1985.59.3.832

Lecumberri, E., Goya, L., Mateos, R., Alía, M., Ramos, S., Izquierdo-Pulido, M., & Bravo, L. (2007). A diet rich in dietary fiber from cocoa improves lipid profile and reduces malondialdehyde in hypercholesterolemic rats. *Nutrition, 23*(4), 332–341. https://doi.org/10.1016/j.nut.2007.01.013

Lecumberri, E., Mateos, R., Izquierdo-Pulido, M., Rupérez, P., Goya, L., & Bravo, L. (2007). Dietary fiber composition, antioxidant capacity and physico-chemical properties of a fiber-rich product from cocoa (*T. cacao* L.). *Food Chemistry, 104*(3), 948–954. https://doi.org/10.1016/j.foodchem.2006.12.054

Lee, H. J., Lee, K. W., Kang, K. S., Kim, D., Park, H. H., Lee, M. J., Kim, H.S., & Kwon, I. B. (2004). Extracts of cacao and cacao bean husk with inhibitory effects on carcinogenesis. *U.S. Patent Application No. 10/202586*, filed February 19, 2004.

Lessens, D. M., & Rakel, D. P. (2011). Does chocolate reduce blood pressure? *Evidence-Based Practice, 14*(9), 12.

Leveille, G. A., & Sauberlich, H. E. (1966). Mechanism of the cholesterol-depressing effect of pectin in the cholesterol-fed rat. *The Journal of Nutrition, 88*(2), 209–214. https://doi.org/10.1093/jn/88.2.209

Lin, H. V., Frassetto, A., Kowalik, E. J., Nawrocki, A. R., Lu, M. M., Kosinski, J. R., . . . Marsh, D. J. (2012). Butyrate and propionate protect against diet-induced obesity and regulate gut hormones via free fatty acid receptor 3-independent mechanisms. *PLoS ONE, 7*(4). https://doi.org/10.1371/journal.pone.0035240

Lin, X., Zhang, I., Li, A., Manson, J. A. E., Sesso, H. D., Wang, L., & Liu, S. (2016). Cocoa flavanol intake and biomarkers for cardiometabolic health: A systematic review and meta-analysis of randomized controlled trials. *Journal of Nutrition, 146*(11), 2325–2333. https://doi.org/10.3945/jn.116.237644

Loopstra-Masters, R. C., Liese, A. D., Haffner, S. M., Wagenknecht, L. E., & Hanley, A. J. (2011). Associations between the intake of caffeinated and decaffeinated coffee and measures of insulin sensitivity and beta cell function. *Diabetologia, 54*(2), 320–328. https://doi.org/10.1007/s00125-010-1957-8

Magistrelli, D., Zanchi, R., Malagutti, L., Galassi, G., Canzi, E., & Rosi, F. (2016). Effects of cocoa husk feeding on the composition of swine intestinal microbiota. *Journal of Agricultural and Food Chemistry, 64*(10), 2046–2052. https://doi.org/10.1021/acs.jafc.5b05732

Makki, K., Deehan, E. C., Walter, J., & Bäckhed, F. (2018). The impact of dietary fiber on gut microbiota in host health and disease. *Cell Host and Microbe, 23*(6), 705–715. https://doi.org/10.1016/J.CHOM.2018.05.012

Mandaliya, D. K., & Seshadri, S. (2019). Short chain fatty acids, pancreatic dysfunction and type 2 diabetes. *Pancreatology, 199*(2), 280–284. https://doi.org/10.1016/j.pan.2019.01.021

Manzano, P., Hernández, J., Quijano-Avilés, M., Barragán, A., Chóez-Guaranda, I., Viteri, R., & Valle, O. (2017). Polyphenols extracted from *T. cacao* waste and its utility as antioxidant. *Emirates Journal of Food and Agriculture, 29*(1), 45–50. https://doi.org/10.9755/ejfa.2016-04-388

Mao, T. K., Van de Water, J., Keen, C. L., Schmitz, H. H., & Gershwin, M. E. (2002). Modulation of TNF-α secretion in peripheral blood mononuclear cells by cocoa flavanols and procyanidins. *Developmental Immunology, 9*(3), 135–141. https://doi.org/10.1080/1044667031000137601

Martin, M. Á., Goya, L., & Ramos, S. (2016). Antidiabetic actions of cocoa flavanols. *Molecular Nutrition and Food Research, 60*(8), 1756–1769.

Martín, M. Á., Ramos, S., Cordero-Herrero, I., Bravo, L., & Goya, L. (2013). Cocoa phenolic extract protects pancreatic beta cells against oxidative stress. *Nutrients, 5*(8), 2955–2968. https://doi.org/10.3390/NU5082955

Martín-Cabrejas, M. A., Valiente, C., Esteban, R. M., Mollá, E., & Waldron, K. (1994). Cocoa hull: A potential source of dietary fiber. *Journal of the Science of Food and Agriculture, 66*(3), 307–311. https://doi.org/10.1002/jsfa.2740660307

Martínez, R., Torres, P., Meneses, M. A. M. A., Figueroa, J. G. J. G., Pérez-Álvarez, J. A. J. A., & Viuda-Martos, M. (2012). Chemical, technological and *in vitro* antioxidant properties of cocoa (*T. cacao* L.) co-products. *Food Research International, 49*(1), 39–45. https://doi.org/10.1016/J.FOODRES.2012.08.005

Martínez-López, S., Sarriá, B., Gómez-Juaristi, M., Goya, L., Mateos, R., & Bravo-Clemente, L. (2014). Theobromine, caffeine, and theophylline metabolites in human plasma and urine after consumption of soluble cocoa products with different methylxanthine contents. *Food Research International, 63*, 446–455. https://doi.org/10.1016/j.foodres.2014.03.009

Martínez-Pinilla, E., Oñatibia-Astibia, A., & Franco, R. (2015). The relevance of theobromine for the beneficial effects of cocoa consumption. *Frontiers in Pharmacology, 6*(FEB), 1–5. https://doi.org/10.3389/fphar.2015.00030

Martín-Peláez, S., Camps-Bossacoma, M., Massot-Cladera, M., Rigo-Adrover, M., Franch, À., Pérez-Cano, F. J., & Castell, M. (2017). Effect of cocoa's theobromine on intestinal microbiota of rats. *Molecular Nutrition and Food Research, 61*(10), 1–31. https://doi.org/10.1002/mnfr.201700238

Massot-Cladera, M., Costabile, A., Childs, C. E., Yaqoob, P., Franch, À., Castell, M., & Pérez-Cano, F. J. (2015). Prebiotic effects of cocoa fiber on rats. *Journal of Functional Foods, 19*, 341–352. https://doi.org/10.1016/j.jff.2015.09.021

Massot-Cladera, M., Mayneris-Perxachs, J., Costabile, A., Swann, J. R., Franch, À., Pérez-Cano, F. J., & Castell, M. (2017). Association between urinary metabolic profile and the intestinal effects of cocoa in rats. *British Journal of Nutrition, 117*(5), 623–634. https://doi.org/10.1017/S0007114517000496

Matsuda, Y., Kobayashi, M., Yamauchi, R., Ojika, M., Hiramitsu, M., Inoue, T., . . . Horio, F. (2011). Coffee and caffeine improve insulin sensitivity and glucose tolerance in C57BL/6J mice fed a high-fat diet. *Bioscience, Biotechnology and Biochemistry, 75*(12), 2309–2315. https://doi.org/10.1271/bbb.110452

Matsui, N., Ito, R., Nishimura, E., Yoshikawa, M., Kato, M., Kamei, M., . . . Hashizume, S. (2005). Ingested cocoa can prevent high-fat diet-induced obesity by regulating the expression of genes for fatty acid metabolism. *Nutrition, 21*(5), 594–601. https://doi.org/10.1016/j.nut.2004.10.008

Matsumoto, C., Petrone, A. B., Sesso, H. D., Gaziano, J. M., & Djoussé, L. (2015). Chocolate consumption and risk of diabetes mellitus in the physicians' health study. *American Journal of Clinical Nutrition, 101*(2), 362–367. https://doi.org/10.3945/ajcn.114.092221

Matsuzawa, Y. (2014). Obesity and metabolic syndrome: the contribution of visceral fat and adiponectin. *Diabetes Management, 4*(4), 391–401. https://doi.org/10.2217/DMT.14.30

Mazor Jolić, S., Radojčic Redovnikovic, I., Marković, K., Ivanec Šipušić, D., & Delonga, K. (2011). Changes of phenolic compounds and antioxidant capacity in cocoa beans processing. *International Journal of Food Science and Technology, 46*(9), 1793–1800. https://doi.org/10.1111/j.1365-2621.2011.02670.x

Mazzutti, S., Rodrigues, L. G. G., Mezzomo, N., Venturi, V., & Ferreira, S. R. S. (2018). Integrated green-based processes using supercritical CO_2 and pressurized ethanol applied to recover antioxidant compouds from cocoa (*T. cacao*) bean hulls. *The Journal of Supercritical Fluids, 135*, 52–59.

Mellinas, A. C., Jiménez, A., & Garrigós, M. C. (2020). Optimization of microwave-assisted extraction of cocoa bean shell waste and evaluation of its antioxidant, physicochemical and functional properties. *LWT, 127*, 109361. https://doi.org/10.1016/j.lwt.2020.109361

Miranda, J. J., Barrientos-Gutiérrez, T., Corvalan, C., Hyder, A. A., Lazo-Porras, M., Oni, T., & Wells, J. C. K. (2019). Understanding the rise of cardiometabolic diseases in low- and middle-income countries. *Nature Medicine, 25*(11), 1667–1679. https://doi.org/10.1038/s41591-019-0644-7

Mitani, T., Watanabe, S., Yoshioka, Y., Katayama, S., Nakamura, S., & Ashida, H. (2017). Theobromine suppresses adipogenesis through enhancement of CCAAT-enhancer-binding protein β degradation by adenosine receptor A1. *Biochimica et Biophysica Acta - Molecular Cell Research, 1864*(12), 2438–2448. https://doi.org/10.1016/j.bbamcr.2017.09.017

Mitchell, E. S., Slettenaar, M., Meer, N., Transler, C., Jans, L., Quadt, F., & Berry, M. (2011). Physiology and Behavior Differential contributions of theobromine and caffeine on mood, psychomotor performance and blood pressure. *Physiology and Behavior, 104*(5), 816–822. https://doi.org/10.1016/j.physbeh.2011.07.027

Morakinyo, A. O., Adekunbi, D. A., Dada, K. A., & Adegoke, O. A. (2013). Coffee consumption attenuates insulin resistance and glucose intolerance in rats fed on high-sucrose diet. *Nigerian Journal of Physiological Sciences, 28*(2), 179–185.

Mozaffarian, D., Benjamin, E. J., Go, A. S., Arnett, D. K., Blaha, M. J., Cushman, M., . . . Turner, M. B. (2016). Heart disease and stroke statistics-2016 update a report from the American Heart Association. *Circulation, 133*(4), e38–e48. https://doi.org/10.1161/CIR.0000000000000350

Norheim, F., Gjelstad, I., Hjorth, M., Vinknes, K., Langleite, T., Holen, T., . . . Drevon, C. (2012). Molecular nutrition research – The modern way of performing nutritional science. *Nutrients, 4*(12), 1898–1944. https://doi.org/10.3390/nu4121898

Nsor-Atindana, J., Zhong, F., & Mothibe, K. J. (2012). *In vitro* hypoglycemic and cholesterol lowering effects of dietary fiber prepared from cocoa (*T. cacao* L.) shells. *Food and Function, 3*(10), 1044. https://doi.org/10.1039/c2fo30091e

Nsor-Atindana, J., Zhong, F., Mothibe, K. J., Bangoura, M. L., & Lagnika, C. (2012). Quantification of total polyphenolic content and antimicrobial activity of cocoa (*T. cacao* L.) bean shells. *Pakistan Journal of Nutrition, 11*(7), 574–579.

Oddoye, E. O. K., Agyente-Badu, C. K., & Gyedu-Akoto, E. (2013). Cocoa and its by-products: identification and utilization. In *Chocolate in health and nutrition* (pp. 23–37). Humana Press. https://doi.org/10.1007/978-1-61779-803-0_3

Oduro-Mensah, D., Ocloo, A., Nortey, T., Antwi, S., Okine, L. K., & Adamafio, N. A. (2020). Nutritional value and safety of animal feed supplemented with *Talaromyces verruculosus*-treated cocoa pod husks. *Scientific Reports, 10*(1), 1–16. https://doi.org/10.1038/s41598-020-69763-9

Okiyama, D. C. G., Navarro, S. L. B., & Rodrigues, C. E. C. (2017). Cocoa shell and its compounds: Applications in the food industry. *Trends in Food Science and Technology, 63*, 103–112. https://doi.org/10.1016/J.TIFS.2017.03.007

Okiyama, D. C. G., Soares, I. D., Cuevas, M. S., Crevelin, E. J., Moraes, L. A. B., Melo, M. P., . . . Rodrigues, C. E. C. (2018). Pressurized liquid extraction of flavanols and alkaloids from cocoa bean shell using ethanol as solvent. *Food Research International, 114*, 20–29. https://doi.org/10.1016/j.foodres.2018.07.055

Okiyama, D. C. G., Soares, I. D., Toda, T. A., Oliveira, A. L., & Rodrigues, C. E. C. (2019). Effect of the temperature on the kinetics of cocoa bean shell fat extraction using pressurized ethanol and evaluation of the lipid fraction and defatted meal. *Industrial Crops and Products, 130*, 96–103. https://doi.org/10.1016/j.indcrop.2018.12.063

Olateju, T., Begley, J., Green, D. J., & Kerr, D. (2015). Physiological and glycemic responses following acute ingestion of a popular functional drink in patients with type 1 diabetes. *Canadian Journal of Diabetes, 39*(1), 78–82. https://doi.org/10.1016/j.jcjd.2014.07.220

Olubamiwa, O., Ikyo, S. M., Adebowale, B. A., Omojola, A. B., & Hamzat, R. A. (2006). Effect of boiling time on the utilization of cocoa bean shell in laying hen feeds. *International Journal of Poultry Science, 5*(12), 1137–1139. https://doi.org/10.3923/ijps.2006.1137.1139

Otero, M. (2018). *La cascarilla, poca y de moda*. La Opinión de A Coruña. https://www.laopinioncoruna.es/coruna/2018/06/24/cascarilla-moda-24030099.html.

Ozdal, T., Sela, D. A., Xiao, J., Boyacioglu, D., Chen, F., & Capanoglu, E. (2016). The reciprocal interactions between polyphenols and gut microbiota and effects on bioaccessibility. *Nutrients*, *8*(2), 78. https://doi.org/10.3390/nu8020078

Paganini-Hill, A. (2011). Lifestyle practices and cardiovascular disease mortality in the elderly: The leisure world cohort study. *Cardiology Research and Practice*, *1*(1). https://doi.org/10.4061/2011/983764

Panak Balentić, J., Ačkar, Đ., Jokić, S., Jozinović, A., Babić, J., Miličević, B., . . . Pavlović, N. (2018). Cocoa shell: A by-product with great potential for wide application. *Molecules (Basel, Switzerland)*, *23*(6), 1404. https://doi.org/10.3390/molecules23061404

Papadimitriou, A., Silva, K. C., Peixoto, E. B. M. I., Borges, C. M., de Faria, J. M. L., & de Faria, J. B. L. (2015). Theobromine increases NAD+/Sirt-1 activity and protects the kidney under diabetic conditions. *American Journal of Physiology—Renal Physiology*, *308*(3), F209–F225. https://doi.org/10.1152/ajprenal.00252.2014

Papillo, V. A., Locatelli, M., Travaglia, F., Bordiga, M., Garino, C., Coïsson, J. D., & Arlorio, M. (2019). Cocoa hulls polyphenols stabilized by microencapsulation as functional ingredient for bakery applications. *Food Research International*, *115*, 511–518. https://doi.org/10.1016/j.foodres.2018.10.004

Pavlović, N., Jokić, S., Jakovljević, M., Blažić, M., & Molnar, M. (2020). Green extraction methods for active compounds from food waste—Cocoa bean shell. *Foods*, *9*(2), 140. https://doi.org/10.3390/FOODS9020140

Pino, J. L., Mujica, V., & Arredondo, M. (2021). Effect of dietary supplementation with oat β-glucan for three months in subjects with type 2 diabetes: A randomized, double-blind, controlled clinical trial. *Journal of Functional Foods*, *77*, 104311. https://doi.org/10.1016/j.jff.2020.104311

Predan, G. M. I., Lazăr, D. A., & Lungu, I. I. (2019). Cocoa industry-from plant cultivation to cocoa drinks production. In *Caffeinated and cocoa based beverages: Volume 8. The science of beverages* (pp. 489–507). https://doi.org/10.1016/B978-0-12-815864-7.00015-5

Queipo-Ortuño, M. I., Seoane, L. M., Murri, M., Pardo, M., Gomez-Zumaquero, J. M., Cardona, F., . . . Tinahones, F. J. (2013). Gut microbiota composition in male rat models under different nutritional status and physical activity and its association with serum leptin and ghrelin levels. *PLoS ONE*, *8*(5). https://doi.org/10.1371/journal.pone.0065465

Quirós-Sauceda, A. E., Palafox-Carlos, H., Sáyago-Ayerdi, S. G., Ayala-Zavala, J. F., Bello-Perez, L. A., Álvarez-Parrilla, E., . . . González-Aguilar, G. A. (2014). Dietary fiber and phenolic compounds as functional ingredients: Interaction and possible effect after ingestion. *Food and Function*, *5*(6), 1063–1072. https://doi.org/10.1039/c4fo00073k

Rabadan-Chávez, G., Quevedo-Corona, L., Garcia, A. M., Reyes-Maldonado, E., & Jaramillo-Flores, M. E. (2016). Cocoa powder, cocoa extract and epicatechin attenuate hypercaloric diet-induced obesity through enhanced β-oxidation and energy expenditure in white adipose tissue. *Journal of Functional Foods*, *20*, 54–67. https://doi.org/10.1016/J.JFF.2015.10.016

Ramos, S., Martín, M. A., & Goya, L. (2017). Effects of cocoa antioxidants in type 2 diabetes mellitus. *Antioxidants*, *6*(4), 84. https://doi.org/10.3390/antiox6040084

Ramos, S., Moulay, L., Granado-Serrano, A. B., Vilanova, O., Muguerza, B., Goya, L., & Bravo, L. (2008). Hypolipidemic effect in cholesterol-fed rats of a soluble fiber-rich product obtained from cocoa husks. *Journal of Agricultural and Food Chemistry*, *56*(16), 6985–6993. https://doi.org/10.1021/jf8009816

Rasouli, M., & Zahraie, M. (2006). Suppression of VLDL associated triacylglycerol secretion by both α- and β-adrenoceptor agonists in isolated rat hepatocytes. *European Journal of Pharmacology*, *545*(2–3), 109–114. https://doi.org/10.1016/j.ejphar.2006.06.066

Ravindran, R., & Jaiswal, A. K. (2016). Exploitation of food industry waste for high-value products. *Trends in Biotechnology*, *34*(1), 58–69. https://doi.org/10.1016/j.tibtech.2015.10.008

Rebollo-Hernanz, M., Aguilera, Y., Martin-Cabrejas, M. A., & Gonzalez de Mejia, E. (2021a). Phytochemicals from cocoa shell protect mitochondrial function and alleviate oxidative stress in hepatocytes via regulation of ERK and PI3K-AKT pathways. *Medical Sciences Forum*, *2*(1), 25. https://doi.org/10.3390/msf2021002025

Rebollo-Hernanz, M., Aguilera, Y., Martin-Cabrejas, M. A., & Gonzalez de Mejia, E. (2021b). Phytochemicals from the cocoa shell modulate mitochondrial function, lipid and glucose metabolism in hepatocytes via activation of FGF21/ERK, AKT, and mTOR pathways. *Antioxidants*, *10*(11), 136.

Rebollo-Hernanz, M., Cañas, S., Aguilera, Y., Benitez, V., Gila-Díaz, A., Rodriguez-Rodriguez, P., . . . Arribas, S. M. (2020). Validation of cocoa shell as a novel antioxidant dietary fiber food ingredient: nutritional value, functional properties, and safety. *Current Developments in Nutrition*, *4*(Supplement_2), 773.

Rebollo-Hernanz, M., Cañas, S., Taladrid, D., Bartolomé, B., Aguilera, Y., & Martin-Cabrejas, M. A. (2021). Extraction of phenolic compounds from cocoa shell: modeling using response surface methodology and artificial neural networks. *Separation and Purification Technology*, *270*, 118779. https://doi.org/10.1016/j.seppur.2021.118779

Rebollo-Hernanz, M., Zhang, Q., Aguilera, Y., Martín-Cabrejas, M. A., & de Mejia, E. G. (2019a). Cocoa shell aqueous phenolic extract preserves mitochondrial function and insulin sensitivity by attenuating inflammation between macrophages and adipocytes *in vitro*. *Molecular Nutrition and Food Research*, *63*(10), 1801413. https://doi.org/10.1002/mnfr.201801413

Rebollo-Hernanz, M., Zhang, Q., Aguilera, Y., Martín-Cabrejas, M. A., & de Mejia, E. G. (2019b). Relationship of the phytochemicals from coffee and cocoa by-products with their potential to modulate biomarkers of metabolic syndrome *in vitro*. *Antioxidants*, *8*(8), 279. https://doi.org/10.3390/antiox8080279

Redgwell, R., Trovato, V., Merinat, S., Curti, D., Hediger, S., & Manez, A. (2003). Dietary fiber in cocoa shell: characterization of component polysaccharides. *Food Chemistry*, *81*(1), 103–112. https://doi.org/10.1016/S0308-8146(02)00385-0

Reimer, R. A., Maurer, A. D., Eller, L. K., Hallam, M. C., Shaykhutdinov, R., Vogel, H. J., & Weljie, A. M. (2012). Satiety hormone and metabolomic response to an intermittent high energy diet differs in rats consuming long-term diets high in protein or prebiotic fiber. *Journal of Proteome Research*, *11*(8), 4065–4074. https://doi.org/10.1021/pr300487s

Richardson, T., Thomas, P., Ryder, J., & Kerr, D. (2005). Influence of caffeine on frequency of hypoglycemia detected by continuous interstitial glucose monitoring system in patients with long-standing type 1 diabetes. *Diabetes Care*, *28*(6), 1316–1320. https://doi.org/10.2337/diacare.28.6.1316

Rojo-Poveda, B.-P., Mateus-Reguengo, B., & Stévigny, Z. (2019). Effects of Particle Size and Extraction Methods on Cocoa Bean Shell Functional Beverage. *Nutrients*, *11*(4), 867. https://doi.org/10.3390/nu11040867

Rojo-Poveda, O., Barbosa-Pereira, L., El Khattabi, C., Youl, E. N. H., Bertolino, M., Delporte, C., . . . Stévigny, C. (2020). Polyphenolic and methylxanthine bioaccessibility of cocoa bean shell functional biscuits: metabolomics approach and intestinal permeability through Caco-2 cell models. *Antioxidants*, *9*(11), 1164. https://doi.org/10.3390/antiox9111164

Rojo-Poveda, O., Barbosa-Pereira, L., Zeppa, G., & Stévigny, C. (2020). Cocoa bean shell—a by-product with nutritional properties and biofunctional potential. *Nutrients*, *12*(4). https://doi.org/10.3390/nu12041123

Rojo-Poveda, O., Zeppa, G., Ferrocino, I., Stévigny, C., & Barbosa-Pereira, L. (2021). Chemometric classification of cocoa bean shells based on their polyphenolic profile determined by rp-hplc-pda analysis and spectrophotometric assays. *Antioxidants*, *10*(10), 1533. https://doi.org/10.3390/antiox10101533

Rossin, D., Barbosa-Pereira, L., Iaia, N., Sottero, B., Danzero, A. C., Poli, G., . . . Biasi, F. (2021). Protective effect of cocoa bean shell against intestinal damage: An example of byproduct valorization. *Antioxidants*, *10*(2), 280. https://doi.org/10.3390/antiox10020280

Rossin, D., Barbosa-Pereira, L., Iaia, N., Testa, G., Sottero, B., Poli, G., . . . Biasi, F. (2019). A dietary mixture of oxysterols induces *in vitro* intestinal inflammation through TLR2/4 activation: The protective effect of cocoa bean shells. *Antioxidants*, *8*(6), 151. https://doi.org/10.3390/antiox8060151

Rowley, T. J., Bitner, B. F., Ray, J. D., Lathen, D. R., Smithson, A. T., Dallon, B. W., . . . Tessem, J. S. (2017). Monomeric cocoa catechins enhance β-cell function by increasing mitochondrial respiration. *The Journal of Nutritional Biochemistry*, *49*, 30–41. https://doi.org/10.1016/J.JNUTBIO.2017.07.015

Ruzaidi, A., Amin, I., Nawalyah, A. G., Hamid, M., & Faizul, H. A. (2005). The effect of Malaysian cocoa extract on glucose levels and lipid profiles in diabetic rats. *Journal of Ethnopharmacology*, *98*(1–2), 55–60. https://doi.org/10.1016/j.jep.2004.12.018

Sánchez, D., Moulay, L., Muguerza, B., Quiñones, M., Miguel, M., & Aleixandre, A. (2010). Effect of a soluble cocoa fiber-enriched diet in zucker fatty rats. *Journal of Medicinal Food*, *13*(3), 621–628. https://doi.org/10.1089/jmf.2009.0127

Schubert, M. M., Hall, S., Leveritt, M., Grant, G., Sabapathy, S., & Desbrow, B. (2014). Caffeine consumption around an exercise bout: Effects on energy expenditure, energy intake, and exercise enjoyment. *Journal of Applied Physiology*, *117*(7), 745–754. https://doi.org/10.1152/japplphysiol.00570.2014

Smolders, L., Mensink, R. P., Boekschoten, M. V., de Ridder, R. J. J., & Plat, J. (2018). Theobromine does not affect postprandial lipid metabolism and duodenal gene expression, but has unfavorable effects on postprandial glucose and insulin responses in humans. *Clinical Nutrition*, *37*(2), 719–727. https://doi.org/10.1016/j.clnu.2017.06.007

Story, J. A. (1985). Dietary Fiber and Lipid Metabolism. *Proceedings of the Society for Experimental Biology and Medicine*, *180*(3), 447–452. https://doi.org/10.3181/00379727-180-42201

Sugimoto, N., Miwa, S., Hitomi, Y., Nakamura, H., Tsuchiya, H., & Yachie, A. (2014). Theobromine, the primary methylxanthine found in theobroma cacao, prevents malignant glioblastoma proliferation by negatively regulating phosphodiesterase-4, extracellular signal-regulated kinase, Akt/mammalian target of rapamycin kinase, and nuclear fact. *Nutrition and Cancer*, *66*(3), 419–423. https://doi.org/10.1080/01635581.2013.877497

Tokede, O. A., Gaziano, J. M., & Djoussé, L. (2011). Effects of cocoa products/dark chocolate on serum lipids: A meta-analysis. *European Journal of Clinical Nutrition*, *65*(8), 879–886. https://doi.org/10.1038/ejcn.2011.64

Tomaru, M., Takano, H., Osakabe, N., Yasuda, A., Inoue, K.-I., Yanagisawa, R., . . . Uematsu, H. (2007). Dietary supplementation with cacao liquor proanthocyanidins prevents elevation of blood glucose levels in diabetic obese mice. *Nutrition*, *23*(4), 351–355. https://doi.org/10.1016/J.NUT.2007.01.007

Turnbull, D., Rodricks, J. V., Mariano, G. F., & Chowdhury, F. (2017). Caffeine and cardiovascular health. *Regulatory Toxicology and Pharmacology*, *89*, 165–185. https://doi.org/10.1016/j.yrtph.2017.07.025

Tzounis, X., Rodriguez-Mateos, A., Vulcevic, J., Gibson, G. R., Kwik-Uribe, C., & Spencer, J. P. (2011). Prebiotic evaluation of cocoa-derived flavanols in healthy humans by using a randomized, controlled, double-blind, crossover intervention study. *American Journal of Clinical Nutrition*, *93*(1), 62–72. https://doi.org/10.3945/ajcn.110.000075

Vahouny, G. V., Tombes, R., Cassidy, M. M., Kritchevsky, D., & Gallo, L. L. (1980). Dietary fibers: V. Binding of bile salts, phospholipids and cholesterol from mixed micelles by bile acid sequestrants and dietary fibers. *Lipids*, *15*(12), 1012–1018. https://doi.org/10.1007/BF02534316

Vásquez, Z. S., de Carvalho Neto, D. P., Pereira, G. V. M., Vandenberghe, L. P. S., de Oliveira, P. Z., Tiburcio, P. B., . . . Soccol, C. R. (2019). Biotechnological approaches for cocoa waste management: A review. *Waste Management*, *90*, 72–93. https://doi.org/10.1016/j.wasman.2019.04.030

Vinson, J. A., Proch, J., Bose, P., Muchler, S., Taffera, P., Shuta, D., . . . Agbor, G. A. (2006). Chocolate is a powerful ex vivo and *in vivo* antioxidant, an antiatherosclerotic agent in an animal model, and a significant contributor to antioxidants in the European and American diets. *Journal of Agricultural and Food Chemistry*, *54*(21), 8071–8076. https://doi.org/10.1021/jf062175j

Vojvodić, A., Komes, D., Vovk, I., Belščak-Cvitanović, A., & Bušić, A. (2016). Compositional evaluation of selected agro-industrial wastes as valuable sources for the recovery of complex carbohydrates. *Food Research International*, *89*, 565–573. https://doi.org/10.1016/j.foodres.2016.07.023

Watson, J. M., Jenkins, E. J. E., Hamilton, P., Lunt, M. J., & Kerr, D. (2000). Influence of caffeine on the frequency and perception of hypoglycemia in free-living patients with type 1 diabetes. *Diabetes Care*, *23*(4), 455–459. https://doi.org/10.2337/diacare.23.4.455

Wenzel, T. J., Gates, E. J., Ranger, A. L., & Klegeris, A. (2020). Short-chain fatty acids (SCFAs) alone or in combination regulate select immune functions of microglia-like cells. *Molecular and Cellular Neuroscience*, *105*, 103493. https://doi.org/10.1016/j.mcn.2020.103493

Williamson, G., & Manach, C. (2005). Bioavailability and bioefficacy of polyphenols in humans. II. Review of 93 intervention studies. *The American Journal of Clinical Nutrition*, *81*(1), 243S–255S. https://doi.org/10.1093/ajcn/81.1.243S

Wollgast, J., & Anklam, E. (2000). Review on polyphenols in *T. cacao*: Changes in composition during the manufacture of chocolate and methodology for identification and quantification. *Food Research International*, *33*(6), 423–447. https://doi.org/10.1016/S0963-9969(00)00068-5

Wu, C., & Rajagopalan, S. (2016). Phosphodiesterase-4 inhibition as a therapeutic strategy for metabolic disorders. *Obesity Reviews*, *17*(5), 429–441. https://doi.org/10.1111/obr.12385

Yamauchi, R., Kobayashi, M., Matsuda, Y., Ojika, M., Shigeoka, S., Yamamoto, Y., . . . Horio, F. (2010). Coffee and caffeine ameliorate hyperglycemia, fatty liver, and inflammatory adipocytokine expression in spontaneously diabetic KK- A^y mice. *Journal of Agricultural and Food Chemistry*, *58*(9), 5597–5603. https://doi.org/10.1021/jf904062c

Yarmolinsky, J., Mueller, N. T., Duncan, B. B., Del Carmen Bisi Molina, M., Goulart, A. C., & Schmidt, M. I. (2015). Coffee consumption, newly diagnosed diabetes, and other alterations in glucose homeostasis: A cross-sectional analysis of the Longitudinal Study of Adult Health (ELSA-Brasil). *PLoS ONE*, *10*(5), 1–15. https://doi.org/10.1371/journal.pone.0126469

Yusof, F., Khanahmadi, S., Amid, A., & Mahmod, S. S. (2016). Cocoa pod husk, a new source of hydrolase enzymes for preparation of cross-linked enzyme aggregate. *SpringerPlus*, *5*(1), 1–18. https://doi.org/10.1186/s40064-015-1621-3

Zainal, B., Abdah, M. A., Taufiq-Yap, Y. H., Roslida, A. H., & Rosmin, K. (2014). Anticancer agents from non-edible parts of *T. cacao*. *Natural Products Chemistry and Research*, *2*(4). https://dx.doi.org/10.4172/2329-6836.1000134

Zapata, F. J. F. J., Rebollo-Hernanz, M., Novakofski, J. E. J. E., Nakamura, M. T. M. T., & Gonzalez de Mejia, E. (2020). Caffeine, but not other phytochemicals, in mate tea (*Ilex paraguariensis* St. Hilaire) attenuates high-fat-high-sucrose-diet-driven lipogenesis and body fat accumulation. *Journal of Functional Foods*, *64*, 103646. https://doi.org/10.1016/j.jff.2019.103646

16

Pseudocereals: Ancient Grains as Sources of Bioactive Compounds for Functional Foods and Modulation of Molecular Markers

Oscar Abel Sánchez-Velázquez[1], *María Lilibeth Manzanilla-Valdez*[2], *Martin Mondor*[3,4], *Guiomar Melgar-Lalanne*[5], *and Alan Javier Hernández-Álvarez*[2]

[1] *Facultad de Ciencias Químico-Biológicas, Universidad Autónoma de Sinaloa, Culiacan, Mexico*
[2] *School of Food Science and Nutrition, University of Leeds, Leeds, United Kingdom*
[3] *St-Hyacinthe Research and Development Centre, Agriculture and Agri-Food Canada, Saint-Hyacinthe, Canada*
[4] *Institute of Nutrition and Functional Foods (INAF), Université Laval, Quebec City, Canada*
[5] *Centro de Investigaciones Biomédicas, Universidad Veracruzana, Xalapa, Mexico*

16.1 Introduction

Pseudocereals are plants that do not belong to the Poaceae family but produce fruits and seeds used as flour for bread and other foods (Mir, Riar, & Singh, 2018; Selma-Garcia, Haros, & Larrapa, 2020). They have been used as food ingredients due to their nutritional characteristics. The best-known pseudocereals are amaranth (*Amaranthus* spp.), quinoa (*Chenopodium quinoa* Willd.), buckwheat (*Fagopyrum* spp.), and chia (*Salvia hispanica*L.) (Grancieri, Duarte Martino, & Gonzalez de Mejia, 2019; Rodríguez, Rahman, Thushar, & Singh, 2020).

Pseudocereals are considered as novel foods or superfoods because of their unique nutritional profiles (Vrenna, Peruccio, Liu, Zhong, & Sun, 2021). “Superfoods” is a marketing term that is used to promote foods associated with high concentrations of vitamins, minerals, fibers, essential fatty acids, and/or antioxidants. These grains are considered a novelty for Western consumers, although they have been consumed in their communities of origin since ancient times (Schiemer, Halloran, Jespersen, & Kaukua, 2018). In the United Nations 2030 Agenda for Sustainable Development, pseudocereals are mentioned as ingredients that can contribute to healthy diets for a growing world population, which is expected to increase between 70% and 100% by 2050 (Melini & Melini, 2021).

Amaranth was a primary food for the ancient Mesoamerican Indian civilizations (Mexico and Maya) and was used in religious ceremonies as a physical representation of the gods (mainly Huitzilopochtli and Tlaloc). The invading Spanish viewed amaranth as a pagan food and prohibited its cultivation and use. For that reason, the crop became marginal. Its current use is limited to making a candy bar called “*alegría*” which means happiness in Spanish (Morales Quezada &del Castilo, 2009). Quinoa originated in the Andean region

Molecular Mechanisms of Functional Food, First Edition. Edited by Rocio Campos-Vega and B. Dave Oomah.

Table 16.1 Proximal composition (g/100 g) of the main pseudocereals.

Pseudocereal	Family	Crude protein	Fat	Dietary fiber	Total carbohydrates	References
Quinoa (*Chenopodium quinoa*)	Amaranthaceae	13.1	5.5	14.7	10.2	Nowak, Du, and Charrondière (2016)
Amaranth (*Amaranthus* spp.)	Amaranthaceae	14.6	5.7	7.0	5.8	Rodríguez, García Fillería, and Tironi (2020)
Buckwheat (*Fagopyrum* spp.)	Polygonaceae	13.3	3.4	10.0	–	Rodríguez, García Fillería, and Tironi (2020)
Chia (*Salvia hispanica*)	Lamiaceae	18.9	31.2	35.3	4.6	da Silva et al. (2017)

and was used as both animal feed and human food until the Spanish conquest, when the crop became marginalized because the Spanish prohibited the indigenous people from using it for medicinal, social, and religious purposes (Martínez, Fuentes, & Bazile, 2015). Today, quinoa is the most widely consumed pseudocereal in the world. Its export volume increased 600-fold from 1995 to 2014 (Bedoya-Perales, Pumi, Mujica, Talamini, & Padula,2018). The consumption of chia has been increasing over the years, given its health benefits related to chronic diseases such as obesity, diabetes, and cancer. Today, chia seeds are used whole or in the form of flour. They can be eaten raw, or added to other foods, such as yogurts and salads, fruits and preparations such as breads, cakes, granola bars, and beverages (Grancieri et al., 2019). Finally, although amaranth can be consumed similarly to chia and quinoa and in the traditional snack bar form, no information appears to be available on the increase in consumption of this grain. In general, pseudocereals are rich in carbohydrates and protein (**Table 16.1**). They are also gluten-free because they do not belong to the Poaceae family.

Pseudocereals are a great source of high-quality proteins and essential amino acids and can potentially be used as a substitute in gluten-free diets because of their lower allergenicity (Prasadi & Joyce, 2020). While leucine is the most abundant essential amino acid in pseudocereals, their lysine and methionine contents are also higher than in rice (Mota et al., 2016). They are also considered a source of unsaturated fatty acids, with linoleic acid being the most abundant fatty acid in amaranth and quinoa, followed by oleic acid. In buckwheat, the linoleic acid and oleic acid contents are almost equal (Alvarez-Jubete, Arendt, & Gallagher, 2009; Alvarez-Jubete, Arendt, & Gallegher, 2010). In chia, the most abundant fatty acid is linoleic acid, followed by linolenic acid and oleic acid (Kulczyński, Kobus-Cisowska, Taczanowski, Kmiecik, & Gramza-Michałowska, 2019).

Unlike cereals, pseudocereals are gluten-free plant-based food sources. Gluten, an allergen, induces intestinal disorders, including coeliac disease, which affects 1% of the global population. This disease has a harsh response to the immune system, resulting in an inflammation response and damage of the upper small intestine with infiltration of intraepithelial lymphocytes, and these includes villous atrophy that leads to reduced nutrients (vitamins A, D, E, and K, iron, calcium, and folate) absorption (Jnawali, Kumar, & Tanwar, 2016: Xhaferaj, Alves, Ferreira, & Scherf, 2020). Gluten-free products are considered by the Codex Alimentarius legislation, which establishes in the Codex Standard 118-1979 that gluten-free products "are considered of/or made only from one or more ingredients which do not contain wheat, rye, barley, oats or their crossbred varieties, and the gluten level does not exceed 20 mg/kg in total" (Xhaferaj et al., 2020). Different studies have shown that pseudocereal could be a safe option for coeliac disease individuals, as they do not contain gluten and are a good source of starch, fiber, protein, minerals vitamins, and phytochemicals (Martínez-Villaluenga, Peñas, & Hernández-Ledesma, 2020).

Pseudocereal grains are high in fiber which is mainly composed of galacturonic acid, arabinose, xylose, glucose, and galactose (Lamothe, Srichuwong, Reuhs, & Hamaker, 2015; Velarde-Salcedo, Bojórquez-Velázquez, & de la Rosa, 2019). They are a good source of minerals, mainly iron, magnesium, and zinc, and contain some minerals in higher concentrations than in cereals and legumes (Rybicka, 2018). Starch is the main carbohydrate found in pseudocereal seeds; it is present in the form of granules of different sizes and shapes. Buckwheat, amaranth, quinoa, and chia are rich sources of compounds such as flavonoids, phenolic acids, trace elements, essential fatty acids, and vitamins, which have beneficial effects on human health, including the prevention and reduction of some degenerative diseases (Campos-Vega, Loarca-Piña,& Oomah, 2010; Thakur & Kumar, 2019).

Many of these phytochemicals are stored in the plant cell structures and fiber; thus, their bioavailability and bioactivity are highly related to their bioaccessibility during gastrointestinal digestion (Delgado, Tironi, & Añón, 2011). Moreover, the microbiota has a crucial role on the biotransformation of nutrients, since innate digestive conditions are not sufficient to release bioactive compounds from the food matrix (Carrillo-Terán, Vilcacundo, & Carpio, 2015; Rowland et al., 2018). It is known that bacteria strains of the species *Bifidobacterium longum*, *Bifidobacterium breve*, *Bifidobacterium lactis*, *Lactobacillus acidophilus*, *Lactobacillus casei*, and *Lactobacillus rhamnosus* promote health benefits by interacting and transforming pseudocereal compounds due to their probiotic characteristics (Ugural & Akyol, 2020). The chemical biotransformation of pseudocereals promoted by microbiota has been effective as a diet therapy to treat obesity, lactose intolerance, inflammatory disorders, and celiac disease, among others in *in vivo* and human trials (Tsafrakidou, Michaelidou, & Biliaderis, 2020).

The biological activity of pseudocereals is related to their protein content, their lipid fraction, and their polysaccharide profile, as well as to the presence of phytochemicals (Martínez-Villaluenga et al., 2020). They have a multitude of health benefits and therefore offer an opportunity to improve public health. They contain several flavonoids, phenolic acids, trace elements, fatty acids, and vitamins with known effects on human health, allowing them to play a role in the prevention and reduction of many degenerative diseases, such as cancer, cardiovascular disease, high blood pressure, and diabetes (Campos-Vega et al., 2010; Lamothe et al., 2015; Thakur, Kumar, & Dhaliwal, 2021; Velarde-Salcedo et al., 2019).

Molecular markers or biomarkers are molecules contained within a sample taken form an organism. There are many types of markers which can be used to identify dominance and co-dominance within the genome. Molecular markers have been successfully assessed through nutritional interventions (*in vitro*, *in vivo*, and clinical trials) in order to elucidate pseudocereals bioactives fate and biological effects (Rodríguez, García Fillería, &

Tironi, 2020). Biomarkers must be capable of being objectively measured and evaluated as an indicator of normal biological processes, pathogenic states, or responses to pharmacological treatment (BDWC, 2001). Biomarkers are among the new tools that are being used in precision medicine and nutrition. They are classified into three groups according to their clinical applications. However, only two of these types of biomarkers—molecular and cellular biomarkers—have begun to be used in applications related to pseudocereals (Pirzadah & Malik, 2020). These markers play a clinical role in improving the precision of or guiding treatment decisions and can be further categorized as predictive, prognostic, or diagnostic. Pseudocereals incorporated into the diet have direct effects on the synthesis, expression, and regulation of specific biomarkers related to chronic degenerative diseases (Lai, Wu & Pan, 2015; Pirzadah & Malik, 2020).

The aim of this chapter is to review the bioactive compounds present in pseudocereals (chia, quinoa, amaranth, and buckwheat) and their modulation of molecular markers.

16.2 Pseudocereals Bioactive Compounds

16.2.1 Phytochemicals

Amaranth, buckwheat, chia, and quinoa seeds are rich sources of bioactive phytochemicals with potential health benefits. More than 125 bioactive phytochemicals have been identified in one or more of these pseudocereals (Dziadek et al., 2016; Huda et al., 2021; Knez Hrnčič, Ivanovski, Cör, & Knez, 2019; Martínez-Cruz & Paredes-López, 2014; Ogrodowska, Zadernowski, Czaplicki, Derewiaka, & Wronowska, 2014; Raguindin et al., 2021; Skwaryło-Bednarz et al., 2020; Tang & Tsao, 2017) (**Table 16.2**). The main group of bioactive phytochemicals identified in these seeds is phenolic compounds, which includes the subgroups of flavonoids, phenolic acids, stilbenoids, tocopherols, and tocotrienols. Phenolic compounds have shown promising bioactivities, such as antioxidant, antimicrobial, anti-inflammatory, and antiproliferative activities (Albuquerque, Heleno, Oliveira, Barros, & Ferreira, 2021; Giménez-Bastida, Hamdi, & Llopis, 2017; Giménez-Batisda G, Zielinski, Piskula, Zielinska, & Szawara-Nowak, 2017; Vega-Galvez et al., 2010). However, they are best known for their antioxidant activity. Serious diseases such as cancer, cardiovascular diseases, hypertension, and diabetes are linked to the generation of free radicals in the body which attack the cells. Owing to their antioxidant activity, phenolic compounds can inhibit free radicals in the body. Their antioxidant activity occurs through transfer of a hydrogen atom and/or an electron to free radicals, which interrupts the chain reaction of oxidation. The antioxidant effect depends on the number and position of the hydroxyl groups (Caleja, Ribeiro, Barreiro, & Ferreira, 2017).

Flavonoids are the most abundant phenolic compounds in pseudocereals generally found in glycoside-bound and free aglycone forms. They consist of 15 carbon atoms (C6–C3–C6) assembled in 2 benzene rings, which are linked by a heterocyclic pyrane ring. They can be divided into six different subgroups: anthocyanins, flavanols, flavones, flavonols, flavonones, and isoflavonoids (Albuquerque et al., 2021). Flavonoids identified in one or more of the aforementioned pseudocereals include apigenin, catechin, (+)-catechin-7-*O*-glucoside, catechin hydrate, cyanidin-3-*O*-galactopyranosyl-rhamnosidedaidzein, cyanidin-3-*O*-galactoside, cyanidin-3-*O*-glucoside, cyanidin-3-*O*-rutinoside, daidzein,

Table 16.2 Major bioactive phytochemicals and their concentrations (mg/kg) in pseudocereals.

Bioactive compounds	Amaranth	Buckwheat	Chia	Quinoa
Phenolic compounds (mg/kg)				
Benzoic acid	n.d.	n.d.	n.d.	19–21
Caffeic acid	6.4–6.6	n.d.	0.33–1,189	6.3
Chlorogenic acid	n.d.	n.d.	2,515	n.d.
***p*-Coumaric acid**	1.2–17.4	n.d.	n.d.	1.1–275
Daidzin	n.d.	n.d.	0.1	n.d
Epigallocatechin	n.d.	n.d.	n.d.	1.6–120
2,4-Dihydroxybenzoic acid	4.7–5.1	n.d.	n.d.	5.4–21.1
3,4-Dihydroxybenzoic acid	4.7–136	n.d.	n.d.	29.8–97
Ferulic acid	120–620	n.d.	n.d.	76–251.5
Ferulic acid 4-*O*-glucoside	n.d.	n.d.	n.d.	14.3–161.4
Gallic acid	11.0–440	n.d.	11	3.7–320
Genistin	n.d.	n.d.	0.1	n.d.
Glycitin	n.d.	n.d.	0.1	n.d.
***p*-Hydroxybenzoic acid**	8.5–20.9	n.d.	n.d.	15.8–275
Isorhamnetin	142–600	n.d.	n.d.	4.0–20.8
Kaempferol	22.4–59.7	n.d.	0.1	0.3–542
Kaempferol dirhamnoside	70	n.d.	n.d.	4.6–14.0
Protocatechuic ethyl ester	n.d.	n.d.	747.1	n.d.
Quercetin	214–843	104.6–1,520	0.1	18.4–680
Quercetin-3-rutinoside	7–592	n.d.	n.d.	50.8–360
Rosmarinic acid	n.d.	n.d.	937	n.d.
Rutin	n.d.	52.5–22,560	n.d.	n.d.
α-Tocopherol	3.0–34.8	n.d.	1.97	8.0–26
β-Tocopherol	5.9–211.8	n.d.	0.88	0.4–2
γ-Tocopherol	1.0–57.1	n.d.	74.8	25.9–53
δ-Tocopherol	0.1–48.8	n.d.	4.06	0.9–3
α-Tocotrienol	10.2–20.6	n.d.	0.32	0.4
β-Tocotrienol	35.4–48.5	n.d.	n.d.	0.8–3
γ-Tocotrienol	2.0–4.0	n.d.	n.d.	n.d.
δ-Tocotrienol	15.5–18.4	n.d.	n.d.	n.d.
Vanillic acid	15.5–69.5	n.d.	n.d.	14–523.9
Vanillic acid glucoside	n.d.	n.d.	n.d.	13.6–60.9
Other bioactive compounds (mg/kg)				
Amaranthine	151.3	n.d.	n.d.	n.d.
β-Carotene	n.d.	463	n.d.	0.3–1.1

Table 16.2 (Continued)

Bioactive compounds	Amaranth	Buckwheat	Chia	Quinoa
20-Hydroxyecdysone	n.d.	n.d.	n.d.	8.6–760
Isoamaranthine	58.7	n.d.	n.d.	n.d.
Lutein	3.6–4.3	666	0.11	4.0–12.4
Squalene	1,970–4,700	n.d.	n.d.	n.d.
Zeaxanthin	0.1–0.3	n.d.	0.46	0.3–5.4

Abbreviations: n.d., no data available. Adapted from: Bojórquez-Velázquez et al. (2018); Bystricka et al. (2011); Knez Hrnčič et al. (2019); Kalinová, Vrchotova, and Triska (2019); Luthar et al. (2020); Martínez-Cruz and Paredes-López (2014); Ogrodowska et al. (2014); Patiño and Robles (2016); Tang and Tsao (2017); Thanh Tien, Dang Trinh, Inoue, Morita, and Van Hung (2018); Tuan et al. (2013); Yıldız, Yalçın, and Güneş (2019); da Silva et al. (2017).

daidzin, epiafzelechin-(4–8)-epiafzelechin-(4–8)-epicatechin, epiafzelechin-(4–8)-epiafzelechin-(4–8)-epicatechin-*O*-(3,4-dimethyl)-gallate, epiafzelechin-(4–6)-epicatechin, epiafzelechin-(4–8)-epicatechin-*p*-OH-benzoate, epiafzelechin-(4–8)-epicatechin(3,4-dimethyl)-gallate, epiafzelechin-(4–8)-epicatechin-methylgallate, epicatechin, epicatechin (4–8)-epicatechin-*O*-(3,4-dimethyl)-gallate, epicatechingallate, (−)-epicatechin-3-*O*-*p*-hydroxybenzoate, epicatechin-3-*O*-(3,4-di-*O*-methyl)-gallate, epigallocatechin, fagopyrins (A, B, C, D, E, and F), formononetin (4′-*O*-methyldaidzein), 3,3-di-*O*-galloyl-procyanidin B-2, 3-*O*-galloyl-procyanidin B-2, genistein, genistin, glycitein, proteglycitin, hesperetin, hesperidin, hesperetin-5-*O*-glucoside, hesperetin-*O*-hexosyl-*O*-hexoside, hesperetin-*O*-malonylhexoside, hesperetin-7-*O*-neohesperidoside, homoeriodictyol, homoorientin, 2′-hydroxydaidzein, 6-hydroxydaidzein, isoorientin, isoquercetin, isoquercitrin, isorhametin, isovitexin, kaempferol, kaempferol dirhamnoside, kaempferol-3-*O*-galactoside, kaempferol-3-*O*-glucoside, kaempferol-3-*O*-glucoside-7-*O*-glucoside, kaempferol-3-*O*-rutinoside, kaempferol-3-*O*-sophoroside, (−)-liquiritigenin, luteolin, 3′,4′-methylenedioxy-7-hydroxy-6-isopentenyl flavone, myricetin, naringenin, naringenin chalcone, naringenin-7-*O*-glucoside orientin, naringenin-*O*-malonylhexoside, naringin, ononin, phloretin, procyanidins (A1, A2, A3, B1, B2, B3, B5, and C1), quercetin, quercetin-3-arabinoside, quercetin-3-*O*-β-D-galactoside, quercetin-3-*O*-glucoside, quercetin-3-rutinoside, quercetin-3-*O*-rutinoside-7-*O*-galactoside, quercetin-3-*O*-rutinoside-3′-β-glucopyranoside, quercetin-3-*O*-rutinoside-3′-*O*-glucoside, quercetin-3-*O*-[β-D-xyloxyl-(1–2)-α-L-rhamnoside], quercitrin, rhamnetin, rutin, sissotrin, tannic acid, tricin 4′-*O*-β-guaiacylglycerol, tricin-7-*O*-β-guaiacylglycerol, tricin-4′-*O*-(β-guaiacylglyceryl) ether-*O*-hexoside, tricin-4′-*O*-syringic acid, tricin-4′-*O*-(syringyl alcohol) ether-5-*O*-hexoside, tricin 4′-*O*-(syringyl alcohol) ether-7-*O*-hexoside, and vitexin.

Another important subgroup of phenolic compounds is the phenolic acids, which have one carboxylic acid group and a phenolic ring. The total phenolic content (12–16 g/kg) of five buckwheat cultivars grown in Western Canada was mostly (70–79%) etherified (Oomah, Campbell, & Mazza, 1996). They can be divided into benzoic acids and cinnamic acids according to their carbon skeleton. Benzoic acids identified in one or more of the aforementioned pseudocereals include *p*-anisic, benzoic, 4-hydroxybenzoic, *p*-hydroxybenzoic, 2,4-dihydroxybenzoic, 2,5-dihydroxybenzoic, and 3,4-dihydroxybenzoic acids, 3,4-dihydroxybenzoic acid-4-*O*-glucoside, gallic acid, *p*-hydroxybenzaldehyde, protocatechuic acid, protocatechuic ethyl ester, syringic acid, vanillic acid, vanillic acid glucoside, and vanillin. Cinnamic acids include caffeic, chlorogenic, *O*-coumaric and *p*-coumaric acids, *p*-coumaric acid 4-glucoside, 2,4-dihyroxycinnamic acid, ferulic acid, ferulic acid 4-*O*-glucoside, isoferulic, neochlorogenic, rosmarinic, and sinapic acids. Stilbenoids are phenolic compounds characterized by the chemical structure C6–C2–C6, with two benzene rings connected by a double bond. One stillbenoid that has been identified in pseudocereals is resveratrol. Finally, tocopherols and tocotrienols which are part of the vitamin E family have also been identified in pseudocereals. Both have a chromanol ring but tocopherols have a saturated phytyl side chain, while tocotrienols have an unsaturated isoprenyl side chain containing three double bonds at C3′, C7′, and C11′. The double bonds of tocotrienols' side chains at C3′ and C7′ have a *trans*-configuration (Niki &Abe, 2019). Both come in four forms: α, β, δ, and γ (Constantinou, Papas, & Constantinou, 2008; Niki & Abe, 2019). The four forms differ with respect to the number and position of methyl groups on the chromanol ring. The α-forms of tocopherol and tocotrienol have three methyl groups at the C5, C7, and C8 positions of the chromanol ring, while the β- and γ-forms have two and the δ-forms have one methyl group (Niki & Abe, 2019).

Other groups of bioactive phytochemicals found in pseudocereals are betalains and carotenoids. Betalains are a class of red and yellow tyrosine-derived pigments which includes molecules such as amaranthine, isoamaranthine, betanin, and isobetanin; they have been identified in one or more of the aforementioned pseudocereals (Asao & Watanabe, 2010). Betalains display antimicrobial, anti-inflammatory, anticancer, antidiabetic, and anti-hepatitis properties (Madadi et al., 2020). Carotenoids are also pigments that can be found in pseudocereals. Lutein, zeaxanthin, and β-carotene have been identified in some of the aforementioned pseudocereals (Julio et al., 2016; Patiño & Robles, 2016). Their primary health benefit relates to their high antioxidant potential (Eggersdorfer & Wyss, 2018; Meléndez-Martinez, 2019). Sapogenins—the aglycones, or non-saccharide, portions of the family of natural products known as saponins—are also present in pseudocereals (Mastebroek, Limburg, Gilles, & Marvin, 2000). Sapogenins contain steroid or other triterpene frameworks as their key organic feature; they include daucosterol, β-daucosterol, ergosterol peroxide, glutinol, glutinone, hecogenin, 6-hydroxy stigmasta-4,22-dien-3-one, 23S-methylcholesterol, olean-12-en-3-ol, peroxidize-ergosterol, β-sitosterol, β-sitosterol palmitate, stigmast-5,24-dien-3-ol, stigmsat-4-ene-3,6-dione, stigmast-5-en-3-ol, α-terpineol, α-thujene, trans-stigmast-5,22-dien-3-ol, urs-12-en-3-ol, and ursolic acid. Phenylpropanoid glycosides which are acylated glycoconjugates carrying a substituted arylalkyl aglycon have also been identified in buckwheat: 3,6-di-*p*-coumaroyl-1,6′-di -feruloyl sucrose, diboside A, lapathoside A, tatarisides (A, B, C, D, E, F, and G), 1,3,6,6′-tetra-feruloyl sucrose, 1,3,6-tri-*p*-coumaroyl-6′-feruloyl sucrose, and 1,3,6′-tri-ferul oyl-6-*p*-coumaroyl sucrose (Raguindin et al., 2021). Phenylpropanoid glycosides demonstrated cytotoxic activity against human cancer cell lines (Zheng et al., 2012). Cyclitols, carbohydrates that contain only hydroxyl groups, were also identified: fagopyritols (A1, A2, A3, B1, B2, and B3) (Raguindin et al., 2021). Cyclitols have antioxidant, anti-inflammatory, and anticancer activities (Owczarczyk-Saczonek, Lahuta, Placek, & Górecki, 2018).

Amaranth is also rich in squalene, a triterpene with bioactive activity (Bojórquez-Velázquez et al., 2018; Ogrodowska et al., 2014).

Other bioactive molecules previously considered antinutritional compounds have been identified in pseudocereals, including saponins, phytic acid, tannins, and protease inhibitors (Vega-Galvez et al., 2010). These are secondary metabolites synthesized by plants mainly as a defense mechanism against adverse environmental conditions, while others are reserve compounds, such as defense proteins (Bowman–Birk inhibitors and Kunitz inhibitors), which are stored in seeds as energy pools in preparation for germination (Muzquiz et al., 2012). These compounds may have beneficial or negative effects on human health or both (Campos-Vega et al., 2010; Muzquiz et al., 2012). These may have a negative impact on protein digestibility and amino acids bioavailability (Gilani, Xiao, & Cockell, 2012). Saponins exhibit anticancer activity and may alleviate hyperlipidemia (Shi et al., 2004; Campos-Vega et al., 2010). In addition, condensed tannins and phytic acid have antioxidant activity and phytic acid exhibits anticancer activity (Campos-Vega et al., 2010).

16.2.2 Peptides

Pseudocereals also contain proteins which play an important nutritional role by providing amino acids for protein synthesis. Proteins are also a source of bioactive peptides which can be released from the parent protein through hydrolysis either during gastrointestinal digestion (*in vivo*) or through the action of enzymes or acids to produce protein hydrolysates (*in vitro*). Bioactive peptides range in size from 2 to 50 amino acid residues. The high diversity of peptides produced offers high potential for bioactivity. They can function as enzyme inhibitors, hormones, or neurotransmitters and can affect body growth, blood pressure, appetite, sleep, learning and stress, among other (Dhaval, Yadav, & Purwar, 2016). Various peptides derived from pseudocereals have been identified.

16.2.2.1 Amaranth

Many studies aimed at generating bioactive peptides from amaranth have been carried out in the last few years (Ayala-Niño, Rodríguez-Serrano, González-Olivares, et al., 2019; Ayala-Niño, Rodríguez-Serrano, Jiménez-Alvarado, et al., 2019; Ayala-Niño, Contreras-López, Castañeda-Ovando, Sánchez-Franco, & González-Olivares, 2020; Barba de la Rosa et al., 2010; Barrio & Añón, 2010; Bojórquez-Velázquez et al., 2018; Delgado et al., 2011; Delgado et al., 2016; Delgado, Galleano, Añón, & Tironi, 2015; Fritz, Vecchi, Rinaldi, & Añón, 2011; Luna-Suárez, Medina-Godoy, Cruz-Hernández, & Paredes-López, 2010; Mendonça, Saldiva, Cruz, & Arêas, 2009; Montoya-Rodríguez & de Mejía, 2015; Montoya-Rodríguez, de Mejía, Dia, Reyes-Moreno, & Milán-Carrillo, 2014; Moronta, Smaldini, Docena, & Añón, 2016); Moronta, Smaldini, Fossati, Añón, & Docena, 2016; Nardo, Suárez, Quiroga, & Añón, 2020; Quiroga, Aphalo, Ventureira, Martínez, & Añón, 2011; Quiroga, Barrio, & Añón, 2015; Rodríguez, García Fillería, & Tironi, 2020; Sabbione, Luna-Vital, Scilingo, Añón, & González de Mejía, 2018; Sabbione, Nardo, Añón, & Scilingo, 2016; Sabbione, Scilingo, & Añón, 2015; Sandoval-Sicairos, Milán-Noris, Luna-Vital, Milán-Carrillo, & Montoya-Rodríguez, 2021; Soares, Mendonça, de Castro, Menezes, & Arêas, 2015; Soriano-Santos & Escalona-Buendía, 2015; Taniya, Reshma, Shanimol, Krishnan, & Priya, 2020; Tironi & Añón, 2010; Tovar-Pérez, Guerrero-Legarreta, Farrés-González, &

Soriano-Santos, 2009; Vecchi & Añón, 2009; Velarde-Salacedo et al., 2013). The peptide structure was not always identified in these studies. Vecchi and Añón (2009) used an *in silico*-based peptide library screening method to identify two peptides (IKP and LEP) from the 11S amaranth globulin fraction; these peptides inhibit angiotensin-converting enzyme (ACE). Four peptides with antioxidant activity were isolated from an *in vitro* gastrointestinal digestion model by Delgado et al. (2016): AWEEREQGSR (IC_{50} = 6.7 μg/mL), YLAGKPQQEH (IC_{50} = 16 μg/mL), IYIEQGNGITGM (IC_{50} = 71 μg/mL), and TEVWDSNEQ (IC_{50} = 20 μg/mL). In another study, Ayala-Niño, Rodríguez-Serrano, González-Olivares, et al. (2019) used Alcalase® and Flavourzyme® to generate amaranth protein hydrolysates and identified two peptides (NIDMLRL and LVRW) with possible antihypertensive activity. Montoya-Rodríguez et al. (2014) identified three peptides with anti-inflammatory activity following *in vitro* digestion: HGSEPFGPR, RPRPWRYT, and RDGPFPWYSH. Lunasin, a 43 amino acids peptide with proven anticancer activity, has also been identified in amaranth proteins (Ayala-Niño et al., 2020). Soares et al. (2015) identified three peptides (GGV, IVG or LVG, and VGVI) with cholesterol-lowering activity that were derived through *in vitro* digestion of amaranth flour.

16.2.2.2 Buckwheat

Several studies similar to amaranth have been carried out to generate bioactive peptides from buckwheat and identify their activities (Fujimura, Minami, Watanabe, & Tadera, 2003; Koyama et al., 2013; Leung & Ng, 2007; Li, Matsui, Matsumoto, Yamasaki, & Kawasaki, 2002; Luo et al., 2020; Ma, In, Hyeon, & Yang, 2006; Ma, Xiong, Zhai, Zhu, & Dziubla, 2010; Oparin et al., 2012; Ruan, Chen, Shao, Wu, & Han, 2011; Yu, Wang, Zhang, & Fan, 2016; Zhou et al., 2015; Zhu, 2021). Li et al. (2002) hydrolyzed solubilized buckwheat powder with pepsin, followed by chymotrypsin and trypsin. The protein hydrolysate contained 11 peptides with ACE inhibitory activity. Their amino acid sequences are FY, AY, LF, YV, VK, YQ, YQY, PSY, LGI, ITF, and INSQ. The first four peptides had already been identified in previous studies, while the remaining seven were identified for the first time. An additional peptide with ACE inhibitory activity (GPP) was also identified by Ma et al. (2006). Fujimura et al. (2003) purified two bioactive peptides (Fa-AMP1 and Fa-AMP2) with antimicrobial activities from buckwheat seeds by gel filtration, ion exchange high performance liquid chromatography (HPLC), and reverse-phase HPLC. The amino acid sequences are AQCGAQGGGATCPGGLCCSQWGWCGSTPKYCGAGCQSNCK for Fa-AMP1 and AQCGAQGGGATCPGGLCCSQWGWCGSTPKYCGAGCQSNCR for Fa-AMP2. Ruan et al. (2011) and Oparin et al. (2012) also identified antimicrobial peptides from buckwheat with the following sequences respectively: FtTI = LIYAKVKCLITGVRTYVGKQSWPELVGTKGKTAAATIDKENAHVTAVLCPPLTTLATCRTFDFRCDR-VRVLINRIGGVVTKTPTVG, and BWI-2c = SEKPQQELEECQNVCRMKRWSTEMVHRCEKKCEEKFERQQR. Koyama et al. (2013) subsequently isolated seven peptides from buckwheat proteins with antihypertensive activities (GHG, DVWY, FDART, FQ, VAE, VVG, and WTFR) using liquid chromatography-mass spectrometry (LC–MS)/MS. Ma et al. (2010) used LC–MS/MS to identify four antioxidant peptides from *in vitro* digests of buckwheat protein. Their amino acid sequences were WPL, VPW, VFPW, and PW. Luo et al. (2020) also identified three antioxidant peptides with the following amino acid sequences: GEVPW, YMENF, and AFYRW. Their respective molecular weights were 586.65, 702.79, and 741.85 Da. AFYRW demonstrated the best antioxidant activity of these three peptides.

16.2.2.3 Chia

Chia is also a good source of bioactive peptides, and many studies have been dedicated to the production and/or identification of these peptides (Aguilar-Toalá, Deering, & Liceaga, 2020; Aguilar-Toalá & Liceaga, 2020; Coelho, de Araujo Aquino, Machado Latorres, & de las Mercedes Salas-Mellado, 2019; Coelho, Soares-Freitas, Gomes Arêas, Avila Gandra, & de las Mercedes Salas-Mellado, 2018; Cotabarren et al., 2019; Grancieri et al., 2019; Grancieri, Duarte Martino, & Gonzalez de Mejia, 2021; Martínez Leo & Segura Campos, 2020; Segura-Campos, Peralta González, Guerrero, & Betancur Ancona, 2013). Grancieri et al. (2019) used the BIOPEP database to identify all the possible bioactive peptides that can be generated from all the protein sequences found in chia seeds; hundreds of peptides with different bioactivities were identified. Aguilar-Toalá and Liceaga (2020) hydrolyzed chia seed meal using sequential enzymatic (Alcalase® followed by Flavourzyme®) microwave-assisted hydrolysis. They identified seven chia peptides in the hydrolysate that exhibited inhibitory activity toward collagenase, hyaluronidase, tyrosinase, and elastase, which are implicated in the degradation of the skin protein matrix. Peptides with the following amino acid sequences were identified: APHWYTN, DQNPRSF, GDAHWAY, GDAHWTY, GDAHWVY, GFEWITF, and KKLKRVYV. The same research group identified 16 peptides from chia exerting antimicrobial activity (Aguilar-Toalá et al., 2020). In decreasing order of antimicrobial potential score, the identified peptides are the following: KADVPGLKK, KGDVIAIR, LKQGDVIAIR, GNIFRGL, FQKGDVIAIR, KGLVLPF, GDKFIQVVNHK, IGTPGKGIL, QKGDVIAIR, KQGDVIAIR, QKGDVIAIRA, KQGDVIAIR, DVPGLKK, GGVLPGIK, HGPIKLH, and NFAVVKQ. Grancieri et al. (2021) identified two peptides with antioxidant activity (NSPGPHDVALDQ and RMVLPEYELLYE) from a digested glutelin fraction.

16.2.2.4 Quinoa

Many studies on quinoa peptides have been carried out in the last few years (Guo, Richel, et al., 2020; Guo, Hao, et al., 2020; Mahdavi Yekta et al., 2020; Mudgil et al., 2020; Vilcacundo, Martínez-Villaluenga, & Hernández-Ledesma, 2017; Vilcacundo, Miralles, Carrillo, & Hernández-Ledesma, 2018; Zheng et al., 2019). Guo, Richel, et al. (2020) also used the BIOPEP database for a predictive analysis of the bioactive peptides that can be released from quinoa protein based on *in silico* enzymolysis with papain. A total of 715 sequences with given activity were predicted. In another study by the same group, three bioactive peptides (FHPFPR, NWFPLPR, and NIFRPF) with ACE inhibitory activity were studied and their inhibitory effects were confirmed with IC_{50} values of 34.92, 16.77, and 32.40 μM, respectively. Mudgil et al. (2020) used enzymes like bromelain, chymotrypsin, and pronase E to produce quinoa protein hydrolysates. They identified 29 peptides (FFE, LYNDWDLR, SGPAR, KPGGTAGSALPRPAHW, HDSF, MFVPVPH, HVQGHPALPGVPAHW, HVASGAGPW, FTVM, ASNLDNPSPEGTVM, CALVGL, PGGAR, AGLR, CDCP, FSAGGLP, MLLL, MYLY, LPLLR, VYML, GARR, CVLSPL, DFTF, CYTF, AHCGGLPY, LLPYH, HMLH, LRL, and MVLP) with antihypertensive activity in the hydrolysates. Vilcacundo et al. (2017) digested a quinoa protein concentrate using the *in vitro* harmonized protocol (Minekus et al., 2014). They identified three novel peptides derived from 11S globulin (IQAEGGLT, DKDYPK, and GEHGSDGNV) which have antidiabetic activity. Zheng et al. (2019)

hydrolyzed the albumin fraction of quinoa protein using Alcalase followed by trypsin and identified one ACE inhibitory and antioxidant peptide, RGQVIYVL, and two antioxidant peptides, ASPKPSSA and QFLLAGR, in the resulting hydrolysates.

16.3 *In Vitro*, *In Vivo*, and Human Bioactivity of Pseudocereals

The literature on the functional and health-promoting effects of pseudocereals is still limited compared to cereals or pulses. However, *in vitro* assays, animal studies, and human trials have demonstrated that the rich bioactive composition of pseudocereals exhibits positive effects against non-transmissible diseases and metabolic disorders. **Table 16.3** displays the bioactivity of four selected pseudocereals.

16.3.1 Cardiovascular

16.3.1.1 Amaranth

One of the main risk factors associated with cardiovascular disease is hypertension (Arnett et al., 2019). Amaranth helps to reduce blood pressure and has an effect on the renin-angiotensin system (RAS) according to several reports.

Suárez et al. (2020) found that amaranth protein isolates and hydrolysates exhibited hypotensive effect with *in vitro* IC_{50} values of 0.7 and 0.14 mg/mL, respectively compared to 39.2 mg/mL for the control (Captopril). For *in vivo* tests, spontaneously hypertensive rats (SHR) were divided into three groups following the administration of a) 50 mg of isolate, b) 1.9 g/kg of hydrolysate, and c) 50 mg of isolate +1.9 g/kg of hydrolysate. The control received 10 mg of captopril and 2.5 mg of aliskiren. Blood pressure was reduced as follows: by 42 mmHg in the group receiving 50 mg of isolate, by 35 mmHg in the group with 50 mg of isolate +1.9 g/kg of hydrolysate, and by 25 mmHg in the group with 1.9 g/kg of hydrolysate. In the control, blood pressure decreased by 32 mmHg. Lastly, for the *ex vivo* model, aorta rings were cut into 2 mm and stabilized for 1 hr with a resting tension of 2–3 g, with changes in saline water every 20 min. The contractile activity of aorta rings was measured in the presence of potassium ions and norepinephrine; after 30 min of treatment, contractile activity was lower in the groups receiving amaranth hydrolysate (0.34 to 0.29 mg/g). In conclusion, amaranth peptides play a key role in regulating blood pressure and ACE-I signaling (Suárez et al., 2020).

Endothelial dysfunction and elevated plasma lipoproteins can lead to strokes. Caselato-Sousa et al. (2014) found that heat-expanded amaranth grains (*Amaranthus caudatus*) had an anticholesterolemic effect. Hypercholesterolemia induced rabbits were fed 5 g of amaranth grains (0.66 g protein) daily for 21 days (Caselato-Sousa et al., 2014). The rabbits on a high-cholesterol diet receiving the amaranth treatment (G2) had significantly ($p < 0.05$) higher endothelial function (83 vs. 55.5%) than those on the same diet with no treatment (G4). Excretion of fecal cholesterol was also significantly ($p < 0.05$) higher in the G2 group (551 vs. 311 mg/kg) than in the G4 group. This finding suggests that digestion residues of heat-expanded amaranth can bind to bile acids by promoting fecal excretion (Caselato-Sousa et al., 2014). In another study, amaranth seeds were fed to hamsters in which hypercholesterolemia was induced through a casein-rich diet (300 g/kg) + cholesterol (0.5 g/kg) for 3 weeks. The animals were divided into three groups and fed ad libitum for 4 weeks, as

Table 16.3 Bioactivity of pseudocereals.

Samples	Amaranth (*Amaranthus spp.*)					
	Experimental conditions	Type of study (*In vitro*, *in vivo*, or clinical trial)	Aim of the study	Biomarkers	Effects/results	Reference
Isolate and hydrolysate	Alcalase® hydrolyzed (5 hr) amaranth protein isolate	*In vitro* ACE-I modulation	ACE-I inhibition activity	Renin and chymase enzymes	Protein hydrolysate inhibited renin (IC_{50} 0.6 mg/mL)	Quiroga, Aphalo, Nardo, and Añón (2017)
Protein hydrolysates	COPRAM™ produced amaranth protein concentrate and Alcalase hydrolyzed (6.16 hr)	*In vivo* spontaneously hypertensive rats (SHR)	Hypotensive activity	ACE-I Blood pressure	Inhibited ACE-I and lowered blood pressure	Ramírez-Torres et al. (2017)
Isolate and hydrolysate	Administration of 1.9 g protein, 50 mg peptide, and 1.9 g protein +50 mg peptide	*In vivo* SHR	Hypotensive activity	ACE-I in plasma	↓ Systolic blood pressure by 19 ± 3 mmHg	Suárez, Aphalo, Rinaldi, Añón, and Quiroga (2020)
Isolate and hydrolysate	Aorta rings of 2 mm, stabilized for 1 hr, resting tension of 2–3 g, changing saline every 20 min	*Ex vivo* SHR	Protection against gastrointestinal digestion	Thoracic aorta	Vasorelaxant effect, depolarization of the membrane, and opening of typeL channels	Suárez et al.(2020)
Seeds	Finely ground heat-expanded amaranth grains suspended in 30 mL of water. A daily dose of 5 g (0.66 g protein) was used.	*In vivo* hypercholesterolemic rabbits	Reverse endothelial dysfunction	Endothelial function, LDL, HDL, and cholesterol	Reverted endothelial dysfunction and ↑ fecal cholesterol excretion	Caselato-Sousa, Ozaki, De Almeida, and Amaya-Farfan (2014)
Leaves	Methanol extract 10 mL (80%), 1 g of leaves in 1% acetic acid	*In vitro*	Antioxidant activity	DPPH and ABTS	Antioxidant capacity for DPPH of 34.38 µg/g, and for ABTS 68.89 µg/g.	Sarkerand Oba (2020)

(*Continued*)

Table 16.3 (Continued)

	Amaranth (*Amaranthus spp.*)					
Samples	**Experimental conditions**	**Type of study (*In vitro*, *in vivo*, or clinical trial)**	**Aim of the study**	**Biomarkers**	**Effects/results**	**Reference**
Flour	Grains were soaked (7 hr), germinated (20 hr, 32 °C, 90% humidity), dried (45 °C, 18 hr), and milled	*In vitro*	Antioxidant activity	DPPH	↑ DPPH activity in germinated seeds	Chauhan, Saxena, and Singh (2015)
Flour	Tempering grains with water for 24 hr before milling	*In vitro*	Antioxidant activity	DPPH and ABTS	16.35 g GAE/kg in DPPH, and 15.3 g GAE/kg in ABTS	Preetham Kumar, Dharmaraj, Sakhare, and Inamdar (2016)
Hydrolyzed protein and peptides	Hexane defatted flour, protein extraction pH 11, and pI 5.7	*In vitro*	Inhibition of HMG-CoA reductase	HMG-CoA reductase	Peptides and hydrolyzed protein inhibited HMG-CoA	Soares et al.(2015)
Peptide fractions	Amaranth protein concentrate hydrolyzed individually and sequentially for 4 hr with Alcalase and Flavourzyme, respectively	*In vitro*	ACE inhibition activity	ACE	IC_{50} 0.158 mg/L	Ayala-Niño et al.(2019)
Peptide fractions	Flour defatted w/hexane (10% w/v) for 24 hr, suspended deionized water (10%) and pH 9 adjusted with NaOH	*In vitro*	Antioxidant activity	ABTS	IC_{50} 1.375 mg/L	Ayala-Niño et al.(2019)
Extract	200 g seeds, macerated (48 hr w/250 mL 70% ethanol). Insulin release batches of 40 or 50 isolated rat islets, perfused w/3.3 mM glucose in KRB for 20 min	*Ex vivo*	Antidiabetic activity	Insulin secretion	↑ Insulin secretion and improved the glucose tolerance	Zambrana et al.(2018)

Extract	200 g seeds, macerated (48 hr w/250 mL 70% ethanol). Daily dose of 2,000, 1,000, 500 mg/kg b.w. was administrated orally for 10–12 hr	*In vivo* Goto-Kakizaki (GK) and Wistar (W) rats	Antidiabetic activity	Bloodstream glucose and HbA1c	↓ HbA1c in GK (19.8%) and in W (10.7%)	Zambrana et al.(2018)
Amaranth's 2-caffeoylisocitric acid and caffeic acid	Synthetic caffeic acid and chlorogenic acid. Macrophages cultured using RPMI 1640 medium w/10% fetal calf serum, 5% CO_2, and 37 °C	*In vitro*	Anti-inflammatory activity	Cell line RAW 264.7 macrophages iNOS, TNF-α, and IL-6	↓ iNOS, TNF-α, and IL-6, by decreasing the translocation of the nuclear factor kB/Rel-like protein 65 into the nucleus	Schröter et al.(2019)
Peptides	Defatted w/hexane and suspended w/distilled water at 4 °C for 1 hr. Pellet suspended NaCl, KH_2PO_4, and EDTA, pH 7.5 to obtain peptides. Concentration of 500 µM substrate, pH 8 for 1 hr at 37 °C	*In vitro*	Antidiabetic activity	DPP-IV	DPP-IV inhibition w/an IC_{50} of 1.1 mg/mL	Velarde- Salcedo et al. (2013)
Peptide fractions	Hydrolysis w/pepsin ratio 1 : 2, pH 2, 37° C, 3 hr. Pancreatin ratio 1 : 2, pH 7.5, 37 °C, 90 min	*In vitro*	Antioxidant activity	ORAC	3–10 kDa fraction with higher ORAC activity of 33.6 mmol TE/100 g	Sandoval-Sicairos et al. (2021)

(*Continued*)

Table 16.3 (Continued)

	Amaranth (*Amaranthus spp.*)					
Samples	**Experimental conditions**	**Type of study (*In vitro*, *in vivo*, or clinical trial)**	**Aim of the study**	**Biomarkers**	**Effects/results**	**Reference**
Peptide fractions	Hydrolysis w/pepsin ratio 1:2, pH 2, 37° C, 3 hr. Pancreatin ratio 1:2, pH 7.5, 37 °C, 90 min. Concentration of 0.25 to 1 mg/mL was used	*In vitro*	Anti-inflammatory activity	Cell RAW 264.7 macrophage iNOS	> 10 and 3–10 kDa fractions with higher NO reduction of 65.2 and 66.6%, respectively	Sandoval-Sicairos et al. (2021)
Isolates and fractions	Defatted flour treated 3X with water:buffer A(0.0325 mol/L K_2HPO_4, 0.0026 mol/L KH_2PO_4, pH 7.5, 0.4 mol/L NaCl), water and NaOH. Hydrophobic fraction treated w/dimethyl sulfoxide: H2O. Synthetic lectins	*In vitro*	Antitumor activity	UMR106 rat osteosarcoma-derived cells	High activity of cell proliferation inhibitors w/ IC_{50} of 0.1 mg/mL. Exhibited cytotoxic effect followed by apoptosis	Quiroga, Barrio, and Añón (2014)
Extracts of *A. lividus* and *A. hybridus* seeds	Seeds dried for 15 days and ground, mixed w/100% methanol, and shaking (24 hr, 37 °C). Concentration of 25, 50, and 100 μg/mL	*In vitro* and *in vivo* albino mice w/EAC	Antioxidant and anticancer activity	DPPH and cell proliferation	DPPH IC_{50} 28.0 μg/mL and 45% growth inhibition of EAC cells at 100 μg/mL concentration	Al-Mamun et al.(2016)
Extract of *Amaranthus virdis* steam and leaves	Air-dried and pulverized. 50 g of powder in 500 mL, of 50% ethanol, cold maceration for 24 hr. Concentration of 10 μg/mL	*In vitro*	Antiproliferative activity	MTT in three cell lines (Jurkat, CEM, and HL-60)	Leaves have high antiproliferative effect	Larbie et al.(2015)

Amaranthus dubius, *A. spinosous*, *A. tricolor*, and *A. viridis* extracts	Shade-dried and ground. Soxhlet extraction w/100% methanol, dissolved in DMSO at 25 mg/mL. Assays of different concentrations from 0 to 100 µg/mL	*In vitro*	Anticancer activity	MCF7 cells	Higher activity in *A. dubius* extract IC_{50} of 48.22 µg/mL *A. spinosous*, *A. tricolor* and *A. viridis* had IC_{50} of 65, 96, and 72 µg/mL, respectively	House, Puthenparampil, Malayil, and Narayanankutty (2020)
Peptides	Heat-denatured amaranth protein. Pepsin (15 min) and pancreatin (5 min) hydrolysis and incubation for 3 min	*In vitro*	Anticancer activity	MDA-MB231 cells	Inhibited cell growth IC_{50} 48.3 µg/mL	Taniya et al.(2020)
Leaf's extract	Freeze-dried plant powder (0.5 g) diluted in 5 mL w/70% ethanol. Concentration for cells 1.4 and 37 µg/mL, for antioxidant 0.31–10 mg/mL	*In vitro*	Antioxidant activity	Cell line HepG2 AFB1 DNA damage	↓ 57 and 81% of AFB1 and DNA damage, respectively	Odongo et al.(2018)
Peptides	Hydrolysis (2 hr), after 1 and 2 hr incubation w/ pepsin and pancreatin.	*In vitro*	Antidiabetic and hypotensive activity	α-Amylase and DPP-IV	DPP-IV IC_{50} of 0.19 µg/mL (KD120, F < 5 KDa), α-amylase IC_{50} of 0.84 µg/mL (KD60, F < 5 KDa), and ACE 39.0 µg/mL (KG120)	Vilcacundo, Martínez-Villaluenga, Miralles, and Hernández-Ledesma (2019)

(*Continued*)

Table 16.3 (Continued)

Samples	Experimental conditions	Type of study (*In vitro*, *in vivo*, or clinical trial)	Aim of the study	Biomarkers	Effects/results	Reference
	Amaranth (*Amaranthus spp.*)					
			Buckwheat (*Fagopyrum* spp.)			
Phenolic compounds	Baked buckwheat flour fermented with lactic acid bacteria and fungi	*In vitro* (bioaccessibility of antiaging compounds)	Antiaging activity	FRAP	Positive correlation ($r = 0.5$) between the antiaging bioaccessibility indexes, antioxidant/reducing bioaccessibility indexes, and total phenolic compounds bioaccessibility	Zieliński, Szawara-Nowak, and Wronkowska (2020)
Phenolic compounds	Methanolic (80%) extract of fermented baked buckwheat flour	*In vitro* (formation of antiaging compounds)	Effect of liquid state fermentation of buckwheat on inhibitory activity against advanced glycation end products formation	Advanced glycation end products	Liquid state fermentation of buckwheat with selected lactic acid bacteria	Zieliński, Szawara-Nowak, Bączek, and Wronkowska (2019)
Enzymatic protein hydrolysate	Alkaline buckwheat protein isolate digested with pepsin and pancreatin	*In vitro* (antioxidant effect)	Antioxidant potential of the buckwheat protein digests was compared by assessing their capacity to scavenge radicals	$ABTS^{+\bullet}$ and $^{\bullet}OH$	Pancreatin digest demonstrated the strongest activity against both radicals; 456 Da and 362 Da fractions showed the highest $ABTS^{+\bullet}$ scavenging activity and were 23–27% superior to mixed buckwheat protein digest; 362 Da fraction was most effective in neutralizing $^{\bullet}OH$ (86%)	Ma et al. (2010)

Not specified products of intake buckwheat	NM	*In vivo* and human trials	Evaluating the impact of buckwheat consumption on cardiovascular risk markers	LDL-cholesterol	24% ↓ concentrations of blood glucose and total cholesterol	Li, Lietz, and Seal (2018)
Peptide from Tartary buckwheat	Water-soluble peptide obtained using a sequential purification with three anion exchange chromatographic gels	*In vitro* (Bcap37 human breast cancer cells)	Antiproliferative effects of peptide on Bcap37 cells	Bcl-2 and Fas	Changes in typical apoptotic morphological, to a round shape, cell shrinkage, the absence of obvious microvilli, plasma membrane blebbing, and formation of apoptotic bodies; peptide treatment arrests G0/G1 cell cycle and prevents cells growth from G0/G1 phase to S phase; expression of Bcl-2 and Fas	Guo, Zhu, Zhang, and Yao (2010)
EBE	Ethanolic (70%) extraction with sonication	*In vivo* (ninety-six 3-month-old male Kunming mice)	Anti-fatigue properties of EBE in an animal model	BLA, BUN, GPx, and SOD	Tartary buckwheat extracts had anti-fatigue properties; ↓ the increase of BLA, ↓ the BUN, ↑ tissue glycogen content, and ↑ the activities of SOD and GPxin mice	Jin and Wei (2011)

(Continued)

Table 16.3 (Continued)

Samples	Amaranth (*Amaranthus spp.*)					
	Experimental conditions	Type of study (*In vitro*, *in vivo*, or clinical trial)	Aim of the study	Biomarkers	Effects/results	Reference
BWPS	Water crude extraction followed by ethanolic (95%) extraction	*In vitro* (THP-1 cells)	Antileukemic activity of BWPS through direct cytotoxic and/or indirect immune-potentiating effects	CD11b and CD14 antigens (indirect induction)	BWPS induce differentiation of human leukemic THP-1 cells; ↑ stimulated cytokine secretion (differentiation inducer); induced THP-1 cell differentiation, and maturity dose-dependently; potential for differentiation therapy in leukemia	Wu and Lee (2011)
Buckwheat seed oil (oleic 33.2%, linoleic 31.9%, and palmitic 13.2% acids) and rutin in a seed ethyl acetate extract	Soxhlet hexane extraction 8 hr	*In vitro* (neuroprotective inhibition of the enzymes: AChE and BChE along with TYRO)	Ethyl acetate extract of the seed, stem, and aerial parts of the plant for their neuroprotective effect through AChE and BChE and TYRO inhibitory and antioxidant activity	$SO^{\bullet}$, DPPH, DMPD, ferric radicals, and $H_2O_2^{\bullet}$; AChE, BChE, and TYRO inhibition linked to neuroprotection	Ethyl acetate seed extract ↓ AChE, BChE, and TYRO; ↓ glutamate release; ↑ scavenging activity against $SO^{\bullet}$, DPPH, DMPD, ferric radicals, and $H_2O_2^{\bullet}$	Gulpinar et al.(2012)
PBF and BPE	PBF obtained by milling and screening seeds (100-mesh screen); BPE obtained by isoelectric precipitation at pH 8.0 and then pH 4.5	*In vivo*, rats	Cardioprotective effects	Cholesterol in serum	Cholesterol in serum ↓ 31–33%; PBF ↓ liver cholesterol, hepatic activity Fas, and adipose tissue weight	Tomotake et al.(2006); Tomotake, Kayashita, and Kato (2015)
CEWP	Whole plant was exposed to increasing polarity solvents	*In vivo*, rats (100 mg/kg)	Anti-stress potential in rats (100 mg/kg) forced to swim	BUN, cholesterol, glucose, plasma cortisol, and triglyceride	Anti-stress potential; ↑ increased swimming time; ↓ lowering BUN, cholesterol, blood glucose, plasma cortisol, and triglyceride levels	Kothiyal and Ratan (2011)

Lectins	Acetate buffer (pH 5) extraction followed by chromatographic purification	*In vitro*, U937 cells (human leukemia cell line), and ex vivo (rabbit and human blood)	Antiproliferation activity	ND	Lectins induced apoptosis at 12.5–100 μg/mL; lectins showed hem-agglutination of both human and rabbit erythrocytes, remarkable preference for blood "type O" in human blood	Bai, Ji, et al. (2015)
Tartary buckwheat lectin	Ethanol (95%) extracted defatted flour	Ex vivo, PBDC	Anti-inflammatory activity	(rh)TNF-α	↑ Pro-inflammatory cytokines (IL-10 and IL-12) in PBDC; ↑ maturation markers CD83 and CD40	Bai, Feng, et al. (2015)
EEB	Ethanolic (95%) extraction by incubation (24 hr, 60 °C, 90 rpm)	*In vitro*, RBL-2H3 cells (rat basophilic leukemia cell line)	Anti-inflammatory activity	FcεRI-mediated activation	↓ Antigen DNP-HSA-induced release of histamine, PGD2, and cysteinyl leukotriene in IgE-sensitized, as well as inhibition in antigen-induced HDC2 and COX-2 and 5-LO mRNA expression in IgE-sensitized cells; suppression of antigen-induced FcεRIβ and FcεRI FCEϒ subunit mRNA expression, antigen-induced phosphorylation of Akt, cPLA2, LAT, Lyn, PI3K, PLCϒ1, and Syk	Kyung-Hwa (2012)

(*Continued*)

Table 16.3 (Continued)

Samples	Amaranth (*Amaranthus spp.*)					
	Experimental conditions	Type of study (*In vitro*, *in vivo*, or clinical trial)	Aim of the study	Biomarkers	Effects/results	Reference
BPS	Water extraction of seeds (10 : 1, v/w) followed by ethanolic (95%) extraction	*In vitro*, THP-1 cells (human leukemic MNC/macrophage like cells)	Anti-inflammatory activity	$SO^{\bullet}$	BPS induced differentiation of THP-1 cells maturity; ↑ exerted phagocytic activity and $SO^{\bullet}$ production in mature cells	Wu and Lee (2011)
MEBBP) (0.76% rutin and 0.001% quercetin)	Ethanol (80%) extraction followed by PBS extraction	*In vivo*, Balb/c mice	Antidiabetic activity	Blood sucrose and glucose	↓ Sucrose *in vitro* activity; ↓ blood glucose; ↓ decreased blood glucose not related to rutin's effect	Hosaka et al.(2011)
Ethanol:water (1 : 1, v/v) extract of KBC	NM	*In vitro*, H4IIE cells (rat hepatoma cell line),	Antidiabetic activity	p42/44 ERK, p70 (S6K) on Thr (421), p38 MAPK	KBC may affect hepatic glucose metabolism (inhibiting glucose flux and uptake); ↑ phosphorylation of p42/44 extracellular-related kinase (p42/44 ERK); ↓ p70 (S6K), on Thr (421), p38 MAPK	Al-Snafi (2017); Curran et al.(2010)
Flavonoids from buckwheat leaves and flowers (FBLF)	Water extraction and drying followed by ethanolic solubilization and filtration	*In vivo*, T2DM diabetic rats	Antidiabetic activity	PTP1B, blood glucose	↓ Fasting blood glucose; ↑ insulin resistance; ↓ on the expression of PTP1B with effect on creatinine clearance; ↓ renal damage;	Chu, Wang, and Han (2011)
Rutin (100 µM) and quercetin (50 µM), from methanolic extracts of common and Tartary buckwheat flours	Methanolic (80%) extraction	*In vitro*, HepG2 cells (human hepatocellular carcinoma cell line)	Antigenotoxic activity	ND	↑ Antioxidant capacity; ↓ 51–67% DNA-damage	Vogrinčič, Kreft, Filipič, and Zegura (2013); Al-Snafi (2017)

Buckwheat fermented with *Lactobacillus delbrueckii* subsp. *lactis* powder	Seeds subjected to hydrothermal treatment (100 °C, 10 min)	*In vitro*	Antioxidant activity	$DPPH^{\bullet}$ and ferric radical	Fermentation ↑ TPC; ↑ twofold $DPPH^{\bullet}$ inhibitory activity; ↑ 1.eightfold ferric reducing antioxidant power;	Gandhi and Dey (2012)
Crude methanolic extract of common buckwheat	Seeds powdered and exposed to methanol	*In vitro*	Antioxidant activity	$DPPH^{\bullet}$	↑ 81.6% $DPPH^{\bullet}$ inhibitory activity	Abbasi, Janjua, Rehman, William, and Khan (2015)
BSMJ and BSNMJ	Methanolic extraction	*In vitro*, 3T3-L1 cells	*in vitro* antioxidant, anti-adiponectin, and antioxidative stress activity	TPRZ-Fe (III) complex, ORAC, GPx and Cu/Zn SOD genes, G6PDH and NOX4 markers	BSMJ induced ↑ 42.9% TPTZ-Fe (II); ORAC values were BSMJ 103.3 μM TE/g and BSNMJ 60.6 μM TE/g; BSMJ downregulated GPx and Cu/Zn SOD genes expression in 3T3-L1 cells; ↓ ROS production	Lee et al. (2013)
			Chia (*Salvia hispanica*)			
Seeds	Whole seeds	Human trial	Cardioprotective effects	Platelet's aggregation and lower systolic blood pressure	37 g of chia seed/day ↓ myocardial infarction risk; ↓ platelets aggregation; ↓ systolic blood pressure by up to 6 mmHg	Vuksan et al. (2010)

(*Continued*)

Table 16.3 (Continued)

Samples	Amaranth (*Amaranthus spp.*)					
	Experimental conditions	Type of study (*In vitro*, *in vivo*, or clinical trial)	Aim of the study	Biomarkers	Effects/results	Reference
Peptides from enzyme hydrolysates	Sequential hydrolysis with Alcalase® and Flavourzyme®	Human trial	Hypotensive effect	ACE-1	↓ ACE-I 58.5–69.3%	Salazar-Vega, Segura-Campos, Chel-Guerrero, and Betancur-Ancona (2012)
Seeds (omega-3 fatty acids, higher iron, and fiber content gluten-free)	Whole seeds	Human trial	Cardioprotective and antidiabetic effects	Blood glucose	37 g of chia seed/day stabilized blood glucose level in diabetic patients	Vuksan et al. (2010)
Chia peptide hydrolysates (1–6 and > 10 kDa)		*In vitro*	Antioxidant and hypotensive effect	ABTS, ACE-1	↓ ACE-1 activity (1–6 kDa 69% and > 10 kDa 54%); 1–6 kDa ↓ TEAC (7.3 mmol/L-mg protein),	Salazar-Vega, Rosado-Rubio, Chel-Guerrero, Betancur-Ancona, and Castellanos-Ruelas (2009); Segura-Campos, Peralta González, Guerrero, and Betancur Ancona (2013)
α-Linolenic acid from chia seeds	Mucilage extracted at different seed: distilled water ratios (1 : 20,40), pH (4–8), and temperature (20–80 °C) conditions	*In vivo*	Cardioprotective activity	Thromboxanes and leukotrienes	Blocking sodium and calcium channels dysfunctions, avoiding possible hypertension and protecting against arrhythmia	Muñoz, Cobos, Diaz, and Aguilera (2012)
EChE	De-oiled seed sample exposed to ethanol (50%)	*In vitro*, THB-1 macrophages	Anti-inflammatory and *in vitro* antioxidant activity	IL-10 and TNF-α; ABTS and FRAP	EChE did not show ↓ in IL-10 expression before nor after exposure to LPS; ↑ TNF-α production; ↑ ABTS (2.4–2.sevenfold) and FRAP (5.1–6.fourfold) scavenging activity	Alemán, Pérez-García, Fernández de Palencia, Montero, and Gómez-Guillén, (2021)

Seeds	Whole seeds	Human trial (hypertensive individuals)	Antihypertensive activity	Blood pressure	35 g of chia seed flour/day for 12 weeks ↓ blood pressure by 5.5 mmHg	Toscano et al. (2014)
Chia seed oil	Cold-pressed filtered oil	*In vivo*, female Wistar rats	Cardioprotective effects	Creatine kinase, creatine kinase-MB, and aspartate antitransferase	Ameliorates doxorubicin-induced cytotoxicity	Ahmed et al. (2021)
Crude extract of chia seeds (high diversity and moderate concentration of alkaloids, cardio-glycosides, flavonoids, phenols, and tannins)	Ethanol (95%) extract	*In vitro*, PC-3 (prostate cancer cell line)	Anticancer and chemoprotective perspectives	ND	Dose-dependent inhibitory effect	Dugganaboyana, Palanisamy, Kumar, Muthusami, and Velliyur (2016)
Chia seed oil	Petroleum ether Soxhlet extraction (6 hr)	*In vivo*	Cardioprotective potential	Serum LDL, lipids, and triglycerides	↓ Body weight and organ index, reduced serum lipid levels of total cholesterol, triglyceride, and low-density lipoprotein cholesterol; ↑dismutase and glutathione peroxidase	Han et al. (2020)

(*Continued*)

Table 16.3 (Continued)

Samples	Experimental conditions	Type of study (*In vitro*, *in vivo*, or clinical trial)	Aim of the study	Biomarkers	Effects/results	Reference
	Amaranth (*Amaranthus spp.*)					
DTP and DPF from chia seeds	DTP mucilage-free and fat-free chia flour adjusted to pH 8; DPF above sample and extracted with isopropanol (70%)	*In vitro*, human monocytes	Anti-inflammatory activity	TNF-α, MCP-1, IL-6, and NF-κB	DPF ↓ the secretion of TNF-α (24.1%), MCP-1 (18.9%), IL-6 (39.6%), and IL-10 (68.7%); ↓ NF-κB translocation to nuclei	Grancieri et al.(2019)
Chia fraction rich in FAFFEFFELLFAFFT peptide	Chia powder exposed to ethanol (50%) 10 min, re-extracted with same solvent	*In vitro* and ex vivo	Antidiabetic effect	Biomarkers related to diabetes	↓ iCAM expression (81%), LOX-1 (36%), iNOS (42%), and NF-κB (71%); ↓ secretion of TNF-α (43%), MCP-1 (47%), and IL-6 (51%), as well as nitric oxide (43%), lipid accumulation (42%), prostaglandins (42%). ↓ glutamic-oxalacetic transaminase, alanine aminotransferase or on associated ones, like (liver fibrolisis and HSP72 upregulation)	Creus, Ferreira, Oliva, and Lombardo (2016); (Creus, Benmelej, Villafañe, and Lombardo, 2017); da Silva et al.(2018); Poudyal, Panchal, Waanders, Ward, and Brown (2012); (Poudyal, Panchal, Ward, Waanders, and Brown, 2012); Poudyal, Panchal, Ward, and Brown (2013); Marineli, Lenquiste, Moraes, and Maróstica (2015); Moura, Lollo, Morato, and Amaya-Farfan (2018)
Fiber flour hydrolyzed extracts	Hexane defatted chia seed cake and ethanol extracted for 48 hr	*In vitro*	Antioxidant capacity	ORAC and FRAP	↑ ORAC (142 μmol Trolox eq./g) and FRAP (> 80 μmol Trolox eq./g); stabilize auto-oxidation of polyunsaturated fatty acids	Azeem, Nadeem, and Ahmad (2015)

Chia seed fatty acid extract	Blended seed extracted with hexane by cold percolation	Ex vivo, human adipocytes	Anti-inflammatory activity	UCP-1, PPARγC1α, PRDM16, and SREBP-1c ARNm	↑ Expression of adipocyte mitochondrial thermogenesis-related genes, such as UCP-1, PPARγC1α, and PRDM16; ↓ expression SREBP-1c	Pandurangan, Al-Maiman, Al-Harbi, and Alshatwi (2020)
Topical formulation containing 4% chia seed oils	NM	Human trials, healthy and diabetic volunteers with xerotic pruritus symptoms	Skin protective effects	Skin pH, hydration; trans-epidermal water loss	↑ Skin hydration, lichen simplex chronicus, prurigo nodularis, and epidermal permeability barrier function	Jeong, Park, Park, and Kim (2010)
Chia seed-rich diet	Chia-rich diet (36.2%, w/w) for 21 days	*In vivo*, Wistar albino rats exposed to aluminum	Hepatorenal protective effect	Alanine aminotransferase, alkaline phosphatase, aspartate aminotransferase, lactate dehydrogenase, creatinine, urea nitrogen, and vitamin B_{12}	Aluminum exposure had no influence on renal function; liver function was disturbed and chia was of no use against it; ↑ concentrations of vitamin B_{12}	Demir and Bilgic (2019)

(*Continued*)

Table 16.3 (Continued)

	Amaranth (*Amaranthus spp.*)					
Samples	**Experimental conditions**	**Type of study (*In vitro*, *in vivo*, or clinical trial)**	**Aim of the study**	**Biomarkers**	**Effects/results**	**Reference**
			Quinoa (*Chenopodium quinoa*)			
Quinoa flour	Commercial quinoa flour incubated w/*Lc. mesenteroides* subsp. *mesenteroides* CRL 2131, *L. plantarum* CRL 1964 and CRL 2107, *Lactobacillus rhamnosus* CRL 1963, CRL 1984, and CRL 1983	*In vivo* BALB/c mice	Prevent vitamins and minerals deficiency	Levels of B_2 and B_9 in blood	↑ B_2, P. Ca^{+2}, Fe^{+2}, Mg^{+2}. ↑ B_9 719 ng/mL, EGRAC value lower 1.3 (no deficiency w/ supplementation)	Carrizo, de Moreno de LeBlanc, LeBlanc, and Rollán (2020)
Food formulated from quinoa	Flour/water slurry (30% solids) cooked (60 °C, 2 hr). 200 g/day	Clinical trials children's	Prevent malnutrition	IGF-1	↑ Plasma levels of IGF-1 Reduced childhood malnutrition	Ruales, de Grijalva, Lopez-Jaramillo, and Nair (2002)
Seeds	50 g seeds/day for 6 weeks	Clinical trials Celiac disease	Prevent gastrointestinal dysfunction	Gastrointestinal parameters, villus height, surface-enterocyte height, and intra-epithelial lymphocytes	Improved ratio of villus height to crypt depth to normal levels, surface-enterocyte cell height from 28.8 to 29.8 μm and intra-epithelial lymphocytes per 100 enterocytes ↓ from 30.3 to 29.7	Zevallos, Ellis, Suligoj, Herencia, and Ciclitira (2012)
Quinoa bread	Enriched bread w/20% of quinoa flour. 160 g/day for 4 week	Clinical trials Cardiovascular disease	Reduced the risk of developing cardiovascular disease	Triglycerides, cholesterol, LDL, and blood glucose	↓ LDL and blood glucose	Li, Lietz, Bal, et al. (2018)

Quinoa flakes	25 g/day for 4 weeks	Clinical trials Postmenopausal women	Reduced metabolic parameters	Triglycerides, cholesterol, LDL, TBARS, and GSH	↓ 3.9% in Tg, 5.2% in CT, 5.5% in TBARS, 6.2% LDL, and ↑ GSH by 0.1 μmol/L	De Carvalho et al. (2014)
Red quinoa seeds	Ground flour. 125 g/day quinoa protein/kg and 40 g casein/kg for 8 weeks	*In vivo* obese-diabetic (db/db) mice	Reduced or prevented hepatic steatosis	Triglycerides, cholesterol, VLDL, and insulin	↓ Tg, CT, VLDL, and ↑ insulin	Noratto, Murphy, and Chew (2019)
Peptide fractions	Gastric digestion (2 hr), after 1 and 2 hr incubation w/pancreatin. Concentration of 50 μL	*In vitro*	Antidiabetic activity	α-Amylase, α-glucosidase, and DPP-IV	↑ DPP-IV inhibition (ID60), α-amylase (ID120), and α-glucosidase (ID60 F < 5 KDa)	Vilcacundo et al. (2017)
Hydrolysates and peptide fractions from quinoa yogurt	Germinated (0, 24, and 72 hr), concentration 6 log CFU/mL, and fermented (8 hr, 42 °C). Inoculated w/*Lactobacillus casei* SY13 and Zhang. Concentration of 10 mg/mL	*In vitro*	Antidiabetic activity	α-Glucosidase	↑ α-glucosidase w/QLCSY13 peptide	Obaroakpo et al. (2019)
Peptides	Gastric digestion (2 hr), after 1 and 2 hr incubation w/pancreatin. Concentration from 4 to 0.0 mg/mL for 24 hr	*In vitro* Caco-2, HT-29, and HCT-116 cells	Anticancer activity	MTT	81.2% inhibitory activity w/4 mg protein/ml concentration	Vilcacundo et al. (2018)
Seeds	20 g/ day of processed quinoa w/80 mL of water ready for consumption, for 28 days	Clinical trials Adults	Antidiabetic activity	BMI, HbA1c, and basal glycemia	↓ BMI by 0.2 kg/m^2, HbA1c by 0.3%, and ↑ basal glycemia by 1.63 mg/dL	Abellán-Ruiz et al. (2017)

(*Continued*)

Table 16.3 (Continued)

Samples	Experimental conditions	Type of study (*In vitro*, *in vivo*, or clinical trial)	Aim of the study	Biomarkers	Effects/results	Reference
	Amaranth (*Amaranthus spp.*)					
Quinoa polysaccharide	Precipitate washed with ethanol, acetone, and ether, dissolved in hot water. Concentration of 20 μL for α-amylase, and 150 μL for α-glucosidase	*In vitro*	Antidiabetic activity	α-Amylase, α-glucosidase, and glycemic index (GI)	α-Glucosidase IC_{50} of 48.7 mg/ml and α-amylase IC_{50} of 61.4 mg/mL, GI of 61.7, w/90% EtOH	Tan et al. (2020)
Quinoa polysaccharide	Precipitate washed with ethanol, acetone, and ether, dissolved in hot water. Concentrations 0–20 mg/mL	*In vitro*	Antioxidant activity	ABTS and DPPH	↑ Scavenging activity w/90% EtOH	Tan et al. (2020)
Polysaccharide fraction	Petroleum ether extraction (60 °C, 2 hr, 3X) and ultrasonic-assisted extraction. Concentration of 2 mL	*In vitro*	Antioxidant activity	ABTS and DPPH	ABTS and DPPH IC_{50} of 1.9 and 1.1 mg/mL, respectively	Hu et al. (2017)
Hydrolysates	Defatted w/hexane (1 : 10 g/mL), dialysis 3.5 kDa membrane for 48 g at 4 °C. Concentration of 50 μL	*In vivo*	Hypotensive activity	ACE-1	$IC_{50} = 38.2\,\mu M$	Zheng et al. (2019)
Seeds	Powder through 100-mesh sieve. Concentration 0, 5, 12.5, 25, and 50 μg mL^{-1}	*In vitro* HEPG2 cell line	Anticancer activity	Cytotoxicity	Cytotoxicity against HEPG2 w/$IC_{50} = 14.6\,\mu g$	Mohamed, Fouda, and Mohamed (2019)
Seeds	Powder through 100-mesh sieve. 25 g/day quinoa for 4 weeks	*In vivo* Sprague–Dawley rats	Hepatoprotection activity	AST, ALAT, triglycerides, cholesterol, and LDL	↓ in Tg 31.2%, CT 27%, VLDL 18.7%, AST 33.2%, and ALT 46.7%	Mohamed et al. (2019)

Quinoa oleanane triterpenoid	Germinated at 0, 24, and 72 hr, concentration 6 log CFU/mL, and fermented (8 hr, 42 °C). Inoculated w/*Streptococcus thermophilus* and *Lactobacillus bulgaricus*. Concentration of 150 µL for DPPH and ABTS, and 200 µL for FRAP	*In vitro* HUVEC$_s$ line	Antioxidant activity	DPPH, ABTS, and FRAP antioxidant capacity	↑ DPPH 132.6% (QY172), ABTS 22.1 µg AAE/g, and FRAP 37.4 µmol Fe^{2+}/g (QY148)	Obaroakpo et al. (2020)
Quinoa oleanane triterpenoid	Germinated at 0, 24, and 72 hr, concentration 6 log CFU/mL, and fermented (8 hr, 42 °C). Inoculated w/*S. thermophilus* and *L. bulgaricus*. Concentration of 50 µL	*In vitro* HUVEC$_s$ line	Antidiabetic activity	α-Glucosidase and DPP-IV	α-Glucosidase IC_{50} of 7.4%, 6.8%, and 10.5% for QY100, QY124, and QY172. DPP-IV highest IC_{50} 12.9% for QY124	Obaroakpo et al. (2020)
Extract	Extraction of ethanol 50% v/v for 1 hr. Concentrations 1.9, 0.9, and 0.19 mg ChL DW/mL	*In vitro* AT-2 and MAT-LyLu cells	Anticancer activity	Cytotoxicity	Inhibitory effect on MAT-LyLu cells motility	Gawlik-Dziki et al. (2013)
Polysaccharide fraction	Petroleum ether extraction (60 °C, 2 hr, 3X) and ultrasonic-assisted extraction. Concentrations 12.5–200 µg/mL	*In vitro* SMMC 7721 and MCF-7 cells	Anticancer activity	Cytotoxicity and proliferation	CQP = IC_{50} values of 121.4 µg/mL (24 hr) and 53.4 µg/mL (48 hr) in SMMC 7721 cells and IC_{50} values of 83.5 µg/mL (24 hr) and 64.7 µg/mL (48 hr) in MCF-7 cells	Hu et al. (2017)

(*Continued*)

Table 16.3 (Continued)

Samples	Experimental conditions	Type of study (*In vitro*, *in vivo*, or clinical trial)	Aim of the study	Biomarkers	Effects/results	Reference
	Amaranth (*Amaranthus spp.*)					
Red quinoa seeds	Ground flour. 125 g/day quinoa protein/kg and 40 g casein/kg for 8 weeks	*In vivo* obese-diabetic (db/db) mice	Anti-inflammatory activity	TNF-α, IL-6, and ROS	↓ TNF-α, IL-6, and ROS	Noratto et al. (2019)
Polysaccharide fraction	Petroleum ether extraction (60 °C, 2 hr, 3X) and ultrasonic-assisted extraction. Concentrations of 12.5, 25, 50, 100, and 200 μg/mL	*In vitro*	Anti-inflammatory activity	LPS and NO synthesis	Significant rise in cell viability w/24 and 48 hr incubation ($p < 0.05$) in LPS, and ↓ NO synthesis. 34–53% inhibition rate of treatment	Hu et al. (2017)

ABTS$^{+\bullet}$, [2,2′-azinobis-(3-ethylbenzothiazoline-6-sulfonate)] radical; ACE, angiotensin-converting enzyme, DMSO, dimethyl sulfoxide; DPPH, 2,2-diphenyl-1-picrylhydrazyl; HMG-CoA, hydroxy methyl glutaryl co-enzyme A; KRB, Krebs buffer; MTT, 3-[4,5-dimethylthiazole-2-yl]-2,5-diphenyltetrazolium bromide; ORAC, oxygen radical absorption capacity; FRAP, ferric-reducing antioxidant power; $^{\bullet}$OH, hydroxyl radical; LDL, low-density lipoprotein; Bcl-2, protein from B-cell lymphoma-2; FAS, fatty acid synthase; BLA, blood lactic acid; BUN, blood urea nitrogen; GPx, glutathione peroxidase; SOD, superoxide dismutase; AChE, acetylcholinesterase; BChE, butyrylcholinesterase; TYRO, tyrosinase; SO$^{\bullet}$, superoxide radical; DMPD, *N,N*-dimethyl-ρ-phenylenediamine; H_2O_2, peroxide radical; PBF, protein buckwheat flour; BPE, buckwheat protein extract; EBE, ethanolic buckwheat extract; CEWP, crude extract of common buckwheat whole plant; BWPS, buckwheat polysaccharides; IL, interleukin, ND, not determined; NM, not mentioned; TNF-α, tumor necrosis factor-α; PBDC, peripheral blood dendritic cells; FcεRI, high-affinity IgE receptor; PGD2, prostaglandin D2; COX-2, cyclooxygenase-2; 5-LO, 5-lipoxygenase; Akt, protein kinase B; cPLA2, cytosolic phospholipase A2; LAT, linker for activation of T-cells; Lyn, tyrosine protein kinase Lyn; PI3K, phosphoinositide 3-kinase; PLC, phospholipase C; Syk, spleen tyrosine kinase; MEBBP, methanolic (80%) extract of buckwheat bran powder; KBC, Koto buckwheat concentrate; ERK, extracellular signal-regulated kinase; MAPK, mitogen-activated protein kinase; PTP1B, tyrosine-protein phosphatase non-receptor type 1; TPC, total phenolic content; BSMJ, buckwheat sprouts treated with methyl jasmonate; BSNMJ, buckwheat sprouts non treated with methyl jasmonate; G6PDH, glucose-6-phosphate dehydrogenase; NOX4, NADPH oxidase 4; MNC, monocyte; TEAC, Trolox equivalence antioxidant capacity; EChE; ethanolic chia extract; IL, interleukin; DTP, digested total protein; DPF, digested protein fractions; MCP-1, monocyte chemoattractant protein 1; NF-κB, nuclear factor kappa-light-chain-enhancer of activated B cells; iCAM, intracellular adhesion molecule 1; LOX-1, lectin-type oxidized LDL receptor 1; iNOS, nitric oxide synthases; HSP72, heat shock 72 kDa protein 1; UCP, uncoupling protein; PPARγC1α, peroxisome proliferator-activated receptor gamma; PRDM16, PR domain containing 16; SREBP-1c, sterol regulatory element-binding protein 1; AST, aspartate aminotransferase; ALT, alanine aminotransferase; EGRAC, erythtocyte glutathione reductase activation coefficient; HMG-CoA, hydroxy methyl glutaryl co-enzyme A; TBARS, thiobarbituric acid reactive substances; VLDL, very low-density lipoprotein.

follows: a) 20% amaranth protein isolate, b) 20% of casein +10% of amaranth protein isolate, and c) 20% of casein as control. Total cholesterol was reduced by 27, 20, and 48% ($p < 0.05$) in groups a, b, and c, respectively. The authors concluded that cholesterol excretion led to hypocholesterolemia, and that amaranth protein has an effect on endogenous cholesterol metabolism (Mendonça et al., 2009).

16.3.1.2 Buckwheat

Serum cholesterol levels decreased (31% and 33%, respectively) in male Sprague–Dawley rats fed with high protein buckwheat flour (PBF) or high protein extract (BPE) (20% protein, 20% sucrose, 2.9% cellulose, 3.5% mineral mixture, 1.0% vitamin mixture, and 3.0% DL-methionine) for 10 days (Tomotake et al., 2006; Tomotake et al., 2015). In addition, consumption of PBF significantly ($p < 0.05$) reduced liver cholesterol, whereas BPE had a minor effect. A diet including PBF also significantly ($p < 0.05$) inhibited hepatic activity of fatty acid synthase (FAS) and slightly decreased adipose tissue weight ($p > 0.05$) in rats fed cholesterol-free diets compared with rats fed buckwheat casein ($p < 0.05$) (Tomotake et al., 2006). Additionally, feeding mice PBF and BPE for 27 days reduced the incidence of cholesterol gallstones and the lithogenic index in mice fed cholesterol-enriched diets. Results also indicated increased fecal excretion of neutral and acidic steroids in individuals fed PBF and BPE compared to those fed casein (control diet) (Tomotake et al., 2006; Tomotake et al., 2015).

The human body during stress synthesizes glucocorticoid which may induce hypertension and other negative cardiovascular effects (Hunter & Bailey, 2015). A crude extract of common buckwheat whole plant exhibited anti-stress potential in rats (100 mg/kg) forced to swim in an endurance test (Kothiyal & Ratan, 2011). The extract significantly ($p < 0.01$) increased swimming time in rats. When compared to the control stress group, rats administered the buckwheat extract showed lower blood urea nitrogen (BUN), cholesterol, blood glucose, plasma cortisol, and triglyceride levels. Phenolics (50–200 mg/mL) from 80% methanolic extract of fermented buckwheat biscuit had high bioaccessibility and dose-response inhibition of anti-advanced glycation end products after gastrointestinal digestion (Zieliński et al., 2019; Zieliński et al., 2020). An ethanolic extract (70%) of Tartary buckwheat flour administered to ninety-six 3-month-old male Kunming mice at 60, 120, and 240 mg/kg body weight inhibited blood lactic acid, decreased levels of BUN, increased glycogen levels in tissues, glutathione peroxidase (GPx), and superoxide dismutase (SOD) during a swimming evaluation compared with the control group (mice not given the extract) (Jin & Wei, 2011). In human studies using random effects models, buckwheat intake decreased total cholesterol by 0.50 mmol/L (95% CI: −0.80, −0.20) and triglycerides by 0.25 mmol/L (95% CI: −0.49, −0.02), respectively. In 14 out of 20 animal studies, total cholesterol decreased (12–54%), along with 2–74% triglycerides reduction (Li, Lietz, & Seal, 2018).

Buckwheat products protect the development of cardiovascular disease by reducing circulating cardiovascular risk factors and oxidative stress markers according to a single-blind crossover trial with 8 weeks washout period (Dinu et al.,2017). In the study, 21 participants (11 F; 10 M; mean age 53 ± 13.4 years) at high cardiovascular disease risk consumed products (bread, pasta, biscuits, and crackers), made from either buckwheat-enriched semi-wholegrain wheat or control semi-wholegrain wheat. Intake of buckwheat products significantly ameliorated total cholesterol, low-density lipoprotein cholesterol, triglycerides, glucose, and insulin from baseline levels. Moreover, thiobarbituric acid reactive substances (TBARs) levels decreased significantly with concomitant increase in plasma oxygen radical absorption capacity (ORAC) levels (Dinu et al.,2017).

16.3.1.3 Chia

Chia proteins display therapeutic activity with antihypertensive effects that can replace conventional drugs. However, there are few reports in the literature on chia proteins and their antihypertensive or cardiovascular effects (Orona-Tamayo, Valverde, & Paredes-López, 2017). In human trials, dietary consumption of 37 g of chia seeds per day prevented myocardial infarction and strokes by inhibiting platelet aggregation and lowering systolic blood pressure (SBP) by up to 6 mmHg (Vuksan et al., 2010). The α-linolenic acid in chia seeds participates in the synthesis of thromboxanes and leukotrienes, which play a role in blocking sodium and calcium channel dysfunctions, preventing possible hypertension and protecting against arrhythmia (Felemban, Attar, & Zeid, 2021; Muñoz et al., 2012). Natural inhibitors of ACE have also been found in chia protein hydrolysates of 1–6 kDa (Salazar-Vega et al., 2009; Segura-Campos, Peralta González, Guerrero, & Betancur Ancona, 2013). Chia short peptides with cardioprotective effects have between 2 and 9 amino acids (Orona-Tamayo et al., 2017). Peptides of ~1 kDa showed 69% ACE inhibition, while the corresponding value for peptides >10 kDa was 54% (Segura-Campos, Peralta González, Guerrero, & Betancur Ancona, 2013). Protein hydrolysates obtained by digestion with proteases (pepsin, trypsin, and chymotrypsin) also exhibited ACE inhibitory activity, with IC_{50} values of 377–339 µg/mL (Orona-Tamayo et al., 2017). Some authors attributed the cardioprotective effects of chia peptides to the high proportion of hydrophobic amino acids, including isoleucine, leucine, phenylalanine, and proline (Alemán, Giménez, Peérez-Santin, Gómez-Guillén, & Montero, 2011; Segura-Campos, Peralta González, Guerrero, & Betancur Ancona, 2013). Dipeptides found in chia seeds, such as alanine-glutamine, have cardioprotective effects related to athletic endurance during high-intensity exercise (McCormack et al., 2015). In treated and untreated hypertensive subjects of both sexes (35 to 65 years of age), Toscano et al. (2014) reported that supplementation of 35 g of chia seed flour per day for 12 weeks reduced blood pressure readings by 5.5 mmHg, a larger decrease than in individuals previously treated with traditional medication and a placebo with medication, respectively. In another study, chia seed oil provided cardioprotection against doxorubicin-induced cytotoxicity in female Wistar rats (Ahmed et al., 2021). A diet with chia seed oil alleviated electrocardiogram changes in rats; it also protected cardiac tissues against doxorubicin-induced rise of serum creatine kinase, creatine kinase-MB, and aspartate anti-transferase levels. Chia seed oil administered to mice fed a high-fat diet was linked to decrease in body weight and organ index, as well as reductions in the serum lipid levels of total cholesterol, triglyceride, and low-density lipoprotein cholesterol (Han et al., 2020). This treatment also increased SOD and GPx activities and reduced malondialdehyde content in serum and liver. The histopathological results demonstrated that chia seed oil improved hepatic steatosis and reduced lipid deposition.

Seven clinical trials (n = 200) were selected in a systematic review (2005–2015) of chia seed (milled or whole) consumption on prevention/control of cardiovascular risk factors in humans (de Souza Ferreira, de Sousa Fomes, Santo da Silva, & Rosa, 2015). Chia seed intervention significantly lowered SBP and inflammatory markers without change in body mass, lipid profile, or blood sugar in only one study. Alpha linolenic acid and eicosapentaenoic acid (EPA) spiked significantly in four of the studies without changes in other parameters. Postprandial blood sugar decreased in two acute randomized trials, and triglycerides, body mass, and inflammatory markers decreased in only one study. Most of the studies, according to the review did not demonstrate statistically significant effects of chia seed consumption on cardiovascular disease risk factors (de Souza Ferreira et al., 2015 and references therein). However, chia seed consumption (40 g/day) for 12 weeks significantly lowered SBP, without changes in lipid profile, body weight, and inflammatory marker (high-sensitivity C-reactive protein) in adults ($n = 42$) with type 2 diabetes and hypertension in a recent randomized controlled trial (Alwosais, Al-Ozairi, Zafar, & Alkandari, 2021).

The estimation of the cardioprotective effect of these pseudocereals (whole, protein concentrate, protein isolate, and seeds fractions) using *in vitro* models can sometimes lead to contradictory and controversial results in human trials, since the ACE inhibitory effects are dependent on the physiological status of the patients, the period of supplementation, and the bioavailability of the bioactive peptides present in these sources (Gostner, Becker, Ueberall, & Fuchs, 2015; Orona-Tamayo et al., 2017).

16.3.1.4 Quinoa

Li, Lietz, Bal, et al. (2018) showed the effects of quinoa consumption on cardiovascular disease markers. Thirty-seven overweight (body mass index, BMI > 25 kg/m^2) men between 35 and 70 years of age were selected for the study and provided with a daily dose of 160 g of quinoa-enriched bread (20% quinoa flour) for 4 weeks. The proximate composition of the quinoa bread (20%) per 100 g was 381 Kcal, 14.04 g protein, 6.52 g fiber, 2.73 g fat, and 72.64 g carbohydrates. After 4 weeks of quinoa consumption, the blood glucose level decreased significantly ($p < 0.05$) by 4.5% and LDL by 5.7%. Thus, quinoa consumption (20 g/day) lowered LDL-cholesterol and improved the reduction of postprandial glycemic response (Li, Lietz, Bal, et al., 2018). Zheng et al. (2019) assessed the *in vivo* antihypertensive activity of peptides derived from quinoa bran albumin using an *in silico* approach. Three selected peptides, RGQVIYVL (946.6 Da), ASPKPSSA (743.8 Da), and QFLLAGR (803.5 Da) were tested in an *in vivo* model of hypertensive rats. Oral administration of the peptides significantly ($p < 0.05$) lowered both diastolic blood pressure (DBP) and SBP from the second week onward (Zheng et al., 2019).

Noratto et al. (2019) studied the effect of daily quinoa flour intake (125 g quinoa protein/kg + 40 g casein/kg) for 8 weeks from the standpoint of preventing hepatic steatosis in obese db/db mice. After 5-week treatment, food intake decreased in the quinoa group, and this was reflected in triglycerides (30.2%) and very low-density lipoprotein (VLDL)-c (29.4%) reduction in the bloodstream. The most beneficial effect of quinoa intake related to total cholesterol (23.4% reduction) was lower than in the obese group and similar to the lean group. Interestingly, the weight of the liver in the quinoa group (5 ± 0.3 mg/g of liver tissue) was lower than in the obese (14 ± 1 mg/g of liver tissue) and lean (6 ± 0.7 mg/g of liver tissue) groups ($p < 0.05$). The study demonstrated that quinoa intake improves hepatic steatosis and plasma lipids (Noratto et al., 2019). In an *in vivo* model, Mohamed, Fouda &Mohamed et al. (2019) found that quinoa seeds prevented against nonalcoholic fatty liver disease (NAFLD). NAFLD was induced in male rats after 4 weeks on a high fructose diet, and one group was treated with 25 g/day of quinoa seeds. The chemical composition of quinoa seeds (per 100 g) was 14.4 g protein, 7.1 g fat, 4 g crude fiber, 72.3 g carbohydrates, 36.4 mg GAE total phenolic content, and 1.03 g phytic acid. Rats fed quinoa seeds had lower levels of Tg (64.6 mg/dL), LDL (58.67 mg/dL), uric acid (1.49 mg/dL), creatinine (0.64 mg/dL), AST (49.4 U L), ALT (24.86 U L), and increased HDL (42.42 mg/dL). The NAFLD group that received no treatment had the following levels: Tg (93.83 mg/dL), LDL (72.17 mg/dL), uric acid (2.25 mg/L), creatinine (0.75 mg/dL), AST (73.91 U L), ALT (46.60 U L), and HDL (27.69 mg/dL). All the results showed significant differences ($p < 0.05$). Quinoa seeds exhibited protective effects against NAFLD, probably due to their high amounts of phenolic compounds, polyunsaturated fatty acids, and phytic acid (Mohamed et al., 2019).

Another study with quinoa flour (Carrizo et al., 2020) evaluated the effectiveness of quinoa pasta fermented by *L. plantarum* and thus producing higher amounts of vitamin B_2 and B_9. The *in vitro* model selected was BALB/c mice which were divided into six groups fed different diets for 42 days: 1) deficient group (DG) with depletion defined diet free of riboflavin and folate (DD), 2) control group (CG), with 7 mg riboflavin/kg and 2 mg folic acid/kg of food (DC), 3) DD + pasta pH control, 4) DD + pasta negative control, 5) DD + pasta with B_2 and B_9 control, and 6) DD + pasta bio-enriched in vitamins and minerals. Group 5 showed increased hemoglobin and hematocrit values of 14.14 g/dL and 47%, respectively, compared with the DG group, which showed an increase in hemoglobin of 12.6 g/dL and an increase in hematocrit of 39.9%. Additionally, histological evaluations showed that quinoa-enriched pasta increased small intestine villi length in the group supplemented with quinoa and B_2 and B_9, compared with the DG (Carrizo et al., 2020).

Quinoa consumption (50 g/day) for 12 weeks lowers serum triglycerides (an independent cardiovascular disease [CVD] risk marker) and prevalence of metabolic syndrome in overweight (42% >25 kg/m^2 BMI) and obese (30 to ≥40 kg/m^2 BMI) participants (average age 40 years; 71% F) according to a dose–response randomized, controlled, single-blind trial with a parallel design (Navarro-Perez, Radcliffe, Tierney, & Jois, 2017). The 36% decrease in serum triglycerides (comparable to pharmacological therapy such as fibrates 35% and statins 20%) was attributed to reduced dietary fat absorption due to the high bile acid-binding capacity of quinoa protein. However, quinoa intake had no effect on anthropometric measures of body composition, circulating hormones, glucose, or total, HDL- or LDL-cholesterol.

16.3.2 Anticarcinogenic

16.3.2.1 Amaranth

To study the antitumor activity of amaranth, Quiroga et al. (2014) used protein isolate from *Amaranthus mantegazzianus* seeds and amaranth lectin purchased from Vector laboratories (Burlingame, CA, USA). After the flour was defatted, protein extraction was carried out at pH 9 followed by pH 5 precipitation. The hydrophobic fraction (MPI-h) was treated with DMSO: H_2O. Amaranth fractions were obtained from defatted flour by successive fractionation with water:buffer A (0.0325 mol/L K_2HPO_4, 0.0026 mol/L KH_2PO_4, pH 7.5, and 0.4 mol/L NaCl) and water:NaOH, to obtain albumins, globulins, globulin-p, and glutelins. For the *in vitro* assays, UMR106 rat osteosarcoma-derived cells were used for their sensitivity to antiproliferative effect. The results showed that *A. caudatus* and *mategazzianus* lectins had high cell proliferation inhibitory activity with IC_{50} values of 0.08 and 0.1 mg/mL, respectively (Quiroga et al., 2014). Another study on the antiproliferative effect of amaranth used *A. viridis* extracts (stem and leaves) on three different human leukemic cell lines: Jurkat, CEM, and HL-60. The leaf extracts showed a lower IC_{50} value than the stem extracts for antiproliferative activity with 111, 123, and > 1,000 μg/mL for Jurkat, HL-60, and CEM, respectively. The stem extracts had higher IC_{50} values of 655, 814, and > 1,000 μg/mL for Jurkat, HL-60, and CEM, respectively, indicating that these extracts had lower antiproliferative activity than amaranth leaf extracts (Larbie et al., 2015). The authors suggested that the antiproliferative effect of *Amaranthus viridis* leaf extracts could be due to their higher phenolic compounds concentrations (tannins) than in the stem extracts (Larbie et al., 2015).

In another *in vitro* study, four methanolic extracts from aerial parts of *Amaranthus dubius* (AD), *Amaranthus spinosus* (AS), *Amaranthus tricolor* (AT), and *A. viridis* (AV) were tested in human breast cancer cells MCF7 and MDAMB231 to assess their antiproliferative activity. The four extracts exhibited different levels of cytotoxicity to the two breast cancer cells: the AS extract showed the highest antiproliferative activity with an $IC_{50} = 82$ μg/mL; the AV and AD extracts showed moderate toxicity with $IC_{50} = 102$ and 107 μg/mL, respectively. Meanwhile, the least active extract was AT with an $IC_{50} = 214$ μg/mL. The high amounts of phenolic compounds and flavonoids such as caffeic acid, ferulic acid, quercetin and kaempferol glycosides, and rutinosides may be responsible for the antiproliferative effect (House et al., 2020).

16.3.2.2 Buckwheat

Tartary buckwheat protein fractions showed cytotoxic effects on the mammary cancer cell line Bcap37. More specifically, the protein called "TBWSP31" showed a time- and dose-dependent ($IC_{50} = 19.75$ μg/mL obtained at 72 hr) antitumor effect on Bcap37 cells, evaluated by the MTT assay (Xiao-na & Hui-yuan, 2007). In U937 cells (human leukemia cell line), lectins purified from Tartary buckwheat seeds induced apoptosis at 12.5, 25, 50, and 100 μg/mL (Zouhir, 2014). In rabbit blood, these lectins also showed hemagglutination in response to both human and rabbit erythrocytes, with a remarkable preference for "type O" human blood (Zouhir, 2014). Bcap37 cells treated with TBWSP31 protein showed typical apoptotic morphological changes (observed by inverted microscopy and scanning electron microscopy), including detachment from the culture plate, change to a round shape, cell shrinkage, the absence of obvious microvilli, plasma membrane blebbing, and formation of apoptotic bodies. Additionally, exposure to TBWSP31 protein interrupted the cell cycle at G0/G1 and prevented the cells from growing from the G0/G1 phase (first growth phase, that "starts" with a cell from a previous division) to the S phase (the period during which DNA duplication takes place), this effect was most pronounced at 48 hr (Guo et al., 2010).

16.3.2.3 Chia

A 95% ethanol chia seeds extract (1 : 5, v/w) exhibited a dose-dependent inhibitory effect (cytotoxic) on prostate cancer cell line PC-3 (Dugganaboyana et al., 2016). This activity was attributed to the high diversity and moderate concentration of alkaloids, cardiac-glycosides, flavonoids, phenols, and tannins in this extract. Thus, chia seed extract can be considered a food source with chemoprotective effects.

16.3.2.4 Quinoa

Reactive oxygen species (ROS) generation overwhelms cellular antioxidant capacity and oxidizes lipids, proteins, and nucleic acids and can lead to cell death or transformation. For this reason, interfering with ROS production can have chemopreventive effects. Gawlik-Dziki et al. (2013) assessed the chemopreventive potential of quinoa ethanolic extract. The results showed significant ($p < 0.05$) and dose-dependent cytostatic effects with concentrations between 0.186 and 1.86 mg of extract DW/mL. AT-2 cells were more sensitive and reacted at a lower concentration (0.186 mg DW/mL). On the other hand, higher concentrations were required to observe the same effect on MAT-LyLu cells (Gawlik-Dziki et al., 2013). Another study on the *in vitro* anticancer activity of quinoa was carried out by Hu et al. (2017). Quinoa polysaccharide fraction (CQP) was evaluated for its anticancer effect in normal liver (L02) cells, normal human breast epithelial (MCF 10A) cells, human liver cancer (SMMC 7721) cells, and human breast cancer (MCF-7) cells. A petroleum ether extraction was carried out for 2 hr at 60 °C three times, then an ultrasonic-assisted extraction was performed to obtain the CQP fraction. CQP significantly inhibited cell proliferation for SMMC 7721 and MCF-7 cells, with IC_{50} values of 121.4 and 83.48 μg/mL at 24 hr, respectively, and 53.4 μg/mL (SMMC 7721) and 64.67 μg/mL (MCF-7) after 48 hr. The authors concluded that no significant inhibition of the proliferation of normal cells was observed, indicating that CQP is biosafe (Hu et al., 2017).

Vilcacundo et al. (2018) investigated the *in vitro* chemopreventive properties of peptides released from quinoa during a simulated gastrointestinal digestion for 120 min, after 60 min and 120 min of pancreatin incubation. Digests were taken at different times, non-digested (Q0), gastric digestate (QG120), and gastroduodenal digestates (QD60 and QD120). Q0 showed no inhibitory effect, even at the highest concentration (4 mg protein/mL). Meanwhile, QG120 exhibited anticancer activity against Caco-2 cells in a dose-dependent manner (4 mg protein/mL), with maximum 81% inhibitory activity (Vilcacundo et al., 2018). The *in vitro* anticancer activity of quinoa seeds was also evaluated by Mohamed et al. (2019): the quinoa seeds were dried and grinded, then phenolic compounds were extracted with ethanol (80%). Human liver tumor-derived cell lines (HepG2) were used and quinoa flour was used at different concentrations (0.5, 12.5, 25, and 50 μg/mL). The cytotoxicity effect of quinoa seeds gave an IC_{50} value of 14.6 μg/mL compared to an IC_{50} of 3.6 μg/mL for the control doxorubicin (Mohamed et al., 2019).

16.3.3 Anti-inflammatory

16.3.3.1 Amaranth

Chronic diseases perpetuate a state of cellular inflammation that worsens those pathologies. Amaranth leaves contain high amounts of phenolic acids such as hydroxycinnamic acid and 2-caffeoylisocitric acid, which are well-known anti-inflammatory compounds (Schröter et al., 2019). Schröter et al. (2019) studied the anti-inflammatory effect of three different secondary metabolites (2-caffeoylisotric acid, chlorogenic acid, and caffeic acid) derived from amaranth, in lipopolysaccharides (LPS)-induced RAW 264.5 cell macrophages.The metabolites significantly ($p < 0.05$) reduced anti-inflammatory parameters (iNOS, TNF-α, and IL-6) (Schröter et al., 2019). Lipoxygenase inhibition is another type of anti-inflammatory activity. House et al. (2020) studied the anti-inflammatory activity of methanolic extracts of four amaranth species (AD, AS, AT, and AV). The IC_{50} values for lipoxygenase inhibition were 74, 72, 66, and 58 μg/mL for AV, AD, AT, and AS extracts, respectively, indicating that the AS extract had the highest anti-inflammatory activity. Recently, Sandoval-Sicairos et al. (2021) studied the anti-inflammatory effect of amaranth peptide fractions, after a 90-min hydrolysis with pancreatin, 3 hr of total hydrolysis. Subsequently, three fractions were produced as follows: F1 (> 10 kDa), F2 (3–10 kDa), and F3 (<3 kDa). F1 and F2 significantly ($p < 0.05$) reduced nitric oxide (NO) production 65 and 67%, respectively; there was no significant difference between F1 and F2. The decrease in NO production can be attributed to the capacity of peptides (*Amaranthus* spp.) formed during hydrolysis to hinder the expression of specific enzymes required for NO synthesis (Sandoval-Sicairos et al., 2021). De la Barca, Rojas-Martínez, Islas-Rubio, and Cabrera-Chávez (2010) assessed the use of gluten-free bread and cookies using raw and popped amaranth; amaranth cookies (raw and popped) exhibited less than 20 ppm gluten content, thus, being a suitable and safe option for celiac patients. Ballabio et al. (2011) evaluated the tolerance of 40 amaranth varieties in coeliac patients; all varieties had negligible cross-reactivity against IgA and IgG, suggesting that amaranth can be safely included in gluten-free foods, but amaranth varieties should be selected carefully due to the possible presence of gluten-like immunoreactivity proteins.

16.3.3.2 Buckwheat

In RBL-2H3 cells (rat basophilic leukemia cell line), an ethanolic (75%) extract of buckwheat mediated inflammatory activity and proximal signal events in FcεRI-mediated activation (Kyung-Hwa, 2012). In a case–control study, an ointment made from honeybee (*Apis mellifera*) and common buckwheat were applied as a dermal treatment on a large dermal wound in male rabbit resulted in its complete resorption of inflammatory exudate at the 4th day, with a wound contraction rate of 5.6%, which progressively reached 32%, 72%, and 99.5% at the 7th, 11th, and 26th days, respectively (Zouhir, 2014). Buckwheat polysaccharides induced differentiation of THP-1 cells (human leukemic monocyte [MNC]/macrophage-like cells) maturity following 3-day and 5-day treatments in a dose-dependent manner and exerted phagocytic activity and superoxide anion production in these mature cells (Wu & Lee, 2011). Buckwheat albumins and glutelins do not react against polyclonal antigliadin antibody (Rajnincova, Galova, Spalekova, Balazova, & Chnapek, 2019).

16.3.3.3 Chia

An ethanolic chia extract did not suppress IL-10 expression of THB-1 macrophages before or after exposure to LPS (Alemán et al., 2021). In the same study, however, chia extract downregulated TNF-α production in THP-1 macrophages exposed to LPS for 4 hr. The authors attributed this effect to the diversity of bioactive compounds found in the extract (salvianolic, rosmarinic, caftaric, and protocatechuic acids). A concentration of 100 μg/mL of silver nanoparticles functionalized with a methanolic extract of chia seeds showed >98% albumin denaturation, which is related to anti-inflammatory activity in *in vivo* trials (Rubavathi et al., 2020). In a monocyte culture (RAW 264.4 macrophages) treated with chia seed total protein and protein fraction digests, reduced NO excretion (65%) during the inflammatory process, with the glutelin-rich fraction also reducing the formation of

ROS (19.7%) and prostaglandins secretion (34.6%) (Grancieri et al., 2019). Chia seeds can benefit childhood obesity due to the anti-inflammatory effects of its fiber content according to results of a double-blind randomized clinical trial (da Silva et al., 2020). Ground chia seeds (25 g/day for 75 days) supplementation reduced fibrinogen levels (an early predictor for MetS individuals at risk of CVD development) and inflammatory biomarkers (TNF-α and NF-κB) that correlated with fiber in obese children (5–10 years old; 24 kg/m^2 average BMI).

16.3.3.4 Quinoa

After optimizing the extraction of polysaccharides from quinoa seeds, Hu et al. (2017) evaluated the immune regulatory effect and anti-inflammatory response of quinoa polysaccharides fractions (CQP). NO is well known for being a mediator and regulator of the inflammatory response. The *in vitro* measurement of NO release was performed with RAW264.7 macrophage cells, one group was treated with CQP, and LPS was used as positive control. The results showed that CQP (50 and 100 μg/mL) can inhibit LPS-stimulated macrophages production and decrease NO production after 24 and 48 hr ($p < 0.05$). The authors concluded that purified CQP exhibited an immune potentiation and anti-inflammatory effect. A petroleum ether extraction (2 hr, 60 °C three times), then an ultrasonic-assisted extraction was performed to obtain the fractions. For the *in vitro* assessment, four different human cell lines were employed: liver (L02) cells, normal breast epithelial (MCF 10A) cells, liver cancer (SMMC 7721) cells, and breast cancer (MCF-7) cells. All the cells exhibited remarkable growth at higher concentrations ranging from 25 to ~100 μg/mL, following the treatment (Hu et al., 2017).

Noratto et al. (2019) investigated the effect of quinoa on obesity and hepatic steatosis biomarkers in obese-diabetic (db/db) mice. Red quinoa seeds were ground into flour and used to prepare pellets to feed the mice. AIN-93G brand was used as a food control. The proximal composition (per kg) of the quinoa pellets was 125 g protein, 541 g carbohydrates, 83 g fiber, and 64.7 g fat. In response to quinoa intake, TNF-α levels were below 12 pg/mL, and IL-6 levels decreased by 35.3% compared to the obese group. There was no statistical difference between the quinoa group and the lean group ($p > 0.05$). Additionally, quinoa intake decreased plasma protein carbonyls by 50% relative to the obese group (Noratto et al., 2019). Berti et al. (2004) showed that quinoa proteins have not immunoreactivity against anti-gliadin antibodies (commercial) and serum from celiac individuals. The study of only 2 quinoa cultivars (Ayacuchana and Pasankalla) out of 15 presented toxic epitopes that activated gliadin-specific T-cell lines (Zevallos et al., 2012). Protein bands with electrophoretic mobility comparable to wheat prolamins were absent in 11 quinoa varieties with low binding affinity to anti-gliadin antibodies and IgAs (Peñas et al., 2014).

16.3.4 Antidiabetic

16.3.4.1 Amaranth

Type 2 diabetes mellitus (T2DM) is a chronic disease, which involves persistent loss of beta pancreatic cells and an increase in glucose (bloodstream), usually associated with insulin resistance (IDF, 2019). T2DM is the second main cause of death worldwide due to the complications that can develop such as chronic kidney disease, neuropathologies, blindness, and diabetic foot (Chigom, 2020). Dipeptidyl peptidase (DPP)-IV is an enzyme known for deactivating incretins hormones, involved in insulin secretions. Amaranth peptides inhibit DPP-IV with IC_{50} of 1.1 mg/mL in a dose-dependent manner (Velarde-Salcedo et al., 2013). Moreover, molecular docking revealed that the possible mechanism to inhibit DPP-IV was by blocking the active dimer formation. A similar study reported the DPP-IV inhibitory activity of amaranth glutelin protein (Soriano-Santos et al., 2015). Glutelin was extracted from defatted flour, albumin was separated from globulin by dialysis after 24 hr, and ethanol (70%) was used to eliminate prolamins. Then, Alcalase hydrolysis (2.4 UA/mL) was used with 0.5 M phosphate buffer (pH 7.4) added to a 5 mg/mL protein solution (incubated for 5 min at 50 °C). The higher inhibition of DPP-IV was achieved with albumin III (IC_{50} 1.98 mg/mL), globulin III (IC_{50} 0.12 mg/mL), and glutelin IV (IC_{50} 0.25 mg/mL) (Soriano-Santos et al., 2015). Zambrana et al. (2018) studied the antidiabetic effect of hydroethanolic extracts from *A. caudatus* (AC) both *in vivo* and *ex vivo*. These extracts were administered (2,000, 1,000, and 500 mg/kg body weight) to two different groups of rats for 21 days: a group of Wistar (W) rats and a group of spontaneously type 2 diabetic Goto-Kakizaki (GK) rats. A glucose tolerance test was performed on fasted rats (10–12 hr), and 1 hr before the experiment; the extracts were administered orally in both groups. The rats' glucose tolerance improved by 24.7% (dose of 1,000 mg/kg). HbA1c levels fell from 7.8 to 6.3% (GK) and from 2.6 to 2.3% (W), a decrease of 19.8 and 10.7%, respectively. The antidiabetic effect of hydroethanolic extract of AC was also evaluated in the GK and W groups, and insulin secretion increased 31-fold and 9-fold ($p < 0.05$), respectively (Zambrana et al., 2018). An *in vitro* study of simulated gastrointestinal digestion of *A. caudatus* flour was performed to assess the release of peptides and evaluate the potential antidiabetic effects (Vilcacundo et al., 2019). Gastric digestion was performed for 2 hr, after 1 hr and 2 hr incubation with pancreatin, and then the inhibition of DPP-IV and α-amylase enzyme was assessed. The 2-hr gastroduodenal digest showed the highest DPP-IV and α-amylase inhibition with IC_{50} values of 0.28 and 2.45 mg peptide/mL, respectively. Meanwhile, the peptide fraction <5 kDa had the lowest IC_{50} for DPP-IV (0.19 mg peptide/mL) and α-amylase (0.84 mg peptide/mL) inhibitory activities. The results of this study demonstrate the role of amaranth proteins as a source of multifunctional peptides for treating hyperglycemia (Vilcacundo et al., 2019).

16.3.4.2 Buckwheat

A methanolic (80%) extract of buckwheat bran powder with 0.76% rutin and 0.001% quercetin showed higher *in vitro* inhibition of sucrose activity than whole buckwheat powder (0.38% rutin and undetectable quercetin). Balb/c mice fed methanolic extract also exhibited a dose-dependent reduction of blood glucose levels within 60 min after oral administration of sucrose (Hosaka et al., 2011). Inositol forms, such as D-*chiro*-inositol (D-CI) and *myo*-inositol (MI), from an ethanol: water (1 : 1, v/v) extract of Koto buckwheat concentrate (BWC) showed insulin-mimetic activities that participate in insulin transduction pathways and glucose uptake in H4IIE cells (rat hepatoma cell line). This suggests that the concentrate may affect hepatic glucose metabolism, possibly by inhibiting glucose flux and uptake (Curran et al., 2010). Diabetic rats treated with 200 mg/kg/day flavonoids extracted from buckwheat leaves and flowers lowered fasting blood glucose, improved insulin resistance, and increased creatinine clearance rate, due to tyrosine-protein phosphatase non-receptor type 1 (PTP1B) suppression (Chu et al., 2011).

16.3.4.3 Chia

The antidiabetic effects of chia proteins have mainly been attributed to storage proteins, which are rich in albumins and globulins and more soluble than similar proteins found in other pseudocereals (Grancieri et al., 2019; Valenzuela-Zamudio & Segura-Campos, 2020). Vuksan et al., 2010 found that daily intake of 37 g of chia seeds stabilized blood glucose levels in diabetic patients. This effect was attributed to the presence of omega-3 fatty acids, high levels of iron and fiber, and elevated calcium and magnesium levels. In an experiment, male Wistar rats received a diet rich in fructose (10 g fructose) combined with pellets supplemented with chia seed extract or ground chia seeds (Mihafu, Kiage, Kimang'a, & Okoth, 2020). Postprandial blood glucose levels were statistically lower ($p < 0.05$) in experimental groups fed ground chia seeds or chia seed extracts compared with the group fed a diet that did not include chia products. The authors considered chia seeds to be a good candidate for lowering postprandial glycemia and improving hematological indicators of physiological status in animals; however, these observations must be validated with human trials. Peptides in chia seeds protein hydrolysates inhibit α-glucosidase, α-amylase, and T2DM biomarker DPP-IV (Valenzuela-Zamudio & Segura-Campos, 2020). Body weight loss was observed in obese and overweight individuals with type 2 diabetes when their calorie-restricted diets (30 g seed/1,000 kcal) were accompanied by chia consumption for 6 months (Ullah et al., 2016; Vuksan et al., 2017). Chia seeds also have an impact on lipogenesis by modulating glucose metabolism, improving the serum lipid profile, reducing visceral adiposity, and decreasing epididymal adipose tissue weight (Chicco, D'Alessandro, Hein, Oliva, & Lombardo, 2009; Creus et al., 2016; Creus et al., 2017; Ferreira, Alvarez, Illesca, Giménez, & Lombardo, 2016; Fortino, Oliva, Rodriguez, Lombardo, & Chicco, 2017; Oliva, Ferreira, Chicco, & Lombardo, 2013; Sierra, Roco, Alarcon, & Medina, 2015). Chia seeds are also a good source of α-linolenic acid, which has physiological relevance since it produces EPA and docosahexaenoic acid (DHA) in rats with metabolic syndrome (Poudyal et al., 2013; Poudyal, Panchal, Waanders, Ward, & Brown, 2012). Chia seed α-linolenic acid was not effective at reducing total body fat, but it induced fat redistribution away from the abdominal area and decreased the risk of cardiovascular diseases (Fonte-Faria et al., 2019; Oliva et al., 2013; Poudyal, Panchal, Waanders, Ward, & Brown, 2012, Poudyal, Panchal, Ward, Waanders, & Brown, 2012; Poudyal et al., 2013). Consumption of chia seed (7 or 14 g) in plain yogurt as a mid-morning snack significantly reduced hunger and increased short-term satiety in 24 healthy women (19–25 years) in a randomized, cross-over trial (Ayaz et al., 2017). The satiating effect of chia seed was attributed to its high content of fiber that can prolong satiety by delaying gastric emptying, increasing gastric distension, and/or attenuating glycemic response.Thus, chia seed consumption can be a useful dietary strategy to prevent overweight and obesity status in healthy individuals according to results from this study.

16.3.4.4 Quinoa

In addition to the potential use of DPP-IV inhibitors to treat hyperglycemia, α-amylase and α-glucosidase inhibitors can be used to delay the release of glucose in the small intestine and subsequently reduce postprandial hyperglycemia. Vilcacundo et al. (2017) studied the potential for a protein concentrate prepared from *C. quinoa* Willd to inhibit DPP-IV, α-amylase, and α-glucosidase. This concentrate was subjected to simulated gastrointestinal digestion for 2 hr, with pepsin and pancreatin, and aliquots were taken at time 0 (ID0), 60 (ID60), and 120 (ID120) min. The resulting samples were fractionated at different molecular weights (< 5 kDa and > 5 kDa). The results showed that the ID60 sample had the lowest DPP-IV inhibition (IC_{50} at 0.23 mg protein/mL), but the <5 kDa fraction of ID60 had the highest α-glucosidase inhibition (IC_{50} = 1.45 mg protein/mL). The ID120 sample had the highest α-amylase inhibitory activity (IC_{50} 0.19 mg protein/mL), highlighting the antidiabetic potential of quinoa seed hydrolysates (Vilcacundo et al., 2017). Abellán-Ruiz et al. (2017) demonstrated the antidiabetic effects of quinoa seeds in a clinical trial. This randomized, placebo-controlled double-blind study with 29 prediabetic volunteers evaluated the effects of processed quinoa (Kuska Active brand) for 28 days, measuring various parameters such as BMI, glycated hemoglobin (HbA1c), fasting plasma glucose (FPG), satiation, and fullness. The nutritional value of the quinoa sample (per 100 g) was 382 kcal, 11.7 g protein, 4.22 g fat, 8.5 g fiber, and 40.2 mg phytosterols. A daily dose of 20 g of quinoa seeds, with 80 mL of water (ready to eat) was provided to volunteers for 28 days. Following the treatment, the quinoa group showed a significant ($p < 0.05$) 0.7% decrease in BMI and a 0.245% decrease in HbA1c ($p < 0.001$), and an increase in satiation and fullness ($p < 0.001$), compared with the placebo group. The authors concluded that quinoa can be used as a coadjutant in the treatment of prediabetic patients (Abellán-Ruiz et al., 2017). Obaroakpo et al. (2019) explored the α-glucosidase inhibitory activity of hydrolysates derived from a quinoa yogurt beverage (QYB) fermented with *L. casei*. Acarbose, used as control, gave an IC_{50} of 3.88 mg/mL. The strongest inhibition of α-glucosidase (IC_{50} = 8.8 mg/mL) was obtained with *L. casei* SY13 (Obaroakpo et al., 2019). Obaroakpo et al. (2020) also evaluated the antidiabetic effect of oleanolic acid (OA) isolated from sprouted QYB, with four different germination times: 0 (QY100), 24 (QY124), 48 (QY148), and 72 (QY172) hr. The maximal inhibitory activity for DPP-IV based on lowest IC_{50} values were 70% (QY100), 13% (QY124), and 16.6% (QY172), while the highest IC_{50} was observed for QY148 (267%). For α-glucosidase enzyme, the IC_{50} values were 7.4%, 6.8%, 17.6%, and 10.5% for QY100, QY124, QY148, and QY172, respectively (Obaroakpo et al., 2020).

Tan et al. (2020) evaluated the antidiabetic effect of polysaccharides from quinoa by analyzing five crude polysaccharides fractionated by gradient with ethanol: 50% (QEP50), 60% (QEP60), 70% (QEP70), 80% (QEP80), and 90% (QEP90). Antidiabetic activity was assessed by measuring the glycemic index (GI), and α-amylase and α-glucosidase enzyme inhibition. Overall, the results showed that QPE90 had the strongest antidiabetic activity, the lowest GI (61.68), and the strongest inhibitory effect on α-amylase (IC_{50} = 61.36 mg/mL) and α-glucosidase (IC_{50} = 48.67 mg/mL) ($p < 0.05$) (Tan et al., 2020).

16.3.5 Antioxidant

16.3.5.1 Amaranth

Al-Mamun et al. (2016) assessed the antioxidant capacity of two species of amaranth, *Amaranthus lividus* (AL) and *Amaranthus hybridus* (AH) by testing methanol extracts for their radical scavenging activity (2,2-diphenyl-1-picrylhydrazyl, DPPH). The AL and AH extracts had IC_{50} values of 93 and 28 μg/mL, respectively, indicating that both amaranth extracts have strong antioxidant capacity due to the presence of other secondary metabolites (Al-Mamun et al., 2016).

Ayala-Niño, Rodríguez-Serrano, González-Olivares, et al. (2019) studied amaranth hydrolysates and their antioxidant activity. The hydrolysis was performed using 5 g of the protein extract diluted in 100 mL of deionized water with Alcalase® enzyme (> 2.4 U/g, pH 10). The Flavourzyme® (> 500 U/g) hydrolysis was performed at pH 7, and both enzymes were used together for 2-hr hydrolysis. The scavenging activities were measured with DPPH, ABTS ([2,2′-azinobis-(3-ethylbenzothiazoline-6-sulfonate)]), and ferric-reducing antioxidant power (FRAP) tests and the following results were obtained: 77 (DPPH), 116 (ABTS), and 63 (FRAP) µmol Trolox E/100 g. However, the fraction prepared with a two-step combined hydrolysis (Alcalase® and Flavourzyme®) had the highest antioxidant activity as indicated by DPPH, ABTS, and FRAP tests which resulted in the following values: 389, 405, and 593 µmol TroloxE/100 g, respectively, showing that amaranth peptides have potential as antioxidant agents (Ayala-Niño, Rodríguez-Serrano, Jiménez-Alvarado, et al., 2019). House et al. (2020) compared four species of amaranth: AD, AS, AT, and AV for the antioxidant capacity of their methanolic extracts. The *in vitro* antioxidant activities were assessed by the DPPH radical scavenging assay, the hydrogen peroxide scavenging assay and the FRAP test. The IC_{50} values for DPPH were 71.06, 92.20, 88.64, and 63.94 µg/mL for AT, AV, AD, and AS, respectively. While the hydrogen peroxide scavenging activity was highest for the AD extract (26 µg/mL), followed by AS (28.55 µg/mL), AV (30.19 µg/mL), and AT (31.13 µg/mL). Similarly, AD (20.44 µg/mL) extract showed the strongest activity in reducing the ferric antioxidant power. The authors concluded that the antioxidant potential was the highest for the AS and AD extracts (House et al., 2020).

In another study, Sandoval-Sicairos et al. (2021) investigated the antioxidant capacity of three fractions of amaranth peptides: F1 (> 10 kDa), F2 (3–10 kDa), and F3 (<3 kDa). Enzymatic hydrolysis was performed with pepsin incubated for 180 min at 37 °C, followed by pancreatin for 90 min at 37 °C, and 30 mL aliquots were collected (10, 25, 60, and 90 min). Antioxidant capacity was evaluated via the ORAC test. F2 had the highest ORAC value with 33.6 mmol Trolox Eq/100 g, followed by F3 and F1, with 14.4 and 13.6 mmol Trolox Eq/100 g, respectively. In addition, F1 and F2 showed the highest peroxyl radical scavenging capacity (64.9%) and the highest activity showed by gastrointestinal digestion at 90 min.

16.3.5.2 Buckwheat

Rutin (100 µM) and quercetin (50 µM) from a methanolic extract of common buckwheat and Tartary buckwheat flour increased antigenotoxic activity 51% and 67%, respectively, in a comet assay on HepG2 cells (human hepatocellular carcinoma cell line) exposed to tert-butyl hyperoxide, which is known to cause DNA damage (Vogrinčič et al., 2013). This ability was attributed to the antioxidant capacity of phenolics, particularly rutin and quercetin (Al-Snafi, 2017; Vogrinčič et al., 2013). A crude methanolic extract of common buckwheat showed high DPPH• inhibitory activity (81.6%) at 1 mg/mL that was attributed to the 85% flavonoids contained in the extract (Abbasi et al., 2015). Fermentation of common buckwheat with *Lactobacillus delbrueckii* subsp. *lactis* increased total phenolic content by 64% compared with unfermented buckwheat powder (Gandhi & Dey, 2012). The increase in total phenolic content impacted the DPPH• inhibitory activity (twofold increase) and the FRAP (1.8-fold increase) after 72-hr fermentation. Buckwheat sprouts treated with methyl jasmonate (BSMJ) reduced TPRZ-Fe (III) complex to TPTZ-Fe (II) with 42.9% activity higher than buckwheat sprouts non-treated with methyl jasmonate (BSNMJ), a phytocompound synthetized for defense and antimicrobial purposes in plants, as well as antioxidant effects and cytochrome C synthesis activity in human cell experiments (Lee et al., 2013). The ORAC values were $103.3 \pm 19.2\,\mu M$ Trolox Eq/g and $60.6 \pm 3.8\,\mu M$ Trolox Eq/g, respectively, for BSMJ and BSNMJ. This effect was attributed to the presence of rutin (Lee et al., 2013). Buckwheat flour fermented with *Lactobacillus plantarum* and *Lactobacillusdelbruecki* subsp. *bulgaricus* increased the reducing power by 36% and 70%, respectively, compared to the unfermented control (Zieliński et al., 2020). Fermentation of buckwheat flour with *Rhizopus oligosporus* also showed an almost two-fold increase in antioxidant/reducing capacity after digestion (Zieliński et al., 2020). Buckwheat protein fractions of 456 Da and 362 Da, isolated using Sephadex G-25, were subjected to pancreatin (2 hr) digestion. The resulting hydrolysate showed an $ABTS^{+\bullet}$ scavenging activity of 23% and 27% greater than non-fractioned buckwheat protein digest ($p < 0.05$) (Ma et al., 2010). The 362 Da fraction was also 86% and 24% more efficient ($p < 0.05$) at neutralizing $^{\bullet}OH$ than mixed buckwheat protein digest and fractions >456 kDa, respectively. Buckwheat seed oil, rich in rutin and oleic, linoleic, and palmitic acids, showed scavenging activity (DPPH, *N,N*-dimethyl-*p*-phenylenediamine [DMPD], superoxide, and hydrogen peroxide radicals), metal-chelating capacity, and FRAP, as well as phosphomolibdenum-reducing antioxidant power (PRAP) (Gulpinar et al., 2012). The authors attributed this *in vitro* antiradical activity to a neuroprotective effect on neuronal damage in a brain ischemia model and to anti-glutamate activity.

16.3.5.3 Chia

Chia seeds are rich sources of phytochemicals with antioxidant potential (carotenoids, peptides, polyphenols, and tocopherols) (Orona-Tamayo et al., 2017). An increased degree of hydrolysis by enzymatic fractionation with Alcalase and Flavourzyme resulted in higher ACE inhibitory activity but lower antioxidant activity (Segura-Campos, Peralta González, Guerrero, & Betancur Ancona, 2013). Antioxidant activity of the chia protein hydrolysates, quantified and calculated as Trolox equivalent antioxidant capacity (TEAC) values (mmol/L-mg), decreased as the degree of hydrolysis increased. The highest TEAC value was obtained for the hydrolysate produced at 90 min (7.31 mmol/L-mg protein), followed by those produced at 120 min (4.66 mmol/L-mg protein) and 150 min (4.49 mmol/L-mg protein) (Segura-Campos, Salazar-Vega, Chel-Guerrero, & Betancur-Ancona, 2013). In another study an omega-3-enriched fraction of chia seeds from Brazil had $ABTS^{\bullet +}$, $DPPH^{\bullet}$, and FRAP antioxidant capacity of 2.56, 1.72, and 2.86 mmol TEAC/g, respectively. These values were attributed to the high proportion of omega-3 fatty acid present in the fraction (Sargi et al., 2013). Brazilian chia seeds grown in Rio Grande do Sul showed a significantly ($p < 0.05$) higher antioxidant capacity (0.48 mmol TEAC/g sample) than chia seeds obtained from Mato Grosso (0.46 mmol TEAC/g sample) (da Silva et al., 2017). The authors attributed these outcomes to the conditions in which the chia was grown. Four solvent extracts of Argentinian chia seeds showed ABTS radical activity ranging between 0.2 and 0.6 mmol TEAC/g (Capitani, Spotorno, Nolasco, & Tomás, 2012). De Falco, Amato, and Lanzotti (2017) as well as Sargi et al. (2013) attributed the difference in antioxidant capacity to the geographical location where the chia seeds were grown, the raw chia material and the fractionation methods used. Alemán et al. (2021) used ABTS (expressed as mg Vit C eq./g) and FRAP (expressed as mM Fe^{2+} eq./g) to evaluate the antioxidant activity of liposomal suspensions of phospholipid-rich fraction (L-PPL) and commercial rapeseed lecithin (L-LEC) as well as an ethanolic extract from chia seeds. The chia seed extract showed antioxidant activity 2.7-fold and 2.4-fold greater than L-PPL and L-LEC, respectively, while for FRAP, the chia seed extract showed antioxidant activity 5.1-fold greater than for L-LEC and 6.4-fold higher than L-PPL. Crude chia seeds, crude fiber flour, and crude oil, as well as hydrolyzed extracts of chia seeds, presented a rich hydroxycinnamic acid profile; the hydroxycinnamic acid was identified as responsible for the high antioxidant activity assessed by ORAC and FRAP (Oliveira-Alves et al., 2017). Azeem et al. (2015) found that the seeds and fiber flour hydrolysate (a methanolic extract followed by alkaline precipitation with H_2SO_4) had high antioxidant activity as determined by the ORAC (> 142 μmol Trolox eq./g) and the FRAP (> 80 μmol Trolox eq./g), with no statistical difference ($p > 0.05$) between them. They also observed that the addition of chia seed oil to cotton oil stabilized the auto-oxidation of polyunsaturated fatty acids. This result was attributed to the high amount of phenolics in the chia oil.

16.3.5.4 Quinoa

Hu et al. (2017) investigated the antioxidant activity of a polysaccharide fraction from *C. quinoa* seeds. A petroleum ether extraction was carried out for 2 hr at 60 °C three times on the quinoa seeds and then an ultrasound-assisted extraction with distilled water was used to obtain the fraction. The scavenging capability of the polysaccharide fraction on DPPH free radicals and ABTS was evaluated. The IC_{50} was 1.9 mg/mL for DPPH and 1.1 mg/mL for ABTS. Both antioxidant effects were dose-dependent (Hu et al., 2017). The antioxidant capacity (ABTS and DPPH) was reported for five crude quinoa polysaccharides (QEP50, QEP60, QEP70, QEP80, and QEP90), fractionated by gradient with ethanol (Tan et al., 2020). QEP90 had the strongest ability to scavenge DPPH and ABTS radicals, with an EC_{50} value of 5.22 and 1.83 mg/mL, respectively. In both assays, QEP50 had the weakest scavenging ability with an EC_{50} of 13.67 mg/mL (DPPH) and 5.21 mg/mL (ABTS) (Tan et al., 2020). Obaroakpo et al. (2020) reported the antioxidant activity of OA extracted from a QYB at four different times of gastric digestion 0 h (QY100), 24 hr (QY124), 48 hr (QY148), and 72 hr (QY172). Samples showed DPPH activity with IC_{50} values of 6.02, 8.31, 17.58, and 132.58 mg/mL for QY100, QY124, QT148, and QY172, respectively. For the FRAP assay, IC_{50} values of 236, 139, 37, and 107 μmol Fe^{2+}/g were obtained for QY100, QY124, QY148, and QY172, respectively. Finally, the scavenging activity of ABTS was quantified producing the following results: IC_{50} values of 146.81 mg AAE/g (QY100), 23.43 mg AAE/g (QY124), 22.93 mg AAE/g (QY148), and 22.07 mg AAE/g (QY172). In conclusion, OAs hold promise for preventing or controlling oxidative stress-related disorders (Obaroakpo et al., 2020).

16.3.6 Renal Effect

16.3.6.1 Buckwheat

Compared to other pseudocereals, research is limited on the beneficial renal effects of buckwheat. The administration of 200 mg/kg/day of flavonoids obtained from buckwheat leaves and flowers produced morphological renal changes in rats with T2DM, including a reduction in atrophies and an enlarged cyst cavity, accompanied by narrow lumens caused by hyperplasia and thickening of the inner lining of renal arteries (Chu et al., 2011). Tartary buckwheat (100 g/day) as a staple food (rice and wheat) replacement can alleviate renal dysfunction in T2DM patients (n = 104; average age 59 years; 27 kg/m^2 BMI; 7.6% HbA_{1c}) according to a parallel, randomized, open-label controlled, 4-week dietary intervention trial (Qiu, Li, Qin, Yue, & Liu, 2016). The tartary buckwheat intervention significantly improved kidney function by reducing urinary albumin to creatinine ratio (risk factor for future kidney disease) and urea nitrogen (essential for removal of residual nitrogen resulting from protein and amino acid metabolism) without changes in blood glucose particularly for patients with microalbuminuria (renal dysfunction stage 1 of diabetic kidney disease). The improvement in urinary albumin excretion is associated with protein metabolism due to the higher protein intake from tartary buckwheat.

16.3.6.2 Chia

As with buckwheat, the renal health-promoting effects of chia have received little attention. Healthy volunteers with xerotic pruritus symptoms and patients with pruritus caused by end-stage renal disease or diabetes were treated with a topical formulation containing 4% chia seed oils in an 8-week-long experiment (Jeong et al., 2010). Significant improvements in skin hydration, lichen simplex chronicus, and prurigo nodularis were observed in all study participants, including healthy volunteers with xerotic pruritus. The chia seed oil topical formulation also improved epidermal permeability barrier function and skin hydration, represented by trans-epidermal water loss and skin capacitance, respectively, and it showed no adverse effects. The authors attributed these effects to the high content of 18 : 3 n−3 fatty acid. In another study, Wistar albino rats exposed to aluminum were fed a chia-rich diet or a standard diet for 21 days to assess the hepatorenal protective effects of chia (Demir & Bilgic, 2019). Liver function (alanine aminotransferase, alkaline phosphatase, aspartate aminotransferase, and lactate dehydrogenase), kidney function (creatinine and urea nitrogen), and vitamin B_{12} and folic acid determined with an automated analyzer showed that aluminum exposure had no influence on renal function and neither did chia supplementation; liver function was disturbed with exposure to aluminum and chia was not effective against it. The animals fed a chia-rich diet displayed higher concentrations of vitamin B_{12}, but this was not the case for folic acid. A chia-rich diet has no effect on renal function and cannot reverse aluminum-induced hepatotoxicity; however, it may help address vitamin B_{12} insufficiency and thus may offer a novel treatment option which is particularly important in the vegan diet.

16.4 Biomarkers

Biological markers are indicators of medical status observed from outside the patient that are commonly used in the medical field for basic and clinical practice and research (Strimbu & Tavel, 2010). Biomarkers have been studied in the context of medical pathway responses to drugs or medical devices. However, biomarkers also represent one of the most powerful tools used in fields using omics technologies, including nutrigenomics, which has a focus on personalized nutrition (Pavlidis et al., 2015). Recent applications in clinical research, including the laboratory-measured biomarkers related to novel foods, still need to be developed further and refined before they can be used for other clinical research end points (Califf, 2018). Pseudocereals have a rich profile of functional phytochemicals and peptides with the capacity to modulate several signaling molecules and metabolic pathways in the human body.

Several compounds from pseudocereals have displayed modulatory effects on biomarkers related to human health and nutrition (**Figure 16.1**). Pure peptides from amaranth (AmPep) may inhibit LOX-1 receptor, which regulates the iCAM mediator for the leukocyte adhesine activity, participate in iNOX induction and the synthesis of endogenous NO, and also modulate the MAPK pathway, associated with the upregulation of transcription factors and several cytokines at the nuclear level (Montoya-Rodríguez & de Mejía, 2015; Schröter et al., 2019). Hydroxycinnamic acids from amaranth (AmHxA), such as caffeic and chlorogenic acids, can upregulate TNF-α receptor activity that can activate the apoptotic pathway via caspase 3, upregulate IL-6, IL-10, and IL-12, the last two related to the activity of CD40 and CD83 membrane receptors, but IL-10 also participates in upregulating CD11b membrane receptor (Taniya et al., 2020). Buckwheat polysaccharides may interact with CD14 membrane receptor and modulate the phosphorylation of ERK1/2, followed by activating c-FAS expression in the nucleus of Ser phosphorylation, activate STAT3 and HIF-1α, which in turn upregulate IL-10 expression, followed by COX-2 suppression (Bai, Feng, et al., 2015). Lectins isolated from buckwheat also upregulated the TNF-α membrane receptor (Bai, Ji, et al., 2015). Furthermore, phytochemicals (mainly phenolic compounds) from an ethanolic extract of buckwheat were able to pierce the cell membrane and interact with the mitochondria, stimulating the expression of mitochondrial P450 oxidase pathway until flavin adenine dinucleotide (FAD) binding with Nox4/p22^{phox} (Lee et al., 2013). This extract also suppressed antigen-induced phosphorylation of Lyn, participating in

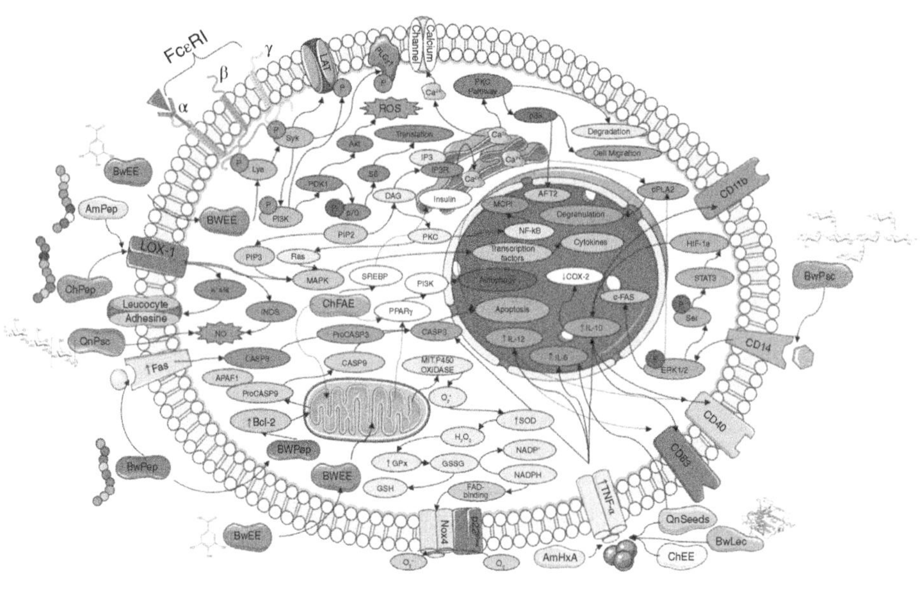

Figure 16.1 Molecular and cellular biomarkers influenced by phytochemicals, polysaccharides, lectins, and peptides from four selected pseudocereals.

modulating membrane molecules, such as FcεRI (α, β and γ), LAT, and PCLγ1, as well as upregulating translation (via p70 phosphorylation) and modulating ROS (via Akt) (Al-Snafi, 2017; Kyung-Hwa, 2012). Buckwheat peptides may participate via upregulating the expression of Bcl-2 and mitochondrial caspase 9 until caspase 3, or via upregulating the Fas membrane receptor, followed by caspase 8 and procaspase 3 until caspase 3 (Bai, Feng, et al., 2015). An ethanolic extract of chia (ChEE) may participate in upregulating TNF-α membrane receptor and apoptosis via regulation of caspase 3, as well as interleukins (Grancieri et al., 2019). Chia peptides (ChPep) can modulate calcium bioavailability and inflammation through the calcium channel from an endogenous regulation of the endoplasmic reticulum and participate in cell degradation and migration, as well as degranulation via cPLA2 (da Silva et al., 2018). A chia fatty acid extract (ChFAE) participated in autophagy induction and pre-insulin peptides packing (Ferreira et al., 2016; Oliva et al., 2013; Pandurangan et al., 2020; Valenzuela-Zamudio & Segura-Campos, 2020). A crude extract from quinoa seeds upregulated IL-6 directly in the nucleus or via TNF-α membrane receptor (Lamothe et al., 2015; Li, Lietz, Bal, et al., 2018). Quinoa polysaccharides may traverse the cell membrane and inhibit NO synthesis (Hu et al., 2017; Melini & Melini, 2021; Tan et al., 2020).

Figure 16.2 summarizes other biological effects of pseudocereal compounds at the systemic level. Pseudocereals bioactive compounds such as peptides, phenolics, fatty acids, polysaccharides, among other compounds exert various biological activities through different mechanisms such as (a) antiproliferative effect by inhibiting cell proliferation or promoting apoptosis in different cancer cell lines (AT-2, MAT-LyLu, H22, EAC, and MDA-MB231) (Gawlik-Dziki et al., 2013; Bai, Feng, et al., 2015; Al-Mamun et al., 2016; Taniya et al., 2020), (b) antioxidant effect by exerting scavenging activity and inhibiting ROS such as SOD, CAT, GSH, GST, and malondialdehyde (MDA) (Gulpinar et al., 2012; Lee et al., 2013; Diwakar et al., 2014; Marineli, Lenquiste, Moraes, & Maróstica, 2015; Al-Snafi, 2017; Odongo et al., 2018; Zheng et al., 2019), (c) antidiabetic effect by inhibiting intestinal enzymes (α-amylase and α-glucosidase), thus reducing the hydrolysis of complex carbohydrates and delaying the absorption of simple sugars; while DPP-IV inhibition decreases glucagon secretion, d) hypolipidemic effect can be observed by HMG-CoA reductase enzyme inhibition, thus decreasing LDL proteins production (Poudyal, Panchal, Waanders, Ward, & Brown, 2012; Poudyal et al., 2013; Marineli, Moura, et al., 2015; Guo, Hao, et al., 2020;Creus et al., 2016; da Silva et al., 2018), e) hypotensive effect occurs by ACE-I inhibition, this prevents the over production of ACE-II (potent vasoconstrictor) which induces aldosterone release resulting in elevated sodium concentration and blood pressure; bioactive compounds lower blood pressure by inhibiting this enzyme (Segura-Campos, Peralta González, Guerrero, & Betancur Ancona, 2013; Quiroga et al., 2017; Ayala-Niño, Rodríguez-Serrano, Jiménez-Alvarado, et al., 2019; Obaroakpo et al., 2019; Zheng et al., 2019; Suárez et al., 2020; Enes et al., 2020; Han et al., 2020), f) anti-inflammatory response via downregulation of pro-inflammatory cytokines activities primarily by interacting with cellular inflammatory and redox signaling pathways attenuating production of pro-inflammatory cytokines such as TNF-α, IL-1β, and IL-6 and other inflammatory markers such as NO and PGE2 via downregulation of iNOS and COX-2 (Wu & Lee, 2011; Bai, Ji, et al., 2015; Creus et al., 2016; da Silva et al., 2018; Schröter et al., 2019; Noratto et al., 2019; House et al., 2020; Enes et al., 2020; Pandurangan et al., 2020; Sandoval-Sicairos et al., 2021; Alemán et al., 2021). An additional mechanism that lowers

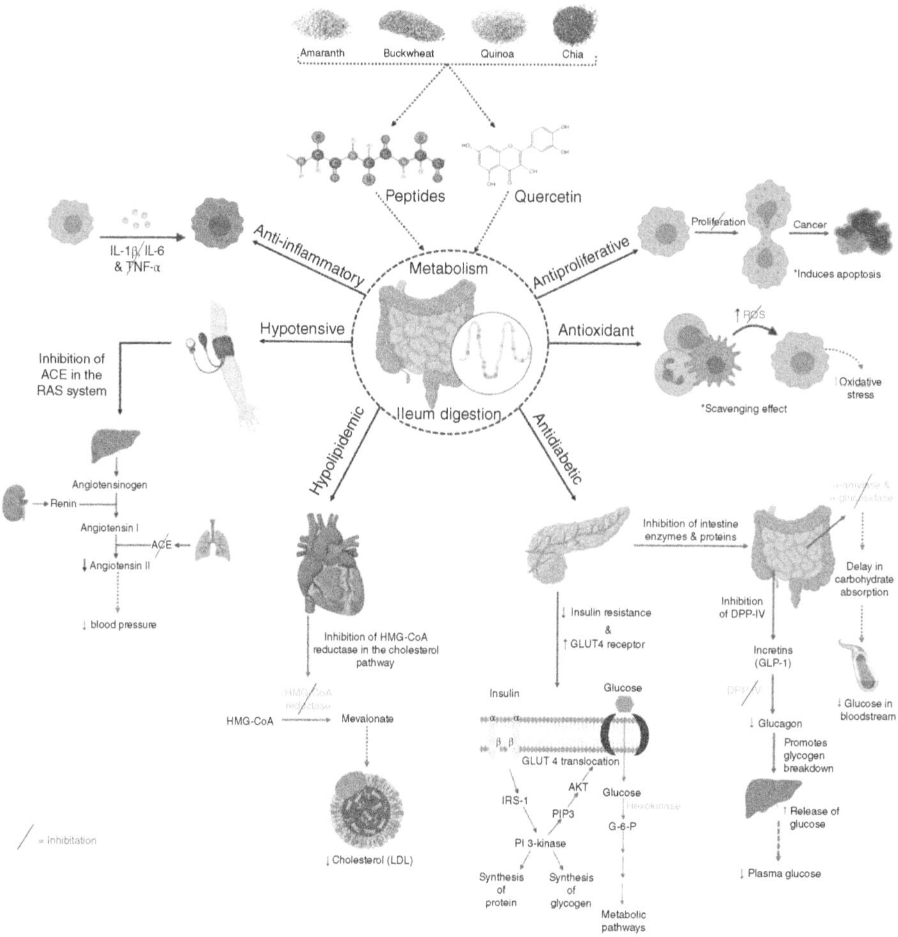

Figure 16.2 Systemic biomarkers influenced by phytochemicals, extracts, fractions, and peptides from four selected pseudocereals.

glucose in the bloodstream is the translocation of GLUT4 receptor (Al-Snafi, 2017; Creus et al., 2016; Curran et al., 2010; da Silva et al., 2018; Enes et al., 2020; Obaroakpo et al., 2020; Zambrana et al., 2018).

16.4.1 Amaranth

Cardiovascular: The bioactive potential of amaranth protein hydrolysates was measured by inhibiting the synthesis of angiotensin-converting enzyme (ACE—I) and thrombin (Ayala-Niño, Rodríguez-Serrano, González-Olivares, et al., 2019). Amaranth protein hydrolysates inhibited ACE-I and thrombin (10.5 and 11.9%), respectively, and the fraction resulting from a two-step combined hydrolysis (Alcalase® and Flavourzyme®) showed higher ACE-I and thrombin inhibition at 58.5% and 92.9%, respectively. Another *in vitro* study of ACE-I inhibition was performed with protein-derived amaranth peptides to identify mechanisms underlying inhibition of the renin-angiotensin-aldosterone system (RAS), inhibition of ACE-I enzyme and inhibition of AT1 receptors (Quiroga et al., 2017). The amaranth hydrolysates did not inhibit chymase activity, but a dose–response effect was found for renin activity with an IC_{50} of 0.6 mg/mL. Suárez et al. (2020) reported results for ACE-I inhibition and the hypotensive effect of amaranth protein isolates and amaranth hydrolysates, specifically *in vitro* with IC_{50} values of 0.7 and 0.14 mg/mL, respectively, compared with 39.2 mg/mL for the control (captopril). For the *in vivo* tests, SHR, amaranth samples and synthetic peptide (VIKP), administered in a single dose (3 hr) lowered blood pressure in SHR rats from 200 to 220 mmHg to 140–170 mmHg. Thus, suggesting amaranth peptides play a key role by acting on plasma ACE, plasma renin, and the vascular system. Another study using amaranth hydrolysates in SHR inhibited ACE-I (93.5%, $p < 0.05$) relative to a control group (captopril). After supplementation, the hydrolysate was bioavailable in rats from 5 to 60 min, and the hypotensive effect started at 4 hr in SHR ($p < 0.05$ vs. water group), while the captopril hypotensive effect started at 3 hr (Ramírez-Torres et al., 2017). The authors concluded that amaranth hydrolysates can be used as ingredients for functional foods to treat hypertension (Ramírez-Torres et al., 2017). Another biomarker used for cardiovascular disease is 3-hydroxy-3-metyl-glutaryl-CoA reductase (HMG-CoA reductase); the inhibition of this enzyme plays a key role in the cholesterol pathway and in reducing cardiovascular risk. Manólio-Soares et al. (2015) studied the *in vitro* inhibition activity of five amaranth peptides (GGV, IVG, LVG, VGVI, and VGVL) on HMG-CoA. All tests were performed with a concentration of 4 μg/mL. The peptides VGVI and LVG showed no significant effect ($p > 0.05$) in terms of HMG-CoA reductase inhibition. In contrast, the GGV, IVG, and VGVL peptides significantly ($p < 0.05$) inhibited the specific activity of HMG-CoA reductase (Soares et al., 2015).

Antioxidant: Odongo et al. (2018) studied the ability of *Amaranthus cruentus* ethanol extract to protect against DNA damage induced by the mycotoxin aflatoxin B1 (AFB1) and oxidative stress. The extract inhibited AFB1 by 57% and oxidative stress-induced DNA damage by 81%, with both activities being concentration-dependent. The amaranth extract also had a protective effect (38%) against oxidative stress (triggered using ferrous sulfate). AFB1 is linked to an increase in ROS production. It is therefore important to inhibit this pathway to prevent cell injury and DNA damage (Odongo et al., 2018).

Anti-inflammatory: Schröter et al. (2019) investigated the anti-inflammatory effect of 2-caffeoylisotric acid, chlorogenic acid, and caffeic acid (extracted from amaranth) in LPS-induced macrophage RAW 264.5 cells. The results showed that 2-caffeoylisocitric acid significantly ($p < 0.05$) reduced mRNA and protein levels of iNOS, TNF-α, and IL-6 by directly decreasing NF-kB translocation (Schröter et al., 2019). Recently, Sandoval-Sicairos et al. (2021) studied the anti-inflammatory effect of amaranth peptide fractions, after a 90-min hydrolysis with pancreatin, and 180 min of total hydrolysis. Subsequently, three fractions were produced as follows: F1 (> 10 kDa), F2 (3–10 kDa), and F3 (< 3 kDa). F1 and F2 significantly ($p < 0.05$) reduced NO production (65.2% and 66.6%), respectively.

Anticancer: Al-Mamun et al. (2016) tested the antiproliferative effect of two edible species of *Amaranthus* from Bangladesh, using the seeds to produce a methanol extract. The extracts of *Amaranthus lividus* (AL) and *Amaranthus hybridus* (AH) were evaluated *in vivo* in 36 Swiss albino mice with Ehrlich's ascites carcinoma (EAC), using three different doses (25, 50, and 100 μg/mL/day) for 6 days. Administration of AL and AH extracts at 100 μg/mL inhibited EAC cells growth by 43% and 45%, respectively, with marked features of apoptosis such as cell shrinkage. The authors concluded that amaranth extracts have strong antiproliferative activity on EAC cells occurring through mitochondria-mediated apoptosis (Al-Mamun et al., 2016). Taniya et al. (2020) used protein hydrolysates from seeds of *Amaranthus caudutus* to measure *in vitro* anticancer activity in breast cancer cells (MDA-MB231). The hydrolysates were obtained through a pepsin and pancreatin hydrolysis for 23 min. The results showed that amaranth protein hydrolysates inhibited cell growth with an IC_{50} of 48.3 μg/mL and induced DNA fragmentation, membrane integrity loss, phosphatidylserine translocation, and caspase 3 activity. In conclusion, amaranth peptides may be a promising anticancer option (Taniya et al., 2020).

Antidiabetic: Zambrana et al. (2018) studied islet insulin secretion variation in Wistar rats and spontaneously type 2 diabetic Goto-Kakizaki (GK) rats, after 21 days of daily consumption of hydroethanolic extracts from *Amaranthus caudatus*. The GK rats increased their insulin secretion 1.7-fold at day 10 and 2.3-fold at day 20 ($p < 0.001$). The authors link the influence of H89 (inhibitor of cAMP, dependent on PKA [protein kinase A] and Cal-C) to GK rat (*ex vivo*). H89 reduced insulin secretion by 55% (3.3 mM glucose) and 77% (16.7 mM glucose), respectively ($p < 0.0001$). In conclusion, the amaranth extract stimulates insulin secretion. This effect is mediated by protein kinase C (PKC) and PKA systems and by L-type calcium channels and insulin G protein-couple exocytosis in the beta pancreatic cells.

16.4.2 Buckwheat

Cardiovascular: The anti-adipogenesis effect of a buckwheat methanolic extract (with high concentration of rutin and other phenolics like flavonoids, *p*-hydroxybenzoic, and hydroxycinnamic acids) was evaluated on 3T3-L1 preadipocytes treated with methyl jasmonate. Results showed statistically ($p < 0.05$) strong inhibition of the adipogenic transcription factor that initiates C/EBPα and PPARγ, which promote a late-stage adipocyte marker (aP2). Bcap37 cells exposed to TBWSP31 protein promoted the expression of bcl-2 and Fas, which in turn induced apoptosis (Guo et al., 2010).

Antioxidant: Buckwheat sprouts treated with methyl jasmonate downregulated the expression of the GPx and Cu/Zn SOD genes in 3T3-L1 cells, supporting the inhibition of ROS production, which induces the synthesis of the adipocyte markers G6PDH and NOX4 (Lee et al., 2013). This suggests a possible relationship between adiponectin and oxidative stress (Al-Snafi, 2017; Lee et al., 2013). Buckwheat extracts showed inhibition activity linked to the enzymes acetylcholinesterase (AChE) and butyrylcholinesterase (BChE) as well as tyrosinase (TYRO) which are involved in neuroprotective activity (Gulpinar et al., 2012).

Anti-inflammatory: Tartary buckwheat lectin (25 μg/mL), combined with 5×10^{-5} M (rh) IL-4, promotes maturation (88.7%) and proliferation (91.5%) of peripheral blood dendritic cells, which were significantly ($p < 0.05$) higher than (rh) TNF-α (20 ng/mL) (Bai, Feng, et al., 2015). Exposure of dendritic cells to 50 μg/mL Tartary buckwheat lectin for 48 hr extensively upregulated maturation markers CD83 and CD40 (Bai, Feng, et al., 2015). This study also revealed a multiple reduction effect of the buckwheat extract on the

RBL-2H3 cells. Antigen (DNP-HSA)-induced release of histamine, prostaglandin D2 (PGD2), and cysteinyl leukotriene in IgE-sensitized, as well as inhibition in antigen-induced HDC2 and COX-2 and 5-lipoxygenase (5-LO) mRNA expression in IgE-sensitized cells. Additionally, antigen-induced FcεRIβ and FcεRI FCEϒ subunit mRNA expression was suppressed, as well as antigen-induced phosphorylation of Akt, cPLA2, LAT, Lyn, PI3K, PLC ϒ1, and Syk. The mechanisms responsible for inhibiting the release of inflammatory mediators were identified, such as histamine, PGD2, cysLT, and the proximal signal events of intracellular FcεRI signaling molecules. In the indirect treatment, buckwheat polysaccharides (BPS) significantly stimulated cytokine secretion (differentiation inducer) in monocytes from peripheral blood mononuclear cells in MNC-conditioned medium (BPS-MNC-CM) following a 24-hr treatment, and THP-1 cell differentiation and maturity were significantly increased after 5 days treatment with the BWPS-MNC-CM (Wu & Lee, 2011).

Anticancer: Recombinant buckwheat trypsin inhibitor induced *in vitro* apoptosis in the H22 cell line (hepatic cancer cells) as evidenced by apoptotic bodies formation, nucleus DNA fragmentation, caspase-3, -8 and -9 activation, and mitochondrial dysfunction leading cytochrome C from the mitochondria to the cytosol (Bai, Feng, et al., 2015).

A BWC containing D-*chiro*-inositol (D-CI) inhibited basal and insulin-simulated glucose uptake. By contrast, insulin and D-CI stimulated glucose uptake. However, insulin-mimetic effects on selected protein phosphorylation events in H4IIE cells were not reported for D-CI or *myo*-inositol (Curran et al., 2010). Furthermore, KBC participates in the phosphorylation of p42/44 extracellular-related kinase (p42/44 ERK) and its downstream target, p70 (S6K), on Thr (421). BWC upregulated p38 mitogen-activated protein kinase (p38 MAPK) phosphorylation of an important insulin target; specific inositol forms and phenolics from BWC were mentioned as being responsible for this effect (Al-Snafi, 2017; Curran et al., 2010).

Renal effect: Flavonoids from buckwheat flowers and leaves downregulated the expression of PTP1B in T2DM rats; they can also lower blood glucose and renal damage (Chu et al., 2011).

16.4.3 Chia

Cardiovascular: Segura-Campos, Peralta González, Guerrero, & Betancur Ancona (2013) observed that the degree of hydrolysis of chia seed significantly ($p < 0.05$) inhibited ACE-I by obstructing angiotensin II generation. The results suggest that the bioactive peptides have the potential to control ACE-I naturally with decreased risk factors. Hydrolyzed chia proteins were added to white bread and carrot cream. The hydrolysates containing peptides with potential ACE inhibitory activities had no significant effect on the quality of these food products (Segura-Campos, Peralta González, Guerrero, & Betancur Ancona, 2013). Altered expression of FAT/CD36 by insulin is a key mechanism for restoring mitochondrial activity and β-oxidation. In rats fed a high-fat diet, chia seeds participated in regulating this mechanism as well as decreasing CTP-1 activity and PPAR-α protein mass, two enzymatic activities related to the balance of lipid oxidation and lipid storage (Creus et al., 2016; da Silva et al., 2018). Chia seeds prevented this condition by increasing PPAR-α, increasing PGC1-α expression, recruiting FAT/CD36 to the sarcolemma through insulin signaling, and decreasing visceral adiposity (Creus et al., 2016; da Silva et al., 2018; Enes et al., 2020; Ferreira et al., 2016; Marineli, Moura, et al., 2015; Poudyal et al., 2013; Poudyal, Panchal, Ward, Waanders, & Brown, 2012). Western blot results for mice fed a diet rich in fat and chia seed oil demonstrated that chia seed oil upregulated the expression of peroxisome proliferator-activated receptor (PPAR) α and carnitine palmitoyltransferase 1a in the liver (Han et al., 2020).

Antioxidant: Diets with high fructose content induce oxidative stress and lipid peroxidation (LPO) in plasma and liver. Marineli, Lenquiste, Moraes, & Maróstica (2015) reported the reduction of the endogenous antioxidant biomarkers glutathione (GHS), catalase (CAT), GPx, and glutathione reductase (GRd) in plasma and liver in obese rats fed a fructose-rich diet. Rats fed chia seed and oil for 6 and 12 weeks showed hepatic and plasmatic antioxidant activity of 47 and 35%, respectively, with increased GHS, CAT, GPx, and GRd activities. Chia oil also suppresses genes related to melanogenesis and skin oxidative damage (Diwakar et al., 2014).

Anti-inflammatory: An ethanolic extract of chia displayed anti-inflammatory response by suppressing TNF-α production on THP-1 macrophages (Alemán et al., 2021). In monocytes treated with digested total protein (DTP) from chia seeds, the secretion of TNF-α (24.1%), MCP-1 (18.9%), IL-6 (39.6%), and IL-10 (68.7%) was reduced, while DTP and a digested protein fraction (DPF) reduced the NF-κB translocation to nuclei (Grancieri et al., 2019). DTP and digested glutelin fraction (rich in the FAFFEFFELLFAFFT peptide) reduced iCAM expression (86.4% and 80.8%), LOX-1 (37.3%, 35.7%), iNOS (67.0%, 42.2%), and NF-κB (57.5% and 71.1%). DTP was also considered effective in reducing TNF-α secretion (43.3%), MCP-1 (47.6%), and IL-6 (50.5%), as well as NO (43.4%), lipid accumulation (41.9%), and prostaglandins (41.9%). Chia seeds have the potential to reduce the inflammation biomarkers (glutamic-oxalacetic transaminase and alanine aminotransferase) or the associated ones, such as liver fibrolisis and HSP72 upregulation (Creus et al., 2016; Creus et al., 2017; da Silva et al., 2018; Marineli, Lenquiste, Moraes, & Maróstica, 2015; Moura et al., 2018; Poudyal et al., 2013; Poudyal, Panchal, Waanders, Ward, & Brown, 2012; Poudyal, Panchal, Ward, Waanders, & Brown, 2012), which are also related to glucose or insulin tolerance as well as lipogenesis control (Enes et al., 2020). A chia seed fatty acid extract increased the expression of adipocyte mitochondrial thermogenesis related genes, such as uncoupling protein-1 (UCP-1), peroxisome proliferator-activated receptor gamma coactivator 1 alpha (PPARγC1α) and PRDM16 (PR domain containing 16); the extract downregulated expression of the lipid synthesis-related gene sterol regulatory element binding of protein-1c (SREBP-1c) (Pandurangan et al., 2020). In addition, the expression of adipogenesis-related genes, such as the proliferator activated receptor γ (PPARγ) and CCAAT/enhancer binding protein (C/EBPα), were downregulated by chia seed consumption. Macrophages treated with chia seeds in adipocyte condition media significantly inhibited obesity-associated inflammatory genes and expression levels of proteins, such as monocyte chemo attractant protein-1 (MCP-1), prostaglandins E2, interleukin-6, plasminogen activator inhibitor-1 (PAI-1), and tumor necrosis factor-α (TNF-α). The authors concluded that a 3 : 1 ratio of omega-3 and omega-6 fatty acid compositions can inhibit lipid accumulation and enhance fatty acid oxidation, via UCP-1 and PRDM16 expression. Macrophage recruitment to adipocytes and the development of obesity associated inflammation was suppressed by chia seeds.

Antidiabetic: Chia seeds have an impact on lipogenesis by modulating glucose metabolism biomarkers which increase PPAR (Creus et al., 2016; da Silva et al., 2018). Rats given a typical "Western diet (unbalanced diet)" after being fed chia showed improved insulin sensitivity and glucose tolerance (Enes et al., 2020). This conclusion was based on: increased adenosine monophosphate-activated protein kinase (AMPK) phosphorylation of gastrocnemius, cardiac muscles, epididymal, and subcutaneous adipose tissues; increased expression of heat shock proteins 25 (HSP25) and HSP70 in soleus muscle; and increased insulin-stimulated phosphorylation of IRS-1 and GLUT-4 translocation on gastrocnemius and heart (Chicco et al., 2009; Creus et al., 2016; Creus et al., 2017; Ferreira et al., 2016; Fonte-Faria et al., 2019; Fortino et al., 2017; Marineli, Lenquiste, Moraes, & Maróstica, 2015,Marineli, Moura, et al., 2015; Oliva et al., 2013; Poudyal, Panchal, Waanders, Ward, & Brown, 2012,Poudyal, Panchal, Ward, Waanders, & Brown, 2012; Rui, Yang, Chen, Qin, & Wan, 2018). A diet including chia seeds stimulated AMPK and HSP expression involved in cellular energy balance (Enes et al., 2020). Chia seeds facilitate AMPK phosphorylation, modulate signaling improvement in mitochondrial activity patterns, fatty acid oxidation, and adipogenesis control, and prevent the overexpression of inflammatory mediators, carnitine palmitoyltransferase 1 (CTP-1), peroxisome proliferator-activated receptor-γ coactivator 1-alpha (PGC-1α), and HSP expression (Creus et al., 2016; Creus et al., 2017; da Silva et al., 2018; Fortino et al., 2017; Marineli, Moura, et al., 2015; Musch, Kapil, & Chang, 2004). Chia seeds and chia seed oil can also restore glucose energy utilization by mitigating obesity-induced insulin sensitivity by regulating AMPK and IRS-1 phosphorylation, which improve GLUT-4 translocation and increase hexokinase and glucose 6-phosphate enzymatic activity, as well as increase CPT-1 activity, PGC-1α expression, and peritoneal glucose tolerance (Chicco et al., 2009; Creus et al., 2017; Ferreira et al., 2016; Fortino et al., 2017; Marineli, Moura, et al., 2015; Oliva et al., 2013; Poudyal, Panchal, Ward, Waanders, & Brown, 2012). Despite all these promising results, the effect of specific compounds found in chia seeds on these biomarkers and their key role in metabolic pathways has not yet been elucidated (Enes et al., 2020).

16.4.3.1 Quinoa

Cardiovascular: Noratto et al. (2019) assessed the lipid-lowering effect of quinoa intake in obese/diabetic mice to avoid fatty liver disease. After 8 weeks, quinoa intake reduced triglycerides (11%) and total cholesterol (26%) compared to control group ($p < 0.05$), and LDL-c (57%) compared to the lean group ($p < 0.05$). Db/db mice is a model used to evaluate dyslipidemia and diabetic dyslipidemia treatments. The results suggest the ability of quinoa to reduce lipoproteins and avoid fatty liver disease. The lipid-lowering effect of quinoa may be linked to its phenolics, which are correlated to their antioxidant properties (Noratto et al., 2019). Mohamed, Fouda & Mohamed et al. (2019) investigated the hypolipidemic effect of quinoa *in vivo*. Male rats (Sprague–Dawley) were induced to fatty liver disease, with a high fructose diet for 4 weeks, one group was supplemented with 20% quinoa powder. In plasma, total cholesterol, triglycerides, HDL, and LDL decreased significantly compared to NAFLD group ($p < 0.05$). The same effect was observed in liver tissue, total fat, triglycerides, and total cholesterol decreased compared to NAFLD ($p < 0.05$). The hypolipidemic effect may be due to the polyunsaturated fatty acids, phenolic compounds, and phytic acid present in quinoa powder. Additionally, vitamin E, omega-3, and omega-6 (available in quinoa powder) are antioxidants and helped to protect membrane function and brain development (Mohamed et al., 2019). Li, Lietz, & Seal, (2018) evaluated the effect of daily consumption of quinoa in cardiovascular risk markers. Quinoa enriched bread (20 g quinoa flour) was given to 37 healthy overweight men for 4 weeks. After the treatment, total cholesterol, LDL, HDL, triglycerides, apoA1, apoB, AST, and ALT were measured, and there were no significant differences between refined wheat bread group and quinoa bread group ($p > 0.05$). In conclusion, quinoa bread does not affect the markers of cardiovascular risk (Li, Lietz, Bal, et al., 2018).

***Antioxidant*:** Cao et al. (2020) evaluated the antioxidant effect of polysaccharides from quinoa in a high-fat diet in rats (Sprague–Dawley). The animals were given a high-fat diet with different doses of quinoa polysaccharides (300 and 600 mg/kg/day) for 8 weeks. MDA measures the degree of cell damage, while SOD measures the ability to eliminate harmful substances, and also inhibits atherosclerosis. The rats fed quinoa polysaccharides had lower serum MDA compared with non-treated rats ($p < 0.05$). The results showed that a high-fat diet reduced serum SOD, while SOD increased in rats treated with quinoa polysaccharide. It is well known that high-fat diets can cause an imbalance of free radicals in the body, and quinoa showed that it can intervene and/or delay oxidative damage.

***Anti-inflammatory*:** The quinoa polysaccharide fraction inhibited NO production in LPS-induced RAW 264.7 macrophages by 34.3% to 52.7% (concentration and time-dependent manner), for 24 and 48 hr, respectively (Hu et al., 2017). Noratto et al., 2019 studied the effect of quinoa intake using an obese-diabetic (db/db) mice model to measure obese and hepatic steatosis biomarkers, showing that reduced ROS levels in blood erythrocytes were similar between groups (lean and obese-diabetic). In conclusion, quinoa intake significantly reduced ($p < 0.05$) protein carbonyls and interleukin (IL-6), along with total-c accumulation in liver.

***Anticancer*:** Gawlik-Dziki et al. (2013) assessed the chemopreventive potential of quinoa ethanolic extract in two cell lines (rat prostate cancer AT-2 and MAT-LyLu cells) and found inhibition of LPO and the ability to inhibit lipoxygenase activity. Phenolic compounds in the quinoa extract prevented LPO (EC_{50} = 3.0 mg DW/mL) and inhibited lipoxygenase activity (EC_{50} = < 1 mg DW/mL). The authors concluded that phenolic compounds in quinoa ethanolic extracts may have anticarcinogenic activity via their effect on oxidative stress and ROS-dependent intracellular signaling (Gawlik-Dziki et al., 2013).

***Antidiabetic*:** Obaroakpo et al. (2020) extracted polyphenol-polysaccharides from sprouted quinoa yogurt and evaluated the antidiabetic effect in human endothelial NCI-H716 cells. Administration of the polyphenol-polysaccharides was associated with upregulation of the stimulation of glucagon-like peptide-1. This hormone is responsible for gut motility and inhibition of gastric acid secretion and has similar activity to glucagon. Lastly, the polyphenol-polysaccharides were tested in the glucagon genes HNF-3 gamma and CCK-2R and showed upregulation in the transcriptional activation of HNF-3 gamma gene receptor and CCK-2R gene receptor compared with the control (untreated) group. The authors concluded that consumption of sprouted quinoa yogurt or the isolated polyphenol-polysaccharides can influence insulin release and be a potential GLP-1 secretagogue for T2DM therapy (Obaroakpo et al., 2020).

16.5 Conclusions and Perspectives

Throughout this chapter, we present an updated review of recent studies focused on the biologically active compounds present in pseudocereals, particularly chia, amaranth, quinoa, and buckwheat, with a focus on their molecular mechanisms related to various degenerative diseases. Although in recent years, these ancient grains have gained popularity around the world and are consumed raw (in the case of chia) or added as ingredients to various food products and traditional dishes, more research is needed to understand the health benefits of their bioactive compounds beyond their contribution to nutrition. Pseudocereals contain many bioactive compounds and the available results from *in vitro*, *in vivo*, and clinical trials shed light on various compounds and provide strong evidence of their beneficial effects on human health. Nonetheless, standardized and validated experimental protocols using *in vitro*, *in vivo* approaches, and human trials to investigate the different types of bioactivity (antioxidant, anti-inflammatory, anticancer, antidiabetic, etc.) would enable more robust and systematic comparisons. Meta-analysis studies conducted between laboratories are also lacking.

Pseudocereals have been used to adapt beneficial lactic acid bacteria (*Lactobacillus fermentum, L. helveticus, Lactobacillus paralimentarius, L. plantarum, Lactobacillus pontis,* and *Lactobacillus spicheri*) and yeasts (*Issatchenkia orientalis* and *Saccharomyces cerevisiae*) and prepare sourdoughs (Giménez-Bastida, Hamdi, & Llopis, 2017). Buckwheat-based diet increases aerobic mesophilic and lactic acid bacteria, particularly *L. plantarum and Bifidobacterium spp.* with slight reduction in enterobacteria and other potentially harmful commensal bacteria in rats (Giménez-Bastida, Hamdi, & Llopis, 2017 and references therein). Moreover, quinoa reduced fermentation time and survival potential of *Lactobacillus casei* LC-1 in the formulation of a symbiotic beverage (Giménez-Bastida, Hamdi, & Llopis, 2017 and references therein).

The gut benefits of pseudocereals are due predominantly to their probiotic effects, since they are good fermentation substrates and maintain bacterial diversity during storage. Pseudocereals exhibit synbiotic effects due to the reduction of *Ruminococcacea, Lachnospiraceae, Helicobacterace,* and *Clostridium, Escherichia* and increase in *Peptoclostridium, Provolellaceae, Lactobacillus, Bifidobacterium, Enterococcus,* and *Eubacteriaceae* (Ugural & Akyol, 2020). Moreover, these effects can be used to develop products and/or formulations to treat dysbiosis, obesity, celiac disease, lactose intolerance, inflammatory bowel diseases, and inflammation-mediated chronic disorders. Amaranth consumption reduces *Ruminococcacea, Lachnospiraceae*, and *Helicobacterace* and increases *Provolellaceae* species in animal intestines. Quinoa supplementation reduces pathogenic microorganisms (*Escherichia/Shigella* and *Peptoclostridiu*) reducing thereby colonic damage in animal models. The total flavonoid content of quinoa presumably contributes to its antimicrobial activity against both gram-negative and gram-positive bacteria. Buckwheat reduces *Erysipelotrichaceae, Ruminococcacea,* and *Lachnospiraceae* and increases *Eubacteriaceae* levels. The gastro protective effects of amaranth seed oil was demonstrated in a randomized split clinical study in duodenal peptic ulcer patients (Cherkas et al., 2018). Supplementation of CO_2 extracted amaranth (*A. cruentus*) seed oil (1 mL/d for 27 days [from day 8–35]) to standard anti-*Helicobacter* treatment (30 mg lansoprazole, 1,000 mg amoxicillin, and 500 mg clarithromycin capsule twice daily for 7 days) significantly reduced histidine adducts accumulation of 4-hydroxynonenal (protein adducts of LPO) and increased heart rate variability (systemic, extragastric pathophysiological alteration parameter) in gastric mucosa in *Helicobacter pylori* positive duodenal peptic ulcer patients ($n = 36$) (Cherkas et al., 2018). Moreover, amaranth seed oil supplementation modulated redox balance and reduced oxidative stress manifestations (due to improved endothelial function and local blood flow that effectively heal ulcer) in gastric mucosa of *H. pylori*-associated duodenal peptic ulcer in clinical conditions and improved heart rate variability in duodenal peptic ulcer patients after *H. pylori* eradication.

With regard to the fate and transformations that nutrients likely undergo during digestion in real food systems, it is crucial to consider the role of the food matrix on the bioaccessibility, bioavailability, and bioactivity of these compounds. Also, consideration must be given to the presence of complex mixtures of bioactives in addition to the purified compounds used in model studies. The recent use of harmonized *in vitro* digestion studies is paving the way toward a better understanding of these phenomena. Furthermore, the increased emphasis on carrying out measurements in model systems, animals, and humans has led to the discovery of several potential biomarkers for states of disease and wellness. This has expanded our understanding of complex mixtures of compounds and purified bioactives and the potential they offer for the development of functional foods. More human trials are urgently needed to understand and elucidate the target sites of these compounds and the metabolites that may be produced during digestion and metabolism of the intact nutrients. This would improve our ability to identify appropriate uses for biomarkers and increase the level of efficiency and precision that can be attained in the development of useful diagnostic and therapeutic strategies incorporating nutrigenomics and personalized nutrition approaches. The ultimate goal of all this work is the development of functional foods formulated with pseudocereals to help prevent non-communicable diseases.

References

Abbasi, R., Janjua, S., Rehman, A., William, K., & Khan, S. W. (2015). Some preliminary studies on phytochemichals and antioxidant potential of *Fagopyrum esculentum* cultivated in Chitral, Pakistan. *The Journal of Animal and Plant Sciences, 25*(3), 576–579.

Abellán-Ruiz, M. S., Barnuevo-Espinosa, M. D., García-Santamaría, C., Contreras-Fernández, C., Aldeguer García, M., Soto-Méndez, F., . . . López-Román, F. J. (2017). Efecto del consumo de quinua (*C. quinoa*) como coadyucante en la intervención nutricional en sujetos prediabéticos. *Nutricion Hospitalaria, 34*(5), 1163–1169. https://doi.org/10.20960/nh.843

Aguilar-Toalá, J. E., Deering, A. J., & Liceaga, A. M. (2020). New insights into the antimicrobial properties of hydrolysates and peptide fractions derived from chia seed (*S. hispanica* L.). *Probiotics and Antimicrobial Proteins, 12*, 1571–1581. https://doi.org/10.1007/s12602-020-09653-8

Aguilar-Toalá, J. E., & Liceaga, A. M. (2020). Identification of chia seed (*S. hispanica* L.) peptides with enzyme inhibition activity toward skin-aging enzymes. *Amino Acids, 52*, 1149–1159. https://doi.org/10.1007/s00726-020-02879-4

Ahmed, A. Z., Mumbrekar, K. D., Satyam, S. M., Shetty, P., D'Souza, M. R., & Singh, V. K. (2021). Chia Seed Oil Ameliorates Doxorubicin-Induced Cardiotoxicity in Female Wistar Rats: An Electrocardiographic, Biochemical and Histopathological Approach. *Cardiovascular Toxicology, 21*(7), 533–542. https://doi.org/10.1007/s12012-021-09644-3

Albuquerque, B. R., Heleno, S. A., Oliveira, M. B. P. P., Barros, L., & Ferreira, I. C. F. R. (2021). Phenolic compounds: current industrial applications, limitations and future challenges. *Food and Function, 12*, 14–29. https://doi.org/10.1039/d0fo02324h

Alemán, A., Giménez, B., Peérez-Santin, E., Gómez-Guillén, M. C., & Montero, P. (2011). Contribution of Leu and Hyp residues to antioxidant and ACE-inhibitory activities of peptide sequences isolated from squid gelatin hydrolysate. *Food Chemistry, 125*(2), 334–341. https://doi.org/10.1016/j.foodchem.2010.08.058

Alemán, A., Pérez-García, S., Fernández de Palencia, P., Montero, M. P., & Gómez-Guillén, M.d.C. (2021). Physicochemical, antioxidant, and anti-Inflammatory properties of rapeseed Lecithin liposomes loading a chia (*S. hispanica* L.) seed extract. *Antioxidants, 10*(5), 693. https://doi.org/10.3390/antiox10050693

Al-Mamun, M. A., Husna, J., Khatun, M., Hasan, R., Kamruzzaman, M., Hoque, K. M. F., . . . Ferdousi, Z. (2016). Assessment of antioxidant, anticancer and antimicrobial activity of two vegetable species of Amaranthus in Bangladesh. *BMC Complementary and Alternative Medicine, 16*(1), 1–11. https://doi.org/10.1186/s12906-016-1130-0

Al-Snafi, A. E. (2017). A review on *F. esculentum*: A potential medicinal plant. *IOSR Journal of Pharmacy, 7*(3), 21–32. https://doi.org/10.9790/3013-0703012132

Alvarez-Jubete, L., Arendt, E. K., & Gallagher, E. (2009). Nutritive value and chemical composition of pseudocereals as gluten-free ingredients. *International Journal of Food Sciences and Nutrition, 60*(4), 240–257. https://doi.org/10.1080/09637480902950597

Alvarez-Jubete, L., Arendt, E. K., & Gallegher, E. (2010). Nutritive value of pseudocereals and their increasing us as functional gluten-free ingredients. *Trends in Food Science and Technology., 21*, 106–113.

Alwosais, E. Z. M., Al-Ozairi, E., Zafar, T. A., & Alkandari, S. (2021). Chia seed (*S. hispanica* L.) supplementation to the diet of adults with type 2 diabetes improved systolic blood pressure: A randomized controlled trial. *Nutrition and Health, 27*(2), 181–189.

Arnett, D. K., Blumenthal, R. S., Albert, M. A., Buroker, A. B., Goldberger, Z. D., Hahn, E. J.,. . .Virani, S. S. (2019). Circulation ACC/AHA clinical practice guideline 2019 ACC/AHA Guideline on the primary prevention of cardiovascular disease: Executive summary association task force on clinical practice guidelines. https://doi.org/10.1161/CIR.0000000000000677.

Asao, M., & Watanabe, K. (2010). Functional and bioactive properties of quinoa and amaranth. *Food Science and Technology Research, 16*(2), 163–168. https://doi.org/10.3136/fstr.16.163

Ayala-Niño, A., Contreras-López, E., Castañeda-Ovando, A., Sánchez-Franco, J. A., & González-Olivares, L. G. (2020). Amaranth proteins as a source of bioactive peptides: A review. *International Food Research Journal, 27*(1), 1–15.

Ayala-Niño, A., Rodríguez-Serrano, G. M., González-Olivares, L. G., Contreras-López, E., Regal-López, P., & Cepeda-Saez, A. (2019). Sequence identification of bioactive peptides from amaranth seed proteins (*Amaranthus hypochondriacus spp.*). *Molecules, 24*(17), 3033. https://doi.org/10.3390/molecules24173033

Ayala-Niño, A., Rodríguez-Serrano, G. M., Jiménez-Alvarado, R., Bautista-Avila, M., Sánchez-Franco, J. A., González-Olivares, L. G., & Cepeda-Saez, A. (2019). Bioactivity of peptides released during lactic fermentation of amaranth proteins with potential cardiovascular protective effect: an *in vitro* study. *Journal of Medicinal Food*, *22*(10), 976–981. https://doi.org/10.1089/jmf.2019.0039

Ayaz, A., Akyol, A., Inan-Eroglu, E., Cetin, A. K., Samur, G., & Akbiyik, F. (2017). Chia seed (*Salvia Hispanica* L.) added yogurt reduces short-term food intake and increases satiety: randomized controlled trial. *Nutrition Research and Practice*, *11*(5), 412–418.

Azeem, W., Nadeem, M., & Ahmad, S. (2015). Stabilization of winterized cottonseed oil with chia (*S. hispanica* l.) seed extract at ambient temperature. *Journal of Food Science and Technology*, *52*(11), 7191–7199. https://doi.org/10.1007/s13197-015-1823-2

Bai, C. Z., Feng, M. L., Hao, X. L., Zhong, Q. M., Tong, L. G., & Wang, Z. H. (2015). Rutin, quercetin, and free amino acid analysis in buckwheat (*Fagopyrum*) seeds from different locations. *Genetics and Molecular Research*, *14*(4), 19040–19048. https://doi.org/10.4238/2015.december.29.11

Bai, C. Z., Ji, H. J., Feng, M. L., Hao, X. L., Zhong, Q. M., Cui, X. D., & Wang, Z. H. (2015). Stimulation of dendritic cell maturation and induction of apoptosis in lymphoma cells by a stable lectin from buckwheat seeds. *Genetics and Molecular Research*, *14*(1), 2162–2175. https://doi.org/10.4238/2015.March.27.3

Ballabio, C., Uberti, F., di Lorenzo, C., Brandolini, A., Peñas, E., & Restani, P. (2011). Biochemical and Immunochemical Characterization of Different Varieties of amaranth (*Amaranthus* L. ssp.) as a safe ingredient for gluten-free products. *Agriculture Food Chemistry*, *59*(24), 12969–12974.

Barba de la Rosa, A. P., Barba-Montoya, A. B., Martínez-Cuevas, P., Hernández-Ledesma, B., León-Galván, M. F., De León-Rodríguez, A., & González, C. (2010). Tryptic amaranth glutelin digests induce endothelial nitric oxide production through inhibition of ACE: antihypertensive role of amaranth peptides. *Nitric Oxide*, *23*(2), 106–111. https://doi.org/10.1016/j.niox.2010.04.006

Barrio, D. A., & Añón, M. A. (2010). Potential antitumor properties of a protein isolated obtained from the seeds of *Amaranthus mentagazzianus*. *European Journal of Nutrition*, *49*, 73–82. https://doi.org/10.1007/s00394-009-0051-9

BDWC. (2001). Biomarkers and surrogate endpoints: Preferred definitions and conceptual framework. *Clinical Pharmacology and Therapeutics*, *69*(3), 89–95. https://doi.org/10.1067/mcp.2001.113989

Bedoya-Perales, N. S., Pumi, G., Mujica, A., Talamini, E., & Padula, A. D. (2018). Quinoa expansion in Peru and its implications for land use management. *Sustainability*, *10*(2), 532. https://doi.org/10.3390/su10020532

Berti, C., Ballabio, C., Restani, P., Porrini, M., Bonomi, F., & Lametti, S. (2004). Immunochemical and molecular properties of proteins in *C. quinoa*. *Cereal Chemistry*, *81*(2), 275–277.

Bojórquez-Velázquez, E., Velarde-Salcedo, A. J., De León-Rodríguez, A., Jimenez-Islas, H., Pérez-Torres, J. L., Herrera-Estrella, A., . . . Barba de la Rosa, A. P. (2018). Morphological, proximal composition, and bioactive compounds characterization of wild and cultivated amaranth (*Amaranthus* spp.) species. *Journal of Cereal Science*, *83*, 222–228. https://doi.org/10.1016/j.jcs.2018.09.004

Bystricka, J., Vollmannova, A., Kupecsek, A., Musilova, J., Polakova, Z., Cicova, I., & Bojnanska, T. (2011). Bioactive compounds in different plant parts of various buckwheat

(*F. esculentum Moench.*) cultivars. *Cereal Research Communications*, *39*(3), 436–444. https://doi.org/10.1556/CRC.39.2011.3.13
Caleja, C., Ribeiro, A., Barreiro, M. F., & Ferreira, I. C. F. R. (2017). Phenolic compounds as nutraceuticals or functional food ingredients. *Current Pharmaceutical Design*, *23*(19), 2787. https://doi.org/10.2174/1381612822666161227153906
Califf, R. M. (2018). Biomarker definitions and their applications. *Experimental Biology and Medicine*, *243*(3), 213–221. https://doi.org/10.1177/1535370217750088
Campos-Vega, R., Loarca-Piña, G., & Oomah, B. D. (2010). Minor components of pulses and their potential impact on human health. *Food Research International*, *43*, 461–482. https://doi.org/10.1016/j.foodres.2009.09.004
Cao Y, Zou L, Li W, Song Y, Zhao G, Hu Y. Dietary quinoa (Chenopodium quinoa Willd.) polysaccharides ameliorate high-fat diet-induced hyperlipidemia and modulate gut microbiota. *International Journal of Biological Macromolecules* 2020 Nov 15;163:55–65. doi: https://doi.org/10.1016/j.ijbiomac.2020.06.241.
Capitani, M. I., Spotorno, V., Nolasco, S. M., & Tomás, M. C. (2012). Physicochemical and functional characterization of by-products from chia (*S. hispanica* L.) seeds of Argentina. *LWT–Food Science and Technology*, *45*(1), 94–102. https://doi.org/10.1016/j.lwt.2011.07.012
Carrillo-Terán, W. I., Vilcacundo, R., & Carpio, C. (2015). Compuestos bioactivos derivados de amaranto y quinoa. *Actualización en Nutrición*, *16*(1), 18–22.
Carrizo, S. L., de Moreno de LeBlanc, A., LeBlanc, J. G., & Rollán, G. C. (2020). Quinoa pasta fermented with lactic acid bacteria prevents nutritional deficiencies in mice. *Food Research International*, *127*, 108735. https://doi.org/10.1016/j.foodres.2019.108735
Caselato-Sousa, V. M., Ozaki, M. R., De Almeida, E. A., & Amaya-Farfan, J. (2014). Intake of heat-expanded amaranth grain reverses endothelial dysfunction in hypercholesterolemic rabbits. *Food and Function*, *5*(12), 3281–3286. https://doi.org/10.1039/c4fo00468j
Chauhan, A., Saxena, D. C., & Singh, S. (2015). Total dietary fiber and antioxidant activity of gluten free cookies made from raw and germinated amaranth (*Amaranthus spp.*) flour. *LWT–Food Science and Technology*, *63*(2), 939–945. https://doi.org/10.1016/j.lwt.2015.03.115
Cherkas, A., Zarkovic, K., Cipak Gasparovic, A., Jaganjac, M., Milkovic, L., Abrahamovych, O., . . . Zarkovic, N. (2018). Amaranth oil reduces accumulation of 4-hydroxynonenal-histidine adducts in gastric mucosa and improves heart rate variability in duodenal peptic ulcer patients undergoing Helicobacter pylorieradication. *Free Radical Research*, *52*(2), 135–149.
Chicco, A. G., D'Alessandro, M. E., Hein, G. J., Oliva, M. E., & Lombardo, Y. B. (2009). Dietary chia seed (*S. hispanica L.*) rich in alpha-linolenic acid improves adiposity and normalizes hypertriacylglycerolaemia and insulin resistance in dyslipaemic rats. *British Journal of Nutrition*, *101*(1), 41–50. https://doi.org/10.1017/S000711450899053X
Chigom, E. (2020). Non-communicable diseases. In *World Health*. doi: https://doi.org/10.5005/jp/books/11410_18.
Chu, J. X., Wang, Z. L., & Han, S. Y. (2011). The effects of total flavonoids from buckwheat flowers and leaves on renal damage and PTP1B expression in type 2 diabetic rats. *Iranian Journal of Pharmaceutical Research:IJPR*, *10*(3), 511–517.
Coelho, M. S., de Araujo Aquino, S., Machado Latorres, J., & de las Mercedes Salas-Mellado, M. (2019). *In vitro* and *in vivo* antioxidant capacity of chia protein hydrolysates and peptides. *Food Hydrocolloids*, *91*(2), 19–25. https://doi.org/10.1016/j.foodhyd.2019.01.018

Coelho, M. S., Soares-Freitas, R. A. M., Gomes Arêas, J. A., Avila Gandra, E., & de las Mercedes Salas-Mellado, M. (2018). Peptides from chia present antibacterial activity and inhibit cholesterol synthesis. *Plant Foods for Human Nutrition, 73*, 101–107. https://doi.org/10.1007/s11130-018-0668-z

Constantinou, C., Papas, A., & Constantinou, A. I. (2008). Vitamin E and cancer: An insight into the anticancer activities of vitamin E isomers and analogs. *International Journal of Cancer, 123*(4), 739–752. https://doi.org/10.1002/ijc.23689

Cotabarren, J., Mabel Rosso, A., Tellechea, M., García-Pardo, J., Lorenzo Rivera, J., Obregón, W. D., & Parisi, M. G. (2019). Adding value to the chia (*S. hispanica L.*) expeller: Production of bioactive peptides with antioxidant properties by enzymatic hydrolysis with Papain. *Food Chemistry, 274*, 848–856. https://doi.org/10.1016/j.foodchem.2018.09.061

Creus, A., Benmelej, A., Villafañe, N., & Lombardo, Y. B. (2017). Dietary Salba (*S. hispanica* L.) improves the altered metabolic fate of glucose and reduces increased collagen deposition in the heart of insulin-resistant rats. *Prostaglandins, Leukotrienes and Essential Fatty Acids, 121*, 30–39. https://doi.org/10.1016/j.plefa.2017.06.002

Creus, A., Ferreira, M. R., Oliva, M. E., & Lombardo, Y. B. (2016). Mechanisms involved in the improvement of lipotoxicity and impaired lipid metabolism by dietary alpha-linolenic acid rich *S. hispanica* L. (Salba) seed in the heart of dyslipemic insulin-resistant rats. *Journal of. Clinical Medicine, 5*(2), 18. https://doi.org/10.3390/jcm5020018

Curran, J. M., Stringer, D. M., Wright, B., Taylor, C. G., Przybylski, R., & Zahradka, P. (2010). Biological response of hepatomas to an extract of *F. esculentum* M. (buckwheat) is not mediated by inositols or rutin. *Journal of Agricultural and Food Chemistry, 58*(5), 3197–3204. https://doi.org/10.1021/jf903890c

da Silva, B. P., Anunciação, P. C., Matyelka da Silva, J. C., Della Lucia, C. M., Martino, H. S. D., & Pinheiro-Sant'Ana, H. M. (2017). Chemical composition of Brazilian chia seeds grown in different places. *Food Chemistry, 221*, 1709–1716. https://doi.org/10.1016/j.foodchem.2016.10.115

da Silva, B. P., Toledo, R. C. L., Grancieri, M., Moreira, M. E. D. C., Medina, N. R., Silva, R. R., & Martino, H. S. D. (2018). Effects of chia (*S. hispanica* L.) on calcium bioavailability and inflammation in Wistar rats. *Food Research International, 116*, 592–599. https://doi.org/10.1016/j.foodres.2018.08.078

da Silva, C. S., Monteiro, C. R. D. A., da Silva, G. H. F., Sarni, R. O. S., Souza, F. I. S., Feder, D., . . . Fonseca, F. L. A. (2020). Assessing the metabolic impact of ground chia seed in overweight and obese prepubescent children: results of a double-blind randomized clinical trial. *Journal of Medicinal Food, 23*(3), 224–232.

De Carvalho, F. G., Ovídio, P. P., Padovan, G. J., Jordão Junior, A. A., Marchini, J. S., & Navarro, A. M. (2014 May). Metabolic parameters of postmenopausal women after quinoa or corn flakes intake – a prospective and double-blind study. *Internation Journal of Food Sciences and Nutrition, 65*(3), 380–385. https://doi.org/10.3109/09637486.2013.866637

De Falco, B., Amato, M., & Lanzotti, V. (2017). Chia seeds products: an overview. *Phytochemistry Reviews, 16*(4), 745–760. https://doi.org/10.1007/s11101-017-9511-7

De la Barca, A. M. C., Rojas-Martínez, M. E., Islas-Rubio, A. R., & Cabrera-Chávez, F. (2010). Gluten-free breads and cookies of raw and popped amaranth flours with attractive technological and nutritional qualities. *Plant Foods for Human Nutrition, 65*, 241–246.

de Souza Ferreira, C., de Sousa Fomes, L. D. F., Santo da Silva, G. E., & Rosa, G. (2015). Effect of chia seed (*S. hispanica* L.) consumption on cardiovascular risk factors in humans: a systematic review. *Nutricion Hospitalaria*, *32*(5), 1909–1918.

Delgado, M. C. O., Galleano, M., Añón, M. C., & Tironi, V. A. (2015). Amaranth peptides from simulated gastrointestinal digestion: antioxidant activity against reactive species. *Plant Foods for Human Nutrition*, *70*(1), 27–34. https://doi.org/10.1007/s11130-014-0457-2

Delgado, M. C. O., Nardo, A., Pavlovic, M., Rogniaux, H., Añón, M. C., & Tironi, V. A. (2016). Identification and characterization of antioxidant peptides obtained by gastrointestinal digestion of amaranth proteins. *Food Chemistry*, *197*, 1160–1167. https://doi.org/10.1016/j.oodchem.2015.11.092

Delgado, M. C. O., Tironi, V. A., & Añón, M. C. (2011). Antioxidant activity of amaranth protein or their hydrolysates under simulated gastrointestinal digestion. *LWT—Food Science and Technology*, *44*(8), 1752–1760. https://doi.org/10.1016/j.lwt.2011.04.002

Demir, E. A., & Bilgic, Y. (2019). Chia seed (*S. hispanica* L.) supplementation may contribute to raising the levels of vitamin B12: An option for the vegan diet. *Revista de Nutrição*, *32*, e180249. https://doi.org/10.1590/1678-9865201932e180249

Dhaval, A., Yadav, N., & Purwar, S. (2016). Potential applications of food derived bioactive peptides in management of health. *International Journal of Peptide Research and Therapeutics*, *22*(3), 377–398. https://doi.org/10.1007/s10989-016-9514-z

Dinu, M., Ghiselli, L., Whittaker, A., Pagliai, G., Cesari, F., Fiorillo, C., . . . Sofi, F. (2017). Consumption of buckwheat products and cardiovascular risk profile: A randomized, single-blinded crossover trial. *Nutrition, Metabolism and Cardiovascular Diseases*, *27*(1), e20–e21.

Diwakar, G., Rana, J., Saito, L., Vredeveld, D., Zemaitis, D., & Scholten, J. (2014). Inhibitory effect of a novel combination of *S. hispanica* (chia) seed and *Punica granatum* (pomegranate) fruit extracts on melanin production. *Fitoterapia*, *97*, 164–171. https://doi.org/10.1016/j.fitote.2014.05.021

Dugganaboyana, G., Palanisamy, C., Kumar, K., Muthusami, S., & Velliyur, K. (2016). Dietary evaluation, antioxidant and cytotoxic activity of crudeextractfromchiaseeds (*S. hispanica* L.) against Human Prostate Cancer Cell Line (PC-3). *International Journal of Pharmacognosy and Phytochemical Research*, *8*(8), 1358–1362.

Dziadek, K., Kopec, A., Pastucha, E., Piatkowska, E., Leszczynska, T., Pisulewska, E., . . . Francik, R. (2016). Basic chemical composition and bioactive compounds content in selected cultivars of buckwheat whole seeds, dehulled seeds and hulls. *Journal of Cereal Science*, *69*, 1–8. https://doi.org/10.1016/j.jcs.2016.02.004

Eggersdorfer, M., & Wyss, A. (2018). Carotenoids in human nutrition and health. *Archives of Biochemistry and Biophysics*, *652*, 18–26. https://doi.org/10.1016/j.abb.2018.06.001

Enes, B. N., Moreira, L. P. D., Silva, B. P., Grancieri, M., Lucio, H. G., Venâncio, V. P., . . . Martino, H. S. D. (2020). Chia seed (*S. hispanica* L.) effects and their molecular mechanisms on unbalanced diet experimental studies: A systematic review. *Journal of Food Science*, *85*(2), 226–239. https://doi.org/10.1111/1750-3841.15003

Felemban, L. F., Attar, A. M. A., & Zeid, I. M. A. (2021). Medicinal and nutraceutical benefits of chia seed (*S. hispanica*). *Journal of Pharmaceutical Research. International*, *32*(41), 15–26. https://doi.org/10.9734/jpri/2020/v32i4131040

Ferreira, M. S., Alvarez, M. R., Illesca, M. S., Giménez, P., & Lombardo, Y. B. (2016). Dietary Salba (*S. hispanica* L.) ameliorates the adipose tissue dysfunction of dyslipemic insulin-resistant rats through mechanisms involving oxidative stress, inflammatory cytokines and peroxisome proliferator-activated receptor γ. *European Journal of Nutrition, 57*(1), 83–94. https://doi.org/10.1007/s00394-016-1299-5

Fonte-Faria, T., Citelli, M., Atella, G. C., Raposo, H. F., Zago, L., de Souza, T., . . . Barja-Fidalgo, C. (2019). Chia oil supplementation changes body composition and activates in- sulin signaling cascade in skeletal muscle tissue of obese animals. *Nutrition, 58*, 167–174. https://doi.org/10.1016/j.nut.2018.08.011

Fortino, M. A., Oliva, M. E., Rodriguez, S., Lombardo, Y. B., & Chicco, A. (2017). Could post-weaning dietary chia seed mitigate the development of dyslipidemia, liver steatosis and altered glucose homeostasis in offspring exposed to a sucrose-rich diet from utero to adulthood? *Prostaglandins Leukotrienes and Essential Fatty Acids, 116*, 19–26. https://doi.org/10.1016/j.plefa.2016.11.003

Fritz, M., Vecchi, B., Rinaldi, G., & Añón, M. C. (2011). Amaranth seed protein hydrolysates have *in vivo and vitro* antihypertensive activity. *Food Chemistry, 126*(3), 878–884. https://doi.org/10.1016/j.foodchem.2010.11.065

Fujimura, M., Minami, Y., Watanabe, K., & Tadera, K. (2003). Purification, characterization, and sequencing of a novel type of antimicrobial peptides, fa-amp1 and fa-amp2, from seeds of buckwheat (*F. esculentum moench*). *Bioscience, Biotechnology and Biochemistry, 67*(8), 1636–1642. https://doi.org/10.1271/bbb.67.1636

Gandhi, A., & Dey, G. (2012). Fermentation responses and *in vitro* radical scavenging activities of*F. esculentum*. *International Journal of Food Sciences and Nutrition, 64*(1), 53–57. https://doi.org/10.3109/09637486.2012.710891

Gawlik-Dziki, U., Świeca, M., Sułkowski, M., Dziki, D., Baraniak, B., & Czyz, J. (2013). Antioxidant and anticancer activities of *C. quinoa* leaves extracts - *In vitro* study. *Food and Chemical Toxicology, 57*, 154–160. https://doi.org/10.1016/j.fct.2013.03.023

Gilani, G. S., Xiao, C. W., & Cockell, K. A. (2012). Impact of antinutritional factors in food proteins on the digestibility of protein and the bioavailability of amino acids and on protein quality. *British Journal of Nutrition, 108*, S315–S332. https://doi.org/10.1017/S0007114512002371

Giménez-Bastida, J. A., Hamdi, S., & Llopis, J. M. L. (2017). Nutritional and health implications of pseudocereal intake. In C. M. Haros & R. Schoenlechner (Eds.), *Pseudocereals: Chemistry and technology* (pp. 217–232). John Wiley & Sons, Ltd.

Giménez-Bastida, J. A., Zielinski, H., Piskula, M., Zielinska, D., & Szawara-Nowak, D. (2017). Buckwheat bioactive compounds, their derived phenolic metabolites and their health benefits. *Molecular Nutrition and Food Research, 61*(7), 1600475. https://doi.org/10.1002/mnfr.201600475

Gostner, J. M., Becker, K., Ueberall, F., & Fuchs, D. (2015). The good and bad of antioxidant foods: An immunological perspective. *Food and Chemical Toxicology, 80*, 72–79. https://doi.org/10.1016/j.fct.2015.02.012

Grancieri, M., Duarte Martino, H. S., & Gonzalez de Mejia, E. (2019). Chia seed (*S. hispanica* L.) as a source of proteins and bioactive peptides with health benefits: A review. *Comprehensive Reviews in Food Science and Food Safety, 18*, 480–499. https://doi.org/10.1111/1541-4337.12423

Grancieri, M., Duarte Martino, H. S., & Gonzalez de Mejia, E. (2021). Protein digests and pure peptides from chia seed prevented adipogenesis and inflammation by inhibiting PPAR and NF-κB pathways in 3T3L-1 adipocytes. *Nutrients, 13*, 176. https://doi.org/10.3390/nu13010176

Gulpinar, A. R., Orhan, I. E., Kan, A., Senol, F. S., Celik, S. A., & Kartal, M. (2012). Estimation of *in vitro* neuroprotective properties and quantification of rutin and fatty acids in buckwheat (*F. esculentum Moench*) cultivated in Turkey. *Food Research International, 46*, 536–543. https://doi.org/10.1016/j.foodres.2011.08.011

Guo, H., Hao, Y., Richel, A., Everaert, N., Chen, Y., Liu, M., . . . Ren, G. (2020). Antihypertensive effect of quinoa protein under simulated gastrointestinal digestion and peptide characterization. *Journal of the Science of Food and Agriculture, 100*, 5569–5576. https://doi.org/10.1002/jsfa.10609

Guo, H., Richel, A., Hao, Y., Fan, X., Everaert, N., Yang, X., & Ren, G. (2020). Novel dipeptidyl peptidase-IV and angiotensin-I-converting enzyme inhibitory peptides released from quinoa protein by in silico proteolysis. *Food Science and Nutrition, 8*, 1415–1422. https://doi.org/10.1002/fsn3.1423

Guo, X., Zhu, K., Zhang, H., & Yao, H. (2010). Anti-tumor activity of a novel protein obtained from tartary buckwheat. *International Journal of Molecular Sciences, 11*(12), 5201–5211. https://doi.org/10.3390/ijms11125201

Han, K., Li, X.-Y., Zhang, Y.-Q., He, Y.-L., Hu, R., Lu, X.-L., & Hui, J. (2020). Chia seed oil prevents high-fat diet-induced hyperlipidemia and oxidative stress in mice. *European Journal of Lipid Science and Technology*, 1900443. https://doi.org/10.1002/ejlt.201900443

Hosaka, T., Nii, Y., Tomotake, H., Itom, T., Tamanaha, A., Yamasaka, Y., & Sakai, T. (2011). Extracts of common buckwheat bran prevent sucrose digestion. *Journal of Nutritional Sciences and Vitaminology, 57*(6), 441–445. https://doi.org/10.3177/jnsv.57.441

House, N. C., Puthenparampil, D., Malayil, D., & Narayanankutty, A. (2020). Variation in the polyphenol composition, antioxidant, and anticancer activity among different Amaranthus species. *South African Journal of Botany, 135*, 408–412. https://doi.org/10.1016/j.sajb.2020.09.026

Hu, Y., Zhang, J., Zou, L., Fu, C., Li, P., & Zhao, G. (2017). Chemical characterization, antioxidant, immune-regulating and anticancer activities of a novel bioactive polysaccharide from *C. quinoa* seeds. *International Journal of Biological Macromolecules, 99*, 622–629. https://doi.org/10.1016/j.ijbiomac.2017.03.019

Huda, M. N., Lu, S., Jahan, T., Ding, M., Jha, R., Zhang, K., . . . Zhou, M. (2021). Treasure from garden: Bioactive compounds of buckwheat. *Food Chemistry, 335*(127), 653. https://doi.org/10.1016/j.foodchem.2020.127653

Hunter, R. W., & Bailey, M. A. (2015). Glucocorticoids and 11β-hydroxysteroid dehydrogenases: Mechanisms for hypertension. *Current Opinion in Pharmacology, 21*, 105–114. https://doi.org/10.1016/j.coph.2015.01.005

IDF, International Diabetes Federation. (2019). *IDF diabetes atlas* (9th ed.). Brussels, Belgium: International Diabetes Federation. In *Atlas de la Diabetes de la FID*. Retrieved from http://www.idf.org/sites/default/files/Atlas-poster-2014_ES.pdf

Jeong, S. K., Park, H. J., Park, B. D., & Kim, I.-H. (2010). Effectiveness of topical chia seed oil on pruritus of end-stage renal disease (ESRD) Patients and Healthy Volunteers. *Annals of Dermatology, 22*(2), 143. https://doi.org/10.5021/ad.2010.22.2.143

Jin, H.-M., & Wei, P. (2011). Anti-fatigue properties of tartary buckwheat extracts in mice. *International Journal of Molecular Sciences, 12*(8), 4770–4780. https://doi.org/10.3390/ijms12084770

Jnawali, P., Kumar, V., & Tanwar, B. (2016). Celiac disease: Overview and considerations for development of gluten-free foods. *Food Science and Human Wellness, 5*, 169–176.

Julio, L. M., Capitani, M. I., Guiotto, E. N., Ixtaina, V. Y., Tomás, M., & Nolasco, S. (2016). Characterization ofchia (*Salvia Hispanica* L.) seed and by-products from Argentina. In D. Betancur-Ancona & M. Segura-Campos (Eds.), *S. hispanica L.: Properties, Applications and health* (pp. 87–105). Nova Science Publishers, Inc.

Kalinová, J. P., Vrchotova, N., & Triska, J. (2019). Phenolics levels in different parts of common buckwheat (*F. esculentum*) achenes. *Journal of Cereal Science, 85*, 243–248. https://doi.org/10.1016/j.jcs.2018.12.012

Knez Hrnčič, M., Ivanovski, M., Cör, D., & Knez, Ž. (2019). Chia seeds (*S. hispanica* L.): An overview—phytochemical profile, isolation methods, and application. *Molecules, 25*(1), 11. https://doi.org/10.3390/molecules25010011

Kothiyal, P., & Ratan, P. (2011). Antistress effect of *F. esculentum* in rats subjected to forced swimming endurance test. *Pharmacologyonline, 3*, 290–296.

Koyama, M., Naramoto, K., Nakajima, T., Aoyama, T., Watanabe, M., & Nakamura, K. (2013). Purification and identification of antihypertensive peptides from fermented buckwheat sprouts. *Journal of Agricultural and Food Chemistry, 61*(12), 3013–3021. https://doi.org/10.1021/jf305157y

Kulczyński, B., Kobus-Cisowska, J., Taczanowski, M., Kmiecik, D., & Gramza-Michałowska, A. (2019). The chemical composition and nutritional value of chia seeds—current state of knowledge. *Nutrients, 11*(6), 1242. https://doi.org/10.3390/nu11061242

Kyung-Hwa, K. (2012). *F. esculentum* extract suppresses the release of inflammatory mediator and proximal signal events inFcεRI-mediated RBL-2H3 cell activation. *The Korean Association of Oriental Medical Physiology, 26*(4), 469–474.

Lai, C.-S., Wu, J.-C., & Pan, M.-H. (2015). Molecular mechanism on functional food bioactives for anti-obesity. *Current Opinion in Food Science, 2*, 9–13. https://doi.org/10.1016/j.cofs.2014.11.008

Lamothe, L. M., Srichuwong, S., Reuhs, B. L., & Hamaker, B. R. (2015). Quinoa (*C. quinoa W.*) and amaranth (*A. caudatus* L.) provide dietary fibers high in pectic substances and xyloglucans. *Food Chemistry, 167*, 490–496. https://doi.org/10.1016/j.foodchem.2014.07.022

Larbie, C., Appiah-Opong, R., Acheampong, F., Tuffour, I., Torkornoo, D., Marfo, E., . . . Abotsi, P. (2015). Anti-proliferative effect of *A. viridis* Linn. on human leukemic cell lines—A preliminary study. *International Journal of Biological and Pharmaceutical Research, 6*(3), 236–243. Retrieved from www.ijbpr.com

Lee, Y. J., Kim, K. J., Park, K. J., Yoon, B. R., Lim, J. H., & Lee, O. H. (2013). Buckwheat (*F. esculentum* M.) sprout treated with methyl jasmonate (MeJA) improved anti-adipogenic activity associated with the oxidative stress system in 3T3-L1 adipocytes. *International Journal of Molecular Sciences, 14*(1), 1428–1442. https://doi.org/10.3390/ijms14011428

Leung, E. H. W., & Ng, T. B. (2007). A relatively stable antifungal peptide from buckwheat seeds with antiproliferative activity toward cancer cells. *Journal of Peptide Science, 13*(11), 762–767. https://doi.org/10.1002/psc.891

Li, C. H., Matsui, T., Matsumoto, K., Yamasaki, R., & Kawasaki, T. (2002). Latent production of angiotensin I-converting enzyme inhibitors from buckwheat protein. *Journal ofPeptide Science*, *8*(6), 267–274. https://doi.org/10.1002/psc.387

Li, L., Lietz, G., Bal, W., Watson, A., Morfey, B., & Seal, C. (2018). Effects of quinoa (*C. quinoa Willd.*) consumption on markers of CVD risk. *Nutrients*, *10*(6), 1–17. https://doi.org/10.3390/nu10060777

Li, L., Lietz, G., & Seal, C. (2018). Buckwheat and CVD risk markers: a systematic review and meta-analysis. *Nutrients*, *10*(5), 619. https://doi.org/10.3390/nu10050619

Luna-Suárez, S., Medina-Godoy, S., Cruz-Hernández, A., & Paredes-López, O. (2010). Modification of the amaranth 11S globulin storage protein to produce an inhibitory peptide of angiotensin I converting enzyme, and its expression in *Escherichia coli*. *Journal of Biotechnology*, *148*(4), 240–247. https://doi.org/10.1016/j.jbiotec.2010.06.009

Luo, X., Fei, Y., Xu, Q., Lei, T., Mo, X., Wang, Z., . . . Li, H. (2020). Isolation and identification of antioxidant peptides from tartary buckwheat albumin (*Fagopyrum tataricumGaertn.*) and their antioxidant activities. *Journal of Food Science*, *85*(3), 611–617. https://doi.org/10.1111/1750-3841.15004

Luthar, Z., Germ, M., Likar, M., Golob, A., Vogel-Mikuš, K., Pongrac, P., . . . Kreft, I. (2020). Breeding buckwheat for increased levels of rutin, quercetin and other bioactive compounds with potential antiviral Effects. *Plants*, *9*, 1638. https://doi.org/10.3390/plants9121638

Ma, M.-S., In, Y. B., Hyeon, G. L., & Yang, C.-B. (2006). Purification and identification of angiotensin I-converting enzyme inhibitory peptide from buckwheat (*F. esculentum Moench*). *Food Chemistry*, *96*(1), 36–42. https://doi.org/10.1007/978-3-642-61625-9_33

Ma, Y., Xiong, Y. L., Zhai, J. J., Zhu, H. N., & Dziubla, T. (2010). Fractionation and evaluation of radical scavenging peptides from *in vitro* digests of buckwheat protein. *Food Chemistry*, *118*, 582–588. https://doi.org/10.1016/j.foodchem.2009.05.024

Madadi, E., Mazloum-Ravasan, S., Yu, J. S., Ha, J. W., Hamishehkar, H., & Kim, K. H. (2020). Therapeutic application of betalains: A review. *Plants*, *9*, 1219. https://doi.org/10.3390/plants9091219

Mahdavi Yekta, M., Nouri, L., Hossein Azizi, M., Karimi Dehkordi, M., Mohammadi, M., Jabbari, M., . . . Mousavi Khaneghah, A. (2020). Peptide extracted from quinoa by pepsin and alcalase enzymes hydrolysis: Evaluation of the antioxidant activity. *Journal of Food Processing and Preservation*, *44*, e14773. https://doi.org/10.1111/jfpp.14773

Marineli, R. S., Lenquiste, S. A., Moraes, É. A., & Maróstica, M. R. (2015). Antioxidant potential of dietary chia seed and oil (*S. hispanica* L.) in diet-induced obese rats. *Food Research International*, *76*, 666674. https://doi.org/10.1016/j.foodres.2015.07.039

Marineli, R. S., Moura, C. S., Moraes, E. A., Lenquiste, S. A., Lollo, P. C. B., Morato, P. N., & Marostica, M. R. J. (2015). Chia (*S. hispanica* L.) enhances HSP, PGC-1a expressions and improves glucose tolerance in diet-induced obese rats. *Nutrition*, *31*(5), 740–748. https://doi.org/10.1016/j.nut.2014.11.009

Martínez, E. A., Fuentes, F., & Bazile, D. (2015). History of Quinoa: its origin, domestication, diversification and cultivation with particular reference to the chilean context. In *Quinoa: Improvement and sustainable production* (pp. 19–24). Hoboken, New Jersey, USA: *John Wiley & Sons Inc*. https://doi.org/10.1002/9781118628041.ch2

Martínez Leo, E. E., & Segura Campos, M. R. (2020). Neuroprotective effect of peptide fractions from chia (*S. hispanica*) on H_2O_2-induced oxidative stress-mediated neuronal damage on N1E-115 cell line. *Neurochemical Research*, *45*, 2278–2285. https://doi.org/10.1007/s11064-020-03085-0

Martínez-Cruz, O., & Paredes-López, O. (2014). Phytochemical profile and nutraceutical potential of chia seeds (*S. hispanica* L.) by ultra high performance liquid chromatography. *Journal of Chromatography A*, *1346*, 43–48. https://doi.org/10.1016/j.chroma.2014.04.007

Martínez-Villaluenga, C., Peñas, E., & Hernández-Ledesma, B. (2020). Pseudocereal grains: Nutritional value, health benefits and current applications for the development of gluten-free foods. *Food and Chemical Toxicology*, *137*, 1–26.

Mastebroek, H. D., Limburg, H., Gilles, T., & Marvin, H. J. P. (2000). Occurrence of sapogenins in leaves and seeds of quinoa (*C. quinoa Willd*). *Journal of the Science of Food and Agriculture*, *80*(1), 152–156. doi:https://doi.org/10.1002/(sici)1097-0010(20000101)80:1<152::aid-jsfa503>3.0.co;2-p.

McCormack, W. P., Hoffman, J. R., Pruna, G. J., Jajtner, A. R., Townsend, J. R., Stout, J. R., . . . Fukuda, D. H. (2015). Effects of L-alanyl- l-glutamine ingestion on one-hour run performance. *Journal of the American College of Nutrition*, *34*(6), 488496. https://doi.org/10.1080/07315724.2015.1009193

Meléndez-Martinez, A. J. (2019). An overview of carotenoids, apocarotenoids, and vitamin A in agro-food, nutrition, health, and disease. *Molecular Nutrition and Food Research*, *63*, 1801045. https://doi.org/10.1002/mnfr.201801045

Melini, V., & Melini, F. (2021). Functional components and anti-nutritional factors in gluten-free grains: A focus on quinoa seeds. *Foods*.. https://doi.org/10.3390/foods10020351

Mendonça, S., Saldiva, P. H., Cruz, R. J., & Arêas, J. A. G. (2009). Amaranth protein presents cholesterol-lowering effect. *Food Chemistry*, *116*(3), 738–742. https://doi.org/10.1016/j.foodchem.2009.03.021

Mihafu, F. D., Kiage, B. N., Kimang'a, A. N., & Okoth, J. K. (2020). Effect of chia seeds (*S. hispanica*) on postprandial glycaemia, body weight and hematological parameters in rats fed a high fat and fructose diet. *International Journal of Biological and Chemical Sciences*, *14*(5), 1752–1762. https://doi.org/10.4314/ijbcs.v14i5.20

Minekus, M., Alminger, M., Alvito, P., Balance, S., Bohn, T., Bourlieu, C., . . . Brodkorb, A. (2014). A standardized static *in vitro* digestion method suitable for food—An international consensus. *Food and Function*, *5*(6), 1113–1124. https://doi.org/10.1039/c3fo60702j

Mir, N. A., Riar, C. S., & Singh, S. (2018). Nutritional constituents of pseudo cereals and their potential use in food systems: A review. *Trends in Food Science and Technology*, *75*, 170–180. https://doi.org/10.1016/j.tifs.2018.03.016

Mohamed, D. A., Fouda, K. A., & Mohamed, R. S. (2019). *in vitro* anticancer activity of quinoa and safflower seeds and their preventive effects on non-alcoholic fatty liver. *Pakistan Journal of Biological Sciences*, *22*(8), 383–392. https://doi.org/10.3923/pjbs.2019.383.392

Montoya-Rodríguez, A., & de Mejía, E. G. (2015). Pure peptides from amaranth (*A. hypochondriacus*) proteins inhibit LOX-1 receptor and cellular markers associated with atherosclerosis development *in vitro*. *Food Research International*, *77*, 204–214. https://doi.org/10.1016/j.foodres.2015.06.032

Montoya-Rodríguez, A., de Mejía, E. G., Dia, V. P., Reyes-Moreno, C., & Milán-Carrillo, J. (2014). Extrusion improved the anti-inflammatory effect of amaranth (*A. hypochondriacus*) hydrolysates in LPS-induced human THP-1 macrophage-like and mouse RAW 264.7 macrophages by preventing activation of NF-κB signaling. *Molecular Nutrition and Food Research*, *58*(5), 1028–1041. https://doi.org/10.1002/mnfr.201300764

Morales Quezada, I., & del Castilo, R. (2009). Amaranthfrom the sacred to the everyday. *Voices of Mexico*. 86 – UNAM

Moronta, J., Smaldini, P. L., Docena, G. H., & Añón, M. C. (2016). Peptides of amaranth were targeted as containing sequences with potential anti-inflammatory properties. *Journal of Functional Foods*, *21*, 463–473. https://doi.org/10.1016/j.jff.2015.12.022

Moronta, J., Smaldini, P. L., Fossati, C. A., Añón, M. C., & Docena, G. H. (2016). The anti-inflammatory SSEDIKE peptide from Amaranth seeds modulates IgE-mediated food allergy. *Journal of Functional Foods*, *25*, 579–587. https://doi.org/10.1016/j.jff.2016.06.031

Mota, C., Santos, M., Mauro, R., Samman, N., Matos, A. S., Torres, D., & Castanheira, I. (2016). Protein content and amino acids profile of pseudocereals. *Food Chemistry*, *193*, 55–61. https://doi.org/10.1016/j.foodchem.2014.11.043

Moura, C. S., Lollo, P. C. B., Morato, P. N., & Amaya-Farfan, J. (2018). Dietary nutrients and bioactive substances modulate heat shock protein (HSP) expression: A review. *Nutrients*, *10*(6), 1–13. https://doi.org/10.3390/nu10060683

Mudgil, P., Kilari, B. P., Kamal, H., Abayomi Olalere, O., FitzGerald, R. J., Gan, C.-Y., & Maqsood, S. (2020). Multifunctional bioactive peptides derived from quinoa protein hydrolysates: Inhibition of α-glucosidase, dipeptidyl peptidase-IV and angiotensin I converting enzymes. *Journal of Cereal Science*, *96*, 103130. https://doi.org/10.1016/j.jcs.2020.103130

Muñoz, L. A., Cobos, A., Diaz, O., & Aguilera, J. M. (2012). Chia seeds: Microstructure, mucilage extraction and hydration. *Journal of Food Engineering*, *108*(1), 216–224. https://doi.org/10.1016/j.jfoodeng.2011.06.037

Musch, M. W., Kapil, A., & Chang, E. B. (2004). Heat shock protein 72 binds and protects dihy- drofolate reductase against oxidative injury. *Biochemical and Biophysical Research Communications*, *313*(1), 185–192. https://doi.org/10.1016/j.bbrc.2003.11.096

Muzquiz, M., Varela, A., Burbano, C., Cuadrado, C., Guillamón, E., & Pedrosa, M. M. (2012). Bioactive compounds in legumes: pronutritive and antinutritive actions. Implications for nutrition and health. *Phytochemistry Reviews*, *11*, 227–244. https://doi.org/10.1007/s11101-012-9233-9

Nardo, A. E., Suárez, S., Quiroga, A. V., & Añón, M. C. (2020). Amaranth as a source of antihypertensive peptides. *Frontiers in Plant Science*, *11*, 578631. https://doi.org/10.3389/fpls.2020.578631

Navarro-Perez, D., Radcliffe, J., Tierney, A., & Jois, M. (2017). Quinoa seed lowers serum triglycerides in overweight and obese subjects: a dose–response randomized controlled clinical trial. *Current Developments in Nutrition*, *1*(9), e001321.

Niki, E., & Abe, K. (2019). Vitamin E: Structure, properties and functions. In E. Niki (Ed.), *Vitamin E: Chemistry and nutritional benefits* (pp. 1–11). Royal Society of Chemistry.

Noratto, G. D., Murphy, K., & Chew, B. P. (2019). Quinoa intake reduces plasma and liver cholesterol, lessens obesity-associated inflammation, and helps to prevent hepatic steatosis in obese db/db mouse. *Food Chemistry*, *287*(November 2018), 107–114. https://doi.org/10.1016/j.foodchem.2019.02.061

Nowak, V., Du, J., & Charrondière, U. R. (2016). Assessment of the nutritional composition of quinoa (*C. quinoa Willd.*). *Food Chemistry*, *193*, 47–54. https://doi.org/10.1016/j.foodchem.2015.02.111

Obaroakpo, J. U., Liu, L., Zhang, S., Jing, L., Liu, L., Pang, X., & Lv, J. (2020). Bioactive assessment of the antioxidative and antidiabetic activities of oleanane triterpenoid isolates of sprouted quinoa yogurt beverages and their anti-angiogenic effects on HUVECS line. *Journal of Functional Foods*, *66*. https://doi.org/10.1016/j.jff.2020.103779

Obaroakpo, J. U., Liu, L., Zhang, S., Lu, J., Pang, X., & Lv, J. (2019). α-Glucosidase and ACE dual inhibitory protein hydrolysates and peptide fractions of sprouted quinoa yogurt beverages inoculated with *L. casei*. *Food Chemistry*, *299*, 1–9. https://doi.org/10.1016/j.foodchem.2019.124985

Odongo, G. A., Schlotz, N., Baldermann, S., Neugart, S., Ngwene, B., Schreiner, M., & Lamy, E. (2018). Effects of *Amaranthus cruentus* L. on aflatoxin B1- and oxidative stress-induced DNA damage in human liver (HepG2) cells. *Food Bioscience*, *26*, 42–48. https://doi.org/10.1016/j.fbio.2018.09.006

Ogrodowska, D., Zadernowski, R., Czaplicki, S., Derewiaka, D., & Wronowska, B. (2014). Amaranth seeds and products—The source of bioactive compounds. *Polish Journal of Food and Nutrition Science*, *64*(3), 165–170. https://doi.org/10.2478/v10222-012-0095-z

Oliva, M. E., Ferreira, M. R., Chicco, A., & Lombardo, Y. B. (2013). Dietary salba (*S. hispanica* L.) seed rich in alpha-linolenic acid improves adipose tissue dysfunction and the altered skeletal muscle glucose and lipid metabolism in dyslipidemic insulin-resistant rats. *Prostaglandins, Leukotrienes, and Essential Fatty Acids*, *89*(5), 279–289. https://doi.org/10.1016/j.plefa.2013.09.010

Oliveira-Alves, S. C., Vendramini-Costa, D. B., Betim Cazarin, C. B., Maróstica Júnior, M. R., Borges Ferreira, J. P., Silva, A. B., & Bronze, M. R. (2017). Characterization of phenolic compounds in chia (*S. hispanica* L.) seeds, fiber flour and oil. *Food Chemistry*, *232*, 295–305. https://doi.org/10.1016/j.foodchem.2017.04.002

Oomah, B. D., Campbell, C. G., & Mazza, G. (1996). Effects of cultivar and environment on phenolic acids in buckwheat. *Euphytica*, *90*, 73–77. https://doi.org/10.1007/BF00025162

Oparin, P. B., Mineev, K. S., Dunaevsky, Y. E., Arseniev, A. S., Belozersky, M. A., Grishin, E. V., . . . Vassilevski, A. A. (2012). Buckwheat trypsin inhibitor with helical hairpin structure belongs to a new family of plant defense peptides. *Biochemistry Journal*, *446*(1), 69–77. https://doi.org/10.1042/bj20120548

Orona-Tamayo, D., Valverde, M. E., & Paredes-López, O. (2017). Chia—The new golden seed for the twenty-first century: Nutraceutical properties and technological uses Chapter 17. In S. Nadathur, W. JPD, & L. Scanlin (Eds.), *Sustainable Protein Sources* (pp. 265–281). Elsevier.

Owczarczyk-Saczonek, A., Lahuta, L. B., Placek, W., & Górecki, R. J. (2018). The potential benefits of plant cyclitols in the treatment of psoriasis. *Polish Annals of Medicine*, *25*(1), 166–171. https://doi.org/10.29089/2017.17.00019

Pandurangan, S.-B., Al-Maiman, S. A., Al-Harbi, L. N., & Alshatwi, A. A. (2020). Beneficial fatty acid ratio of *S. hispanica* L. (Chia Seed) potentially inhibits adipocyte hypertrophy, and decreases adipokines expression and inflammation in macrophage. *Foods*, *9*(3), 368. https://doi.org/10.3390/foods9030368

Patiño, P. J., & Robles, V. Q. (2016). Chia seed: Dietary fiber, bioactive compounds and its application in functional foods. In D. Betancur-Ancona & M. Segura-Campos (Eds.), *S. hispanica L.: Properties, applications and health* (pp. 1–14). Nova Science Publishers, Inc.

Pavlidis, C., Lanara, Z., Balasopoulou, A., Nebel, J. C., Katsila, T., & Patrinos, G. P. (2015). Meta-analysis of genes in commercially available nutrigenomic tests denotes lack of association with dietary intake and nutrient-related pathologies. *OMICS*, *19*(9), 512–520. https://doi.org/10.1089/omi.2015.0109

Peñas, E., Uberti, F., di Lorenzo, C., Ballabio, C., Brandolini, A., & Restani, P. (2014). Biochemical and immunochemical evidences supporting the inclusion of quinoa (*C. quinoa Willd*) as a gluten-free ingredient. *Plants Foods for Human Nutrition.*, *69*, 297–303.

Pirzadah, T. B., & Malik, B. (2020). Pseudocereals as super foods of twenty-first-century: Recent technological interventions. *Journal of Agriculture and Food Research*, *2*(100), 052. https://doi.org/10.1016/j.jafr.2020.100052

Poudyal, H., Panchal, S. K., Waanders, J., Ward, L., & Brown, L. (2012). Lipid redistribution by alpha-linolenic acid-rich chia seed inhibits stearoyl-CoA desaturase-1 and induces cardiac and hepatic protection in diet-induced obese rats. *Journal of Nutritional Biochemistry*, *23*(2), 153–162. https://doi.org/10.1016/j.jnutbio.2010.11.011

Poudyal, H., Panchal, S. K., Ward, L. C., & Brown, L. (2013). Effects of ALA, EPA and DHA in high-carbohydrate, high-fat diet-induced metabolic syndrome in rats. *Journal of Nutritional Biochemistry*, *24*(6), 1041–1052. https://doi.org/10.1016/j.jnutbio.2012.07.014

Poudyal, H., Panchal, S. K., Ward, L. C., Waanders, J., & Brown, L. (2012). Chronic high-carbohydrate, high-fat feeding in rats induces reversible metabolic, cardiovascular, and liver changes. *American Journal of Physiology, Endocrinology and Metabolism*, *302*(12), E1472–E1482. https://doi.org/10.1152/ajpendo.00102.2012

Prasadi, N. V. P., & Joyce, I. J. (2020). Dietary fiber from whole grains and their benefits on metabolic health. *Nutrients*, *12*, 3045. https://doi.org/10.3390/nu12103045

Preetham Kumar, K. V., Dharmaraj, U., Sakhare, S. D., & Inamdar, A. A. (2016). Preparation of protein and mineral rich fraction from grain amaranth and evaluation of its functional characteristics. *Journal of Cereal Science*, *69*, 358–362. https://doi.org/10.1016/j.jcs.2016.05.002

Qiu, J., Li, Z., Qin, Y., Yue, Y., & Liu, Y. (2016). Protective effect of tartary buckwheat on renal function in type 2 diabetics: a randomized controlled trial. *Therapeutics and Clinical Risk Management*, *12*, 1721–1727.

Quiroga, A. V., Aphalo, P., Nardo, A. E., & Añón, M. C. (2017). *In vitro* modulation of renin-angiotensin system enzymes by amaranth (*A. hypochondriacus*) protein-derived peptides: alternative mechanisms different from ACE inhibition. *Journal of Agricultural and Food Chemistry*, *65*(34), 7415–7423. https://doi.org/10.1021/acs.jafc.7b02240

Quiroga, A. V., Aphalo, P., Ventureira, J. L., Martínez, E. N., & Añón, M. C. (2011). Physicochemical, functional and angiotensin converting enzyme inhibitory properties of amaranth (*A. hypochondriacus*) 7S globulin. *Journal of the Science of Food and Agriculture*, *92*(2), 397–403. https://doi.org/10.1002/jsfa.4590

Quiroga, A. V., Barrio, D. A., & Añón, M. C. (2014). Amaranth lectin presents potential antitumor properties. *LWT*, *60*(1), 478–485. https://doi.org/10.1016/j.lwt.2014.07.035

Quiroga, A. V., Barrio, D. A., & Añón, M. C. (2015). Amaranth lectin presents potential antitumor properties. *LWT—Food Science and Technology*, *60*(1), 478–485. https://doi.org/10.1016/j.lwt.2014.07.035

Raguindin, P. F., Itodo, O. A., Stoyanov, J., Dejanovic, G. M., Gamba, M., Asllanaj, E., . . . Kern, H. (2021). A systematic review of phytochemicals in oat and buckwheat. *Food Chemistry*, *338*, 127982. https://doi.org/10.1016/j.foodchem.2020.127982

Rajnincova, D., Galova, Z., Spalekova, A., Balazova, Z., & Chnapek, M. (2019). Analysis of wheat, oat, and buckwheat proteome from aspect of celiac active polypeptides presence. *Journal of Biotechnology.*, *305*, s33–s88.

Ramírez-Torres, G., Ontiveros, N., Lopez-Teros, V., Ibarra-DIarte, J. A., Reyes-Moreno, C., Cuevas-Rodríguez, E. O., & Cabrera-Chávez, F. (2017). Amaranth protein hydrolysates efficiently reduce systolic blood pressure in spontaneously hypertensive rats. *Molecules*, *22*(11). https://doi.org/10.3390/molecules22111905

Rodríguez, J. P., Rahman, H., Thushar, S., & Singh, R. K. (2020). Healthy and resilient cereals and pseudo-cereals for marginal agriculture: Molecular advances for improving nutrient bioavailability. *Frontiers in Genetics*, *11*, 49. https://doi.org/10.3389/fgene.2020.00049

Rodríguez, M., García Fillería, S. F., & Tironi, V. A. (2020). Simulated gastrointestinal digestion of amaranth flour and protein isolate: Comparison of methodologies and release of antioxidant peptides. *Food Research International*, *138*(109), 735. https://doi.org/10.1016/j.foodres.2020.109735

Rowland, I., Gibson, G., Heinken, A., Scott, K., Swann, J., Thiele, I., & Tuohy, K. (2018). Gut microbiota functions: Metabolism of nutrients and other food components. *European Journal of Nutrition*, *57*(1), 1–24. https://doi.org/10.1007/s00394-017-1445-8

Ruales, J., de Grijalva, Y., Lopez-Jaramillo, P., & Nair, B. M. (2002 Mar). The nutritional quality of an infant food from quinoa and its effect on the plasma level of insulin-like growth factor-1 (IGF-1) in undernourished children. *Internation Journal of Food Sciences and Nutrition*, *53*(2), 143–154. https://doi.org/10.1080/09637480220132157

Ruan, J. J., Chen, H., Shao, J. R., Wu, Q., & Han, X. Y. (2011). An antifungal peptide from *F. tataricum* seeds. *Peptides*, *32*(6), 1151–1158. https://doi.org/10.1016/j.peptides.2011.03.015

Rubavathi, S., Ayyappadasan, G., Sangeetha, N., Harini, T., Saranya, D., & Harshapradha, P. (2020). Studies on antioxidant and anti-obesity activity of *salvia hispanica* (Chia) seeds extracts. *Journal of Drug Delivery and Therapeutics*, *10*(3-s), 98–106. https://doi.org/10.22270/jddt.v10i3-s.4169

Rui, Y., Yang, S., Chen, L. H., Qin, L. Q., & Wan, Z. (2018 Aug). Chia Seed supplementation reduces senescence markers in epididymal adipose tissue of high-fat diet-fed SAMP8 Mice. *Journal of Medicinal Food*, *21*(8), 755–760. https://doi.org/10.1089/jmf.2017.4129

Rybicka, I. (2018). The handbook of minerals on a gluten-free diet. *Nutrients*. https://doi.org/10.3390/nu10111683

Sabbione, A. C., Luna-Vital, D., Scilingo, A., Añón, M. C., & González de Mejía, E. (2018). Amaranth peptides decreased the activity and expression of cellular tissue factor on LPS activated THP-1 human monocytes. *Food and Function*, *9*(7), 3823–3834. https://doi.org/10.1039/c8fo00323h

Sabbione, A. C., Nardo, A. E., Añón, M. C., & Scilingo, A. (2016). Amaranth peptides with antithrombotic activity released by simulated gastrointestinal digestion. *Journal of Functional Foods*, *20*, 204–214. https://doi.org/10.1016/j.jff.2015.10.015

Sabbione, A. C., Scilingo, A., & Añón, M. C. (2015). Potential antithrombotic activity detected in amaranth proteins and its hydrolysates. *LWT—Food Science and Technology*, *60*(1), 171–177. https://doi.org/10.1016/j.lwt.2014.07.015

Salazar-Vega, I. M., Rosado-Rubio, G., Chel-Guerrero, L. A., Betancur-Ancona, D. A., & Castellanos-Ruelas, A. F. (2009). Composición en ácido graso alfa linolénico (w3) en el huevo y carne de aves empleando chia (*S. hispanica*) en el alimento. *Interciencia*, *34*, 209e213.

Salazar-Vega, I. M., Segura-Campos, M. R., Chel-Guerrero, L. A., & Betancur-Ancona, D. A. (2012). Antihypertensive and antioxidant effects of functional foods containing chia (Salvia hispanica) protein hydrolysates. In *Scientific, Health and Social Aspects of the Food Industry*. IntechOpen. https://doi.org/10.5772/30993

Sandoval-Sicairos, E. S., Milán-Noris, A. K., Luna-Vital, D. A., Milán-Carrillo, J., & Montoya-Rodríguez, A. (2021). Anti-inflammatory and antioxidant effects of peptides released from germinated amaranth during *in vitro* simulated gastrointestinal digestion. *Food Chemistry, 343*, 128394. https://doi.org/10.1016/j.foodchem.2020.128394

Sargi, S. C., Silva, B. C., Santos, H. M. C., Montanher, P. F., Boeing, J. S., Santos Júnior, O. O., & Visentainer, J. V. (2013). Antioxidant capacity and chemical composition in seeds rich in omega-3: Chia, flax, and perilla. *Food Science and Technology, 33*(3), 541–548. https://doi.org/10.1590/s0101-20612013005000057

Sarker, U., & Oba, S. (2020). Polyphenol and flavonoid profiles and radical scavenging activity in leafy vegetable *Amranthus Gangeticus*. *BMC Plant Biology, 20*(1). https://doi.org/10.1186/s12870-020-02700-0

Schiemer, C., Halloran, A., Jespersen, K., & Kaukua, P. (2018). Marketing Insects: Superfood or Solution-Food? *Edible Insects in Sustainable Food Systems*, 213–236. https://doi.org/10.1007/978-3-319-74011-9_14

Schröter, D., Neugart, S., Schreiner, M., Grune, T., Rohn, S., & Ott, C. (2019). Amaranth's 2-caffeoylisocitric acid—an anti-inflammatory caffeic acid derivative that impairs NF-κB signaling in LPS-Challenged RAW 264.7 Macrophages. *Nutrients, 11*(3). https://doi.org/10.3390/nu11030571

Segura-Campos, M. R., Peralta González, F., Guerrero, L. C., & Betancur Ancona, D. (2013). Angiotensin I-converting enzyme inhibitory peptides of chia (*S. hispanica*) produced by enzymatic hydrolysis. *International Journal of Food Science, 2013*, 158482. https://doi.org/10.1155/2013/158482

Segura-Campos, M. R., Salazar-Vega, I. M., Chel-Guerrero, L. A., & Betancur-Ancona, D. A. (2013). Biological potential of chia (*S. hispanica* L.) protein hydrolysates and their incorporation into functional foods. *LWT—Food Science and Technology, 50*(2), 723–731. https://doi.org/10.1016/j.lwt.2012.07.017

Selma-Gracia, R., Haros, C. M., & Laparra, J. (2020). Inclusion of *S. hispanica* L. and *C. quinoa* into bread formulations improve metabolic imbalances derived from a high fat intake in hyperglycaemic mice. *Food and Function, 11*, 7994–8002. https://doi.org/10.1039/d0fo01453b

Shi, J., Arunasalam, K., Yeung, D., Kakuda, Y., Mittal, G., & Jiang, Y. (2004). Saponins from edible legumes: Chemistry, processing, and health benefits. *Journal of Medicinal Food*, 7, 67–78. https://doi.org/10.1089/109662004322984734

Sierra, L., Roco, J., Alarcon, G., & Medina, M. (2015). Dietary intervention with *Salvia hispan- ica* (Chia) oil improves vascular function in rabbits under hypercholesterolaemic conditions. *Journal of Functional Foods, 14*, 641–649. https://doi.org/10.1016/j.jff.2015.02.042

Skwaryło-Bednarz, B., Stępniak, P. M., Jamiołkowska, A., Kopacki, M., Krzepiłko, A., & Klikocka, H. (2020). Amaranth seeds as a source of nutrients and bioactive substances in human diet. *Acta Scientiarum Polonorum Hortorum Cultus, 19*(6), 153–164. https://doi.org/10.24326/asphc.2020.6.13

Soares, R. A. M., Mendonça, S., de Castro, L. Í. A., Menezes, A. C. C. C. C., & Arêas, J. A. G. (2015). Major peptides from Amaranth (*A. cruentus*) protein inhibit HMG-CoA reductase activity. *International Journal of Molecular Sciences, 16*(2), 4150–4160. https://doi.org/10.3390/ijms16024150

Soriano-Santos, J., & Escalona-Buendía, H. (2015). Angiotensin I-converting inhibitory and antioxidant activities and surfactant properties of protein hydrolysates as obtained of *A. hypochondriacus* L. grain. *Journal of Food Science and Technology, 52*(4), 2073–2082. https://doi.org/10.1007/s13197-013-1223-4

Soriano-Santos, J., Reyes-Bautista, R., Guerrero-Legarreta, I., Ponce-Alquicira, E., Escalona-Buendía, H. B., Almaza-Pérez, J. C., . . . Román-Ramos, R. (2015). Dipeptidyl peptidase iv inhibitory activity of protein hydrolyzates from *A. hypochondriacus* L. Grain and their influence on postprandial glycemia in streptozotocin-induced diabetic mice. *African Journal of Traditional, Complementary and Alternative Medicines, 12*(1), 90–98. https://doi.org/10.4314/ajtcam.v12i1.13

Strimbu, K., & Tavel, J. A. (2010). What are biomarkers? *Current Opinion in HIV and AIDS, 5*(6), 463–466. https://doi.org/10.1097/coh.0b013e32833ed1

Suárez, S., Aphalo, P., Rinaldi, G., Añón, M. C., & Quiroga, A. (2020). Effect of amaranth proteins on the RAS system. *In vitro, in vivo* and ex vivo assays. *Food Chemistry, 308*, 125601. https://doi.org/10.1016/j.foodchem.2019.125601

Tan, M., Chang, S., Liu, J., Li, H., Xu, P., Wang, P., . . . Zhao, Q. (2020). Physicochemical Properties, Antioxidant and Antidiabetic Activities of Polysaccharides from Quinoa (*C. quinoa Willd.*) Seeds. *Molecules, 25*(17). https://doi.org/10.3390/molecules25173840

Tang, Y., & Tsao, R. (2017). Phytochemicals in quinoa and amaranth grains and their antioxidant, anti-inflammatory, and potential health beneficial effects: A review. *Molecular Nutrition and Food Research, 61*(7), 1600767. https://doi.org/10.1002/mnfr.201600767

Taniya, M. S., Reshma, M. V. R., Shanimol, P. S., Krishnan, G., & Priya, S. (2020). Bioactive peptides from amaranth seed protein hydrolysates induced apoptosis and antimigratory effects in breast cancer cells. *Food Bioscience, 35*, 100588. https://doi.org/10.1016/j.fbio.2020.100588

Thakur, P., & Kumar, K. (2019). Nutritional importance and processing aspects of Pseudo-cereals. *Journal of Agricultural Engineering and Food Technology, 6*, 155–160.

Thakur, P., Kumar, K., & Dhaliwal, H. S. (2021). Nutritional facts, bio-active components, and processing aspects of pseudocereals: A comprehensive review. *Food Bioscience, 42*, 1–13.

Thanh Tien, N. N., Dang Trinh, L. N., Inoue, N., Morita, N., & Van Hung, P. (2018). Nutritional composition, bioactive compounds, and diabetic enzyme inhibition capacity of three varieties of buckwheat in Japan. *Cereal Chemistry, 95*, 615–624. https://doi.org/10.1002/cche.10069

Tironi, V. A., & Añón, M. C. (2010). Amaranth proteins as a source of antioxidant peptides: effect of proteolysis. *Food Research International, 43*(1), 315–322. https://doi.org/10.1016/j.foodres.2009.10.001

Tomotake, H., Kayashita, J., & Kato, N. (2015). Hypolipidemic activity of common (*F. esculentum*Moench) and tartary (*F. tataricum Gaertn.*) buckwheat. *Journal of Science of Food and Agriculture, 95*(10), 1963–1967. https://doi.org/10.1002/jsfa.6981

Tomotake, H., Yamamoto, N., Yanaka, N., Ohinata, H., Yamazaki, R., Kayashita, J., & Kato, N. (2006). High protein buckwheat flour suppresses hypercholesterolemia in rats and gallstone formation in mice by hypercholesterolemic diet and body fat in rats because of its low protein digestibility. *Nutrition, 22*(2), 166–173. https://doi.org/10.1016/j.nut.2005.01.012

Toscano, L. T., da Silva, C. S. O., Toscano, L. T., de Almeida, A. E. M., da Cruz Santos, A., & Silva, A. S. (2014). Chia flour supplementation reduces blood pressure in hypertensive subjects. *Plant Foods for Human Nutrition*, *69*(4), 392–398. https://doi.org/10.1007/s11130-014-0452-7

Tovar-Pérez, E. G., Guerrero-Legarreta, I., Farrés-González, A., & Soriano-Santos, J. (2009). Angiotensin I-converting enzyme-inhibitory peptide fractions from albumin 1 and globulin as obtained of amaranth grain. *Food Chemistry, 116*(2), 437–444. https://doi.org/10.1016/j.foodchem.2009.02.062

Tsafrakidou, P., Michaelidou, A. M., & Biliaderis, G. C. (2020). Fermented cereal-based products: Nutritional aspects, possible impact on Gut Microbiota and health implications. *Foods (Basel, Switzerland)*, *9*(6), 734. https://doi.org/10.3390/foods9060734

Tuan, P. A., Thwe, A. A., Kim, Y. B., Kim, J. K., Kim, S. J., Lee, S., . . . Park, S. U. (2013). Effects of white, blue, and red light-emitting diodes on carotenoid biosynthetic gene expression levels and carotenoid accumulation in sprouts of tartary buckwheat (*F. tataricum Gaertn.*). *Journal of Agricultural and Food Chemistry*, *61*, 12356–12361. https://doi.org/10.1021/jf4039937

Ugural, A., & Akyol, A. (2020). Can pseudocereals modulate microbiota by functioning as probiotics or prebiotics? *Critical Reviews in Food Science and Nutrition*, 1–15. https://doi.org/10.1080/10408398.2020.1846493

Ullah, R., Nadeem, M., Khalique, A., Imran, M., Mehmood, S., & Javid, A. (2016). Nutritional and therapeutic perspectives of Chia (*S. hispanica* L.): A review. *Journal of Food Science and Technology*, *53*(4), 1750–1758. https://doi.org/10.1007/s13197-015-1967-0

Valenzuela-Zamudio, F., & Segura-Campos, M. R. (2020). Amaranth, quinoa and chia bioactive peptides: a comprehensive review on three ancient grains and their potential role in management and prevention of Type 2 diabetes. *Critical Reviews in Food Science and Nutrition*. https://doi.org/10.1080/10408398.2020.1857683

Vecchi, B., & Añón, M. C. (2009). ACE inhibitory tetrapeptides from *A. hypochondriacus* 11S globulin. *Phytochemistry*, *70*(7), 864–870. https://doi.org/10.1016/j.phytochem.2009.04.006

Vega-Galvez, A., Miranda, M., Vergara, J., Uribe, E., Puente, L., & Martinez, E. A. (2010). Nutrition facts and functional potential of quinoa (*C. quinoa willd.*), an ancient Andean grain: a review. *Journal of the Science of Food and Agriculture*, *90*, 2541–2547. https://doi.org/10.1002/jsfa.4158

Velarde-Salcedo, A. J., Barrera-Pacheco, A., Lara-González, S., Montero-Morán, G. M., Díaz-Gois, A., González De Mejia, E., & Barba De La Rosa, A. P. (2013). *in vitro* inhibition of dipeptidyl peptidase IV by peptides derived from the hydrolysis of amaranth (*A. hypochondriacus* L.) proteins. *Food Chemistry*, *136*(2), 758–764. https://doi.org/10.1016/j.foodchem.2012.08.032

Velarde-Salcedo, A. J.,Bojórquez-Velázquez, E.,&de la Rosa, A. P. B. (2019). Amaranth. In: *Whole Grains and their Bioactives: composition and Health*, pp. 209–250.

Vilcacundo, R., Martínez-Villaluenga, C., & Hernández-Ledesma, B. (2017). Release of dipeptidyl peptidase IV, α-amylase and α-glucosidase inhibitory peptides from quinoa (*C. quinoa Willd.*) during *in vitro* simulated gastrointestinal digestion. *Journal of Functional Foods*, *35*, 531–539. https://doi.org/10.1016/j.jff.2017.06.024

Vilcacundo, R., Martínez-Villaluenga, C., Miralles, B., & Hernández-Ledesma, B. (2019). Release of multifunctional peptides from kiwicha (*A. caudatus*) protein under *in vitro* gastrointestinal digestion. *Journal of the Science of Food and Agriculture*, *99*(3), 1225–1232. https://doi.org/10.1002/jsfa.9294

Vilcacundo, R., Miralles, B., Carrillo, W., & Hernández-Ledesma, B. (2018). *in vitro* chemopreventive properties of peptides released from quinoa (*C. quinoa* Willd.) protein under simulated gastrointestinal digestion. *Food Research International*, *105*, 403–411. https://doi.org/10.1016/j.foodres.2017.11.036

Vogrinčič, M., Kreft, I., Filipič, M., & Zegura, B. (2013). Antigenotoxic effect of tartary (*F. tataricum*) and common (*F. esculentum*) buckwheat flour. *Journal of Medical Food*, *16*(10), 944–952.

Vrenna, M., Peruccio, P. P., Liu, X., Zhong, F., & Sun, Y. (2021). Microalgae as future superfoods: fostering adoption through practice-based design research. *Sustainability*, *13*(5), 2848. https://doi.org/10.3390/su13052848

Vuksan, V., Jenkins, A. L., Brissette, C., Choleva, L., Jovanovski, E., & Gibbs, A. L. (2017). Salba-chia (*S. hispanica* L.) in the treatment of overweight and obese patients with type 2 diabetes: A double- blind randomized controlled trial. *Nutrition, Metabolism and Cardiovascular Diseases*, *27*(2), 138–146. https://doi.org/10.1016/j.numecd.2016.11.124

Vuksan, V., Jenkins, A. L., Dias, A. G., Lee, A. S., Jovanovski, E., Rogovik, A. L., & Hanna, A. (2010). Reduction in postprandial glucose excursion and prolongation of satiety: possible explanation of the long-term effects of whole grain Salba (*S. hispanica* L.). *European Journal of Clinical Nutrition*, *64*(4), 436–438. https://doi.org/10.1038/ejcn.2009.159

Wu, S.-C., & Lee, B.-H. (2011). Buckwheat polysaccharide exerts antiproliferative effects in THP-1 human leukemia cells by inducing differentiation. *Journal of Medicinal Food*, *14*(1–2), 26–33. https://doi.org/10.1089/jmf.2010.1252

Xhaferaj, M., Alves, T. O., Ferreira, M., & Scherf, K. A. (2020). Recent progress in analytical method development to ensure the safety of gluten-free foods for celiac disease patients. *Journal of Cereal Science*, 1–11.

Xiao-na, G., & Hui-yuan, Y. (2007). Isolation, purification and structure analysis of antitumor protein from tartary buckwheat. *Food Science*, *07*. https://doi.org/10.1021/jf071032+

Yıldız, N., Yalçın, E., & Güneş, A. (2019). Chemical, nutritional and bioactive properties of common buckwheat cultivars bred in Turkey. *Quality Assurance and Safety of Crops & Foods*, *11*(2), 137–143. https://doi.org/10.3920/QAS2018.1391

Yu, G., Wang, F., Zhang, B., & Fan, J. (2016). *in vitro* inhibition of platelet aggregation by peptides derived from oat (*Avena sativa* L.), highland barley (*Hordeum vulgare* Linn. var. *nudum* Hook. f.), and buckwheat (*F. esculentum* Moench) proteins. *Food Chemistry*, *194*, 577–586. https://doi.org/10.1016/j.foodchem.2015.08.058

Zambrana, S., Lundqvist, L. C. E., Veliz, V., Catrina, S. B., Gonzales, E., & Östenson, C. G. (2018). *A. caudatus* stimulates insulin secretion in goto-kakizaki rats, a model of diabetes mellitus type 2. *Nutrients*, *10*(1), 1–17. https://doi.org/10.3390/nu10010094

Zevallos, V. F., Ellis, H. J., Suligoj, T., Herencia, L. I., & Ciclitira, P. J. (2012). Variable activation of immune response by (*C. quinoa Willd.*) prolamins in celiac disease. *American Journal of Clinical Nutrition*, *96*(2), 337–344.

Zheng, C., Hu, C., Ma, X., Peng, C., Zhang, H., & Qin, L. (2012). Cytotoxic phenylpropanoid glycosides from *F. tataricum* (L.) Gaertn. *Food Chemistry*, *132*, 433–438. https://doi.org/10.1016/j.foodchem.2011.11.017

Zheng, Y., Wang, X., Zhuang, Y., Li, Y., Tian, H., Shi, P., & Li, G. (2019). Isolation of novel ACE-inhibitory and antioxidant peptides from quinoa bran albumin assisted with an in silico approach: Characterization, *in vivo* antihypertension, and molecular docking. *Molecules*, *24*(24), 1–17. https://doi.org/10.3390/molecules24244562

Zhou, X., Wen, L., Li, Z., Zhou, Y., Chen, Y., & Lu, Y. (2015). Advance on the benefits of bioactive peptides from buckwheat. *Phytochemistry Reviews*, *14*, 381–388. https://doi.org/10.1007/s11101-014-9390-0

Zhu, F. (2021). Buckwheat proteins and peptides: Biological functions and food applications. *Trends in Food Science and Technology*, *110*, 155. https://doi.org/10.1016/j.tifs.2021.01.081

Zieliński, H., Szawara-Nowak, D., Bączek, N., & Wronkowska, M. (2019). Effect of li- quid-state fermentation on the antioxidant and functional properties of raw and roasted buckwheat flours. *Food Chemistry*, *271*, 291–297. https://doi.org/10.1016/j.foodchem.2018.07.182

Zieliński, H., Szawara-Nowak, D., & Wronkowska, M. (2020). Bioaccessibility of anti-AGEs activity, antioxidant capacity and phenolics from water biscuits prepared from fermented buckwheat flours. *LWT—Food Science and Technology*, *123*, 109051. https://doi.org/10.1016/j.lwt.2020.109051

Zouhir, D. (2014). Efficacy of honey bee and *F. esculentum* Moench ointment in the treatment of sub chronic wound in rabbits: A case control study. *American Journal of Animal and Veterinary Sciences*, *9*(1), 14–18. https://doi.org/10.3844/ajavsp.2014.14.18

17

Berries as Functional Foods: Molecular Mechanisms Promoting Human Health

Shazia Parveen[1], Irshad Ul Haq Bhat[2], and Rajeev Bhat[3]

[1] Chemistry Department, Faculty of Science, Taibah University, Yanbu, Saudi Arabia,
[2] Department of Applied Sciences, Chemistry Section, University of Technology and Applied Sciences, P. O. Box 74, Al-Khuwair, 133 Muscat, Oman
[3] ERA-Chair for Food (By-) Products Valorisation Technologies (VALORTECH), Estonian University of Life Sciences, Kreutzwaldi 1a, 51014, Tartu, Estonia, EU

17.1 Introduction

Recently, studies have increased considerably particularly pertaining to plant foods derived natural ingredients that can promote human health. Subsequently, food researches have focused on recognizing various bioactive components, which are efficacious in delivering physiological benefits beyond providing the basic nutritional value (Kareb & Aïder, 2019). The term "functional foods," first coined during 1980s (in Japan), is frequently misinterpreted as they remain regulated but not officially acknowledged in many countries (Ye, Georges, & Selomulya, 2018). A food is considered as "functional" if it is capable of establishing a beneficial effect on one or more targeted functions inside the human body, over and above providing satisfactory nutritional properties, consequently improving overall health and well-being, preventing nutrient deficiencies, and aiding in prevention from disease risks (Birch & Bonwick, 2019; Granato et al., 2020). The functional food can be either natural or from a processed product containing one or more bioactive components. Functional foods includes (a) foods having natural bioactive components, (b) foods supplemented with bioactive substances, (c) food enhanced or derived with ingredients (Kareb & Aïder, 2019), and foods (d) produced in food industries by utilizing potential functional components either recovered from industrial wastes or by-products (Dalle Zotte & Szendrő, 2011) or by implementing certain technologies (Gutiérrez, 2018). The investigations undertaken on "functional foods" has been considerably augmented over the past decade. However, only a few selected studies have addressed the bioavailability of active ingredients for their clinical potential. Functional foods are not medicine as they can neither cure nor prevent diseases or any human ailments.

Recently, epidemiological studies have demonstrated that regular intake of fruits, mainly berries, can impart effective health benefits (Baby, Antony, & Vijayan, 2018; Miller, Feucht, & Schmid, 2019; Nile & Park, 2014; Yang & Kortesniemi, 2015). Berries such as blueberry,

Molecular Mechanisms of Functional Food, First Edition. Edited by Rocio Campos-Vega and B. Dave Oomah.

cranberry, raspberry, strawberry, grapes, blackcurrant, goji berry, and chokeberry are excellent sources of diverse bioactive components consisting of phenolic acids (viz., benzoic acid and its derivatives), stilbenes, lignans, flavonoids, anthocyanins, flavonols, carotenoids, tannins, vitamins, alkaloids, xanthones, and polysaccharides (Afrin et al., 2016) (Figure 17.1). Berries are considered to be the best dietary sources of functional food due to their rich health benefits linked with their explicit biochemical properties (Ma, Sun, Zeng, Luo, & Yang, 2018; Paredes-López, Cervantes-Ceja, Vigna-Pérez, & Hernández-Pérez, 2010; Wu, Wang, Prior, & Pehrsson, 2018). They are known to contain not only fundamental dietary constituents but also a broad range of phytonutrients that play synergistic and cumulative role in human health promotion (Yeung et al., 2019). The presence of various berry bioactive components testify their human health benefits, and studies have shown them to exhibit anticancer (Aqil et al., 2016; Dienaitė et al., 2020; Gill et al., 2020), antioxidant (Sytařová et al., 2020; Tang, Dunshea, & Suleria, 2020; Umdale, Ahire, Aiwale, Jadhav, & Mundada, 2020; Wu et al., 2019; Zorzi et al., 2020), antidiabetic (De Silva & Rupasinghe, 2020; Liu, Yu, Guo, Fang, & Chang, 2020; Nazir, Zahoor, Ullah, Ezzeldin, & Mostafa, 2020; Zhang et al., 2020), cardioprotective (Basu, Rhone, & Lyons, 2010; Najjar, Schwartz, Wong, Mehta, & Feresin, 2021), and neurodegenerative activities (Flanagan, Müller, Hornberger, & Vauzour, 2018; Shabani, Rabiei, & Amini-Khoei, 2020; Tan & Ismail, 2020), which contributes to moderating pathological effect on various human ailments. The polyphenolic components are recognized as the main antioxidants or free radical inhibitory agents, owing to their ability to donate hydrogen atoms or electrons (Singh, Singh, Kaur, & Singh, 2016). Furthermore, the phytochemicals present in the fruit berries play a significant role in blood pressure modulation, platelet accumulation, atherosclerosis, and full-body metabolism. Moreover, anthocyanins and proanthocyanidin (PA) rich berries have been known to inhibit elevated blood glucose level to improve diabetes and other associated metabolic disorders. Several recent studies highlight the importance of exploring berry pomace, which occurs as a waste/by-product post-processing in food industries (Nour et al., 2021; Oreopoulou & Tzia, 2007; Rohm et al., 2015). Therefore, this chapter summarizes various underlying molecular mechanisms involved in different biological activities exerted by berries and their bioactive compounds (e.g. as an antioxidant, antidiabetic, cardioprotective, neuroprotective, anticancer, and anti-inflammation component).

17.2 Bioactives in Berries and Mechanism of Action

The aim to understand the consumption of berries is essential to elucidate their molecular mechanism. The secondary metabolites found in berries are primarily responsible for providing health benefits to humans. Based on various biological activities and types of bioactive components, a range of molecular mechanisms are involved, some of which includes free radicals scavenging activity, metabolic, oxidative stress and signaling pathways, and anti-inflammatory and enzymatic inhibition activities summarized in Table 17.1.

Further, Table 17.2 summarizes on the *in vitro*, *in vivo*, and human intervention studies along with their biological activities, undertaken in selected common berries.

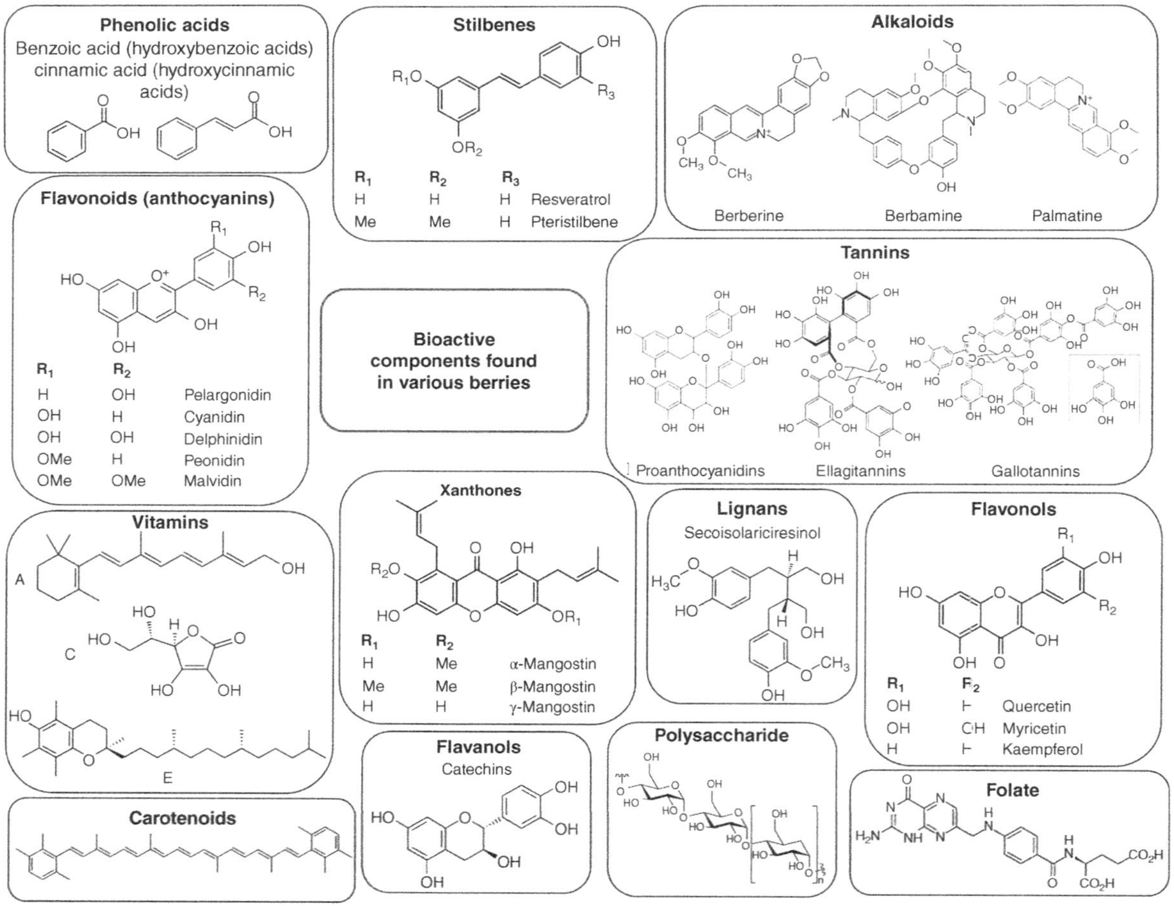

Figure 17.1 Various bioactive components in edible berries. *Source:* Afrin et al. (2016)/MDPI.

Table 17.1 Bioactive components of berries responsible for biological activity and their molecular mechanisms.

Bioactivities	Bioactive components	Molecular mechanism	References
Antioxidant	Anthocyanins, phenolic acids, ellagic acid, and resveratrol	i) Free radicals scavenging ii) Prevention of free radical formation and lipid peroxidation iii) Inhibition of oxidative enzyme activity	Sytařová et al. (2020); Tang et al. (2020); Umdale et al. (2020); Zorzi et al. (2020)
Antidiabetic	Anthocyanins and polyphenolic compounds	i) Carbohydrate-metabolizing enzyme inhibition ii) Glucose transporters inhibition iii) Glycogenolysis inhibition iv) Modulation of gut microbiota by anthocyanin breakdown products	Nazir et al. (2020); Yazdankhah, Hojjati, and Azizi (2019)
Antimicrobial	Phenolic compounds	Cranberry fraction FC111 affected PG synthesis of *Staphylococcus aureus* and acted in synergy with β-lactam antibiotics	Diarra et al. (2013)
Prevention of cardiovascular diseases	Phenolic acids and flavanoids	i) Countering free radicals generation ii) Suppressing inflammatory gene expression iii) Upregulating eNOS expression associated with NO generation iv) Oxidative stress reduction v) Apoptotic inhibition	Cassidy (2018); Najjar et al. (2021); Oak et al. (2018); Olas (2018); Wu et al. (2018)
Prevent neurodegenerative diseases	Anthocyanins and polyphenols	i) Providing significant cellular protection against ROS damage ii) Substantial reduction in β-amyloid (Aβ) peptide production iii) Upregulating cerebral antioxidant enzymatic activities iv) Improvement in cognitive performance v) Neuroregeneration by reversing damage	Granzotto and Zatta (2014); Kelly, Vyas, and Weber (2017); Shabani et al. (2020); Tan and Ismail (2020)
Anticancer	Anthocyanins, phenolic acids, and stilbenes	i) Inflammation suppression ii) Angiogenesis inhibition iii) DNA damage protection iv) Effects on apoptosis or proliferation rates of malignant cells	Dienaitė et al. (2020); Kristo, Klimis-Zacas, and Sikalidis (2016)
Anti-inflammatory	Anthocyanins, phenolic acids, and flavonoids	i) Arachidonic acid (AA) metabolism inhibition ii) Cyclooxygenase (COX) inhibition iii) Lipoxygenase (LOX) inhibition iv) Cytokines interleukin and TNF inhibition v) Nuclear factor kappa B (NF-κB) inhibition	Banach, Wiloch, Zawada, Cyplik, and Kujawski (2020); Gasparrini et al. (2021); Seo et al. (2019); Xiang, Wang, Yi, and He (2018)

Table 17.2 *In vitro*, *in vivo*, and human intervention studies of most common berries along with their biological activities

Berry	Bioactivity	Status	References
Açaí berry	COVID-19 anti-inflammation trial	Human interventional (clinical trial)	Farkouh (2021)
Bay laurel berries	SARS-CoV	*In vitro*	Duke, Godwin, Cellier, and Duke (2002) Silveira et al. (2020)
Blackberry	Anticancer	*In vitro* and *In vivo*	Shi et al. (2019)
Blackcurrant	Antidiabetic	*In vivo*	Iizuka, Ozeki, Tani, and Tsuda (2018)
Blueberry	Antidiabetic	*In vivo*	Elks, Terrebonne, Ingram, and Stephens (2015)
Blueberry	Antidiabetic		Takikawa, Inoue, Horio, and Tsuda (2010)
Blueberry	Anticancer activity	*In vitro*	Kostin, McDonald, and McFadden (2012)
Blueberry	Antioxidant	*In vitro* and *In vivo*	Newsome et al. (2014)
Blueberry	Antidiabetic	Human trial and human intervention	Rebello, Burton, Heiman, and Greenway (2015); Stote et al. (2017)
Blueberry	Neuro-preventive	Human trial	Bowtell, Aboo-Bakkar, Conway, Adlam, and Fulford (2017)
Cranberry, chokeberry, blackberry, gooseberry, black grape, bilberry, blueberry, red raspberry, blackcurrant, redcurrant, and strawberry	Antidiabetic	*In vivo* and human studies	Císedas, Les, Gómez-Serranillos, Smith, and López (2016)
Red berry	Anti-inflammatory	*In vivo*	Nabavi et al. (2018)
Strawberry	Cardiovascular prevention	Human intervention studies	Park et al. (2016)
Strawberry	Anti-inflammatory	Human intervention studies	Schell et al. (2017)
Strawberry	Antioxidant	Human intervention studies	Basu et al. (2016)
Mixture of berries (blueberries, blackcurrant, elderberry, lingonberries, strawberry, and tomatoes)	Cardio-prevention	Human intervention	Nilsson, Salo, Plaza, and Björck (2017)
Strawberry	Neuro-prevention and cardio protection	*In vitro*	Cianciosi, Ansary, Regolo, Giampieri, and Forbes-Hernandez (2021)

17.2.1 Antioxidant Properties of Berries

Antioxidants play a crucial role in preventing or delaying the oxidation process. Oxidation induces numerous reactions, viz., lipid peroxidation, causing degradation in the overall quality food products generating unwanted chemical species (Umdale et al., 2020; Uttara, Singh, Zamboni, & Mahajan, 2009). These reactions can harm vital tissues in the human body as well as interfere with several physiological routes leading to several diseases. The reactive oxygen species (ROS) or free radicals which occur as a result of these reactions inside the body weaken the autoimmune system leading to various chronic diseases such as cancer, cardiovascular diseases (CVDs), diabetes, and neurological disorders. The free radicals formed in the human body possess an unpaired electron in the outer orbit resulting in instability and high reactivity. The human body has natural defense mechanisms against free radical-induced damage, like "oxidative stress," but collective oxidative damage can lead to numerous diseases. Furthermore, dietary antioxidants can reduce the occurrence of oxidative stress-induced DNA damage. Various berries possess antioxidant property by different molecular mechanisms: (a) scavenging or quenching of the free radicals formed in the human body, (b) preventing the formation of free radicals, (c) preventing lipid peroxidation, and (d) the inhibition of oxidative enzymes.

Extracts of 11 different berries were recently evaluated for their antioxidant activity, fatty acid profile, total anthocyanins (TACs), total flavonoids (TFCs) contents, ferric-reducing antioxidant power (FRAP), and radical scavenging activities determined by 2,2-diphenyl-1-picrylhydrazyl (DPPH)• radical assays (Zorzi et al., 2020). The results revealed that black currants, blackberries, and blueberries displayed the highest TAC and TFC contents accounting for their enhanced antioxidant activity. Additionally, anthocyanidins structure can affect antioxidant capacity; however, cyanidin and delphinidin generally exhibit higher antioxidant activity compared to others (Cao, Sofic, & Prior, 1996). Moreover, the flavonol content enhances the berry antioxidant capacity as they are more powerful antioxidants than anthocyanins. The antioxidant activity in sea buckthorn (*Hippophaë rhamnoides* L.) was evaluated by determining the total polyphenols, TFCs, vitamins C and E, and carotenoids (Sytařová et al., 2020). The antioxidant activity has been estimated by free radical scavenging (DPPH) and photo-chemiluminescence assays (Orsavová, Hlaváčová, Mlček, Snopek, & Mišurcová, 2019). For example, sea buckthorn berries are an important source of vitamin C (4–100 times) higher amounts than any other fruits. The presence of carotenoids and phenolic compounds of these berries exhibited high antioxidant activity, where vitamin C exhibited higher antioxidant effect compared to carotenoids. Recently, Tang et al. (2020) attributed the antioxidant activities of Hops (*Humulus lupulus* L.) and juniper berries (*Juniperus communis* L.) to the presence of polyphenols. TPC, flavonoids, and tannins were determined for polyphenol estimation; however, antioxidant activity was ascertained by three methods such as the DPPH˙, FRAP, and 2,2'-azino-bis(3-ethyl benzothiazoline-6-sulfonic acid) ($ABTS^+$) assays. Although "Hops" displayed higher antioxidant activity than juniper berries, the polyphenolic content in juniper berries largely comprised of flavonoids and bioflavonoids exhibited antioxidant activity by free radical scavenging, prevention of free radical formation, and lipid peroxidation. The antioxidant efficacy and the phytoconstituents of wild berries of Indian Sandalwood (*Santalum album* L.) have been evaluated by DPPH, FRAP, $ABTS^+$, and phosphomolybdenum assays

(Umdale et al., 2020). The results indicated that the wild berry pulp extract is a rich source of phytoconstituents (viz., phenolics, flavonoids, alkaloids, and tannins) and antioxidant potential; hence it can be utilized as a functional food. Goji berries (*Lycium* sp.) as an excellent source of macronutrients are obtained from *Lycium chinense* and *Lycium barbarum* (Ma et al., 2019). They consist mostly of polysaccharides but also have other bioactive compounds viz., carotenoids and phenolics that relates to the health-promoting properties of goji berries. Goji berries have high antioxidant potential due to the presence of polyphenolic components which helps in alleviating oxidative stress by preventing DNA damage by the free radicals, lipids, and proteins. High amounts of polyphenolic compounds (phenolic acids and tannins flavonoids such as anthocyanins and flavonols) and vitamin C-exhibiting antioxidant activities are reported for berries belonging to Rosaceae (strawberry, raspberry, and blackberry) and Ericaceae (blueberry and cranberry) family. Water-soluble vitamin C is an established potent antioxidant compound capable of scavenging free radicals and other ROS formed during normal cell metabolism.

17.2.2 Anti-Inflammatory Properties of Berries

Inflammation is a biological pattern of response to injury, which triggers the accumulation of cells in irritated tissues protecting further damage. Chronic inflammation plays a vital role in the instigation and advancement of numerous chronic disorders that include cancer, arteriosclerosis, CVDs, diabetes, obesity, and neurodegenerative disorders. Inflammation process has been linked with oxidative stress. Besides, nitric oxide (NO) is a signaling molecule associated with the pathogenesis of inflammation. The molecules that inhibit the production of NO qualify as probable anti-inflammatory therapeutic agents. The pro-inflammatory cytokines viz., interleukin (IL)-6, IL-1β, and tumor necrosis factor (TNF) are triggered inducing ROS production. These pro-inflammatory genes are initiated by the nuclear factor kappa B (NF-κB) leading to inflammation (Seo et al., 2019). The bioactive components of several berries exert their anti-inflammatory activity by inhibiting either: (a) arachidonic acid metabolism, (b) cyclooxygenase (COX), (c) lipoxygenase (LOX), (d) cytokines and IL, and/or (e) NF-κB. Wang, Xiang, Yi, & He (2017), reported the anti-inflammatory of steroidal saponins extracted from the berries of *Solanum nigrum* L. (European black nightshade), through cell viability evaluation by 3-(4,5-dimethylthiazol-2-yl)-2,5-diphenyltetrazolium bromide (MTT) assay. Their probable inhibitory effects on NO and IL-6 and IL-1β production by lipopolysaccharide (LPS)-induced macrophages cell line RAW 264.7 were assessed to determine whether inhibitory activities of NO production of these compounds were due to their cytotoxicity. One steroidal saponin extract substantially inhibited NO production ($IC_{50} = 9.7\,\mu M$) compared to the positive control (indomethacin, $IC_{50} = 47.4\,\mu M$). Other extracts inhibited LPS-induced IL-6 and IL-1β production considerably. However, no cytotoxicity was observed against RAW 264.7 macrophage-treated cells suggesting potential anti-inflammatory activity of *S. nigrum* berries. The molecular mechanism involved in the anti-inflammatory activity exhibited by *S. nigrum* is through the inhibition of NO production, as well as the pro-inflammatory cytokines (IL-6 and IL-1β) induced by LPS (in RAW 264.7 macrophages). In another study, Xiang et al. (2018) reported the anti-inflammatory activity of steroidal glycosides having cholestane 16, 22-dione skeleton extracted from *S. nigrum* L. berries by evaluating their inhibitory effects on LPS induced NO production (RAW 264.7 cell line). All extracts inhibited NO production (IC_{50} = 11.33–49.35 mM), suggesting their anti-inflammatory potential. The anti-inflammatory effects of anthocyanin-rich chokeberry extracts using LPS-stimulated RAW 264.7 cells have been investigated (Banach et al., 2020). IL-6 production was inhibited by myocytes a pro-inflammatory cytokinine that inhibits the production of pro-inflammatory TNF-α. TNF-α and IL-1β are mediators of inflammation. Moreover, the production of inflammation markers decreased considerably in RAW 264.7 cells, with no cytotoxicity. The anti-inflammatory activity has been reported for seven phenolics in ethanolic extracts of *Crataegus orientalis* berries by HPLC-DAD (diode-array detection) (Šavikin et al., 2017). These berries predominantly contained hyperoside, isoquercitrin, and chlorogenic acid. The extracts exhibited an excellent anti-inflammatory effect based on their COX-1 and 12-LOX inhibitory activities. However, COX-1 and 12-LOX inhibition potential of *C. orientalis* extracts was considerably lower than that of aspirin and quercetin, well-known potent inhibitor of COX-1 and 12-LOX, respectively. The anti-inflammatory activity of these berries was attributed to the presence of phenolics that can disrupt the synthesis of compounds in the COX-1 and 12-LOX pathways. The bioactive components of *Hippophae rhamnoides* L. (sea buckthorn) have been evaluated for their anti-inflammatory activities via UHPLC-ESI-QTOF-MS (ultra-high-performance liquid chromatography accurate-mass quadrupole time-of-flight mass spectrometry with electrospray ionization) (Zheng et al., 2019). The activity was measured as inhibitory potential of NO production in macrophage RAW 264.7 cell lines. Sea buckthorn extracts (10 μg/mL) displayed satisfactory anti-inflammatory effects having NO inhibition values in the range of 73–98%. Further, the anti-inflammatory properties of *Rubus coreanus* (Korean black raspberry) were investigated by quantifying NO production, expression of inflammatory proteins (inducible NO synthase, iNOS), COX-2, NF-κB, and inhibitor of NF-κB (IκB), and inflammatory cytokines via LPS-stimulated RAW 264.7 cells (Seo et al., 2019). Results showed that immature berry extracts (200 μg/mL) significantly reduced NO production (~80%). The extracts also suppressed iNOS, COX-2, phosphorylated (p)-IκB, and p-NF-κB activation compared with mature berry extract. Besides, immature berry extract attenuated the inflammatory cytokines TNF-α, IL-1β, and IL-6 compared to the mature berry extract, via NF-κB signaling pathways. Figure 17.2 depicts the mechanism of antioxidant and anti-inflammatory activities as elaborated by Gasparrini et al. (2021).

17.2.3 Antidiabetic Properties of Berries

Diabetes mellitus (DM) is considered a chronic and noncommunicable disease due to either insulin deficiency (type 1 diabetes, T1D) or damage of insulin action (type 2 diabetes, T2D), both resulting in hyperglycemia in humans (Liu et al., 2020). T2D is the most widespread diabetes accounting for ˃90% of all DM. This type is characterized by excessive blood glucose accumulation, resulting in poor response to insulin followed by insulin resistance developed in the body. Flavonoids, especially anthocyanins which are bioactive components of several berries like cranberry, chokeberry, blackberry, gooseberry, black grape, bilberry, blueberry, red raspberry, blackcurrant, redcurrant, or strawberry possess antidiabetic activity via carbohydrate-hydrolyzing enzyme (alpha-amylase and alpha-glucosidase) inhibition, by regulating the prompt release of glucose into the bloodstream

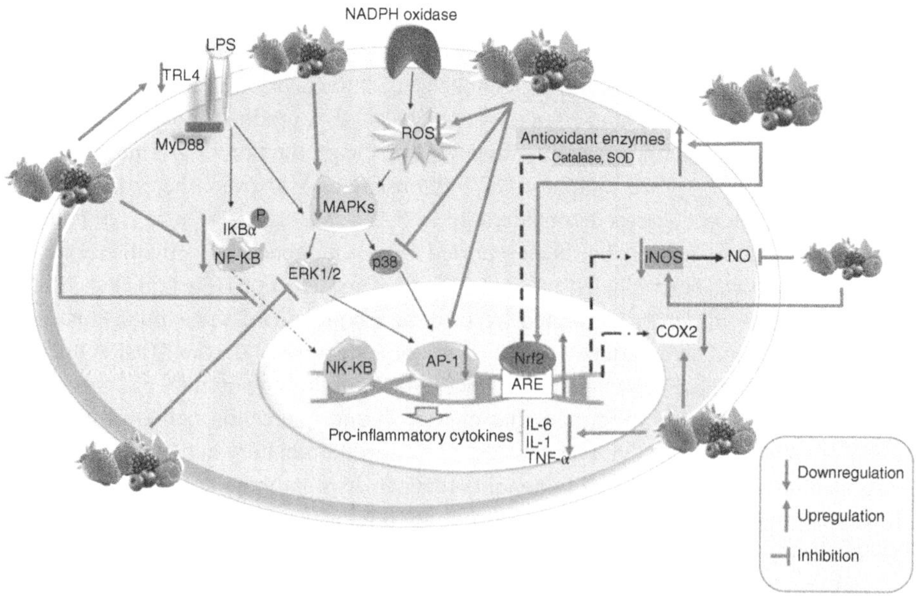

Figure 17.2 Antioxidant and anti-inflammatory effect of berries (*Source:* Gasparrini et al. (2021)/ Reprinted with permission from Elsevier License Number: 5062920212931; dt. 06 May, 2021).

(Rupasinghe, Balasuriya, & Wang, 2017; Yazdankhah et al., 2019). Inhibition of dipeptidyl peptidase-4 (DPP-4) is known to convert the active glucose-dependent insulin tropic polypeptide (GIP) and glucagon-like peptide-1 (GLP-1) to an inactive form and is considered a novel pharmacological target for T2D treatment. This limits the functioning of insulin secretion from the beta-cells of the pancreas (Wu et al., 2019). Thus, the molecular mechanisms involved to counter the metabolic variations related to T2D include: (a) carbohydrate-metabolizing enzyme inhibition, (b) glucose transporters inhibition, (c) glycogenolysis inhibition, and (d) gut microbiota manipulation by anthocyanin breakdown products. The polyphenolic content and *in vitro* antidiabetic properties have been reported for haskap (*Lonicera caerulea* L.) berries also known as blue honeysuckles (De Silva & Rupasinghe, 2020). Accordingly, cyanidin-3-*O*-glucoside content is nearly 80% of TACs present in haskap berries, higher than other berries. The TAC was assessed by a nondestructive technique that displayed a solid association for calculated anthocyanin by liquid chromatography-mass spectrometry (UHPLC-ESI-MS). The berries extracts exhibited antidiabetic properties inhibiting alpha-amylase (IC_{50} = 2.38–5.08 mg/mL), alpha-glucosidase (IC_{50} = 1.13–2.12 mg/mL), DPP-4 (IC_{50} = 2.15–11.6 mg/mL), and formation of advanced glycation end products (AGE, IC_{50} = 1.2–4.79 mg/mL) *in vitro*. The polyphenolic components and *in vivo* antidiabetic properties of Chinese bayberry "Shui Jing" have been investigated by liquid chromatography quadrupole time-of-flight mass spectrometry (LC-Q-TOF-MS) (Liu et al., 2020). Apart from anthocyanins, PAs, epigallocatechin gallate, flavonols, myricitrin, and quercetin were the principal constituents of these berries. Fasting blood glucose, elevated glucose tolerance, and insulin sensitivity decreased substantially in diabetic KK-A^y mice suggesting that these white bayberries rich in PAs and flavonols can be beneficial in diabetes regulation. The *in vitro* and *in vivo* antidiabetic activities were investigated by inhibiting carbohydrate digestive enzymes in catechin isolated from *Elaeagnus umbellata* berries (autumn olive or Japanese silverberry) (Nazir et al., 2020). Catechin from *E. umbellata* berries considerably reduced blood glucose levels in diabetic rats by (a) inhibiting α-amylase and α-glucosidase enzymes and (b) increasing the secretion of healthy pancreatic β-cells. Hence, *E. umbellata* Thunb. berries can efficiently be used as a useful candidate in T2D management. The inhibitory effects of *L. caerulea* berry polyphenols (LCBPs) on α-amylase activity and the involved mechanisms were evaluated using an enzyme inhibition assay, enzyme kinetics analysis, fluorescence quenching, and molecular docking (Liu et al., 2020). The results indicated that LCBP inhibited α-amylase activity ($IC_{50} = 301.5\,\mu g/mL$) and displayed higher inhibitory effect than acarbose which is a positive control for carbohydrate digestion. The main bioactive components are cyanidin-3-*O*-glucoside, catechins, and chlorogenic acid, which exhibited reversible competitive inhibitory effects on α-amylase activity, thereby managing T2D. Along with anthocyanins, flavan-3-ols and flavonols are also present in berries. The role of berry flavonoids on blood glucose management in diabetes has been studied. Their antidiabetic effects included regulating glucose metabolism, elevating insulin production, reducing apoptosis, and stimulating pancreatic β-cells proliferation (Vinayagam & Xu, 2015).

Dietary supplement containing sea buckthorn and blueberry extract (Diavit®, Plantarom Microproducţie, Cluj-Napoca, Romania) administered (3/day for 2 months) improved blood sugar and lipid levels in 30 type 1 diabetic children (average age 12.6 years, range 8–17 years; ≥7.5% glycated hemoglobin [HbA1C] receiving insulin 4–5 injections/day) (Nemes-Nagy et al., 2008). Moreover, the supplement significantly reduced glycated hemoglobin (HbA1C) and oxidative tissue damage by erythrocyte superoxide dismutase activity and increased level of endogenous insulin production (C peptide concentration). This phytotherapeutic can therefore be considered a treatment in the fight against DM.

17.2.4 Cardioprotective Effects of Berries

CVDs are linked with the heart and blood vessels and include coronary and congenital heart diseases, rheumatic heart disease (caused by *Streptococcal* bacteria), peripheral arterial disease, cerebrovascular disease, deep vein thrombosis, and pulmonary embolism (WHO, 2021). More than half of all CVDs can be prevented just by altering daily lifestyle and improved diet. An enhanced vascular oxidative stress due to elevated ROS is caused by an imbalance between ROS-generating and -eliminating enzyme systems. An increase in ROS levels can lead to thiol oxidation, comparatively higher in ischemic heart disease patients (Najjar et al., 2021). Blueberries contain high levels of polyphenolic components and are known to play an important role in atherosclerosis prevention. The mechanisms through which blueberries might act include oxidative stress and inflammation reduction, endothelial dysfunction control, regulating cholesterol accumulation, along with possible gut microbiota modulation (Wu et al., 2018). Oxidative stress and inflammation are the key players in the commencement and growth of atherosclerosis (plaque builds up inside the arteries) and CVD (Real et al., 2010). The bioactive components of berries mainly encompasses the polyphenols (including flavonoids, phenolic acids, lignans, and stilbenes) along with appreciable amounts of flavonoids particularly anthocyanins. These compounds play an important role not only in suppressing oxidative or inflammatory damages (in cellular and animal models) but also significantly decreases the cytotoxicity generated by the inducers of oxidative stress (Cassidy, 2018; Olas, 2018). Blueberries, chokeberries, cranberries, and strawberries anthocyanin extracts have been validated for substantial enhancement in low-density lipoprotein (LDL) oxidation and lipid peroxidation.

The cardioprotective effects of berries consumption presumably involve scavenging of the free radicals generation, suppressing inflammatory genes expression, reducing oxidative stress and regulating eNOS (endothelial nitric oxide synthase 3) expression, which is associated with NO generation. Atherosclerosis progress is reduced by these effects, thus helping maintain vascular functions (Basu et al., 2010). Foods rich in anthocyanins, largely obtained from berries, are readily integrated into the diet and simple dietary modification can decrease CVD risk. Another important component, quercetin, present in ample amounts in berries exhibited antihypertensive effects in experimental animals and humans cardiovascular system. There are several underlying mechanisms for antihypertensive effect including the suppression of oxidative stress. This affects intracellular protein kinase cascades via transformation of extracellular matrix inside the vasculature (Ferenczyova, Kalocayova, & Bartekova, 2020). Besides the vascular effects, quercetin also shows excellent heart-defensive effects in several types of heart injury, viz., doxorubicin-induced cardiotoxicity, ischemia–reperfusion injury, and diabetic cardiomyopathy. Quercetin's cardioprotective effects can be attributed to various diversified signaling pathways such as apoptotic inhibition, decreased oxidative stress, and effects on cardiac inflammatory proteins (Castillo et al., 2018). Figure 17.3 presents an overview of the proposed scheme for anti-atherogenicity effects of anthocyanins on cardiometabolic risk factors and platelet functions (Aboonabi et al., 2020). Furthermore, a schematic presentation (Figure 17.4) highlights the potential mechanisms that can contribute to the beneficial effects of polyphenols/polyphenol-rich products by protecting endothelial cells against oxidative-stress-induced premature senescence.

The cardiovascular benefits of some berries may be due to reduced blood clotting effects of their oil. For example, supercritical CO_2-extracted sea buckthorn (5 g/day for 4 weeks) decreased the rate and extent of adenosine-5′-diphosphate-induced platelet aggregation in 12 healthy normolipidemic men (Johansson, Korte, Yang, Stanley, & Kallio, 2000).

17.2.5 Neuroprotective Berries

Neurodegenerative diseases such as Parkinson's disease (PD), Alzheimer's disease (AD), and Huntington's disease (HD) are incurable. Nerve degeneration is a common incident even in the absence of any neurodegenerative disease. As humans age, the body becomes incapable of maintaining and producing new nerve tissues, thus leading to many cognitive and motor impairments. The prolonged effects of oxidative stress in central nervous system (CNS) are well-defined primary mechanism of neurodegeneration. It is caused when ROS generation is much faster than their elimination by endogenous redox reactions (Ataie, Ataie, & Shadifar, 2016). Anthocyanins present in various berries have neuroprotective potentials in the CNS. One of the crucial mechanisms for the neuroprotective effects of anthocyanins is the suppression of neuro-inflammation and oxidative stress. Anthocyanins exert multiple effects, such as (a) providing significant cellular protection against ROS-damaging effects,

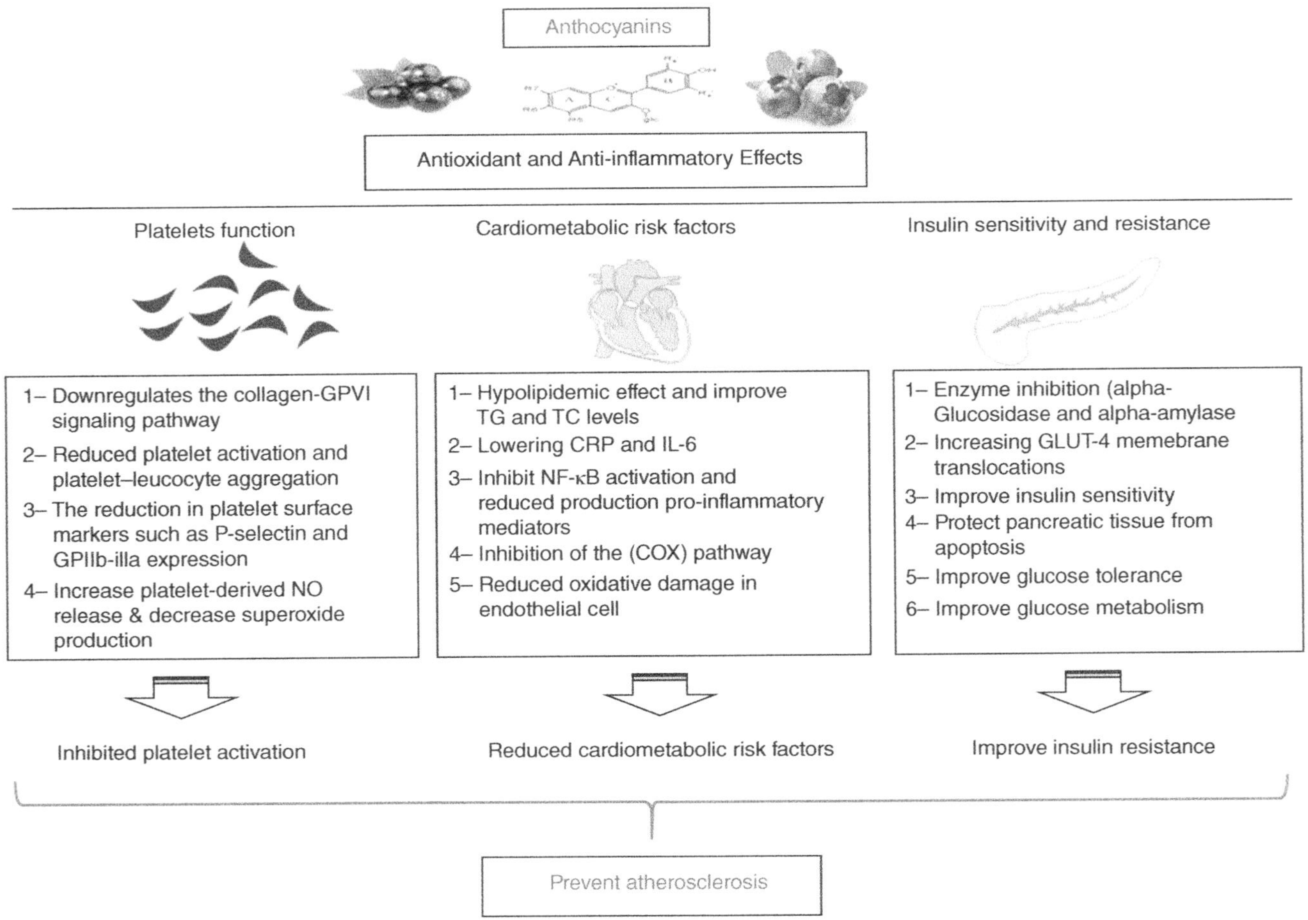

Figure 17.3 Proposed scheme for anti-atherogenicity effects of anthocyanins on the cardiometabolic risk factors and platelet functions (*Source:* Aboonabi, Meyer, Gaiz, and Singh (2020); Reprinted with permission from Elsevier License Number: 5062930027182; dt. 06 May, 2021).

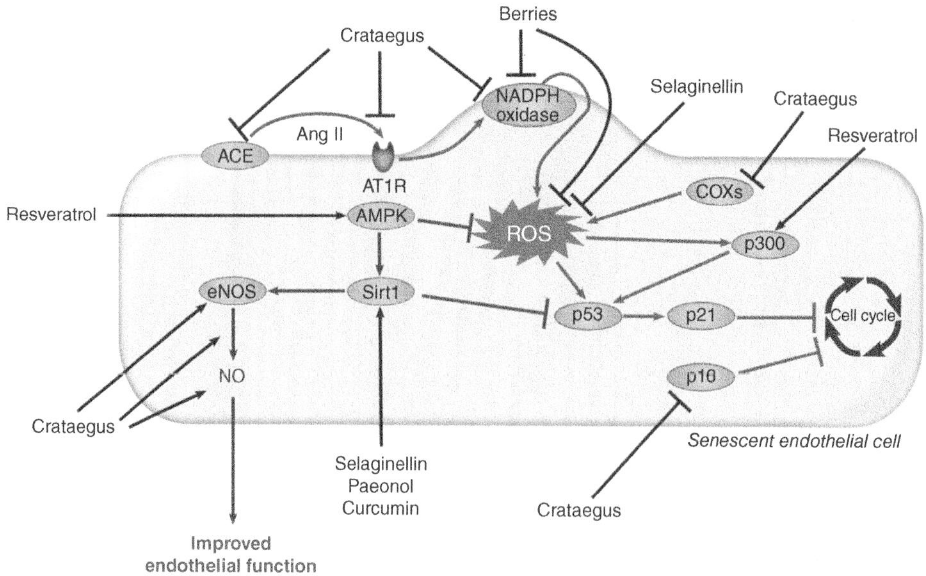

Figure 17.4 Schematic representation summarizing potential mechanisms contributing to the beneficial effect of several polyphenols and polyphenol-rich products by protecting endothelial cells against oxidative-stress-induced premature senescence. (*Source:* Oak et al. (2018), Reprinted with permission from Elsevier; License no. 5062930839658; dt. 06 May 2021).

(b) substantial reduction in β-amyloid (Aβ) peptide production, (c) upregulation of cerebral antioxidant enzymatic activities, (d) improvement in cognitive performance, and (e) neuroregeneration by reversing some of the damage (Kelly et al., 2017). Resveratrol is known to exert neuroprotective effects by reducing microglia-induced neuro-inflammation, hence protecting the brain against hypoxic–ischemic damage and upgrading cognitive function in the AD model (Granzotto & Zatta, 2014). Anti-glycation and neuroprotective effects have been evaluated for six polyphenol-rich edible berries (blackberry, black raspberry, blueberry, cranberry, red raspberry, and strawberry), for their neuroprotective properties via anti-Aβ fibrillation and murine microglia bioassays (Ma et al., 2018). The anthocyanins-enriched (ACEs) berry extracts displayed excellent free radical scavenging, entrapping reactive carbonyl species and anti-glycation effects compared to their corresponding anthocyanins-free (ACFs) extracts. Hence, these edible berries may have the potential to exert microglial neuroprotective effects attributable to their anthocyanin content. The neuroprotective and anti-amnesic activity of *E. umbellate* extracts on scopolamine-induced memory impairment in mice has been studied (Nazir et al., 2020). The results revealed promising *in vitro* anticholinesterase (AChE and BChE) activity. Further, the extracts were investigated for behavioral tests (learning and memory behavior) in normal and scopolamine-induced cognitive impaired mice. The extracts exhibited maximal inhibition of tested cholinesterase as well as for the free radicals. In conclusion, this investigation showed *E. umbellata* to possess substantial memory enhancing potency along with imparting anti-amnesic effects. Hence, *E. umbellate* having potential neuroprotective activities can be recommended for further investigation to control memory impairments and neurological disorders.

Neuroprotective effect of *Aronia melanocarpa* (black chokeberry) extract was investigated against glutamate-induced oxidative stress in HT22 (mouse hippocampal) cells by MTT assay (Lee et al., 2017). *Aronia melanocarpa* has high polyphenolic content that include anthocyanins, flavanols, flavonoids, chlorogenic acids, triterpenes, and caffeic acid derivatives. The results revealed that the berry extract demonstrated neuroprotective properties by lowering glutamate-induced cell death to ~16.8% (relative protection at 100 μg/mL). Moreover, another bioactive component, cyanidin-3-*O*-galactoside, considerably protected HT22 cells against glutamate-induced neurotoxicity, with relative protection of ~35.4% at 10 μg/mL, suggesting its therapeutic potential against neurodegenerative diseases, for example, AD. The prospective neuroprotective benefits of berry anthocyanins can deliver promising unconventional approach to fight age-related cognitive deterioration and pathological neurodegeneration. Oxidative stress inhibition and neuro-inflammation are the two key mechanisms by which anthocyanins exert protective effects in preventing and/or treatment of CNS disorder.

17.2.6 Anticancer Properties of Berries

Berry components that may be responsible for exerting cancer-protective effects in cells are phenolic acids, stilbenes (resveratrol), flavonoids (anthocyanins, flavonols, and catechins), lignans, and tannins (e.g. PAs and ellagitannins) (Forester & Waterhouse, 2010; Kristo et al., 2016). In addition, resveratrol and anethole inhibited AP-1 (activator protein 1), which is related to growth regulation and cell transformation. Resveratrol exhibits antioxidant activity and has been effective at molecular targets that are linked with tumor initiation, promotion, and progression (Du, Yang, Hu, Meng, & Lin, 2020). It can also induce apoptosis as well as diminish angiogenesis by attenuating FGF-2 (fibroblast growth factor) and vascular endothelial growth factor enzymes (De Filippis et al., 2017). Polyphenols modulate the redox state of the cancer microenvironment, increasing its cytotoxicity that leads to apoptosis (Pan et al., 2018). Most berries are known to extend effects on the proliferation rates of both premalignant and malignant cells (Kristo et al., 2016). The mechanisms involved are (a) inflammation suppression, (b) angiogenesis inhibition, (c) DNA damage protection, and (d) apoptotic effects or proliferation rates of malignant cells.

Anticancer activity of three chokeberries [*Aronia arbutifolia, prunifolia*, and *melanocarpa* (red, purple, and black, respectively)] rich in phytochemicals particularly anthocyanins and phenolic acids have been studied for growth inhibition in HT-29 human colon cancer cells by the cell viability MTT assay (Gill et al., 2020). The results revealed that black chokeberry among the three chokeberries extract exhibited the greatest reduction in cell proliferation. The phenolic acids present were responsible for the growth inhibition of cancer cells. Another study demonstrated the antiproliferative activity of sea buckthorn (*Hippophaë rhamnoides* L.) berry pomace extracts (Dienaitė et al., 2020). The extracts inhibited HT-29 cell proliferation at noncytotoxic concentrations quantified in non-tumoral Caco-2 cells. The antitumor activity can be attributed to phenolic components viz., flavonoids catechin, kaempferol, quercetin, and isorhamnetin that protect cells from oxidative damage. Overall, the antiproliferative effect of these components was due to their ability to target different molecular switches in carcinogen metabolism steps that include inflammation, cell proliferation, cell cycle, apoptosis, and angiogenesis. The antitumor activities of blueberry (alone and in combination) with black raspberry has been examined in animal model (nude mice) for lung tumor xenograft effects (Aqil et al., 2016). The results showed individual anthocyanidins (aglycones) exhibit significant effects (with low IC_{50} value) in inhibiting non-small-cell lung cancer cell growth. Similarly, delphinidin (a bioactive component of blueberry) and punicalagin (a bioactive component of black raspberry that is converted to ellagic acid *in vivo*) are reported to exhibit high tumor growth inhibition. The antitumor activity of blueberry is attributable to the polyphenolics present, especially anthocyanins. The anticancer activity of three wild *Rubus* berries (*Rubus ellipticus, Rubus niveus*, and *R. fairholmianus*) against human colon cancer cell line (Caco-2) has been evaluated via MTT assay (Muniyandi et al., 2019). These berries contain high amount of polyphenolic components that can be efficient in scavenging ROS and preventing the cells from oxidative damage that can lead to cancer and CVDs. The studies suggested that the anticancer activity of *Rubus* extracts is due to the presence of kaempferol, and Caco-2 cells viability was efficiently controlled. The effects of blueberry fermented with a tannase-producing lactic acid bacteria (*Lactobacillus plantarum*) on the anticancer activities in human cancer cell lines, including cervical, pancreatic, gastric, and human liver cancer cells have been determined (Ryu, Kang, & Cho, 2019). The *L. plantarum* fermentation converted the blueberry polyphenols into active phenol metabolites exhibiting higher antiproliferative activities at inhibiting HeLa cells proliferation via apoptosis (DNA cell damage), with very low cytotoxicity to normal fibroblast cells. The *in vitro* cytotoxic activity of *Juniperus oxycedrus* subsp. *oxycedrus* berries against two human breast cancer cell lines (MDA-MB-468 and MCF-7) has been reported (Ben Mrid et al., 2019). The results revealed potent cytotoxic effects in a dose-dependent manner against both breast tumor cells lines with high selectivity for both tumor cell lines without affecting normal cells. The *in vitro* cytotoxicity in terms of IC_{50} was calculated and the results showed that the lowest IC_{50} value (5.23 μg/mL) for the MCF-7-treated cells compared to the MDA-MB-468 cell line, ($IC_{50} = 6.43$ μg/mL). The bioactive components in berries can exert chemotherapeutic and chemo-preventive effects by suppressing inflammation, attenuating oxidative stress, and reducing angiogenesis through manipulation of multiple signaling pathways (e.g. NF-κ-light-chain-enhancer of activated B cells) that control DNA transcription, and Wnt/β-catenin (Afrin et al., 2016). The mode of action of berry polyphenols on anticancer activity has been discussed with a schematic representation (May, Parry, & Parry, 2020).

17.3 Conclusion and Outlook

Functional foods impart the much required health effects on certain targeted functionalities inside the human body. Functional foods promote the overall well-being beyond the normal nutritional effects by contributing toward minimizing the occurrence of several chronic and degenerative diseases. Consumption of various types of edible berries as functional foods are attaining global eminence and have been the subject of several studies aimed to explore their role in promoting human health. ROS generated in the human body is responsible for various pathological conditions that include aging, atherosclerosis, cancer, diabetes, microbial infections, inflammation, Parkinsonism, and others. Berries (and their extracts) which comprises various bioactive components such as polyphenols, anthocyanins, stilbenes, flavonoids, and lignans have been individually or synergistically shown to impart health benefits such as oxidative stress regulation, retarding CVDs, protecting against DNA damage, maintaining cellular metabolic processes, and apoptosis. Though the exact mechanism on the mode of action at the molecular level remains obscure up to a certain extent, the contribution of certain mechanistic pathways is well established in the available databases. These pathways account for the attenuation of diversified cellular processes, demonstrating some common signaling events that include oxidative stress management, cell cycle arrest, apoptotic induction, and inflammation, together with reduction of angiogenesis and metastasis.

Several studies demonstrate that berry consumption benefit human health, particularly those with metabolic diseases. For example, regular berry intake improves lipid profile in humans (Pap et al., 2021). Berry (blueberry, cranberry, and chokeberry) juice consumption (21 days) reduced LDL cholesterol levels (30%) and increased high-density lipoprotein (HDL) cholesterol (by up to 20%) and total antioxidant status in healthy human volunteers. Moreover, intake of blueberries (150 g/day) for 6 months improved endothelial function, elevated HDL cholesterol and total concentrations of anthocyanin-derived phenolic acid metabolites in serum and 24-h urine of overweight and obese adults. Therefore, daily blueberry consumption can reduce (12–15%) CVD risk. Whole berry-puree (mixture of equal amounts of blackcurrants, bilberries, cranberries, and strawberries) significantly reduced postprandial insulin response in healthy women (Pap et al., 2021 and references therein). Cranberry beverage (450 mL/day) consumption for 8 weeks significantly reduced serum insulin and fasting C-reactive protein (CRP) and increased HDL cholesterol in overweight or obese adults with abdominal adiposity in a randomized, double-blind, placebo-controlled, parallel design trial (Chew et al., 2019). In a prospective, population-based cohort study of middle-aged and older Finnish men (n = 2,332, average age 53 years, mean 26.6 kg/m^2 BMI) higher intake of berries was related with significant (35%) reduced T2D risk similar to an earlier health survey (Finnish Mobile Clinic Health Survey of 4,304 men and women 40–69 years of age) (Mursu, Virtanen, Tuomainen, Nurmi, & Voutilainen, 2014). This long 20-year follow-up study demonstrated that berry intake can reduce metabolic syndrome (by improving glucose metabolism) up to 35% independent of body weight.

Berry consumption improves postprandial response and profile of circulatory inflammatory markers and increases plasma antioxidant capacity according to published human studies (2011–2014) (Yang & Kortesniemi, 2015). Moreover, prolonged intake of berries and berry products can improve plasma lipid profile, reduce chronic inflammation, and maintain cardiovascular health, particularly for people with metabolic syndrome. In a meta-analysis of randomized controlled trials, berries (including juice of barberry, cranberry, blueberry and raspberry powder, and freeze-dried strawberry) significantly reduced systolic blood pressure (SBP) (by 3.68 mmHg) and diastolic blood pressure (DBP) (by 1.52 mmHg) and significantly elevated vascular adhesion molecules (sVCAM-1) levels (by 14.6 ng/mL) (Wang, Gallegos, Haskell-Ramsay, & Lodge 2021). However, in a subgroup analysis, these associations were limited to cranberry juice (SBP and DBP by 1.52 and 1.78 mmHg, respectively) indicating that cranberry supplementation can improve blood pressure. The quantitative analysis showed that only cranberry juice improves CVD outcomes, although blueberry juice and its freeze-dried form have been frequently studied for their cardio and vascular protective effects. Another systematic review of seven randomized clinical trials, involving 270 adults with T2D showed that daily cranberry juice (240 mL) consumption for 12 weeks and blueberry extract or powder supplementation (9.1–9.8 mg of anthocyanins, respectively) for 8–12 weeks had a beneficial effect on glucose control in T2DM subjects. Those results indicate a promising use of these berries in T2DM management, although more studies are required to better understand the mechanisms involved (Rocha, Caldas, da Silva, Hermsdorff, & Alfenas, 2019).

New Zealand blackcurrant extract (300 mg active cassis containing 105 mg anthocyanins [35–50% delphinidin-3-rutinoside, 5–20% delphinidin-3-glucoside, 30–45% cyanidin-3-rutinoside, and 3–10% cyanidin-3-glucoside]) intake for 8 days improved recovery from exercise-induced muscle damage in nonresistance trained adults in a double-blind, randomized, placebo-controlled, parallel study (Hunt et al., 2021). Consumption of anthocyanin-rich blackcurrant extract can alleviate oxidative stress and reduce plasma protein carbonyl levels and plasma oxidative capacity. This plasma anthocyanin increase activates cellular redox-sensitive signaling pathways (Nrf2/ARE transcription), upregulating cellular antioxidant systems thereby reducing oxidative stress indices during exercise recovery and promoting tissue repair (Hunt et al., 2021 and references therein). The blackcurrant extract significantly reduced serum creatine kinase (marker of muscle damage) post-exercise indicating decreased secondary muscle damage caused by dampened inflammatory response.

High flavonoid-rich berry intake/consumption (equivalent to 1.6 portions/day or 695–792 mg flavonoid/day) was associated with lower (up to 4 mmHg) SBP and pulse pressure, greater microbial diversity, and higher abundance of *Parabacteroides* and *Ruminococcaceae* in North German population (PopGen cohort; n = 904 [516 M+ 388F]; average age 61 years) (Jennings et al., 2021). Moreover, the gut microbiome accounted for 15% of the bidirectional flavonoid–gut microbiome interaction associated with blood pressure. The study highlights protective associations between dietary achievable intakes of flavonoid-rich berries/foods and their moderating effects on clinically relevant blood pressure improvements partly explained by gut microbiome characteristics. This SBP reduction is known to lower stroke risk (by 13% over 295,652 patient-years), provide cardiovascular protection and can save 45,000 quality life years as well as decrease related health and social care costs (Jennings et al., 2021 and references therein).

Sea buckthorn oil protects skin and mucous membranes due to its anti-inflammatory effects providing positive wound healing benefits. For example, sea buckthorn oil (3 g/day for 3 months) administered orally benefited vaginal health (improved mucosal integrity of vaginal epithelium) in postmenopausal women (n = 116, 57–69 years old), unable to use treatment for vaginal atrophy in a randomized, double-blind, placebo-controlled study (Larmo, Yang, Hyssälä, Kallio, & Erkkola, 2014). These effects were associated with the anti-inflammatory activity of the sea buckthorn oil attributed to carotenoids, tocopherols, and other lipid components (e.g. low LA to ALA ratio that favors production of low pro-inflammatory eicosanoids). Sea buckthorn oil (2 g/day for 3 months) positively affected dry eye symptoms by attenuating the increase in tear film osmolarity in adults (n = 100, average age 45 years) in a randomized, double-blind, parallel, placebo-controlled trial (Larmo et al., 2010).

There is also ample scope to explore some of the edible berries wastes and by-products (pomace) to obtain value-added functional components and study their effects and mechanisms under *in vitro* and *in vivo* conditions. In conclusion, additional investigations on the beneficial effects of berry phytochemicals need to emphasize the enhanced understanding of the underlying molecular mechanisms in an integrative way recognizing the *in vivo* effects of the bioactive components, development of robust and well-controlled clinical studies and thereby interpreting the plausible overall synergistic effects.

Acknowledgments

This chapter is based on our ongoing project VALORTECH, which has received funding from the European Union's Horizon 2020 research and innovation program under grant agreement No. 810630.

References

Aboonabi, A., Meyer, R. R., Gaiz, A., & Singh, I. (2020). Anthocyanins in berries exhibited anti-atherogenicity and antiplatelet activities in a metabolic syndrome population. *Nutrition Research*, *76*, 82–93. https://doi.org/10.1016/j.nutres.2020.02.011

Afrin, S., Giampieri, F., Gasparrini, M., Forbes-Hernandez, T. Y., Varela-López, A., Quiles, J. L., . . . Battino, M. (2016). Chemopreventive and therapeutic effects of edible berries: A focus on colon cancer prevention and treatment. *Molecules*, *21*, 169. https://doi.org/10.3390/molecules21020169

Aqil, F., Jeyabalan, J., Kausar, H., Munagala, R., Singh, I. P., & Gupta, R. (2016). Lung cancer inhibitory activity of dietary berries and berry polyphenolics. *Journal of Berry Research*, *6*, 105–114. https://doi.org/10.3233/JBR-160120

Ataie, A., Ataie, R., & Shadifar, M. (2016). Polyphenolic antioxidants and neuronal regeneration. *Basic Clinical and Neuroscience Journal*, *7*. https://doi.org/10.15412/J.BCN.03070201

Baby, B., Antony, P., & Vijayan, R. (2018). Antioxidant and anticancer properties of berries. *Critical Reviews in Food Science and Nutrition*, *58*(15), 2491–2507. https://doi.org/10.1080/10408398.2017.132919

Banach, M., Wiloch, M., Zawada, K., Cyplik, W., & Kujawski, W. (2020). Evaluation of antioxidant and anti-Inflammatory activity of anthocyanin-rich water-soluble aronia dry extracts. *Molecules*, *25*(4), 055. https://doi.org/10.3390/molecules25184055

Basu, A., Morris, S., Nguyen, A., Betts, N. M., Fu, D., & Lyons, T. J. (2016). Effects of dietary strawberry supplementation on antioxidant biomarkers in obese adults with above optimal serum lipids. *Journal of Nutrition and Metabolism*, *1–9*. https://doi.org/10.1155/2016/3910630

Basu, A., Rhone, M., & Lyons, T. J. (2010). Berries: emerging impact on cardiovascular health. *Nutrition Reviews*, *68*, 168–177. https://doi.org/10.1111/j.1753-4887.2010.00273.x

Ben Mrid, R., Bouchmaa, N., Bouargalne, Y., Ramdan, B., Karrouchi, K., Kabach, I., . . . Nhiri, M. (2019). Phytochemical characterization, antioxidant and *in vitro* cytotoxic activity evaluation of *Juniperus oxycedrus* Subsp. oxycedrus needles and berries. *Molecules*, *24*, 502. https://doi.org/10.3390/molecules24030502

Birch, C. S., & Bonwick, G. A. (2019). Ensuring the future of functional foods. *International Journal of Food Science and Technolology, 54*, 1467–1485. https://doi.org/10.1111/ijfs.14060

Bowtell, J. L., Aboo-Bakkar, Z., Conway, M. E., Adlam, A.-L. R., & Fulford, J. (2017). Enhanced task-related brain activation and resting perfusion in healthy older adults after chronic blueberry supplementation. *Applied Physiology, Nutrition, and Metabolism, 42*(7), 773–779. https://doi.org/10.1139/apnm-2016-0550

Cao, G., Sofic, E., & Prior, R. L. (1996). Antioxidant capacity of tea and common vegetables. *Journal of Agricultural and Food Chemistry, 44*(11), 3426–3431. https://doi.org/10.1021/jf9602535

Cásedas, G., Les, F., Gómez-Serranillos, M. P., Smith, C., & López, V. (2016). Bioactive and functional properties of sour cherry juice (Prunus cerasus). *Food and Function, 7*, 4675–4682. https://doi.org/10.1039/C6FO01295G

Cassidy, A. (2018). Berry anthocyanin intake and cardiovascular health. *Molecular Aspects of Medicine, 61*, 76–82. https://doi.org/10.1016/j.mam.2017.05.002

Castillo, R. L., Herrera, E. A., Gonzalez-Candia, A., Reyes-Farias, M., De La Jara, N., Peña, J. P., & Carrasco-Pozo, C. (2018). Quercetin prevents diastolic dysfunction induced by a high-cholesterol diet: Role of oxidative stress and bioenergetics in hyperglycemic rats. *Oxididative Medicine and Cellular Longevity, 1–14.* https://doi.org/10.1155/2018/7239123

Chew, B., Mathison, B., Kimble, L., McKay, D., Kaspar, K., Khoo, C., . . . Blumberg, J. (2019). Chronic consumption of a low calorie, high polyphenol cranberry beverage attenuates inflammation and improves glucoregulation and HDL cholesterol in healthy overweight humans: A randomized controlled trial. *European Journal of Nutrition, 58*, 1223–1235.

Cianciosi, D., Ansary, J., Regolo, L., Giampieri, F., & Forbes-Hernandez, T. Y., (2021). The preventive effect of strawberries against the onset of common human diseases induced by oxidative stress. *Acta Horticulturae 1309: IX International Strawberry Symposium*, 917–924. https://doi.org/10.17660/ActaHortic.2021.1309.131.

Dalle Zotte, A., & Szendrő, Z. (2011). The role of rabbit meat as functional food. *Meat Science, 88*, 319–331. https://doi.org/10.1016/j.meatsci.2011.02.017

De Filippis, B., Ammazzalorso, A., Fantacuzzi, M., Giampietro, L., Maccallini, C., & Amoroso, R. (2017). Anticancer activity of stilbene-based derivatives. *ChemMedChem, 12*(8), 558–570. https://doi.org/10.1002/cmdc.201700045

De Silva, A. B. K. H., & Rupasinghe, H. P. V. (2020). Polyphenols composition and anti-diabetic properties *in vitro* of haskap (*Lonicera caerulea* L.) berries in relation to cultivar and harvesting date. *Journal of Food Composition and Analysis, 88*, 103402. https://doi.org/10.1016/j.jfca.2019.103402

Diarra, M. S., Block, G., Rempel, H., Oomah, B. D., Harrison, J., McCallum, J., ... Malouin, F. (2013). *In vitro* and *in vivo* antibacterial activities of cranberry press cake extracts alone or in combination with β-lactams against *Staphylococcus aureus. BMC Complementary and Alternative Medicine,* ***13***, 90. https://doi.org/10.1186/1472-6882-13-90

Dienaitė, L., Pukalskas, A., Pukalskienė, M., Pereira, C. V., Matias, A. A., & Venskutonis, P. R. (2020). Phytochemical composition, antioxidant and antiproliferative activities of defatted sea buckthorn (*Hippophaë rhamnoides* L.) berry pomace fractions consecutively recovered by pressurized ethanol and water. *Antioxidants, 9*, 274. https://doi.org/10.3390/antiox9040274

Du, J., Yang, D., Hu, W., Meng, F., & Lin, Z., 2020. Study on the anticancer biological mechanism of resveratrol. *E3S Web of Conferences,* 218, EDPSciences. https://doi.org/10.1051/e3sconf/202021803002.

Duke, J. A., Godwin, M. J., Cellier, D. U. J., & Duke, P. A. K. (2002). *Handbook of Medicinal Herbs.* Boca Ratón: CRC Press.

Elks, C. M., Terrebonne, J. D., Ingram, D. K., & Stephens, J. M. (2015). Blueberries improve glucose tolerance without altering body composition in obese postmenopausal mice. *Obesity, 23,* 573–580. https://doi.org/10.1002/oby.20926

Farkouh, M. (2021). *The Açai Berry COVID-19 Anti-Inflammation Trial (ACAI).* Toronto: University Health Network. Available from https://clinicaltrials.gov/ct2/show/NCT04404218

Ferenczyova, K., Kalocayova, B., & Bartekova, M. (2020). Potential implications of quercetin and its derivatives in cardioprotection. *International Journal of Molecular Sciences, 21,* 1585. https://doi.org/10.3390/ijms21051585

Flanagan, E., Müller, M., Hornberger, M., & Vauzour, D. (2018). Impact of flavonoids on cellular and molecular mechanisms underlying age-related cognitive decline and neurodegeneration. *Current Nutrition Reports, 7,* 49–57. https://doi.org/10.1007/s13668-018-0226-1

Forester, S. C., & Waterhouse, A. L. (2010). Gut metabolites of anthocyanins, gallic acid, 3- O -methylgallic acid, and 2,4,6-trihydroxybenzaldehyde, inhibit cell proliferation of Caco-2 cells. *Journal of Agricultural and Food Chemistry, 58,* 5320–5327. https://doi.org/10.1021/jf9040172

Gasparrini, M., Forbes-Hernandez, T. Y., Cianciosi, D., Quiles, J. L., Mezzetti, B., Xiao, J., . . . Battino, M. (2021). The efficacy of berries against lipopolysaccharide-induced inflammation: A review. *Trends in Food Science and Technology.* https://doi.org/10.1016/j.tifs.2021.01.015

Gill, N. K., Rios, D., Osorio-Camacena, E., Mojica, B. E., Kaur, B., Soderstrom, M. A., . . . Forester, S. C. (2020). Anticancer effects of extracts from three different chokeberry species. *Nutrition and Cancer, 73*(7), 1168–1174. https://doi.org/10.1080/01635581.2020.1789679

Granato, D., Barba, F. J., Bursać Kovačević, D., Lorenzo, J. M., Cruz, A. G., & Putnik, P. (2020). Functional foods: product development, technological trends, efficacy testing, and safety. *Annual Review of Food Science and Technolology, 11,* 93–118. https://doi.org/10.1146/annurev-food-032519-051708

Granzotto, A., & Zatta, P. (2014). Resveratrol and Alzheimer's disease: Message in a bottle on red wine and cognition. *Frontiers in Aging and Neuroscience, 6,* 95. https://doi.org/10.3389/fnagi.2014.00095

Gutiérrez, T. J. (2018). Plantain flours as potential raw materials for the development of gluten-free functional foods. *Carbohydrate Polymers, 202,* 265–279. https://doi.org/10.1016/j.carbpol.2018.08.121

Hunt, J. E., Coelho, M. O., Buxton, S., Butcher, R., Foran, D., Rowland, D., . . . King, D. G. (2021). Consumption of New Zealand blackcurrant extract improves recovery from exercise-induced muscle damage in non-resistance trained men and women: A double-blind randomized trial. *Nutrients, 13*(8), 2875. https://doi.org/10.3390/nu13082875

Iizuka, Y., Ozeki, A., Tani, T., & Tsuda, T. (2018). Blackcurrant extract ameliorates hyperglycemia in type 2 diabetic mice in association with increased basal secretion of glucagon-like peptide-1 and activation of AMP-activated protein kinase. *Journal of Nutritional Science and Vitaminology (Tokyo), 64*, 258–264. https://doi.org/10.3177/jnsv.64.258

Jennings, A., Koch, M., Bang, C., Franke, A., Lieb, W., & Cassidy, A. (2021). Microbial diversity and abundance of *Parabacteroides* mediate the associations between higher intake of flavonoid-rich Foods and lower blood pressure. *Hypertension*. https://doi.org/10.1161/HYPERTENSIONHA.121.17441

Johansson, A. K., Korte, H., Yang, B., Stanley, J. C., & Kallio, H. P. (2000). Sea buckthorn berry oil inhibits platelet aggregation. *The Journal of Nutritional Biochemistry, 11*(10), 491–495.

Kareb, O., & Aïder, M. (2019). Whey and its derivatives for probiotics, prebiotics, synbiotics, and functional foods: a critical review. *Probiotics and Antimicrobial Proteins, 11*, 348–369. https://doi.org/10.1007/s12602-018-9427-6

Kelly, E., Vyas, P., & Weber, J. (2017). Biochemical properties and neuroprotective effects of compounds in various species of berries. *Molecules, 23*, 26. https://doi.org/10.3390/molecules23010026

Kostin, S. F., McDonald, D. E., & McFadden, D. W. (2012). Inhibitory effects of (−)-epigallocatechin-3-gallate and pterostilbene on pancreatic cancer growth *in vitro*. *Journal of Surgical Research, 177*(2), 255–262. https://doi.org/10.1016/j.jss.2012.04.023

Kristo, A. S., Klimis-Zacas, D., & Sikalidis, A. K. (2016). Protective role of dietary berries in cancer. *Antioxidants, 5*, 37. https://doi.org/10.3390/antiox5040037

Larmo, P. S., Järvinen, R. L., Setälä, N. L., Yang, B., Viitanen, M. H., Engblom, J. R. K., . . . Kallio, H. P. (2010). Oral sea buckthorn oil attenuates tear film osmolarity and symptoms in individuals with dry eye. *Journal of Nutrition, 140*, 1462–1468.

Larmo, P. S., Yang, B., Hyssälä, J., Kallio, H. P., & Erkkola, R. (2014). Effects of sea buckthorn oil intake on vaginal atrophy in postmenopausal women: A randomized, double-blind, placebo-controlled study. *Maturitas, 79*, 316–321.

Lee, H. Y., Weon, J. B., Ryu, G., Yang, W. S., Kim, N. Y., Kim, M. K., & Ma, C. J. (2017). Neuroprotective effect of *aronia melanocarpa* extract against glutamate-induced oxidative stress in HT22 cells. *BMC Complementary and Alternative Medicine, 17*(1), 207. https://doi.org/10.1186/s12906-017-1716-1

Liu, S., Yu, J., Guo, S., Fang, H., & Chang, X. (2020). Inhibition of pancreatic α-amylase by *Lonicera caerulea* berry polyphenols *in vitro* and their potential as hyperglycemic agents. *LWT, 126*(109), 288. https://doi.org/10.1016/j.lwt.2020.109288

Liu, Y., Zhang, X., Zhan, L., Xu, C., Sun, L., Jiang, H., . . . Li, X. (2020). LC-Q-TOF-MS Characterization of polyphenols from white bayberry fruit and its antidiabetic effect in KK-Ay mice. *ACS Omega, 5*, 17839–17849. https://doi.org/10.1021/acsomega.0c02759

Ma, H., Johnson, S., Liu, W., DaSilva, N., Meschwitz, S., Dain, J., & Seeram, N. (2018). Evaluation of polyphenol anthocyanin-enriched extracts of blackberry, black raspberry, blueberry, cranberry, red raspberry, and strawberry for free radical scavenging, reactive carbonyl species trapping, anti-Glycation, anti-β-amyloid aggregation, and microglial neuroprotective effects. *International Journal of Molecular Sciences, 19*(2), 461. https://doi.org/10.3390/ijms19020461

Ma, L., Sun, Z., Zeng, Y., Luo, M., & Yang, J. (2018). Molecular mechanism and health role of functional ingredients in blueberry for chronic disease in human beings. *International Journal of Molecular Sciences*, *19*, 2785. https://doi.org/10.3390/ijms19092785

Ma, Z. F., Zhang, H., Teh, S. S., Wang, C. W., Zhang, Y., Hayford, F., . . . Zhu, Y. (2019). Goji berries as a potential natural antioxidant medicine: An insight into their molecular mechanisms of action. *Oxididative Medicine and Cellular Longevity*, *2019*, 1–9. https://doi.org/10.1155/2019/2437397

May, S., Parry, C., & Parry, L. (2020). Berry chemoprevention: Do berries decrease the window of opportunity for tumorigenesis. *Food Frontiers*, *1*(3), 260–275. https://doi.org/10.1002/fft2.32

Miller, K., Feucht, W., & Schmid, M. (2019). Bioactive compounds of strawberry and blueberry and their potential health effects based on human intervention studies: A brief overview. *Nutrients*, *11*(7), 1510. https://doi.org/10.3390/nu11071510

Muniyandi, K., George, E., Sathyanarayanan, S., George, B. P., Abrahamse, H., Thamburaj, S., & Thangaraj, P. (2019). Phenolics, tannins, flavonoids and anthocyanins contents influenced antioxidant and anticancer activities of Rubus fruits from Western Ghats, India. *Food Science and Human Wellness*, *8*, 73–81. https://doi.org/10.1016/j.fshw.2019.03.005

Mursu, J., Virtanen, J. K., Tuomainen, T.-P., Nurmi, T., & Voutilainen, S. (2014). Intake of fruit, berries, and vegetables and risk of type 2 diabetes in Finnish men: the Kuopio Ischaemic Heart Disease Risk Factor Study. *American Journal of Clinical Nutrition*, *99*, 328–333.

Nabavi, S. M., Nabavi, S. F., Sureda, A., Caprioli, G., Iannarelli, R., Sokeng, A. J. T., . . . Maggi, F. (2018). The water extract of tutsan (*Hypericum androsaemum* L.) red berries exerts antidepressive-like effects and *in vivo* antioxidant activity in a mouse model of post-stroke depression. *Biomedicine and Pharmacotherapy*, *99*, 290–298. https://doi.org/10.1016/j.biopha.2018.01.073

Najjar, R. S., Schwartz, A. M., Wong, B. J., Mehta, P. K., & Feresin, R. G. (2021). Berries and their polyphenols as a potential therapy for coronary microvascular dysfunction: A mini-review. *International Journal of Molecular Sciences*, *22*(3), 373. https://doi.org/10.3390/ijms22073373

Nazir, N., Zahoor, M., Nisar, M., Karim, N., Latif, A., Ahmad, S., & Uddin, Z. (2020). Evaluation of neuroprotective and anti-amnesic effects of *Elaeagnus umbellata* Thunb. on scopolamine-induced memory impairment in mice. *BMC Complementary Medicine and Therapies*, *20*, 143. https://doi.org/10.1186/s12906-020-02942-3

Nazir, N., Zahoor, M., Ullah, R., Ezzeldin, E., & Mostafa, G. A. E. (2020). Curative effect of catechin isolated from *Elaeagnus Umbellata* Thunb. berries for iabetes and related complications in streptozotocin-induced diabetic rats model. *Molecules*, *26*, 137. https://doi.org/10.3390/molecules26010137

Nemes-Nagy, E., Szocs-Molnár, T., Dunca, I., Balogh- Sămărghiţan, V., Hobai, S., Morar, R., . . . Crăciun, E. C. (2008). Effect of a dietary supplement containing blueberry and sea buckthorn concentrate on antioxidant capacity in type 1 diabetic children. *Acta Physiologica Hungarica*, *95*(4), 383–393.

Newsome, B. J., Petriello, M. C., Han, S. G., Murphy, M. O., Eske, K. E., Sunkara, M., . . . Hennig, B. (2014). Green tea diet decreases PCB 126-induced oxidative stress in mice by up-regulating antioxidant enzymes. *Journal of Nutritional Biochemistry*, *25*, 126–135. https://doi.org/10.1016/j.jnutbio.2013.10.003

Nile, S. H., & Park, S. W. (2014). Edible berries: Bioactive components and their effect on human health. *Nutrition, 30*, 134–144. https://doi.org/10.1016/j.nut.2013.04.007

Nilsson, A., Salo, I., Plaza, M., & Björck, I. (2017). Effects of a mixed berry beverage on cognitive functions and cardiometabolic risk markers; A randomized cross-over study in healthy older adults. *PLoS One, 12*, e0188173. https://doi.org/10.1371/journal.pone.0188173

Nour, V., Panaite, T. D., Corbu, A. R., Ropota, M., & Turcu, R. P. (2021). Nutritional and bioactive compounds in dried Sea-Buckthorn Pomace. *Erwerbs-Obstbau*, ***63***, 91–98. https://doi.org/10.1007/s10341-020-00539-1

Oak, M.-H., Auger, C., Belcastro, E., Park, S.-H., Lee, H.-H., & Schini-Kerth, V. B. (2018). Potential mechanisms underlying cardiovascular protection by polyphenols: Role of the endothelium. *Free Radical in Biology and Medicine, 122*, 161–170. https://doi.org/10.1016/j.freeradbiomed.2018.03.018

Olas, B. (2018). Berry phenolic antioxidants – implications for human health? *Frontiers in Pharmacology, 9*. https://doi.org/10.3389/fphar.2018.00078

Oreopoulou, V., & Tzia, C. (2007). Utilization of plant by-products for the recovery of proteins, dietary fibers, antioxidants, and colorants. In V. Oreopoulou & W. Russ (Eds.), *Utilization of By-Products and Treatment of Waste in the Food Industry* (pp. 209–232). US: Springer. https://doi.org/10.1007/978-0-387-35766-9_11

Orsavová, J., Hlaváčová, I., Mlček, J., Snopek, L., & Mišurcová, L. (2019). Contribution of phenolic compounds, ascorbic acid and vitamin E to antioxidant activity of currant (Ribes L.) and gooseberry (*Ribes uva-crispa* L.) fruits. *Food Chemistry, 284*, 323–333. https://doi.org/10.1016/j.foodchem.2019.01.072

Pan, P., Huang, Y. W., Oshima, K., Yearsley, M., Zhang, J., Yu, J., . . . Wang, L. S. (2018). An immunological perspective for preventing cancer with berries. *Journal of Berry Research, 8*, 163–175. https://doi.org/10.3233/JBR-180305

Pap, N., Fidelis, M., Azevedo, L., Carmo, M. A. V., Wang, D., Mocan, A., . . . Granato, D. (2021). Berry polyphenols and human health: evidence of antioxidant, anti-inflammatory, microbiota modulation, and cell-protecting effects. *Current Opinion in Food Science*. https://doi.org/10.1016/j.cofs.2021.06.003

Paredes-López, O., Cervantes-Ceja, M. L., Vigna-Pérez, M., & Hernández-Pérez, T. (2010). Berries: Improving human health and healthy aging, and promoting quality life-A review. *Plant Foods for Human Nutrition, 65*, 299–308. https://doi.org/10.1007/s11130-010-0177-1

Park, E., Edirisinghe, I., Wei, H., Vijayakumar, L. P., Banaszewski, K., Cappozzo, J. C., & Burton-Freeman, B. (2016). A dose–response evaluation of freeze-dried strawberries independent of fiber content on metabolic indices in abdominally obese individuals with insulin resistance in a randomized, single-blinded, diet-controlled crossover trial. *Molecular Nutrition and Food Research, 60*(5), 1099–1109. https://doi.org/10.1002/mnfr.201500845

Real, J. T., Martínez-Hervás, S., Tormos, M. C., Domenech, E., Pallardó, F. V., Sáez-Tormo, G., . . . García-García, A.-B. (2010). Increased oxidative stress levels and normal antioxidant enzyme activity in circulating mononuclear cells from patients of familial hypercholesterolemia. *Metabolism, 59*, 293–298. https://doi.org/10.1016/j.metabol.2009.07.026

Rebello, C. J., Burton, J., Heiman, M., & Greenway, F. L. (2015). Gastrointestinal microbiome modulator improves glucose tolerance in overweight and obese subjects: A randomized controlled pilot trial. *Journal of Diabetes Complications*, *29*, 1272–1276. https://doi.org/10.1016/j.jdiacomp.2015.08.023

Rocha, D. M. U. P., Caldas, A. P. S., da Silva, B. P., Hermsdorff, H. H. M., & Alfenas, R. D. C. G. (2019). Effects of blueberry and cranberry consumption on type 2 diabetes glycemic control: A systematic review. *Critical Reviews in Food Science and Nutrition*, *59*(11), 1816–1828.

Rohm, H., Brennan, C., Turner, C., Günther, E., Campbell, G., Hernando, I., . . . Kontogiorgos, V. (2015). Adding value to fruit processing waste: Innovative ways to incorporate fibers from berry pomace in baked and extruded cereal-based foods – a susfood project. *Foods*, *4*, 690–697. https://doi.org/10.3390/foods4040690

Rupasinghe, H. P. V., Balasuriya, N., & Wang, Y. (2017). Prevention of type 2 diabetes by polyphenols of fruits. In K. Al-Gubory & I. Laher (Eds.), *Nutritional Antioxidant Therapies: Treatments and Perspectives* (pp. 447–466). Cham: Springer International Publishing. https://doi.org/10.1007/978-3-319-67625-8_17

Ryu, J., Kang, H. R., & Cho, S. K. (2019). Changes over the fermentation period in phenolic compounds and antioxidant and anticancer activities of blueberries fermented by *Lactobacillus plantarum*. *Journal of Food Science*, *84*, 2347–2356. https://doi.org/10.1111/1750-3841.14731

Šavikin, K. P., Krstić-Milošević, D. B., Menković, N. R., Beara, I. N., Mrkonjić, Z. O., & Pljevljakušić, D. S. (2017). Crataegus orientalis leaves and berries: phenolic profiles, antioxidant and anti-inflammatory activity. *Natural Products. Communication*, *12*, 1934578X1701200. https://doi.org/10.1177/1934578X1701200204

Schell, J., Scofield, R., Barrett, J., Kurien, B., Betts, N., Lyons, T., . . . Basu, A. (2017). Strawberries improve pain and inflammation in obese adults with radiographic evidence of knee osteoarthritis. *Nutrients*, *9*, 949. https://doi.org/10.3390/nu9090949

Seo, K. H., Lee, J. Y., Park, J.-Y., Jang, G. Y., Kim, H. D., Lee, Y.-S., & Kim, D. H. (2019). Differences in anti-inflammatory effect of immature and mature of *Rubus coreanus* fruits on LPS-induced RAW 264.7 macrophages via NF-κB signal pathways. *BMC Complementary and Alternative Medicine*, *19*, 89. https://doi.org/10.1186/s12906-019-2496-6

Shabani, S., Rabiei, Z., & Amini-Khoei, H. (2020). Exploring the multifaceted neuroprotective actions of gallic acid: A review. *International Journal of Food Properties*, *23*, 736–752. https://doi.org/10.1080/10942912.2020.1753769

Shi, F., Wang, C., Wang, L., Song, X., Yang, H., Fu, Q., & Zhao, W. (2019). Preparative isolation and purification of steroidal glycoalkaloid from the ripe berries of *Solanum nigrum* L. by preparative HPLC–MS and UHPLC–TOF-MS/MS and its anti-non-small cell lung tumors effects *in vitro* and *in vivo*. *Journal of Separation Science*, *42*, 2471–2481. https://doi.org/10.1002/jssc.201801165

Silveira, D., Prieto-Garcia, J. M., Boylan, F., Estrada, O., Fonseca-Bazzo, Y. M., Jamal, C. M., . . . Heinrich, M. (2020). COVID-19: is there evidence for the use of herbal medicines as adjuvant symptomatic therapy? *Frontiers in Pharmacology*, *1479*. https://doi.org/10.3389/fphar.2020.581840

Singh, B., Singh, J. P., Kaur, A., & Singh, N. (2016). Bioactive compounds in banana and their associated health benefits—A review. *Food Chemistry*, *206*, 1–11. https://doi.org/10.1016/j.foodchem.2016.03.033

Stote, K. S., Sweeney, M. I., Kean, T., Baer, D. J., Novotny, J. A., Shakerley, N. L., . . . Gottschall-Pass, K. T. (2017). The effects of 100% wild blueberry (*Vaccinium angustifolium*) juice consumption on cardiometablic biomarkers: a randomized, placebo-controlled, crossover trial in adults with increased risk for type 2 diabetes. *BMC Nutrition, 3*, 45. https://doi.org/10.1186/s40795-017-0164-0

Sytařová, I., Orsavová, J., Snopek, L., Mlček, J., Byczyński, Ł., & Mišurcová, L. (2020). Impact of phenolic compounds and vitamins C and E on antioxidant activity of sea buckthorn (Hippophaë rhamnoides L.) berries and leaves of diverse ripening times. *Food Chemistry, 310*, 125784. https://doi.org/10.1016/j.foodchem.2019.125784

Takikawa, M., Inoue, S., Horio, F., & Tsuda, T. (2010). Dietary anthocyanin-rich bilberry extract ameliorates hyperglycemia and insulin sensitivity via activation of AMP-activated protein kinase in diabetic mice. *Journal of Nutrition, 140*, 527–533. https://doi.org/10.3945/jn.109.118216

Tan, S. J., & Ismail, I. S. (2020). Potency of selected berries, grapes, and citrus fruit as neuroprotective agents. *Evidence-based Complementary and Alternative Medicine, 2020*, 1–12. https://doi.org/10.1155/2020/3582947

Tang, J., Dunshea, F. R., & Suleria, H. A. R. (2020). LC-ESI-QTOF/MS characterization of phenolic compounds from medicinal plants (Hops and Juniper Berries) and their antioxidant activity. *Foods, 9*, 7. https://doi.org/10.3390/foods9010007

Umdale, S., Ahire, M., Aiwale, V., Jadhav, A., & Mundada, P. (2020). Phytochemical investigation and antioxidant efficacy of wild, underutilized berries of economically important Indian Sandalwood (*Santalum album* L.). *Biocatalysis and AgriculturalBiotechnology, 27*, 101705. https://doi.org/10.1016/j.bcab.2020.101705

Uttara, B., Singh, A., Zamboni, P., & Mahajan, R. (2009). Oxidative stress and neurodegenerative diseases: A review of upstream and ownstream antioxidant therapeutic options. *Current Neuropharmacology, 7*(1), 65–74. https://doi.org/10.2174/157015909787602823

Vinayagam, R., & Xu, B. (2015). Antidiabetic properties of dietary flavonoids: a cellular mechanism review. *Nutrition Metabolism (London), 12*, 60. https://doi.org/10.1186/s12986-015-0057-7

Wang, Y., Gallegos, J. L., Haskell-Ramsay, C., & Lodge, J. K. (2021). Effects of chronic consumption of specific fruit (berries, citrus and cherries) on CVD risk factors: a systematic review and meta-analysis of randomized controlled trials. *European Journal of Nutrition, 60*, 615–639.

Wang, Y., Xiang, L., Yi, X., & He, X. (2017). Potential anti-inflammatory steroidal saponins from the berries of Solanum nigrum L. (European Black Nightshade). *Journal of Agricultural and Food Chemistry, 65*(21), 4262–4272. https://doi.org/10.1021/acs.jafc.7b00985

WHO. (2021). https://www.who.int/news-room/fact-sheets/detail/cardiovascular-diseases-(cvds).

Wu, S., Gao, L., Cipriani, A., Huang, Y., Yang, Z., Yang, J., . . . Zhan, S. (2019). The effects of incretin-based therapies on β-cell function and insulin resistance in type 2 diabetes: A systematic review and network meta-analysis combining 360 trials. *Diabetes, Obesity and Metabolism, 21*(4), 975–983. https://doi.org/10.1111/dom.13613

Wu, X., Wang, T. T. Y., Prior, R. L., & Pehrsson, P. R. (2018). Prevention of atherosclerosis by berries: The case of blueberries. *Journal of Agricultual and Food Chemistry, 66*, 9172–9188. https://doi.org/10.1021/acs.jafc.8b03201

Wu, Y. G., Wang, D. B., Hu, J. J., Song, X. Q., Xie, C. Z., Ma, Z. Y., & Xu, J. Y. (2019). An iron(iii) complex selectively mediated cancer cell death: Crystal structure, DNA targeting and: *In vitro* antitumor activities. *Inorganic Chemistry Frontiers*, *6*(4), 1040–1049. https://doi.org/10.1039/c9qi00030e

Xiang, L., Wang, Y., Yi, X., & He, X. (2018). Anti-inflammatory steroidal glycosides from the berries of *Solanum nigrum* L. (European black nightshade). *Phytochemistry*, *148*, 87–96. https://doi.org/10.1016/j.phytochem.2018.01.019

Yang, B., & Kortesniemi, M. (2015). Clinical evidence on potential health benefits of berries. *Current Opinion in Food Science*, *2*, 36–42. https://doi.org/10.1016/j.cofs.2015.01.002

Yazdankhah, S., Hojjati, M., & Azizi, M. H. (2019). The antidiabetic potential of black mulberry extract-enriched pasta through inhibition of enzymes and glycemic index. *Plant Foods for Human Nutrition*, *74*(1), 149–155. https://doi.org/10.1007/s11130-018-0711-0

Ye, Q., Georges, N., & Selomulya, C. (2018). Microencapsulation of active ingredients in functional foods: From research stage to commercial food products. *Trends in Food Science andTechnology*, *78*, 167–179. https://doi.org/10.1016/j.tifs.2018.05.025

Yeung, A. W. K., Tzvetkov, N. T., Zengin, G., Wang, D., Xu, S., Mitrović, G., . . . Atanasov, A. G. (2019). The berries on the top. *Journal of Berry Research*, *9*, 125–139. https://doi.org/10.3233/JBR-180357

Zhang, Y., Zhang, M., Wang, Y., Fan, Y., Chen, X., Yang, Y., . . . Liu, H. (2020). Protein–ligand interaction-guided discovery of novel VEGFR-2 inhibitors. *Journal of Biomolecular Structure and Dynamics*, *38*(9), 2559–2574. https://doi.org/10.1080/07391102.2019.1635915

Zheng, W.-H., Bai, H.-Y., Han, S., Bao, F., Zhang, K.-X., Sun, L.-L., . . . Yang, Z.-G. (2019). Analysis on the constituents of branches, berries, and leaves of *Hippophae rhamnoides* L. by UHPLC-ESI-QTOF-MS and their anti-inflammatory activities. *Natural Product Communications*, *14*(8), 1934578X1987140. https://doi.org/10.1177/1934578X19871404

Zorzi, M., Gai, F., Medana, C., Aigotti, R., Morello, S., & Peiretti, P. G. (2020). Bioactive compounds and antioxidant capacity of small berries. *Foods*, *9*, 623. https://doi.org/10.3390/foods9050623

Index

Page locators in **bold** indicate tables. Page locators in *italics* indicate figures. This index uses letter-by-letter alphabetization.

d

n

o

t